THE SOCIETY OF NAVAL ARCHITECTS AND MARINE ENGINEERS

TRANSACTIONS

Volume 115
2007

Published in 2008 by

The Society of Naval Architects and Marine Engineers
601 Pavonia Avenue, Jersey City, New Jersey 07306
www.sname.org

The volumes of Transactions contain professional information of permanent value. Information on the various functions and activities of the Society and a list of members may be found in the SNAME Membership Directory online at www.sname.org.

ISSN 0081 1661

ISBN (Hard Cover) 0-939773-62-7
(Soft Cover) 0-939773-63-5
(CD) 0-93977-3-64-3

Table of Contents

2007 Annual Meeting Activities

Reports

Refereed Papers

The following papers were selected for presentation at the 2007 Annual Meeting by the Society's Papers Committee

Commemoratives

2007 EXECUTIVE COMMITTEE

2007 MEMBERS OF COUNCIL

President: Robert E. Kramek

Executive Director: Philip B. Kimball Treasurer: Joseph J. Cuneo

Roderick J. Allan............SNAME Representative to the OTC Board of Directors
Glenn M. Ashe...............................V. P., Atlantic South Region
Robert Axelrod...Hon.Vice President
Christopher D. Barry........................Chair, Small Craft Comm.
Robert F. Beck...........Chair, Journal of Ship Research Comm.
Suzanne M. Beckstoffer................V. P., Atlantic South Region
Gerald E. Bellows.......................................V. P., Pacific Region
Willaim T. Bennett, Jr....................Co-Chair, Offshore Comm.
Donald M. Birney............................Hon.Vice President,Chair, Pension Committee
*Brent Blackburn..Section Chair, Gulf
*Nathan Bossett.................Section Chair, Northern California
James N. Brekke..............................Co-Chair, Offshore Comm.
*Alex Brovkin..Section Chair, Arctic
*Richard Burns.................................. Section Chair, Southwest
*Donald Burris.. Section Chair, Texas
Edward J. Campbell..Past President
Edward W.H. Clendenning........................Hon. Vice President
Roger H. Compton..Past President
Donald P. Courtsal......................................Hon. Vice President
Joseph J. Cuneo............Past President and SNAME Treasurer
Lauren G. Daly.........................Chair, Student Steering Comm.
*Lee Dana...Section Chair, Southeast
William G. Day, Jr.....................Chair, Hydrodynamics Comm.
Richard D. Delpizzo..............Chair, Ships' Machinery Comm.
Jose Femenia...Past President, Chair, Nominating and Public Policy Comm.
C. L. French..Past President
*Gregory J. Grigoropoulos..............V. P. International Region, Section Chair, Greece
Arthur J. Haskell..Past President
*Glen Hauptmann...Section Chair, Gulf
*Brig Henry............................Section Chair, Pacific Northwest
*David Herbein.............................Section Chair, New England
John H. Higginbotham............Chair, Annual Meeting Comm.
*Albert W. Horsman......................Section Chair, Great Lakes/ Great Rivers
William L. Hurley...............................Functional V. P. Planning and Public Policy, and Chair, Planning Comm.
Robert G. Keane, Jr.........................Chair, Ship Design Comm.
Philip B. Kimball.........................Executive Director, SNAME
Ronald K. Kiss..Past President
David H. Klinges..Hon. Vice President
Robert E. Kramek..President
Petros A. Lalangas.............................V. P., International Region
Kenneth R. Lane..................................Chair, Sections Comm.
Norman Lemley.............................Chair, Public Policy Comm.
*Albert P. Levesque.........................Section Chair, Philadelphia
Robert A. Levine..........................Chair, Development Comm.
Edward M. Lewandowski.................Chair, Marine Technology Comm.

Section Chairs from 9/1/06-8/31/07

Thomas P. Mackey..................................Hon. Vice President,
Walter M. Maclean......................Functional V. P., Education, Chair, Professional Engineering Licensure Comm.
Monroe D. Macpherson..........................Hon. Vice President
John A. Malone..V. P., Pacific Region
Naresh M. Maniar................................Chair, Fellows Comm.
Stephen P. Markle.......................Chair, Joint ASNE/SNAME Environmental Eng. Comm.
Scott C. McClure.........................Chair, Scholarships Comm.
Robert G. Mende......................................Hon. Vice President
John A. Mercier.......................V. P., Central and Gulf Region
R. Keith Michel.......................... Hon. Vice President,Chair, Investments Comm.
*Paul H. Miller.............................Section Chair, Chesapeake
Richards T. Miller...................................Hon. Vice President
Arnold P. Moore..................................Chair, Awards Comm.
William B. Morgan.................................Hon. Vice President
Perry W. Nelson...........Past President, Chair, Audit Comm.
Peter G. Noble..............................Hon. Vice President, V. P. Central and Gulf Region, Chair International Activities Comm.
Samuel J. Oncea.................Chair, Electronic Media Comm.
Ellsworth L. Peterson..............................Hon.Vice President
*Harilaos Psaraftis...................................Hon. Vice President
Paul J. Roden...........................Functional V. P., Membership, Chair, Membership Comm.
Bruce S. Rosenblatt...Past President
Philip G. Rynn.........................Chair, Hull Structure Comm.
Robert J. Scott..Hon. Vice President
Jaideep Sirkar.............................Functional V. P., Technology, Chair, T&R Steering Comm.
Rick S. Spaulding.................Chair, Ship Production Comm. Section Chair, Hampton Roads
Arnold M. Stein..Hon. Vice President
Richard L. Storch.............Chair, Journal of Ship Production Comm.
Robert D. Tagg..........Functional V. P., Knowledge Mngmnt, Chair, Publications Comm.
William duBarry Thomas...................Chair, History Comm.
Martin Toyen...................................Functional V. P., Finance, Chair, Finance Comm.
*Yvonne Traynhamq.............. Section Chair, New York Met
*Jocelyn Turgeon............... Section Chair, Eastern Canadian
Hendrik F. Van Hemmen...................Chair, Ship Repair and Conversion Comm.
*John Volk...........................Section Chair, Canadian Atlantic
Daniel Walker............................V. P., Atlantic North Region
Xiaozhi (Christina) Wang....................Chair, Papers Comm.
Jennifer J. Zeien...Chair, Ship Technical Operations Comm.
David R. Zoller...Hon.Vice President

Ernest V. Luzzatto,Legal Advisor - Non-Member

Note: Hawaii Section - no chairman at present time.

Section Officers

September 1, 2006 to August 31, 2007

ARCTIC
Alex Brovkin

CANADIAN ATLANTIC
John Volc

CHESAPEAKE
Paul H. Miller

EASTERN CANADIAN
Jocelyn Turgeon

GREAT LAKES/GREAT RIVERS
Al Horsmon

GREEK
Harilaos Psaraftis

GULF
Brent Blackburn

HAMPTON ROADS
Rick Spaulding

NEW ENGLAND
David Herbein

NEW YORK METROPOLITAN
Yvonne Traynham

NORTHERN CALIFORNIA
Nathan Bossett

PACIFIC NORTHWEST
Brig Henry

PHILADELPHIA
Al Levesque

SOUTHEAST
Lee Dana

SOUTHWEST
Richard Burns

TEXAS
Donald Burris

SNAME Administrative Staff

Philip B. Kimball, Executive Director

Joseph Caggiano	Information Systems Manager
Connie Cali-Poutre	Web Site Manager
Linda Davis	Membership Representative, Member Services Dept.
Brenda Elmore-Feliciano	Manager, Member Records, Membership Services Dept.
Susan Evans Grove	Publications Director
Tommie-Anne Faix	Publications Sales Associate
Erlinda Faustino	Administrative Assistant/Scholarships and Education Coordinator
Susan Giver	Director of Outreach and Strategic Development
Allan Gray	Associate Executive Director (Finance and Administration)
Carolyn Marus	Annual Meeting Manager
Anne Mikaelian	Accounts Payable/ Receivable
Rich Mouk	Technical Assistant
Edith Nowak	Manager of Sections and Student Sections
Nancy Ostin	Director of Membership & Marketing
Raymond Rivera	Mailroom Coordinator
Alan Rowen	Technical Director
Barbara Trentham	Assistant to the Executive Director

2007 Annual Meeting

The Society's 2007 SNAME Maritime Technology Conference & Exposition and Ship Production Symposium (SMTC&E/SPS) was held again for the second year in Ft. Lauderdale, convenient to all types and sizes of recreational water craft, power and sailing yachts, cruise ships and ocean-going commercial vessels. Our host hotel, the Hyatt Regency Pier 66, was just over the bridge and on the other side of the Intracoastal Waterway, but still very near the Greater Fort Lauderdale/Broward County Convention Center, where the technical program, the Cruise Ship and Shipbuilding Panels, numerous committee meetings and the Exposition were located.

The President's Reception

The Annual Meeting's pre-events began with the Executive Committee meeting on Tuesday night. The following day there were numerous standing committee meetings, three accredited technical courses and the Council meeting, all conducted prior to the opening gala, the President's Reception, located this year in the Expo Hall. Members roamed the exhibit hall to the delight of exhibitors while affording expanded socializing and networking opportunities with friends and colleagues. President Kramek both thanked Lloyd's Register for sponsoring the reception and announced that Lloyd's Educational Trust will be funding a new SNAME Student-of-the-Year Award for the 2008-09 academic years. In fact, it was hard not to run into a student this year, as they attended in record numbers (201) from colleges and universities located all over the Americas and Europe.

The President's Address and General Business Meeting

The first plenary session, which started promptly on Thursday morning, was opened by President Kramek with the delivery of his President's Address which emphasized the diversity of our profession and the significant opportunities for naval architects and marine engineers at this particular time. He then introduced the Cruise Ship Panel session and its moderator, VADM James C. Card, USCG (Ret.). Distinguished panelists included RDML James A. Watson, USCG, Director of Prevention Policy, who at the last minute replaced RADM Brian M. Salerno, USCG, Assistant Commandant for Marine Safety, Security and Stewardship; Harris Kulovaara, Executive Vice President, Royal Caribbean Cruise Lines; Captain Thomas E. (Ted) Thompson, USCG (Ret.); Alan Gavin, Marine Director, Lloyd's Register and Thomas Weigand, Director, Sales and Design Department, Meyer Werft Shipyard. Clearly, this industry has grown rapidly in the last three decades, including the tonnage of its ships, with RCCL about to deliver its 232,000 ton, mega cruise ship, GENESIS OF THE SEAS. Technical advances in materials, operational safety, environmental conservation and energy efficiency were documented by the panelists and validated by over 200 members in attendance.

The 50-year Golden Membership Certificates were awarded to those members in attendance, Donald L. Caldera, Joseph J. Cuneo, John H. Klose, Raymond Pearlson and James G. Price. President Kramek then introduced the Chairman of the Papers Committee, Dr. Xiaozhi (Christina) Wang, who presented the American Bureau of Shipping – Captain Joseph H. Linnard Prize for the best paper presented at the previous annual meeting to Hans G. Payer and Raul Brostella for their paper entitled, "The Panama Canal Expansion and the Panamax Vessels of the Future."

President Kramek then called on Dr. Edward M. Lewandowski to present the Vice Admiral E.L. Cochrane awards to Daniel Savitsky, Michael F. DeLorme and Raju Datla for their paper entitled, "Inclusion of Whisker Spray Drag in Performance Prediction Method for High-Speed Planing Hulls," published in the January 2007 issue of Marine Technology, Volume 44, Number 1.

President Kramek announced that the recipients of the Distinguished Service Awards, Barbara L. Lamb, Southwest Section, and Dawson Miller, Canadian Atlantic Section, were receiving the SNAME Distinguished Service Awards for their dedicated personal service and technical contribution to the Society.

Executive Director Kimball called on recipients of Certificates of Appreciation, Thomas P. Mackey, Matthew L. Unger and John H. Higginbotham, to approach the podium whereupon President Kramek presented the award to each in turn, and certificates (in absentia) to Thomas Lamb and Frank G. Bercha.

President Kramek called upon Jose Femenia, Chairman of the Nominating Committee, to present the Honorary Member Certificates to Thomas P. Mackey and to Michael G. Parsons. The Chairs of the ten T&R Technical Committees were then recognized for their individual contributions to the program over the past year.

The Society's Business Session was then called to order by President Kramek who asked Prof. Jose Femenia, Chair of the Nominating Committee, to announce the nomination of R. Keith Michel for President. Following seconding speeches by Joseph J. Cuneo, Peter G. Noble and Bruce S. Rosenblatt, and a vote by the membership, newly elected President-Elect Michel addressed the membership. The business Session concluded with the passage of numerous Bylaw changes, a report on elections naming William DuBarry Thomas as Honorary Vice President, and the naming of Roderick A. Barr, Norman D. Birrell, Dietmar R. Fales, Carlos Guedes Soares, Richard W. Thorpe and Johannes E. Wichers as Fellows of the Society.

The SNAME Expo

The SNAME Expo numbered 107 booths and afforded event-goers an opportunity to network, meet potential business partners or clients, and to examine new products and emerging technologies in the ship and boat design, construction, technology and production and operations segments of the industry. For the third consecutive year, the schedule for the Innovation Sessions, short informative presentations on emerging technologies in the marine industry, have been filled and well attended. These have proven to be popular for members, students and visitors and the exhibitors have found a way to reach an even larger targeted audience.

The Technical Program

The technical program included 13 papers on hydrodynamics, ship design, structure, propulsion and operations, and the ever-popular Ship Production Symposium with 13, production-oriented papers. Also included were a Cruise Ship panel and a Shipbuilding panel held as plenary sessions, two T&R presentations, three student papers and a student technical poster session. All technical papers and courses were accredited for those desirous of earning continuing education credits.

Three, day-long accredited technical courses were delivered simultaneously on Wednesday and presented in the hotel including "Seakeeping" by Dr. Edward M. Lewandowski, "Ship Arrangements" by H. David Kaysen, and "Sailing Yacht Performance Analysis" by Dr. Robert F. Beck and Dr. Kevin Maki.

Friday morning's Shipbuilding Panel and second plenary session, moderated by Bruce Rosenblatt, President, ROH Inc., was comprised of Sean Connaughton, Maritime Administrator, US Maritime Administration; Fred Harris, President, General Dynamics/NASSCO; Ron McAlear, President and COO, Alabama Shipyards and Spencer Schilling, President, Herbert Engineering. Although all agreed that commercial shipbuilding in the US may never approach the affordability of ships currently being built in South Korean shipyards, the US yards can significantly close the gap by completing the design definition prior to detail design and by following the design/build concept. We can further close the gap through proven designs, design for production, modular construction, standardized parts, economies of scale, building in series as an "assembler," and with US government assistance and the strategic deployment of capital.

The SNAME Annual Banquet

Following a lively pre-Banquet Reception on Thursday evening, the Crystal Ballroom was the venue for the 2007 SNAME Banquet. This year, in a departure from tradition, Susan Giver, SNAME's Director of Outreach and Strategic Development, joined past president, Dr. Roger Compton and his wife, Jill, along with 9 of 28 members of the Webb Family Singers, the WOOFS, in the singing of the American and Canadian National Anthems much to the delight of an overflow audience. President Kramek followed by thanking the Banquet's sponsor, ABS, and acknowledging the Southeast Section Chairman, Lee Dana, for his support of the 2007 SMTC&E/SPS. Rick Spaulding, Chair of the Ship Production Committee, was called upon to present the 2007 Elmer L. Hann Award to Tyau-da Huang, William Patrick Keene (in absentia), Lee G. Kvidahl, Christopher Conrardy, Pingsha Dong and Lawrence Decan. He continued with the presentation of the 2007 William M. Kennedy Award to Jay P. Carson.

Following dinner, President Kramek called upon Ronald K. Kiss to present the William H. Webb Medal to Dr. Roger H. Compton, F. Tim Pease to present the Blakeley Smith Medal to Howard L. Shatto, R. Keith Michel to present the "Vice Admiral 'Jerry' Land Medal to Bruce S. Rosenblatt. President Kramek then presented the David W. Taylor Medal to Hans G. Payer.

President Kramek then introduced our principal speaker for the evening, Vice Admiral Paul E. Sullivan, USN, Commander, Naval Sea Systems Command, who delivered the Annual Banquet address. VADM Sullivan described the current capabilities of the Navy's 278- vessel fleet, and his challenge to build our country's fighting naval capability to 313 in the next decade. He specifically addressed the students in the audience, and made it very clear to them that ample opportunities abound today for naval architects and marine engineers interested in a career in naval ship design and construction.

The Student Program

On Thursday morning over 120 students assembled to participate in the design/build portion of the Student Congress, and to attend the Student Steering Committee meeting and elections for 2008/2009. Immediately following the SSC was the award ceremony to announce the prizewinners of the various SNAME-sponsored student design competitions with President Kramek and Executive Director Kimball presiding.

Matthew Werner, Chair of the SNAME Education Committee, awarded the 2007 Graduate Paper Honor Prize to James P. Morash (in absentia) for his paper, "A Modular High Voltage Lithium-Ion Battery for Unmanned Marine Vehicles" presented to the New England Section, February 8, 2007. Gregory R. Mitchell was awarded the Graduate Paper Award (in absentia) for his paper, "A Concept Design Tool for Small Planing Craft" presented to the New England Section, February 8, 2007.

The 2007 Undergraduate Paper Honor Prize was presented to Peter Bryn for his paper, "A Systematic Approach for Modifying the Stern of the 0.65 Block Series 60 Hull," presented to the New York Metropolitan Section, June 2007. And lastly, the 2007 Undergraduate Paper Award was presented to James Sannino, Evan Karlik, Christina Gomez, Seth E. Clark, Nina Young and Nathaniel Chan for their paper, "RoBoat: Design and Field Testing of an Autonomous Surface Vessel and Acoustic Tracking System" presented to the New England Section, Fall 2006.

President Kramek and David Stephenson, Technical Director for ASNE, made the presentation of the First Place prize for the 2006 Dr. James A. Lisnyk Ship Design Competition to the student team from The Norwegian University of Science and Technology for its winning entry, "Rapid Response Vessel." A team from Virginia Polytechnic Institute and State University received the Second Prize for its winning design, "Large Ocean Interface Submarine SSLOI." Third Prize was awarded to the team from University of Newcastle upon Tyne for its winning entry, "Double Acting Container Carrier."

The 2007 International Student Offshore Design Competition (ISODC) First Place prize was awarded by Dr. Jon Mikkelsen, Senior Instructor/Capstone Design Coordinator, Department of Mechanical Engineering, University of British Columbia, to a student team from Texas A&M for its winning design, "Conceptual Design of a Semi-Submersible Floating Oil and Gas Production System for Offshore Malaysia." Dr. Mikkelsen announced the awards for Second, Third, Fourth and Fifth Place winning designs to Texas A&M Team 1 for its winning entry, "Design of a Semi-Submersible Production and Drilling Facility for the Gulf of Mexico," to the Federal University for it winning entry, "Preliminary Design of a Dry Completion System for a Mono-Column Platform," to Texas A&M Team 2 for its winning entry, "Design of a Semi-Submersible Floating Production System for the Brazil Offshore Environment," and to Texas A&M Team 4 for its winning entry, "Design of a Semi-Submersible Floating Production System for Offshore West Africa," respectively.

Dr. Alan Brown, NAVSEA Professor of Ship Design at Virginia Tech, presented the 2007 International Student Workboat/Small Craft/Yacht Design Competition Awards. First Prize was awarded to Webb Institute for its winning design, "Modular Express Cruiser." Second Prize was awarded to a team from Webb Institute for its winning entry, "Hybrid Solar Electric Sailboat," and Third Prize was awarded to a team from the University of British Columbia for its winning design, "Fast Oil Spill Response Vessel."

The Student Congress Competition

A highlight for the registrants and visitors in attendance was the continuing focus on SNAME Students who participated again in their own "icebreaker" social, the popular Job Fair, and the Student Steering Committee meeting to elect new officers for 2008. On Thursday, over 120 students took part in the Student Congress, the Awards presentations, and the "design and build" competition created by members of the Student Steering Committee. This year's competition, again held in the Expo Hall, called for the construction of a hull made from basic materials such as aluminum foil, wire coat hangers, Popsicle sticks and packaging tape. The finished designs were put to the test in a pool of water by being filled with soda cans. Each hull design was strategically loaded by students to maintain its stability until it submerged and sank from the weight. The winning design held an incredible 185 soda cans, about 100 pounds, much to the surprise and delight of the large and attentive audience.

EXECUTIVE DIRECTOR'S REPORT TO COUNCIL, 2007

I am pleased to submit the following report for the fiscal year ending August 31, 2007.

MEMBERSHIP

The Society's total membership increased by 1.5 % during fiscal year 2007. A comparison of the membership of the Society as of August 31, 2007, and August 31, 2006, is as follows:

	FY 2007	FY 2006
Members (under age 32)	38	29
Members (aged 32 and older)	3,614	3,655
Affiliates (under age 32)	8	8
Affiliates (aged 32 and older)	65	72
Associate Members (under age 32)	576	501
Associate Members (aged 32 and older)	2,651	2,528
Reciprocal Members	4	4
Student Members	923	979
Honorary Members	9	10
Honorary Fellows	20	24
Fellows	186	175
International Electronic Members	38	29
Total	8,132	8,014

With last year's membership increase of 5.1% this year's 1.5% increase continues the positive change, reversing the trend of the prior seven years. There was a dues increase of $25 in 2007, the first increase in eight years. We had predicted a 4% decrease in overall membership because of the dues increase; therefore the result was actually significantly better than expected. The decreases in membership can be seen in the full dues categories, and dues were not increased in the under 32 year categories primarily to continue our focus on young member retention. Overall, membership dues income increased to $891,636 in 2007 from $766,145 in 2006.

Membership Retention

Membership retention for fiscal year 2007 stands at 87% (94% for members, 37% for students). This, too, exceeds similar society retention trends, compared to 88% in 2006. Of the 8,014 members at the beginning of fiscal year 2007, 886 to-date, have not renewed their SNAME membership.

A postcard was developed and mailed to all domestic members reminding them of the early pay contest "Renew in November and win an iPod." Membership renewals ran ahead of budget with members paying earlier than in prior years. In conjunction with our Sections, a series of reminder letters, e-mail blasts and contacts from Section volunteers assisted in the timely renewal of memberships.

The Staff Telethon of 2007 realized a 49% success rate, resulting in over $68,000 in dues collected. For comparison, 2006 had a 66% success rate and over $106,000 collected. An important note: this year there were 331 fewer members to call in the staff telethon. Factors that influenced these results were: membership renewals mailed earlier than in the prior year, an additional mailing of a reminder postcard for membership renewal, a coordinated effort with Sections to contact and renew members along with an assertive blast email campaign. Those left to be called were the more difficult to reach (information no longer valid) and less likely to renew.

Particular emphasis was placed on the retention of Student Members to Associate Members post graduation. An electronic reminder was sent to over 300 graduating seniors with a promotional rate and graduation gift offered. We asked student section leaders and faculty advisors to provide us with after-school addresses. Student retention decreased, but this was only due the re-categorizing of graduating students to students in transition (296 in '07). This will allow us to better count and follow these first year associate members and track their retention.

Recruitment

An electronic recruitment campaign was run in collaboration with Maritime Reporter with over 40 new members recruited through this effort. A planned membership campaign using the OTC list did not happen because the list could not be made available to us. A campaign in 2008 is being pursued in conjunction with the SNAME OTC Conference Steering Committee.

Under the International Activities Committee the European Office opened in Greece. Plans to open an office in Shanghai were unsuccessful. The goal for these offices was to recruit 100 new members each and was not achieved. In the spring of 2007 Spyros Volonakis joined SNAME as the manager of the European Regional Office.

As noted above, overall membership grew by 118 members over the prior year.

Applications

The Applications Committee, chaired by Neil Miller, reviewed 918 applications for new membership and upgrade requests during fiscal year 2007 (including 590 new Student Members). By comparison, applications reviewed in fiscal year 2005 were 855 (including 531 Student Members).

In response to challenges of convening the Applications Committee in a bi-monthly meeting at headquarters, an online "virtual" committee system (MARS – Membership Application Review System) was developed. The MARS system allows members of the committee to log into a secure site, review applications, submit comments or approvals and allow for a second committee member to agree or ask further questions. This system allows the Membership Department to process and approve members on a more timely basis, generally in a week or two. New Member certificates and kits are sent to new members on a monthly basis, rather than the two-month delay experienced previously. This better serves the committees' needs, allowing members who are no longer in the HQ local area to continue to serve as volunteers, and assist the Society in this important area on their schedule. In addition, it improves service to our new members and welcomes them into our Society on a more timely basis.

Member Benefits

The 'Member Benefits' pages on the SNAME website were redesigned and updated. Membership has a whole new look and there is an interactive member benefits chart that connects to the detailed information regarding SNAME member and discount programs.

In addition to discounted rates for publications, conferences, and continuing education, SNAME currently offers the following benefits to members: Life Insurance; High-Limit Accident Insurance; health products including Catastrophic Major Medical, Dental, Medicare Supplement, Cancer Protection, and Long-Term Care Insurance; Car and Equipment Leasing; discounts from UPS, Avis Car rental, and Dell Computers; and the personalized SNAME MBNA credit card.

An improved Career Center was designed and launched with our partner, Job Target, in August of 2006, generating $16,919 in royalty revenue. In addition, this enhanced offering provides a valuable service to members and non-members. Over 140 candidates have posted resumes on the site and there are on average 20-25 job opportunities each month.

New Member Benefits negotiated during the year include a partnership with AON Affinity Partners to offer Professional Liability Insurance. This was one of the most requested items under the benefits area in our 2006 Member Value Survey. This unique program will offer a comprehensive insurance program at a discount to SNAME members.

International Electronic Membership

As a pilot program under the International Activities Committee, this program entered its second year. Membership in this category rose from 29 to 38 members. Efforts to promote this program have been limited and the committee will review and recommend on the status for the future.

An inventory of publications has been shipped and is onsite and available for sale in the European Regional office in Greece. Member prospect lists have been established and a recruitment campaign is ongoing. Future plans involve working with the International Activities Committee on web content to advertise the ERO as well as promote SNAME's International programs.

Member Communications

As a follow up to the Member Value survey, all **SNAM***eNewsletters* of the past 12 months included an article about membership and member benefits. The Monthly SNAME Opinion Survey (SOS) produced limited results, but did give us an opportunity to educate members on different SNAME offerings, programs and issues. A webpage was developed to showcase the results.

Section Update

Significant changes were made to the Chairman Elect Seminar designed to engage and educate the Section leadership including the introduction of the SNAME VOYAGE and the Best Ideas Dinner. Overall, our Regional Vice Presidents are working closely with our Sections and HQ staff to assist Sections in membership. Super Section criteria were developed and were distributed. Sections were more involved in membership renewals, goal setting, planning. Attendance in Membership/Section teleconferences was strong and growing.

Member Contact Update

A blast e-mail was designed and sent to all members in the spring and summer of 2007, asking for them to update their contact information. The ease of updating through this technology resulted in nearly 2,000 updates to member records each time. Much work has been done to verify and correct member records and data. The online system by which members can change their contact information was re-engineered and now matches our HQ database and provides an easier updating to the master HQ database.

NECROLOGY

James L. Albright
Capt. Merle W. Allen
Theodore C. Ballew
CAPT Theodore Banvard
Robert J. Bosnak
Ralph R. Brown
Joseph D. Bryan
Roger Alan Burger
Wilson Lyall Creighton
Capt. Cecil Douglas Davies
Robert P. Giblon
John T. Gilbride
Jon R. Gribskov
Capt. William E. Heronemus
C. William Lapworth
Ed MacCutcheon
James W. Mavor Jr.
William R. Mayhew
William L. McCarthy
Capt. Wm B. Mitchell USNR
Edward K. Moll
Waldemar S. Nelson
Kris Ramakrishna
Robert H. Riddervold
CAPT Virgil Rinehart
Capt. David R. Smith
Fred George Speidel
Edward V. Stewart Jr.
R. Tamboowalla
Robert J. Tapscott
Larry P. Toups
Dr. Jan Dirk VanManen
Russell J. Ware
CAPT SV Wroblewski

SNAME'S FIVE-YEAR STRATEGIC PLAN

The Planning Committee, under the leadership of Bill Hurley, continued to implement the Society's Strategic Plan by facilitating the initiatives for changing the future course of the annual meeting and developing a new web-based association management system, both of which were results of the ad hoc committees to study Knowledge Management and the Annual Meeting and other Symposia in 2006.

Concurrently, throughout the year the committee worked diligently to review the entire plan for years 2007 through 2011. The committee members debated the benefits and liabilities of proposed changes with the common goal of completing the Strategic Plan's update before year end. The main focus of the monthly meetings was to review and update elements of the plan and make final edits, a process that was completed in June and overwhelmingly approved by the Executive Committee at its meeting in July.

The Strategic Plan, originally issued in July 2005, is our guide for meeting the mission and goals for the Society over the next five years. It is a living document, subject to change, revision and evaluation for effectiveness, and will continue to be reviewed on a regular basis.

SNAME/ASNE STRATEGIC ALLIANCE

The leadership and senior staff of both societies agreed to establish an Oversight Committee consisting of their respective presidents, executive directors, finance directors and volunteer finance committee chairs to discuss the business aspects of the strategic alliance. This committee met in January of 2007, and again in June, with the intent of meeting twice a year going forward.

The permanent SNAME/ASNE Joint Strategic Alliance Committee continued its meetings throughout the year and met jointly with the Oversight Committee in June. The strategic alliance committee, chaired by Paul Cojeen, continued to promote the concepts of joint Education and Ship Design Committees and the joint Student Mentoring Program. The Oversight Committee introduced the concepts of a joint Ship Design Conference for 2009, a joint electronic daily marine news service and further agreement to provide the digital versions of our respective flagship journals to students rather than mailing hard copies. The savings in publication and postage could result in a reduction of future student dues.

INTERNATIONAL ACTIVITIES COMMITTEE

The International Activities Committee, chaired by Peter Noble, planned and hosted two Collegia, one in Brazil and one in Norway, both on October 8, 2007. The Brazil event, held jointly by SNAME and SOBENA (Sociedade Brasileira de Enginharia Naval) featured presentations and panel discussions on topics including activities related to natural gas in Brazil, advances in the use of loading computers for offshore product units, CNG as a solution for offshore gas utilization, LNG transportation developments and design and construction and ballast water issues. The event was organized by Prof. Sergio Hamilton Sphaier, Department of Ocean Engineering, COPPE - UFRJ, Federal University of Rio de Janeiro, Rio de Janeiro, RJ Brazil and will be hosted by Petrobras.

The Norway event, hosted by Det Norske Veritas, featured discussions and presentations on a particular area of interest: Naval Architecture for Harsh Environments with presentations on classification concerns, energy transport shipping, oil and gas exploration, and deepwater development innovations, all with a harsh environment focus.

The format of the collegia includes a focused technical seminar on a topic of interest to the particular region, but in addition, provides networking opportunities and the ability to exchange ideas on the future of the profession and on the professional societies. Perhaps the most important benefit will be the chance to discuss topics of interest with like-minded professionals and the opportunity to network in a stimulating and exciting environment.

The response was strong both in attendance and in support from individual members and the maritime community. Additional Collegia are being planned for the UK, China and Arabian Gulf. Discussions are underway with the Texas Section to establish an affiliate section relationship with members in the Arabian Gulf.

2006 SNAME MARITIME TECHNOLOGY CONFERENCE & EXPO AND SHIP PRODUCTION SYMPOSIUM

The Society's 2006 SNAME Maritime Technology Conference & Exposition and Ship Production Symposium (SMTC&E/SPS) was held in Ft. Lauderdale, convenient to all types and sizes of recreational water craft, power and sailing yachts, cruise ships and ocean-going commercial vessels. There was something for everyone, and this was abundantly clear when viewing from the windows of our host hotel, the Ft. Lauderdale Grande Hotel and Yacht Club. It was just across the street from the Broward County Convention Center where the technical program, committee meetings and the Exposition were located. This was our first of two years in Ft. Lauderdale as a trial to attempt to ease Headquarters' planning burden and to "brand" the location and time of year, thus improving industry awareness and potential for attendance.

The President's Reception

The Annual Meeting began with the Executive Committee meeting Tuesday night. The following morning there were numerous standing committee meetings, two accredited technical courses and the Council meeting, all conducted prior to the President's Reception, located that evening outside under the stars overlooking the ever-changing boat traffic navigating the Intra-coastal Waterway. There was ample opportunity for socializing and networking with fellow members, friends, colleagues, students and young members at this event, and throughout the week at the Expo, the SNAME Banquet and the alumni dinners. In fact, it was hard not to run into a student, as they attended in record numbers (153) this year from colleges and universities all over North and South America and Europe.

The Keynote Session

The Keynote Session, which started promptly on Thursday morning, was opened by President Roger H. Compton who introduced Roger J. Marshall, our Keynote Speaker and President of Roger Marshall, Inc. Mr. Marshall provided us with a fascinating account of the development of "super" and "mega" yachts and their place in the world today from design to materials to construction supplemented by slides of yachts in various stages of construction and operation.

The 50-year Golden Membership Certificates were then awarded to those members in attendance, Naresh M. Maniar and James R. Wilkins, Jr., followed by the Distinguished Service Awards presented to Dr. Richard C. Boutwell and Michael T. Martin. President Compton then introduced the Chairman of the Papers Committee, Dr. Christina Wang, who presented the American Bureau of Shipping – Captain Joseph H. Linnard Prize for the best paper presented at the previous annual meeting. Award certificates were presented to Peter G. Noble, Lars Ronning, John R. Paulling, III, Rong Zhao, and Hoseong Lee for their paper entitled, "A Novel LNG Tank Containment Design for Large LNG Carriers."

President Compton then called on Jaideep Sirkar, Functional Vice President for Technology and Chairman of the Technical & Research Committee, to present the Admiral E. L. Cochrane Award for the best paper published in Marine Technology to Wilhelm Magelssen for his paper "Operation of Ships in Cold Climates with Emphasis on Tankers and New Requirements." Mr. Sirkar then presented T&R Certificates of Recognition to Thomas Lamb and Charles H. Piersall for their outstanding service in the Society's Technical and Research Program. Chairman Sirkar also recognized the Chairs of the ten T&R Technical Committees for their individual contributions to the program over the past year.
President Compton announced the names of the newly elected Fellows of the Society in 2006: Cengiz R. Ertekin, Alexander C. Landsburg, Apostolos Papanikolaou, David R. Pedrick, Dracos Vassalos and

Preben T. Pedersen.

The session was concluded with Past President Jose Femenia, Chairman of the Nominating Committee, stepping forward to present the Honorary Member Certificate to President Compton. Immediately following, Dr. Compton delivered the President's Annual Address citing the progress made over the last year, and the new and exciting initiatives to modernize and expand the Society's website, the annual meeting and the T&R Program to provide enhanced relevance to our expanding membership both domestically and internationally. Upon completion of the Society's Annual Business Meeting the Keynote Session concluded and all delegates were escorted to the first floor for the grand opening of the SNAME Expo by a spirited and lively mariachi band.

The SNAME Expo

The SNAME Expo numbered 147 booths and afforded event-goers an opportunity to network, meet potential partners or clients and examine new products and emerging technology in the boat construction, shipbuilding, design, technology and production segments of the industry. For the second year in a row the schedule for the Innovation Sessions, short informative sessions on emerging technology in the marine industry, has been filled and the sessions have been well attended. These have proven to be popular for members, students and visitors and the exhibitors have found a way to reach an even larger targeted audience.

Results	2006	2005	2004	2003
Number of Booths	152	185	163	153
Number of Barters	2	6	6	11
Attendees	510	857	745	1,259

The Technical Program

The technical program was developed with small craft and yachts in mind, and included the popular Ship Production Symposium, technical sessions incorporating energy transportation, environmental effects, high-speed craft and ferries, LNG ship design and technology, and small craft development. Specific events drawing considerable attention featured a Super Yacht panel, a Small-Craft Technical Session, a panel on International Ship and Port Security (ISPS) in Southeast Florida, the Alternative Energy Session and a presentation on the planned expansion of the Panama Canal.

Three, day-long accredited technical courses were prepared for Wednesday and Friday: "A Review of Basic Naval Architecture" by Bill Garzke, "Introduction to the Design of Floating Offshore Oil and Gas Facilities" by Dr. Robert Randall, and "Seakeeping" by Dr. Edward Lewandowski [cancelled on Friday due to a medical emergency].

The SNAME Annual Banquet

Following a lively pre-banquet reception on Thursday night, the Grand Ballroom was the site for the 2006 SNAME Banquet. Again this year, Susan Giver, SNAME's Director of Marketing and Communications, ably led us in the singing of the American and Canadian National Anthems. Dr. Roger Compton followed by thanking the Banquet's sponsors and acknowledging the Southeast Section Chairman, Lee Dana, for his support of the 2006 SMTC&E/SPS. President Compton called on Functional V.P., Jaideep Sirkar, to present the 2006 Elmer L. Hann Award to Robert G. Keane, Jr., Howard Fireman and Daniel W. Billingsley. He continued with the presentation of the 2006 William M. Kennedy Award to Tim Colton.

Following dinner, Dr. Compton called on Dr. Alaa Mansour to present the Davidson Medal for outstanding scientific accomplishment in ship research to Dr. Preben Terndrup Pedersen. R. Keith Michel then presented the Vice Admiral "Jerry" Land Medal for outstanding accomplishment in the marine field to Mr. Peter G. Noble. Lastly, President Compton presented the David W. Taylor Medal for notable achievement in naval architecture and/or marine engineering to Mr. Edward N. Comstock.

President Compton then introduced our principal speaker for the evening, Mr. Halsey C. Herreshoff, President of Herreshoff Designs. Mr. Herreshoff treated the filled-to-capacity banquet hall guests with a slide-supported anecdotal account of notable events in the history of the America's Cup as well as the historical significance of his illustrious family, the Herreshoff's of Bristol, RI, early yacht designers and boat builders. His entertaining comments highlighted the brilliant career of his grandfather, Capt. Nathanael G. Herreshoff, one of this country's foremost and innovative power and sailing yacht designers and builders from the last two centuries.

The Student Program

On Thursday during the Student Congress in the Expo Hall, a number of awards were made prior to the judging of the Student Design Competition. Gathered for the award presentations were members, students and numerous observers.

First, the 2006 Graduate Paper Honor Prize was awarded to Robert K. M. Seah for his paper, "Resonant Modes of Rectangular Twin Hulls," presented to the Northern California Section on February 15, 2006. The Graduate Paper Award was presented to Matthew L. Unger for his paper, "An Upgrade of the MIT Towing Tank," presented to the New England Section on February 9, 2006.

The 2006 Undergraduate Paper Honor Prize was presented to Michael J. Stanway, Peteris Augenbergs, Heather Brundage, Lauren Cooney, Bridget Downey, Timothy Pennington, Taaddeus Stefanov-Wagner and David Tobias for their paper, "Design of an Acoustic Autonomous Surface Vessel," presented to the New England Section on February 9, 2006. And lastly, the 2006 Undergraduate Paper Award was presented to Seth R. Kruger for his paper, "Developing a Safe and Efficient Surface Maneuvering Envelope for Submarines," presented to the Chesapeake Section on April 30, 2005.

First and third prizes for the 2006 Dr. James A. Lisnyk Ship Design Competition went to student teams from the University of Newcastle upon Tyne for their winning entries, "Design of a Trimaran Liquid Hydrogen Tanker" and "SeQuest ERV" (Liquified CO_2 Tanker), respectively. Second prize was awarded to a team from the Norwegian University of Science and Technology for its entry, "Angier Cruises (Cruise Ship Design)." The 2006 International Student Offshore Design Competition (ISODC), first prize was awarded by Tom Johnson, co-chairman of the design competition, to a student team from the University of Michigan for it winning entry, "Singapura FONGPSO." Mr. Johnson announced the second through fifth place awards to teams from Texas A&M University with entries, "Design of a Floating Production, Storage and Offloading Vessel for Operation in the South China Sea," "Design of a Floating Production, Storage and Offloading Vessel for Offshore Indonesia," "Design of a Floating Production, Storage and Offloading Vessel for Offshore Qatar" and "Design of a Floating Production, Storage and Offloading System Design for Offshore Nigeria," respectively.

Lastly, first and second prizes were awarded for the 2006 International Workboat/Small Craft/Yacht Design Competition to student teams from the University of Michigan for their winning entries, "220' Trimaran" and "Multi-Mission Patrol Boat," respectively.

Third prize was awarded to a team from SUNY Maritime College for its winning entry, "LNG Assist/Escort/Emergency Response Tug."

A highlight for the registrants and visitors in attendance was the focus on students who participated in a first time "icebreaker" social event, the popular Job Fair, and the Student Steering Committee meeting to elect new officers for 2007. Over 130 students took part in the Student Congress and Awards Lunch, held on Thursday in the Expo Hall, with a "design and build" competition created by young members from the Student Steering Committee, which called for the construction of a "Transpacific" sailboat by each of the nine teams in a little over an hour. The 12" model Trans-Pac sailboats, built from the simple materials provided, were tested for stability and righting moment, as well as their ability to sail downwind both in a straight line and on a broad reach in the "official" test pool. A panel of four judges, made up of industry professionals, scored the teams based on performance, design, presentation and teamwork. When all of the scores were tallied, Team Ram Rod, dominated with an outstanding point total of 84.3 out of 100, earning the first prize consisting of $500 in cash divided among the team members.

Friday was designated exclusively for the Student Technical Paper presentation track, and, new this year, the Student Technical Poster Session, which was well-received and impressed members and visitors alike.

The Student Summit

The Student Summit, in its second year, was held on Saturday, with over 80 members and students in attendance to discuss changes to benefit student participation in SNAME and new ideas to promote more student interest for future conferences. One idea, suggested by the students called for their own student forum on the SNAME website. Within two weeks following the SMTC&E, the forum was "up and running" for students and members alike to begin using.

Comparative statistical data of note for the Annual Meeting:

	2006	2005	2004
Total Registrations	482	591	631
Banquet Attendance	320	438*	260
President's Reception	380	450	420

*First time Student registration included attendance at the Banquet.

MARKETING

The strategic partnership with The Maritime Group afforded SNAME the opportunity to reach over 60,000 marine professionals per month with advertising featuring the new SNAME Career Center, SNAME membership, and the SNAME Maritime Technology Conference & Expo and Ship Production Symposium. Editorial content featuring incoming SNAME President, ADM Bob Kramek, and a monthly SNAME column in Maritime Reporter was created. SNAME exhibited at five trade shows throughout the year, marketing memberships, publications, events, and generating awareness of programs.

OUTREACH

Based upon SNAME's commitment to K-12 student outreach, SNAME has been awarded a grant by the Office of Naval Research as part of its National Naval Responsibility Initiative. The initiative focuses on bringing academia, government and industry to work together to ensure that the talent needed to design the Navy's next generation of ships and submarines will be there when needed, and it utilizes Sea Perch, an underwater ROV, to achieve its goals. SNAME has been tasked with expanding the SEAPerch Program into a national program and has designated one staff person to manage this effort.

FINANCES

The Fiscal Year 2007 operating fund financials show that total revenue at the conclusion of the year was $2,830,768, a $41,373 increase from Fiscal Year 2006. Dues revenue increased $125,491 from FY2006 as a result of the 2007 dues increase while functions income decreased $108,565 due to lower attendance at the Annual Meeting and Exposition in Ft. Lauderdale.

Actual expenses in Fiscal 2007 were $2,964,687, which was $37,402 lower than in the previous fiscal year. Overall, the Fiscal Year 2007 Operating Fund completed the year with an excess of $115,996 compared to $4,337 in FY2006.

The budgetary process is monitored by the Finance Committee chaired by Martin Toyen. A breakeven budget was approved for Fiscal Year 2008 by the Finance Committee.

A statement of the Society's assets, liabilities, and fund balances is contained in the Report of the Treasurer.

The Audit Committee report for Fiscal Year 2007 will be submitted to Council not later than January 31, 2008, and the independent auditor's full detailed report will be available for inspection at Society Headquarters.

Finally, the performance of the Society's investments continues to be monitored by Chairman Keith Michel and the Investments Committee. The value of the portfolio increased from $9,040,216 on August 31, 2006, to $9,996,933 on August 31, 2007.

SECTIONS

2007 Chair Elect Seminar

The Society held its 17th Chair/Vice Chair-Elect Seminar on May 18 and 19. Incoming Student Section Officers also joined us for their second annual Student Chair-elect Seminar.

The Seminar began with a "Best Ideas" dinner on Friday night. Incoming Section and Student Section Chair's/Vice Chair's were asked to provide their best idea or the most effective idea that has worked for their Section to headquarters prior to the event. The "best ideas" were presented at the dinner on Friday night, and each participating officer took home the pre-compiled list of "best ideas" as a resource.

The seminar continued on Saturday with presentations, afternoon breakout sessions, and discussions. This year was a first for the "SNAME Voyage," a scavenger hunt format that took place at the seminar. Section and Student Officers were divided into their appropriate Regions, led by their Regional Vice Presidents. Dr. John Mercier (Black Pearl) led the Central and Gulf Region, Mr. Glenn Ashe (The Rising Tide) headed the Atlantic South Region, Mr. Dan Walker (SS Blowdown) worked with the Atlantic North Region, and Peter Noble (N/V Ambiguous) stood in for the Pacific Region. Each Region was sent on a voyage to develop a plan for each of the Sections, and to find answers to questions by visiting a number of "ports" representing all of the various areas of SNAME. All participants agreed that it was very useful to have the Regional Vice Presidents involved as well as to learn more about the Society by having to find the answers to questions in a friendly competition with the other regions.

Sections Tool Box

A Sections Tool Box was created to serve as a reference tool for Section Officers to answer frequently asked questions (FAQs) at the click of a button.

SNAME Mentor Program Update

In 2006, SNAME launched its trial of the SNAME Student Mentor Program, providing an opportunity for SNAME Student Members to have one-on-one interaction with a SNAME member industry contact providing academic and professional coaching in preparation for their career.

In 2007, SNAME continued the program by assigning a mentor to students members in their Junior year. Assignments were made based on the student's career goal and mentor's experience in the industry. Monthly topics were sent to facilitate communication and much positive feedback was received. SNAME will continue the program in 2008 by matching students from the class of 2009 with mentors. Monthly topics will continue and assignments will take place on a first come first serve basis, where mentors will be provided with a list of student members with their career goals, and will choose the student they feel best fit to mentor.

SUPER SECTIONS:

We will be introducing "Super Sections" for the fiscal year of 2007-2008. At the end of the year, the Section that meets the most criteria will win a prize along with the title of "Super Section" for that year. The criteria for which that each Section will receive points are:
1. Meeting Schedules submitted by the deadline (September 1st),
2. Audit submitted by the deadline (September 30th),
3. Number of Section Meeting Reports published in MT/Newsletter,
4. Average number of days the meeting notice was sent in advance to HQ/Members,
5. Number of Council Meetings the Section Chair attended,
6. If the Section has a Membership Chair and
7. If the Section is represented at the Chair/Vice Chair-Elect Seminar.

Student Sections

A new Student Section at the University of Florida was approved by the Chair of the Sections Committee on August 21. The Student funding formula is being reviewed for possible revision.

PUBLICATIONS

The position of chair of the Publications Committee was vacated during the year and a search is underway for a new chair. The last meeting of the Committee was held at the SMTC&E in Fort Lauderdale, Florida. During a meeting of the PNA Committee at the SMTC&E, there was a great deal of discussion regarding the status various chapters and their schedule for completion, and the need for an introductory chapter.

Journals

The Journal of Ship Research continues to benefit from a healthy backlog. We have been publishing extra pages per issue to try to have papers published in a timely fashion.

Marine Technology's SNAME News continued to publish one general interest article or paper and more color was added in the SNAME News portion for each issue. An ad-hoc committee continued to analyze ways to improve this publication and make it more valuable to the membership.

Journal of Ship Production has made great strides in having been published on a more-timely basis. The journal continues to be primarily papers provided by the Ship Production Symposium, but the number of overseas submissions outside of that symposium seems to be increasing.

Publications Director Susan Evans Grove and Publication Sales Associate Tommie-Anne Faix began the development of a marketing plan to increase subscriptions to all the journals, particularly in the overseas market.

Reference Publications

The revisions to Principles of Naval Architecture and Ship Structural Design continue to move slowly forward. The role of the Control Committee for Principles of Naval Architecture has been revitalized in hopes of moving each individual chapter along more quickly. Roger Compton and Randy Paulling began work on a Chapter 0 that will be published as a forward to each individual chapter. The second completed chapter of Ship Structural Design, authored by Dr. Domique Beghin, was released in August as Technical and Research Bulletin 2-32 and is available on CD. These chapters will soon be sold as individual supplements to the book. Due to chapter re-numbering the revised chapters could not be inserted into the original book.

<u>Stability and the Safety of Ships</u>

This book authored by SNAME member, Vadim Belenky, was previously published by Elsevier. Mr. Belenky subsequently offered the copyright to SNAME, and with aggressive marketing, we are optimistic about sales for FY 2008.

Ship Production
The copyright for this book, previously published by Cornell Maritime Press, has been obtained by the authors and the book was re-released by SNAME with a new cover.

Naval Architecture for Non-Naval Architects
This book was reprinted with updates made by the author Harry Benford. A new cover has also increased interest in this classic text.

Transactions
The 2006 Transactions were mailed in June. Options for hardcover, soft cover and CD ROM continue to be available.

Publications Sales

The Publication team promoted SNAME publications through postcard and catalog mailings to targeted groups during the year, and a marketing survey was developed for authors of new books in order to direct more targeted marketing. While direct mail was the focus this fiscal year, the team plans to expand to advertisements in relevant trade publications early in the next fiscal year.

TECHNICAL AND RESEARCH

Twenty-three funded T&R projects were in progress or pending approval. Four other proposals are known to be under development by the panels. One T&R bulletin and one updated and supplemented report were released. An updated edition of another bulletin and one new bulletin will be released shortly.

The Joint ASNE/SNAME Environmental Engineering Committee began to organize itself with nine panels, activated its web page, and developed proposals for funded projects. The Joint ASNE/SNAME Ship Design Committee organized itself with nine panels and met three times during 2007. A proposal to establish on a trial basis a Ship Design Committee Collaborative web site was submitted for approval. A proposal for an electronic journal is being developed. The Ship Repair and Conversion Committee was disbanded. Its mission will be absorbed by the Ship Technical Operations Committee. Four Ad Hoc Panels transitioned into the standing T&R Panel structure or disbanded after completing their tasks. One more panel will follow shortly, leaving three active Ad Hoc Panels.

All SNAME Annual Meeting papers are being added to the database with abstracts and pdf copies. The job will be completed by or shortly after the end of 2007. This work is being done by interns at SNAME HQ.

The Technical Program at SMTC&E/SPS 2007 in Fort Lauderdale will include, in addition to two plenary-session panel discussions, thirteen peer-reviewed SNAME papers and eleven to thirteen SPS papers and presentations. In addition, three hours have been set aside for T&R Sessions, and two hours have been reserved for student papers. All of the SNAME papers and at least ten of the SPS papers and presentations will be approved for PE Continuing Education Credit.

SNAME T&R staff absorbed the role of secretariat for the Ship Production Symposium Papers Committee. Twenty papers or presentations were processed for SMTC&E/SPS 2007.

For the third time, the Expo at SMTC&E/SPS 2007 will include Innovation Sessions. This very popular feature has been expanded to fourteen 30-minute slots.

SNAME will soon complete its third year as a Sponsor of PE Continuing Education Courses. Section papers and presentations are routinely evaluated and approved for meetings of NY Metropolitan and Philadelphia Sections. Other Sections and Regions have been invited to submit papers and presentations for evaluation, and several have done so. In addition to Annual Meeting Papers and SPS papers and presentations, papers at other conferences have been accredited. To date, 16 classroom courses have been approved for PE Continuing Education Credit, some for the second or third time.

SNAME is now involved regularly as a sponsor and/or an active participant in ICMES, MEETS, INEC, and IMDC.

The Indian edition of the SNAME textbook Introduction to Practical Marine Engineering, printed under license by the Institute of Marine Engineers (India), is in its second printing, with fees totaling some $3000 to date.

EDUCATION

During the 2007 Fiscal Year, the Functional V.P. for Education has overseen activities of the Scholarships Committee, chaired by Mr. Scott C. McClure, the SNAME Education Committee, chaired by Matthew R. Werner along with the ASNE-SNAME Joint Education Committee, currently chaired by Captain Alan J. Brown and the Professional Engineer Licensure Committee, chaired by the undersigned. The activities are described as follows:

Scholarships Committee

The Scholarships Committee held its regular meeting as a teleconference on Tuesday, February 27. Eight of the committee's members participated and ten of the committee's eleven members submitted applicant evaluations for committee member discussions. Alan Gray informed the committee that the approved budget was $120,000, including $20,000 that the Executuve Committee specified should be used to increase the amount of funds allocated for undergraduate scholarship support at the various universities in the program. In view of the $6,000 needed for the new, M&L Rosenblatt undergraduate scholarship, this left $114,000 for graduate/undergraduate allocation by the committee.

After reviewing the expected tuition costs of the universities during the 2007-8 academic year, the committee allocated $70,000 for undergraduate scholarships and $50,000 for graduate scholarships. The undergraduate scholarship funds were allocated as follows:

Webb Institute	$ 10,600
University of Michigan	$ 10,600
University of New Orleans	$ 8,600
SUNY Maritime College	$ 3,800
Florida Atlantic University	$ 3,800
Virginia Polytechnic University	$ 3,800
Memorial University of Newfoundland	$ 3,800
Massachusetts Institute of Technology	$ 3,800
Texas A&M University	$ 3,800
Maine Maritime Academy	$ 2,800
University of British Columbia	$ 2,800
University of California at Berkeley	$ 3,800
Florida Institute of Technology	$ 1,000
California Maritime Institute	$ 1,000
Undergraduate University Total	$ 64,000
Mandell and Lester Rosenblatt	$ 6,000
Total Undergraduate Scholarships	$ 70,000

Undergraduate funds to the listed institutions were distributed in August.

Eight graduate scholarship applications were reviewed and ranked. Applicants included six U.S./Canadian Citizens (including one woman) and two International students, one a citizen of Germany and the other a citizen of Cyprus. One application was dismssed owing to missing GRE scores.

The Committee selected the following individuals to receive graduate scholarships:

Name Scholarship Amount

Robert E. Dvorak Landers Scholar $ 1,000
2007 Webb Institute graduate, Hydrodynamics (CFD) at MIT
The award was reduced as Mr. Dvorak also received a full scholarship with tuition and stipend.

Candance P. Brown Kennedy Scholar $ 12,000
2004 Webb Institute graduate, Ocean Engineering at MIT, FIT, or FAU

Robert T. Madsen Richards Scholar $ 11,000
2007 University of Michigan graduate, Concurrent Marine Design at Michigan

Thomas J. Matlak McClure Scholar $ 11,000
2006 Univ. of California, Berkeley graduate, Ocean Engineering at UC Berkeley

Johan Kemnitz $ 10,000
2007 University of Michigan graduate, Hydrodynamics at Michigan

Marios Dias $ 0
2007 University of Newcastle graduate, was unable to pursue graduate study at this time

Alternate Miguel R. Quintero $ 5,000
2007 Florida Atlantic University graduate, Ocean Engineering at University of Hawaii, Manoa

$50,000

Ten Mandell and Lester Rosenblatt Undergraduate Scholarship Applications for the 2007-8 academic year were received in accordance with the published deadlines. Applications were received from six men and four women, including one citizen of Turkey, one citizen of Canada and eight citizens of the U.S. All received materials were loaded onto the Scholarships Committee Intranet Site for committee member access. The Scholarships Committee met by teleconference on June 27 to review and act upon them. Evaluations were submitted and discussed by seven of the committee members under the guidance of Scott McClure, the committee chairman. The committee recommended that a scholarship award in the amount of $6,000 be made to Ms. Heather Caroline Brown, currently a student in the Ocean and Naval Architecture Engineering program at the Memorial University of Newfoundland.

Funding of a new Robert N. Herbert undergraduate scholarship was received in the amount of $80,000. It is expected that additional funding will be received to allow awarding of a new, Robert N. Herbert $5,000 undergraduate scholarship in June of 2008. It is intended that the criteria for award will be the same as for the Rosenblatt scholarship.

Fundraising continues for the new John V. Wehausen graduate scholarship with total funds allocated to that account amounting to $53,435 as of the end of the 2007 fiscal year. It is expected that additional funraising will allow the first of this new graduate scholarship to be awarded by the Scholarships Committee in February of 2008. It is estimated that the 2007 contributions to the scholarship program amount to $135,230.

In anticipation of the 2008 scholarship program, the SNAME web site pages for the undergraduate and graduate scholarships applications have been revised, pending final approval by the committee chairman.
It was agreed by the committee that a young member should be appointed to the committee. "A recent graduate scholarship recipient with a minimum of four years beyond graduation is to be added to the Scholarships Committee. The Committee chairman is to request the appointment."

Scholarships Campaign
Preparations have been made to initiate the Society's two-year, Million Dollar Scholarship fund raising campaign. Matching funds are being offered for contributions ranging from $100 to $50,000 with rights to name a scholarship for contributions in excess of $50,000.

Education Committee
Matthew Werner has been appointed Chairman of the Education Committee. In his role as Chairman of the Education Committee, he will also Co-Chair with Alan Brown the Joint ASNE/SNAME Education Committee. Committee meetings were held during the 2006 Annual Meeting, on March 5, 2007, and on June 27, 2007. Operating under its new Joint Committee Charter, the Committee's work continued with the development of the student competitions (Lisnyk, ISODC, ISWSYDC and MISCSDC), continuing education courses, and development of ABET training for SNAME accreditation teams. The committee wishes to encourage university faculty to urge greater student participation in the design competitions and wishes to have more faculty membership on the committee. In the future, the Lisnyk competition announcement will be published earlier to encourage greater participation.

Matt Werner and Bill Garzke have worked on the task of refining the Continuing Education Business Plan, shortening the presentation, updating the cash flow analysis, and updating the list of courses offered 2002 to present. Some of the work previously planned, such as making a market survey, was never carried out. Interest in online distance learning courses was discusssed, but conducting a survey has not yet been done. Professor Neu is to work with Kelley Clarkson to develop an online course inventory.

A proposal is being developed for establishing a visiting professorship for engineering design for producibility. The visiting professor would be able to visit the universities to provide guidance on designing for producibility when the students are working on their Capstone Projects. Video conferencing is being considered as a possible vehicle for the conduct of the professor's guidance.

P.E. Licensure Committee
The work of the committee continues as before. There were 29 registrants for the 2007 P.E. Review Course (PERC), a number of them registering early so as to stretch their study time. There were also 29 examinees for the April 20 NAME examination. The pass rate for first-time takers was 89% while the pass rate for repeat takers was 100%. The PERC continues to be a valuable aide for those preparing for the exam.

The new NAME exam specification will be implemented for the 2008 exam, the draft of which has been formed and awaits critique by a panel of Subject Matter Experts (SME). The SNAME web site P.E. Licensure page has been updated with the new specification. The new specification has also been put onto the NCEES website. The NAME Study Guide has had the exam specification updated as well, and a copy of the Study Guide in its present state of development has been sent to NCEES for their record.

With a new exam specification, the Study Guide Sample Examination is in need of restructuring and that was a current project. In addition, there is a need for problems in subject areas not covered in the previous specification.

Announcements have been placed in the **SNAM*eNewsLetter*** noting that those wishing to get an early start on their NAME Exam study can do so by registering early.

AWARDS

The following awards were presented at the Annual Banquet on Thursday, October 12, 2006:

The Vice Admiral "Jerry" Land Medal for "Outstanding Accomplishment in the Marine Field" was presented to Peter G. Noble by R. Keith Michel.
The David W. Taylor Medal for "Notable Achievement in Naval Architecture and/or Marine Engineering" was presented to Edward N. Comstock by Roger H. Compton.

The Davidson Medal for "Outstanding Scientific Accomplishment in Ship Research" was presented to Preben T. Pedersen by Alaa Mansour.

The William M. Kennedy Award for "Outstanding Service and Contribution in the Development of Systems and Planning Applying to Shipbuilding and Repair" was presented to Tim Colton by Jaideep Sirkar, and

The Elmer L. Hann Award for the "Best Paper on Ship Production Delivered at the 2005 Ship Production Symposium" was presented to Robert G. Keane, Jr., Howard Fireman and Daniel W. Billingsley by Jaideep Sirkar.

The following awards were presented at the Keynote Address on Thursday, October 12, 2006:

50-Year Membership Certificates
Membership Certificates representing 50 years of membership were presented by President Roger H. Compton to: William B. Alexander (in absentia), Anthony C. Antoniou (in absentia), Grant W. Bauer (in absentia), James A. Bortner (in absentia), William G. Bullock (in absentia), Reid Schell Byers (in absentia), David W. Carreau (in absentia), Albert Stanley Carter (in absentia), J.W. Cuthbert (in absentia), Theodore C. Dawson (in absentia), Clifford Gaddess (in absentia), Norman O. Hammer (in absentia), Halsey Chase Herreshoff (in absentia), Thomas J. Kingston (in absentia), George H. Levine (in absentia), Michael J. Markey (in absentia), Justin Huntley McCarthy (in absentia), Naresh M. Maniar, J. Cameron McKernan (in absentia), James A. McQuilling (in absentia), A. (Gus) Prasciunas (in absentia), Francis John Slyker (in absentia), Thomas John Stewart (in absentia), Daniel H. Thompson, Jr. (in absentia), James R. Wilkins, Jr., John H. Van Arken (in absentia), Luther B. Ward (in absentia), and, Trevor White (in absentia).

Distinguished Service Awards
President Roger H. Compton read and presented the Distinguished Service Award for dedicated personal service and/or technical contributions to the Society to Richard C. Boutwell and Michael T. Martin.

Certificates of Appreciation
There were no Certificates of Appreciation awarded in 2006.

American Bureau of Shipping - Captain Joseph H. Linnard Prize
Dr. Xiaozhi (Christina) Wang presented the American Bureau of Shipping - Captain Joseph H. Linnard Prize for "the best paper contributed to the Annual Meeting and appearing in the Transactions of the Society" (2005) to Peter G. Noble, Lars Ronning, John R. Paulling, III (in absentia), Rong Zhao (in absentia) and Hoseong Lee (in absentia).

Vice Admiral E. L. Cochrane Award
Edward M. Lewandowski presented The Vice Admiral E. L. Cochrane Award for "the best paper published in Marine Technology," to Wilhelm Magelssen.

T&R Certificates of Recognition
Jaideep Sirkar presented a Technical and Research Certificate of Recognition for "Outstanding Service in the Technical and Research Program" to Thomas Lamb (in absentia) and Charles H. Piersall (in absentia).

Honorary Member Certificate
Jose Femenia, Chairman of the Nominating Committee, presented an Honorary Member Certificate to President Roger H. Compton.

The following awards were presented at the Student Congress on Thursday, October 12, 2007:

Student Paper Awards
Student Paper Awards for "Outstanding Student Papers Presented at a Section during the 2005-2006 Program Year," were awarded as follows: the Graduate Paper Honor Prize to Militiadis Kotinis; the Undergraduate Paper Honor Prize to Evan A. Karlik; the Undergraduate Paper Award to Robert Dickinson, Peter K. T. Burger and Anthony Cote; and a 2nd Undergraduate Paper Award to Brendan M. Abbott and Daniela T. Hanson.

Dr. James A. Lisnyk Ship Design Competition
The 2006 Dr. James A. Lisnyk Ship Design Competition Awards were presented by President SNAME Roger H. Compton and ASNE President Richard D. Hepburn, as follows: The University of Newcastle Upon Tyne was awarded First and Third Prize headed by Faculty Advisor, Richard Birmingham (in absentia) and Peter N.H. Wright, Senior Lecturer in Naval Architecture, School of Marine Science and Technology, representing The University of Newcastle Upon Tyne. The members of the First Prize winning team were Michalis Christoforou, Georgios Georgiou, Shaw Chuan Ivan Lim, Laura Phillips and Philip Herbert Wykeham. The Third Prize winning team members were Benedict A. Graat, James Maclean, Michalis Pouris, Kyle R. Pritchard and Konstantinos Thalassetis. Virginia Polytechnic and State University won Second Prize, headed by Faculty Advisor, Alan J.

Brown. The Second Prize winning team members were Morgan C. Baldwin, Aaron J. Cox, Nathan A. Good, Nicholas E. Marickovich, Travis S. Smith and Ryan Webster.

International Student Offshore Design Competition

The 2006 International Student Offshore Design Competition Award for First Place was presented by Thomas Johnson and Ray Fales to the Federal University of Rio de Janeiro. The members of the winning team were Diogo Do Amaral Mercedo Amante, Tiago Pace Estefan, Joao Paulo Carrijo Jorge, Leandro Cerqueira Trovoado and Daniel Santos Werneck, headed by Faculty Advisor Sergen F. Estefen. Announcements were made as follows: Second Place, Texas A&M University Team 3, Third Place, Texas A&M University Team 1, Fourth Place, University of Michigan and Fifth Place, Texas A&M University Team 2.

2006 International Workboat/Small Craft/Yacht Design Competition

The 2006 International Workboat/Small Craft/Yacht Design Competition Award for 1st Prize was presented to the University of Michigan by Thomas Lamb. The members of the winning team headed by Faculty Advisor Thomas Lamb were David Elwood, Benjamin van Dam and Christopher Amory. The 2nd Prize was presented to the University of British Columbia, headed by Faculty Advisor Dr. Sander M. Calisal. The members of the winning team were William Rawlings, Joel Atwater and Kolya Harpe. The 3rd Prize was presented by the University of New Orleans headed by Faculty Advisor Professor Cliff Whitcomb. The members of the winning team were John Huber, Justin Iserhardt, Eric Lord, Glen Grogan and Sattam al Showler.

2005 SNAME/MarAd Commercial Ship Design Competition

The competition did not take place in 2006.

APPOINTMENTS

SOCIETY REPRESENTATIVES TO OTHER ORGANIZATIONS AND OTHER APPOINTMENTS

Accreditation Board for Engineering and Technology (ABET)

Board of Directors — Dr. Edwin G. Wiggins

Engineering Accreditation Commission — Dr. Michael Fleahman

Engineering Accreditation Commission Alternate — Open at present

Technology Accreditation Commission — Prof. Paul C. Jackson

Technology Accreditation Commission Alternate — Open at present

Program Evaluators:

Prof. Harold C. Alexander
Mr. Robert E. Apple
Mr. Edward Barrett
Mr. Elwood Baumgart
Dr. Rameswar Bhattacharyya
Mr. Norman Birrell
Prof. Neil Bose
Prof. Boris S. Butman
Dr. Tyne-Hsien Chang
Prof. Chrys Chrystomidis
Cdr. Kurt J. Colella, USCG
Dean Roger Compton
Mr. Alexander Delli Paoli
Dr. Jeffrey Falzarano
Prof. Jose Femenia
Mr. Michael Fleahman
Prof. James Harbach
Dr. Groves E. Herrick
CDR Dwight G. Hutchinson, USCG (Ret.)
Prof. Bahadir Inozu
Prof. Paul C. Jackson
Dr. Robert G. Latorre
Dr. Alaa E. Mansour
Prof. Charles J. Munsch
Dr. Wayne Neu
Dr. Vijay Panchang
Dr. Michael G. Parsons
Dr. Robert E. Randall
CDR Paul Roden, USCG
Prof. William J. Sembler
Dr. Herman E. Sheets
Prof. James R. Sherrard
Dr. John Spencer
Dr. Edwin G. Wiggins

Council of Engineering and Scientific Society (CESSE):

Mr. Joseph Caggiano
Mrs. Susan Evans
Ms. Susan Giver
Mr. Allan Gray
Mr. Philip B. Kimball
Mrs. Carolyn Marus
Mrs. Nancy Ostin
Mrs. Barbara Trentham

Marine Engineering Council Underwriters Laboratories	Open
Maritime Technical Information Facility	Dr. Charles R. Cushing
SPE Liaison Society Executive Directors	Open
National Inventors Hall of Fame Foundation Selection	Philip B. Kimball
Offshore Technology Conference Board of Directors (to April 30, 2006)	Mr. Roderick J. Allan
(from May 1, 2006)	Mr. Edward G. Stokes
Program Committee 2005/2006	Gregory G. Thory Craig Kolby
Program Committee 2006/2007 Pan American Institute of Naval Engineering	ADM Robert E. Kramek, USCG (ret.)
Ship Structures Subcommittee	Dr. Walter M. Maclean
Sperry Board of Award	Mr. Naresh M. Maniar Mr. William duB. Thomas Mr. Philip B. Kimball (Alternate)
U.S. Committee on Theoretical and Applied Mechanics	Dr. Dale Karr

GEORG P. WEINBLUM LECTURE

The Twenty-Ninth Weinblum Memorial Lecture was given by Professor Gunther F. Clauss, on November 22, 2006, in Hamburg, Germany.

OFFSHORE TECHNOLOGY CONFERENCE

The Offshore Technology Conference was held at Houston's Reliant Center in May and attended by 67,155 registrants. The 39th year of the Exposition recorded 2,400 booths for a total of 529,700 net square feet of space in Reliant Hall. The Society's share of the 2006 Offshore Technology Conference was $615,955.

WEB ENABLEMENT AND REDESIGN PROJECT

In the fall of 2006, the Council approved the web redesign project. Ellipsis Partners, a web project consultant, was selected to assist with the preparation of an RFP, as well as the selection of the provider. Ellipsis performed an in-depth analysis of SNAME requirements with the staff and with the volunteer advisory committee. In addition, a web user survey was created and the results analyzed, as well as the existing member value survey. The RFP was generated, reviewed by staff and advisory committee, and was sent to over 30 potential vendors by the consultant. The provider will be selected by year-end.

MANAGEMENT INFORMATION SERVICES

Headquarters owns 20 PC's, one MAC PC and five file servers running a wide variety of Microsoft and Computer Associate products. The five file servers consist of two data servers, a mail server a terminal server and a backup terminal server. Over the last year, hq has upgraded all their Computer Associates Pproducts and their Spam software.

All PC's have Microsoft Office XP Professional and IncoculateIT anti virus software from Computer Associates installed. SNAME runs an Association Management software package from GoMembers.com that consists of Dues Processing, Events Management, Exhibitor Management, Subscriptions and Order Entry modules.

In the winter of 2007, Headquarters installed direct inward dialing phone numbers, which eliminated the need to have a full-time phone receptionist. Also in the winter of 2007, Headquarters replaced two leased printers/copiers providing the capability to print, copy and scan in color.

SNAME PROPERTIES INC.

The building was fully rented for the entire year. Revenues in Fiscal Year 2007 were $537,884, which exceeded Fiscal Year 2006 by $10,427. Expenses were $19,687 (3.8%) lower than budget. Overall, revenues exceeded expenses by $42,451 in FY2007, a 17% improvement over FY2006.

HEADQUARTERS STAFF

Last year the Membership Department reported an overall increase in membership as well as Member retention. Due to the efforts of the entire staff membership increased again this year to the highest in over five years, o and retention hit a record high. The department continues its focus on enhancing the member experience with SNAME by increasing member retention, by improving responsiveness and communication with members and by paying greater attention to detail.

The new SNAME Regional Offices for membership opened on September 1, with a Regional Office Manager (ROM) managing the European office in Athens, Greece and the Herbert Engineering staff managing membership applicatioms in Shanghai, China.

The annual Membership Renewal Telethon conducted in April by the staff successfully re-signed 502 of the 1024 un-renewed members recorded at the end of March.

As a result of the Headquarters reorganization in 2006, this year with the new program in place, we have adopted a strong commitment to service for our members by empowering the staff with the flexibility to address and resolve concerns, and to adapt and improve our programs and services to meet the ever-changing needs of our members.

The Education Department at Headquarters continued to work with its partner, High Caliber Solutions, to develop Pod-casts and mini-courses for members, with special emphasis on serving Young Professionals' needs. We also established baseline protocols for e-Journals with Guidelines for Editorial and Administrative Committees and Guidelines for Acceptability.

In late March the Society welcomed 43 NA&ME Engineering students from the National Technical University of Athens (NTUA) to SNAME Headquarters. The students toured Headquarters and were introduced to the Society's programs, its technical program, membership, publications and events by department heads. We recruited a number of new Student members and sold books and accessories.

On May 1, Headquarters received notification of a three-year grant award of $167,000 from the Office of Naval Research (ONR) to develop and promote interest by middle school students in engineering and naval architecture and marine/ocean engineering with the Sea Perch, a remotely operated underwater vehicle (ROV), which they assemble and operate to learn the first principles of engineering and technology.

On June 1, we introduced the **SNAMe***MarineNewsDaily*, a customizable marine news service, which allows members to customize their content by entering selected key words. An archival service is also available from a resource of 23,000 news articles available from Maritime Reporter, our project partners. The digital versions of Marine Technology continue be made available to the members both in "flash" and "non-flash" formats.

Later in June we engaged our human resources consultant to conduct non-harassment training for all staff at Society Headquarters. Additionally, we are updating the Employee Manual and various forms for staff employment and annual review and evaluation.

The Executive Director and certain staff members attended conferences and trade shows during the year, and managed the SNAME booth to promote membership, book and accessory sales. The Society exhibited this year at the Connecticut Maritime Association and Annual Meeting in Stamford, CT, in March, the Offshore Technology Conference in Houston, TX, in May, and at the ASNE Day Conference in Arlington, VA.

In conclusion, I would like to personally thank our president, the Society officers, the Council and Executive Committee members, the Chairs of standing committees, as well as the numerous volunteers who provide the spirit, motivation and expertise that make the Society a standout in the industry. Also, my personal thanks to our dedicated staff, which works tirelessly to insure that our members continue to receive value for their membership along with relevant activities, benefits and services.

Respectfully submitted,

Philip B. Kimball
Executive Director

TREASURER'S REPORT TO COUNCIL, 2007

Balance Sheet at August 31, 2007 and August 31, 2006

	2007	2006
ASSETS :		
Cash & Cash Equivalents	$586,096	$578,192
Accounts Receivable	194,378	188,441
Prepaid Expenses		
Section Advances	251,101	236,423
Other Prepaid Expenses	284,981	352,755
Total Prepaid Expenses	536,082	589,178
Intangible Pension Asset	0	64,273
Total Current Assets	1,316,556	1,420,084
Investments	9,996,933	9,040,216
Property & Equipment	302,732	293,895
Less Depreciation	(224,195)	(218,649)
Net Property & Equipment	78,537	75,246
Due from SNAME Properties	3,282,032	3,220,115
TOTAL ASSETS	14,674,058	13,755,661
LIABILITIES & NET ASSETS		
Accounts Payable	118,751	112,955
Deferred Revenue	300,197	454,212
Deferred Pension Contributions	200,820	338,272
Other Liabilities	211,285	181,174
Total - Current Liabilities	831,053	1,086,613
Total Liabilities	831,053	1,086,613
Net Assets	13,843,005	12,669,048
Total Liabilities & Net Assets	$14,674,058	13,755,661

	8/31/07	8/31/06
Revenues		
Member Dues	$891,636	$766,145
Publication Sales	537,472	532,365
Royalties	40,944	19,500
Contributions	163,309	80,161
Outreach Program	8,759	7,701
Annual Meeting & Expo.	657,775	863,595
OTC	615,955	520,433
Advertising	44,792	39,586
P.E. Exam	13,500	12,101
Educational Courses	6,243	8,845
Interest & Dividends	249,915	216,831
Total Revenue	3,230,300	3,067,26
Expenses		
Program Services		
Publications	381,969	429,027
Functions	373,180	418,356
Scholarships,Grants	132,618	93,571
Total Program Expenses	887,767	940,954
Management & General		
Salaries and Fees	1,077,784	1,067,658
Payroll Taxes, Benefits, Insurance	357,799	355,686
Subtotal	1,435,583	1,423,344
Facility Expenses	152,842	153,343
Membership Services	133,222	139,204
Professional Fees	82,377	48,735
Headquarter & Section Exp.	141,168	126,064
T & R Expenses	15,925	38,448
Depreciation and Other	22,305	14,950
Sundries	41,966	42,012
P.E. Exam	51,167	58,506
Educational Courses	23,269	24,148
Advertising	26,584	23,978
Outreach Program	6,220	4,159
Website	41,627	42,689
Subtotal (Lines19-28)	738,672	716,236
Total MGT & General (Lines 18 + 29)	2,174,255	2,139,580
Total Expenses (Lines 14 + 30)	3,062,022	3,080,534
From Operations Increase in Net Assets	$168,278	$(13,271)

	2007	2006
Assets		
Cash & Cash Equivalents	$93,414	$120,302
Accounts Receivable	0	0
Prepaid Expenses	0	6,044
Total Current Assets	93,414	126,346
Property & Equipment at Cost	3,667,304	3,595,364
Less Depreciation	(1,903,472)	(1,785,030)
Net Property & Equipment	1,763,832	1,810,334
Total Assets	1,857,246	1,936,680
Liabilities & Net Assets (Net Worth)		
Accounts Payable	13,539	17,804
Mortgage Current Portion	107,964	102,564
Total Current Liabilities	121,503	120,368
Mortgage Payable	505,110	604,982
Due to SNAME	3,282,032	3,220,115
Other Liabilities	23,852	23,582
Total Liabilities	3,810,994	3,848,679
Net Assets (Net Worth)	(1,953,748)	(1,911,999)
Total Liabilities & Net Assets	$1,857,246	$1,936,680

	8/31/07	8/31/06
Revenue		
Rental Income	$501,956	$492,322
Parking and Other Income	35,928	35,460
Total Revenue	537,884	527,782
Expenses		
Cleaning & Rubbish Removal	39,175	34,881
Utilities	62,512	65,781
Maintenance & Repairs	48,885	64,006
Real Estate Taxes	73,675	71,000
Property Mtg. Fees	12,400	11,200
Mortgage Interest	34,532	38,506
Professional Fees	1,245	1,236
Depreciation	114,475	103,750
Insurance	6,950	6,900
Leasing Commissions	0	8,000
Security	7,924	9,270
Administrative Expenses	78,660	77,000
Total Expenses	480,433	491,530
Increase Net Assets	$57,451	$36,252

Reviewed, approved and respectfully submitted by
Joseph J. Cuneo, Treasurer

Awards and Citations Presented at the 2006 Annual Meeting

FIFTY-YEAR GOLDEN MEMBERSHIP CERTIFICATES WERE AWARDED TO THE RECIPIENTS LISTED BELOW BY PRESIDENT ROGER H. COMPTON

Egil Abrahamsen
Eugene L. Aronne
Carlie Baxter, Jr.
John H. Boyd, Jr.
John G. Brown
Donald L. Caldera
Robert T. Church
Denis D.A. Csupor
Richard A. Cumming
Joseph J. Cuneo
Alfred E. Deeble
Harold L. Essex
Water R. Gerich
Herman M. Gilson
Douglas M. Gluntz
John Grinstead
Herbert C. Hammond
John F. Hillmann
Robert L. Jack, Jr.
Peter A. Johnson
Roger G. Kline
John H. Klose
Eric E. Lithen
Charles W. Lofft, Jr.
Richard M. McCabe
John Meghrian
Neil M. Miller
Francis X. Nicastro
Raymond Pearlson
James G. Price
William H. Russ, III
Lawrence A. Schultz
Feridum K. Serim
Andrew G. Spyrou
Arthur E. Sturtevant
Donald J. Szostak
D.P. Tzoannos
Richard P. Voelker
William R. Walsh
James R. Wittmeyer
Bruce J. Wooden
John B. Woodward
John Woolston
Howard F. Yaffe
George A. Zahn
Andrs Ziedins

ABS – CAPTAIN JOSEPH H. LINNARD PRIZE
Hans G. Payer
Raul Brostella

DISTINGUISHED SERVICE AWARDS
Barbara L. Lamb
Dawson Miller

CERTIFICATES OF APPRECIATION
Frank Bercha
Thomas Lamb
Thomas P. Mackey
Matthew L. Unger
John H. Higginbotham

HONORARY MEMBER CERTIFICATE
Thomas P. Mackey
Michael G. Parsons

STUDENT PAPER AWARDS

Graduate Paper Honor Prize
James P. Morash

Graduate Paper Award
Gregory R. Mitchell

Undergraduate Paper Honor Prize
Peter Bryn

Undergraduate Paper Award
James Sannino
Evan Karlik
Christina Gomez
Seth E. Clark
Nina Young
Nathaniel Chan

TECHNICAL AND RESEARCH PROGRAM CHAIRMEN
Introduced by: Mr. Jaideep Sirkar

William G. Day, Jr.
William T. Bennett, Jr.
Robert G. Keane, Jr.,
Philip G. Rynn
Richard Delprizzo
Rick S. Spaulding
Jennifer J. Zeien
Hendrik F. van Hemmen
Stephen P. Markle
William duBarry Thomas

THE VICE ADMIRAL E.L. COCHRANE AWARD
Daniel Savitsky
Michael F. DeLorme
Raju Datla

DR. JAMES A. LISNYK SHIP DESIGN COMPETITION AWARDS

First Prize: Norwegian University of Science and Technology

Students: Erlend Vågsholm
Trygve Espeland Halvorsen
Thomas Petter Gram
Anders Martin Moe
Torstein Alexander Pettersen
Even Høgset Olsen

Faculty Advisors: Professor Kai Levander
Professor Stian Erichsen

Second Prize: Virginia Polytechnic Institute and State University

Students: Andrea Maines
Matthew Martz
J. Spencer Ovren
Michael Palmer
Amy Sloan
W. Rob Story

Faculty Advisor: Dr. Alan Brown

Third Prize: The University of Newcastle upon Tyne

Students: Marios Agathocleous
Patrick Carey
Claire McKee
Dale Smith
Robin Steward
Richard Whitaker

Faculty Advisors: Dr. Peter N. H. Wright
Prof. Richard Birmingham
Prof. Martin J. Downie

INTERNATIONAL STUDENT OFFSHORE DESIGN COMPETITION AWARDS

First Place: Texas A&M Team 3

Students: HuiDong Bea
Aaron Gregg
Milton Hooks
Brian Riordan
Justin Russell
Miles Williams

Seccond Place: Texas A&M University Team 1

Students: Christopher Aurich
Tyson Breedlove
Eric Brown
Sean Evans
Ryan Koska
Bryce Read

Third Place: The Federal University of Rio de Janeiro

Students: Bernardo Barreto Xavier
Guilherme Manso Haddad
José Da Costa Pereira Júnior
Luiz Felipe De Paula Carvalho
Márcio Valente Scultori Da Silva
Renata Mendonça Grabowsky Nunes

Fourth Place: Texas A&M University Team 2

Students: Blake Allen
Jieh Larson
Jaime Lazcano
Miguel Urias
Dustin Young

Fifth Place: Texas A&M Team 4

Students: Kingsley Ewetuya
Dane Nelson
B.J. Turner
Travis Wallace
Casey Williams

INTERNATIONAL WORKBOAT/SMALL CRAFT/YACHT DESIGN COMPETITION AWARDS

First Place: The Webb Institute

Students: Nathan Pratt
Mark D'Angelo
Jonathan Leivo
Julie Rotramel

Faculty Advisor: Professor George Petrie

Second Prize : The Webb Institute

Students: Robert I. Bara
Patrick Fass

Faculty Advisor: Prof. Professor George Petrie

Third Prize: The University of British Columbia

Students: Lauren Best
Jen Alloway
Silas Spence
Paal Jetlund Thorsagen

Faculty Advisor: Dr. Jon Mikkelsen

William M. Kennedy Award

The 2007 William M. Kennedy Award for "outstanding service and contribution in the development of systems and planning applying to shipbuilding and ship repair" was presented to **Jay P. Carson** for outstanding service and contribution in the development of systems and planning applying to shipbuilding and ship repair.

Elmer L. Hann Award

The 2007 Elmer L. Hann Award for the best paper on ship production delivered at the 2006 Ship Production Symposium was presented to **Tyau-da Huang, William Patrick Keene (in absentia), Lee G. Kvidahl, Christopher Conrardy, Pingsha Dong, and Lawrence Decan.**

Vice Admiral "Jerry" Land Medal

The 2007 Vice Admiral "Jerry" Land Medal for "outstanding accomplishments in the marine field" was presented to **Bruce Rosenblatt** by R. Keith Michel.

Land Medal Presentation

It is my pleasure and honor to introduce Bruce S. Rosenblatt as this year's recipient of the Vice Admiral "Jerry" Land medal. A pleasure because Bruce is a very close friend; an honor because I truly admire his leadership and commitment to the maritime industry.

Jerry Land had an illustrious career with the U.S. Navy and in public office. He was a naval architect who specialized in the design and construction of submarines during World War I, and later served as Chief of the Bureau of Construction and Repair. In his post-naval career as chairman of the U.S. Maritime Commission, Jerry Land was responsible for overseeing the design and construction of more than 5,000 merchant ships including some 4,000 Liberty and Victory ships. Bruce Rosenblatt is very deserving of this award, which is granted in honor of a man who gave so much to his country and his profession. I have known Bruce for many years and worked closely with him at SNAME and on the board of trustees of Webb Institute. I know of no one who gives as much time and personal energy into the betterment of our profession and the mentoring of students and young graduates.

This "giving" to the profession began with his involvement in SNAME student sections during his college years. Bruce was one of the youngest Section chairs of the NY Met section, one of the youngest members of the Executive Committee, and the youngest president in SNAME's history. During Bruce's tenure as SNAME president, SNAME streamlined its governance structure, expanded its international presence and led the first World Maritime Technical Conference. Bruce visited every SNAME Section, including the newly formed Section in Greece, seeking advice from the membership on how to improve our Society. Bruce took the comments seriously, and did much to enhance member benefits and improve communications within the Society. Following his service as president, he has continued his commitment to SNAME and remains a vibrant member of many of its committees. Bruce is also an active member of ASNE, having served as a vice president and member of ASNE's council.

Perhaps Bruce's most ardent volunteer efforts have been in support of students and young professionals. Bruce always finds time in his busy schedule to mentor anyone with interest in the marine industries. He is an active member of the Advisory Committee of SUNY Maritime, he chairs Michigan's Industry Advisory Group, and is a member of Webb's Board of Trustees and chairs its Nominating Committee.

Bruce Rosenblatt is a third generation naval architect. His grandfather, Mandell Rosenblatt, was a remarkable man – a self-taught naval architect who together with his son Lester would establish the very successful design firm, M. Rosenblatt & Son. Many of you knew Lester well, as he was a mainstay of our Society and a leader in the maritime industry for so many years. Bruce followed Mandell and Lester Rosenblatt as president of M. Rosenblatt & Son. Bruce has also continued and expanded upon the family tradition of giving back to their chosen profession. As did his father before him, Bruce was a dynamic and effective president of our Society. The growth in membership that SNAME experienced during the 1990's under Bruce's direction was second only to the growth experienced when Lester Rosenblatt chaired the Membership Committee some twenty years earlier. With the presentation of this award, Bruce and Lester become the first father and son to be awarded "Jerry Land" medals.

Bruce was previously awarded SNAME's Centennial Medallion and its Distinguished Service Award, and is an Honorary Member and Fellow of the Society. He now becomes one of the youngest persons to be awarded one of SNAME's prestigious medals. Fortunately, it is not in Bruce's nature to rest on his laurels. He will no doubt continue to provide creative leadership and his tireless support to our Society and to the education institutions that are so important to the future of the maritime industry. Please join me in acknowledging Bruce Rosenblatt as the 2007 recipient of the "Jerry" Land medal.

Land Medal Acceptance

Thank you, Keith, for those very kind words.

President Kramek, Admiral Sullivan, many Honored Guests, Ladies and Gentlemen on the dais and throughout the ballroom, I am honored to receive the Vice Admiral "Jerry" Land Medal. As Keith Michel discussed, Admiral Land's distinguished career was quite remarkable. Just look at the list of past recipients of the Land Medal, so many great marine industry professionals, from Admiral Land, to many others who I respect so much, how could one not be humbled?

I have so many people to thank:
My family - First and foremost my wife, Jan, my daughter, Melissa, and my son, Daniel, are the ones who paid the most. For the times I worked when on vacation, I guess most of the time, and for the days and nights I missed sharing with them, and for their unending support, thank you. A few too many days out of school this year kept Daniel and therefore Jan away tonight, but I'm fortunate to have Melissa here as well as a surprise visit by my sister, Roberta and my niece, Shanna.

My teachers - So many come to mind. They all had the unenviable task of trying to teach me. Sometimes it worked and other times ...... OK, let's forget about the other times. In a speech like this you always hate to mention anyone as there will be others you forget, but how can I not mention Harry Benford? He made my Naval Architecture classes magic for me ..... Why could he not have taught me differential equations?

My co-workers at M. Rosenblatt & Son, Inc. - These fine men and women, some here this evening, had the task of teaching "the boss' son" the "practical" aspects of being a naval architect and manager. Luckily most didn't mind that I'd likely end up being their boss some day and their caring, guidance and desire to help train me will never be forgotten. Whether it was the guys who gave me the dirty jobs in the bilge on shipchecks, or made me realize, generally the hard way, the "right" ways for things to be done, they made sure I learned how everything is different when you get "beyond the theoretical."

I even learned not to push certain buttons on ships. Yes, it's true, one compartment of a destroyer got "partially" flooded, but it was a "fairly" old destroyer, and a "fairly" small compartment and in this case "partially" flooded might better be called just "slightly" flooded and anyway, I proved that the pumps were working and the drains were clear......My apologies, Admiral. But yet I digress. Another very important part of design that I learned from these "practical teachers" at M. Rosenblatt & Son, was that functionality and cost should go hand-in-hand with the technical part of a solution.

My mother - She passed away when I was just starting high school, but I know I have a special part of her in me.

My grandfather – Some of you here tonight knew him better than I did, at least from a professional standpoint. I knew him more as a grandfather than a business or engineering mentor, but I've heard so many stories of how he was the consummate professional. I'll never have his innate, great engineering skills, but hopefully I've picked up a few of his other good traits.

My father – I know many of you here tonight knew him; he was an inspiration to me. My father and I had a bit of a unique relationship. In addition to his full time job of being my father, he also did yeoman's work as a mentor in sailing and the marine industry. We spent weeks together each summer, sailing on our sailboat the Rosa II, which he and my grandfather, Mandell, designed. Those annual sails to Maine, were educational in many ways and the memories are countless and wonderful. But he was not always complimentary. In fact, when I showed him my first Lines Drawing done in my first semester at Michigan, after looking it over he said "if your grandfather could see this he'd roll over in his grave." I made sure that the one I did for my final design project, just a couple years later, was top notch, and yes, I did get his approval. Many of you knew that he loved telling stories, even a "few" clean ones, but I could tell he cherished it whenever, in his eyes, I made it to the next plateau.

Whether it was:

- When I was younger and I'd be on the foredeck of the Rosa II in some pretty stormy situations,
- To when he could introduce me as "his son, the Naval Architect",
- To when I became the President of our firm,
- To when he seconded my nomination as President of this society,
- To when he called me on my official first day as the President of SNAME and said "Mr. President, how may I help you?"

I was elected President of SNAME 23 years after my father was elected President, and now I am the recipient of the Land Medal 23 years after he did. Perhaps 23 is my new favorite number! When my father won his Land Medal, he mentioned that he deeply regretted his dad, my grandfather, not being here. For me, the same is so true regarding both of them.

I feel much professional satisfaction when I help the marine industry students and our young professionals. For the over 200 students and other young professionals in the audience, what a great time to be heading into our business.

My message to you is "get involved" and "stay involved."

There are lots of ways of doing it. First of all,

- Keep your membership in SNAME current.
- Use our Mentor Program – there are lots of us that are

happy to answer your questions and hopefully lead you in the right direction. o

- Become a leader in your Student Section of SNAME.
- Become a leader in the SNAME Student Steering Committee.

Upon graduation:

- Join a T&R Committee
- Get involved in one of our Young Professionals programs.
- Get involved in your Local Section of SNAME.

Yes, at first the work may not be the most exciting, but you will be working with other professionals and I guarantee you that work will hone your organizational and managerial skills. My father told me many times that "all experience is good experience" and throughout my career I have learned that is another way he was spot on!

Some of you here may know Matt Unger. He is the consummate Young Professional. I want to thank him not only for nominating me for this medal, but also for being a role model for all of you students and young professionals. In a rather short time he has worked his way to senior levels in SNAME's governance. I'm counting on many of you to do the same. Please do, as the marine industry needs you to be our future leaders.

Ladies and gentlemen, please know that it is with deep gratitude that I thank the Society for bestowing this great honor upon me. I will always cherish this moment and this medal.

Blakeley Smith Medal

The 2007 Blakely Smith Medal Medal for "Outstanding Accomplishment in Ocean Engineering" was presented to **Howard L. Shatto** by F. Tim Pease.

Blakeley Smith Medal Presentation

Ladies and Gentlemen, Howard Shatto is this year's recipient of the Blakely Smith Medal and your program briefly describes some of his many significant achievements related to ocean engineering, so I do not intend to repeat that information. I would like to mention though that the remotely operated vehicles described in the 2nd paragraph of his citation are the forerunners of the ROVs discussed further down in the citation.

What I would like to do is to place into historical context the time frame when his early achievements occurred.

By 1946 the shipbuilding industry in the United States was well established, however, the offshore oil industry was in its infancy. For example:

- In 1946 diesel-electric powered drilling rigs were just beginning to be used on inland swamp barges.
- In October 1947 the first offshore oil well out of the sight of land was discovered in water that was only 18 feet deep, and this event is considered to signify the beginning of the offshore oil industry.
- In 1949 the first offshore open water pipeline was laid. This Gulf of Mexico 10-inch diameter pipeline was only 10 miles long and in water depths ranging from 14 to 30 feet. Today, by contrast, there are over 20,000 miles of underwater pipelines in the Gulf of Mexico alone.

In 1946, during this offshore start-up period, Howard received his Electrical Engineering degree and immediately joined Shell Oil Company and began his career in the offshore oil industry. There were no codes, rules, regulations or prior examples to guide engineers like Howard as they set out to develop that industry. However, by 1949, Howard engineered Shell's first offshore diesel-electric powered drilling rig (an accomplishment significant enough that it was included in the 1951 World Petroleum Conference). A quick literature search found that Howard's early 1960's patents have been cited by others more than 30 times in more recently issued patents, which assists our understanding of the true creativeness Howard has brought to the offshore industry.

Because of Howard's willingness to step into this emerging industry, we now routinely operate in water depths over 10,000 feet deep. There have been more than 1,300 DP systems built using Howard's concepts. ROVs have multiplied into the thousands. Only recently two 600 million dollar shipyard construction contracts were signed for dynamically positioned drillships capable of operating in 12,000 foot water depths.

Howard is a true offshore industry pioneer, and it is my privilege to introduce him to you this evening.

Blakeley Smith Medal Acceptance

Thank you, Tim, for those kind words......It is hard for me to describe how much fun I have had in this career of almost 63 years...but I want to give it a quick try.

My first job with Shell was to help build our first diesel-electric rig. It was designed to work in swamps in dredged canals. But its first location was to drill a wildcat several miles offshore, off the Mississippi Delta. That was deep water...ten feet deep.

I guess my biggest challenge came in January 1960. It was to begin working with a small group of engineers to develop subsea well completions using remotely operated vehicles in place of divers. This was an idea that had been put forth by my Division Manager Bill Bates.

I thought it was a great idea then, and I still do. When I came into that project the group was also building a small core drill ship called Eureka. In order to hold position while drilling, it was to have two 200 HP azimuthing thrusters, each with a manual speed control and contacts to turn each thruster clockwise or counter-clockwise. I knew that was going to be a monster to control, and I thought long and hard about how to do it all automatically.

In late summer of 1960, in a hot traffic-jam on the Ventura freeway, all of a sudden the solution came to me and I got goose bumps on my arms. We got Hughes Aircraft to bid to build the new DP control system for $50,000. Within five months we put it into full operation in May 1961. It was almost miraculously successful. That was exciting.

The Eureka led the way for thousands of DP vessels to follow. DP controls are now all done digitally, where the Eureka was purely analog. But they still control surge, sway and yaw independently in the same way. It still gives me a big thrill to watch a 100,000 ton ship hold still in the face of wind, waves and current and to watch eight 7000 HP thrusters maneuvered by computers with measurements from satellites and sonar. How could that not be exciting?

We had worked in secret for over two years, and that was fun...then had to go for maximum publicity, which was also fun. We had to do that because no leases would be allowed until other companies were given the technology so they could compete. In January 1963 we taught a school for other companies about all of Shell's deep water technology at $100,000 tuition.

We built the world's first DP oil well drill ship, Sedco 445, in Japan in 1970 and put it to work off Borneo. Later we reworked Transocean's Discoverer Seven Seas for rougher seas and deeper water and set many new water depth records in the Atlantic and GOM.

After working for one company, Shell, for 42 years, it was a remarkable and exciting change to consult for all of the majors and many of the drilling and construction contractors.

Over the years since 1961 we have made huge improvements in the performance and reliability of both Dynamic Positioning and ROVs. Both have seen their MTBFs increase several fold.

DP and ROVs are a marriage made in heaven. Each enables the other to do far more than it could do alone. Our belief, forty five years ago, that this combination would be the key to deep water exploration has been borne out in spades. And that is very satisfying, indeed.

The people I have worked with all these years have been superb...some of the finest, brightest and kindest people in the world.

My thanks to my wife, Ann, who is here and who kept our children flourishing while being shuttled from our three coasts to Holland and back. Traveling has been a great pleasure to us all.

Altogether this has been a fantastic lot of fun...probably more than one person should be allowed.

Thanks again to the Society of Naval Architects and Marine Engineers for this prestigious award. The list of prior recipients is humbling but I have to say this is yet another source of joy.

William H. Webb Medal

The 2007 William H. Webb Medal for "outstanding contributions to education in naval architecture, marine or ocean engineering" was presented to **Roger H. Compton** by Ronald K. Kiss.

William H. Webb Medal Presentation

I am extraordinarily privileged to have the opportunity to present the Webb Medal to my friend, colleague and Dean of Webb Institute, Dr. Roger H. Compton.

Having just completed his term as President of our Society he is well known to many of you, but holding that office was not what made Roger deserving of this award. The award is given for "Outstanding Contributions to Education in Naval Architecture, Marine or Ocean Engineering." Roger's contributions came through scholarship, engineering and teaching.

For scholarship he has written over 39 publications during his career. Five were published in Marine Technology (one earned SNAME's Cochrane Prize); two were published in the ASNE Journal; and two were presented before the Annual Meeting of our Society. For engineering he was a key part of the team that developed the conceptual design for the United States Naval Academy's 120-foot and the 380-foot towing tanks including their data acquisition systems. He then served as the Director of the Ship Hydromechanics Laboratory, which serves as a vital part of the experimental education for midshipmen. The American Society of Naval Engineers honored him by awarding him the Solberg Award for excellence in research in 1996. But to me it is in teaching that one best exemplifies the spirit of William H. Webb for who this award is named. It was Mr. Webb who made Roger's undergraduate education possible. He was taught by Jens Holm and Cedric Nevitt, two of the best classroom professors there were.

With such a great foundation Roger has been in the classroom for over 41 years and he has set the standard for mentoring, motivating and educating countless young men and women over this time period. During his service at the U.S. Naval Academy, he directed and supported at least 68 midshipmen in research projects and served as faculty advisor for another five midshipmen in their preparation of Trident Scholar projects, which are the equivalent of an undergraduate thesis. Thirteen of the research projects were presented as papers before technical groups including ASNE and SNAME. Four of these papers were awarded Undergraduate Paper Honor Prizes by SNAME. Three of the Trident Scholar projects were presented as technical papers at annual meetings of our Society and several were recognized as the most outstanding projects developed at the Academy among all departments. Dean Compton continues this leadership as a thesis advisor on numerous undergraduate theses at Webb Institute.

While at Webb Roger introduced the concept of having freshman teams design AND construct a vessel starting during orientation and culminating with a performance contest during parents' weekend in late September. Get'em interested while they're young was the idea, and Webb's enrollment now stands near a record, which demonstrates the wisdom of this concept. As many of you know I served as President of Webb for six years and presidents are always concerned about their legacy. I can honestly say that the legacy I'm most proud of is bringing Roger to Webb as Dean and seeing him and his students thrive. There is a story here – Roger had been pursued off and on over the years to leave his post at Annapolis and come to Webb. For various reasons he elected to stay put in Annapolis. But in November 1997, he heard about my appointment to Webb and sent me a fateful email of congratulations wherein he said "If there's anything I can do to help, let me know." And the rest is history – Roger arrived at Webb with me and he continues to earn the Webb Medal everyday – as a superb teacher and educator.

Join me in congratulating Roger Compton, the 2007 Webb Medalist.

William H. Webb Acceptance

President Kramek, ADM Sullivan, SNAME Colleagues, and my true friend, Ron Kiss:

I can think of no greater or more humbling honor than to receive from you, my professional colleagues an award named for William H. Webb. To have had the privilege of receiving a "free and gratuitous" education because of a man whose accumulated fortune was dedicated at the end of the 19th century to establishing two pillars of our industry – Webb's Academy and Home and the Society of Naval Architects and Marine Engineers; to have been mentored and taught by true giants of our discipline including Cedric Ridgely-Nevitt, Jens Holm, and Tom Curran (the first two of whom would be truly confounded by my being here tonight); to have developed so may life-long friends and professional colleagues like Ron Kiss; to have been offered the opportunity to have a career where my love of naval architecture could be shared with eager young minds at two of our country's finest institutions of higher learning – the U.S. Naval Academy and Webb Institute (of Naval Architecture!); to have enjoyed seeing the professional success of many former students; and, probably most importantly on a personal level, to have looked forward to going to work virtually every day of my career – I have been truly blessed.

Even being considered for this most prestigious award for academics who specialize in naval architecture, marine or ocean engineering has exceeded my fondest hopes of professional achievement. When I stumbled across an ad in The Chronicle for an assistant professor to teach naval architecture at the U.S. Naval Academy, I was strongly encouraged to apply by Dean Tom Curran and 2001 Webb medalist, Prof. Bob Zubaly. As I look into the faces of current Webb students, I'm envious of the challenges and opportunities that lay ahead of them as they enter an international marine industry that's robust, expanding, and hungry for their talents. It's our job as senior faculty members to ensure that we're preparing them optimally for careers that will stretch beyond the midpoint of the twenty-first century. While the fundamental principles of ship performance and design must continue to be the foundation of our curricula, we've got to be aggressive in introducing rigorous coverage of advanced and nascent technologies with promise to revolutionize structures, energy conversion and delivery systems, automation, robotics, fabrication and environmental stewardship. We must do that while improving the leadership potential, the communication skills, and the recognition of the importance of life-long learning of our students. A tall order, but a national priority.

Of course, my ability to focus on my career and students is completely dependant on having a partner who not only supports me, but is actively involved with student and faculty activities – on and off campus. This she does in addition to pursuing her own professional activities. Thank you, Jill, for putting up with the time I spend in my office and for sharing the joys and pain of maintaining our personal fleet of small boats.

Thanksgiving has come early for me this year. Thank you Mr. Webb, and thank you all – very much.

David W. Taylor Medal

The 2007 David W. Taylor Medal for "Notable Achievement in Naval Architecture and/or Marine Engineering" was presented to **Hans G. Payer** by Robert Kramek.

David W. Taylor Medal Presentation

I have know Hans Payer for many years.....not only has he been a leader in our industry and teacher as well, but a Leader of the World Classification Societies, having served as Chairman of IACS (International Association of Class Societies), and represented GL as their Council Member for many years....he is a friend and valued member of SNAME...let me recall his achievments for you.

Dr. Payer joined the German Ship Classification Society, Germanischer Lloyd, or GL, in 1971 after completing his university studies. Early in his career at GL he was made head of the Structural Research Department and later head of the entire Research Department. Under his direction, there was a thorough upgrading of structural analysis capabilities including the then new finite element technique that he had learned at Berkeley. As head of research, he continued to improve GL's full range of capabilities for both structural and hydrodynamic analysis.

In 1988, Dr. Payer was appointed Managing Director of the Hamburg Ship Model Basin, or HSVA, a post that he held for five years before returning to GL. During his tenure at HSVA, he oversaw a thorough modernization of the facilities and technical capability of the laboratory until it now ranks with the finest in Europe. After returning to GL from HSVA, he was made a member of the GL Executive Board and CEO of GL Maritime Services, where he remained until his retirement in 2003. He continues today in an advisory capacity to GL as Consultant and Foreign Affairs Advisor. Both the GL and the HSVA positions made use of his outstanding managerial and organizational skills while enabling him to exercise his technical knowledge and abilities at the highest levels in both ship structures and ship hydrodynamics. There is no doubt that he played a key role in guiding GL to its dominant world position in the design, analysis and classification of large container ships.

Dr. Payer's international activities in the ship field are most extensive, including both the International Ship Structures Congress and the International Towing Tank Conference. He has served as head of the International Association of Classification Societies, the Tanker Structures Cooperative Forum and advisor to the International Registry of the Marshall Islands. He has held academic appointments as Visiting Professor at Berkeley and at the Technical University of Rio de Janeiro.

Dr. Payer's publications include over 100 technical papers on subjects as diverse as finite element analysis, post buckling behavior of structures, propeller-induced hull vibration, computational fluid dynamics, cavitation, ship operations, and human factors in ship safety. Virtually all of his publications have been technically substantial, not merely descriptive, and comprise important contributions to the technical literature of the field. A significant number of them have been published in the SNAME Transactions and Journals including four Transactions papers since 2001. He lectures extensively on contemporary topics such as Green Ships, the Panama Canal Expansion, Global Integration of Transport, and Human Factors in Ship Design and Operation.

Dr. Payer was born in Munich, Germany and spent his early years in Innsbruck, Austria. After graduation from high school in Innsbruck, he pursued undergraduate studies at the Technical University of Vienna. He then moved to the University of California at Berkeley for graduate study in the Department of Naval Architecture which awarded him the MS and PhD degrees. Dr. Payer joined SNAME in 1967 and was elected a Fellow in 2005. He is also a member of the Schiffbautechnische Gesellschaft of Germany which awarded him its silver medal in 1982, and, in 2004, he received national recognition when he was awarded the Cross of the Order of Merit of the German Federal Republic.

It is with great pleasure and an honor to present this award to Doctor Hans Payer.

David W. Taylor Medal Acceptance

Dear Mr. President, dear Ladies and Gentlemen,

There are two affiliations of particular importance in my professional life, the University of California, Berkeley and the Society of Naval Architects and Marine Engineers, SNAME.

I came to Berkeley in the late sixties, the time of flower-power. Many of the students as well as other affiliates of UCB with long hair were walking about the campus with balloons, reciting poetry. One could get the impression this was all play and no study. But the reality was not so. Life at UCB was inspirational but also hard work, particularly in engineering.

I joined SNAME and had my first experience with monthly meetings of the SNAME – Northern California Section in the top floor of the Engineer's Club in downtown San Francisco. Here I remember two aspects which impressed me most at these meetings:

- The view of the city
- The contents of the papers

The late sixties were exciting times in shipbuilding; tankers were rapidly getting bigger, the 240,000 dwt designs followed soon after the first 160,000 dwt tanker had been launched and they in turn were followed rapidly by tankers of then unbelievable 300,000 all the way to 500,000 dwt. Containerships emerged with all their challenges of open-top design. For both ship types most of the new problems were structural. At the graduate naval architecture department of UCB you had the choice between hydrodynamics as your main subject, which was arguably more glamorous, and ship structures, which was important in the background but also more down to earth or hands-on. I decided for structures.

Naval architects and marine engineers are lucky people. Most can do what they really like: to find the best solutions to the challenges of efficient and reliable maritime transportation.

The growth of tankers stopped abruptly with the oil crisis of 1973. Growth in the demand for oil stagnated and suddenly there was a severe overcapacity of tanker tonnage. Containerships, however, really took off. Designs of the first, second and third generation of containerships followed one another within three to five years. The designs at that time came mostly from U.S. naval architecture firms. But many of these ships were built in Europe, particularly in Germany. That is where I went from Berkeley. And I took my Berkeley experience in the finite element technique with me. Together with a young engineering team at the GL Research Department we developed in Hamburg analysis techniques particularly for containerships. Over the years GL became the leading Class Society for this type of vessels.

In our complicated world most accomplishments do not come from one person alone but from a good team and in co-operation with other teams. The development of containerships was all about team work; a close co-operation between designers, owners, yard and class was a prerequisite for the successful evolution of this type of ship.
At this point I would like to address the many naval architecture and marine engineering students and the young SNAME members here this evening: You are in a very exciting field. Shipping, and particularly container shipping today, is the backbone of world trade and of the world economy. There would not be the present network of world manufacturing and world trade, there would not be globalization to the degree we see it today without the reliable, dependable and efficient container transport system. There is a strong need worldwide for good naval architects and marine engineers as well as for mariners. My experience has shown that if you take a positive attitude with all the challenges you encounter you'll be able to really move things.

The shipping world is relatively small and very international. You make many friends and you'll mostly work with friends. I have had a wonderful time so far.

I want to thank first my teachers at UC Berkeley, and I am very happy and appreciate it very much that Professor J.R. Paulling is here tonight. I have learned a lot from him, and he invited me to write my first SNAME paper with him in 1968. I want to thank GL for all the freedom I had during the more than 30 years of affiliation and all the support I received. My wife has accompanied me to Berkeley – we got married in San Francisco – and has been at my side with understanding and support throughout my career. I thank you from my heart, Aiga. Finally I thank SNAME, thank you Robert Kramek, thank you Naresh Maniar.

I am very happy and I am proud to receive the David Taylor Medal. I am in good company. Professor Paulling and Professor Schade are also recipients of this prestigious honor. "I have the simplest of taste" Oscar Wilde wrote, "I am always satisfied with the best" Thank you.

Election of New President

Prof. Jose Femenia, Chair of the Nominating Committee, presented the nomination for the next President of the Society. Prof. Femenia announced the committee's nomination was R. Keith Michel. The nomination was seconded by Bruce S. Rosenblatt, Peter G. Noble and Joseph J. Cuneo.

Bruce S. Rosenblatt
It is really a true pleasure to second the nomination of my great friend, Keith Michel, to be the next President of our great society.

I have to thank SNAME for introducing the two of us. Fifteen years ago, we both sat on SNAME's Executive Committee, and I guarantee you we were not the typical Executive Committee members. Dave O'Neil dubbed us the "two young bucks" as

- we were far below the average age of other Executive Committee members,
- we were anxious (in other words, not very patient) for great things and
- we wanted to change the world

Soon it will be Keith's turn to try to lasso in today's "young bucks."

The friendship that developed quickly led to annual, or even more frequent, vacations with Keith, his wife Peggy and various Michel family members. These vacations generally centered on scuba diving, but when not diving, we'd often discuss just about every aspect of SNAME, including who we thought should be the next President of SNAME. Had Keith been less active running Hebert Engineering and in other marine industry related work such as the Marine Board and Webb Institute, perhaps I could have been doing the same thing I am doing at this very moment, just ten years ago.

But they say good things are worth waiting for and that saying applies perfectly when we think of Keith as SNAME's next President. Keith is a leader, often a quiet leader, but a leader no doubt. He has done just about anything one can with SNAME with perhaps his efforts chairing SNAME Planning Committee standing out. His overall ability to always think strategically will make him the perfect President. Keith is a man of great integrity, modesty and loyalty. I consider it an honor to be called his friend and to be able to endorse his nomination.

Accpetance Speech R. Keith Michel SNAME President-elect, November 15, 2007

President Kramek and fellow members:

It is an honor to be nominated as your next president. I would like to express my thanks to Joe and Peter and Bruce for their kind words.

As you know, SNAME functions through the efforts of a large group of dedicated volunteers and the support provided by a very capable headquarters' staff. I look forward to working with all of you over the next several years.

Those of us involved with the governance of our Society recognize the importance of enhancing the benefits SNAME provides to its members. A number of new initiatives are underway. These include:

- Expanding SNAME's scholarship program
- Strengthening the Technical & Research programs
- Re-tooling the SNAME website
- Changes to the structure of the Annual Meeting
- Holding collegia round the world, as we seek to enhance our outreach to international members

These are exciting times for the maritime and offshore industries, and SNAME must keep pace with the changing environment. In the interest of continued improvement, President Kramek has asked me to spend the next year working with Bill Hurley and his Planning Committee, to further refine SNAME's strategic plan. I look forward to participating in the planning efforts, and to serving as your president.

2007 Annual Meeting Photographs

Following are selected formal photographs from the Annual Meeting.
Some events and awardees are not pictured because formal photographs were not available.

50 YEAR GOLDEN MEMBERS

Fifty-Year Golden Membership Certificate awarded to Donald Caldera (r) by Bob Kramek (l)

Fifty-Year Golden Membership Certificate awarded to Joe Cuneo (r) by Bob Kramek (l)

Fifty-Year Golden Membership Certificate awarded to James Price (l) by Bob Kramek (r)

Fifty-Year Golden Membership Certificate awarded to Ray Pearlson (r) by Bob Kramek (l)

AWARDS

Honorary Membership Certificates went to (l to r): Michael Parsons and Thomas Mackey, presented by Jose Femenia and Bob Kramek

Bob Kramek (l) presented Certificates of Appreciation to (l to r): Tom Mackey, John Higginbotham, and Matt Unger

The Cochrane Prize was awarded to (l to r): Raju Datla and Dan Savitsky by Ed Lewandowski

The ABS Linn ard Prize was awarded by Xiaozhi Wang (c) to Raul Brostella (l) and Hans Payer (r)

AWARDS

The Elmer L. Hann Award presented by Rick Spaulding to (l to r): Larry DeCan, T.D. Huang, Chris Conrardy, and Lee Kvidahl

The William Kennedy Award presented by Rick Spaulding (l) to Jay Carson (r)

MEDAL RECIPIENTS

Bob Kramek (l) presented the David W. Taylor Medal to Hans Payer (r)

Roger Compton (l) was presented with the William H. Webb Medal by Ron Kiss (r)

Howard Shatto (l) was presented with the Blakely Smith Medal by Tim Pease (r)

MEDAL RECIPIENTS

Bruce Rosenblatt (l) was presented with the Vice Admiral "Jerry" Land Medal by Keith Michel (r)

Past, Present and Future Presidents (l to r): Jose Femenia, Bob Kramek, Roger Compton, Ron Kiss, Joe Cuneo, Bruce Rosenblatt, and Keith Michel

Papers Presented at the 2007 Annual Meeting

Review of Structural Design of Aluminum Ships and Craft

Robert A. Sielski (AM)

ABSTRACT

Design procedures used for aluminum structure are compared with those used for steel. The paper highlights basic areas of loads, analysis procedures, and allowable stress levels in the context of materials and fabrication to benefit the designer who is well versed in design of steel structure when designing for aluminum.

KEY WORDS: aluminum; ship structure; structural design.

NOMENCLATURE

ASTM	- American Society for Testing and Materials
AWS	- American Welding Society
b_e	- Effective breadth of plating in plate-beam
C	- Coefficient used in plate design
D	- Depth of a structural member
da/dN	- Fatigue crack growth rate in distance per stress cycle
E	- Elastic modulus
Eurocode 9	- European structural design standard
Extrusion	- Structural profile formed by pressing heated metal through a die
GRP	- Glass reinforced plastic
HAZ	- Heat affected zone
k	- Factor based on panel aspect ratio
K	- Stress intensity factor
l	- Length of a welded panel
M	- Bending moment
p	- Pressure
s	- Width of a stiffened panel
SM	- Section modulus
t	- Thickness of plating
g	- Specific weight of seawater
s	- Stress
σ_y	- Yield strength, proof stress
σ_a	- Allowable design stress

INTRODUCTION

Aluminum was unknown until 1808 when its existence was established by Sir Humphry Davy in Britain, who named it alumium. It did not become commercially viable until 1886 when the Hall-Héroult electrolytic process was developed independently in the United States and in France. After that, the price dropped dramatically, and aluminum became an increasingly useful metal for a multitude of purposes, from cookware to boats.

The development of aluminum in marine use is chronicled in many places, including papers read before the Society of Naval Architects and Marine Engineers. McGuire (1895) described the use of aluminum for the America's Cup yacht *Defender*. Faragher (1931) described aluminum's use up till 1930, and Rock (1931) told of the alloys in use at that time, with Efield (1937) providing an update six years later. Ferris (1931) told of the use of aluminum in passenger ships, as did Sharp (1937), who described comparative fire tests between wooden and aluminum joiner work. McComb and Archibald (1950) illustrated further fire testing on staterooms. Mears and Brown (1944) told of early testing to prove the corrosion resistance of aluminum alloys developed for marine use. Bates (1945) advocated the increased use of aluminum for the superstructure of large vessels. Forrest (1947) described the use of aluminum in naval and commercial applications as did Boykin and Sellers (1953). Vasta (1949) described structural tests on the interaction between the steel hull and the aluminum superstructure of a passenger ship. Holtyn (1966) brought an update on the use of aluminum, and Holtyn et al. (1972) described the design and construction of the largest aluminum ship built up to that time.

Aluminum began to be used in the 1940s for pleasure craft and for workboats, which have greatly increased in size over the years. The use of aluminum for the hulls of high-speed merchant vessels began in the 1990s with increased construction of high-speed ferries. These vessels have become so technologically advanced that they have surpassed the capabilities of many naval vessels; many navies today are adapting derivatives of these high-speed vessels to combatant craft. Two recent papers were read before this society discussing the structural design of such craft. Kramer (2005) demonstrated the application of classification society rules to the design of a high-speed aluminum catamaran, and Stone and Novak (2006) compared the requirements of several classification societies for the design of aluminum high-speed craft. Because of this recent interest in aluminum for marine construction, the interagency Ship Structure Committee has sponsored and continues to sponsor work in aluminum structures, including the development of the recently completed Aluminum Marine Structure Design and Fabrication Guide, on which much of this paper is based.

DESIGNING WITH ALUMINUM

Boykin and Sellers (1953) pointed out that aluminum is not just a substitute for steel; it is an entirely different material. Therefore, even though most of the design procedures used for steel structure can be used for aluminum, the designer should be aware of differences. Those differences should begin with an overall awareness of the structural design process. The process is illustrated by Fig. 1, which was first illustrated to me by Malcolm Dick of Gibbs & Cox in the mid 1970s. The structural designer must consider structural loads, analysis procedures, and allowable stress levels, but all in the context of the materials used and the fabrication methods employed.

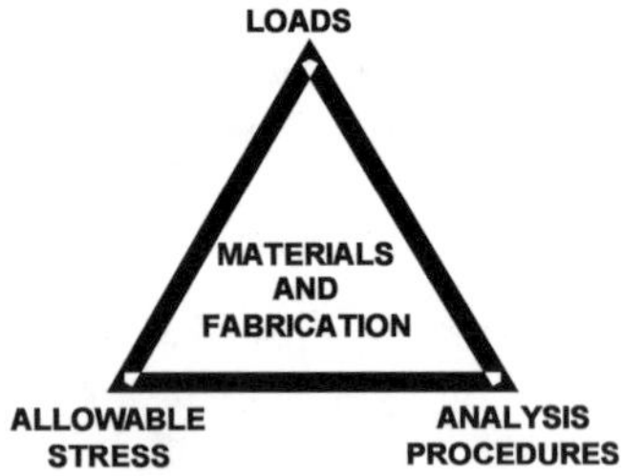

Fig. 1: The structural design triangle.

An important aspect of the design triangle is that one must not mix-and-match between different criteria, but must approach all in an integrated fashion. A particular design load that has worked well for steel structures may not be directly applicable to aluminum. As an example, the equivalent hydrostatic head used by the U.S. Navy for the bow structure of cruisers and destroyers has always been augmented by the undocumented practice of placing intercostal stiffeners to limit plate deformation. Therefore, these design loads would not be applicable to different structural configurations, and may not be applicable for aluminum, which deforms differently from steel, as will be explained below.

Structural analysis procedures can range from the use of simple design formulas such as used in classification society rules to detailed analysis using methods such as finite element analysis or even through testing of prototype structures. The design allowable stresses and the applicable loads for one form of analysis do not necessarily apply to the others.

Reliability Methods

The above discussion on closing the design triangle of loads, analysis, and allowable stress levels has been based on deterministic methods of developing a design procedure with inherent factors of safety that are in line with operational experience of ships designed to the same criteria. However, there is much uncertainty in the analysis, particularly in the loads. Material properties not only have their inherent range of properties due to differences in manufacture and in fabrication of the final product, but uncertainty in areas such as the effect of welding and imperfections on strength. Even when structural analysis is performed by the finite element method, there are still uncertainties associated with choices made in modeling, and compromises made, such as the amount of plate to be included in plate-beam combinations.

Much progress has been made in developing load and resistance factor design methods for steel ships, but there is little progress in the assessment of aluminum craft. In principle, the method developed for steel ships can be applied directly to aluminum craft, but there are calibration issues that should be considered, particularly because the response of high-speed craft with unusual hull forms that characterize aluminum vessels were not considered in code development to date.

Materials and Fabrication

Aluminum Product Forms. The principal product forms in which aluminum is used are plate and extrusions, which along with forgings are called wrought products. With aluminum, a four-digit number describes the chemical composition of the alloy, and this number is followed by a letter and several digits that describe the heat treatment and work hardening that the alloy receives. There are two basic chemical compositions of aluminum used for marine purposes, the 5xxx-series and the 6xxx-series. The alloys that have 5 for the first digit, the 5xxx-series, have magnesium as the principal alloying agent, and many also have a significant amount of manganese. The 5xxx-series are not heat-treatable, but obtain additional strengthening by work hardening. The alloys with a 6 as the first digit, the 6xxx series, have magnesium and silicon as principal alloying agents. These form magnesium silicide, which makes the alloys heat-treatable. The remaining three digits represent the specific alloy composition. These alloys have good strength and corrosion resistance, and are easily welded, although the strength of both series diminishes significantly with welding, particularly the 6xxx-series. Other alloys should be avoided for marine use, especially the 2xxx and 7xxx-series, which are used extensively for aerospace structures. The chemical compositions of some of the commonly used alloys are provided in Table 1. Following the 4-digit number in the alloy designation there is a letter followed by up to three digits. The letter H indicates strain hardening, and T indicates heat treatment. If the alloy is produced in the annealed condition, the letter O is used.

A variety of alloys are available within these series, but the 5xxx-series plate alloys for marine use should be restricted to those listed in ASTM B 928-04, which are the following: 5059-H116, 5059-H321, 5083-H116, 5083-H321, 5086-H116, 5383-H116, 5383-H321, 5456-H116, and 5456-H321. The exceptions are the 5xxx alloys containing less than 3 percent Mg or in tempers that are not susceptible to sensitization, such as the annealed or O temper. An example is alloy 5454-H32, which has a specified magnesium range of 2.4 to 3.0 percent. This alloy / temper product can be used in applications where the operating temperature is in excess of 65 ^{0}C (150 ^{0}F) to avoid problems of stress corrosion cracking. The International Association of Classification Societies in their Requirements Concerning Materials and Welding (IACS, 2005) specify, "… the alloy grades 6005A, 6061 of the 6xxx series should not be used in direct contact with sea water unless protected by anodes and/or paint system." Alloy 6061 has 0.15 to 0.40 percent copper, 6005A has 0.30 percent copper, but the 6082 alloy has only 0.10 percent copper, and has no such restriction on its use. Up to recently, the U.S. Navy has completely prohibited the use of 6xxx-series alloys, but has begun to accept them for some applications.

Table 1: Chemical Composition of Aluminum Alloys (Percentage by weight) (The Aluminum Association, 2006a)

Alloy	Si	Fe	Cu	Mn	Mg	Cr	Zn	Ti
5059	0.45	0.50	0.25	0.60-1.2	5.0-6.0	0.25	0.40-0.90	0.20
5083	0.40	0.40	0.10	0.40-1.0	4.0-4.9	0.05-0.25	0.25	0.15
5086	0.40	0.50	0.10	0.20-0.70	3.5-4.5	0.05-0.25	0.25	0.15
5383	0.25	0.25	0.20	0.7-1.0	4.0-5.2	0.25	0.40	0.15
5454	0.25	0.40	0.10	0.50-1.00	2.4-3.0	0.05-0.20	0.25	0.20
5456	0.25	0.40	0.10	0.50-1.00	4.7-5.5	0.05-0.20	0.25	0.20
6005A	.50-.90	0.35	0.30	0.50	.40-.70	0.30	0.20	0.10
6061	.40-.80	0.7	.15-.40	0.15	.80-1.20	0.04-0.35	0.25	0.15
6063	0.20-0.60	0.35	0.10	0.10	0.45-0.90	0.10	0.10	0.10
6082	0.7-1.3	0.50	0.10	0.40-1.0	0.6-1.2	0.25	0.20	0.10

The alloys listed under ASTM B 928 are produced in plate and sheet product forms. With aluminum, structural shapes are produced by the extrusion process, which is similar to dough being pushed through a cookie press. A billet of aluminum is heated to near the melting point, and is pressed through a die to form the desired structural profile. The process is versatile, and a new design for a die is relatively inexpensive to produce, so a wide variety of extrusions are available, or can be specially developed to meet a need in a new vessel design.

For extrusions, the alloys most used are 5059-H111, 5083-H111, 5086-H111, 5383-H112, 5454-H111, 6005A-T61, 6061-T6, 6063-T6, and 6082-T6. The 6xxx-series alloys extrude at a rate that is about four times faster than a similar shape in a 5xxx-series, and can be produced in much thinner sections, as little as 2-mm (0.08-inch) thick sections, whereas the 5xxx-series alloys have a limitation of about 3.3 mm (0.128 inches) of thickness in extruded shapes. The 6xxx-series extrusions are therefore less expensive and more versatile than the 5xxx-series, in spite of the limitation on their use.

Extrusions are commonly produced in the conventional flat bar, bulb angle, angle, and tee sections such as are used in steel construction. However, a wide variety of other shapes are used. The rounded tee section shown in Fig. 2 is more compact than a conventional tee section, and can be welded more closely together. Integrally stiffened panels as shown in Fig. 3 and sandwich panels shown in Fig. 4 are less expensive to produce than welded panels. They weigh less than conventional structure because of the close stiffener spacing, but have less distortion than welded structure. Special extrusions such as those shown in Fig. 5 can designed and produced at a small cost.

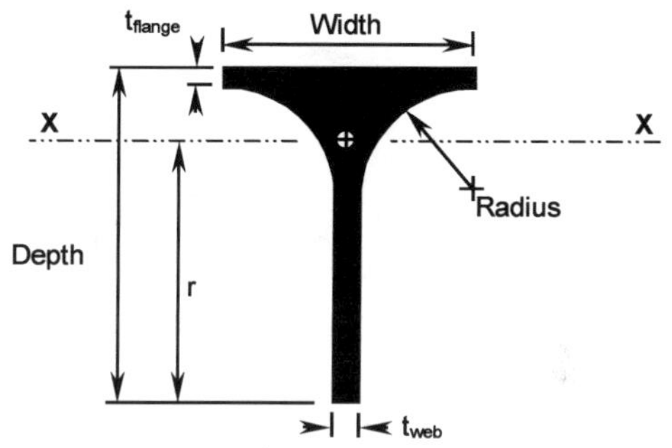

Fig. 2: Rounded tee profile.

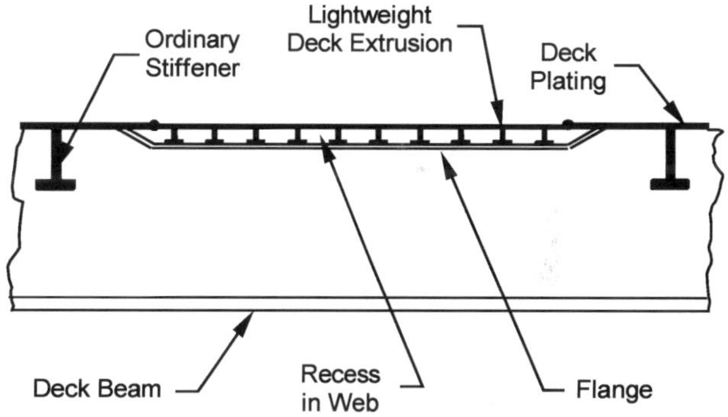

Fig. 3: Recess in deck transverse in way of lightweight deck extrusion.

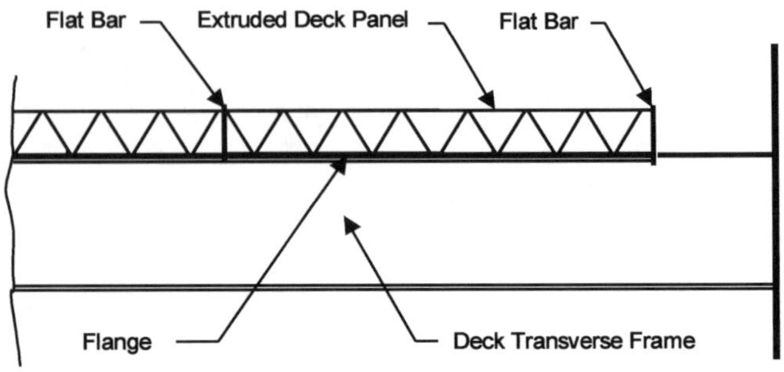

Fig. 4: Corrugated deck extrusion.

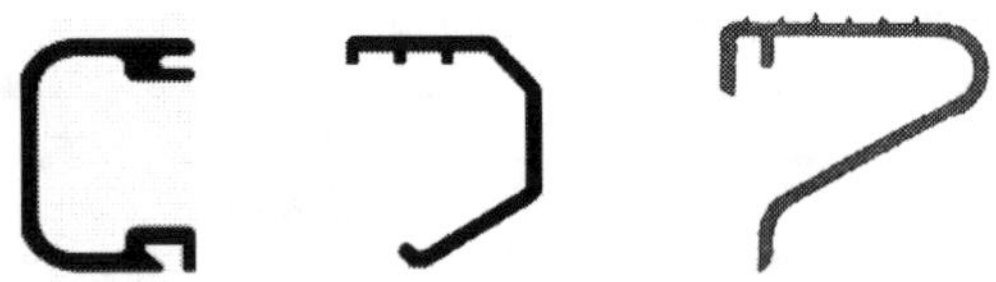

Fig. 5: Special extrusions for gunwale.

The friction stir welding process, which has made the lightweight extruded panels even more viable, is illustrated in Fig. 6. This process has proven to be economical for joining several extrusions to form a single panel.

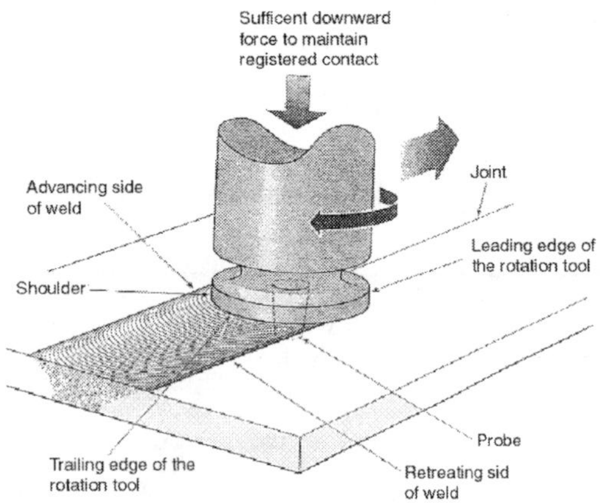

Fig. 6: Schematic illustration of friction stir welding (The Welding Institute Web Site).

In this process a steel tool is rotated at high speed to soften the aluminum to a temperature that is less than the melting point, but softens it so that the tool can pass through with the aluminum forming a bond after the tool passes. With this process, extruded panels that are about one foot wide are joined to form wider panels, reducing the amount of welding done in the shipyard, and producing a panel that has very little distortion.

Mechanical properties. The mechanical properties of several alloys are given in Table 2. The properties given for the tensile strength and yield strength are for base metal, not for welded material. The welded strength of aluminum is significantly less than the strength of the base metal, particularly in the 6xxx-series. The welded strength is not a specified value used in material forming and in procurement, but is an experimental value determined through testing. In general, the strength of weld metal is about the same strength as the heat affected zone surrounding a weld. Historically, there have been two standards for determining the welded properties of aluminum, tensile testing using either a 50-mm (2-inch) gauge length specimen or a 250-mm (10-inch) gauge length specimen. The longer specimen includes significant base metal, but the shorter specimen generally contains only the weld metal and heat affected zone of the weld. Therefore, the shorter specimens indicate a lower strength than do the longer specimens. Different authorities have used one or the other standard over the years, although The Aluminum Association has recently changed to the shorter specimen.

The variation in the testing standards as well as differences in test data make for a significant difference in the welded strength, as shown in Table 3. These differences can leave a designer in a quandary. When designing in accordance with a particular design code, the value of welded strength used should be the value specified in that design code because that welded strength should be consistent with the analysis methods and allowable strength levels that is used in the design code.

Fig. 7, taken from Boykin and Sellers (1953), illustrates some of the differences in the behavior of aluminum and steel under load. The comparison is between mild steel and 6061-T6, which is used for structural shapes. Unlike most structural steel alloys, aluminum does not have a defined yield point. Instead, the material begins to gradually soften as stress increases. To define the yield strength, an offset to the initial stress-strain curve is taken at 0.2 percent strain, and the intersection of this line with the stress-strain curve is defined as the yield strength or proof stress.

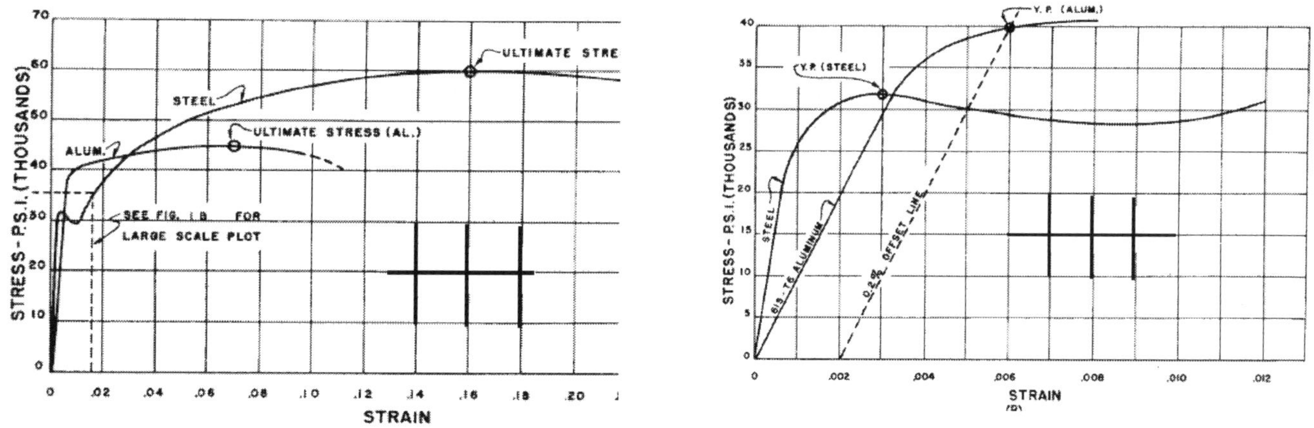

Fig. 7: Typical stress-strain curves for aluminum and steel (from Boykin and Sellers, 1953).

Table 2: Mechanical Properties Marine Aluminum Alloys

Alloy and Temper	Thickness Range		Ultimate Strength [5]		Yield Strength [5] (0.2% Offset)		Elastic Modulus		Density	
	in	mm	ksi	MPa	ksi	MPa	ksi $x10^3$	MPa x 10^3	lbs/in^3	g/cm^3
5059-H111 (E)[1]	0.114-1.968	3.0-50	47.7	329	23.2	160			0.096	2.66
5059-H116 (S&P)[1]	0.114-0.787	3.0-20	53.5	438	39.1	270			0.096	2.66
5059-H116 (P)[1]	0.788-1.968	20.1-50	52.1	359	37.6	259				
5059-H321 (P)[1]	0.114-0.787	3.0-20	53.5	369	39.1	270				
5059-H321 (S&P)[1]	0.788-1.968	20.1-50	52.1	359	37.6	259				
5083-H111 (E)	<=5.0	<=130	40.0	275	24.0	165	10.3	71.0	0.096	2.66
5083-H116 (S&P)	0.188-1.5	4.0-40	44.0	305	31.0	215	10.3	71.0	0.096	2.66
5083-H116 (P)	1.5-3.0	40-80	41.0	285	29.0	200	10.3	71.0	0.096	2.66
5083-H321 (S&P)[1]	0.063-1.5	1.6-38	44.0	303	31.0	214				
5083-H321 (P)[1]	1.501-3.0	38.1-76.5	41.0	283	29.0	200				
5086-H111 (E)	<=5.0	<=130	36.0	250	21.0	145	10.3	71.0	0.096	2.66
5086-H116 (S&P)	All	All	40.0	275	28.0	195	10.3	71.0	0.096	2.66
5383-H-112 (E)[2]			45.0	310	27.6	190	10.2	70.0	0.096	2.66
5383-H116 (P)[2]	<0.79	<20	44.2	305	31.2	215	10.2	70.0	0.096	2.66
5454-H111 (E)	<=5.0	<=130	33.0	230	19.0	130	10.3	71.0	0.097	2.69
5454-H32 (S&P)	0.02-2.0	0.5-50	36.0	250	26.0	180	10.3	71.0	0.097	2.69
5456-H116 (S&P)	0.188-1.25	4.0-12.5	46.0	315	33.0	230	10.3	71.0	0.096	2.66
5456-H116 (P)	1.251-1.5	12.51-0.0	44.0	305	31.0	215	10.3	71.0	0.096	2.66
5456-H116 (P)	1.501-3.0	40.01-80	41.0	285	29.0	200	10.3	71.0	0.096	2.66
5456-H321 (S&P)[1]	0.188-0.499		46.0 59.0		33.0 46.0				0.096	2.66
6005A-T61 (E)[3]			38.0	260	35.0	240	10.0	68.9	0.098	2.70
6061-T6 (E)	All	All	38.0	260	35.0	240	10.0	68.9	0.098	2.70
6063-T6 (E)	All	All	30.0	205	25.0	170	10.0	68.9	0.097	2.70
6082-T6 (E)[1]	All	All	45.0	310	38.0	262			0.098	2.70

(E) Extrusions, (S&P) Sheet and Plate, (P) Plate
Notes
1. ABS, Rules for Materials and Welding, 2006
2. ALCAN, Aluminium and the Sea, 2004.
3. Data supplied by Tower Extrusions, Olney, Texas.
4. Aluminum Association, Aluminum Standards & Data – 2006b
5. Where two values are given; the first is the minimum allowable stress and the second is the maximum allowable stress.

Table 3: Yield Strength of Welded Aluminum Alloys (MPa, ksi)

Alloy	Type[1]	Source					
		ABS[2]	DNV[3]	Aluminum Association[4]	AWS Hull Welding[5]	ALCAN[6]	US Navy[7]
5086-H0	E		92 (13)	95 (14)			97 (14)
5086-H32	P	131 (19)	92	95 (14)	131 (19)		152 (22)
5086-H111	E	124 (18)	92	95 (14)	124 (18)		110 (16)
5086-H116	P	131 (19)	92	95 (14)	131 (19)		152 (22)
5083-H111	E	145 (21)		110 (16)	145 (21)		
5083-H116	P	165 (24)	116[8] (17)	115 (18)	165 (24)[9]	125 (18)	
5383-H111	E	145 (21)				145 (21)	
5383-H116	P	145 (21)	140 (20)			145 (21)	
5454-H111	E	110 (16)	76 (11)	85 (12)	110 (16)		110 (16)
5454-H34	P	110 (16)	76 (11)	85 (12)	110 (16)		110 (16)
5454-H32	P	110 (16)	76 (11)	85 (12)	110 (16)		
5456-H111	E	165 (24)			165 (24)		145 (21)
5456-H116	P	179 (26)		125 (19)	179 (26)		179 (26)
6061-T6[10]	E, P	138 (20)	105 (15)	105 (15)	138 (20)		
6061-T6	E, P	103 (15)	105 (15)	80 (11)	103 (15)		

1. E = Extrusion, P = Plate
2. ABS, Rules for Materials and Welding, Part 2, Aluminum and Fiber Reinforced Plastics, Chapter 5, Appendix 1, Table 2, American Bureau of Shipping
3. Det Norske Veritas, Rules for Classification of High Speed, Light Craft and Naval Surface Craft, Part 3, Chapter 3, Section 2, Table B4. Yield strength determined from the values of f_1 published by the equation $\sigma_1 = f_1 \times 240 / 1.1$.
4. Aluminum Design Manual, 2005, Table 3.3-2, The Aluminum Association, Arlington, Virginia. Based on 50 mm (2 in.) gauge length
5. Guide for Aluminum Hull Welding, AWS D3.7, American Welding Society, 2004. Based on 250 mm (10 in.) gauge length
6. Aluminium and the Sea, ALCAN, Paris, 2004.
7. A Guide for the Use of Aluminum Alloys in Naval Ship Construction and Design, Volume II, Table 4.1, David W. Taylor Naval Ship Research and Development Center, DTNSRDC 84/015, 1984.
8. Thickness $\leq$ 9.5 mm (0.275 in.)
9. 5083-H 321 Plate, thickness $\leq$ 38 mm (1.5 in.)
10. Welded with 5356 filler.

As a consequence of the difference in the stress-strain properties of aluminum and steel, some design methods, particularly analysis for compressive strength, which have been validated through testing in steel structure need to be reevaluated by testing with aluminum structure. An example is the testing performed by Paik et al. (2005) on welded aluminum panels.

Another attribute of aluminum illustrated by Fig. 7 is the difference in elastic modulus between aluminum and steel. For the same stress, aluminum will elastically deform three times greater than steel. This affects design in several ways, including consideration of deflection of structure under load, buckling strength, and vibration. The design codes of most classification societies now include some calculations for minimum inertia for structural sections and for the hull girder, and these calculations include the elastic modulus. However, serviceability considerations should always be made, such as the "springiness" of deck structure when walked upon, or the ability of plating to support deck coverings without the materials cracking under load.

The reduced elastic modulus means that aluminum has much less buckling strength than steel, so that the performance of buckling strength calculations during design is very important. Fortunately, aluminum has about one-third the density of steel, so that equivalent structural panels of aluminum and steel have about the same natural frequency of vibration. However, when there is a significant amount of nonstructural mass associated with the structure, the aluminum structure will have a lower vibration frequency, which could lead to resonance problems. Additionally, the greater susceptibility of aluminum to fatigue damage limits the magnitude of allowable vibration levels in aluminum structure.

The difference between vibration of steel and aluminum structure is illustrated in Table 4. A bottom plate panel in the bow of the 60-meter 50-knot craft examined by Stone and Novak (2006) was designed in aluminum and in steel using the criteria of the ABS high-speed craft guide (ABS, 2001). In this case the natural frequency of vibration of the aluminum plate as well as the plate-stiffener is higher than corresponding frequencies of the steel structure. However, with the addition of a nominal dead load of 200 kg/m² (40 lb / ft²), the frequencies of the plate alone become about equal, and the frequency of the aluminum plate-stiffener become lower than the equivalent steel structure. The energy required to maintain 1 mm of amplitude of vibration in the stiffener vibrating in its of the first mode is slightly less for the aluminum stiffener than for the equivalent steel stiffener in the absence of significant damping. Because the energy is proportional to the square of the amplitude, the amplitude of vibration of the aluminum stiffener is only 6 percent greater than the amplitude of the steel stiffener for the same energy of excitation.

Table 4: Comparison of Vibration Frequencies of Equivalent Aluminum and Steel Panels

Characteristic	Aluminum	Steel
Alloy	Plate 5086-H116 stiffeners 6061-T6	HS 36
Yield Strength	Plate 165 MPa (24 ksi) Stiffeners 138 MPa (20 ksi)	355 MPa (51 ksi)
Allowable stress	Plate 148 MPa (21 ksi) Stiffeners 76 MPa (11 ksi)	Plate 320 MPa (46 ksi) Stiffeners 195 MPa (28 ksi)
Panel size	800 mm x 260 mm (31.5 in x 10.2) in	800 mm x 260 mm (31.5 in x 10.2) in
Design pressure	222 kN/m² (32 psi)	222 kN/m² (32 psi)
Scantlings Selected	76mm x 76 mm x 2.2 kg T (3″ x 3″ x 1.5# T) 7 mm plate	63 mm x 51 mm x 4.8 mm L (2.5″ x 2″ x 3/16″ L) 5 mm plate
Plate Frequency (Hz)	281	200
Plate Frequency with 200 kg/m² (40 psf) dead load (Hz)	82	81
Stiffener Frequency (Hz)	395	295
Stiffener Frequency with 200 kg/m² (40 psf) dead load (Hz)	136	148
Energy for 1 mm (0.04 in) amplitude vibration in first mode	8.64 Joules (6.4 ft-lbs)	9.80 Joules (7.2 ft-lbs)

Fatigue. Fatigue is failure of a structure under repeated loading, which for a typical ship can be as many as 100,000,000 wave loading cycles of varying amplitude over the lifetime of the ship. The most common source of fatigue loading comes from repeated encounters of the hull with waves, but there can be other sources such as machinery vibration or waterjet propulsors that cause repeated loading cycles of several orders of magnitude greater than from wave loading, and must be considered in design.

The long-term distribution of loads from wave encounter can be represented by the Weibull distribution, which is given by the formula:

$$N = N_{max} \times \exp\left(-\left(\frac{S}{S_{max}} \right)^{\xi} \times \ln(N_{max}) \right) \quad (1)$$

where:

N is the number of loading cycles at stress range S,
N_{max} is total number of loading cycles,
S_{max} is the maximum stress range,
ζ is the Weibull parameter.

In the Weibull distribution, if $\zeta = 1.0$, the distribution is an exponential distribution. To determine the parameter ζ for the Weibull distribution and the anticipated numbers of loading cycles that will be encountered for monohulls of various lengths, the computer program SPECTRA 8.2 (Michaelson, 2000) developed by NSWCCD was used assuming a frigate hull form operating 80 percent of the time over 20 years in the North Atlantic. The results are shown in Fig. 8 and in Fig. 9. Fig. 8 also shows the Weibull coefficients developed by the International Association of Classification Societies (IACS, 1999) for a draft fatigue guidance document. The numbers differ probably because the IACS document was developer for full-formed bulk carriers, and the current analysis is based on fine-hulled combatant ships.

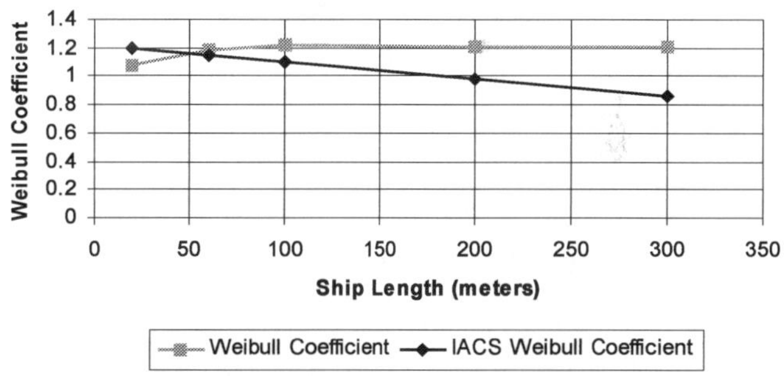

Fig. 8 Weibull coefficient for fatigue spectra versus ship length.

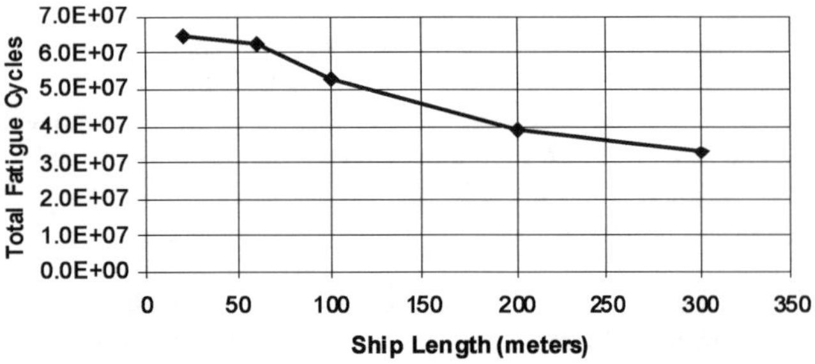

Fig. 9 Total fatigue loading cycles for operation 80 percent over 20 years versus ship length.

If a crack develops in a structure that is subjected to fatigue loading, the crack will propagate at a rate that is a function of the range of stress intensity factor, ΔK, as shown in Fig. 10 for a typical aluminum alloy, 5083, and in Fig. 11 for ABS Grade HS-36 steel. The stress intensity factor is proportional to the applied stress and the square root of the crack size. In the figures, the crack growth per cycle of alternating stress, da/dN is plotted against the range of stress intensity factor, ΔK. In comparing the two figures, it can be seen that for the same ΔK, the crack growth rate in aluminum is about 30 times greater than in steel.

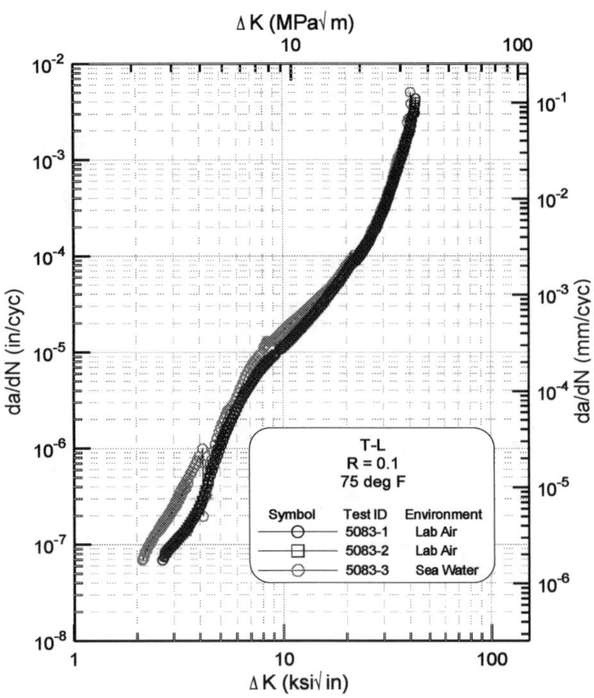

Fig. 10: Fatigue crack growth da/dN curve for 5083 aluminum (Donald and Blair, 2007).

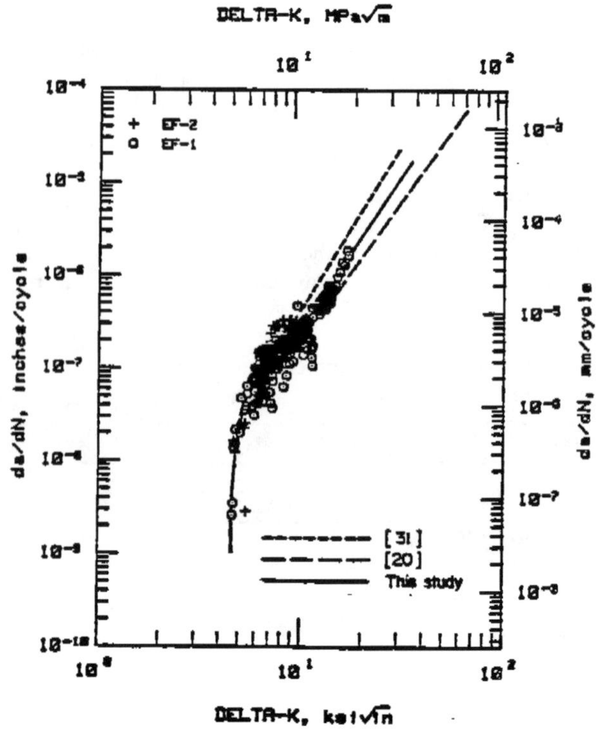

Fig. 11: Fatigue crack growth da/dN curve for ABS Grade CH 36 steel (Leis, 1990).

A computation was made of crack growth rates in the deck of a 41.39-meter 32-knot crew boat for which a comparative design was made in aluminum and in steel. The maximum hull girder bending moments were calculated using the ABS high-speed craft guide (ABS, 2001) and were 19,580 kN-m (6,450 ft-ton) sag and 4,210 kN-m 1,390 ft-ton) hog. These maximum moments and the resulting stresses represent a probability of exceedance of 10^{-8}. As will be explained below, the scantlings in the deck of the aluminum design were increased above the rule minimum to provide additional section modulus to prevent fatigue crack initiation. The resulting section modulus to the deck of the aluminum design was 3,510 cm^2-m, and the resulting maximum stresses in the main deck for the aluminum design were 55.7 MPa (8.1 ksi) compression and 12.0 MPa (1.7 ksi) tension, a maximum stress range of 67.7 MPa (9.8 ksi). The exceedance curve for the fatigue-loading spectrum was assumed to follow a Weibull distribution with a coefficient of 1.13 as determined using Fig. 8, and 6.3×10^7 fatigue loading cycles for 20 years of service at 80 percent operability, determined using Fig. 9. For the HS-36 steel version of the crew boat, the hull girder section modulus was not increased above the amount provided by rule minimum scantlings, and the stresses in the deck were 118 MPa (17.1 ksi) compression and 25 MPa (3.60 ksi) tension, a stress range of 143 MPa (20.7 ksi).

The results for these computations are shown in Fig. 12, where an initial crack size of 24 mm (1 in) was assumed. Under those assumptions, the crack in the aluminum craft would have grown by 26 mm (1.0 in) to be 50 mm (2 in) in length after 2 years of service, whereas the corresponding crack in the steel craft would have grown by only 6 mm to be 30 mm (1.2 in) long. If the aluminum craft continued to operate for another two years without repairs made to the 50-mm crack, the crack would have grown to be about 600 mm (24 in) long, after which it would become unstable and begin to propagate at an extremely rapid rate. For the steel craft, the 30-mm (1.2-in) crack would grow to 38 mm (1.5 in) in the same time, with continued propagation at a fairly stable rate.

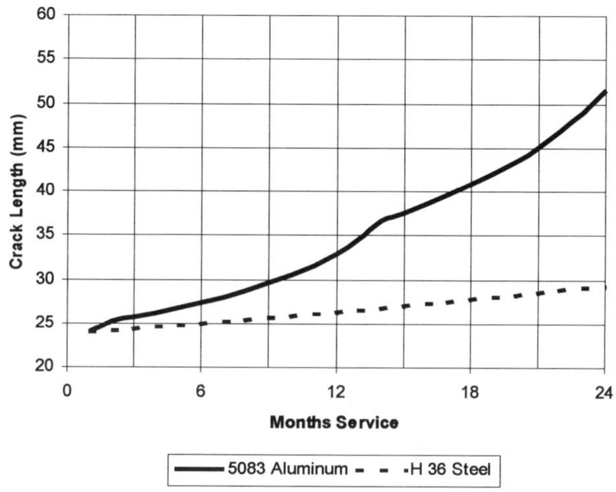

Fig. 12: Predicted crack growth for a 43.9-m 32-knot craft.

The above analysis demonstrates how susceptible aluminum hull structures are to fatigue damage. Furthermore, no methods have been devised to effectively stop or significantly retard fatigue crack growth in aluminum structure. To avoid hull girder failure, the structure must be designed to avoid fatigue crack initiation, and an inspection program is needed to discover and repair cracks that do occur before they become too large.

Design to prevent fatigue crack initiation uses the S-N approach, where N is the number of loading cycles to cause crack initiation at an the alternating applied stress level of S. Fig. 13 is a typical structural detail with its S-N curve illustrated in Fig. 14 (Beach et al., 1981). The data show that for this detail subjected to a constant alternating stress range of 50 MPa (7.3 ksi), crack initiation will occur at anywhere between 1.5 x 10^5 cycles to 1.5 x 10^6 cycles. This order of magnitude range in fatigue testing data is typical for welded aluminum and steel, and for design purposes, a lower limit is set to the data. In Fig. 14, the dotted line represents such a lower limit, with a probability of less than 2.3 percent that failure would occur at a lower stress range. The figure also shows the corresponding S-N curve for this type of detail as required by a new design code, Eurocode 9, which is more conservative than the data would suggest

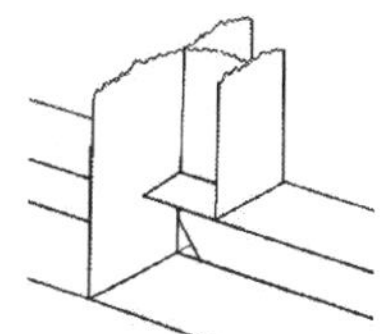

Fig. 13: Typical structural detail for which fatigue testing was conducted (Beach et al., 1981).

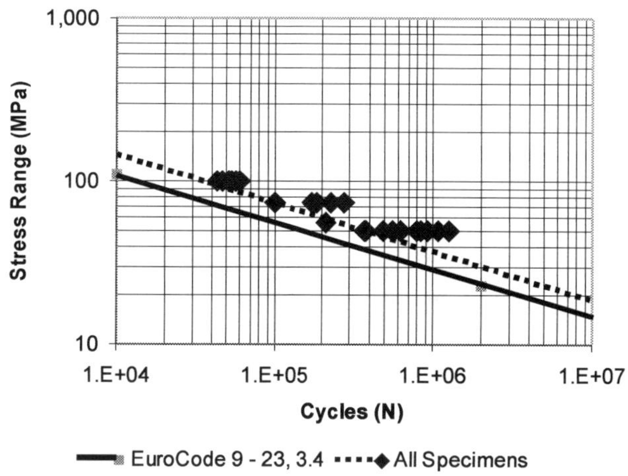

Fig. 14: Fatigue data for stiffener intersection compared to Eurocode 9. Data points from (Beach et al., 1981).

For the high-speed crew boat considered, the scantlings were first determined using the ABS high-speed craft guide (ABS, 2001) for local load requirements. It was found that for both the aluminum and steel craft, the resulting hull girder section modulus exceeded the minimum section modulus required by the ABS guide. The hull girder bending moments calculated using the ABS high-speed craft guide were 19,580 kN-m (6,450 ft-ton) sag and 4,210 kN-m (1,390 ft-ton) hog, as mentioned above. The computed section modulus to the deck of the aluminum design was 2,150 cm^2-m (107 in^2-ft), so the corresponding stresses were 91.0 MPa (13.2 ksi) sag and 20.0 MPa (2.9 ksi) hog, a maximum stress range of 111.0 MPa (16.1 ksi). Using this maximum stress range in a cumulative fatigue damage analysis showed that the fatigue life would be only a little more than 3 years. A fatigue crack growth calculation with that stress range and an assumed crack size of 24 mm (0.94 in) showed that the crack would propagate to about 200 mm (7.9 in) in 10 months of operation, after which it would become unstable and propagate across the deck, a definitely unsatisfactory situation. To improve the fatigue life, the hull girder section modulus to deck was increased to 3,510 cm^2-m 1,780 in^2-ft), at a cost of an approximately 12 percent increase in structural weight. With this increase, the fatigue crack growth shown in Fig. 12 resulted, and the computed fatigue life for crack initiation was 21 years.

It is possible that the above fatigue life computation was overly conservative. The various classification societies all offer formulas for estimating hull girder bending moments and for determining minimum hull girder section moduli. However, these societies caution that for high-speed craft and for unusual hull configurations, the actual loading should be based upon hydrodynamic analysis, model testing, or testing of a prototype or similar vessel.

Loads

The first leg of the design triangle is loads, which can come from many sources. A listing of loads is provided in the U.S. Navy's Structural Design Manual for Surface Ships (NAVSEA, 1976)

1. Basic Loads
    a. Standard Live Loads
    b. Dead Loads
    c. Liquid/Tank loads
    d. Equipment and Cargo Loads
2. Sea Environment
    a. Hull Girder Loads
    b. Sea Loads
    c. Weather Loads
    d. Ship Motion Loads
3. Operational Environment
    a. Slamming
    b. Flooding
    c. Aircraft Landing
    d. Tank Overfill
    e. Docking
    f. Underway Replenishment
4. Combat Environment
    a. Shock
    b. Airblast
    c. Fragments
    d. Gun Blast and Reaction
    e. Missile Blast and Accidental Ignition

Not all of these loads apply to all designs, and there are other specific loads that must be considered for other designs.
Determination of the loads to be applied to a structure is an important part of the structural analysis procedure. Often, when designing to a particular design code, loads will be explicitly specified, such as nominal deck loading. Even in those cases, it is incumbent on the designer to consider the use to which the structure will be placed, and if higher loads are anticipated, ensure that they can be safely withstood. For unusual hull forms and high-speed vessels, nominal design loads may be specified, but most classification societies require that the final loads used in design be verified through hydrodynamic analysis, model testing, or testing of prototypes or other similar vessels.

Hull girder bending moments are no longer determined by using a standard wave, such as the 0.606 √L (1.1 √L in feet) wave and conducting a hydrostatic balance, which used to the be the mainstay of structural naval architectural calculations. Today the hydrostatic balance is mainly used to determining the still water bending moment. Classification societies offer rule bending moments, but as mentioned above, these are generally used only during early design stages, and must be supplemented by other means that are more specific to the hull being analyzed.

Just as important as the hull girder bending moments are the slam-induced pressures on the hull of high-speed craft. The classification societies offer equations for estimating these pressures, but generally require analytical or experimental verification of the loads. However, analytic or experimental determination of slam pressures is extremely difficult because the peak pressures are confined to a small area, and the average pressure acting over a larger area is considerably reduced. This effect is recognized by the classification societies, as shown in Fig. 15, where the pressure reductions calculated for a 61-meter and a 12.35-meter craft are shown using the ABS high-speed craft guide (ABS, 2001) and the DNV high-speed light craft rules (DNV, 2000). The two societies take a slightly different approach, with ABS reducing the peak pressure by the ratio of the actual area being designed to a reference area, which is a function of the ratio of the displacement to the draft. The DNV factor is inherent in design pressure computations, with a minimum design area considered, which places an upper cutoff on the pressure. ABS limits the pressure reduction to 0.4, which is a lower cutoff.

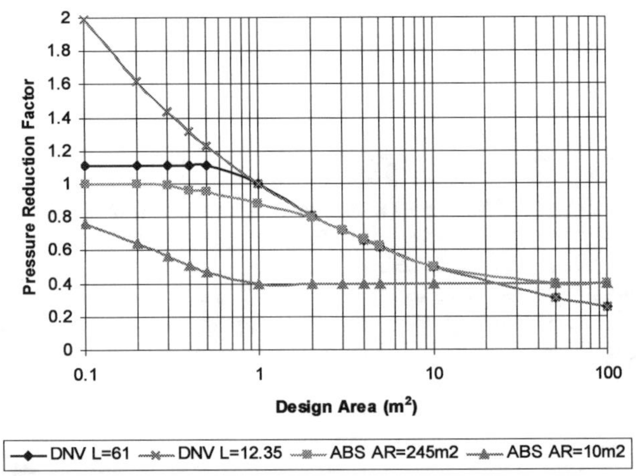

Fig. 15: Comparison of ABS and DNV slam pressure reduction factors.

It would be very difficult to reproduce the logic inherent in the classification society rules through an analytic or experimental approach during design. However, one of the principal variables in the slam pressure calculations is the hull acceleration during a slam event. Because the acceleration is the integrated result of the slam pressure loading and wave action, it is easier to determine in a consistent manner. Therefore, the best approach during design is to determine the acceleration and use this in the corresponding slam pressure computations for the classification society rules being used.

Loads that result from ship motion are computed by using the accelerations in all six degrees of freedom to translate a mass into a structural load. Classification societies offer equations for vertical accelerations that can be used in early design stages, and Section 301 of MIL-STD 1399 provides guidance on the other motions. These should be verified for final design by analytic or experimental means.

Using cargo and other variable mass with accelerations from ship motions is not trivial; especially as design procedures call for inclusion of that mass in an overall finite element model of the ship. The difficulty comes from the trying to apply the maximum loading for which all deck should be designed to all areas of the ship simultaneously, which will exceed the deadweight capacity of the hull. Judicious use must be made of maximum loads in certain areas, and other areas designed in other load cases.

Structural Analysis Procedures

The second leg of the design triangle is structural analysis. The analysis considers the first leg; design loads, and computes a stress or some other parameter that is consistent with the third leg, allowable stresses. There are many different analyses that must be conducted in the design of a ship, and often the loads and allowable stress levels will be different for the different types of analysis. Hull girder analysis, which will be discussed below, is necessary for ensuring adequate longitudinal, and sometimes torsional and transverse, strength. Fatigue analysis, which was discussed above, in integrally connected with a specialized load histogram and a design life, rather than a design stress level. Buckling analysis, which is discussed below, is especially important for aluminum because of the reduced elastic modulus of the material. Other types of analysis, such as hull girder collapse, deflection of the hull girder and structural members, vibration, and reliability may be appropriate for a particular ship design, and it is incumbent on the designer to consider all types of structural analysis and their appropriateness for the specific ship design, and not rely on a cookbook approach.

Hull Girder Analysis. The longitudinal strength of the hull girder in bending and in torsion was traditionally evaluated by computing bending moments and shears through static balance on a wave of standard height, such as 0.606 times the length in meters. For this procedure, standard allowable stress levels were established as the maximum permissible stress computed using a standardized approach to computation of the hull girder section modulus at established positions along the length of the ship. As an example, the U.S. Navy (NAVSEA, 1976) based longitudinal strength on hydrostatic balance on a wave of height 0.606 times the square root of the length of the hull in meters (1.1 $\sqrt{L}$ in feet). The allowable stress level determined by dividing the calculated bending moment at a section by the section modulus at that section was 147 MPa (9.5 tsi) for high tensile steel, and 70 MPa (4.5 tsi) for all alloys of aluminum. In combining the calculated longitudinal hull girder bending stress with stresses from local loads, a margin of 15 MPa (1.0 tsi) was added to the calculated stress, with the same margin used for all grades of steel and for all aluminum alloys. However, the largest aluminum ships built to this standard were a class of 58-meter (190-foot) patrol craft, which were analyzed for fatigue (Beach et al., 1981), although that analysis was limited to bottom longitudinals and for operations in a more benign area than for operation in the North Atlantic, on which most fatigue-life calculations are based. Subsequent evaluation of fatigue life of aluminum deckhouses for the U.S. Navy (Sielski, 1987) placed a limiting stress of 76 MPa (11 ksi) for hull girder bending stress in deckhouses, but an appropriate stress level for aluminum hull girders was not determined.

Current hull girder strength standards are based on standard bending moment and section modulus formulations implemented by classification societies following guidance from the International Association of Classification Societies. For many ships, especially high-speed vessels and ships of unusual hull form, the standard requirements are supplemented by a more rigorous determination of loads, such as discussed above. When the rule-based hull girder bending moments are used, standardized section moduli are used to compute stress, but when the more refined loads are used, hull girder analysis is conducted using finite element analysis, combining global loads with local loading to determine the maximum stress levels from all conditions of loading. The allowable stress levels for such analyses are generally set at some fraction of the yield strength of the material. For example, the ABS high-speed craft guide (ABS, 2001) specifies that for all locations and members the allowable stress for finite element analysis is 0.833 of the welded yield strength for aluminum and for steel. In the ABS high-speed naval craft guide (ABS, 2003), the allowable stresses are 0.85 of the welded yield strength for aluminum and 0.95 of the welded yield strength for steel. These allowable stress levels are subject to further analysis for buckling and for fatigue.

Frames and Stiffeners. In general, the intention of most historic rule-based structural design procedures for structure other than plating has been to calculate stresses as accurately as possible on a first-principles basis. For example, whenever a stiffener can be considered as having well-defined end conditions, the bending moment, m, at the ends is determined by some form of the equation $m = wl^2/k$, where w is the load per unit length over the span, l, of the stiffener, and k is a factor that depends on end fixity, equal to 12 for fixed ends. Factors that are based on mechanics of materials analysis are also introduced for other than full fixity at the ends of the stiffeners. The rules of the classification societies follow this method, except that allowable stress levels are sometimes combined with factors for end conditions, and when the loading is expressed as a head of water, a factor for density is also introduced. Such formulations are independent of the elastic modulus, and are therefore the same for steel and aluminum.

This is an exact solution if all of the assumptions regarding end fixity and load distribution are met, which is generally not the case. The span is generally defined as the length to some point along an end bracket, if fitted, and loads are not always uniform, and some averaging is required. For principal members such as transverse frames and longitudinal girders, end fixity in not easily determined and is a function of loading on adjacent members. In the past, other attempts at exact solutions including grillage analysis, ring frame analysis, and moment distribution methods were used to improve the accuracy of calculated stresses, even if the stresses were the response to arbitrary design loads. For this reason, finite element analysis has become readily accepted for analysis of structure, with as much detail used in analysis as can be practically done to model structural response to defined loads. The above considerations are the same for both steel and aluminum structures, and because finite element analysis is generally conducted as a linear elastic analysis, there is no essential difference in application of the methodology.

Effective Plating. When a structural shape is welded to plating, this plate-stiffener combination forms the full section that bends in accordance with beam theory. For a tee-stiffener, the flange of the tee forms one flange of the plate-stiffener, and the plate forms the other flange. In general, when there are parallel stiffeners, half of the plate between the stiffeners is credited to each stiffener in the determination of the section modulus and inertia of the plate-stiffener combination. However, shear lag and plate buckling considerations can make less than the full amount of plating effective in stiffener bending. The reduced elastic modulus of aluminum compared to steel can make less plating effective in an aluminum plate-stiffener combination when buckling strength governs.

Shear lag is a function of end support condition and the ratio of span to width of plating, as shown in Fig. 16, which was developed using an equation developed by Timoshenko (1956). Neither the elastic modulus nor the yield strength affects effective breadth based on shear lag so the factor is the same for both aluminum and steel. Most classification societies as well as the U.S. Navy usually take the effective breadth for shear lag to be no greater than one-third the span, a sort of average value from Fig. 16.

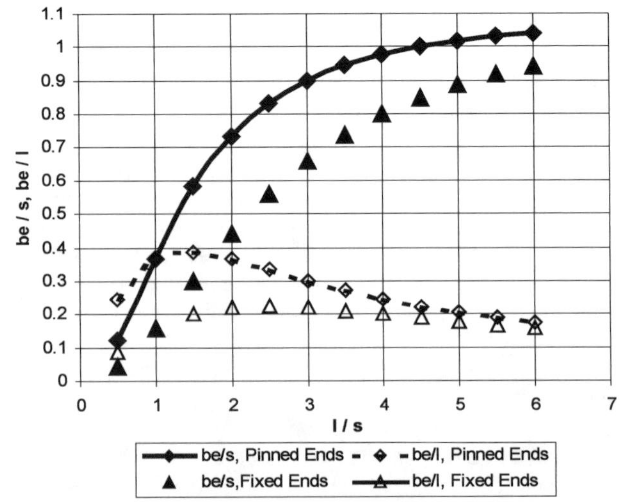

Fig. 16: Effective breadth of plating based on shear lag. Based on (Timoshenko, 1956).

Effective plating is also limited by buckling considerations. Faulkner (1975) attributes eq. 2 to von Kármán. The U.S. Navy uses this equation with the coefficient rounded to 2.0.

$$\frac{b_e}{t} = 1.9\sqrt{\frac{E}{\sigma_y}}$$

(2)

In eq. 2 σ_y should be taken as the yield strength that is consistent with the value that is used for the design of the plate, either of base metal or welded. The effective width of plate calculated in accordance with eq. 2 is shown in Table 5 for several aluminum alloys. The higher the yield strength, the less plate that is effective. In the ABS high-speed craft guide (ABS, 2001) and the ABS high-speed naval craft guide (ABS, 2003) the effective width of plating for secondary structure is limited to $b_e / t \leq 60$. For $b_e / t = 60$ in eq. 2, the yield strength is 80 MPa (11.6 ksi), which is 0.53 of the welded yield strength of 5086-H116. However, secondary structure, such as tank bulkheads and watertight bulkheads, have design allowable stresses of 0.60 σ_y and 0.90 σ_y respectively, which means that they are designed for a higher stress than the buckling strength of the plating.

Table 5: Effective Width of Plate based on Buckling Strength. Calculations based on (Faulkner, 1975).

Alloy	Base Metal		Welded Properties	
	Yield Strength (MPa/ksi)	b_e/t	Yield Strength (MPa/ksi)	b_e/t
5083-H116	215 (31.2)	36	165 (23.9)	41
5086-H116	195 (28.3)	38	152 (22.0)	43
5383-H116	215 (31.2)	36	145 (21.0)	44
5456-H116	230 (33.4)	35	179 (26.0)	40

Design of Plating under Lateral Load. When plating is under a lateral load such as hydrostatic pressure, the plate develops both bending stresses and in-plane membrane stresses. This is particularly true for thin aluminum plates, which initially deform from bending strains, but as the plate deforms more, the membrane strains develop, the stress from which reduce the bending stresses, and the plate is capable of absorbing a greater lateral load than the bending analysis alone would predict. The classic closed-forms of plate analysis reflect only the bending stresses, and must be augmented by membrane stress analysis or by empirical methods based on experimentation. The experimental data also includes permanent deformation of the plate, which permits a much greater load to be carried than would be indicated by a closed-form linear-elastic analysis.

A simple analysis can be made by analyzing a strip of unit width across the breadth, s, of a plate of infinite aspect ratio l/s and clamped edges, and loaded by a uniform pressure, p. The bending moment per unit width is $M = pb^2/12$. The section modulus for that strip is $SM = t^2/6$. Therefore, the bending stress is $\sigma = M/SM = pb^2/2t^2$. Solving for t, with an allowable stress σ_a:

$$t = s\sqrt{\frac{0.5p}{\sigma_a}} \qquad (3)$$

The ABS high-speed craft guide (ABS, 2001) and the ABS high-speed naval craft guide (ABS, 2003) both use the following equation to determine plate thickness under lateral loading:

$$t = s\sqrt{\frac{pk}{1000\,\sigma_a}} \qquad (4)$$

where k is a factor based on panel aspect ratio, equal to 0.5 for $l/s \geq 2.0$.

DNV has the following formula for design of aluminum plating (DNV, 2000):

$$t = s\sqrt{\frac{Cp}{\sigma}} \qquad (5)$$

where C is a coefficient that depends on edge conditions and aspect ratio of the plate. For a plate panel with aspect ratio $l/s \geq 2.0$ and fixed edges, C = 500.

The ABS formula defines s in millimeters, but DNV defines s in meters, a factor of 10^3, which becomes 10^6 when taken inside the radical, explaining the difference between the coefficients k and C. Otherwise the two equations are identical and are the same as eq. 3 for aspect ratios greater than or equal to 2, the usual situation for most designs. Therefore the methods of the classification societies are linear elastic in nature, considering only the plate bending stress. Any other considerations are included in the allowable stress levels.

The U.S. Navy design equation for aluminum plating is the same as is used for steel plating:

$$\frac{s}{t} \leq \frac{C}{k\sqrt{h}} \qquad (6)$$

where

k is a factor dependant on the aspect ratio of the plate panel,

h is the design head of water in feet, and

C is a coefficient.

Table 6 shows the coefficients C, which are based on the material and the location on the ship, which determines the amount of permanent set that is permitted from loading. The derivation of the coefficients, C, is uncertain, but they have appeared in U.S. Navy specifications for many decades. Several analyses have been made to derive the coefficients on a consistent basis, including that of Heller (1974), who considered plastic strain and work hardening of most of the materials in his analysis. Heller's analysis was based on plastic bending analysis of a strip of plating of infinite aspect ratio, but did not include in-plane membrane stresses. Heller concluded that the values of k were determined from experiments conducted in Germany in 1876 by an individual named Bach (Heller, 1974). Bach used mild steel plates, and they were loaded until they had considerable plastic deformation.

Table 6: Values of Coefficient C to be used in U.S. Navy Equation for Plating

Alloy	σ_y [1] (ksi)	σ_u [2] (ksi)	σ_b [3] (ksi)	Calculated C Values			Specified C Values		
				No Set [4]	Some Set [5]	More Set [6]	No Set [4]	Some Set [5]	More Set [6]
5083-H116	24.0 [6]	44.0	19.8	298	467	597			
5086-H116	22.0	40.0	18.0	285	445	569	250 [8]	400 [8]	500 [8]
5383-H116	21.0 [7]	44.2	18.6	289	468	598			
5456-H116	26.0	46.0	21.0	308	477	610	300	470	600
6005A-T61	16.7 [9]	38.0	15.5	264	434	555			
6061-T6	15.0	38.0	14.8	258	434	555			

6063-T6	9.4[9]	30.0	10.7	220	385	493			
6082-T6	16.7[9]	45.0	17.1	277	472	604			

1. Welded yield strength
2. Ultimate strength of unwelded base metal.
3. $\sigma_b = 0.5\,[\,(\boldsymbol{\sigma_y} / 1.26) + (\boldsymbol{\sigma_u} / 2.15)\,]$
4. For locations where no permanent deformation is permitted.
5. For locations where some deformation is permitted, including shell plating below a line 2 feet lower than the waterline, and tank boundaries.
6. For locations where considerable plastic deformation is permitted, including watertight bulkheads.
7. ABS
8. Specified values erroneously developed from properties of 5086-H112 extrusions.
9. DNV

Heller determined that the values for C are equal to:

$$C = \sqrt{\frac{2\sigma_a}{\gamma}} \tag{7}$$

where

σ_a is the allowable stress in psi, and

γ = the specific weight of seawater = 4/9 psi/ft.

The allowable stress is taken as the U.S. Navy allowable bending stress, σ_b, which is based on a combination of yield strength and ultimate strength. σ_a = 1.0 σ_u if no permanent deformation is permitted, 1.1 σ_u if slight deformation is permitted, and 1.8 σ_u if significant permanent deformation is permitted. The equation for allowable stress is the formula:

$$\sigma_b = 0.5\left(\frac{\sigma_y}{1.26} + \frac{\sigma_u}{2.15}\right) \tag{8}$$

where:

σ_b = the allowable bending stress,

σ_y = the yield stress, which for aluminum is the welded yield stress, and

σ_u = the ultimate strength, which is for the unwelded condition of aluminum.

Table 6 provides the appropriate values of C for several aluminum alloys that have been calculated from the above assumptions. Note that the C values for 5086-H116 that have been contained in U.S. Navy specifications were erroneously derived from the properties of 5086-H112 extrusions, and should be corrected.

Although the U.S. Navy method is based to a certain degree on experimental data, that data was for riveted steel panels, and should be verified for design of welded aluminum panels.

Buckling Analysis. Because the elastic modulus of aluminum is one-third the modulus of steel, buckling and instability issues are more important. The equations used in most marine design methods for computing buckling strength are the same for aluminum and steel, with differences reflected in the inclusion of the material's elastic modulus and yield strength. There are differences reflected in some civil engineering design codes for aluminum in the stress-strain behavior of aluminum compared to steel and between different aluminum alloys. The 5xxx-series aluminum alloys have a more rounded stress-strain curve than most steels and the 6xxx-series alloys, and will buckle earlier than those materials because the tangent modulus is reduced before the yield stress is reached. The 5xxx-series alloys receive their strength through work hardening, with extrusions stretched in tension. Therefore, their yield strength in tension is different from the strength in compression, a fact often overlooked in design.

Stiffener Tripping. Torsional instability of stiffeners, commonly called stiffener tripping, is a particularly important failure mode for asymmetric sections such as angles and for members with little or no flange, such as bulb flats and flat bar stiffeners. Vara et al. (2003) made a review of various methods of analysis of stiffener tripping and design criteria to prevent tripping. The analysis methods are based on linear elastic analysis, which is then corrected for plasticity by differing methods. Most analysis methods are based on axial compression, although there are several methods developed for analyzing behavior under lateral loading, and some consider combined lateral and compressive loads. However, the solutions for lateral loads have been shown to be accurate only when the ends of the stiffeners are free to rotate, and combined loading methods have been developed only to analyze symmetric stiffeners. Furthermore, there is little experimental data to validate the analysis methods, particularly for aluminum. Vara et al. made a comparative analysis of the methods of the ABS high-speed naval craft guide, Germanischer Lloyd, Lloyd's Register Rules for Classification of Naval Ships, the method of Adamchak (1979), and the U.S. Navy Design Data Sheet DDS 100-4, and the results, shown in Table 7, agreed within 10 percent. This comparison was made for steel stiffeners, but because the equations are the same for steel and aluminum, results should be similar for aluminum.

Table 7: Comparison of Tripping Equations, Maximum Compressive Force [kilo Newtons (kilo pounds)] (Vara et al., 2003)

Criterion	Tee Section 1	Tee Section 2	Angle 1	Angle 2	Bulb Flat
ABS	144.2 (32.4)	135.218 0 (30.4)	147.5 (33.1)	139.5 (31.4)	119.3 (26.8)
GL	135.4 (30.4)	140.9 (31,7)	140.5 (31.6)	134.4 (30.2)	123.1 (27.7)
USN DDS	150.3 (33.8)	149.0 (33.5)	-	-	-
Adamchak	145.5 (32.7)	148.6 (33.4)	-	-	-
LR Navy	150.3 (33.8)	148.6 (33.4)	150.8 (33.9)	150.8 (33.9)	128.6 (28.9)

Stability of Stiffener and Frame Flanges and Webs. In addition to overall buckling strength and tripping strength, stiffeners and frames and other structural members must be proportioned to avoid local instability. Typical guidance for this is provided in the ABS high-speed naval craft guide (ABS, 2003) for flat bars, outstanding face bars, and flanges as:

$$D/t \leq 0.5\ (E/\sigma_y)^{1/2} \qquad (9)$$

where d and t are the depth and thickness of the member, and σ_y is the yield strength of the unwelded aluminum in tension. The same equations are used for steel and aluminum. For higher strength steel with σ_y = 350 MPa (51 ksi) and E = 207 x 10^3 MPa (30 x 10^3 ksi), the limiting b/t for flanges is 12. For 5083-H116 aluminum with σ_y = 215 MPa (31 ksi) and E = 71 x 10^3 MPa (10.3 x 10^3 ksi), the limiting b/t for flanges is 9. Therefore, in general, the thickness of aluminum structural members should be about 12/9 = 4/3 times thicker than equivalent steel members.

Allowable Stress Levels

There are two types of loads, nominal design loads and estimates of maximum lifetime loads. There are also different methods of structural analysis, design equations and estimates of stress levels resulting from defined loads. Because there are different types of loads and analysis methods, one allowable stress level cannot be used for all design situations.

In general, most allowable stress levels make implicit or explicit recognition of the stresses arising from different loads. In addition, there are implicit factors of safety that will vary depending on the certainty of the analysis and the type of load applied. An example is the ASB HSC guide, which has different allowable stress levels that vary from 0.30 to 0.80 times the welded yield strength, σ_y, of the material. For example, the allowable stress for slam pressures on bottom longitudinals of craft less than 50 m in length is 0.65 σ_y , but for sea pressure on bottom longitudinals, the allowable stress level is 0.50 σ_y . The lower allowable for sea pressure is an implicit inclusion of hull girder bending, and an assumption that the maximum wave loading will occur at the same time as the maximum hull girder bending moment. For slam loading, the phasing between maximum slam pressures and the response of the hull to that pressure being different, so greater stresses are permitted.

In the above, the allowable stress is the same fraction of the welded yield strength of both steel and aluminum. When the design calculations are based on finite element calculations that consider the interaction of all applied loads, the allowable stress levels in the ABS high-speed craft guide (ABS, 2001) are 0.833 of the welded yield strength for aluminum and for steel, but in the ABS high-speed naval craft guide (ABS, 2003) the allowable stresses are 0.85 of the welded yield strength for aluminum and 0.95 of the welded yield strength for steel. Otherwise, most allowable stress levels, including those of the U.S. Navy make no distinction between steel and aluminum other than in the material yield strength.

The allowable strength levels have implicit recognition of initial imperfections and residual stresses from fabrication to the extent that design authorities have established allowable tolerances for construction. For ABS, those tolerances for aluminum structure are given a supplement to the rules for aluminum (ABS, 2006). For the U.S. Navy, tolerances are specified in MIL-STD 1689. Both the ABS and U.S. Navy tolerances for aluminum structure are similar.

Paik et al. (2006) investigated the effect of tolerances on the compressive strength of stiffened plate panels. In their measurements of imperfections of the welded aluminum panels, the published tolerances were mostly met, and no significant degradation in compressive strength was found.

The disparity between different authorities on the welded yield strength of aluminum was mentioned above. Those differences are related to uncertainty as to way in which the welded strength is to be used in design. In most cases, allowable stress levels are based on the welded yield strength, but the base metal yield strength or tensile strength are sometimes used. An example of this is the allowable stress used to determine the required section modulus in the ABS high-speed craft and high-speed naval craft guides, which is a factor of the both the unwelded yield strength and the welded ultimate strength.

A larger issue is the extent to which a relatively narrow strip of weld metal and heat-affected zone can weaken the strength of the larger unwelded structure. A simplified analysis was conducted by Collette (2005), who analyzed a welded plate in tension as a grouping of springs in series and in parallel. He analyzed panels of 5083-H116 and 6082-T6 considering the stress-strain curves of the base metals and of the heat-affected zone (HAZ) of the welds, assuming two different widths for the HAZ. The results are shown in Fig. 17, where the initial stress-strain curve has been offset by 0.2 percent to suit the definition of yield strength. The strength of the welded panel is shown to be greater than the strength of the weld alone, and in the case of the 5083, almost equal to the strength of the base metal.

Paik and Duran (2004) have proposed another approach to developing an effective yield strength using a weighted average of yield strengths based on the volume of base metal and HAZ. This method has been validated through use in comparing computed strengths of welded panels in compression to experimental results. For the geometry analyzed by Collette, the area–weighted yield strength in the 5083-H116 is 207 MPa (30 ksi) for the 12.5-mm (0.5-in) HAZ, and 200 MPa (29.0 ksi) for the 25-mm (1.0-in) HAZ, values only slightly greater than those from the results shown in Fig. 17. In the 6082-T6, the effective yield strength is 247 MPa (35.8 ksi) for the 12.5-mm (0.5-in) HAZ and 235 MPa (34.1 ksi) for the 25-mm (1.0-in) HAZ; values significantly lower than determined by the analysis of Collette. More detailed analysis validated by experiment is needed to determine an appropriate methodology. However, these analyses show that the effective yield strength in tension is greater than that of the HAZ alone, so the allowable stress levels used in current design codes are conservative.

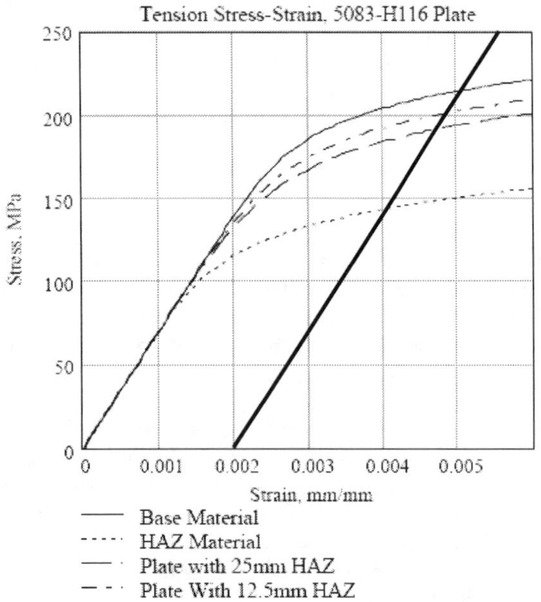

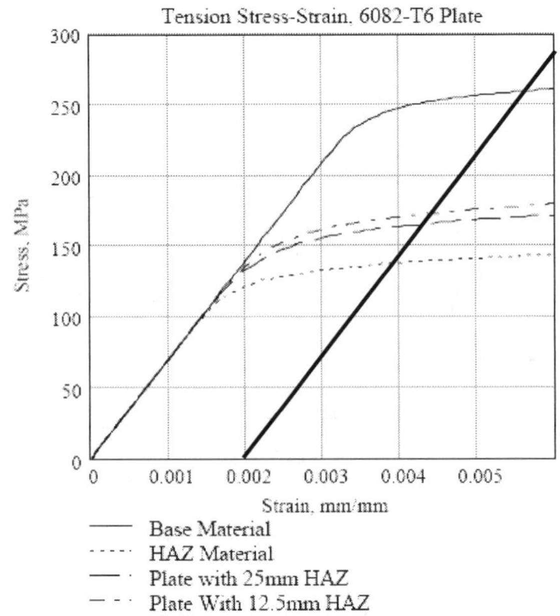

Fig. 17: Analysis of welded plate in tension (Collette, 2005).

CORROSION

The discussion above on different alloys implied that some alloys should be considered and others not used at all because of corrosion issues. On the whole, aluminum has excellent corrosion resistance, and requires no painting for resistance to corrosion. There are many unpainted aluminum vessels that have had more than 30 years of ocean service with no signs of corrosion. However, there are circumstances under which aluminum will corrode or otherwise show reduced strength because of the attack of seawater. In general, there are seven types of corrosion of concern (Czyryca and Vassilaros, 1972):

- General corrosion and pitting
- Exfoliation
- Intergranular corrosion
- Stress-corrosion cracking
- Corrosion-erosion, impingement, and cavitation
- Galvanic corrosion
- Crevice corrosion

General Corrosion

General corrosion is the wastage of material that generally occurs over a broad area. In aluminum an unpainted surface becomes covered with a gray oxide film, which is essential to the corrosion resistance of the alloy, as it can prevent further corrosion as long as the oxide film remains stable and unbroken. Associated with general corrosion is pitting, where small holes appear scattered over the surface. The 5xxx-series alloys are noted for good resistance to general corrosion. Although the 6xxx-series are not considered to be as good, and have prohibition on their use mentioned above, the data that is available on corrosion of those alloys do not support that assertion.

Exfoliation

The process of exfoliation is typified by corrosion along the grain boundaries of rolled plates that results in the material separating into flakes. The problem is associated with alloys in the 5xxx-series with magnesium contents greater than 3 percent. It was first discovered to occur in many vessels in the late 1960s, particularly in various patrol craft used by the U.S. Navy. The aluminum industry developed exfoliation-resistant heat treatments, particularly the H116 temper, which is resistant to exfoliation. The ASTM G 66 (ASSET) test is used to provide visual assessment of exfoliation corrosion susceptibility, and is required for all alloys procured to the ASTM B 928 specification.

Intergranular Corrosion

Intergranular corrosion is similar to exfoliation in that it affects the higher-magnesium content 5xxx-series alloys through precipitation of magnesium to grain boundaries. This type of corrosion recently became a particular problem on a number of ships in the Pacific Northwest that were fabricated with 5083-H321 plate from a particular manufacturer who had changed the processing of this alloy. To address the problem of intergranular corrosion, the American Society of Testing and Materials (ASTM) formed the ASTM B07.03 Task Group on Marine Alloys, which in turn developed the ASTM B 928 specification. That specification requires the ASTM G 67 (NAMLT) test to determine the material's susceptibility to intergranular corrosion, a test that should be performed for all alloys with a magnesium content greater than 3 percent. The ASTM B 928 specification also requires testing to ASTM G 66 (ASSET).

Stress-Corrosion Cracking

The mechanism of stress corrosion involves both electrochemical and mechanical processes. Local galvanic cell action creates a sharp pit, which may induce mechanical tearing at the root, thus exposing a fresh metal surface to accelerated corrosion. Further corrosion deepens the crack and continues the process. Stress corrosion in aluminum alloys is characterized by intergranular cracks normal to the metal surface and stress direction. The electrochemical processes involved in stress-corrosion cracking of the alloys are the same as those in intergranular corrosion; i.e., selective attack of grain boundaries sensitized by precipitates or of areas adjacent to grain boundaries which are highly anodic due to depletion of certain elements. However, not all alloys subject to intergranular corrosion will undergo stress corrosion. Stress corrosion causes a brittle-type failure in an otherwise ductile material. Both the 5xxx-series and 6xxx-series alloys are considered to be resistant to stress-corrosion cracking. However, when the higher magnesium 5xxx-series alloys become sensitized, stress-corrosion cracking can result.

Corrosion-Erosion, Impingement, and Cavitation

The corrosion resistance of aluminum comes from the oxide film that forms on the surface and protects the underling metal from further oxidation. If some process such as high velocity fluid flow continuously disrupts that film, corrosion will continue to occur. The 5xxx-series alloys generally have better corrosion-erosion resistance than the 6xxx-series. Impingement attack occurs from a stream of liquid striking a surface, and it is accelerated by the presence of air bubbles in the liquid. Cavitation erosion is caused by the collapse of voids at high speed, and is the most severe form of combined mechanical and corrosion action. Resistance to cavitation is proportional to the hardness of the material, so the higher strength alloys generally have higher cavitation erosion resistance. However, aluminum alloys have very poor cavitation erosion resistance compared to most other metals and should not be used where cavitation is anticipated, such as high-speed propellers.

Galvanic Corrosion

When two dissimilar metals are coupled in an electrolyte, galvanic corrosion results, with the metal that has the more positive electrical potential being the cathode that attacks the metal with the lower electrical potential, the anode. Magnesium and zinc are the only two metals that are anodic to aluminum and will protect it from corrosion. All other metals, including steel, lead, tin, brass, copper, stainless steel, Monel, Inconel, and nickel are cathodic towards aluminum, and when placed in electrical contact with aluminum in seawater will cause corrosion of the aluminum. This is a particular problem with propellers, propeller shafts and rudder shafts, which are commonly made with these metals. To prevent galvanic corrosion of the aluminum hull, either anodes of zinc or magnesium need to be installed on the hull near these dissimilar metals, or an impressed current system installed, maintained, and operated. Such cathodic protection of the bottom is particularly important if the bottom is painted, as it usually is for antifouling. When a scratch develops in the paint, then a very small area of aluminum exists compared to the area of the metal acting as a cathode, and the local rate of corrosion can be significantly high. Different aluminum alloys have different galvanic potential, and so using two dissimilar alloys can cause a galvanic corrosion problem.

Crevice Corrosion

Crevice corrosion is caused by a difference in electrical potential, similar to galvanic corrosion. In this case, the difference in potential is caused by a greater concentration of the electrolytes in different areas of the solution in which the metal is immersed. A difference in oxygen content can occur in locations such as lapped joints in riveted construction, and crevice corrosion can result. Therefore, faying surfaces should be primed before assembly, or a caulking compound can protect the metals. The 5xxx-series alloys are generally resistant to crevice corrosion, as are the 6xxx-series, but the presence of dissimilar metals can accelerate the tendency for crevice corrosion.

Sensitization

The 5xxx-series alloys with magnesium contents greater than 3 percent are subject to exfoliation, intergranular corrosion, and stress corrosion cracking. Processing the alloys to meet the requirements of ASTM B-928 results in alloys that are resistant to these forms of corrosion. However, exposure to elevated temperatures for extended periods of time can cause the magnesium within the alloys to precipitate to the grain boundaries and make the alloy more prone to one of these forms of corrosion. This change over time is called sensitization. For service at temperatures greater than of 65 °C (150 °F), alloy 5454-H32, which has a magnesium range of 2.4 to 3.0 percent, should be used. The service temperature limit of 65 °C is currently under review because some ships have been found to have 5456-H116 in the weather decks of the superstructure that has become sensitized.

Sensitization can occur during fabrication from heating or deforming the metal if they are not done properly. For this reason, the use of methods such as flame-straightening to remove distortion is generally prohibited. Such heat treatment can be used if it is done carefully, ensuring that maximum temperature limits are met and are held for a maximum amount of time but controlled procedures should be developed and used (Hay and Holtyn, 1980). Certain limits on distortion, such as minimum cold-forming radii (AWS, 2004) are developed to preclude cracking during working and sensitization afterward.

FIRE PROTECTION

Because aluminum has a relatively low melting point, fire protection insulation is required if it is to maintain its strength during a fire. Requirements have been developed by most regulatory agencies to provide protection, including the International Maritime Organization in its High Speed Vessel Code, the U.S. Coast Guard, and the U.S. Navy. These requirements are all developed to prevent the temperature of the aluminum from exceeding 230 °C (446 °F) during the time of a standardized test. Actual test requirements state that the temperature rise of the aluminum be no greater than 200 °C (360 °F) above the ambient, which assumes that the ambient temperature is 30 °C (86 °F). None of the regulations specifically mention the additional insulation requirements that may be necessary if the ambient temperature is significantly greater than 30 °C (86 °F), but special consideration should be made if those conditions are anticipated.

There is a major difference between the requirements of the U.S. Navy and of the various requirements for commercial ships. The commercial requirements are developed to contain a wood fire (or equivalent materials) but the U.S. Navy requirements are developed to withstand an oil fuel fire, which is significantly hotter. A typical fire protection insulation that will provide Class A-60 protection for a bulkhead of a commercial vessel will weigh about 8.64 kg / m^2 (1.76 lb / ft^2). A system to provide the U.S. Navy N-30 protection for a bulkhead will weigh about 14.6 kg / m^2 (3.0 lb / ft^2).

Guidance for protection of aluminum is provided by SNAME T&R Bulletin 2-21 (SNAME 1974). This document was developed following an extensive testing program, and is referred to in U.S. Coast Guard regulations. The actual design configuration of a bulkhead or deck will govern the amount of fire protection insulation required. For example, if the opposite side of a bulkhead is not insulated because there is no fire threat on that side, such as the side of the ship, then the amount of insulation required on the side exposed to a fire is reduced, because radiation will help conduct heat away. Conversely, if both sides are insulated, radiation is significantly reduced, and the amount of required insulation is greater. Other factors also govern insulation requirements, including the presence of stiffeners on one side or the other, and an air space provided by non-flammable sheathing will reduce the amount of insulation required.

The testing conducted to develop the SNAME T&R bulletin did not cover all situations, including fire protection insulation for structure surrounded by fire, such as stanchions. Also, the commercial standards do not specifically pertain to protection from oil and similar higher-temperature fires. There is an informal IMO fire protection level H-60 for resistance to hydrocarbon fires. Because the manner in which the IMO requirements are stated and the way the U.S. Navy N-30 standard is stated, meeting the H-60 standard requires slightly less insulation than does N-30. However, the weight impact is nearly as high, and many builders choose to use steel stanchions and girders to support the deck above vehicles decks, as the weight is about the same as H-60 insulated aluminum and the cost is less.

DESIGN STUDIES

Over the years, many comparisons of comparable designs in aluminum and steel have been made, the first of which was probably reported in 1897 (Hobson, 1897). That study showed the weight of aluminum plates to be 0.43 that of steel plates designed for the same service, and the weight of aluminum shapes to be 0.48 of the weight of comparable steel shapes. Over the years, many other such comparisons were made, with the weight of aluminum structure being about one-half the weight of steel structure. To evaluate the effect of current design criteria on the relative weight of aluminum and steel structure, four comparative designs were made. In one of the studies, a glass reinforced plastic (GRP) variant was also included.

The first two studies are extensions of recent SNAME papers on relative design criteria. In those studies, only limited portions of the hull were analyzed for the effect of differences between various classification societies. The original studies addressed only changes in scantlings in aluminum structure, and did not address weight differences. Although it might have been interesting to compare the effects of the different classification society rules on weight, the current study focused on the difference between the weight of the original aluminum structure, using one of the classification society aluminum scantlings, and redesigning the structure in steel, using the same criteria. In these studies, only the initial design formulas were used, although in a complete design additional calculations would undoubtedly be required, including buckling and fatigue.

The third study addressed the complete midship hull structure of a high-speed craft, including longitudinal strength criteria, developing structure for aluminum, steel, and GRP alternatives. As with the previous studies, the design calculations were not complete to the extent that would be required for a final design. Rule equations were used for longitudinal strength criteria, rather than hydrodynamic analysis or model testing, as would be required by most classification societies. Furthermore, no detailed finite element analysis or buckling calculations were performed, although an initial fatigue assessment was made. In this study, the relatively recent requirement for fire protection insulation reduced the weight benefits of aluminum compared to previous comparative studies.

The final design study was for the deckhouse of a naval combatant that was designed with steel in the 1980s. This deckhouse was redesigned with aluminum and with the inclusion of fire protection insulation showed a significantly reduced weight advantage compared to steel.

Study 1, Comparative Design of a 61-meter, 50-knot High Speed Craft

The vessel selected for this study was the craft used by Stone and Novak (2006) in a comparison of rules of classification societies for high-speed craft. Their analysis was for aluminum only, and has been expanded to include steel and GRP. The Stone and Novak study was performed assuming 5083-H116 plating and 6061-T6 extrusions, with welded yield strengths of 165 MPa (24 ksi) and 138 MPa (20 ksi), respectively. The same aluminum will be used in this analysis; the steel will be HS-36, with a yield strength of 355 MPa (51.5 ksi). For a ship of this length, composite construction is possible, but sufficiently unusual that should such a vessel be designed and constructed, there would probably be unusual structural features or special materials used, so a GRP alternative will not be considered in this study. Although the Stone and Novak study included the entire hull girder section modulus calculations, the entire midship section was not given, with the report focusing on structure subjected to slamming pressures. Therefore, this continuation of that study is limited to comparison of the structure of bottom and side panels amidships. The calculations were conducted using the ABS high-speed craft guide (ABS, 2001).

Principal Dimensions

L_{WL} = 61 meters (200 ft)
B = 12.9 m (42.3 ft)
B_{WL} = 11.7 m (38.4 ft)
V = 50 knots
$H_{1/3}$ = L/12 = 5.08 m (16.7 ft)
Δ = 950 tonnes (935 long tons)
β_{LCG} (deadrise angle) = 17°
C_B = 0.451
τ (trim angle at speed) = 3°

Table 8: Summary of Design for Study 1

Location		Aluminum		Steel	
		Scantlings	Weight of Panel [kg (lb)]	Scantlings	Weight of Panel [kg (lb)]
Bottom	Plate Stiffeners Frames	9.1 mm (0.358 in) 120 x 50 x 3 kg/m T (4.7x2.0x2#) T 300 x 8 /100 x 15 T (11.8 x 0.3 / 3.9 x 0.31 T)	68 (150)	6.2 mm (0.24 in) 90 x 9.1 (3.54 x 0.36) flat bar 200 x 5 / 100 x 12 T (7.9 x 0.2 / 3.9 x 0.47 T)	133 (293)
Side	Plate Stiffeners Frames	5.8 mm (0.23 in) 76.2 x 21 x 1.397 kg/m (3 x 0.83 x 0.94#) bulb flat 100 x 5 / 100 x 8 T (3.9 x 0.2 / 3.9 x 0.31 T)	44 (97)	4.7 mm 50 x 5.2 (2.0 x 0.2) flat bar 80 x 5 / 80 x 6 T 3.1 x 0.2 / 3.1 x 0.24 T)	96 (212)
Total Weight [kg (lb)]			112 (247)		229 (505)

Weight of aluminum / weight of steel = 112 / 229 = 0.489

Study 2: Comparative Design of a High-Speed Catamaran

The second study was for the 96-meter high-speed aluminum catamaran ferry for which Kramer (2005) had made a comparative study of aluminum alternatives. The vessel on which the design study was based was constructed for restricted service in accordance with DNV +1A1 HSLC R4 (enclosed waters 20,20). The vessel was redesigned in steel to DNV Class R4, the same DNV classification as the original design. The vessel was built with 5083-H116 plate and 6061-T6 extrusions. The redesign is in NV-36 steel, which has a 355 MPa (51 ksi) yield strength. Kramer provided detailed calculations for only the bottom, side, and wet deck structure, so this study is limited to those areas.

Principal Dimensions
Length Overall: 122.5 m (402 ft)
Length Waterline: 96.0m (315 ft)
Beam: 25.8m 84.6 ft)
Beam of Hulls: 6.0m (19.7 ft)
Draft: 3.76m (12.33 ft)
Δ = 1,822 tonnes (1,793 long tons)
$C_B = \Delta / (1.025\ L\ B_{WL}\ T)$ = 1,882 / (1.025 x 96 x (2 x 6) x 3.76) = 0.403
Speed: 34 knots

Table 9: Summary of Design for Study 2

Location	Aluminum		Steel	
	Scantlings	Weight of Panel [kg (lb)]	Scantlings	Weight of Panel [kg (lb)]
Bottom	8 mm (0.31 in) plate 140 x 50 IT (5.5 x 2.0 IT)	11.01 (24.3)	6 mm plate HP 100 x 8 (4.0 x 0.31) bulb flat	24.42 (53.8)
Side	6 mm plate 70 x 40 IT	6.83 (15.1)	4.5 mm (0.18 in) plate HP 80 x 6 (3.15 x 0.24) bulb flat	16.02 (35.3)
Wet Deck	7 mm (0.28) plate 120 x 50 (4.7 x 2.0) IT	8.57 (18.9)	6 mm (0.24 in) plate HP 100 x 6 (4.0 x 0.24) bulb flat	20.89 (46.1)
Total Weight [kg (lbs)])		26.41 (58.2)		61.33 (135.2)

Weight of aluminum / weight of steel = 26.41 / 61.33 = 0.43

Study 3: Comparative Design of a 42.67 meter High-Speed Craft

The third study was for a notional 42.67-meter, 32-knot crew boat. The study was made using the ABS high-speed craft guide (ABS, 2001). The steel craft was designed using a higher-strength steel, HS-36, which has a yield strength of 355 MPa (51.5 ksi). The aluminum craft was designed using 5083-H116 plate and 5083-H111 stiffeners, with welded yield strengths of 165 MPa (23.9 ksi) and 145 MPa 21.0 ksi), respectively. These alloys are both of higher strength than the weakest alloy, but not the highest strength available. For the GRP alternative, orthotropic properties were assumed with 207 MPa (30.0 ksi) ultimate tensile strength, and 71.0 x 10^3 MPA (10.3 x 10^3 ksi) elastic modulus. These properties are near the high end for hand lay-up, but easily achieved with lay-up methods such as vacuum-assisted resin transfer that are extensively used in GRP fabrication today. A full design of the midship section was made, including longitudinal strength and fatigue analysis.

The result of the fatigue analysis, which was described above, was an increase of hull girder section modulus to deck. The initial calculations using only the minimum scantlings needed to meet local strength requirements resulted in section moduli to deck and to keel that ex ceded the ABS minimum hull girder section modulus requirement, with the section modulus to keel greatly in excess of the minimum. The fatigue analysis for a typical deck structural detail showed a fatigue initiation life for the aluminum variant of slightly more than 3 years. The steel variant analyzed by the same method had a fatigue life of 54 years for the same type of structural detail used in the aluminum variant.

In order to increase the calculated fatigue life of the aluminum deck to 20 years, the thickness of the strength deck plating, and the sheer and stringer strakes was increased, resulting in an increase of the total weight of structure from 46.2 tons to 52.0 tons, an increase of 5.8 tons or 12 percent.

Aluminum structure requires insulation to protect it from fire. The degree of protection depends upon the specific type of ship (cargo, passenger, etc.), the general arrangements, and the service for which it used, including U.S. Navy requirements. For this study, the assumption was made that the transverse bulkheads are required to meet Class A-60, and the underside of the main deck must also meet class A-60 requirements. The requirements for fire protection of GRP are not as well defined as for aluminum, but the same weight of insulation was assumed for the GRP variant as for aluminum.

Principal Dimensions
Length on Waterline: L_{WL} = 42.67 meters (140 ft)
Length for Rules: L = 41.39 meters (135.8 ft)
Beam at Deck: B = 7.62 meters (25.0 ft)
Beam at waterline: B_{WL} = 7.27 meters (23.9 ft)
Depth midships: D = 3.73 meters (12.2 ft)
Draft midships: d = 2.057 meters (6.75 ft)
Displacement: 416.6 metric tonnes (410.0 long tons)
Block Coefficient on waterline length: C_B =0.637
Speed: V = 32 knots
Design sea state: $h_{1/3}$ = 3.96 meters (13.0 ft)
Trim: τ = 4 degrees
Deadrise at LCG: β_{CG} = 19 degrees
Transverse frame spacing = 1.219 m (4.0 ft)

Transverse bulkhead spacing = 5 x 1.219 = 6.095 m (20.0 ft)

The midship sections for the variants are shown in Fig. 18 and Fig. 19 and the designs are summarized in Table 10.

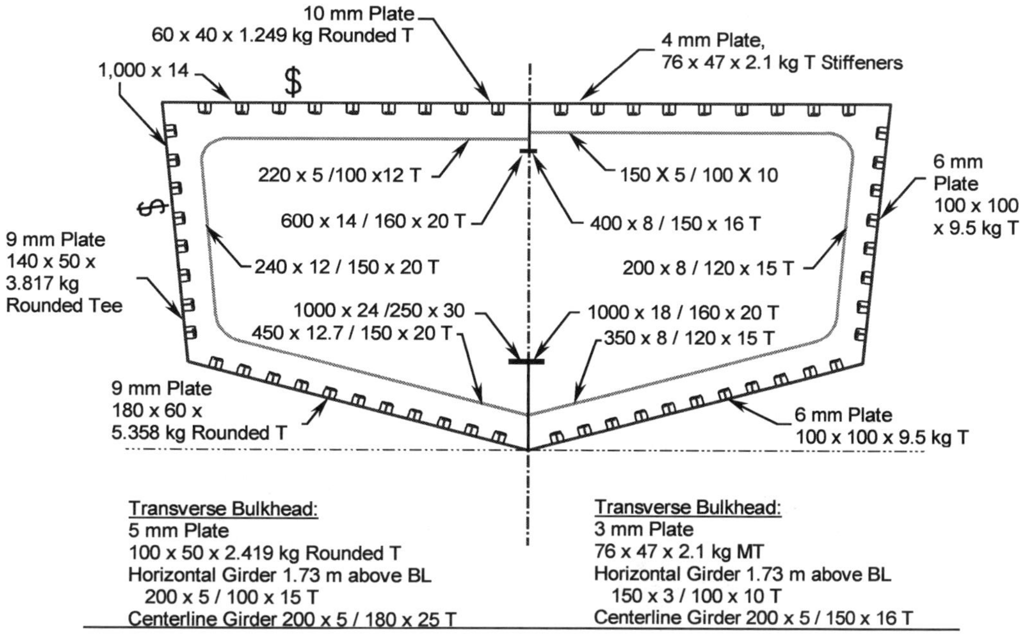

Aluminum Steel

Fig. 18: Midship section of aluminum and steel designs.

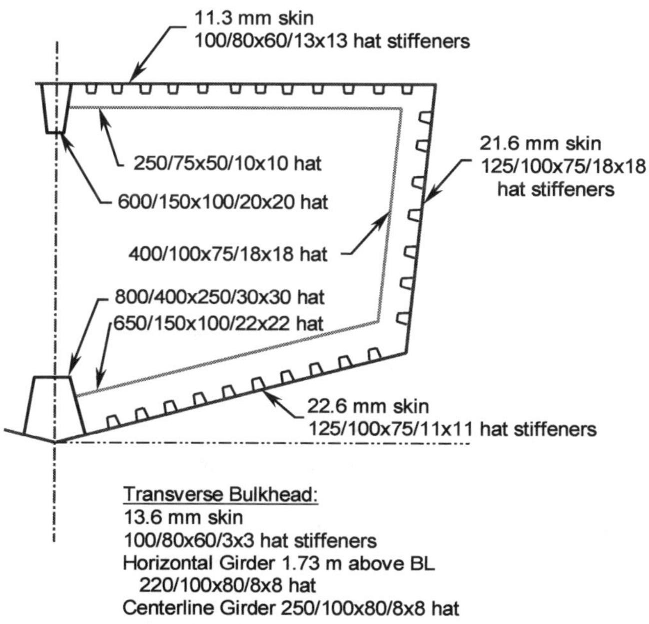

Fig. 19: GRP variant midship section.

Table 10: Summary of Design for Study 3

Item	Weight [kg/m (lb/ft)]		
	Aluminum Variant	Steel Variant	GRP Variant
Longitudinally continuous structure	954 (641)	1,726 (1,160)	1,031 (693)
Transverse frames	184 (124)	289 (194)	271 (182)
Transverse frames	87 (58)	165 (111)	105 (71)
Subtotal	1,225 (823)	2,180 (1,465)	1,407 (945)
Miscellaneous structure (10 percent)	123 (83)	218 (146)	141 (95)
Welding (2.5 %) (5 % for GRP laps, etc.)	31 (21)	55 (37)	70 (47)
Fire Protection Insulation	132 (89)		132 (89)
Total kg/meter (lb/ft)	1,511 (1,015)	2,453 (1,648)	1,750 (1,176)
Total weight of structure (kg/m x 43m x 0.8) [tonnes (l. tons)]	52.0 (51.2)	84.4 (83.1)	60.2 (59.3)
Weight of structure / displacement	12.5 %	20.3%	14.5%

Weight of aluminum structure and insulation / weight of steel = 52.0 / 84.4 = 0.62

Weight of GRP structure and insulation / weight of steel =
60.2 / 84.4 = 0.71

The weight of the aluminum structure is 86 percent of the weight of the GRP structure. There are undoubtedly some items for which the weight was not estimated, such as the deckhouse and any intermediate deck. Therefore, the total weight of structure will be a higher percentage of total displacement. However, because the calculations were conducted in a consistent manner, the comparison between steel and aluminum is valid.

Study 4: Redesign of the Steel Deckhouse of a Destroyer in Aluminum

Starting with early experimentation in the 1930s, the U.S. Navy eventually came to the point where the standard material for the deckhouse structure of all frigates, destroyers, and cruisers was 5456-H116 and 5456-H111 aluminum. In the 1980s that practice changed, and a steel deckhouse appeared on the destroyer class designed at that time. This study will look at the design of the deckhouse of that destroyer using aluminum. Unlike the earlier U.S. Navy aluminum designs, fatigue and fire containment issues will be explicitly addressed. The redesign included only the first two levels of the forward deckhouse, but the results should be representative of the remainder of the deckhouse. Calculations were made using the U.S. Navy design criteria used in the 1980s. The original steel design was made with Higher Strength Steel, which has a yield strength of 350 MPa (51 ksi). The aluminum alloy chosen for this study is 5083-H116 plate and 5083-H111 extrusions following current practice rather than the 5456 aluminum preferred by the U.S. Navy for ships designed in the 1970s. No special lightweight integrally stiffened aluminum panels were used in the redesign to maintain a similar level of construction technology as the steel design.

Because the ship is a naval combatant, all the external structure and major transverse bulkheads in the superstructure were designed to withstand nuclear air blast. Using "reverse engineering" an appropriate design criteria for nuclear blast for the original steel structure was developed using the method of Dye and Langford (1966). Those same criteria were then applied to the aluminum design to develop scantlings. For internal structure, a local design criterion of a deck loading of 366 kg/m^2 (75 lbs/ft^2) was used. In examining some members, principally the deck transverse frames, it was found that the existing steel members were capable of a much greater load, implying that perhaps considerations brought out by the finite element analysis used in the original design required heavier members. Such aluminum members were given a section modulus equal to the section modulus of the steel member multiplied by the ration of the allowable stress for steel divided by the allowable stress level for aluminum. Because of the classified nature on nuclear air blast calculations, they have not been presented here. However, some of the calculations for ordinary design loads have been included below. Because some of the U.S. Navy design equations are based on empirical factors defined in U.S. customary units of measure, the calculations are made in those units.

Design of enclosed deck structure.
Although weather structure of decks and house sides are designed for nuclear air blast, internal structure is not so designed other than for overall in-plane strength. The following calculations are for a typical enclosed deck.

Plating
The US Navy design equation for plating subject to lateral loads is given by the formula:

$$t = \frac{bK\sqrt{H}}{C}$$

where:

b = stiffener spacing (inches)
K = 1.0 for aspect ratio l/b ≥ 2.0
H = design head in feet salt water, 64 pounds per cubic feet
C = 400 for HSS (no permanent set under load)
290 for 5083-H116 (no permanent set under load)

The stiffener spacing for all decks is 27 inches

$$t = \frac{bK\sqrt{H}}{C} = \frac{27 \times 1.0\sqrt{\frac{75}{64}}}{C}$$
= 0.07 in. for steel
= 0.10 in. for aluminum

The existing steel plating is 0.188 inch thick, a thickness selected as a minimum for fabrication purposes. For fabrication of aluminum structure using conventional welding methods and avoiding excessive distortion, a minimum thickness is generally 0.25 inches, which will be used.

Deck Longitudinals
Steel structure
0.188-inch plate
Longitudinal Stiffeners
l = 96 in
s = 27 in
Design load = 75 psf
Moment = $w\, l^2 / 12 = (75/144)\,(27)\,(96)^2 / 12 = 10.8$ in-kips
Assume there is a 10.5 tsi = 23.5 ksi primary stress.
Existing 4 x 4 x 5# T on 50T 0.188-in plate:
$SM_f = 3.6$ in^3
r = 1.73 in
l / r = 96 / 1.73 = 56, Design Data Sheet 100-4 gives F_C = 39.0 ksi
bending stress $f_b = 10.8 / 3.6 = 3.0$ ksi

The acceptance criteria $= (f_c / 0.8\, F_c) + (f_b / F_b)$

$= (23.5 / 0.8 \times 39) + (3.0 / 40) = 0.82 < 1.0$

Aluminum Structure (5083-H111 stiffeners)
Now assume the design primary stress is 11.0 ksi (NAVSEA Design Standard)
Using 4 x 4 x 2.0# T, on 35 T 0.25" plate:

$SM_f = 4.57\ in^3$
$r = 1.78$ in
$l / r = 96 / 1.78 = 54$, Design Data Sheet 100-4 gives $F_C = 17.2$ ksi

bending stress $f_b = 10.8 / 4.57 = 2.36$ ksi

The acceptance criteria $= (f_c / 0.8\ F_c) + (f_b / F_b)$
$= (11.0 / 0.8 \times 17.2) + (2.36 / 18.0) = 0.93 < 1.0$

The acceptance criteria for steel and aluminum are similar, therefore, the aluminum 4 x 4 x 2.0# T are comparable to the existing steel stiffeners.

Deck Transverse Frames
Typical frame 12 x 4 x 22# I-T

$l = 177$ in
$b = 96$ in
$M = w\,l^2 / 12 = (75/144)\,(96)\,(177)^2 / 12 = 130.5$ in-kips
$SM_r = 130.5 / 40.0 = 3.2\ in^3$
Actual SM for 12 x 4 x 22# I-T on 50t 0.188-in plate = 26 in^3

Therefore, the deck transverses are considerably oversized for the nominal design load. The actual design was based on a finite element analysis that considers continuity of structure and the total load situation. For the aluminum structure, adjust the required section modulus on the ratio of allowable bending stress.

$SM_{Al} = 40/18\ SM_{steel} = 2.2\ SM_{steel}$

Aluminum Transverse Frames

$SM_R = 2.2\ SM_{steel} = 2.2 \times 26 = 57\ in^3$
Use 16 x 6 / 0.375 x 0.75 T
On 8.75" 0.25" plate, $SM = 56.7\ in^3$, $I = 557\ in^4$, $Ax = 10.5\ in^2$

The thickness of deck plating and side plating of the steel deckhouse structure was found to be far greater than would be required by design calculations. It was assumed that this was done for fabrication purposes, with a minimum thickness of 0.188 inches (4.8 mm) used. The thickness of corresponding aluminum plate was made 0.25 inches (6.4 mm), which is an appropriate minimum thickness for the fabrication methods generally applied for the deckhouse structure of naval combatants. In other places, particularly the deckhouse side, thicker steel plate was used, presumably to provide a transition to the sheer strake. The aluminum house side plating was increased to be 0.125 inches (3.2 mm) thicker than the corresponding steel plate.

A fatigue analysis had previously been made of the existing steel ship, including the deckhouse. A detailed finite element analysis was used for that study but was not available for this redesign. For the aluminum variant the following assumptions were made to determine the stress in the deckhouse structure.

1. If all other factors remained the same, for the same strain, aluminum has 1/3 the stress of steel, based on the ratio of elastic moduli.
2. The redesign changed deck thickness from 0.188 inches to 0.25 inches thick, an increase of 4/3. Because there is one-third the elastic modulus in aluminum, the effective area changed by the ratio 1/3 x 4/3 = 0.44.
3. Studies of the effectiveness of deckhouse structure show that including the effect of a continuous 2-level deckhouse that exists over a significant length of the ship in longitudinal strength calculations will decrease strength deck stress by 10 percent. Because the deckhouse of this ship is discontinuous, it should not make as great a contribution, although the fact that it is steel does increase its effectiveness.
4. Therefore, it is conservative to assume that changing from a steel to an aluminum deckhouse will increase strength deck stress by 10 percent.
5. Accordingly, the stress in the 03 Level of the deckhouse will be considered to change by a factor of 1.1 x 0.44 = 0.48.

Using that stress with the fatigue-loading spectrum used for the original fatigue analysis, the fatigue life for an aluminum structural detail having a Eurocode 9 classification of 23-3.4 was found to have a fatigue life of 38 years. Eurocode 9 is representative of many typical structural details used in aluminum ship structure, and as a very conservative classification for the detail of the intersection of an aluminum longitudinal deck stiffener with a transverse bulkhead vertical stiffener. Consequently, the aluminum deckhouse should have a very good fatigue life with no increases in scantlings for reduction of stress. The final scantlings are as shown in Table 11.

Table 11: Scantlings for Design Study 4

Location	Steel (HSS)		Aluminum (5083)	
	Member Size	Section Modulus (in^3)	Member Size	Section Modulus (in^3)
House Side, 01–02 Level				

Longitudinals	172 x 102 x 8.9 T (5 x 4 x 6#T)	(5.3)	178 x 121 x 5.2T (7 x 4.75 x 3.5#T)	(12.2)
Vertical Frames	203 x 102 x 14.9 I-T (8 x 4 x 10# I-T)	(9.2)	203 x 102 x 7.4T (8 x 4.75 x 5.0#T)	(19.0)
	203 x 102 x 19.36 I-T (8 x 4 x 13# I-T)	(11.6)	254 x 172 x8.5T (10 x 5 x 5.75#T)	(23.0)
House Side, 02–03 Level				
Longitudinals	102 x 102 x 7.4 T (4 x 4 x 5#T)	(3.8)	152 x 102 x 4.5T (6 x 4.5 x 3.0#T)	(9.2)
Vertical Frames	254 x 102 x22.3 T (10 x 4 x 15#T)	(15.8)	305 x 102 x 14.9T (12 x 4 x 10.0#T)	(36.7)
	305 x 102 x 20.8 I-T (12 x 4 x 14# I-T)	(17)	305 x 172 x 11.9T (12 x 5 x 8# T)	(34.4)
House Front, 01–03 Level, vertical stiffeners	102 x 102 x 7.4T (4 x 4 x 5#T)	(3.6)	172 x 114 x 3.7T (5 x 4.5 x 2.5#T)	(7.0)
02 Level				
Longitudinals	152 x 102 x 10.4T (6 x 4 x 7#T)	(6.6)	178 x 121 x 5.2T (7 x 4.75 x 3.5#T)	(12.2)
Transverse Frames	254 x 102 x 17.9 I-T (10 x 4 x 12# I-T)	(11.9)	254 x 102 x 9.7T (10 x 5 x 6.5#T)	(26.3)

Fire Protection Insulation

If a U.S. Navy ship were built today with an aluminum deckhouse, fire protection insulation would be required for the aluminum structure, both to provide fire barriers and to protect the aluminum structure from melting. For this study, Structo-Guard insulation produced by Thermal Ceramics, Inc. will be used. According to the manufacturer, this product has a NAVSEA rating of N-30 if a bulkhead is insulated with two 1.5-inch thick layers on each side, a total of 6 inches of insulation on a bulkhead. It will be assumed that house sides and ends will be insulated internally, and will require only 3 inches of this insulation. The underside of decks will be insulated with 3 inches of insulation on the underside, and all stiffeners, frames, and girders on decks, bulkheads, house sides and ends, and stanchions will be surrounded with 3 inches of insulation. The insulation weighs 6 pounds per cubic foot, and the following weights result:

Table 12: Weight of Structural Fire Protection Insulation on Aluminum for Design Study 4

Location	Weight [kg (lbs)]
Decks (including stiffening and stanchions below)	15,470 (34,107)
Major Transverse Bulkheads	3,757 (8,282)
Miscellaneous Transverse and Longitudinal Bulkheads	11,077 (24,421)
House Sides and Ends	5,441 (11,995)
Total Weight of Insulation	35,746 (78,805)

Table 13: Final Weight of Structure and Insulation for Design Study 4 [kg (lbs)]

Item	Steel	Aluminum
02 Level		
Plate	241,215 (46,770)	9,595 (21,154)
Longitudinal Stiffeners and Girders	6,214 (13,700)	4,055 (8,940)
Transverse Frames	4,878 (10,754)	3,096 (6,826)
03 Level		
Plate	16,375 (36,100)	7,408 (16,332)
Longitudinal Stiffeners and Girders	3,842 (8,470)	2,265 (4,994)
Transverse Frames	4,326 (9,536)	2,860 (6,306)
House Side 01-03 Level		
Plate	14,827 (32,688)	7,070 (15,586)
Longitudinal Stiffeners	4,717 (10,400)	2,827 (6,232)
Vertical Frames	1,297 (2,860)	903 (1,990)
Stanchions	507 (1,118)	357 (787)
Transverse Bulkheads	16998 (37,474)	7,562 (16,671)
Longitudinal Bulkheads	24,709 (54,473)	10,840 (23,897)
Subtotal Structure	**119,905 (264,343)**	**58,834 (129,705)**
Miscellaneous Structure (10%)	11,990 (26,434)	5,883 (12,970)

Welding, mill tolerance (2.5%)	2,997 (6,608)	1,471 (3,243)
Total Structure	**134,900 (297,400)**	**66,180 (145,900)**
Fire Protection Insulation		35,746 (78,805)
Total Weight of Structure and Insulation	**134,324 (296,130)**	**101,923 (224,700)**

Weight of aluminum / weight of steel:
With fire protection insulation: 224,700 / 296,139 = 0.76
Without fire protection insulation: 145,900 / 297,400 = 0.49

SUMMARY

Aluminum is a versatile metal that has become the material of choice for most larger high-speed vessels in use today, including offshore crew boats and high-speed ferries. If the correct alloy is selected and properly handled during fabrication, the vessels will have little problems with corrosion. Design methods may have questionable areas that have been treated conservatively, but are well proven to develop scantlings that provide good service over time. Fatigue must be considered during the design of an aluminum vessel, and scantling increases may be necessary, which will reduce the weight advantage of aluminum over steel and GRP. Fire protection insulation is required if aluminum is to have the same resistance to fire as steel, and depending on the standards used, can seriously impact the weight of the vessel.

ACKNOWLEDGEMENTS

This work was sponsored by the interagency Ship Structure Committee. I would like to thank the members of the SSC Project Technical Committee, which had technical oversight for this project, especially committee chair Derek S. Novak of the American Bureau of Shipping, Sergei Petinov of St. Petersburg Polytechnic University, Matthew Collette of SAIC, Jeom Kee Paik of Pusan National University, and Paul Hess of the Office of Naval Research, all of whom made many helpful suggestions and provided much reference material.

Many producers and suppliers of aluminum were helpful, including Michael Skillingberg of The Aluminum Association, Stan Guess of Tower Extrusions, Larry Moffett of Taber Extrusions, Joseph B. Wolf and Rob Menard of Aluminum and Stainless, Inc., Douglas J Waldron of Advanced Joining Technologies, Harold Bushfield of Alcoa, and Mike Farrell of G. James Australia Pty. Ltd. I was kindly received at the Alcoa Technical Center, and received helpful information from Nathaniel Beavers, Francine Bovard, Israel Stol, Michael Brandt, James Burg, Roger Kaufold, James Marinelli, Jean Ann Skiles, and Rebecca Wyss.

Catherine Wong of NSWCCD provided much useful advice and information on corrosion, Edward Devine of NSWCCD reviewed the material on design, and Raymond H. Kramer of Raytheon provided information on the INCAT design.

I received help from many shipyard personnel, especially Grant Pecoraro of Gulfcraft, Al Dodson of Swiftships, Jim Towers of Kvichak Marine Industries, Michael Duquesnoy of Austal USA, Paul Kotzebue of Swath Ocean Systems, W. Philip Nuss of Trinity Yachts, and George Lundgren of Workskiff.

REFERENCES

ABS, 2001. Guide for Building and Classing High-Speed Craft, American Bureau of Shipping, October 2001.

ABS, 2003. Guide for Building and Classing High Speed Naval Craft, American Bureau of Shipping.

ABS, 2006. Rules for Materials and Welding, Part 2, Aluminum and Fiber Reinforced Plastics (Chapters 5–6).

Adamchak, John C., Design Equations for Tripping of Stiffeners under Inplane and Lateral Loads, David Taylor Naval Ship Research and Development Center report DTNSRDC 79/064, October 1979.

ALCAN, 2004, Aluminum and the Sea, ALCAN Aerospace Transportation and Industry, Pechiney Rhenalu ALCAN Group, www.alcan-marine.com, Paris, December, 2004.

Aluminum Association, 2006a. International Alloy Designations and Chemical Composition Limits for Wrought Aluminum and Wrought Aluminum Alloys, The Aluminum Association, Arlington, Virginia, April 2006.

Aluminum Association, 2006b, Aluminum Standards and Data, 2006, The Aluminum Association, Arlington, Virginia.

Aluminum Association, 2005, Aluminum Design Manual, The Aluminum Association, Arlington, Virginia.

ASTM B 928. High Magnesium Aluminum-Alloy Sheet & Plate for Marine Service, 2004 Annual Book of ASTM Standards, American Society for Testing and Materials, Philadelphia, Pennsylvania.

AWS. Guide for Aluminum Hull Welding, AWS D3.7, American Welding Society, 2004.

Bates, James L. Large Passenger-Carrying Ships for Certain Essential Trade Routes, Transactions of the Society of Naval Architects and Marine Engineers, Vol. 53, 1945.

Beach, Jeffrey E., Robert E. Johnson, Natale S. Nappi, Sr., Robert A. Sielski, William A. Palko, Thomas W. Montemarano, and William E. Lukens. A Guide for the Use of Aluminum Alloys in Naval Ship Construction and Design, David W. Taylor Naval Ship Research and Development Center, DTNSRDC 84/015, 1984.

Beach, Jeffrey E., Lloyd C. Dye, Joseph A Hauser, Robert E. Johnson, Robert R. Jones, and Jerome P. Sikora. Fatigue Life Assessment of PGG-511-Class Bottom Longitudinals. David W. Taylor Naval Ship Research and Development Center, June 1981.

Boykin, C.V. and M.L. Sellers. Practical Problems Relating to the Use of Aluminum Alloys in ship Construction, Transactions of the Society of Naval Architects and Marine Engineers, 1953, pp. 358–409.

Collette, Matthew. Strength and Reliability of Aluminium Stiffened Panels, submitted for the degree of Doctor of Philosophy, School of Marine Science and Technology, Faculty of Science, Agriculture and Engineering, University of Newcastle, Newcastle upon Tyne, 2005.

Czyryca, Earnest J. and Michael G. Vassilaros. Engineering Properties of Structural Aluminum Alloys: A Handbook, Naval Ship Research and Development Center Report 3571, June 1972.

DNV, Rules for Classification of High Speed, Light Craft and Naval Surface Craft, Det Norske Veritas, July, 2000.

Donald, J. Keith and Amy Blair. Fracture Mechanics Characterization of Aluminum Alloys for Marine Structural Applications, Ship Structure Committee report SSC-448, 2007.

Dye, Lloyd C. and Thomas Dawson. Ship Hull Structure Maintenance and Repair, Naval Sea Systems Command, NAVSEA 0900-LP-082-3010, March 1974.

Efield, P. Alloys in Shipbuilding, Transactions of the Society of Naval Architects and Marine Engineers, 1937, pp. 174–200.

Eurocode 9: Design of Aluminium Structures, Part 2: Structures Susceptible to Fatigue, CEN/TC250/SC 9, Draft for Eurocode 9 Part 2, December 1996,

Faulkner, Douglas. A Review of Effective Plating for Use in the Analysis of Stiffened Plating in Bending and Compression, Journal of Ship Research, Vol. 19, No. 1, pp. 1–17, March 1975.

Faragher, P.V. Uses of Aluminum in Shipbuilding, Transactions of the Society of Naval Architects and Marine Engineers, Vol. 39, 1931, pp. 27–42.

Ferris, Theodore. Design of American Superliners, Transactions of the Society of Naval Architects and Marine Engineers, Vol. 39, 1931.

Forrest, Matthew G. Applications and Uses of Aluminum Alloys in Ship Construction, Transactions of the Society of Naval Architects and Marine Engineers, 1947, pp. 305–331.

Hay, Robert A. and Chester H. Holtyn. The Effect of Thermal Fabrication Practices on Aluminum, Naval Engineers Journal, pp. 37–43, 1980.

Heller, S. Reese Jr. Structural Design of Ship Plating Subjected to Uniform Lateral Load, International Shipbuilding Progress, Vol. 21, No. 244, December 1974.

Hobson, R.P. Notes on the Yacht Defender and the Use of Aluminum in Marine Construction, U.S. Naval Institute Proceedings, Vol. 23, No. 82, 1897, pp. 249-282; Vol. 23, ,No. 83, 1897, pp. 523-562.

Holtyn, Chester, C.H. Aluminum—The Age of Ships, Transactions of the Society of Naval Architects and Marine Engineers, 1966.

Holtyn, Chester. H., G.S. Gresham, J. Snodgrasss, R.A. Hay, and P. Hawner. The Construction and Service Record of a 306-Foot Aluminum Trailership, Transactions of the Society of Naval Architects and Marine Engineers, 1972.

IACS. Requirements Concerning Materials and Welding, International Association of Classification Societies, 2005.

IACS, 1999. Fatigue Assessment of the Ship Structures, Report No. 56, International Association of Classification Societies, 1999.

Kramer, Raymond H. U.S. Navy High Speed Craft – Comparison of ABS and DNV Structural Requirements, Transactions of the Society of Naval Architects and Marine Engineers, Vol. 113, 2005.

Leis, Brian N. Structural Behavior after Fatigue, Ship Structure Committee Report SSC-358, 1990.

McComb, Archibald H. and Edwin E. Benzenberg. Stateroom Fire Test, Transactions of the Society of Naval Architects and Marine Engineers, Vol. 58, 1950.

McGuire, S.C., Aluminum: Its Alloys and Their Use in Ship Construction, Transactions of the Society of Naval Architects and Marine Engineers, 1895, pp. 69-97.

Mears, R.B. and R.H. Brown. Resistance of Marine Alloys to Marine Exposure, Transactions of the Society of Naval Architects and Marine engineers, 1944, pp. 91–113.

Michaelson, Robert W., User's Guide for SPECTRA: Version 8.3. Naval Surface Warfare Center, Carderock Division Report NSWCCD-65-TR-2000/07, 2000.

MIL-STD 1399, Section 301. Interface Standard for Shipboard Systems, Ship Motion Environment, Department of Defense.

MIL-STD 1689. Fabrication, Welding and Inspection of Ship Structure, Department of Defense

NAVSEA 0900-LP-097-4010. Structural Design Manual for Surface Ships, 1976.

Paik, Jeom Kee, Anil K. Thayamballi, Jung Yong Ryu, Joon Ho Jang, Jung Kwan Seo, Sang Wook Park, Sun Kee Seo, Celine Renaud, H. Paul Cojeen, and Nam In Kim (2006b). The Statistics of Weld Induced Initial Imperfections in Aluminum Stiffened Plate Structures for Marine Applications, International Journal of Maritime Engineering, June 2006.

Paik, Jeom Kee, Owen F. Hughes, Paul E. Hess III, and Celine Renaud. Ultimate Limit State Design Technology for Aluminum Multi-Hull Ship Structures, Transactions of the Society of Naval Architects and Marine Engineers, Vol. 113, 2005, pp. 270–305.

Paik, Jeom Kee and Alexandre Duran. Ultimate Strength of Aluminum Plates and Stiffened Panels for Marine Applications, Marine Technology, Vol. 41, No. 3, July 2004.

Rock, G.H. Notes On the Development of Certain Materials used in Ships of the U.S. Navy, Transactions of the Society of Naval Architects and Marine Engineers, 1931, pp. 215–249.

Sharp, George G. Fire Control for Passenger Vessels, Transactions of the Society of Naval Architects and Marine Engineers, Vol. 45, 1937.

Sielski, Robert A., 1987. The History of Aluminum as a Deckhouse Material, Naval Engineers Journal, Vol. 99, No. 3, pp. 165–172, May 1987.

SNAME, 1974. Aluminum Fire Protection Guidelines, Society of Naval Architects and Marine Engineers, Technical and Research Bulletin 2-21.

Stone, Kevin F. and Derek S. Novak, Comparative Structural Requirements for High Speed Craft, Transactions of the Society of Naval Architects and Marine Engineers, 2006.
Timoshenko, S. Theory of Materials, Part II, Van Nostrand, 1956.
Vara, Roger, Christopher M. Potter, Robert A. Sielski, Jerome P. Sikora, L. Robert Hill, John C. Adamchak, David P. Kihl, Jay Herbert, Roger I. Basu, Larrie Ferreiro, Joan Watts, Paul D. Herrington. Adaptation of Commercial Structural Criteria to Military Needs, Gulf Coast Region Maritime Technology Center, University of New Orleans, GCRMTC Project No. AMTC 00-5161, May 2003. NTIS No. PB2004-101323.
Vasta, John. Structural Tests on the Passenger Ship *S.S. President Wilson*—Interaction between Superstructure and Main Hull Girder, Transactions of the Society of Naval Architects and Marine Engineers, Vol. 57, 1949.

Discussion

Matthew Collette, Member

The author is to be congratulated on presenting a clear and concise review of aluminum structural design, the tradeoffs and unique considerations of this material, and many of the lessons learned in the past. With continued interest in aluminum for passenger and military applications, there is increasing demand for improved engineering analysis for aluminum structures. The author's expertise and previous experience with such structures makes this paper a valuable reference for those more recently acquainted with aluminum as a material for vessel construction. The inclusion of the point made by Boykin and Sellers that aluminum "is not just a substitute for steel; it is an entirely different material" is an especially important observation to keep in mind when designing vessels in aluminum.

The increased use of the 6xxx-series alloys in modern construction is a departure from past experience, especially in naval applications, which have been restricted to 5xxx-series alloys in the past. The 5xxx and 6xxx are have quite distinct metallurgic properties as the author mentions; one result of this difference is that the welded ultimate stress of the 6xxx series is reduced proportionally further from base material properties than the welded ultimate stress of the 5xxx series, a situation which is not clear if the comparison is limited to welded proof stresses presented in Table 3. I mention this as the U.S. Navy plating formulae given in eq. 6-8 for lateral loads make use of a combination of the welded proof stress and un-welded ultimate stress of the material for a welded plate. Is the author aware of any experimental validation to confirm these equations for 6xxx-series material, especially equation eight?

I was interested in the design studies presented in the paper, particularly in the analysis of the deckhouse for the destroyer. I understand that the purpose of the comparison was to reflect 1980s construction approaches, but do feel that it is worth mentioning that using modern aluminum construction techniques, especially the use of extrusions joined by friction stir welding, may have resulted in a more favorable weight comparison for the aluminum structure. I base this observation on the fact that the plate in the deckhouse, which appears to be over 50% of the total structural weight, was sized based on minimum thicknesses for traditional fabrication (fusion-welded stiffeners) techniques, not strength requirements. Extrusions joined by friction stir welding would most likely have permitted thinner aluminum plate to be used for the aluminum alloy while still avoiding excessive deformation. I concur with the author's conclusion that more research is required into the structural fire protection of aluminum, especially given the weight impact of such insulation on weight-critical applications.

In weight-critical structures, the advantages of going to a limit-state design formulation, where the ultimate strength of the structure is explicitly considered, is becoming increasingly accepted even if a full reliability-based design procedure is still some time away. In light of the author's comments on the difficulty of ensuring consistent definition of welded proof stress between codes that use different gauge lengths (as shown clearly in Table 3), I would suggest that aluminum structures would benefit from a limit-based design approach. One clear advantage is that the ultimate strength of a welded joint is constant regardless of the gauge length, while the proof stress is not. Such limit-state formulations are starting to appear now for aluminum, but are still behind those formulated for steel. In completing his review for this work, has the author received any feedback from other industries which have moved to a more limit-state oriented design approach for aluminum?

In closing, I would like to thank the author for this contribution to the field of aluminum structures and for providing me the opportunity to comment on the paper.

Thomas J. Summerson, Visitor

My review is focused on the corrosion, fatigue and alloy properties of the paper. Comments are provided in sequence.

In the section on materials and fabrication it would be better to say, "Other high strength alloys should be avoided for marine use…." For clarity, I think that the term "sensitization" should be defined. Why does the International Association of Classification Societies specify that 6005A and 6061 should not be used in direct contact with seawater unless protected by anodes and/or paint system? Also, why did US Navy completely prohibit use of 6XXX alloys until recently? This seems contrary to later statements that the 6XXX alloys have shown good service performance.

Shouldn't 5456-H112 and 5383-H116 extrusions be included in the list of extrusions used? Are the limitations cited for 6XXX series related to strength or to thickness?

In the section "mechanical properties" I think that the word "yield" is a better description than the word "soften." The sentence should read "Instead the material begins to gradually yield as stress increases."

Why aren't the ASTM Standards referenced in Table 2 — for example, B209, B221 and B928? In Table 3, alloy 5086-HO should be 5086-O. Again, I think that ASTM standards should be referenced.

I believe the three paragraphs on exfoliation, intergranular and stress-corrosion cracking could be enhanced by changes in wording. It is most important for the reader to understand that both exfoliation and intergranular corrosion are caused by selective attack at the grain boundaries and is dependent on the grain morphology. Corrosion is the result of selective precipitation of an Mg2Al3 phase along the grain boundaries.

Exfoliation corrosion is intergranular corrosion in a grain structure that is highly elongated and is largely unrecrystallized while intergranular pitting occurs with a recrystallized, equi-axial grain structure. Thus, the H116 temper is not only resistant to exfoliation corrosion but to intergranular corrosion as well. Reference could be made to the Aluminum Association Technical Committee report T1, issued in 1972 for specific background.

Intergranular corrosion is the result of selective attack along the grain boundaries of a recrystallized grain structure.

In the description of stress corrosion cracking, you should delete the first three sentences and begin with the fourth sentence because it is not necessary to have a sharp pit and mechanical tearing at the root of the pit or notch. While the sharp pit or notch creates a stress raiser, it is the presence of a steady tensional load that will lead to SCC when the material is susceptible to IGC. The fourth sentence reads, " Stress corrosion in aluminum alloys is characterized by IG cracks normal to the metal surface and stress direction."

In the section on galvanic corrosion, it seems rather strange to consider the cathode attacking the metal with the lower electrical potential, the anode. Usual phraseology indicates that the lower potential metal (anode) is corroded while sacrificially protecting the more positive potential material (cathode). That is useful in explaining cathodic protection. The last sentence of this section states that using two dissimilar alloys with different galvanic potentials can cause a galvanic corrosion problem. It might be better to replace "can cause" with "may cause" because the potential differences may not be too great.

In the section on sensitization, change "The 5xxx-series alloys with magnesium contents greater than 3 percent are subject to exfoliation…" to "…may be subject to exfoliation…depending upon fabrication processes." Also, why not define sensitizing temperature range as being from 100C to 200C (212 F to 400 F)?

Is there a reference that can be cited in the reported case of 5456-H116 on weather decks of a superstructure becoming sensitized by service temperature above the 65 C limit? Also, what form of failure did the "sensitization" cause? For example, I know that exposing stressed test samples to sensitization treatments does not result in SCC because the sensitizing treatment causes thermal stress relief and the resulting stress was below the threshold level.

When discussing flame straightening, replace the word "heat" with "thermal" to describe the process as a thermal treatment.

I don't understand what the relationship between certain limits on distortion, such as minimum cold-forming radii (AWS 2004), has to do with sensitization afterward. Please explain.

I am surprised that the paper by Harold Bushfield, et al. on the establishment of B928 is not cited.

Authors' Closure

I thank Mr. Summerson for adding to the paper from the perspective of a metallurgist. His comments and clarifications are welcome to this naval architect, who has only a basic understanding of metallurgy. I offer the following responses.

Sensitization is the process where excess magnesium that is dispersed throughout individual grains precipitates to the grain boundaries. The standard laboratory process for inducing sensitization is to hold the material at a temperature of 100°C for one week. However, sensitization can occur at temperatures as low as 50°C if the material is at that temperature for at least 36 weeks (Vassilaros and Czyryca).

I, along with many metallurgists with whom I have spoken, am baffled by the restrictions on the use of 6xxx-series aluminum for marine structures, as there is no experimental evidence to show them to be a concern. Perhaps there were some early poor experiences, such a mixing aluminum with other metals with resulting corrosion, and the restrictions were developed and continue to this day with the reasons lost in history.

There are many different alloys that may be suited for marine use in addition to those listed in the tables of the paper. One of the difficulties facing the designer is to determine if an alloy suggested by a supplier is suitable, a problem exacerbated by a lack of standards, particularly for long-term corrosion. Classification societies provide additional guidance, as alloys that have proved unsuitable have generally been sorted out. Specifications of the ASTM are useful for procurement, but should be used with caution, as there are sometimes optional testing requirements to which all heats have not been subjected. Background for the development of ASTM B 928 is given by Bushfield, et al. (2003).

The 1979 report by Vassilaros and Czyryca helped set standards for use of higher magnesium-content 5xxx-series alloys at elevated service temperatures. The cases of sensitization that I have cited in this paper are from ongoing studies for which no reports have been made in the open literature. However, most of the failures have occurred near welds.

References

Vassilaros, Michael G. and Czyryca, Ernest J., 1979. Natural aging and sensitization of 5000-series marine aluminum alloys, Report DTNSRDC/SME-78/106, David W. Taylor Naval Research and Development Center, June.

Bushfield, Harold Sr., Cruder, Marc, Farley, Rendall, and Towers, Jim, 2003. Marine Aluminum Plate — ASTM Standard Specification B 928 and the Events Leading to its Adoption, *Proceedings*, Society of Naval Architects and Marine Engineers, San Francisco, California.

Fatigue Assessment of Aluminum Ship Details by Hot-Spot Stress Approach

Bård W. Tveiten (V), Xiaozhi Wang (M), Stig Berge (V)

ABSTRACT

Fatigue tests of full scale models of an aluminum ship structural detail were carried out to obtain a design S-N curve. The test detail was also analyzed by the finite element method, using several modeling techniques and element types. The results from both experimental tests and finite element analysis are discussed. Recommendations on the procedure of fatigue assessment of aluminum ships including S-N curve to be used are also presented.

KEY WORDS:

Fatigue strength, fatigue testing, S-N data, hot-spot stress, finite element analysis, welded joints, aluminum structures, fatigue assessment, ship structures.

NOMENCLATURE

a	Weld throat section
E	Young's modulus
K_s	Hot-spot (structural) stress concentration factor
log(a)	Intercept of S-N curve (log-log format)
m	Slope of S-N curve (log-log format)
N	Cycles to failure
R-ratio	$R = S_{min}/S_{max}$ (min. stress/max. stress)
S33	Longitudinal stress
SCF	Stress concentration factor
SP_{Max}	Maximum principal stress perpendicular to the weld toe
t	Flange plate thickness
t_w	Shell thickness used in shell weld idealization
$\Delta\sigma_{nominal}$	Nominal stress range
$\Delta\sigma_s$	Hot-spot stress range
$\varepsilon_{nominal}$	Nominal longitudinal strain (elastic)
ε_{22}	Longitudinal strain, shell element model
ε_{33}	Longitudinal strain, finite solid element model
ν	Poisson ratio
$\sigma_{nominal,\ elastic\text{-}beam}$	Nominal stress obtained from elastic beam theory
σ_s ($K_s \cdot \sigma_{nominal}$)	Hot-spot (structural) stress
$\sigma_{0.4\cdot t\,/0.5\cdot t\,/1.0\cdot t\,/1.5\cdot t}$	Stress value at extrapolation points
ϕ	Weld flank angle
λ	Weld leg length
ρ	Weld toe radius

INTRODUCTION

Unlike large commercial ships, high speed vessels are often constructed of aluminum. Compared with steel structures, aluminum alloys structures provide more flexibility in the design of complex weld details such as aluminum alloy extrusions which offer designers a large freedom from restrictions often given by the "standard" shapes. However, this freedom brings more difficulties in standardizing the fatigue design and analysis methodology of welded aluminum structures.

Some work has been done to assess the fatigue strength of aircraft and civil engineering structures. However, limited work has been performed with regard to welded joints typically used in aluminum ships. The main objective of the study presented in this paper is to investigate the existing design and analysis methodologies and thereby develop a reliable fatigue design and analysis procedure for aluminum ship structures through numerical evaluation and experimental verification.

The study consists of three main parts; a review of existing methodologies for fatigue assessment of welded aluminum ship structures based on the hot-spot stress approach, numerical analyses using the finite element method (FEM) to determine and validate the hot-spot stress assessment procedures, and full scale fatigue tests of a typical welded aluminum ship detail for assessment of design S-N curves to be used together with the hot-spot stress results to be derived from the numerical stress analyses.

FATIGUE OF WELDED ALUMINIUM STRUCTURES

The Hot-Spot Stress Concept

A key issue addressed in this paper is the definition and the calculation of the hot-spot stress, σ_s, used in fatigue design of welded structures.

Fatigue design codes for welded aluminum structures are in general based on a nominal stress range S-N curve approach where the design stress is the local nominal stress range, $\Delta\sigma_{nominal}$, neglecting the effect of any stress raisers due to local weld geometry. However, the nominal stress approach is not practical for welded structures characterized by relatively complex geometry, or combinations of loads. As an alternative, a design principle based on a hot-spot stress range S-N curve has been proposed. This reduces the number of design S-N curves since the geometrical features of the structural details are included in the design stress range rather than in the design S-N curves. However, the local notch effects are still embedded in the design S-N curves. Note that the hot-spot stress range approach applies only to fatigue failures initiating at the weld toe location.

The hot-spot stress, σ_s in Fig. 1, is defined as the surface stress at the weld toe location that includes the stress raising effects due to structural (geometrical) discontinuities, but excludes the stress concentration due to the presence of the weld notch. The non-linear stress peak due to the presence of the weld notch may be excluded by means of an extrapolation procedure of stress from outside the region that is influenced by the local notch.

Generally, all hot-spot stress extrapolation procedures proposed in the open literature are based on surface stress extrapolation, with the assumption that the hot-spot stress varies linearly through the plate thickness and the effect of the weld notch is localized within a distance close to the weld, where the distance is normally expressed as a function of the main plate thickness. It has been shown that the extrapolation methods may depend on joint type, see Tveiten (1999), Tveiten and Moan (2000). That is, the method may be calibrated to a given joint geometry but may fail to get an accurate assessment of different joint configurations and dimensions.

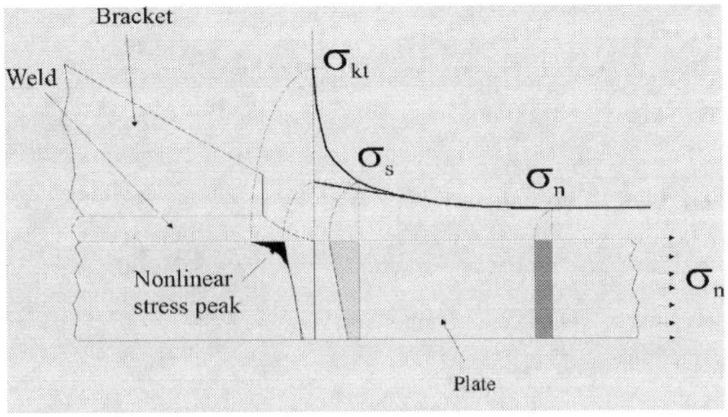

Fig. 1. Definitions of stress used in fatigue analysis.

The through-thickness stress field will also govern crack growth rate. Different details with the same hot-spot stress value may show different stress fields through the plate, and the crack growth life is affected. Thus, the concept of a uniquely determined design S-N curve based on a hot-spot stress is an approximation. For tubular joints a two-parameter hot spot stress has been proposed, including a degree-of-bending component, Berge, Haswell and Engesvik (1994).

Several methods have been proposed for assessment of the hot-spot stress on the basis of the stress distribution through the thickness. It has been shown that these methods could be promising in order to cope with problems related to mesh sensitivity and geometry dependency that have been experienced with surface extrapolation methods: Lotsberg (2003), Doerk, Fricke and Weissenborn (2003), Dong (2004), Poutiainen, Tanskanen and Marquis (2004). Dong (2004) has proposed a method that has been claimed to be mesh-size insensitive and can be used in combination with both solid and shell element models. However, it has been demonstrated that the method in some cases is mesh-sensitive or even sensitive to how element stress values are determined during the post-processing: Doerk, Fricke and Weissenborn (2003), Wang, Sun and Cheng (2004), Poutiainen, Tanskanen and Marquis (2004).

Modeling Principles for FE Analysis

Several possibilities are available for modeling of the structural geometry, ranging from solid elements, thick or thin shell elements, or a combination of these. Shell element models as well as coarse solid element models are characterized by a linear stress distribution over the plate thickness. Therefore, both types of finite element modeling are suitable for the calculation of the hot-spot stress since the nonlinear stress distribution due to the presence of the weld notch is excluded. The main problem that arises when shell elements are used is that the shell element formulation only provides a model for the mid-plane of the plates (the element thickness is given as an element property) and thus the local change of the stiffness associated with the weld shape can not be modeled, see Sumi (1997), Tveiten and Moan (2000).

Existing surface extrapolation methods require mesh refinement, to the point where further refinement does not result in significant change of the stress distribution between the extrapolation points. Alternatively, element mesh requirements can be specified with respect to size and order (linear or quadratic shape functions). Hence, the stress extrapolation procedures are linked to specific requirements for different finite element types and finite element meshes. For practical design there is a conflict of interests between the required accuracy of the resolved hot-spot stress and model complexity and computing time.

Design S-N Curves for Welded Aluminum Structures - Hot-Spot Stress

Limited S-N data for validation of the hot-spot stress approach for welded aluminum ship structures is available. Partanen and Niemi (1999) compiled fatigue test results for a variety of aluminum test specimens with moderate thickness (up to 6mm) subjected to pulsating tension (stress ratio $R > 0$). The hot-spot stress was based on the extrapolation procedure proposed by IIW (2005). The authors concluded that the hot-spot stress range approach can safely be used with an S-N curve of fatigue class FAT40 (FAT40 refers to the characteristic stress range at $2 \cdot 10^6$ cycles) for butt and fillet welded aluminum joints of relatively thin plates and extrusions (up to 6mm) with crack initiation from the weld toe (Fig. 2).

Tveiten and Moan (2000) published S-N data on flat bars with fillet welded in-plane brackets. The hot-spot stress was calculated using the extrapolation methods proposed by Tveiten and Moan (2000) and Niemi (1993). The S-N data obtained for the flat bar/bracket connections seemed to correspond well with the findings of Partanen and Niemi (1999) with respect to an appropriate hot-spot stress design S-N curve of fatigue class FAT 40 (Fig. 3). Aluminum box-stiffener/lap joints with wall thickness 3mm were studied by Ye, Moan and Tveiten (2001). The results showed that using a hot-spot stress range approach, design S-N curve of fatigue class 44 proposed by CEN (1999) seems to give conservative fatigue assessments, Fig. 3. Additional experimental studies by Macdonald, Haagensen and Søvik (1998) and Tveiten (2003) have proposed a fatigue design methodology for welded aluminum space frames made of rectangular hollow section joints using a linear (or quadratic) extrapolation of the hot-spot stress according to recommendations given by the IIW (2005) and a design S-N curve with a fatigue class of FAT 40 as shown in Fig. 3.

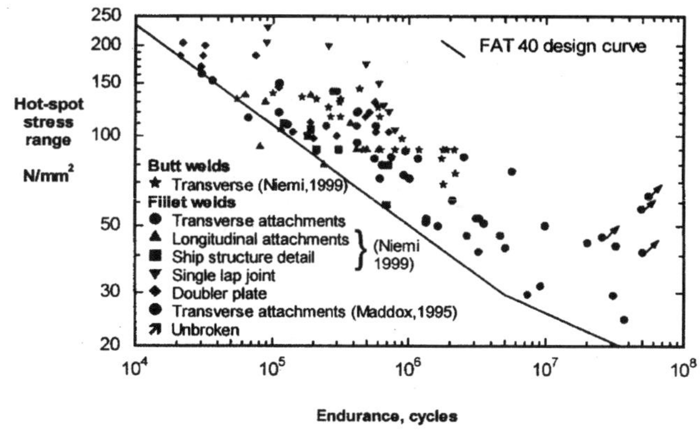

Fig. 2. S-N data expressed in terms of IIW (2005) hot-spot stress range definition, FAT 40 design curve (Figure from Maddox, 2001).

IIW (2005) has recently included a design S-N curve of fatigue class 40 for use with extrapolated hot-spot stress values for a large range of welded aluminum details. However, for some specific details including cruciform joints with load-carrying fillet welds the use of a design S-N curve of fatigue class 36 is proposed, IIW(2005).

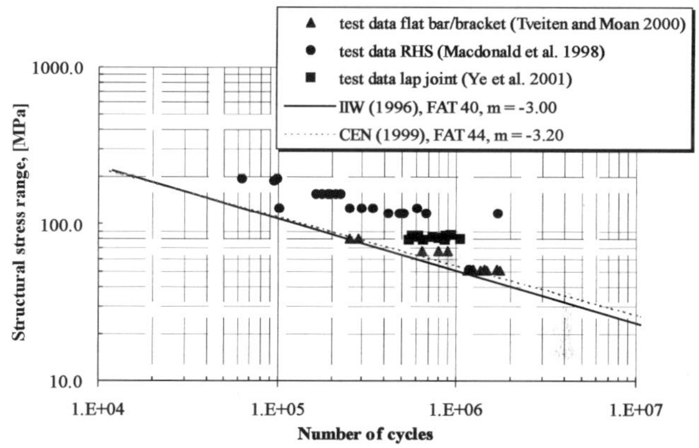

Fig. 3. S-N data expressed in terms of the hot-spot stress range definition proposed by Tveiten and Moan (2000) and IIW (2005) (Figure from Tveiten, Naiquan and Moan, 2002).

STRESS ANALYSIS

Test Model

A schematic outline of the specimen geometry is given in Fig. 4. The transverse web plate and end plates were made of aluminum alloy 5083- H22/32, while the extruded longitudinal stiffener was made of aluminum alloy 6082-T6. The test specimens were designed and manufactured at a Norwegian shipyard according to normal workmanship standards. All the fillet welds were produced manually and tested in the as-welded condition.

Test Rig and Boundary Conditions

The test rig used for the fatigue tests was designed in order to simulate the effect of lateral load transfer from a longitudinal stiffener into a transverse web. At the end of the models a 12mm aluminum plate was welded and fitted with a fixed cylindrical shaft that acted as a bearing against the test ramp. On one side a ball bearing was fitted, to eliminate any frictional forces from the contact with the test ramp. The loading was provided by a servo-hydraulic actuator with a load capacity of ±50 kN, operated in load control.

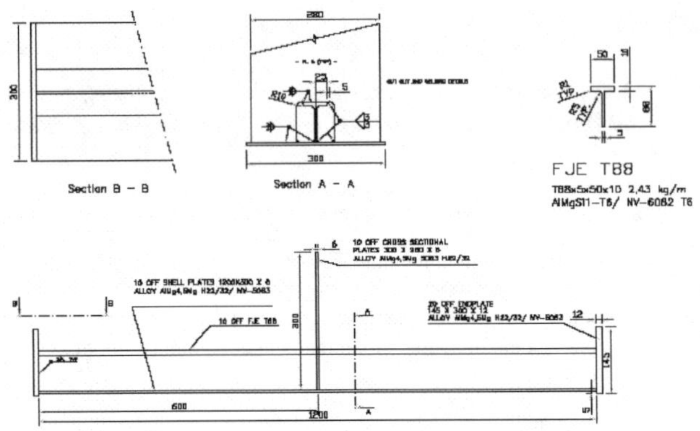

Fig. 4. Schematic outline of test specimen, nominal dimensions

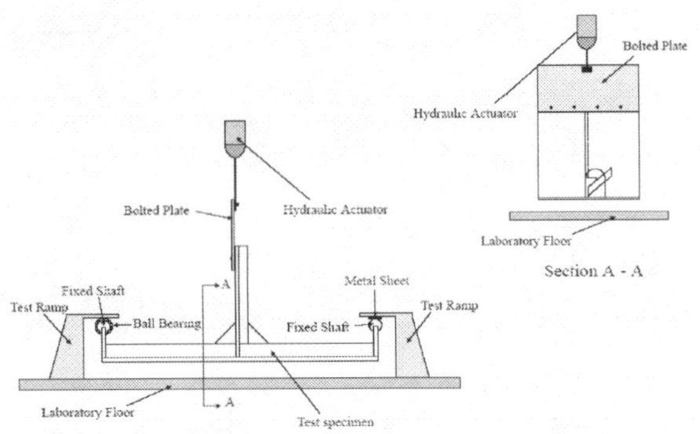

Figure 5. Test rig, loading and boundary conditions for models.

Finite Element Models

In this study, the numerical stress analyses were carried out using a commercial finite element program (ABAQUS 2004) and using a Dell LATITUDE X300 Lap-top with an Intel Pentium M processor of 1.40GHz, and 1.11GB of RAM. All analyses were linear elastic.

The finite elements used in the solid models were 20-node quadratic elements (C3D20/R) with full and reduced integration and 8-node linear brick elements with full and reduced integration (C3D8/R) for all main elements. Note that the weld was modeled without a radius fillet at the weld toe in the solid element models. Hence, a numerical singularity will exist at the weld toe location.

The finite elements used in the shell element models were 4-node double curved general-purpose shell elements with full and reduced integration (S4/R) with five integration points through the shell thickness, and 8-node doubly curved thick shell with reduced integration (S8R).

Pinned-end boundary conditions were assumed, corresponding to the test conditions (Fig. 5). The finite element models had two symmetry planes (the [1-2]-plane and the [2-3]-plane), thus, only ¼ of the finite element model was modeled (Fig. 6).

The finite element analyses were linear-elastic assuming:

- Young modulus of $E = 0.70 \cdot 10^5$ MPa
- Poisson ratio of $\nu = 0.3$

All dimensions are given in millimeter (mm), and loads in Newton (N).

In the case of the shell element models, some of the analysis models used an 'idealized weld' in order to include the stiffness of the weld. It has been proposed by IIW (2005) and Fricke (2001) to include the weld stiffness when using shell elements, by e.g. inclined shell elements or constraint equations.

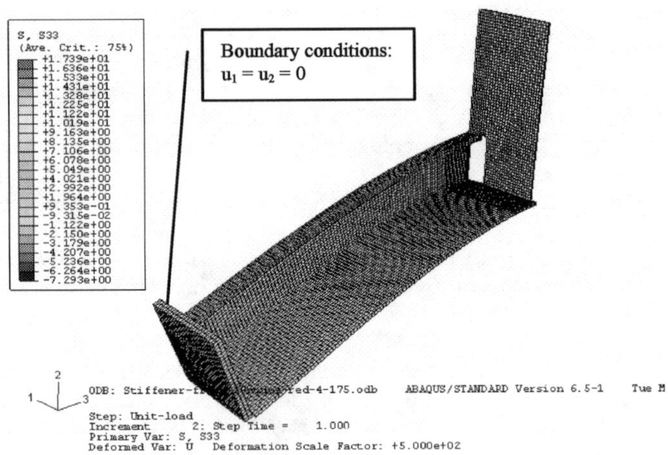

Fig. 6. Solid FE model (deformations scaled a factor 2000). Stress in longitudinal direction, S33 (global co-ordinates).

In the shell element analyses presented in this paper, the weld was idealized by means of:

a) Tied constraints between the surfaces located "inside" the weld, where the flange is the master and the transverse plate is the slave. Tie constraints tie together two surfaces where each node on the slave surface is constrained to have the same displacements as the point on the master surface.

b) Inclined shell elements with increased thickness, $t_w = 10 \cdot t$ (100mm).

c) Solid elements connected to the shell surfaces at the flange and the transverse plate, using a surface-based option in the FEM code that allows for a transition from shell element modeling to solid element modeling.

Finite Element Analysis Results - Comparison of FE Models

Fig. 7 shows the distribution of the normalized longitudinal stress ($S33/\sigma_{nominal,\ elastic\text{-}beam}$) results along the centre line of the stiffener for a set of different solid and shell element analyses using full integration elements and reduced-integration solid and shell elements. In the subsequent figures presented in this paper the notation "red" means 'reduced integration', while the notation "full" means "full integration".

The normalized stresses are based on the derived nodal stress, which is an averaged nodal stress that is extrapolated to nodes from integration points using element shape functions and averaged over all elements that contribute to the node. The nominal stress, $\sigma_{nominal,\ elastic\text{-}beam}$, was calculated using elastic beam theory, neglecting effects of shear lag and the small fillet of the stiffener cross section.

For ship structures characterized by relatively complex geometry subjected to combinations of loads, defining a nominal stress becomes rather cumbersome or very often impossible. For the structural detail subjected to 3-point bending investigated in this study, the nominal stress was easily defined and could also be calculated using elastic beam theory. Hence, for the discussion regarding the hot-spot stress calculations, the hot-spot stress has been normalized for two main reasons; a) to verify the validity and applicability of the finite element models compared to elastic beam theory, and b) to obtain a stress concentration factor that indicates the stress increase caused by the structural geometry of the intersection of the transverse frame/longitudinal stiffener. For the subsequent discussion of the hot-spot stress range approach, stress concentrations relative to beam theory are given.

IIW (2005) presented rather comprehensive recommendations (see Section 2.2.3.4 in IIW 2005) with respect to recommended meshing of coarse and fine element models, element formulation and weld characterization (for shell element models) together with recommended extrapolation procedures. For relatively fine element models, IIW (2005) proposes that the distance of the centre point of the first integration point of the elements to the weld is no more than $0.4 \cdot t$.

ABS (2005) and IACS (2006) propose an element mesh density approximately equal to the representative net thickness in the assessed area for oil tankers (applies to hopper knuckle connections only) and bulk carriers.

Fig. 7 shows that with respect to the global stress distribution, there is a good correspondence between the solid element and the shell element analyses, and the calculations based on elastic beam theory. As expected, the 8-node solid element model is stiffer, thus giving around 3% lower nominal stresses compared to the 20-node solid element model. Note that for the solid element model, the stress is plotted from the weld toe, while for the shell element models (without the weld) the stress is plotted from the element intersection line.

Fig. 8 shows the normalized longitudinal stress along the centre line of the stiffener obtained from various solid element analyses using full-integration and reduced-integration elements.

As shown in Fig. 8, a continued refinement of the element mesh of the finite solid element model using 20-node quadratic brick elements with full integration close to the weld will confine the effect of the singularity at the notch (no weld toe radius was modeled) to an increasingly smaller region. Convergence in the region where it is assumed that the effect of the hot-spot stress can be derived from ($0.4 \cdot t$ - $1.5 \cdot t$) was obtained for element sizes of the order of approximately $0.2 \cdot t$ (1.92x3.75x3.00 (mm)). However, for the analyses using 20-node quadratic brick elements with reduced integration it is seen that convergence in the region between $0.4 \cdot t$ and $1.5 \cdot t$ was obtained for element sizes in the order of approximately $0.4 \cdot t$ (4.175x3.75x5.00 (mm)). The 8-node brick elements with reduced integration provide results similar to the 20-node reduced-elements for same mesh densities.

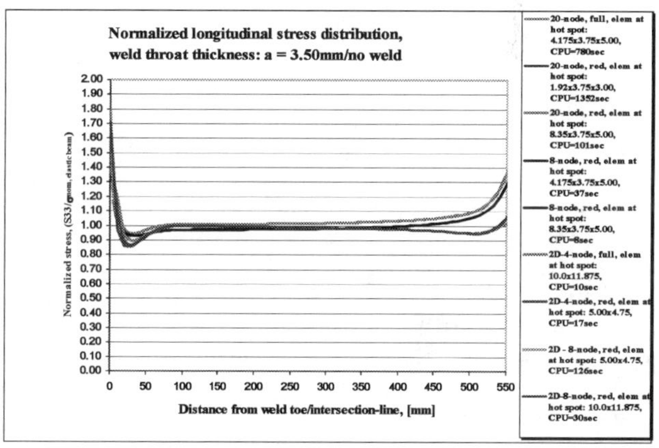

Figure 7. Normalized longitudinal stress distribution along centre line of stiffener for various finite element models.

The results presented in Fig. 9 show that by reducing the integration and the order of the element (from quadratic to linear), the effect of the mesh size with respect to the hot-spot stress distribution is somewhat eliminated and the finite element models with much less refined mesh give approximately equal stress results within the extrapolation zone (0.4·t - 1.5·t). It is important to bear in mind that the definition of hot-spot stress is the linear through-thickness stress, and by using lower order elements or reduced-integration elements, the stresses become linearized, which effectively reduces the effect of the non-linear stress peak at the weld toe.

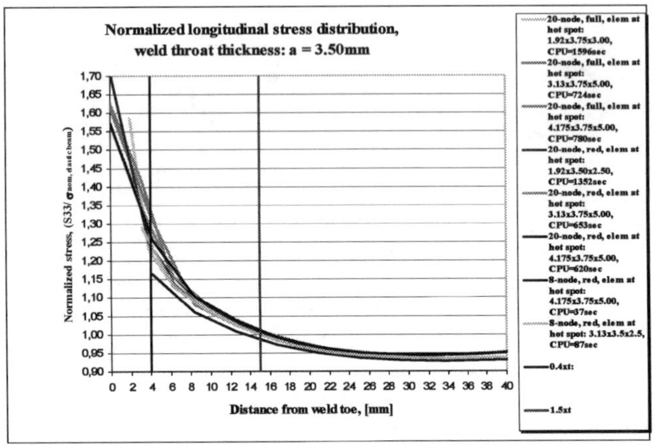

Figure 8. Normalized longitudinal stress distribution along centre line of stiffener close to weld toe for various solid element models. Stress distribution is plotted from the weld toe.

Even though the stress distribution obtained from the 20-node and 8-node solid element models are rather coincident, the difference in CPU time between the 20-node finite solid element analysis using full-integration and the 8-node reduced-integration analyses are quite large, from 1596 seconds to 87 seconds (a factor of ≈18). Hence, tremendous computing time can be saved by using lower order, reduced finite element models. However, it is important to note that at the weld toe, the higher order elements describe the stress concentration better as the stress is derived using higher order polynomials. The linear 8-node reduced-integration elements provide good results for many applications if the mesh density is sufficiently refined, which is particularly important when the structure is carrying bending loads. For instance a single linear element with reduced integration through the thickness of a beam, plate or pipe section is unable to resist bending loads as all the integration points are laying on the neutral axis.

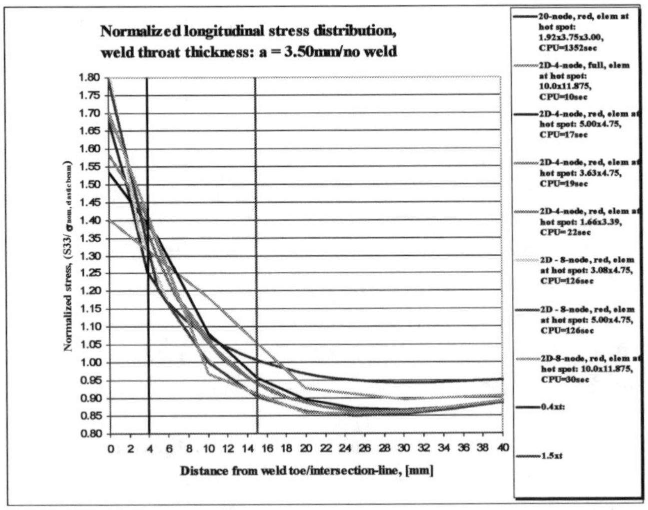

Figure 9. Normalized longitudinal stress distribution along centre line of stiffener close to weld toe for various solid and shell models. The shell element models (2D) do not include a model of weld (stress distribution plotted from the element intersection lines).

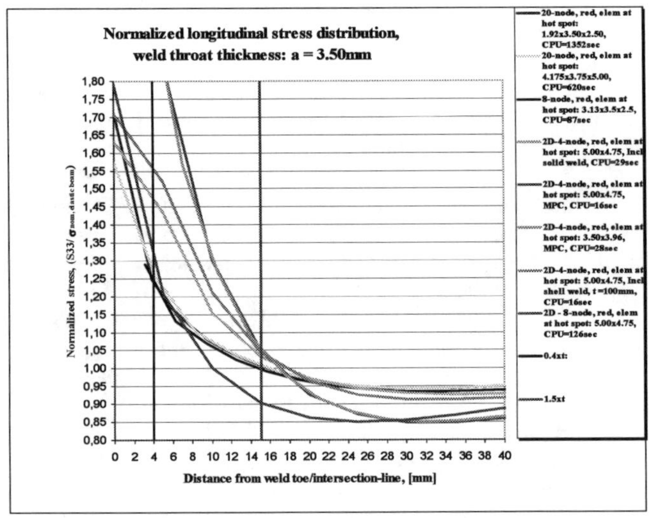

Fig. 10. Normalized longitudinal stress distribution along centre line of stiffener close to weld toe location for various solid and shell element models. The shell element models (2D) include different models of weld (tied contact, inclined shell element with thickness $t_w = 10 \cdot t_{flange} = 100mm$, and solid weld connected to shell). Stress distribution is plotted from toe of the weld.

Although the solid and shell element models show excellent agreement for the global nominal stress distribution, the stress distribution at the hot-spot location is somewhat different resulting in higher nodal stresses for the shell element models than for the solid element models, Fig. 9. This is most likely due to the absence of the weld in the shell models. It has been proposed by Fricke (2001) that vertical or inclined elements with appropriate stiffness, or constrained equations to couple node displacements to model, the weld stiffness need to be considered for cases where the results are affected by a high degree of local bending. However, the various methods for including the weld stiffness should be used with caution. Tveiten (1999) has shown by means of shell and solid element analyses combined with strain gage measurements of longitudinal stiffener/bracket connections that extrapolated hot-spot stress values were rather sensitive to how the weld stiffness had been accounted for in the shell element model. Fig. 10 shows that for this particular detail, various models for including the weld in the shell models actually lead to less accurate stress distributions close to the weld toe when compared to the results obtained with the solid element models. The tied constraint model gave the largest deviation.

The use of 4-node full integration shell elements with rather dense mesh in the order of 1.0·t is not able to properly describe the stress increase at the notch (element intersection line). However, by using 8-node reduced integration shell elements the stress results are comparable to the much denser 4-node shell element models. As shown in Figure 9, reducing the mesh size of the shell 8-node reduced-integration elements, does not change the stress distribution significantly.

Figs. 11 and 12 present the finite element results from the "converged" models, showing no significant change in the stress distribution with increased mesh density.

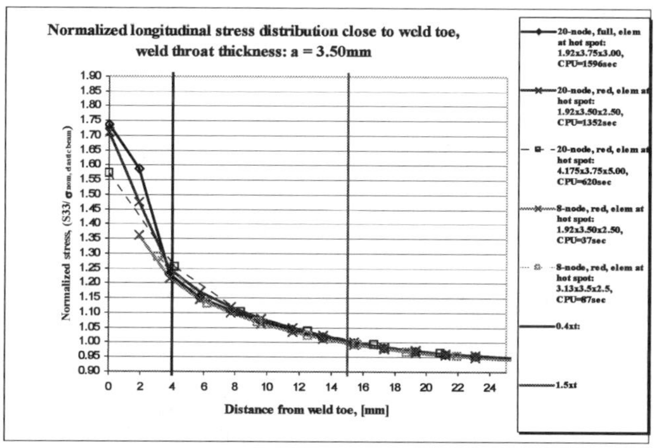

Fig. 11. Normalized longitudinal stress distribution along centre line of stiffener close to weld toe location for the "converged" solid finite element models.

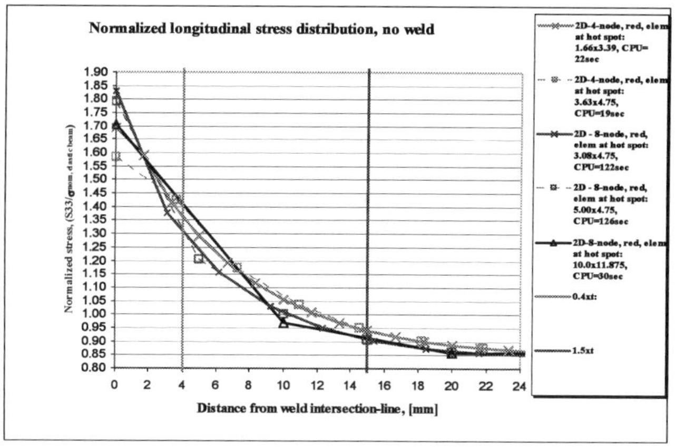

Fig. 12. Normalized longitudinal stress distribution along centre line of stiffener close to weld toe location for the "converged" shell element models

Based on the recommendations given by: Niemi (1993), Tveiten and Moan (2000), Wang, Sun and Cheng (2004), ABS (2005), IIW (2005) and IACS (2006) with respect to mesh refinement and alternatives with respect to modeling, the following finite element models have proved to be the most applicable in order to derive the hot-spot stress:

1. Solid, 20-node, red, element at hot-spot: 4.175x3.75x5.0 (mm), longitudinal element length ≈ 0.4·t
2. Solid, 8-node, red, element at hot-spot: 3.13x3.5x2.5 (mm), longitudinal element length ≈ 0.3·t
3. Shell, 8-node, red, element at hot-spot: 10.0x11.875 (mm), longitudinal element length = 1.0 t.
4. Shell, 4-node, red, element at hot-spot: 3.63x4.75 (mm), longitudinal element length ≈ 0.3·t
5. Shell, 8-node, red, element at hot-spot: 5.0x4.75 (mm) longitudinal element length = 0.5·t

Model 3 corresponds to the guidelines given by ABS (2005) for oil tankers and IACS (2006) (element size approximately equal to the representative net thickness in the assessed area), while the models 1, 2, 4 and 5 are selected on the basis of recommendations given by Niemi (1993), Tveiten and Moan (2000), and IIW (2005) where it is observed that further mesh refinement does not significantly change the stress distribution close to the weld toe, Figs. 11 and 12. Note that model 5 deviates from the requirements given by IIW (2005), which states that the first element should not be larger than 0.4·t. However, as seen in Fig. 12, further refinement of the mesh from 0.5·t does not significantly change the stress distribution.

Some aspects influencing the hot-spot stress calculations

In tubular joints, for which the hot-spot stress extrapolation methods were originally developed, the crown or saddle points are normally the locations of the hot-spot under most loading conditions. As opposed to tubular joints the extrapolation path is not obvious for welded plate connections. In the case of the specimen geometry used in this study the maximum hot-spot stress is not at the centre line but closer to the longitudinal stiffener's edge, Fig. 13.

In the FE analyses that were discussed above the weld was idealized as a 'wedged cylinder' at the intersection of the longitudinal stiffener/web-frame connection. In a design process this is the most likely choice due to its simplicity with respect to modeling. For this model the variation in longitudinal nominal stress across the width of the flange was insignificant (less than 4% in the extrapolation region). However, as shown in Fig. 14 the actual welds were highly irregular across the stiffener and had a fillet also at the side of the flange. Hence, in order to calculate the effect of a more realistic weld shape, the solid element models were modified, adding a 'wedge shaped cylinder cell' at the side of the upper flange, Fig. 13. The introduction of a more realistic weld geometry lead to a significant increase in the stress close to the weld toe towards the edge of the stiffener. Fig. 15 shows stress profiles at the centre line and at the edge of the stiffener flange from the 20-node solid element model. The difference between the stress distribution along the centre line and the edge of the longitudinal stiffener range from 5% to 15% in the 'extrapolation' region. At the centre line the effect of this modification of the weld is insignificant.

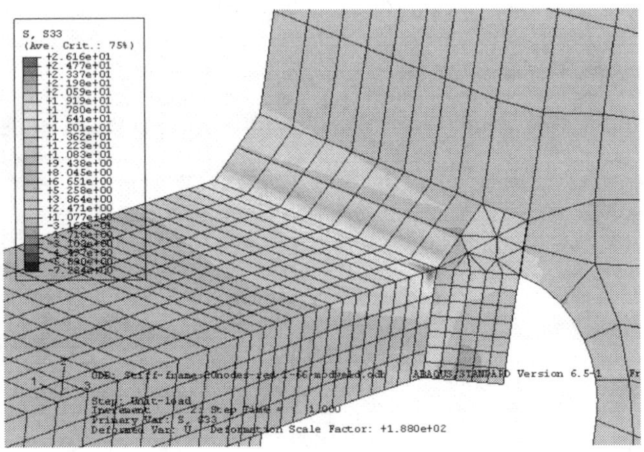

Fig. 13. Solid element model, including a weld shape modeling the as-built geometry shown in Fig. 14.

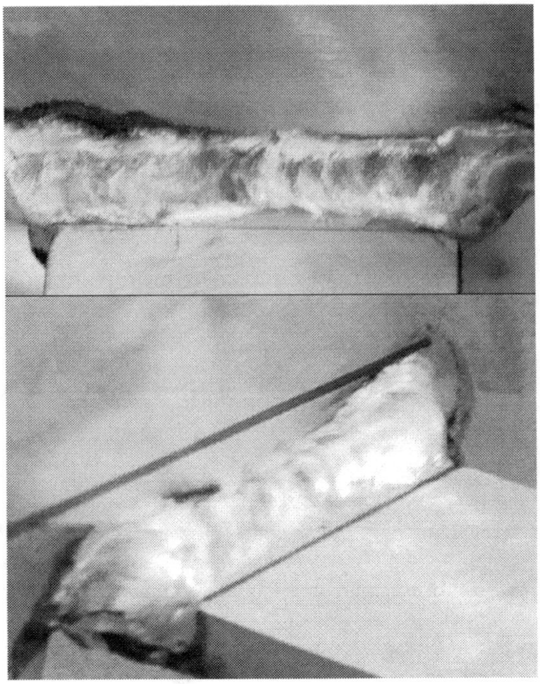

Fig. 14. Sample of weld shape at hot-spot.

For the shell element models (with no weld included), the variations were somewhat larger, with around 15% to 20% difference, Fig. 16. Obviously, this is not due to a modification of the weld shape as the weld is not included in the model. The stress increase seen in the shell element model is most likely related to the rather complex interaction of the web frame/stiffener and the cut-out in the transverse web plate.

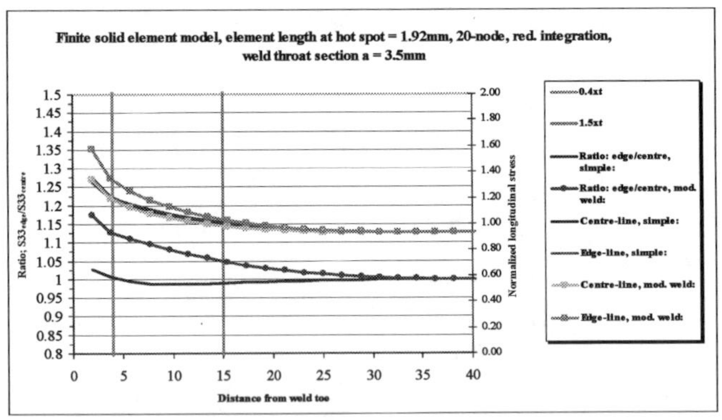

Fig. 15. Normalized longitudinal stress distribution along the centre line and at the edge of the upper flange close to weld toe location

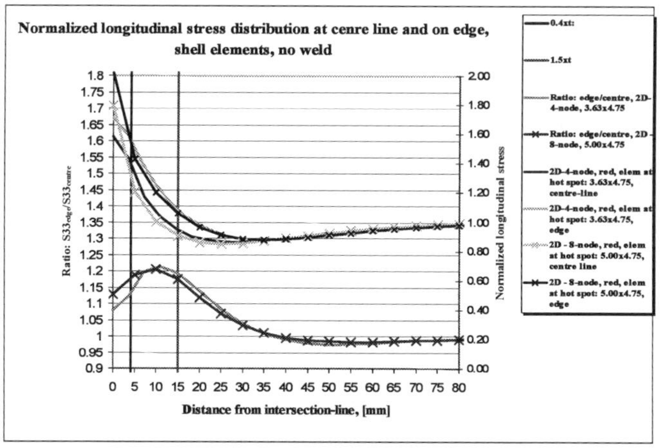

Fig. 16. Normalized longitudinal stress distribution along the centre line and at the edge of the upper flange close to weld toe location - shell element models.

The maximum stress is not located exactly at the edge, but at some distance from the edge corresponding to approximately one element size. Hence, the extrapolated maximum values of the hot-spot stress presented in Table 2 are based on the stresses derived at one element length from the edge of the longitudinal stiffener.

Hot-Spot Stress Calculations Based on FE Analysis

The hot-spot stress extrapolation schemes applied in this study for deriving the hot-spot stress were:

1. Linear extrapolation, using points at 0.4·t and 1.0·t, (IIW (2005)
2. Linear extrapolation, using points at 0.5·t and 1.5·t, IACS (2006) for Bulk Carriers
3. Hot-spot stress derived using one single point at 0.5·t, IACS (2006) for tankers

Note that for post-processing purposes when basing the extrapolation schemes on recommendations given by 1-3 above, the linear extrapolated hot-spot stress can be derived directly from the averaged nodal surface stresses at the 'reference points', using extrapolation equations given below as:

a) Linear extrapolation, using points at 0.4·t and 1.0·t: $\sigma_S = 1.67 \cdot \sigma_{0.4t} - 0.67 \cdot \sigma_{1.0t}$
b) Linear extrapolation, using points at 0.5·t and 1.5·t: $\sigma_S = 1.50 \cdot \sigma_{0.5t} - 0.50 \cdot \sigma_{1.5t}$

The subscript refers to the specific location of the 'reference points.'

Normal practice in hot-spot stress extrapolation procedures is to carry out the hot-spot stress extrapolation to the weld toe when using solid finite elements and to the element intersection line when using shell elements. However, ABS (2005) has proposed to carry out the extrapolation to the weld toe, even though the weld has not been included in the shell element model. In this study, a weld leg length of 4.95mm has been used (which corresponds to a weld throat thickness of 3.5mm). (The applicable surface stress used in the extrapolation scheme proposed by ABS (2005) is the stress derived at the centroid of the elements, Fig. 17.

Tables 1 and 2 present the calculated hot-spot stress concentration factors, K_s, based on the stresses derived from the finite element models. The calculated hot-spot stress concentration factors obtained from the finite element analyses are based on the normalized longitudinal maximum principal stress ($SP_{Max}/\sigma_{nominal,\ elastic\text{-}beam}$) results. The nominal stress, $\sigma_{nominal,\ elastic\text{-}beam}$, was calculated using elastic beam theory. For finite element models where the reference stress can not be derived directly at nodes, linear interpolation was used in order to derive the stress value.

Independent of the finite element model, element type and extrapolation method, a difference of approximately a factor 1.3 (1.75/1.31) to 1.4 (1.63/1.18) between the highest and lowest value is observed, Tables 1 and 2. Note that even though a thorough evaluation of the various finite element models was carried out initially, large differences in hot-spot stress results were experienced. For instance, in an initial design process this will have resulted in a difference in predicted fatigue lives by a factor of approximately 2.2 to 2.7.

The hot-spot stress concentration factors derived from the extrapolation points at 0.4·t and 1.0·t gives the largest values for all models since the extrapolation is performed in a region close to the weld, where the stress gradient is steepest. Except for the 8-node shell element model with element length at hot-spot of 5.0mm where a difference of 1.13 (1.54/1.36) at the centre location is seen, the difference is rather small, i.e. between 0.99 (1.75/1.76) to 1.07 (1.37/1.28). This is due to a rather linear trend of the stress distribution as seen in Figs. 11, 12, 20 and 21.

It should be noted that the "single point" definition of the hot-spot stress (at 0.5·t) shows the smallest variability. Increasing the element size to 1.0·t (recommended by IACS 2006) and using quadratic 8-node shell elements do not significantly change the derived hot-spot stress compared to the denser 4-node element model. It is worth noting that the hot-spot stress concentration factor derived from the 8 node solid element model gives the overall lowest calculated hot-spot stress values.

The values obtained by the ABS (2005) extrapolation method using shell elements are close to the values found by solid elements with extrapolation to the weld toe.

Table 3 presents the ratio between the extrapolated hot-spot stress derived at the edge of the stiffener (one element length from the edge) and the extrapolated hot-spot stress derived at the centre line. It is seen that 'edge-based' extrapolation of the hot-spot stress gives a range 1.11 - 1.18 for the solid element models, while for the shell element models the range is somewhat larger (1.05 - 1.30).

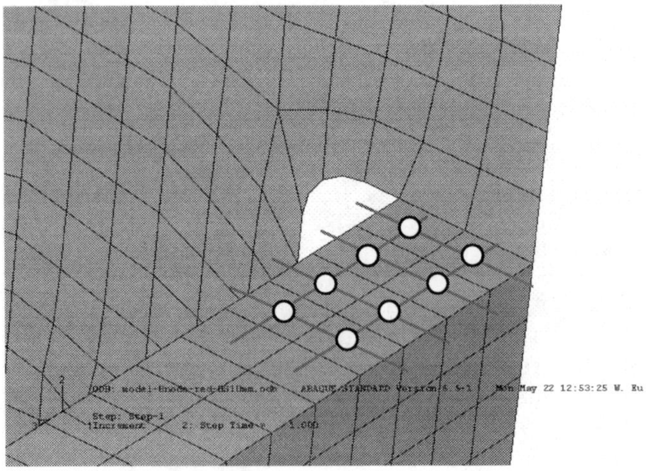

Fig. 17. Reference points at centroid of elements used for the assessment of hot-spot stress according to ABS (2005), element size in hot-spot region ~1.0 t.

Experimental Stress Analysis - Verification of FE Analysis

For measurement of strain close to the hot-spot, strip gauges were used. Along each line, four measurements of strain was obtained covering the region 0.5t - 1.5t. The strips were located to measure strain at the centre line and at the two edges of the flange, corresponding to the extrapolation paths used in the FE analysis. The instrumentation is shown in Figure 18.

In the solid element analyses, the weld was idealized with a weld flank angle, ϕ = 45degrees, a weld throat section, a = 3.5mm (leg lengths, λ = 4.95mm), and with a weld toe radius, ρ = 0mm. As shown in Fig. 14, the welds were highly irregular across the stiffener with much larger weld leg lengths at the edges of the stiffener. In the subsequent plots of the measured and calculated strains, the distance from the weld toe are taken as the actual distance from the weld along a line perpendicular to the weld.

Fig. 19 shows a comparison between the strain gage measurements and the finite element results at the centre line, based on the same finite element models used for the derivation of the hot-spot stress, Tables 1 and 2. As seen from Fig. 19, there is an excellent agreement between the nominal stresses, defined as the stress at a distance more than approximately 80mm from the weld toe. The difference between the finite element models is around 3%. It is also seen that the nominal strains measured at the centre line at each side of the transverse web plate (named L and R) are in excellent agreement. Thus, the measured strains show that the rig arrangement provided a symmetrical 3-point bending loading condition.

In Fig. 20 data from the region close to the weld is shown, at a larger scale. Data for two models are shown, from both sides of the transverse web plate, in effect for four models of the weld. The measured data are seen to fall in a relatively narrow scatter-band. Considering the relatively large variability of the weld profile (Fig. 14) this may be taken as a confirmation that at a distance of 0.4t or more from the weld toe, the stress distribution is essentially unaffected by the local notch of the weld.

The shell models are overpredicting the stress in this region, most likely due to the extrapolation to the shell intersection line. For the solid element models the agreement with measured stress is very good. For these models the definition of the weld toe location is the same as in the experiments.

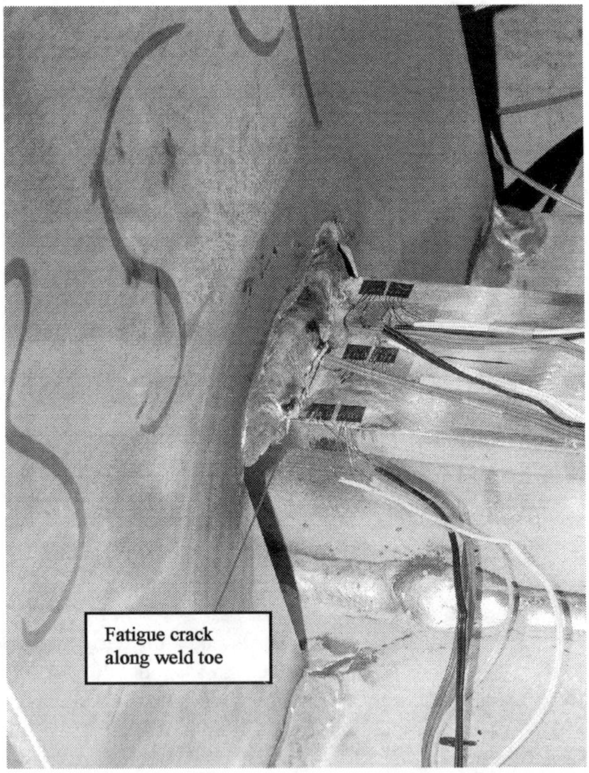

Fig. 18. Fatigue cracking was initiated along weld toe on top of stiffener, in most cases close to one of the edges of the stiffener. the fatigue crack would grow along the weld toe and through the flange of the stiffener (end of fatigue life).

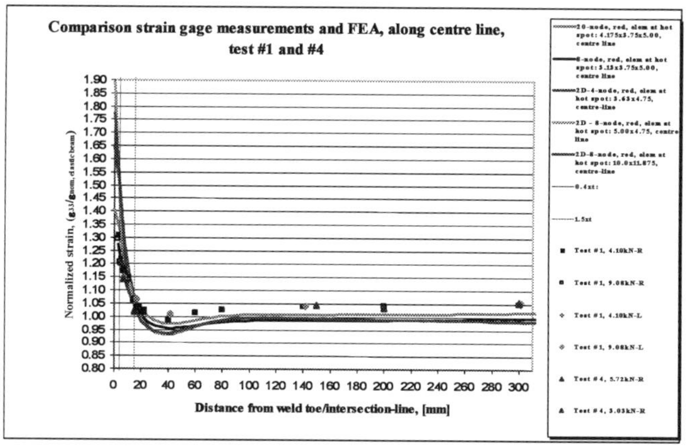

Fig. 19. Normalized longitudinal stress distribution from FE analysis and strain gauge measurements, at centre line of stiffener. Stress is normalized to nominal stress from beam theory.

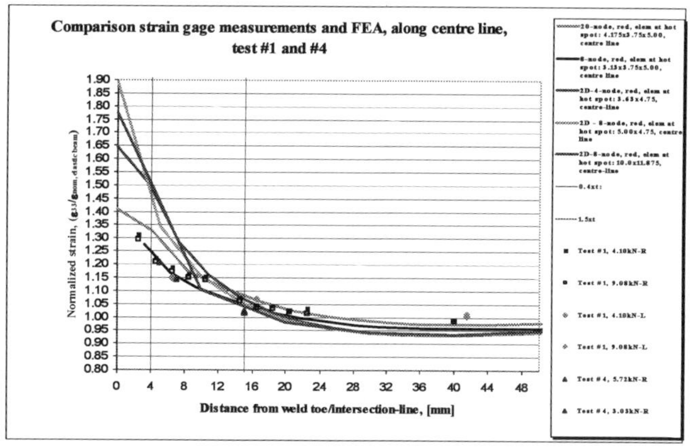

Fig. 20. Normalized longitudinal stress distribution from FE analysis and strain gauge measurements, at centre line of stiffener, close to weld toe. Stress is normalized to nominal stress from beam theory.

Effect of fabrication tolerances and weld distortions

The test models were instrumented to measure strain at the centre-line and at the edges of the stiffener, Fig. 18. The results for test #5 are shown in Fig. 21. The difference between edge and centre line for the measured strains is much larger than the difference that was calculated from the model shown in Figure 13. In the tests a bending gradient across the width of the flange is apparent. It is also seen that the stress at the edges is nonlinear, i.e. the SCF varies with load level. The stress at the centre-line is unaffected by the load level, and in good agreement with the FE results. Bending component occurred due to weld distortions that had lead to some warping of the models, and subsequently an uneven contact between the models and the fixtures that were used in the testing.

Shown in Fig. 22 is an FE model that represents the warping as a torsional displacement of the test models. Non-linear analysis was carried out to simulate the load-dependent contact between the model and the fixture. Two different finite element analyses were carried out with contact between the test model and the ramp of respectively 25% and 100% of the total width of the model, Fig. 23. The first case corresponds to low load levels with little contact between the test ramp and the end plate, while the second case corresponds to high load levels with full contact between the test ramp and the end plate.

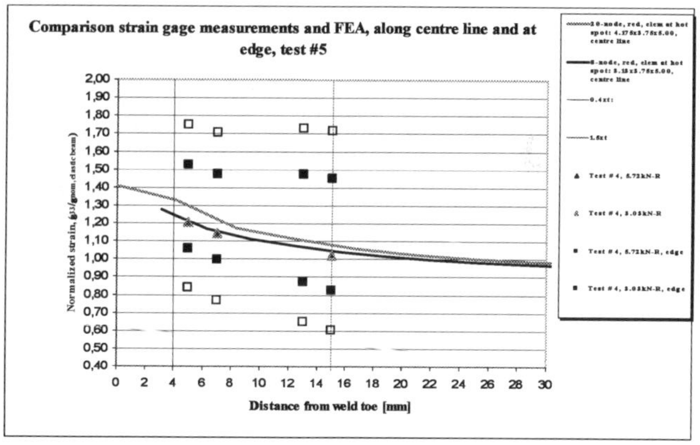

Fig. 21. Normalized longitudinal stress distribution obtained from FE analysis (centre line) and strain gage measurements (centre line and edges), at various load levels.

Fig. 24 shows the finite element results using a combination of the two models. The model using 25% contact was applied to loads up to 5kN, while the model using 100% contact was applied to loads larger than 5kN. The plot in Fig. 24 clearly shows the effect of the load dependency with respect to uneven contact between the test ramp and the end plate.

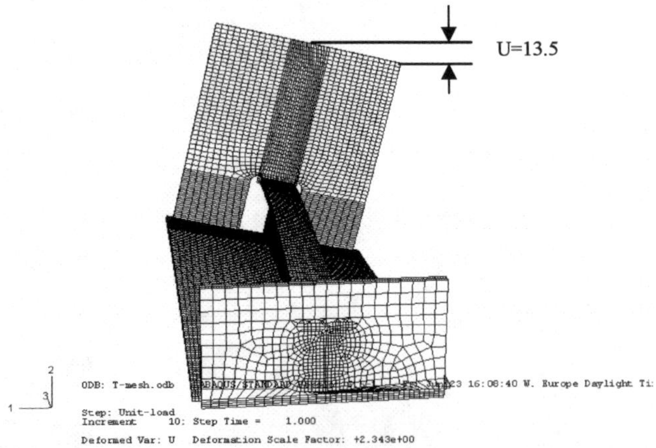

Fig. 22. Initial displaced mesh, angular displacement caused by a torsional moment (end plate fixed). Two levels of angular displacement, displacement of corner nodes: U = 13.5mm.

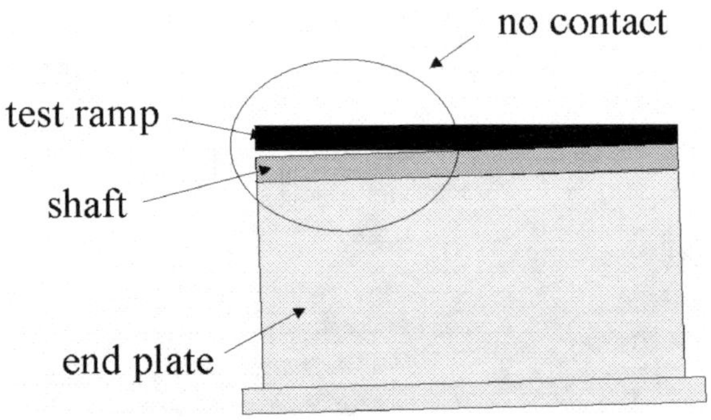

Fig. 23. Contact of test specimen and test ramp for warped specimen, at low loads.

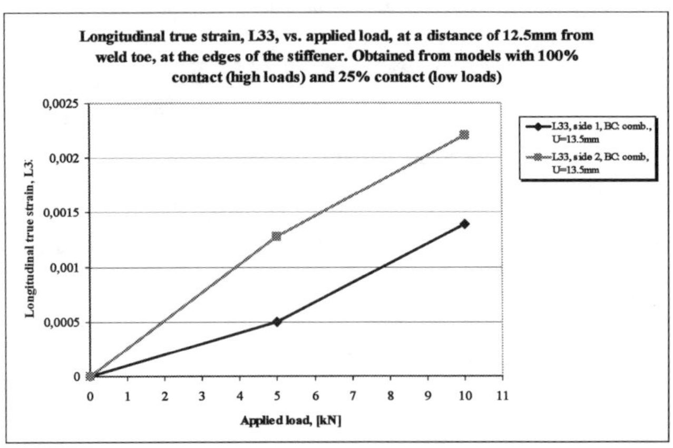

Fig. 24. Calculated strain at edges of flange, using nonlinear FE model to represent warping deformations (cf. Figs. 22 and 23).

It should be noted that along the centre line of the stiffener the measured and calculated stresses are linear, and apparently unaffected by the warping (Fig. 21). At the edges there is an additional stress concentration due to fabrication factors, in this case welding distortions. Although the test models would have different boundary conditions in a ship structure, similar effects may be expected, and need to be taken into account.

Welded structures are in general subject to fabrication tolerances that may lead to significant secondary stresses. FE analysis with idealized geometry may in many cases give nonconservative results, due to the neglect of effects of as-built geometry.

Discussion of the FE Analyses and Assessment of Hot Spot Stress

The results presented in Tables 1 and 2 demonstrate the variability in stress results close to the hot-spot locations for different finite element models and extrapolation paths. The results are consistent with previous studies on structural details (plate joints) presented by e.g. Sumi (1997), Tveiten (1999), Tveiten and Moan (2000) and Tveiten (2003), in that the stress distribution close to the weld toe (extrapolated stresses) are highly dependent on whether a shell element model or a solid element model have been used in the stress analysis, and in particular how the weld has been modeled.

Analysis using a solid element model provides stress results that are most consistent with strain gage measurements. However, the use of finite element tools in structural design is a trade-off with respect to the desired accuracy and time (or actually cost). Hence, in practical design the use of simplified shell element analyses with rather dense element meshes is preferred for analysis of complicated geometries. A reliable hot-spot stress analysis procedure is dependent on a well-defined link between the calculation of the hot-spot stress and the stress implicitly used as the design stress in the S-N curve. Assuming that the hot-spot stress is consistently predicted and reproducible regardless of detail configuration, the same calculation procedure should be used to calculate the hot-spot stress at an actual detail that was used when the hot-spot stress design S-N curve was derived from test data.

Assessment of an extrapolated hot-spot stress using FE models is subject to significant uncertainties. The large variability in hot-spot stress demonstrated here, using the recommendations given in literature, demonstrates the possible pitfalls involved in the process of deriving hot-spot stress values used in design. New methods for hot-spot stress extrapolation which have been claimed to be mesh-insensitive have been proposed, Dong (2004), Poutiainen and Marquis (2004). These methods may substantially improve the methodology for hot-spot stress calculations. However, there are still open questions related to these new methods (see e.g.: Doerk, Fricke and Weissenborn 2003, Wang, Sun and Cheng 2004, Poutiainen, Tanskanen and Marquis 2004), and they are indeed highly laborious and complicated when it comes to post-processing and result derivation.

S-N Curves

Introduction

Based on the work presented in the previous section, full scale fatigue tests were carried out. The objective of the tests was to verify appropriate design S-N curves to be used together with the hot-spot stress derived from the numerical stress analyses and strain gage measurements.

Fatigue test results

A total number of nine models were tested at a constant amplitude loading (load control) with an R-ratio equal to 0.1. Cycles to failure, N, was defined as a complete loss of the load carrying capacity of the longitudinal stiffener. The testing was performed in laboratory air at ambient temperature with a loading frequency of 2.5 Hz. The frequency of 2.5 Hz was used as this was the maximum frequency possible in order to obtain a stable and well controlled sinusoidal loading with defined maximum and minimum. It has been reported in literature that no frequency effects on crack growth can be detected in ambient air, Davidson, Griffiths and Machin (2002).

The crack growth of test specimen #1 was closely monitored to determine the residual fatigue strength of the detail after a visible fatigue crack had been observed. A visible "engineering sized" surface crack of about 2mm length was first observed close to the edge of the longitudinal stiffener, at the point of maximum stress range. Thereafter, the observed crack grew through the thickness of the flange within three thousand. After that, within one thousand cycles, a complete loss of the load carrying capacity of the longitudinal stiffener occurred. Fatigue life from a visible crack to failure was thus negligible.

The fatigue test results based on a nominal stress range are presented in Table 4. The stress range, $\Delta\sigma_{nominal,\ elastic\text{-}beam}$, is the nominal stress value at the weld toe location calculated by means of elastic beam theory

The S-N data were analyzed on the basis of a log-linear S-N curve

$$\log N = \log a - m \cdot \log S$$

The test results are shown in Fig. 25, with a mean life curve and a parallel curve at minus 2 standard deviations (notional 97.7% lower confidence band). The slope parameter of the S-N curve from regression analysis is *m = 3.42.*

Comparison With Proposed S-N Curve

In Fig. 26 the fatigue test data are plotted on the basis of the measured hot spot stress, using linear extrapolation 0.5t-1.5t. The stress profile for the most highly stressed edge of the stiffener was used, including the effect of secondary stress due to the warping. Also plotted are regression lines, and IIW design curves. The regression analysis was carried out with a slope parameter m = 3.0 in agreement with IIW S-N curves.

The FAT 40 S-N curve has been recommended as a design curve for hot spot stress, Partanen and Niemi (1999), IIW (2005). Studies by Tveiten and Moan (1997) on an aluminum stiffener/girder connection have shown that the use of a FAT 30 curve together with a linear procedure to derive the structural stress concentration provides a safe fatigue design.

The data shown in Fig. 25 indicate that the FAT 40 curve may be unconservative for design of this type of detail, and that FAT 32 or 36 may be more appropriate.

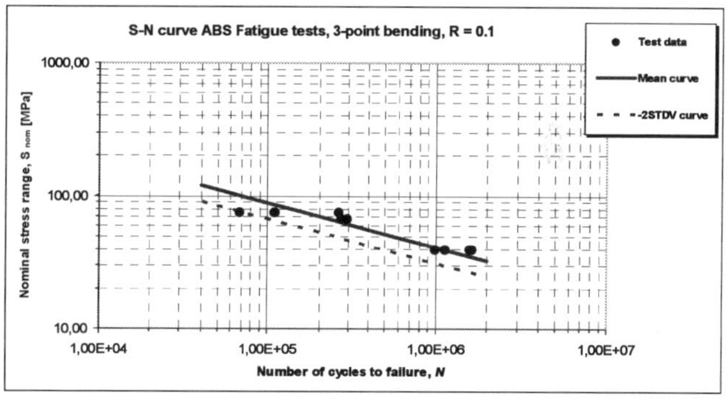

Fig. 25. S-N data based on nominal stress range calculated from elastic beam theory, with regression lines.

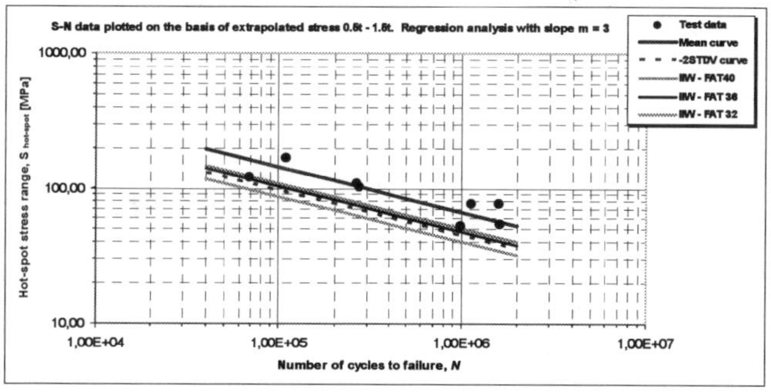

Fig. 26. S-N data plotted on the basis of extrapolated hot spot stress that was measured for each model (0.5t - 1.5·t). Regression lines and IIW design curves are shown.

In Fig. 27 the same data are plotted on the basis of stress measured 0.5t from the weld toe. As shown, the scatter is somewhat decreased compared to the plot based on extrapolated hot spot stress. In this case the FAT 32 curve appears to be most relevant for design.

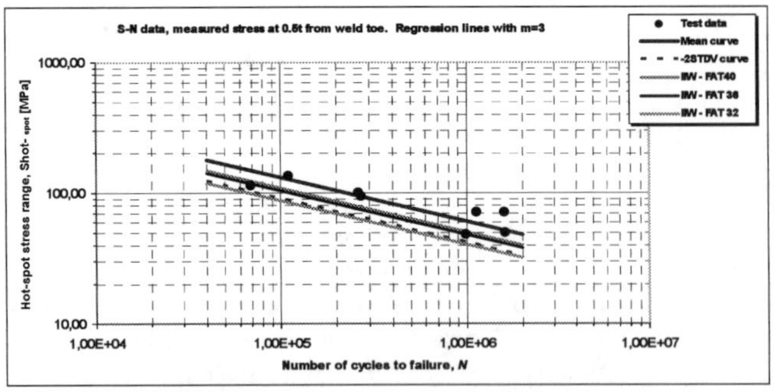

Fig. 27. S-N data plotted on the basis of hot spot stress 0.5·t from the weld toe, which was measured for each model. Regression lines and IIW design curves are shown.

CONCLUSIONS

Fatigue tests of full scale models of an aluminum ship detail were carried out to obtain an S-N curve. The test detail was also analyzed by the finite element method, using several modeling techniques and element types. The results from both experimental tests and finite element analysis were discussed. Recommendations on the procedure of fatigue assessment of aluminum ships including S-N curve to be used were presented.

The following recommendations based on this study are given:

- Preferably shell element models should be used. However, no weld idealization should be included and the extrapolation should be carried out to the element intersection line.
- Solid element models can also be used. In that case the weld should be modeled with a flank angle of 45 degrees, no weld toe radius, and with a weld throat thickness (and overall weld geometry) according to expected sizes seen in the structure. The extrapolation should be carried out to the weld toe.
- The element size in the hot-spot region should be sufficiently small to ensure convergence.
- If an extrapolated hot spot stress is used, the extrapolation path should be verified by an initial evaluation of possible locations of the hot spot stress, since the extrapolation path is not obvious for welded plate connections.
- The hot spot stress from FE analysis could be taken at a point 0.5·t from the weld toe, or by linear extrapolation 0.5t - 1.5·t. The former is simpler, and is somewhat less sensitive to details of the FE model.
- Due to fabrication tolerances significant secondary stresses may be present in a welded structure. In a design based on idealized FE models these stresses need to be taken into account.
- On the basis of measured hot spot stress using linear extrapolation 0.5·t - 1.5·t the fatigue test data indicate a design curve in the range FAT 32 to FAT 36.
- On the basis of measured hot spot stress using the stress at 0.5·t the fatigue test data indicate a design curve corresponding to FAT 32.
- The overall recommendation from this work is to use a hot spot stress defined at 0.5·t from the weld toe and a FAT 32 design curve for this type of welded aluminum details.

ACKNOWLEDGEMENT

This work was carried out as commissioned work for American Bureau of Shipping (ABS). Financial support from ABS is gratefully acknowledged.

Mr. Sheng Wei, National University of Singapore and Mr. Asmund Saelevik, Norwegian University of Science and Technology (NTNU) are acknowledged for their significant contributions to the experimental program.

REFERENCES

ABAQUS, version 6.5.1, http://www.abaqus.com/, 2004.

ABS, "Rules for Building and Classing Steel Vessels", Part 5 – Specific vessel Types, Chapter 1 –Vessels Intended to Carry Oil in Bulk (150m (492ft) or more in length), Appendix 1 – Guide for Fatigue Assessment of Tankers, Chapter 13, USA, 2005.

Berge, S., Haswell, J. and Engesvik, K. "Fracture mechanics analysis of tubular joint tests and degree of bending effects", 13th International Conference on Offshore Mechanics and Arctic Engineering (OMAE), Houston, Vol. III, pp. 189-194, 1994.

CEN, "Design of aluminium structures-Part 2: Structures susceptible to fatigue", Eurocode 9, ENV 1999-2, CEN, Brussels, Belgium, 1999.

Davidson, C.J., Griffiths, J.R., Machin A.S., "The effect of solution heat-treatment time on the fatigue properties of an Al-Si-Mg casting alloy", Fatigue Fract Engng Struct, **25**, pp. 223-230, 2002.

Doerk, O., Fricke, W. and Weissenborn, C., "Comparison of different calculation methods for hot-spot stresses at welded joints", International Journal of Fatigue, Vol. 25, 2003, pp. 359-369.

Dong, P., "The Mesh-insensitive hot-spot stress and Master S-N Curve Method for Ship Structures", In: Proceedings of OMAE Specialty Conference on Integrity of Floating

Production, Storage & Offloading (FPSO) Systems, Houston, USA, August 30 - September 2, 2004

Fricke, W., "Recommended Hot-spot Analysis Procedure for Structural Details of FPSOs and Ships Based Round-Robin FE Analyses", In: Proceedings of the Eleventh International Offshore and Polar Conference (ISOPE'2001), Stavanger, Norway, 2001.

IACS, "IACS Common Structural Rules", http://www.iacs.org.uk/csr/index.html, 2006.

IIW, "Recommendations for fatigue design of welded joints and components", Technical Report IIW-XIII-1965-03/XV-1127-03, IIW, June, 2005.

Lotsberg, I., "Derivation of hot-spot stress from finite element analysis for fatigue assessment of structures", Computational Mechanics – Theory and Practice, K.M. Mathisen, T. Kvamsdal and K.M. Okstad (Eds.) CIMNE, Barcelona, Spain, 2003

MacDonald, K., Haagensen, P. J. and Søvik, O. P., "Progress in fatigue assessment of welded hollow-section aluminium structures", Proc. of the Int. Conf. of Aluminium Structures - INALCO'98, United Kingdom, 1998.

Maddox, S. J., "Hot-spot fatigue data for welded steel and aluminium as a basis for design", Doc. IIW-XIII-1900-01, IIW, 2001.

Niemi, E., "Stress Determination for Fatigue Analysis of Welded Components", IIS/IIW-1221-93 (ex doc XIII-1458-92, XV-797-92), International Institute of Welding (IIW), 1993.

Partanen, T. and Niemi, E., "Hot-spot S-N curves based on fatigue tests of small MIG welded aluminium specimens", Welding in the World, Vol. 43, No. 1, 1999, pp. 16-22.

Poutiainen, I., Tanskanen, P., Marquis, G, "Finite element methods for structural hot-spot stress determination – a comparison of procedures", International Journal of Fatigue, Vol. 26, 2004, pp. 1147-1157.

Poutiainen, I., Marquis, G, "A single-point hot-spot stress assessment procedure for load-carrying fillet welds", IIW document - IIW-XIII-2012-04/XV1174-04, IIW - The International Institute of Welding, Cambridge, United Kingdom, 2004.

Sumi, Y., "Finite element comparative study of ship structural details", Proc. of the 13th Int. Conf. of Ship and Offshore Structures Congress, ISSC'97 - Volume 1, Edited by T. Moan and S. Berge, Trondheim, Norway, 1997.

Tveiten, B.W. and Moan, T. (1997). "Fatigue of Aluminium Stiffener-Girder under Constant Amplitude Loading". Proc. 4th Int. Conf. Fast Sea Transportation, FAST'97, July 21-23, 1997, Sydney, Australia.

Tveiten, B.W. (1999), "Fatigue Assessment of Welded Aluminium Ship Details", Dr. Ing. thesis, Norwegian University of Science and Technology (NTNU), Trondheim, Norway, 1999.

Tveiten, B.W. and Moan, T., "Determination of hot-spot stress for Fatigue Assessment of Welded Aluminium Ship Details", Journal of Marine Structures, Vol. 13, No. 3, 2000, pp. 189-212.

Tveiten, B. W., "The Fatigue Strength of RHS T-joints - NFR Kompetanseprosjekt – KMB 144004/213 – Fatigue and Durability", SINTEF Technical report, STF24 A03220, SINTEF, Trondheim, Norway, 2003.

Ye, N., Moan, T. and Tveiten, B. W., "Fatigue Analysis of Aluminium Box-Stiffener Lap Joints by Nominal", Structural and Notch Stress Range Approaches, Proc. of PRADS'2001, Shanghai, China, 2001.

Wang, X., Sun, H. and Cheng, Z., "Methods for Fatigue Assessment of Critical Ship Details", In: Proceeding of 23rd International Conference on Offshore Mechanics and Artic Engineering, OMAE 2004 – 51630, Vancouver, Canada, 2004.

Table 1. Calculated hot-spot stress concentration factors, K_s, based on normalized longitudinal stress, $SP_{Max}/\sigma_{nominal,\ elastic\text{-}beam}$ at the centre line of the upper flange obtained from finite element models

Extrapolation points	**Shell, 4-node, red, element at hot-spot: 3.33x4.75[2)]**	**Shell, 8-node, red, element at hot-spot: 10x10[2)]**	**Shell, 8-node, red, element at hot-spot: 5x4.75[2)]**	**Solid, 20-node, red, element at hot-spot: 4.175x3.75x5.00[3)]**	**Solid, 8-node, red, element at hot-spot: 3.13x3.75x5.00[3)]**
0.4·t	1.41	-[1)]	1.32	1.26	1.24
1.0·t	1.07	-	1.00	1.06	1.04
K_s	**1.63**	-	**1.54**	**1.40**	**1.37**
0.5·t	1.34	1.34	1.21	1.21	1.18
1.5·t	0.96	0.91	0.90	1.00	0.99
K_s	**1.53**	**1.55**	**1.36**	**1.31**	**1.28**
0.5·t	1.34	1.34	1.21	1.21	1.18
K_s	**1.34**	**1.34**	**1.21**	**1.21**	**1.18**
X_L = 0.5·t	-	1.19	-	-	-
X_R =1.5·t	-	0.89	-	-	-
$K_{s,\ ABS}$	-	**1.34**	-	-	-

[1)]Element sizes too large to derive nodal stresses accurately
[2)]Hot-spot stress has been derived carrying out the stress extrapolation to the element intersection line, except for the extrapolation scheme proposed by ABS (2004) where the extrapolation has been carried out to a 'fictitious' weld toe location
[3)]Hot-spot stress has been derived carrying out the stress extrapolation to the weld toe

Table 2. Calculated hot-spot stress concentration factors, K_s, based on normalized longitudinal stress, $SP_{Max}/\sigma_{nominal,\ elastic\text{-}beam}$ at the edge (at one element from the edge) of the upper flange obtained from finite element models

Extrapolation points	**Shell, 4-node, red, element at hot-spot: 3.33x4.75[2)]**	**Shell, 8-node, red, element at hot-spot: 10x10[2)]**	**Shell, 8-node, red, element at hot-spot: 5x4.75[2)]**	**Solid, 20-node, red, element at hot-spot: 4.175x3.75x5.00[3)]**	**Solid, 8-node, red, element at hot-spot: 3.13x3.75x5.00[3)]**
0.4·t	1.56	-[1)]	1.56	1.45	1.37
1.0·t	1.26	-	1.28	1.21	1.14
K_s	**1.75**	-	**1.75**	**1.60**	**1.52**
0.5·t	1.50	1.45	1.54	1.40	1.31
1.5·t	1.10	1.10	1.10	1.09	1.04
K_s	**1.71**	**1.63**	**1.76**	**1.55**	**1.45**
0.5·t	1.50	1.45	1.54	1.40	1.31
K_s	**1.50**	**1.45**	**1.54**	**1.40**	**1.31**
X_L = 0.5·t	-	1.32	-	-	-
X_R =1.5·t	-	0.92	-	-	-
$K_{s,\ ABS}$	-	**1.53**	-	-	-

[1)]Element sizes too large to derive nodal stresses accurately
[2)]Hot-spot stress has been derived carrying out the stress extrapolation to the element intersection line, except for the extrapolation scheme proposed by ABS (2004) where the extrapolation has been carried out to a 'fictitious' weld toe location
[3)]Hot-spot stress has been derived carrying out the stress extrapolation to the weld toe

Table 3. Ratio - extrapolated hot-spot SCF, K_s, derived at the edge of the stiffener and at the centre line of the stiffener

Extrapolation scheme	Shell, 4-node, red, element at hot-spot: 3.33x4.75[2)]	Shell, 8-node, red, element at hot-spot: 10x10[2)]	Shell, 8-node, red, element at hot-spot: 5x4.75[2)]	Solid, 20-node, red, element at hot-spot: 4.175x3.75x5.00[3)]	Solid, 8-node, red, element at hot-spot: 3.13x3.75x5.00[3)]
Linear extrapolation, 0.4·t and 1.0·t	1.05	-[1)]	1.14	1.15	1.12
Linear extrapolation, 0.5·t and 1.5·t	1.12	1.05	1.30	1.18	1.13
Hot-spot stress, one single point at 0.5·t	1.13	1.09	1.28	1.16	1.11
Linear extrapolation, ABS (2004)	-	1.14	-	-	-

[1)]Element sizes too large to derive stresses accurately

[2)]Hot-spot stress has been derived carrying out the stress extrapolation to the element intersection line, except for the extrapolation scheme proposed by ABS (2004) where the extrapolation has been carried out to a 'fictitious' weld toe location

[3)]Hot-spot stress has been derived carrying out the stress extrapolation to the weld toe

Table 4. Test data, nominal stress range calculated from beam theory.
SCF, K_s, from strain gauge measurements, extrapolated (0.5·t-1.5·t) and at 0.5·t from weld toe, at edge of stiffener flange.

Test spec. no.	$\Delta\sigma_{nominal, elastic\text{-}beam}$ [MPa]	N	K_s 0.5·t-1.5·t	K_s 0.5·t
1	74.65	264,012	1.47	1.34
2	74.65	109,917	2.25	1.80
3	74.65	68,609	1.61	1.51
4	39.19	1,625,397	1.39	1.26
5	39.19	1,132,858	1.95	1.80
6	39.19	Overloaded	-	-
7	39.19	990,897	1.33	1.22
8	39.19	1,596,500	1.95	1.81
9	67.19	271,863	1.52	1.40
10	67.19	297,170	1.33	1.24

Discussion

Robert A. Sielski, Member

The authors are to be congratulated on adding to and perhaps refining the confusing literature concerning the use of the hot-spot stress approach for determining the fatigue initiation strength of aluminum structural details. Unfortunately, they have confined their extensive study to only one structural detail, and as they point out, specifically in the case of tubular joints, such analyses are geometry-dependent. Will the authors please comment on how applicable they feel their conclusions and recommendations are to structural details other than the one that they studied, and what limitations on geometry of details should be considered when applying their method?

Numerous finite element analyses have been made of steel structural details to determine the hot-spot stress. As the approach is a linear elastic approach, what differences between the elastic behavior of steel and aluminum would prohibit the extension of such analyses to aluminum details, or do the authors believe such an extension can be made?

In their test program, the stress ratio of $R = 0.1$ was used, meaning that the specimens were always under tension. However, ship structures are generally loaded under near stress reversal, with the alternating tensile and compressive stresses helping to reduce the effect of initial residual stresses at the detail. Will the authors please indicate why stress reversal was not used in the testing?

In light of the uncertainty remaining in hot-spot stress analysis, I believe that the primary conclusion of this study should be that extensive large-scale testing of a wide variety of aluminum structural details is needed. Such a study could form both a basis for design and a possible means for confirming or calibrating analysis methods such as the hot-spot stress approach for use with structural details not tested.

Authors' Closure

We agree that our study is limited in scope, the primary reason being — alas — the state of funding for this type of research.

The geometry of the detail selected for this study is assumed to be typical for aluminum ship structures and could be referred to as a "base-line" detail. It is correct and also pointed out in the paper, that the analyses to some extent are geometry dependent and might even be load-dependent. Hence, as pointed out later there is a need for further studies on more details to verify the method. The geometries used in such studies should be selected in close cooperation with ship yards and naval architects so to represent structural details that are normally used in ship design.

However, from studies carried out in Norway and Finland on a number of aluminum details (butt welded plate connections, cruciform joints, RHS connections, box stiffener/lap-joints, flat bar/stiffener connections, stiffener/girder connections), it has been proposed that the surface-stress based methods used in this study apply to a large range of typical ship-like details. The authors believe that the robustness of the method is dependent on a robust finite element scheme and a calibrated design S-N curve. The use of finite elements in a hot-spot stress evaluation should be based on a simple and well-defined method with respect to element formulation, stress extrapolation method, and modeling principles. It should be borne in mind that the method should be within reach of the practical design engineer. The large variability in the assessment of hot-spot stress, which is demonstrated in this study using the recommendations given in literature, clearly demonstrates the possible pitfalls involved in the process of deriving hot-spot stress values to be used in design.

Dr. Sielsky makes a reference to tubular joints. The extrapolated "hot spot stress" concept was indeed conceived for fatigue design of tubular joints, Radenkovic (1981). Using this fictitious stress, fatigue test data (S-N-data) for tubular joints turned out to fall within the same scatter-band, regardless of joint geometry and loading mode. For this reason, tubular joints are fatigue designed on the basis of one S-N-curve only, the T-curve. The hot spot stress for tubular joints is thus assumed to be a geometry-independent stress for assessment of fatigue strength. It remains to see if a hot spot stress for ship structural details could be applied in the same way.

The hot-spot stress for fatigue assessment is an elastic stress. Under load-controlled conditions the stress distribution for a given geometry — and indeed the hot spot stress — will be the same, irrespective of the elastic modulus.

Material properties (e.g., steel versus aluminum) are important for the empirical assessment of an S-N-curve, owing to different crack growth characteristics, fatigue notch factor of a weld, etc.

The purpose of the test program was to provide data for discussion of the hot-spot stress concept in fatigue design of ship structural details. Neither the mean stress effect nor the merits of the "stress range" approach were targeted; hence, all tests were run at the same R-ratio. The stress ratio of R = 0.1 was selected mainly for consistency with the small scale test data that are underlying the S-N-curves found in standards and used as reference curves in the present work.

We also believe that R = 0.1 is a realistic load condition for the type of detail that was tested, that is, pressure load transferred from longitudinal stiffeners into a transverse girder web.

The authors would like to propose that a continuation of this work is initiated including a study of a large range of typical ship structural details subjected to fatigue testing and finite element verification work.

In addition, it has been shown in this study that fabrication tolerances play a significant role in determining the hot-spot stress levels. Although the test models would have different boundary conditions in a ship structure, similar effects may be expected and need to be taken into account. Welded structures are in general subject to fabrication tolerances that may lead to significant secondary stresses. FE analysis with idealized geometry may in many cases give non-conservative results, owing to the neglect of effects of as-built geometry. Hence, the effects of fabrication effects should also be subject for further investigations.

Reference

Radenkovic, Dragos 1981 Stress analysis in tubular joints, Plenary Session, International Conference on Steel in Marine Structures, Institut de Recherches de la Sidérurgie Française, Paris.

Hydrodynamic Investigation of the Ballast-Free Ship Concept

Miltiadis Kotinis, (AM)
Michael G. Parsons, (FL)

The hydrodynamic aspect of the Ballast-Free Ship concept is further investigated in order to analyze the water discharge effect on the resistance and propulsion of a vessel. For this purpose, a Seaway-sized bulk carrier model was tested in the towing tank of the University of Michigan Marine Hydrodynamics Laboratory. Additional results were obtained with the aid of Computational Fluid Dynamics. The cost effectiveness of the concept is demonstrated through a detailed economic analysis.

NOMENCLATURE

ACC - annual cargo capacity
B - bias error
C_P - pressure drag coefficient
C_F - friction drag coefficient
CRF - Capital Recovery Factor
C_T - total (friction plus pressure) drag coefficient
D - experimental form factor
E - comparison error
k - form factor
k-ε - turbulence closure model
k-ω - turbulence closure model
P - precision error
Q_m - model-scale ballast trunk flow rate
Q_s - full-scale ballast trunk flow rate
r_G - grid refinement ratio
U_D - experimental data uncertainty
U_G - grid uncertainty
U_I - iterative uncertainty
U_{SN} - simulation numerical uncertainty
U_T - total experimental uncertainty
U_V - validation uncertainty
y^+ - non-dimensional distance to centroid of cell nearest wall
ΔA - change in annual cost
ΔP - change in capital cost
ΔRFR - change in Required Freight Rate
λ - geometric scale ratio

INTRODUCTION

The detrimental environmental and socio-economic effects of the introduction of nonindigenous aquatic species into coastal waters have motivated the international maritime regulatory bodies to adopt strict regulations in order to control and eventually eliminate this phenomenon. The International Maritime Organization (IMO) has recently adopted new regulations (IMO 2004) applicable to both new and existing vessels that conduct ballast operations. These regulations mandate the utilization of a ballast water treatment system and prescribe strict limits regarding the concentration of viable organisms in the discharged ballast water. The new IMO ballast water convention extends the biota control to sizes between 50 and 10 microns and adds specific requirements for selected smaller pathogens (i.e., toxicogenic *Vibrio cholerae*, *Escherichia coli*, and intestinal *enterococci*). Even though the introduction of the requirements will not be in full effect until 2014 for existing vessels; starting in 2009, ships without an approved ballast treatment system will have to conduct ballast water exchange on the high seas, a method that has been criticized regarding its effectiveness (Forsberg et al. 2005, Kent and Parsons 2004).

A combination of a primary treatment system (e.g., filtration with backflush) to remove the larger organisms and a secondary treatment system (UV irradiation, chemical treatment) has proven to be effective in many cases. The installation of a ballast water treatment system, though, significantly increases the capital and operating costs for trans-oceanic vessels. An alternative to ballast treatment systems is the Ballast-Free Ship concept (U.S. Patent #6,694,908 2004), a new paradigm that approaches ballast operation as the reduction of buoyancy rather than the addition of weight to get the vessel to its required ballast drafts. The feasibility and effectiveness of the Ballast-Free Ship concept has been demonstrated (Kotinis et al. 2004). This is a sequel to that earlier SNAME paper.

In the Ballast-Free Ship concept, the traditional ballast tanks are replaced by longitudinal trunks that extend beneath the cargo region of the ship below the ballast draft. The trunks are connected to the sea through symmetrical with respect to the ship centerplane plena; one at the bow and one at the stern. The pressure developed on a moving ship is positive close to the bow stagnation point and negative (suction) near the stern. Therefore, the hydrodynamic pressure differential between the bow and the stern region of a ship is theoretically able to induce a slow water flow inside the ballast trunks without the need for pumps. While the ship is in the ballast condition, these trunks are constantly flooded with "local seawater," thus, reducing the buoyancy of the vessel. The net result is the elimination of the transportation of foreign ballast water and the nonindigenous aquatic species it may contain.

In this paper, the hydrodynamic aspect of the concept is further investigated in an attempt to better understand the effect of the water discharge on the flow around the ship hull and the propeller operation. The current research focuses on ships deballasting in the Great Lakes; however, this does not limit the validity or applicability of the results. A scaled model of a typical Seaway-sized bulk carrier was constructed and utilized during resistance and propulsion tests performed at the University of Michigan Marine Hydrodynamics Laboratory (MHL). The experimental results were supplemented by theoretical (numerical) results obtained through Computational Fluid Dynamics (CFD) simulations performed using the commercial CFD software FLUENT® (Fluent 2006). The numerical investigation was performed before the experimental investigation. First, the flow around the hull without any water suction or discharge was analyzed (initial case). Subsequently, an analysis was attempted for the effect of the water suction on the flow at the bow and the effect of the water discharge on the flow at the stern (modified case). Finally, the numerical results were compared with the experimental results, and a validation procedure was performed for the integral variables of the initial case.

BALLAST-FREE BULK CARRIER MODEL

The initial investigation of the Ballast-Free Ship concept demonstrated the general feasibility of the concept (Kotinis et al. 2004, Kotinis 2005), with an emphasis on the invasive aquatic species problem as encountered in the Great Lakes. The majority of the transoceanic vessels entering the Great Lakes through the St. Lawrence Seaway are Handymax bulk carriers (Farley 1996). The current research attempts to analyze the impact of the ballast trunk discharge on the flow around the hull of a typical Seaway-sized bulk carrier in addition to its effect on the propeller operation.

For this purpose, a typical Seaway-sized Handymax bulk carrier was designed and utilized in towing tank experiments and a numerical hydrodynamic investigation. The bulk carrier was designed using data from similar vessels. These vessels were the *Isa* owned by the Polish Steamship Company and a bulk carrier designed by the Jiangnan shipyard in China. The vessel was designed using the commercial ship design software Maxsurf® (Formation Design Systems 2006). The main particulars of the vessel are presented in Table 1.

Table 1. Main Particulars of the Ballast-Free Bulk Carrier

Waterline length (m)	195.5
Maximum beam (m)	23.76
Depth to main deck	16.00
Full-load draft (m)	10.70
Block coefficient	0.835

The service speed is assumed to be 14.5 knots and the speed of the vessel while at ballast drafts 15.5 knots. During ballast operation, the ship drafts at the forward and aft perpendiculars are 40% and 70% of the design draft, respectively. This condition corresponds to heavy ballast used under extreme weather conditions.

A significant parameter in determining the geometric scale factor of the Ballast-Free bulk carrier model was the size of the available stock propellers. The No. 23 stock propeller at the MHL was the available propeller providing the highest propulsive efficiency and, at the same time, satisfying the hull clearance requirements, assuming a full-scale propeller diameter of 6.0 m. The corresponding geometric scale ratio $\lambda = 37.92$. Constraints relative to the avoidance of the blockage and shallow water effects were also taken into account during this selection process. The propeller characteristics for the No. 23 model propeller are shown in Table 2. The non-dimensional thrust and torque coefficients plotted versus the coefficient of advance (Kt, Kq – J) of the No. 23 model propeller are shown in Fig.1. The main particulars of the Ballast-Free bulk carrier model in the ballast condition are listed in Table 3. The bow and the stern of the constructed model are shown in Figs. 2 and 3, respectively.

Table 2. Characteristics of the MHL No. 23 Propeller

Number of blades	4
Diameter (m)	0.158
Hub diameter (m)	0.031
Pitch-diameter ratio	1.08
Expanded area ratio	0.55

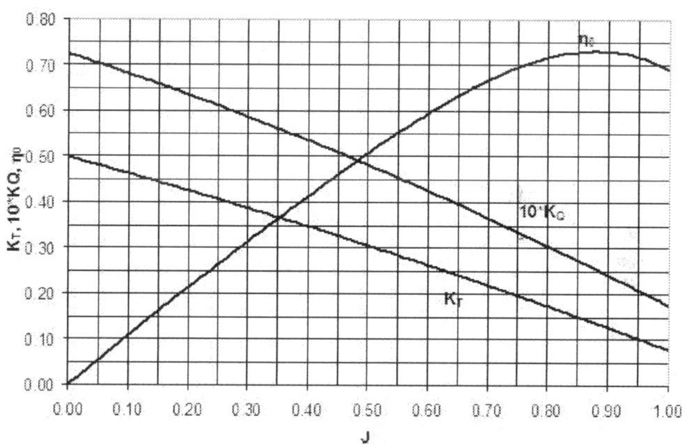

Figure 1. Propeller Coefficients versus Advance Coefficient

Table 3. Characteristics of the Ballast-Free Bulk Carrier Model in Ballast Condition

Waterline length (m)	5.00
Maximum beam (m)	0.627
F.P. draft @ 40% DWL (m)	0.113
A.P. draft @ 70% DWL (m)	0.198
Wetted surface area (m^2)	5.34
Geometric scale ratio λ	37.92

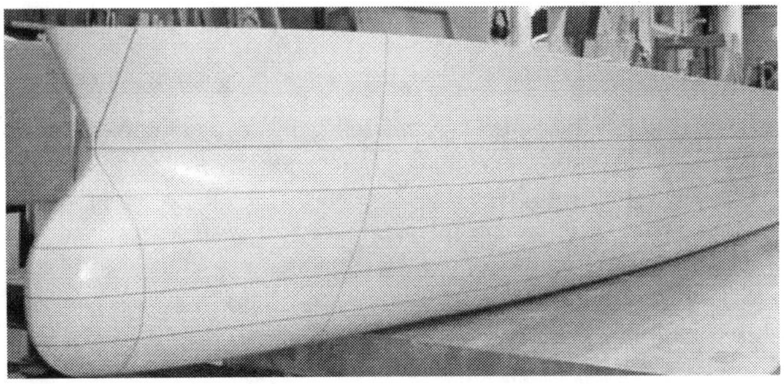

Figure 2. Bow View of the Seaway-sized Bulk Carrier Model

Figure 3. Stern View of the Seaway-sized Bulk Carrier Model

NUMERICAL HYDRODYNAMIC INVESTIGATION

The commercial CFD software FLUENT® (Fluent 2006) was utilized to study the external flow around the bulk carrier model. An IGES file of the model-scale design was imported into Gridgen® (Gridgen 2007) and the flow domain was meshed using a multi-block hybrid grid, which is shown in Fig. 4. A 'double-body' flow model, which does not take the free-surface flow into account, was adopted considering the free surface as a plane of symmetry. In addition to this, only half of the hull surface was considered assuming the flow to be symmetric with respect to the centerplane. The shape of the flow domain was a quarter-cylinder; the outer cylindrical boundary, which was treated as a slip wall, has a radius equal to half a ship length. At the velocity inlet, which was placed at a distance equal to half a ship length upstream of the forward stagnation point, a uniform velocity boundary condition was imposed. The flow domain was truncated downstream of the hull stern at a distance approximately equal to one ship length. At this location, the flow had essentially regained its parallel character; thus, it was being unaffected by the presence of the ship hull. At the outlet of the flow domain, a Neumann boundary condition was imposed and the flow properties were obtained by extrapolating the values from the interior of the domain.

In order to obtain a converged solution for the initial case, three different grids were created using a systematic refinement method and a verification procedure was applied to the results. The friction and pressure drag coefficients on the hull and the nominal wake fraction at the propeller plane were monitored to ensure the convergence of the integral variables. The finest grid was utilized for the hydrodynamic investigation and the selection of the inlet/outlet positions. The post-processing of the CFD results was aided by the utilization of Tecplot 360® (Tecplot 2006).

In the modified case, the water inlet and discharge locations were modeled and their effect on the hull resistance was investigated. An attempt was also made to analyze the effect on propulsion in a qualitative manner.

Description of the Numerical Solver

The numerical solver of FLUENT is based on a finite volume method with the flow properties calculated at the cell centers. The mesh, in 3-D problems, can be either structured or unstructured, single-block or multi-block, with cells of various topologies (hexahedral, tetrahedral, pyramidal, and prismatic). For incompressible flows, a pressure-based segregated solver is typically utilized. The fluid velocity is obtained by the solution of the Reynolds-Averaged Navier-Stokes (RANS) equations. The pressure-velocity coupling is achieved through an iterative procedure where a pressure correction equation is utilized to ensure conservation of mass and eventually compute the pressure field. For steady-state flows, either the SIMPLE or SIMPLEC algorithms are usually employed. The diffusion terms in the RANS equations are discretized with a central differencing scheme. The convection terms are discretized using a higher-order upwind scheme to minimize numerical diffusion; for unstructured meshes with tetrahedral elements a second-order upwind scheme is typically utilized. The discretized equations are solved using the Gauss-Seidel iterative algorithm. The solution convergence is accelerated through the utilization of an algebraic multi-grid method. Further details of the numerical methods can be found in (Mathur et al. 1997) and (Kim et al. 1998).

Turbulence and Near-Wall Modeling

The turbulence model utilized in the present computations was the shear-stress transport (SST) model (Menter 1994). This model implements a blending function in order to apply the standard k - ω model close to solid boundaries (ship hull) and a transformed version of the k - ε model in the far field. The SST model has been shown to provide quite accurate results for ship flows (Kim and Rhee 2002, Duvigneau et al. 2002).

A mesh capable of fully resolving the viscosity-affected near-wall region was constructed. This was deemed necessary, especially in the modified case where the interaction of the water discharge with the boundary layer flow needs to be adequately modeled. In addition to this, a low-Reynolds-number version of the k - ω model, which applies a damping coefficient to the turbulent viscosity, was employed.

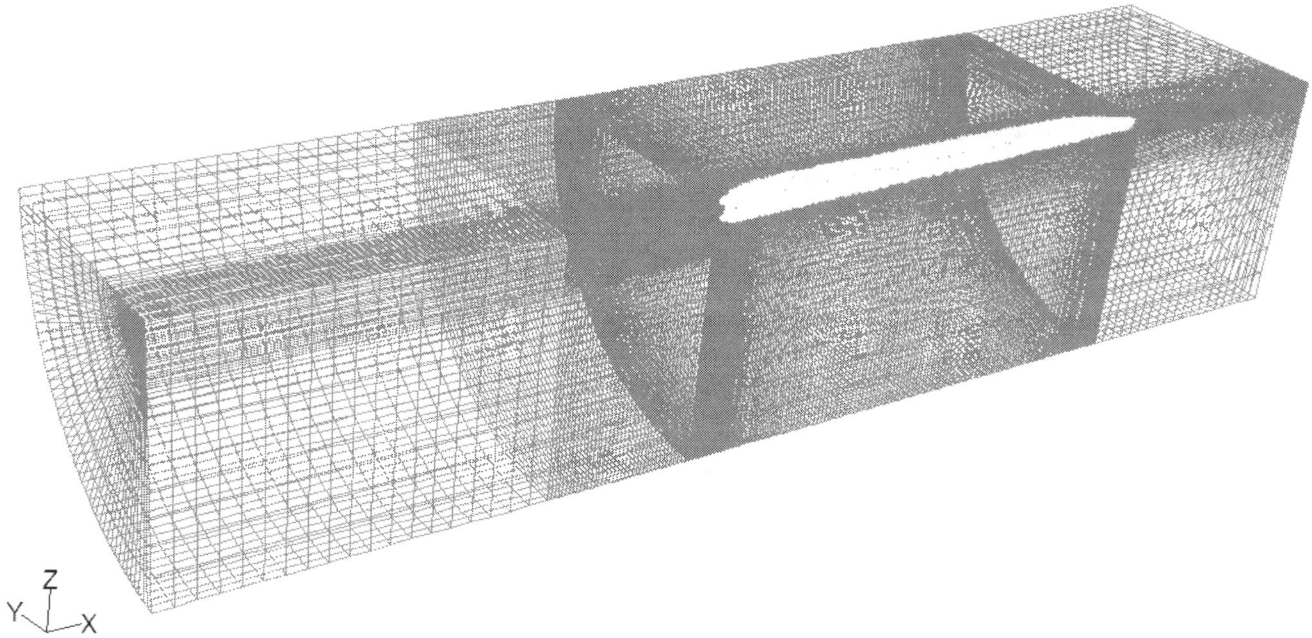

Figure 4. Computational Domain for the Ballast-Free Bulk Carrier

Numerical Results

Computations were performed for the Ballast-Free bulk carrier model in the ballast condition, where the model-scale speed is 1.295 m/s and the corresponding Reynolds number (in fresh water at 15°C) is 6.10e+6. The corresponding friction coefficient based on the ITTC model-ship correlation line is 3.28e-3. The stopping criterion for the simulations was the convergence of the total resistance coefficient and the reduction of the solution residuals by four orders of magnitude. Prior to analyzing the numerical results, a grid convergence study was performed. The dimensions of the three grids utilized for this grid study are listed in Table 4 along with the y+ values. The y+ value is the non-dimensional distance between the centroid of the cell adjacent to the solid boundary and the solid boundary. A systematic refinement was achieved by adjusting the grid spacing on the hull surface and in the flow domain around the hull so that the resulting grid refinement ratio between consecutive grids was $r_G = \sqrt{2}$.

Verification Procedure. A verification study was performed using the methodology developed in (Stern et al. 2001) in order to assess the simulation numerical uncertainty. The computed values of the pressure, friction, and total drag coefficients along with the corresponding form factor are listed in Table 5.

Table 4. Grid Dimensions and y+ Values

Grid	Total Number of Cells	y+
Coarse	510,613	4.6
Medium	715,410	3.3
Fine	1,019,973	2.4

Table 5. Grid Convergence Study

Grid	Coarse	Medium	Fine
C_P	1.167e-3	1.155e-3	1.154e-3
Change (%)		-1.0	-0.1
C_F	2.932e-3	2.975e-3	2.999e-3
Change (%)		1.5	0.8
C_T	4.099e-3	4.130e-3	4.153e-3
Change (%)		0.8	0.6
Form Factor k	0.250	0.259	0.266
Change (%)		3.8	2.6

The total drag coefficient listed above is not the overall drag coefficient as it does not include the contribution of the wave drag nor its interaction with the viscous drag. On the other hand, the 'double body' flow model is ideal for the estimation of the form factor. This is why the latter was selected as the integral variable to utilize in the verification and validation procedure.

The results in Table 5 show that the friction drag coefficient C_F is monotonically convergent and the pressure drag coefficient C_P is grid-independent. The physics behind the friction and pressure drag coefficients are different; a fact that partially justifies the different convergence behavior. The total drag coefficient C_T was monotonically convergent with a change of only 0.6% between the medium and fine grid. The convergence ratio is defined as the solution change between the medium and fine grid over the solution change between the coarse and medium grid. The convergence ratio for the total drag coefficient is 0.74, a value that demonstrates monotonic convergence. The form factor k is computed by dividing the computed total 'double-body' drag coefficient by the ITTC friction coefficient value. The value of the form factor and its corresponding uncertainty was also determined during the towing tank experiments.

The simulation numerical uncertainty, U_{SN}, for steady-state simulations consists of iterative uncertainty, U_I, and grid uncertainty, U_G. Iterative uncertainty, which is defined as half the range of maximum and minimum values of the integral variable over the last two periods of oscillation (before the stopping criterion is reached), was monitored and found to be negligible compared to grid uncertainty. The estimated order of accuracy based on the solution change between grids and the refinement ratio value is 0.86. This value is relatively far from the theoretical value of 2.0, which is the formal order of accuracy of the utilized spatial discretization formulation. However, this result is not uncommon for non-orthogonal, non-uniform grids (Wilson et al. 2001). The grid uncertainty of the uncorrected solution is estimated using a generalized Richardson Extrapolation. The estimated value is equal to about 7.6% of the computed value using the fine grid.

Analysis of Flow around the Hull of the Ballast-Free Bulk Carrier Model and Investigation of Nominal Wake. The pressure coefficient contours and the velocity vectors at the bow of the Ballast-Free bulk carrier model in the ballast condition are shown in Figs. 5 and 6, respectively. Large pressure gradients exist in a small region around the stagnation point, with a significant pressure relief occurring immediately after that region. The positive pressure area at the bow extends up to approximately 7% of the ship length aft of the F.P.; thus, the available locations for the inlet of the bow plenum are limited. In order to ensure an adequate pressure differential to drive the ballast trunk flow, it was decided to place the water inlet at the center of the bulbous bow and close to the keel to take full advantage of the stagnation pressure in that area and to avoid potential problems due to bow emergence. The velocity vectors depicted in Fig. 6 show the downward fluid flow induced by the bulbous bow, even though the ballast condition is not the operating condition for which the bulbous bow has been optimized. This downward flow is expected to reduce the height of the bow wave; a low bow wave height was observed, indeed, during the towing tank model tests.

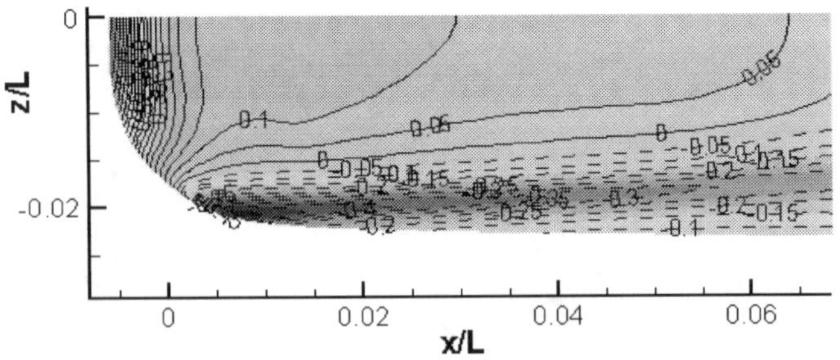

Figure 5. Pressure Coefficient Contours at the Bow of the Ballast-Free Bulk Carrier Model – Initial Case

The corresponding pressure coefficient contours at the stern are presented in Fig.7. Suction pressure exists over the parallel section and most of the ship stern, with a peak contour value close to the keel at Station 18 (x/L = 0.90). Between Stations 17 and 18 (0.85≤x/L≤0.90) and close to the free surface, a lower suction pressure region exists, which combined with the suction pressure peak produces a considerable girthwise pressure gradient. The latter results in the formation of a streamwise vortex in that area; a typical phenomenon for ships with full hull shapes.

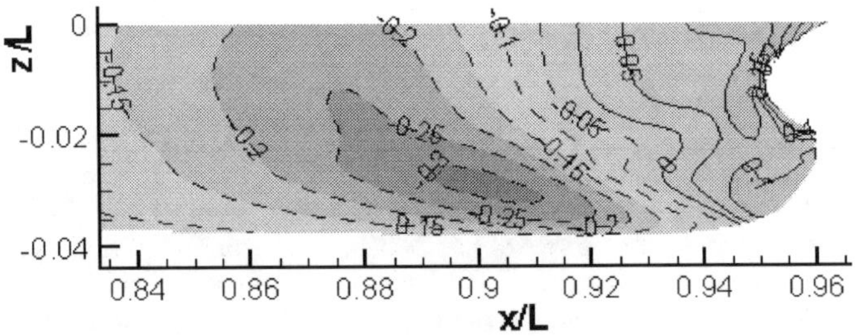

Figure 7. Pressure Coefficient Contours at the Stern of the Ballast-Free Bulk Carrier Model – Initial Case

Axial vorticity contours at x/L = 0.93 and x/L = 0.95 are displayed in Figs. 8 and 9, respectively. At x/L = 0.93, the vorticity contours reveal the existence of a streamwise bilge vortex. This vortex crosses the propeller plane, as demonstrated in Fig. 9. In the same figure, the existence of a small counter-rotating vortex at the base of the bulbous stern is also displayed.

The nominal wake at the propeller plane was investigated in order to help understand and quantify the effect of the longitudinal bilge vortex. Axial velocity contours (non-dimensionalized by the parallel flow velocity) and axial vorticity contours in the propeller disk are shown in Figs. 10 and 11, respectively. In order to investigate the uniformity of the flow in the propeller disk, the simplistic method of computing the standard deviation of the axial velocity was applied. The computed value is 0.205 m/s. The area-weighted average value of the axial vorticity is 10.7 s^{-1}.

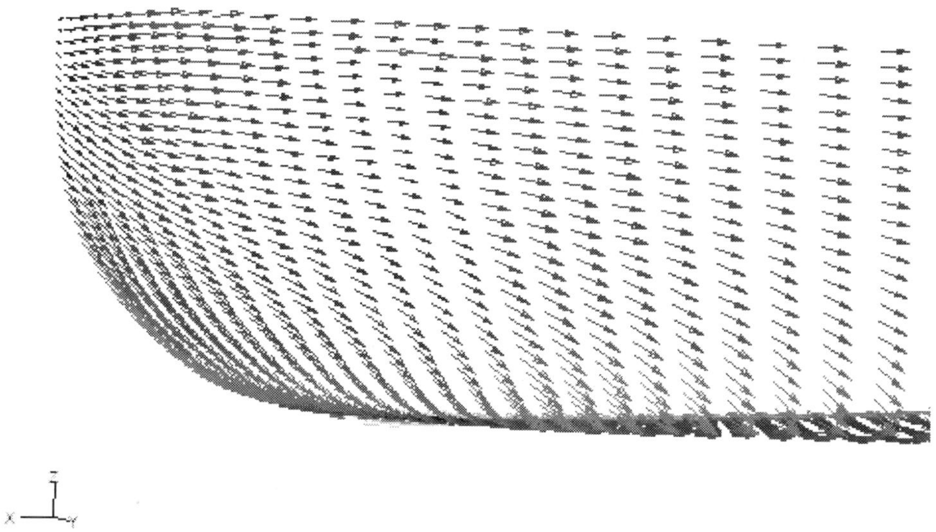

Figure 6. Velocity Vectors at the Bow of the Ballast-Free Bulk Carrier Model – Initial Case

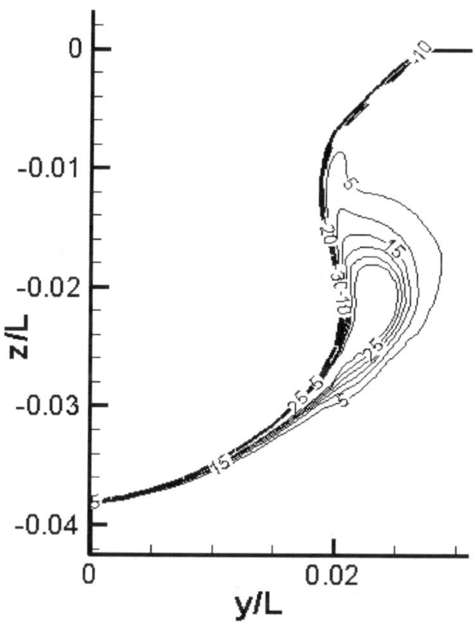

Figure 8. Axial Vorticity Contours at x/L = 0.93 – Initial Case

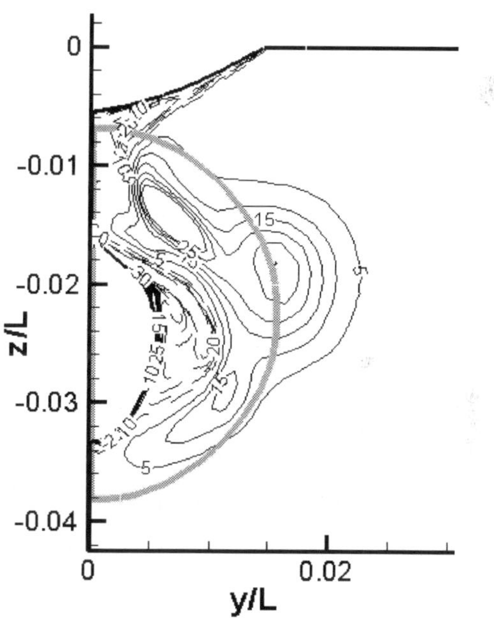

Figure 9. Axial Vorticity Contours at x/L = 0.95 – Initial Case (the propeller disk is also shown in the figure)

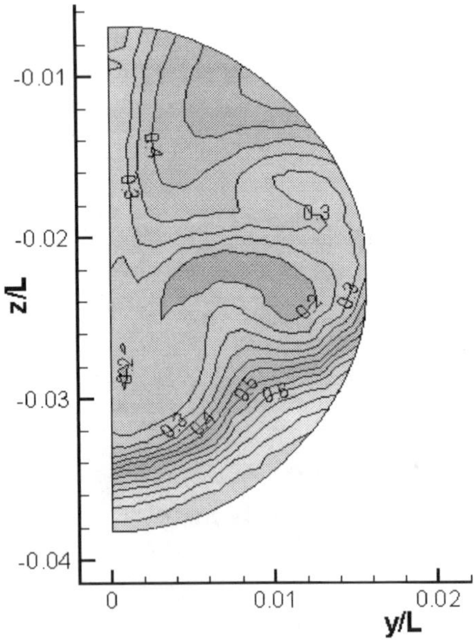

Figure 10. Axial Velocity Contours in the Propeller Disk – Initial Case

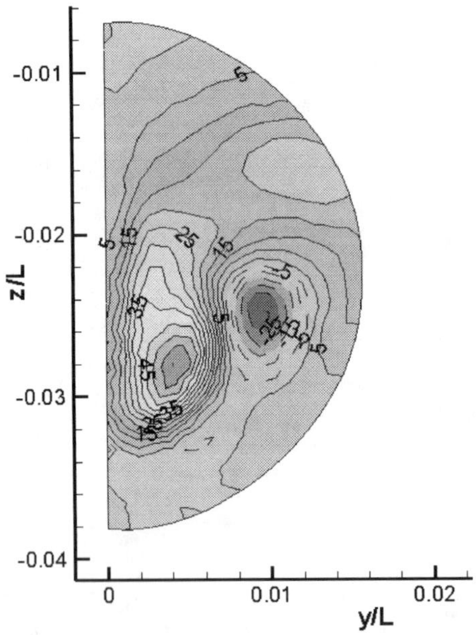

Figure 11. Axial Vorticity Contours in the Propeller Disk – Initial Case

Selection of Inlet and Outlet Plena Locations

Based on the aforementioned flow analysis, it was decided to locate the water inlet right on the face of the bulbous bow in the area around the stagnation point to take advantage of the high positive pressure in this region. In this way, the water exchange goal of 99% in less than two hours can be reached, or even exceeded (Kotinis 2005). For the utilized ballast speed of 15.5 knots (assuming no voluntary speed reduction due to heavy weather), the flow exchange will be achieved in a distance less than 30 nautical miles. A similar approach regarding the water inlet location selection has been followed successfully by Teekay Shipping for a source of pressure to drive flow-through ballast exchange, at a reduced ballast tank level, without the use of pumps (BWN 2002).

A full-scale diameter of approximately 1 m was chosen for the plena inlet and outlet to ensure a smooth inflow and outflow without imposing severe constraints on the structural arrangements. The corresponding inlet/outlet diameter value in model scale is approximately 2.6 cm. The flow rate in the longitudinal trunks was calculated assuming a full-scale volume of ballast water equal to 18,500 m^3. This value, which was obtained from similar ships, is based upon flooding both the normal ballast tanks and a central cargo hold for a heavy weather ballast condition. Assuming an exchange time of 90 min and utilizing Froude scaling, the internal flow rate in model scale is $Q_m = Q_s\ \lambda^{-5/2} = 3.9 \cdot 10^{-4}\ m^3/s$. Using the continuity equation and assuming a symmetrical plenum about the centerplane, the discharge fluid speed is 0.382 m/s.

The centroid of the water inlet was placed a distance approximately 25% DWL above the keel as shown in Fig. 12. The relatively low forward draft in ballast condition (40% DWL) imposes the controlling constraint regarding this selection. The selected position was a trade-off between structural arrangement feasibility and avoidance of water inlet emergence.

The location of the water discharge can only be determined after a careful consideration of the pressure coefficient contours at the stern and the axial velocity contours in the propeller disk. If trunk flow rate maximization were the only criterion, the water outlet should be located in an area with high suction pressure to maximize the pressure differential. On the other hand, when the propeller operation is taken into account, the objective is to minimize the power requirements subject to achieving adequate flow. The discharge location then can only be determined through a propeller flow analysis and propeller performance optimization procedure. It is generally accepted that a more uniform wake and, thus, a more uniform loading on the propeller blades results in less required power for a given delivered thrust. This criterion, though simplistic, can be evaluated initially by merely computing the standard deviation of the axial velocity in the propeller disk.

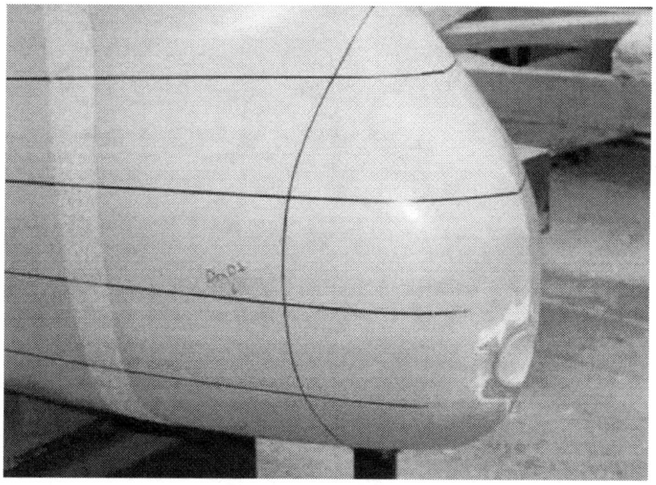

Figure 12. Location of Forward Ballast Trunk Inlet

The primary difficulty in the discharge location selection process stems from the fact that required power minimization and trunk flow rate maximization constitute conflicting criteria in the system optimization. This fact was demonstrated in the initial investigation and development of the Ballast-Free Ship concept (Kotinis 2005). Observing the axial velocity contours in Fig. 10, the upper half-plane is dominated by the primary longitudinal vortex, which provides a significant level of homogenization, but also low fluid velocity, which reduces the propeller loading in this region. On the other hand, a strong velocity gradient dominates the lower half-plane. Even though the secondary vortex covers a smaller region in the propeller disk, it has considerable vorticity and opposes the action of the primary vortex.

If the water outlet is placed in a position where it can supply the fluid particles in the inner boundary layer with additional momentum to delay the separation point and also provide additional momentum to the fluid that flows through the upper half-plane in the propeller disk, then a more uniform wake might occur. On the other hand, given that the water suction is located at a high positive pressure region and the longitudinal pressure gradient at the stern is not steep, the trunk flow rate is not expected to be highly sensitive to the longitudinal position of the water outlet.

In order to investigate the effect of the water discharge location on the flow at the stern, it was decided to test two different discharge locations during the towing tank experiments; one close to Station 17 and one close to Station 19 as shown in Fig. 13. Station 17 is approximately at the location of the forward engineroom bulkhead in the full-scale ship; Station 19 is approximately at the aft engineroom bulkhead. The discharge at Station 17 was located about 45% DWL and the discharge at Station 19 about 30% DWL. In this way, the effect on the flow in the upper half-plane of the propeller disk could be investigated, in addition to the effect of discharging water before and after the formation of the primary longitudinal vortex. The exact positioning of the discharge locations could only be determined after all the model construction and arrangement constraints in the Ballast-Free bulk carrier model were taken into account. The major constraints are the available interior space of the model for the testing equipment (i.e. propeller motor, dynamometer, heave staff) and the feasibility of drilling holes and installing the pipes required for the trunk flow modeling at certain locations inside the model.

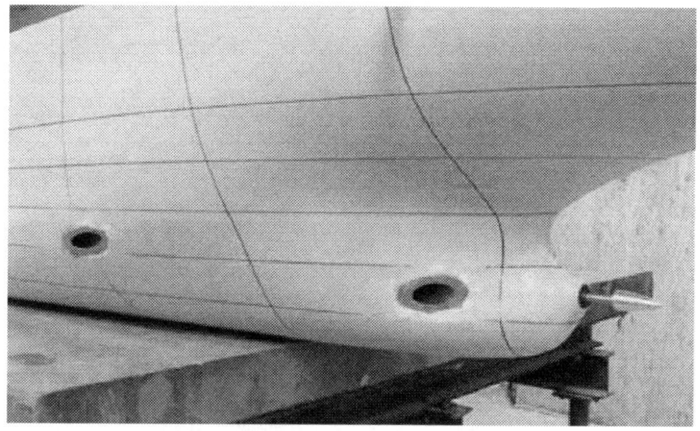

Figure 13. Location of Ballast Trunk Discharges

Modeling and Numerical Investigation of the Water Inlet at the Bow and the Water Discharge at the Stern

The effect of the water suction at the bow and the water discharge close to Station 17 on the flow around the model was investigated numerically with the aid of FLUENT®. The discharge flow direction was set to 10° with respect to the surface tangent. Apparently the lower the angle value, the less obstructive the water discharge would be to the boundary layer flow. This was suspected to be the reason behind the significant increase in power requirements (+ 7.4% for a faster hull form) observed during the initial investigation of the Ballast-Free Ship concept (Kotinis et al. 2004). A smaller angle, even though feasible in the numerical computations, would probably be an unattainable goal during the towing tank experiments or a full-scale implementation.

The major modeling requirement was to provide adequate grid resolution to account for the interaction between the boundary layer flow and the trunk inflow and outflow. This was accomplished by creating a mesh fine enough to resolve the flow in the viscous sublayer, which is the portion of the boundary layer adjacent to the hull solid boundary, in the region around the inlet and discharge locations. For this purpose, a slightly modified version of the fine grid created for the initial case was utilized. The modifications were limited to the modeling of the region close to the inlet and discharge locations. In this case, the total number of cells is 1,074,580.

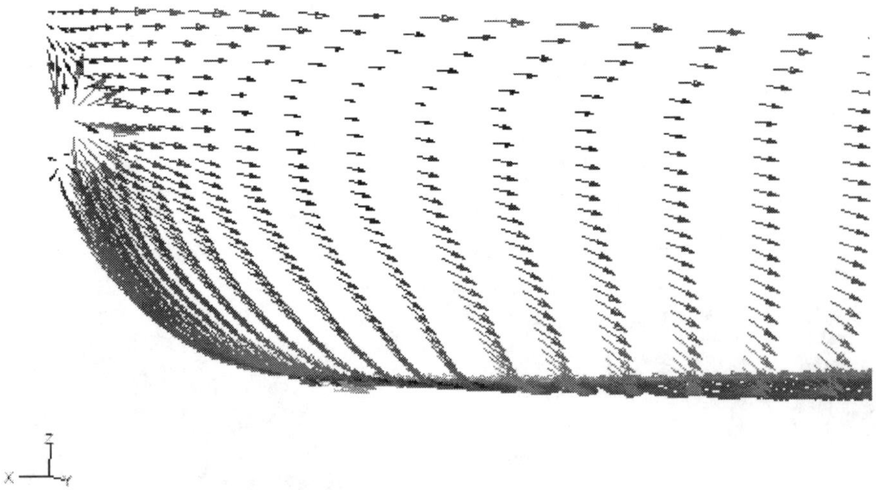

Figure 14. Velocity Vectors at the Bow of the Ballast-Free Bulk Carrier Model – Modified Case

The pressure coefficient contours at the bow are shown in Fig. 15. A comparison with the pressure coefficient contours in Fig. 5 reveals that the positive pressure levels increase in the vicinity of the water inlet, even though this effect vanishes downstream, of x/L = 0.02. This observation is corroborated by the velocity vectors displayed in Fig. 14. In the same figure, a slightly accelerated fluid flow on the edge of the water inlet is displayed, which can be attributed to the suction effect.

The pressure coefficient contours at the stern are displayed in Fig. 16. The effect of the water discharge on the pressure distribution seems to be limited to the vicinity of the discharge location. A small reduction in suction pressure, relative to the pressure distribution of the initial case shown in Fig. 7, is observed slightly upstream of the discharge location. The opposite effect is observed slightly downstream. Therefore, the net effect on the pressure force is expected to be minimal.

The computed values of the friction and pressure drag coefficients are 2.995e-3 and 1.195e-3, respectively. The total drag coefficient is equal to 4.190e-3, an increased of 0.9% compared to the initial case. The corresponding form factor value is 0.277, approximately 4.1% higher, than the initial case. This increase can be attributed mainly to the existence of the water inlet at the bow.

Axial vorticity contours at x/L = 0.93 and x/L = 0.95 in the initial case are demonstrated in Figs. 17 and 18, respectively. At x/L = 0.93, the vorticity contours reveal that the flow discharge close to Station 17 causes a slight stretching of the longitudinal vortex in the transverse direction. Further downstream, at x/L = 0.95, the interaction between the two counter-rotating vortices is significantly affected by the stretching of the primary vortex, as the secondary vortex generated at the bulbous stern, now affects a larger region of the lower half-plane in the propeller disk. The axial vorticity contours in the propeller disk shown in Fig. 19 corroborate the aforementioned conclusion. The area-weighted average value of axial vorticity is 10.5 s^{-1}.

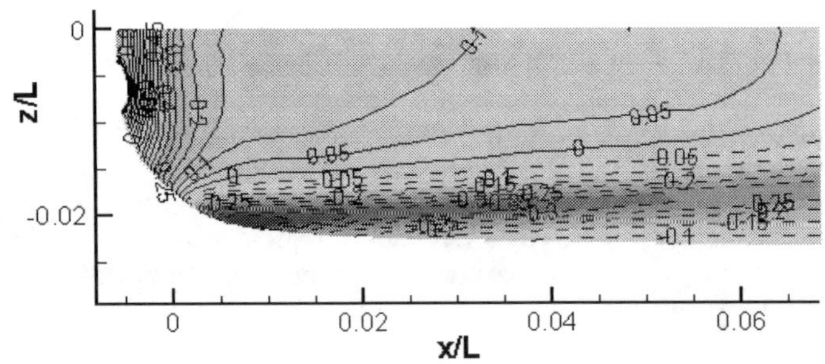

Figure 15. Pressure Coefficient Contours at the Bow of the Ballast-Free Bulk Carrier Model – Modified Case

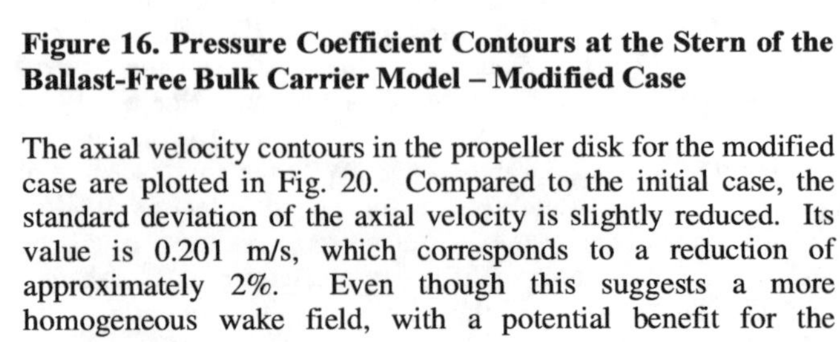

Figure 16. Pressure Coefficient Contours at the Stern of the Ballast-Free Bulk Carrier Model – Modified Case

The axial velocity contours in the propeller disk for the modified case are plotted in Fig. 20. Compared to the initial case, the standard deviation of the axial velocity is slightly reduced. Its value is 0.201 m/s, which corresponds to a reduction of approximately 2%. Even though this suggests a more homogeneous wake field, with a potential benefit for the propeller operation, the changes in the wake distribution necessitate the direct investigation of their effect on the operating propeller.

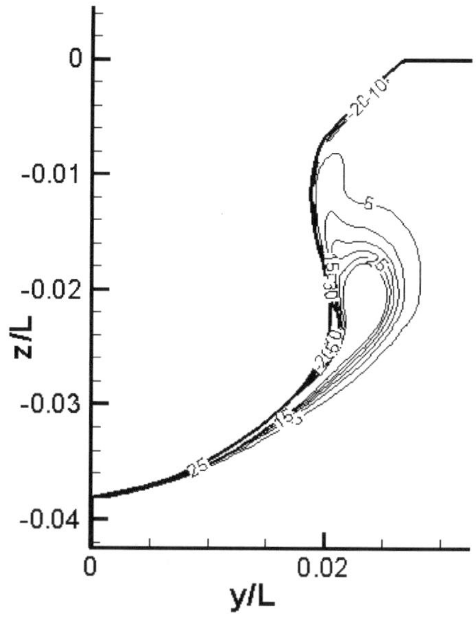

Figure 17. Axial Vorticity Contours at x/L = 0.93 – Modified Case

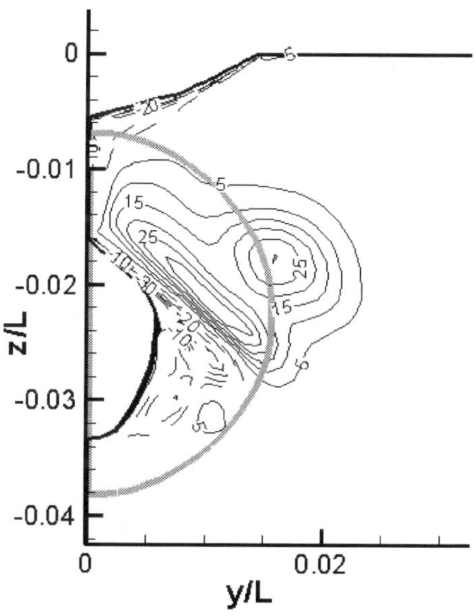

Figure 18. Axial Vorticity Contours at x/L = 0.95 – Modified Case (the propeller disk is also shown in the figure)

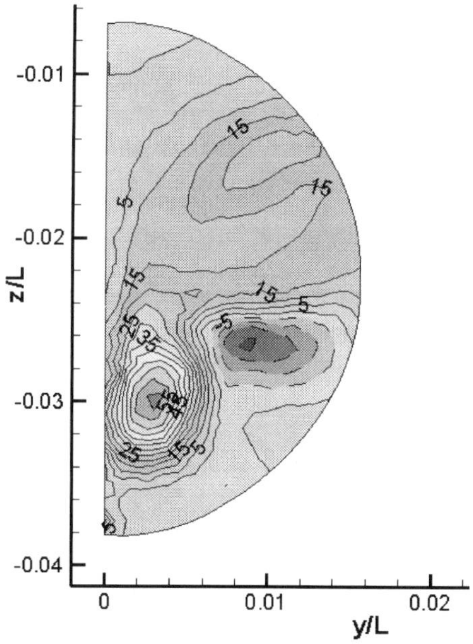

Figure 19. Axial Vorticity Contours in the Propeller Disk – Modified Case

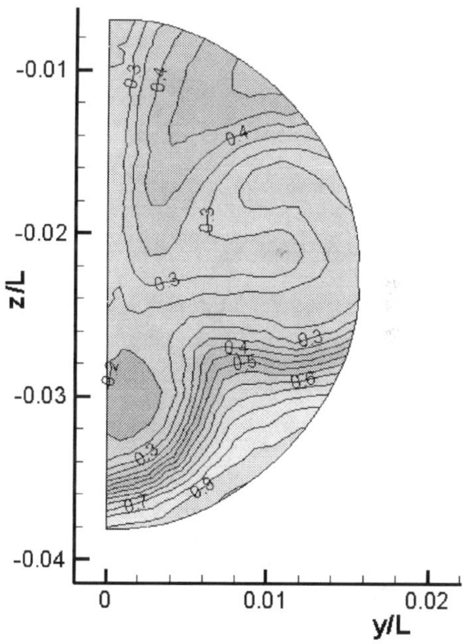

Figure 20. Axial Velocity Contours in the Propeller Disk – Modified Case

EXPERIMENTAL HYDRODYNAMIC INVESTIGATION

The insight gained through the numerical hydrodynamic investigation of the Ballast-Free Ship concept described above was utilized as the starting point for the experimental towing tank model test investigation. The experimental investigation also provided the opportunity for a qualitative and quantitative validation of the numerical results.

The experiments were performed at the University of Michigan Marine Hydrodynamics Laboratory (MHL) towing tank using the Ballast-Free bulk carrier model. The total resistance and the propulsion requirements of the model were determined for three different speeds in the ballast condition where the ballast trunks would be used. All the tests were carried out at the ballast drafts. The experimental test plan is shown in Table 6.

Table 6. Ballast Condition Experimental Test Plan

	Ship speed (knots)	Model speed (m/s)	Froude number
1st speed	14.50	1.210	0.173
2nd speed	15.50	1.295	0.185
3rd speed	16.50	1.378	0.197

Four different measurements were obtained at each speed in an attempt to minimize the precision error. In addition, a randomized testing sequence was utilized to allow for minimization of the effect of extraneous variables. Prior to the model testing, a static calibration of the load cell was performed in order to estimate the instrument error. The unmodified hull, shown in Figs. 2 and 3, was tested first. Subsequently, the hull was modified, as shown in Figs. 12 and 13, and the tests were repeated, first, for the water discharge high close to Station 17 and then for the low discharge close to Station 19.

Because the modeling of the internal flow trunks could not be reliably scaled at the small model scale used, the scaled total trunk flow was pumped rather than using natural flow. The internal flow was modeled primarily by a 1-inch internal diameter flexible pipe, which was connected to the water suction at the bow and the water discharge at the stern. The internal flow was aided by a flexible-impeller pump, which was used to achieve the scaled flow rate. The flow rate was controlled by a high-precision needle valve and monitored by a flow meter. The flow was diverted to the selected discharge location and subsequently split to provide a symmetric water discharge at the stern of the model. Details of the internal pipe connections and system are displayed in Figs. 21, 22 and 23. As already mentioned, the model-scale internal flow rate was set to $3.9 \cdot 10^{-4}$ m^3/s or 6.1 gallons per minute.

Figure 21. Internal Flow Details in the Bow Region

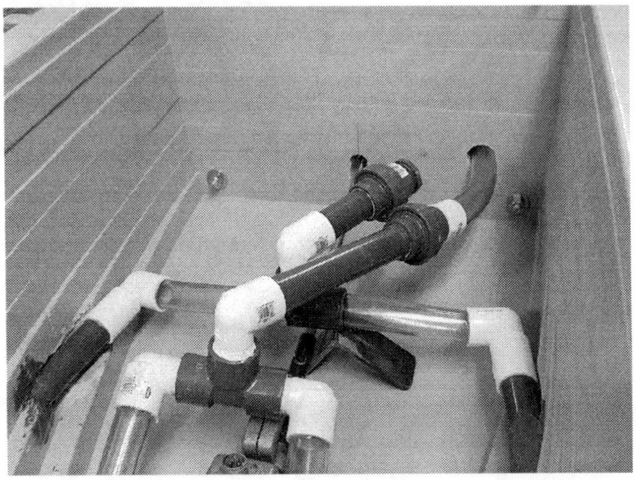

Figure 22. Internal Flow Details in the Stern Region – Looking Forward

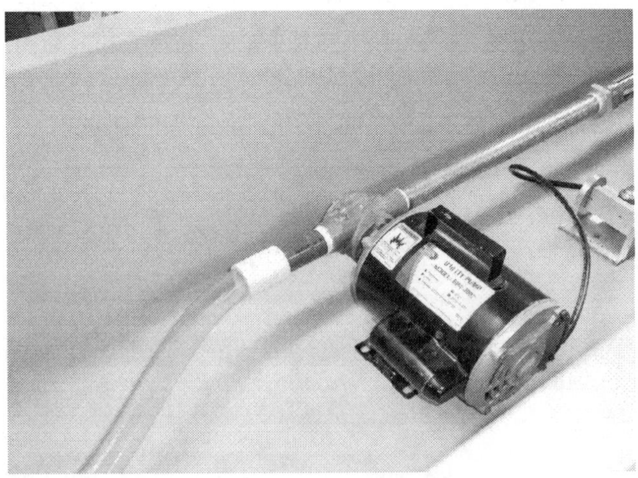

Figure 23. Internal Flow Details – Pump and Flow Meter

Resistance Tests

The resistance of the Ballast-Free bulk carrier model was determined for the three speeds listed in Table 6. The measured total resistance was corrected using the calibration results and then extrapolated to full scale by utilizing the ITTC-recommended method (ITTC 1978). For all testing conditions, the results are reported at a standard temperature of 15°C. The uncertainty in the full-scale resistance and effective power was estimated by taking into account both bias and precision errors. Errors related to the static calibration and the form factor derivation were considered as sources of bias error. During the uncertainty analysis, a 95% level of confidence was assumed. In general, the propagation of the bias limit from a single error source to the result (i.e. full-scale resistance) was calculated by utilizing sensitivity indices and assuming that the error sources are statistically independent. The sensitivity index relates how changes in each of the error sources affect the result. The total bias limit includes the contribution of all the bias error sources. The total precision limit is a measure of the random variation in the results. It is equal to the standard deviation of the average result value at each speed multiplied by the appropriate Student-t distribution value based on the confidence interval and the number of measurements. The total uncertainty is calculated as the root sum square of the total bias limit and the total precision limit:

$$U_T = \sqrt{B^2 + P^2} \qquad (1)$$

Static Calibration. A static calibration of the load cell was performed by utilizing a sequential test. The results are shown in Table 7. It needs to be mentioned that the hysteresis error observed during the calibration procedure was negligible. The bias limit reported in Table 7 corresponds to the standard error due to the scatter in the load cell calibration dataset relative to a linear least-squares regression curve fit assuming a 95% confidence interval. Other contributions to the bias limit of the load cell measurement can be considered negligible (Longo and Stern 2005).

Table 7. Static Calibration Results

Calibration curve slope	1.0003
Zero drift	-0.012 N
Bias limit	0.036 N

Form Factor Calculation and Validation Procedure. The form factor k was obtained by measuring the model resistance at low Froude numbers and then utilizing a linear least-squares regression curve fit. The intercept of the regression model corresponds to the form factor value plus 1. The form factor uncertainty is also reported in Table 8.

Table 8. Form Factor and Uncertainty

	Form factor k	Uncertainty (%D)
Initial case	0.291	6.0
Discharge at St.17	0.259	9.4
Discharge at St.19	0.282	6.3

The experimental form factor value (D) of the baseline (initial) case listed in Table 8 can be utilized to validate the CFD results using the methodology derived in Stern et al. (2001). The experimental data uncertainty, U_D, combined with the computed simulation numerical uncertainty U_{SN} = 7.6%D, provides the validation uncertainty, U_V:

$$U_V = \sqrt{U_{SN}^2 + U_D^2} = 9.2\%D \qquad (2)$$

This value is compared to the absolute value of the comparison error, |E|, which is equal to the difference between the numerical result at the fine grid and the experimental result. In this case, the absolute value of the comparison error is equal to 8.4%D; thus, the form factor value is validated at the uncertainty level of U_V = 9.2%D. Further reducing the levels of validation uncertainty would require reduction in both experimental and simulation numerical uncertainty.

Ship Resistance and Effective Power Results. The full-scale ship resistance and the corresponding effective power were calculated for each of the three cases. The results are shown in Figs. 24 and 25. The water discharge close to Station 17 appears to have a negative effect on ship resistance and, to a lesser extent, the same is true for the discharge close to Station 19. This conclusion, regarding the effect of the water discharge close to Station 17 on the hull resistance, is supported,by the numerical results. The computed form factor value for the modified case is greater than the computed value for the initial case. The corresponding values based on the experimental results showed a different trend; however, the experimental uncertainty was rather high, especially when discharging close to Station 17. The analysis of the numerical results provided a reasonable explanation for the relative increase of the form factor value. Despite the fact that the average resistance values plotted in Fig. 24 reveal an increase in resistance, the difference with respect to the baseline case is not statistically significant as seen by the overlapping error bands.

Propulsion Tests

The resistance tests were followed by a series of propulsion tests using the MHL stock model propeller No. 23. The thrust and torque measurements at the self-propulsion condition at each speed were analyzed using the ITTC-recommended method (ITTC 1978). The calculated delivered power is shown in Fig. 26. An uncertainty analysis was also performed for the propulsion test results.

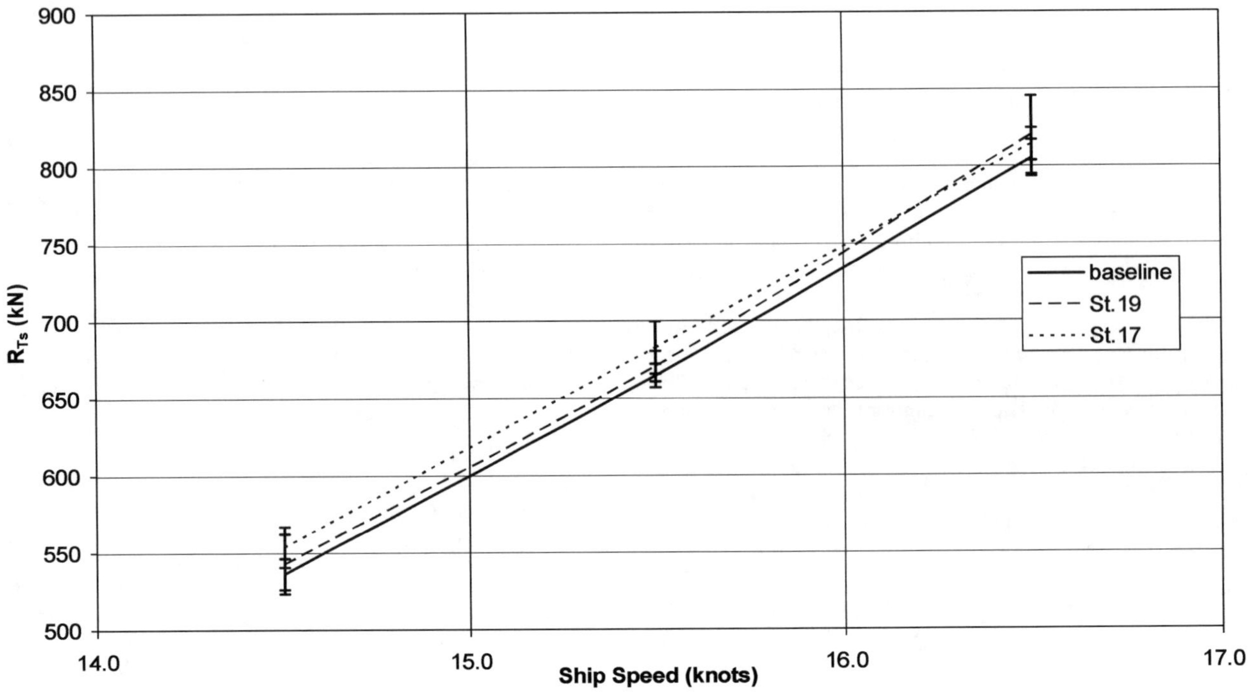

Figure 24. Calculated Ship Total Resistance

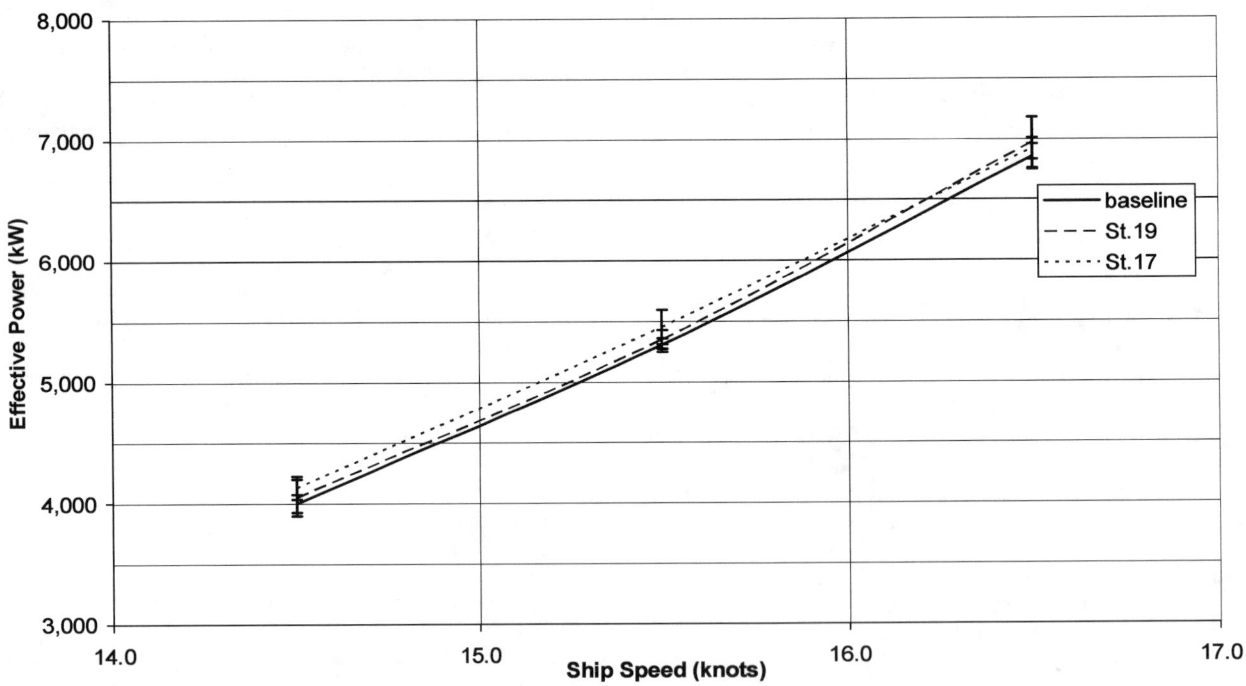

Figure 25. Calculated Effective Power

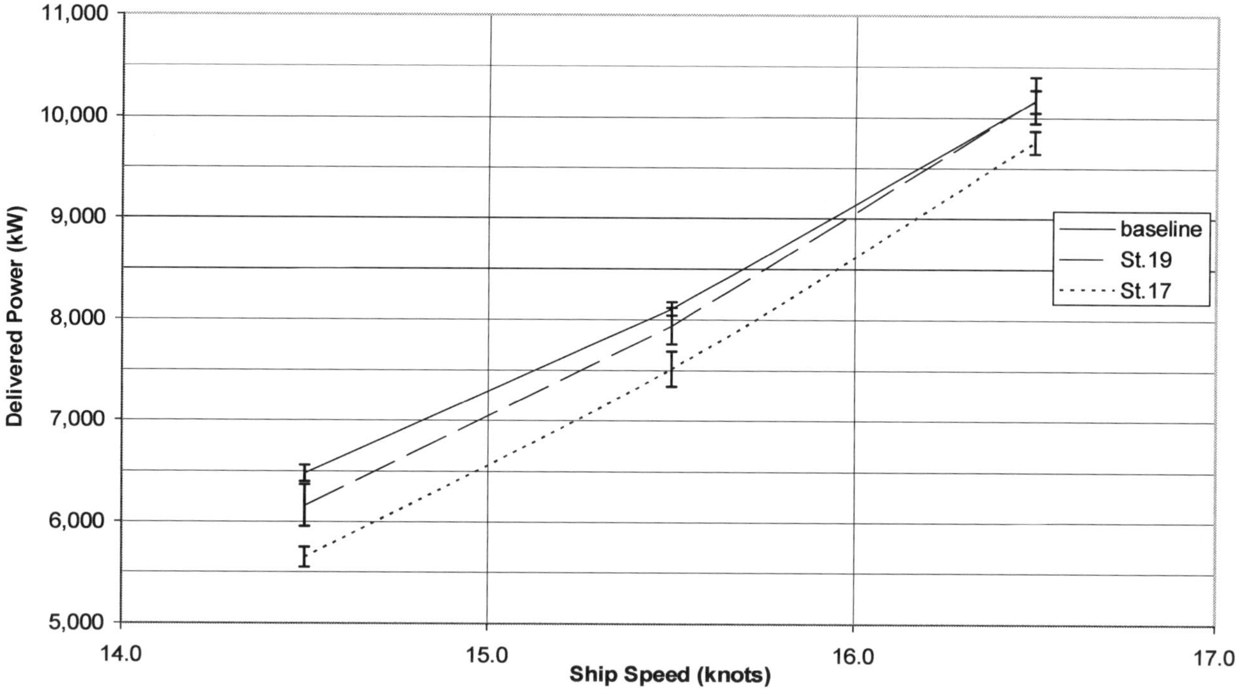

Figure 26. Calculated Delivered Power

The plotted results reveal a significant reduction in the powering requirements caused by the water discharge at the stern. It needs to be emphasized that these results were obtained with a non wake-adapted stock propeller. This reduction can be attributed to increased homogenization of the wake field in the propeller disk, as demonstrated in the numerical investigation, but also to changes in the inflow to the propeller, as shown in Figs. 10 and 20. The physics behind the significant reduction in powering requirements due to the water discharge close to Station 17 cannot be explained further without a detailed analysis of the effective wake. However, judging from the numerical results obtained for the discharge close to Station 17, it can be speculated that the interaction between the water discharge and the longitudinal vortices contributes to the considerable change in delivered power.

At the ballast speed of 15.5 knots, the reduction in delivered power is 7.3% for the discharge close to Station 17 and 2.1% for the discharge close to Station 19. The location near Station 17 would also be preferred from an engineroom arrangements viewpoint since the ballast trunks would not have to be carried through the engineroom.

ECONOMIC ANALYSIS

The economic impact of the Ballast-Free Ship concept on the capital and operating cost of a typical Seaway-sized bulk carrier was estimated in a manner similar to that used in the initial investigation of the concept (Kotinis et al. 2004). Because the Ballast-Free bulk carrier was designed to maintain the same grain capacity, its annual cargo capacity (ACC) would be unchanged. Thus, an appropriate measure of merit is the change in the Required Freight Rate (ΔRFR) using the equation (Mackey et al. 2000):

$$\Delta RFR = (CRF(i, n) \cdot \Delta P + \Delta A)/ACC \qquad (3)$$

where ΔP is the change in the capital cost, ΔA is the change in the annual operating cost, ACC is the constant annual cargo capacity, and CRF(i, n) is the Capital Recovery Factor for an i return on investment over a ship life of n years.

A realistic scenario was adopted for the economic analysis: a Handymax bulk carrier transporting grain from the upper Great Lakes (e.g. Duluth, Thunder Bay) to ports in Northern Europe and occasionally transporting steel into the Great Lakes. A North Atlantic voyage route between Rotterdam and Montreal, entering the Great Lakes through the St. Lawrence Seaway while in a ballast condition, is assumed.

A major, conservative assumption is that the 2.1% and 7.3 % reductions in the required power of the Ballast Free bulk carrier will not be enough to permit a change in the main engine; thus, no propulsion machinery capital cost reduction is included. For the specific vessel type investigated in this study, the reduction in the delivered power indicated by the propulsion results would lead to decreased fuel use, assuming constant ship speed, but would probably not actually permit a change in number of cylinders or a move to a smaller alternative engine model. The results for the water discharge close to Station 17 and close to Station 19 are presented in Fig. 27.

Based on the results of the initial investigation (Kotinis 2005), an increase in the block coefficient value from 0.835 to 0.841 was required in order to permit the transportation of equal annual cargo capacity and offset the lost buoyancy of the flooded plena at constant Seaway draft while fully loaded. The increase in ship powering requirements due to the higher block coefficient value was factored in the analysis input data. The increase in the hull steel weight results in a moderate increase in the hull steel cost. Foreign new construction, typical of Korea, was assumed for the calculation of the hull steel and other construction costs. The eliminated ballast water treatment system was assumed to consist of automatic backflush filtration as a primary treatment combined with UV irradiation for a secondary treatment. The estimated cost of this treatment equipment was based upon a study commissioned by the Great Lakes Ballast Technology Demonstration Project (Hurley et al. 2001). Several other costs relative to changes in arrangements and outfitting are also taken into account. A detailed description of these can be found in (Kotinis 2005).

It needs to be emphasized that no account was taken for the power consumption, crew workload, and maintenance savings associated with the elimination of the ballast water treatment system. This would further reduce the annual operating cost and favor the use of the Ballast-Free Ship concept. The lifecycle cost of the aforementioned ballast treatment system would probably exceed $800,000 (Hurley et al. 2001).

The net savings in terms of the ΔRFR with the ballast trunk water discharge close to Station 17 is estimated to be about $0.93 per ton of cargo. The corresponding savings with the water discharge close to Station 19 is estimated to be about $0.44 per ton of cargo. These savings are relative to the use of filtration primary and UV secondary ballast water treatment when ballast water exchange is no longer permitted in the future.

SUMMARY AND CONCLUSIONS

The initial development and investigation of the Ballast-Free Ship concept demonstrated its feasibility and efficiency, but failed to show its full cost-effectiveness. The main reason was the significant fuel penalty, resulting from increased power requirements found in the initial preliminary testing of an existing, higher-speed vessel with a non-optimum propeller (Kotinis et al. 2004).

The current paper focused on a more detailed hydrodynamic investigation of the concept for a Seaway-sized bulk carrier in order to obtain a more realistic assessment of these effects. For this purpose, a new Seaway-sized bulk carrier model was designed and built.

The hydrodynamic investigation was performed in two parts. In the first part, the numerical investigation using CFD provided useful insight regarding the physics of the flow and, to a certain extent, provided quantitative results to compare with the experimental outcomes.

The selection of the trunk flow inlet and outlet locations utilized in the towing tank experiments was guided by these numerical results. The trunk flow inlet was located in the center of the bulbous bow. Two different locations were tested for the water discharge: one high close to Station 17 and one lower close to Station 19.

The experimental hydrodynamic investigation consisted of resistance and propulsion experiments in the towing tank. The analysis of the model test data revealed that the experimental results were in good agreement with the numerical results.

In general, the water discharge at the stern was shown to increase ship resistance. On the other hand, it has a very favorable effect on the propulsion characteristics for this Seaway-sized bulk carrier design. The reduction in powering requirements, compared to the baseline case, was 7.3% for water discharge close to Station 17 and 2.1% for water discharge close to Station 19 at the assumed ballast speed of 15.5 knots.

It must be cautioned that since some of the effect of the water discharge may be related to the location of the separation near the stern, and this is known to not scale effectively for models of the size used in this testing, the exact magnitude of the full-scale power improvement cannot be stated absolutely. The 5-m model used in this experimental work was the largest that could be tested in the University of Michigan MHL so further confirmation would have to be made in a larger facility that would permit the use of a larger model.

A realistic operating scenario was adopted in order to investigate the economic impact of the Ballast-Free Ship concept. Based on the aforementioned reduction in powering requirements, the net savings with respect to a ballast treatment system would be $0.93 per ton of cargo for the water discharge close to Station 17 and $0.44 per ton of cargo for the water discharge close to Station 19.

The Ballast-Free Ship concept <u>essentially eliminates the transport of foreign ballast water</u> from ships operating in the ballast condition. The elimination of the fuel penalty found earlier makes the Ballast-Free Ship concept a very attractive alternative to more costly ballast water treatment systems.

Further numerical hydrodynamic investigation that includes the propeller effect is expected to provide an optimum location for the water discharge position. This will be confirmed by additional MHL testing planned in 2008.

ACKNOWLEDGEMENTS

This paper is a result of work sponsored by the Great Lakes Maritime Research Institute (GLMRI), a Consortium of the University of Wisconsin, Superior, and University of Minnesota,

Duluth. The GLMRI is funded by the U. S. Department of Transportation, Maritime Administration, under contract DTOS59-05-G-00019 and agreement DTMA1G0605. This financial support is gratefully acknowledged.

REFERENCES

BWN 2002 "Alternative Exchange Design Proposed," *Ballast Water News*, Global Ballast Water Management Program, IMO, **8**, January-March.

DUVIGNEAU, R., VISONNEAU, M. and DENG, G.B. 2002 "On the Role Played by Turbulence Closures in Hull Shape Optimization at Model and Full Scale," 24th Symposium on Naval Hydrodynamics, Fukuoka, Japan, August 8-13.

FARLEY, R. 1996 "Analysis of Overseas Vessel Transits Into the Great Lakes and Resultant Distribution of Ballast Water," Report No. 331, University of Michigan, Department of Naval Architecture and Marine Engineering.

FLUENT 2006 "FLUENT 6.3 User's Manual," Fluent Inc., Lebanon, NH.

FORMATION DESIGN SYSTEMS Pty 2006 "MAXSURF Windows Version 12 User's Manual."

FORSBERG, R., BAIER, R., MEYER, A., DOBLIN, M., and STROM, M. 2005 "Fine Particle Persistence in Ballast Water Sediments and Ballast Tank Biofilms," 28th Annual Meeting of The Adhesive Society, Mobile, Alabama, February 13-16.

GRIDGEN 2007 "Gridgen 15.1 User Manual," Pointwise Inc., Fort Worth, Texas.

HURLEY, W. L. JR., SCHILLING, S. S. JR. and MACKEY, T. P. 2001 "Contract Designs for Ballast Water Treatment Systems on Containership *R. J. Pfeiffer* and Tanker *Polar Endeavor*," SNAME/ ASNE Marine Environmental Engineering Technical Symposium, Arlington, VA, May 31-June 1 (CD).

IMO 2004 "International Convention for the Control and Management of Ships' Ballast Water & Sediments," Diplomatic Conference, February, London.

ITTC 1978 "15th International Towing Tank Conference, 3-10 September 1978, The Hague, The Netherlands", Netherlands Ship Model Basin, Wageningen.

KENT, C.P. and PARSONS M.G. 2004 "Computational Fluid Dynamics Study of the Effectiveness of Flow-Though Ballast Exchange," *Transactions* SNAME, **112**.

KIM, S.E., MATHUR, S.R., MURTHY, J.Y., and CHOUDHURY, D. 1998 "A Reynolds-Averaged Navier-Stokes Solver using Unstructured Mesh-Based Finite Volume Scheme," AIAA Paper 98-0231.

KIM, S.E., and RHEE, S.H. 2002 "Assessment of Eight Turbulence Models for a Three-Dimensional Boundary Layer involving Crossflow and Streamwise Vortices," AIAA 2002-0852, 40th AIAA Aerospace Sciences Meeting and Exhibit, Reno, NV, Jan.

KOTINIS, M. 2005 "Development and Investigation of the Ballast-Free Ship Concept," Ph.D. Dissertation, University of Michigan, Department of Naval Architecture and Marine Engineering.

KOTINIS, M., PARSONS, M. G., LAMB, T. and SIRVIENTE, A. 2004 "Development and Investigation of the Ballast-Free Ship Concept," *Transactions* SNAME, **112**, pp. 206-240.

LONGO, J. and STERN, F. 2005 "Uncertainty Assessment for Towing Tank Tests with Example for Surface Combatant DTMB Model 5415," *Journal of Ship Research*, **49**(1), pp. 55-68.

MACKEY, T. P., TAGG, R. D. and PARSONS, M. G. 2000 "Technologies for Ballast Water Management," 8th ICMES2000/SNAME New York Metropolitan Section Symposium, May 22-23.

MATHUR, S.R. and MURTHY, J.Y. 1997 "A Pressure-Based Method for Unstructured Meshes," *Numerical Heat Transfer*, **31**, pp. 195-215.

MENTER, F.R. 1994 Two-Equation Eddy-Viscosity Turbulence Models for Engineering Applications," *AIAA Journal*, **32**(8):1598-1605, August.

STERN, F., WILSON, R.V., COLEMAN, H.W., and PATERSON, E.G. 2001 "Comprehensive Approach to Verification and Validation of CFD Simulations – Part 1: Methodology and Procedures," ASME *Journal of Fluids Engineering*, **123**, pp. 793-802.

TECPLOT 2006 "Tecplot 360 User's Manual," Tecplot Inc., Bellevue, WA.

U. S. PATENT #6,694,908 2004 "Ballast Free Ship System," U. S. Patent and Trademark Office, Washington, DC.

WILSON, R.V., STERN, F., COLEMAN, H.W., and PATERSON E.G. 2001 "Comprehensive Approach to Verification and Validation of CFD Simulations – Part 2: Applications of RANS Simulation of a Cargo/Container Ship," ASME *Journal of Fluids Engineering*, **123**, pp. 803-810.

Vessel data and trip scenario	Typical bulk carrier	Ballast-free bulk carrier		Comments
		Discharge at St.17	Discharge at St.19	
Round-trip distance (nautical miles)	6,280			Montreal (CAN) to Rotterdam (NL) through the Seaway
Service speed (kts)	14.5			Typical data for an ocean-going Handymax bulk carrier transporting grain cargo from the Great Lakes (Duluth, Thunder Bay) to ports in Northern Europe and occasionally transporting steel into the Great Lakes.
Speed in ballast condition (kts)	15.5			
Proportion of miles in ballast (%)	35			
Average loaded cargo / maximum cargo (%)	90			
Load factor (%)	58.5			
Days of navigation through the Great Lakes	8			Passage up through the Great Lakes towards the western end
Port days per round trip	14			Includes loading/unloading time, bunkering time, and time waiting for berth
Round trips per annum	7			
Maximum payload (metric tons)	32,000			
Cargo carried per annum (metric tons)	131,000			
Engine nominal MCR (kW)	8,580			Data for the MAN B&W 6S50MC two-stroke engine
Block coefficient	0.835	0.841		Compensate for increased hull steel weight and lost buoyancy at plena
Hull steel weight (metric tons)	5,550	5,770		
Hull steel cost ($)	2,220,000	2,308,000		Assuming a steel price of $400/metric ton
Continuous service rating in ballast condition (kW)	7,700	7,140	7,540	Includes 15% sea margin and effect of change in C_B value
Continuous service rating in full load condition (kW)	7,700	7,155	7,555	Includes 15% sea margin and effect of inlet/outlet hull openings and change in C_B value
Specific fuel consumption (g/(kW*hr))	168.7	166.4	168.0	Data for the MAN B&W 6S50MC engine, ISO ambient conditions
Annual heavy fuel cost ($)	1,039,000	951,000	1,014,000	Fuel price (IFO380) of $270/metric ton, transatlantic part of trip only
	Changes in capital cost			
Additional hull steel cost ($)	88,000			
Sluice gates cost ($)	260,000			Acquisition cost plus labor for 52 450x600 mm sluice gates (@ $5,000 each)
Elimination of ballast tank valves ($)	-14,000			14 @ 1,000 each
Reduction in ballast piping cost ($)	-314,000			Removal of main ballast headers (material plus labor)
Reduction in welding cost ($)	-9,500			Reduced welding at the bottom of solid floors (material plus labor)
Additional ballast piping cost ($)	79,000			Addition of ballast piping for F.P. tank (material plus labor)
Additional welding cost ($)	2,600			Additional welding due to raise of inner bottom (material plus labor)
Elimination of ballast water treatment system ($)	-375,000			Assuming automatic backflush filtration combined with UV irradiation
	Changes in operating cost			
	Discharge at St.17	Discharge at St.19		
Change in heavy fuel oil cost ($)	-88,000	-25,000		
Net capital cost change ($)	-282,900			
Net operating cost change per annum ($)	-88,000	-25,000		
Capital recovery factor	0.1175			i = 10%, n = 20 years
Change in required freight rate ($/metric ton)	-0.93	-0.44		savings

Figure 27. Economics Summary Comparing a Typical Bulk Carrier with Filtration and UV Treatment with Ballast-Free Bulk Carrier with Two Discharge Locations

Discussion

Richard D. Stewart, Member

Control of the introduction of exotic species by ballast water from vessels entering the Great Lakes is one of the most critical issues facing that portion of our nation's marine transportation system. The paper by Kotinis and Parsons is well written, cited, organized and very topical. The power savings discovered when testing the new hull form is remarkable, and this information needs to be shared with a wide audience.

I do have some suggestions for the authors. Table 1 lists the full-load draft of the vessel as 10.7 meters. The Seaway system only allows an 8-meter draft (http://www.tc.gc.ca/TDC/publication/updates/v12n3.htm) and in special circumstances 8.08 meters. (http://www.greatlakes-seaway.com/en/navigation/notice20060922.pdf) The authors need to make it clear what the advantages observed in their testing will be at the 8-meter draft and if there is an operational step where the vessel reduces from an salt water draft of 10.7 meters to a seaway maximum draft of 8 meters.

The tremendous potential benefit of this design concept could be clouded if the issue of ballast water sediment is not discussed. The rate of flow of the system is slow enough that there will be settling of sediment in the ballast tanks. The authors state "the elimination of the foreign ballast water and the nonindigenous aquatic species it may contain" (right hand column, last paragraph, page 1). While this statement is true, the sediment from the ballast water that flowed through may not be. The advantage of this design is the ease of entrance into the tanks for cleaning of sediment when the vessel is in non-ballast condition. Also the constant flow of the water should not result in as much sediment as water that remains in a tank allowing for settling. Ship owners may accrue two economic benefits from reduced sediment: 1) lower sediment removal costs; and 2) increased cargo DWT from less sediment accumulating in the tanks.

The last item I would like to address is that I feel that the authors are too conservative in their estimates of the life cycle costs of alternative ballast water treatment systems. Their citation (Hurley, et al., 2001) reflects energy, commodity, and labor costs that have in some cases tripled since Hurley's 2001 data. Recent studies in Finland put the cost higher and to leave out cost factors, such as crew workload, while conservative, does not accurately reflect the true value of this new design. I feel that their paper would benefit from refining the costs of alternative ballast water systems.

Authors' Closure

The authors would like to thank Dr. Stewart for his very useful comments. Dr. Stewart's remarks regarding the ballast water sediment is an issue that was addressed by the authors in the initial investigation of the concept (Kotinis, et al., 2004). Specifically, it was stated that the area of the floors just above the bottom shell plating should be as open as possible to enhance the water flow and minimize the accumulation of sediment. The feasibility of this modification and its effect on the structural strength of the vessel was investigated by performing a detailed structural analysis. Further, to accommodate the required storm ballast volume below the ballast waterline the Ballast-Free Concept Seaway-sized bulk carrier utilizes a 2.4m high inner bottom instead of a more typical 1.6m. This greater height of the double bottom and the cutaway of the floors at the bottom shell will make the cleaning of sediment from the ballast trunks much easier for the crew or contractor Tiger teams. The authors agree with Dr. Stewart regarding the potential economic benefits of the minimization of ballast water sediment. However, a quantification of the cost savings was not possible as part of the presented research project.

An updated economic analysis is provided below to demonstrate changes in the RFR assuming that the vessel operates, when loaded with cargo, at an 8-meter draft. These changes are reflected mainly in the cargo carried per annum. The speed in the loaded condition was also adjusted to reflect operation at an 8-meter draft. An updated cost estimate regarding the capital cost and operating cost associated with the eliminated filtration/UV treatment system has also been included. This estimate is based on the Finnish study (Sassi et al. 2005) recommended by Dr. Stewart. The net savings with the ballast trunk water discharge close to Station 17 are estimated to be about $2.55 per ton of cargo. For water discharge close to Station 19, the savings are estimated to be about $1.57 per ton of cargo.

Reference

Sassi, J., Viitasalo, S., Rytkonen, J. and Leppakoski, E. 2005. Experiments with ultraviolet light, ultrasound and ozone technologies for onboard ballast water treatment, VTT Research Notes, 2313, Finland.

Methodology for the Optimization of Commonality in Multiple Ship Classes

Michael J. Corl, (M)
Michael G. Parsons, (FL)
Michael Kokkolaras, (V)

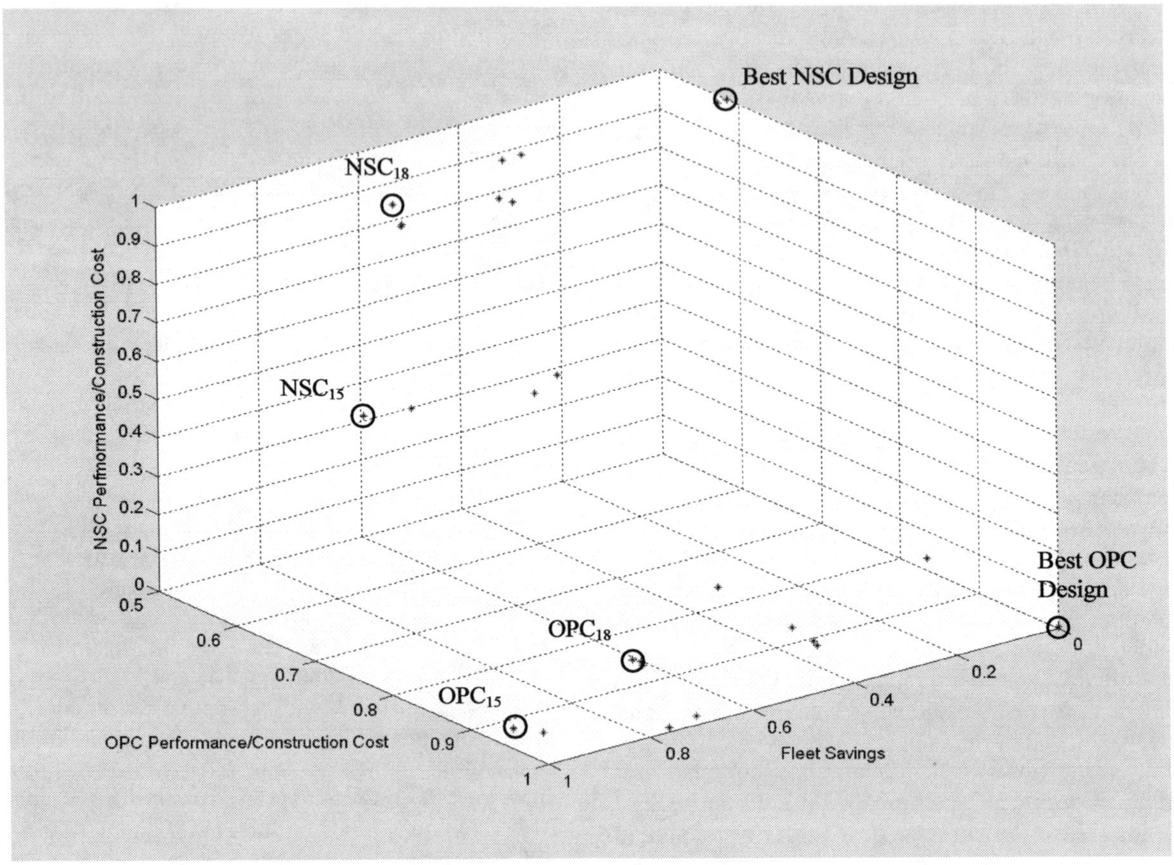

A methodology is presented for the optimization of the use of commonality in multiple classes of ships using mission effectiveness, cost and fleet savings models to evaluate design alternatives. A multicriterion evolutionary algorithm is used to produce the discrete Pareto surface for ships' mission/cost and fleet savings. The methodology is demonstrated and tested using the missions for the U.S. Coast Guard Deepwater High and Medium Endurance Fleets.

KEY WORDS: commonality; ship design; missions; platforms; multicriterion; evolutionary optimization; savings.

NOMENCLATURE

AMIO – Alien Migration Interdiction Operations
C_I – Transverse Inertia Coefficient
CODAD – Combined Diesel and Diesel
CODOG – Combined Diesel or Gas Turbine
d_{ij} – Distance between two data points
D – Normalized distance between data points
D_{RAW} – Average distance between data points
GDO – General Defense Operations
GM_T – Transverse Metacentric Height
k – Number of children generated in each generation
KG – Vertical Center of Gravity

LMR – Living Marine Resources
M_{mut} – Mutation Magnitude
MP_{ij} – Percent of operational time spent on mission
NSC – National Security Cutter
n – Number of independent variables
OPC – Offshore Patrol Cutter
R_{mut} – Mutation Rate
WHEC – High Endurance Cutter
WMEC – Medium Endurance Cutter
WMSM – Maritime Security Cutter Medium (OPC)
WMSL – Maritime Security Cutter Large (NSC)

INTRODUCTION

The overall goal of this research was to apply multicriterion optimization methods to platform decisions for families of ship variants while explicitly taking into account fleet-wide savings. In this context, a platform is the set of common elements used in more than one ship class. The Optimal Design Laboratory of the Department of Mechanical Engineering at the University of Michigan has developed analytical methods for making optimal platform decisions in consumer products and the automotive industry. This research has adapted these methods and extended them to utilize the multicriterion optimization approach necessary to effectively treat naval fleet design problems. The methodology was then tested through modeling and application to determine the optimal common platform and ship designs to use for the missions of the U.S. Coast Guard's Deepwater Fleet.

Motivation

A common practice within the automotive industry is to use the same frame, engine, etc. for perhaps a light sport utility vehicle and other variants within an automobile manufacture's line of vehicles. If this practice is so widely used in the automotive industry, why is it not utilized in naval ship design? In ship design, common hull blocks, main engines, engine rooms, ship service generators, sensors and weapons could be used to provide commonality and savings across multiple ship class variants. Savings can be obtained in training of personnel, logistical support, procurement, detailed design development, and construction costs. The use of optimally determined platforms may offer part of the solution for an affordable fleet in the future.

Traditionally naval ship design has been performed on a ship class by ship class basis. Ships are generally designed in order to maximize their mission effectiveness without consideration of the detailed design of other ships in the fleet; an exception being compatible communications and weapons control. As ship designs have progressed, ships are being developed with more systematic commonality in mind. Different ship classes will have certain systems that are common to each other. The motivation behind this is to improve interoperability and decrease costs associated with the design, development, construction and operation of the ships. Despite the use of commonality, the shipbuilding industry has yet to utilize platform design techniques as standard practice. This research develops a logical methodology to establish the optimal platforms within a family, or fleet, of ships.

The strategic design question is how many and which elements should be included in the platform definition to maximize savings without excessive degradation of the performance of the variants in the family. The use of commonality in design often comes with compromises in mission effectiveness of individual designs. A multicriterion design optimization decision results – how to maximize the savings through the use of the platform while also maximizing the performance of each of the variants within the family. In order to use platforms in ship design, one must develop a way to measure the effects of commonality decisions on each variant's performance. This research develops a system of measuring the change in performance associated with the use of a platform and comparing this to the resulting fleet-wide cost savings.

Previous Product Family Work

The development of rational, analytical methods for the definition of platforms has been the subject of a number of recent research efforts. Simpson provides an extensive survey on these efforts (Simpson 2004). Here, we only review some basic relevant work.

At the University of Michigan, the Optimal Design Laboratory of the Department of Mechanical Engineering has developed optimization-based methodologies for making platform decisions. Fellini et al. used information from the optimization of individual product variants to determine optimal platforms (Fellini et al. 2004). Further work by the same authors evaluated commonality decisions while controlling performance losses. Sharing of product components was based on designer-specified loss tolerances (Fellini et al. 2005). By combining two previous approaches, Fellini, Kokkolaras, and Papalambros developed a methodology that identifies an initial set of shared components for a platform and then evaluates that platform using performance loss standards (Fellini et al. 2006) under the tacit assumption that more platform content is better. Sensitivity measures were used to establish the platform content to consider without requiring detailed cost savings estimation.

At M.I.T.'s Engineering Design Laboratory, Gonzalez-Zugasti, Otto, and Baker used a general optimization formulation that balances the advantages of sharing components with the constraints of individual product variants to form an interactive, team-based negotiation model for designing a product family based on a common platform (Gonzalez-Zugasti et al. 2000). In addition, Gonzalez-Zugasti and Otto explored methodologies for designing families of products built onto modular platforms (Gonzalez-Zugasti and Otto 2000). These authors rely heavily on design team input throughout their product platform design methodology. In this methodology, the design team meets to decide which portions of the individual designs should be used as platforms. From this decision, the product variants are developed. The variant designs are optimized with regard to performance and cost constraints. The optimization is performed on the variants on a "one at a time" basis rather than optimizing them concurrently.

Simpson, Maier, and Mistree have also focused their attention on product family design (Simpson et al. 2001). Their primary focus was on scale-based product families derived from product platforms that can be derived from functional and manufacturing considerations. This methodology has been extended several times (Nayak et al. 2002, Messac et al. 2002, Simpson et al. 2005). Nayak, Chen, and Simpson developed robust design concepts to formulate a variation-based platform design methodology. Messac, Martinez, and Simpson used a physical programming approach to optimize designs. The authors considered performance and production, but did not explicitly mention cost savings considerations.

Fujita and Yoshida proposed a simultaneous optimization method for module combination and module attributes of multiple products. Their work optimized the combinatorial pattern of commonality and similarity, optimized similarities on scale-based variety, and optimized the continuous module attributes (Fujita and Yoshida 2004). Considerations were made for performance, cost and profit of the design variants based on a fixed modular architecture. Design trends in the optimization were used to help narrow the number of design variants.

Related Ship Design Optimization Work

In 1992, the U.S. Navy began an initiative titled "Affordability through Commonality" (Bosworth and Hough 1993, Cecere et al. 1995). The goal of this initiative was to lower the cost of fleet ownership through the use of increased commonality. The Navy defined commonality as using modularity, equipment standardization and process simplification. The authors argue that the use of commonality would ultimately lower all life cycle costs associated with design, construction and operation of the Navy's ships. Although the Navy is using this fleet ownership strategy, it does not appear that a formal methodology has been developed to aid cost-effective commonality decision-making.

Brown and Salcedo presented a ship design optimization methodology based on life cycle cost and mission effectiveness (Brown and Salcedo 2003). They developed a methodology for exploring the many variations that are possible in a given ship design. By using various combinations of combat systems, engine selections, hull form parameters, manning, endurance, and mobility, they efficiently explored the design space for non-dominated designs. The designs are compared using life cycle costs and a measure of mission effectiveness. Multiple-Objective Genetic Algorithm (MOGA) optimization was used to search the design space. Their work did not consider the design of a family of ship variants, but rather gives the methodology for the optimal design of one ship class.

Zalek, Parsons and Papalambros used a multicriterion evolutionary algorithm to search the design space for monohull forms optimized with respect to calm water powering and seakeeping (Zalek et al. 2006a, Zalek et al. 2006b, Zalek 2007). Their experience in developing a multicriterion evolutionary algorithm was a starting point for this research.

Multicriterion Optimization

In most real world applications, optimization problems involve multiple competing objectives. "When an optimization problem involves more than one objective function, the task of finding one or more optimum solutions is known as multicriterion optimization" (Deb 2001). The multicriterion or multi-objective optimization problem can be formulated as:

$$\begin{aligned}
&\text{minimize } \mathbf{F}(\mathbf{x}) = [f_1(\mathbf{x}), f_2(\mathbf{x}), f_3(\mathbf{x}), \ldots, f_K(\mathbf{x})]^T \\
&\qquad\qquad \mathbf{x} = [x_1, x_2, \ldots, x_n]^T \\
&\text{subject to } h_i(\mathbf{x}) = 0, \qquad i = 1, \ldots, I \\
&\qquad\qquad g_j(\mathbf{x}) \geq 0, \qquad j = 1, \ldots, J
\end{aligned} \tag{1}$$

where $f_1, f_2, f_3, \ldots, f_K$ represent the multiple objective functions and the vector $\mathbf{x}$ contains the design independent variables.

Some of the multiple objective functions usually conflict with one another. As one objective function is improved some of the other objective functions often suffer. Therefore, compromises must take place in order to reach acceptable levels of satisfaction among each of the objective functions. Because of these compromises, the results of a multicriterion optimization problem differ from that of a single criterion optimization problem. A single criterion optimization will have one globally optimal solution. A multicriterion solution will have many possible solutions. A given solution may be optimal for one of the objective functions, but not the others. No solution will be the best for all the objectives. The result of a multicriterion optimization is a Pareto optimal set.

The Pareto optimal set is composed of the non-dominated set of solutions to the problem. Each design on the Pareto front is such that no criterion can be improved without sacrificing another. An example of a convex two criterion optimization solution is shown Fig. 1.

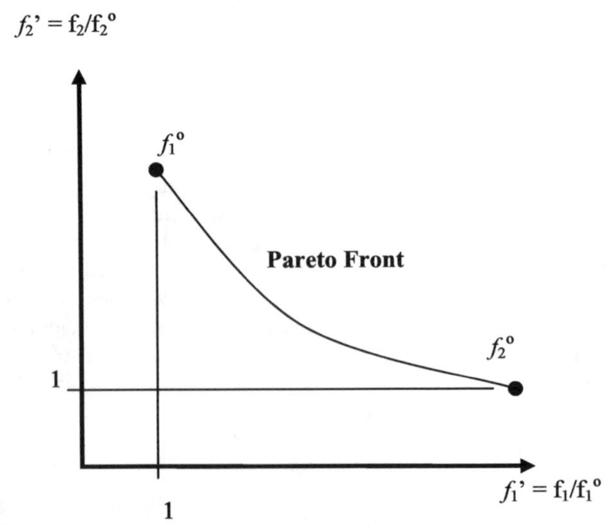

Figure 1. Multicriterion Optimization Solution

In Fig. 1, f_1 and f_2 represent the two objective functions to be minimized. The Pareto front is usually bounded by the optimum solutions for f_1 and f_2 obtained considering only their objective functions one at a time, f_1^o and f_2^o, respectively.

Problem Formulation

For the purposes of this research, the objective functions, $f_1, f_2, f_3, \ldots, f_K$, will be limited to three. The objective functions are:

$f_1(\mathbf{x_1}, \mathbf{x_C})$ = Ship A Mission Effectiveness/Average Ship Cost
$f_2(\mathbf{x_2}, \mathbf{x_C})$ = Ship B Mission Effectiveness/Average Ship Cost
$f_3(\mathbf{x_1}, \mathbf{x_2}, \mathbf{x_C})$ = Net Fleet Savings from the Commonality

where f_1 and f_2 are subject to the ship design constraints for their respective design as specified by naval architecture practice and the customer. The objective function f_3 is the total fleet savings realized through the use of the commonality. The design independent variable vector **x** is composed of the following:

$\mathbf{x_1}$ = Ship A design independent variables
$\mathbf{x_2}$ = Ship B design independent variables
$\mathbf{x_C}$ = Commonality components.

The performance or mission effectiveness of the two ship classes ($i = 1, 2$) is related to their specific missions ($j = 1,\ldots,n$). The ability of each ship i to successfully accomplish each mission j is assumed to depend upon k performance characteristics, y_k. The contribution of each of these performance characteristics k to the success in each mission j is characterized by a fuzzy membership function or fuzzy utility $0 \le U_{ijk}(y_k) \le 1$. The overall mission effectiveness or performance per average ship cost is then obtained by minimum correlation inference (Kosko 1992) as follows:

$$\left[\frac{Performance}{Cost}\right]_i = \sum_{j=1}^{n} \frac{MP_{ij} \min\left[U_{ijk}(y_k)\right]}{Cost_i} \tag{2}$$

where the MP_{ij} are the percent of time that vessel i is engaged in mission j and $Cost_i$ is the average acquisition cost of vessel i. The cost of the ship will be estimated using common ship construction cost estimating methods. The cost used in the objective functions will be the average cost of all the ships in the fleet.

The net fleet savings function $f_3(x)$ will take into consideration all fleet-wide costs directly associated with the use of common components. To demonstrate the methodology, these are limited to bulk purchase savings and construction learning curve savings. The savings function will consider all ships in each design class as contributing to the total fleet savings. The global effect of commonality on the cost of the entire fleet of ships involving the two classes A and B is used.

The commonality components will be comprised of a set of integers that specify which ship components will be common between both ship classes. If a given component is designated as common, both ships will be constrained to use that component. Each commonality component will have two or three component choices. By varying the number and combinations of the commonality components, the design space will be populated. Common components in the design optimization will consist of weapon systems, ship service generators, cruise engines, superstructure and midship section hull blocks. The various combinations of these commonality components will be used to determine which set of common components will result in Pareto optimal designs for Ship A and Ship B.

As the various combinations of commonality are applied to the designs, the optimization will begin to fill out the three objective Pareto front or Pareto surface. Figure 2 shows a schematic of the expected discrete Pareto front that will be obtained for this multicriterion optimization.

Every set of commonality components ℓ will result in a solution for Ship A_ℓ and Ship B_ℓ that will be located on a line of commonality. If a single ship were being considered for both missions, this line would be the two-objective Pareto Front for Ship A_ℓ performance/cost and Ship B_ℓ performance/cost. For specific commonalities, Ships A_1 and B_1 might share Ship A's midship section hull blocks, Ships A_2 and B_2 might share Ship A's midship section and cruise engines, and so on. As more things become common amongst the ships, the savings can increase and the ship designs will tend toward each other on the Pareto surface as more effectiveness is sacrificed for commonality. Once every item on the ship is determined to be common, the result will be one design for both missions. This design is shown as point C in Fig 2. Once every combination of common components is used in the optimization, the discrete Pareto front will be fully populated. The Pareto Front will not be continuous because of the discrete nature of the commonality variable. Rather, the Pareto front will be a collection of discrete points as shown in Fig. 2.

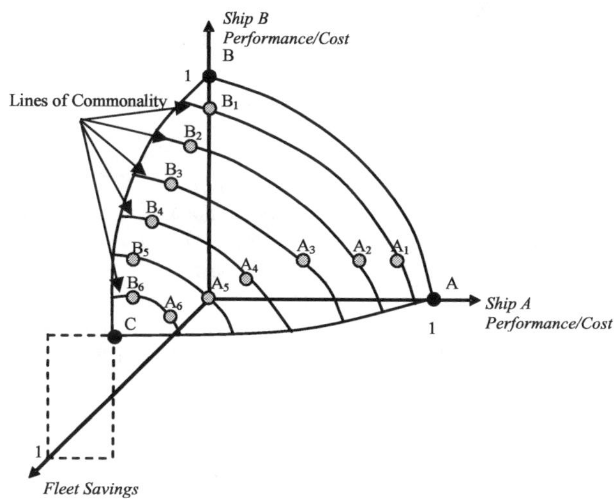

Figure 2. Expected Discrete Pareto Front

SHIP SYNTHESIS MODEL

For the purposes of this research, a ship synthesis model was needed. Since the goal of the optimization was to generate scores of ship variations with minimal input, the synthesis model had to be relatively simple in nature. A simple ship synthesis model would be adequate to provide the initial point design characteristics needed for the basic cost estimates and performance evaluations.

The ship synthesis model used was adapted from the Performance Based Cost Model used by the U.S. Coast Guard Engineering Logistic Center. The model was developed by the Naval Surface Warfare Center Carderock Division as a means to do comparative ship studies (Naval Surface Warfare Center 1998). The model is capable of synthesizing frigate-sized, deep-water, white-hull cutters and reporting acquisition as well as operational and support costs.

The model was developed using previously developed models for relevant ship types. The ship synthesis algorithms are based on a combination of SHOP 5 and ASSET algorithms. SHOP 5 is a Canadian developed model for monohull frigates and destroyers based on NATO frigates. ASSET is the Advanced Surface Ship Evaluation Tool used extensively within NAVSEA and represents a mixture of first principle algorithms as well as regression analyses of historical U.S. combatant ship data, including the U.S. Coast Guard's WMEC 270 class of ships (ASSET 2005). The model provides reasonable results for deep-water cutters with displacements of 1500 tonnes or greater. The Cost Estimating Relationships (CERs) for the basic construction costs were developed by SPAR Associates Inc. and are based on the U.S. Coast Guard's WHEC 378, WMEC 270, and WMEC 210 classes of ships. Additional CERs were adapted from the CERs that were developed for the U.S. Coast Guard's Great Lakes Icebreaker estimate. All costs are reported in constant Year 1998 U.S. dollars.

The inputs to the U.S. Coast Guard model are design and performance-based and allow the user to examine the effects of 21 independent variables. From the inputs, the program calculates ship dimensions, powering requirements, electrical load, auxiliary systems weight estimate, outfit weight estimate, variable loads, and habitability/personnel space volumes. This information is used to determine the weight of each Ship Weight Breakdown Structure (SWBS) group, lightship displacement, growth margins, ship loads, and full load displacement. A volumetric check is performed to ensure adequate space is allotted for necessary compartment volumes. Once the ship is balanced and has adequate volume, the program calculates the procurement costs for the lead ship and follow-on ships as well as operating and support costs for the life of the ships.

In order to make the ship synthesis model more suited to the purpose here, a number of changes were made. The original Microsoft Excel model was reprogrammed in C++. The adapted synthesis model was also changed to require fewer inputs than the Coast Guard model. Table 1 shows the independent variables used in the two models. The adapted synthesis model limited the number of independent variables to the eight listed. The remaining variables are made dependent on the eight input variables and follow logical ship design practice.

The ship synthesis was also extended to ensure that its outputs satisfied a few naval architecture constraints. These include a more refined weight-displacement check, a basic initial intact stability check, and a more robust volume check.

The iterative process within the synthesis starts with initial guesses for beam, draft, and superstructure volume. As the synthesis model conducts calculations for weights, the initial guesses are updated and the iterative process continues until the calculated displacements of successive iterations are within one half of a percent of each other. As the vessel weight changes, the buoyancy requirement is met by modifying the draft.

The beam of the ship is dependent on the required initial intact stability of the ship. The required GM_T is estimated using the parametric equations in Chapter 11 of *Ship Design and Construction* (Parsons 2003). The vertical center of buoyancy (KB) is calculated here using Wobig's equation, the waterplane transverse inertia coefficient, C_I, is estimated using the formula proposed by D'Arcangelo, a 3% free surface allowance on KG is utilized, and the required GM_T is set at 7% of the beam.

The hull volume check is performed at each design iteration. The required volume of each space is calculated and compared to the total volume of the ship. If the required volume is less than that of the actual ship's volume, the superstructure volume is decreased and vice versa. This calculation ensures that the ship is sized correctly and there is no extra volume on the ship.

Occasionally the design inputs, as a set, may be unrealistic and not able to produce a working design. There are a few safeguards that prevent these sets of inputs from being considered in future analysis. These safeguards are designed to stop calculations when it is found that the given inputs will not generate a practical design. Two occurrences have been found when unrealistic ship requirements are made. The first occurs when the required horsepower is greater than the available horsepower of the engines in the database. The engine database consists of over forty gas turbines and diesel engines. The engines range in horsepower from 447 to over 42,500 kW. If the required engine rating for maximum speed is greater than 42,500 kW, the synthesis stops and that design is aborted. Given the discrete nature of the engine sizes, the synthesis will at times iterate between two designs without converging to within one half of a percent of successive displacements. If a synthesis conducts 100 iterations without convergence, the calculations are stopped and the design is aborted. Through repeated runs, it was found that most designs converge in less than 10 iterations.

Once a ship's displacement converges, the ship's characteristics are fully defined. At this point the cost model is used to determine the costs associated with each ship in the class.

Table 1. Ship Synthesis Independent Variables

U. S. Coast Guard Synthesis Model	Adapted Synthesis Model
Power plant type	Power plant type
Prismatic coefficient	Midship section coefficient
Block coefficient	Block coefficient
Froude length constant (circle M)	Length
Beam/draft ratio	
Maximum speed or shaft horsepower	Maximum speed
Number of main engines	
Number of cruise engines	
Number of diesel-generator sets	
Total accommodations	
Range @ cruising speed	Range @ cruising speed
Cruising speed	
Endurance dry stores	
Endurance chilled stores	
Endurance frozen stores	
Endurance general stores	
Helicopter hangars (1=yes, 0=no)	Number of helicopter hangars
	Weapon system type
Combat system weight input	
Combat system variables loads input	
WG700 weapons weight	
Ratio of superstructure volume to hull volume	

MULTICRITERION EVOLUTIONARY OPTIMIZATION FORMULATION

Problem Formulation without Commonality

In this initial design problem without any commonality, the following multicriterion design optimization will be used.

- Maximize the mission performance of the OPC mission ship relative to the average ship cost for the entire fleet
- Maximize the mission performance of the NSC mission ship relative to the average ship cost for the entire fleet

The problem can then be formulated as

$$\text{maximize } \mathbf{F}(\mathbf{x}) = [f_1(\mathbf{x}_1), f_2(\mathbf{x}_2)]^T$$
$$\mathbf{x} = [\mathbf{x}_1, \mathbf{x}_2]^T = [x_1, x_2, \ldots, x_n]^T \quad (3)$$

with

$f_1(\mathbf{x}_1)$ = OPC Mission Ship Effectiveness / Average Ship Cost
$f_2(\mathbf{x}_2)$ = NSC Mission Ship Effectiveness / Average Ship Cost,

where f_1 and f_2 are subject to the ship design constraints for their respective designs as specified by naval architecture practice and the customer. For this application, the constraints have been included in the synthesis model and are, therefore, not explicitly stated in the problem formulation. As mentioned previously, the independent variables have been limited to eight as shown in Table 1.

Mission effectiveness measures how well a design meets the mission requirements for the ship. The effectiveness of each design is modeled using fuzzy utility functions. It is important to note that if a design exceeds its design requirements, it will not receive more credit and since the cost of exceeding requirements will be higher, the value of its objective function will decrease due to its over design for its intended mission.

Evolutionary Optimization Process

An evolutionary (real-coded Genetic Algorithm) optimization process (Goldberg 1989, Michalewicz 1996, Deb 2001, Osyczka 2002) was designed to provide the Pareto front with a diverse set of solutions. The basic optimization process used in this research is illustrated in Fig. 3. The details of each portion of this algorithm are described in the following subsections.

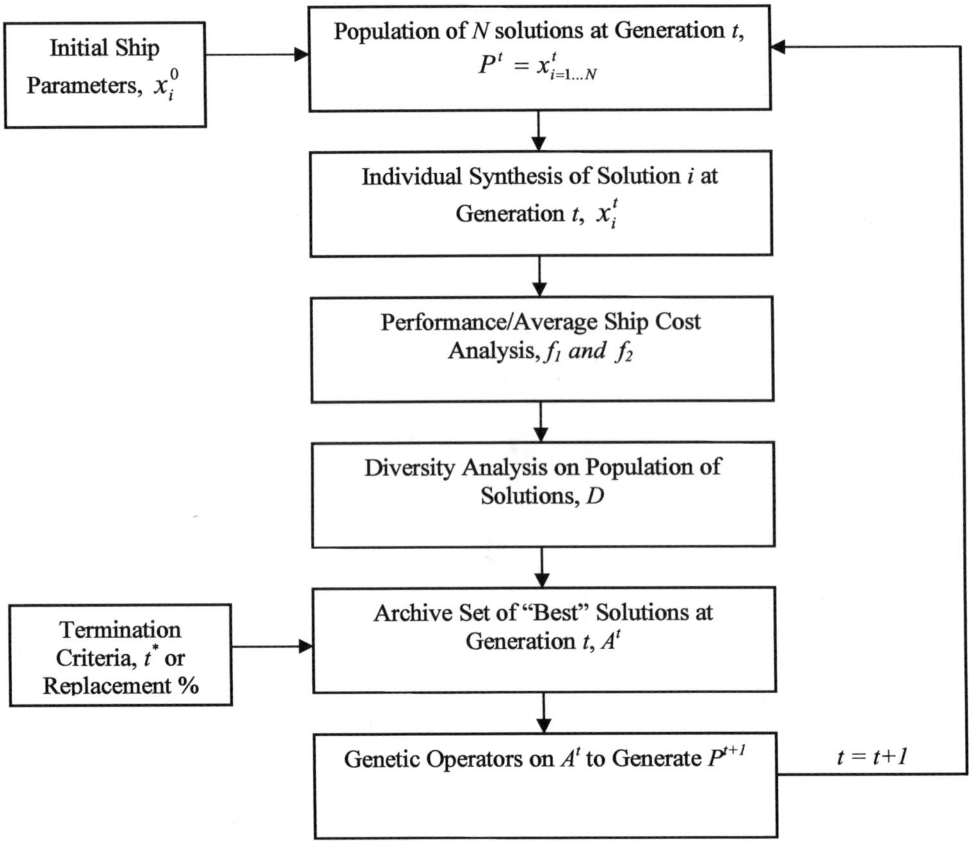

Figure 3. Basic Optimization Process

The optimization process is a multicriterion evolutionary algorithm based primarily on Zalek's work (Zalek et al. 2006a). His experience and lessons learned were very helpful in the initial formulation of this work. Zalek's algorithm was based primarily on Deb (Deb 2001) and Zitzler (Zitzler et al. 2003) with some original concepts that were developed for the specific nature of his work. Many of the procedures used are standard methodologies in evolutionary algorithms. However, Deb's influence can be seen in the nondominance sorting algorithm and in the tournament selection method used. The use of an archive as an elitism operator was taken from the Strength Pareto Evolutionary Algorithm (SPEA) work of Zitzler.

Initial Ship Parameters. The initial population of ships, P^0, consists of randomly generated combinations of the n independent variables, x_i^j. The random parameters are generated using the following standard equation:

$$x_i^j = x_{lower}^j + random(0,1) \cdot (x_{upper}^j - x_{lower}^j) \cdot \qquad (4)$$

Each set of parameters is input into the ship synthesis model and a ship is developed. Not all combinations of inputs will generate a feasible ship. If a ship is not feasible, a new set of parameters is developed and synthesized. The process continues until the minimum population of ships, N, has been created.

Population at Generation *t*. The population at any given generation t is set to have a minimum number of ship variants, N. There is no maximum on the number of solutions in the archive. This was done to ensure that nondominated solutions were not inadvertently left out of the solution set. By allowing the population to grow without a maximum, the user is able to search the variable space more efficiently and effectively. If an artificial criteria were used to limit the size of the population, as done by Zitzler, nondominated solutions may be lost and never recovered. In order to maintain elitism, the population at $t>0$ consists of the previous generation's archive and the offspring that are created in the current generation.

Individual Synthesis. Each set of solution parameters is input into the ship synthesis model during each generation.

Performance/Average Ship Cost. The goal of the optimization is to maximize the performance over average ship cost. The performance measurement of a given solution is calculated using fuzzy utility functions. The utility functions represent how effective ship i will be in performing its assigned missions j given its individual capabilities. Each mission has performance attributes y_k that contribute to the successful performance of the mission. The performance attributes are assigned a fuzzy utility $0 \leq U_{ijk}(y_k) \leq 1$ that represents what percent of a given mission the ship can perform with that performance attribute y_k. Fuzzy minimum correlation inference is used to assign an overall effectiveness for each mission (Kosko 1992). The percent of time that each ship i spends performing mission j (MP_{ij}) is multiplied by the minimum utility for that mission. The values used for the MP were obtained from the projected operational profiles of the OPC and NSC (USCG Internal). The j mission utilities are then summed to yield the overall mission performance and this is divided by the average cost of each ship.

$$\left[\frac{Performance}{Cost}\right]_i = \sum_{j=1}^{n} \frac{MP_{ij} \min\left[U_{ijk}(y_k)\right]}{Cost_i} \tag{5}$$

A ship that has the capability to perform all of its missions well will have a high measure of performance approaching one. However, the ability to perform these missions comes at a price. The more capable a ship is, the higher the cost of the ship. By dividing the performance by cost, the algorithm penalizes ships for being overly capable. There is a fine balance between capabilities and cost. Overly capable ships will prove costly and the performance over average ship cost will suffer as a result. On the other hand, inexpensive ships will not be as mission capable and they too will have poor performance over cost.

Diversity. In order to ensure that a wide range of solutions is generated along the entire Pareto front, diversity calculations are performed. The diversity calculations measure a given solution's distance from its nearest neighbor solutions.

There are three common methods of ensuring solution diversity. They are the kernel method, nearest neighbor method, and histogram methods as described by Zitzler (Zitzler et al. 2003). Kernel methods take the distances between solutions as an argument. The distances (d_i) are calculated from each individual solution to all the other solutions. Then they are put through a kernel function, K, and summed. The sums of the kernel functions, $K(d_i)$, represent the solution density. The nearest neighbor technique used here takes the distance of a given point to its k nearest neighbors in order to estimate the solution density. The histogram method uses a grid to define the density of solutions. The density is determined by calculating the number of solutions in a given box of the grid.

Using a method similar to that utilized by Zalek (2007), the distance to the nearest three neighbors is calculated. The diversity operator $D(x_i)$ can be calculated in either objective function space or independent variable space. Depending on the nature of the optimization there may be a difference in the performance of these two methods. Distances, d_{ij}, between solutions in n-dimensional independent variable space are calculated using,

$$d_{ij} = \sqrt{(x_i^1 - x_j^1)^2 + \ldots + (x_i^n - x_j^n)^2}\,. \tag{6}$$

The average of the three closest solutions, D_{RAW}, is calculated and then normalized by dividing by the maximum value of D_{RAW} for all solutions in the population.

$$D_{RAW}(x_i) = (d_i^1 + d_i^2 + d_i^3)/3 \tag{7}$$

$$D(x_i) = D_{RAW}(x_i) / \max\{D_{RAW}(x_1), \ldots, D_{RAW}(x_N)\} \tag{8}$$

Using the three nearest neighbors ensures that pairs of solutions that are each other's neighbors will not be penalized for having a single close neighbor. By only using the three nearest neighbors a localized diversity is calculated.

Identical solutions are not considered in the calculation of diversity. By eliminating duplicate solutions, the diversity of the population is more easily maintained. Duplicate solutions run the risk of dominating the genetic processes and creating additional duplicates.

Archiving. The archiving of best solutions serves three important purposes. First, it creates the pool of potential parents for tournament selection and the evolutionary generation of offspring. Second, it allows for the measurement of how much the Pareto front is progressing from one generation to the next. Finally, it serves as the elitism operator for the algorithm.

The archive is developed using standard dominance sorting techniques. Each solution in the population is compared to every other solution to check for dominance. Dominance occurs when each of the objective criterion values of the solution are no worse than those of the other solutions and at least one objective criterion is better than that of the other solutions. When this occurs a solution is said to dominate the other solutions. The set of solutions that are not dominated by any other solutions are called the nondominated set of solutions.

The archive is generally made up of the nondominated set of solutions. However, during the early generations it may consist of lower ranked solutions in order to reach a minimum number of solutions. As the solutions become more refined, the number of nondominated solutions increases and eliminates the need to carry lower ranked solutions in the archive. There is no maximum to the size of the archive. If the number of solutions on the Pareto front were limited in size, the fine details of the Pareto front might not become apparent. These might include important knuckles or gaps in the front.

Termination Criteria. The optimization process has two conditions under which it will terminate. The user can set a maximum number of generations t^*. When the program completes t^* generations, it will stop and output a final archive, A^*. By selecting a maximum number of generations, the user prevents the optimization from running indefinitely. A second stopping condition exists when the archive becomes stagnant. The solutions in the archive carry markers which indicate if they have been carried over from previous generations or if they are newly generated offspring. If 99% of the solutions from one archive to the next are the same, the program stops and outputs the final archive. This saves computation time and only allows the optimization to continue as long as it is creating new solutions along the Pareto front.

Genetic Operators. The archived solutions in A make up the potential parent solutions in the mating pool. Once the archive has been created, those solutions are compared in a tournament selection process (Michalewicz 1996, Li and Parsons 1998). In tournament selection, archived solutions are randomly paired together. Each pair of solutions is compared using two tests. The first test asks if either solution dominates the other. If one solution dominates the other, the dominant solution is placed in the mating pool. If neither solution dominates the other, the objective functions are added together with the solution's diversity value. The sum of these values is compared to the sum of the values for the other solution in the pair and the solution with the higher sum is selected for the mating pool. In earlier generations, the first test is more likely to determine which solution in a pair will become a parent. As the archive is filled entirely with nondominated solutions, the second test tends to distinguish between the two solutions. In essence, early in the optimization process, the goal is to generate more nondominated solutions. Later in the process the goal shifts to creating more diverse solutions.

The optimization process creates a minimum of k child solutions per generation. As the archive grows beyond the minimum, more child solutions are developed in proportion to the size of the archive. For every child that is created two parents are needed. The crossover operator is used on the parents to create an offspring. Arithmetic crossover (Michalewicz 1996) is applied to the real number variables using

$$x^j_{child} = \alpha \cdot x^j_{parent1} + (1-\alpha) \cdot x^j_{parent2} \cdot \tag{9}$$

The weighted parent blending factor α is randomly selected between 0 and 1. This serves as a means to weigh which parent more heavily influences the characteristics of the offspring. For discrete variables, α is used again. In this case if α is less than 0.50, the value of the first parent is chosen; if α is greater than or equal to 0.50, the value of the second parent is chosen.

The optimization process also utilizes a mutation operator that allows for a more global search of the variable space. There are two parts to the mutation operator. The first part is the mutation rate R_{mut} and the second is the mutation magnitude M_{mut}:

$$R_{mut}(t) = 0.15 \cdot e^{1.2 \cdot (t/t^*)} \tag{10}$$

$$M^j_{mut}(t) = (2 \cdot random(0,1) - 1) \cdot \frac{1}{2 \cdot e^{(t/t^*)}} \cdot (x^j_{upper} - x^j_{lower}) \cdot \tag{11}$$

The mutation is set up similar to Zalek's work in that the rate of mutation starts out small with a large mutation magnitude in the earlier generations. This allows for a broad search of the variable space. As the generations progress toward the termination condition, the rate of mutation is increased exponentially, while the mutation magnitude is decreased exponentially. This allows for a more local search of new design solutions. By searching closer to existing solutions, the optimization process looks to fill in the gaps between existing solutions creating a more refined Pareto front.

The mutation operation is randomly applied to the entire set of child solutions with replacement. Thus, some solutions may be mutated multiple times, while others may not be mutated at all. The mutation of real variables is performed using

$$(x^j_i)^{t+} = (x^j_i)^t + M^j_{mut}(t) \text{ if } x^j_{lower} \le (x^j_i)^{t+} \le x^j_{upper} \cdot \tag{12}$$

If the mutated value falls outside of the selected limits for that variable, it is discarded and another mutation is attempted. Failed mutations do not count toward the total number of mutations

$$Number\ of\ Mutations = R_{mut} \cdot n \cdot k \tag{13}$$

Case Study

A two-objective case study was performed to test the methodology incorporated into the optimization program. The study utilized the mission requirements for the U.S. Coast Guard's Deepwater Fleet. Specifically, the Operational Requirements of the Maritime Security Cutter Large (WMSL), formerly known as the National Security Cutter (NSC), and the Maritime Security Cutter Medium (WMSM), formerly known as the Offshore Patrol Cutter (OPC), were used to create a multicriterion optimization problem. (The first NSC was actually launched in September 2007 and the OPC is currently being designed.) The mission requirements for these two real classes of ships were used extensively in this research to examine the validity of the optimization methodology. Table 2 shows the actual design characteristics of both ships (USCG Website 2006).

The goal of the study was to find the Pareto optimal set of solutions that would satisfy the mission needs of both ships. In other words, if only one class of ships that maximized its ability to satisfy both ships' missions were to be built, what would the Pareto front look like?

For the purposes of this case study, the design parameters for the eight independent variables were set to approximately +/- 10% of the actual ship design characteristics (USCG Internal). The design ranges can be seen in Table 3. The power plant type inputs 1 or 2 represent a four (two cruise, two sprint) diesel engine (CODAD) plant or a two cruise diesel engine and one sprint gas turbine (CODOG) plant, respectively. The propulsion configuration is limited to twin screw, geared mechanical with controllable pitch propellers. The weapons system type inputs 1, 2, or 3 represent a 46 mm gun, a 57 mm gun, or both a 57 mm gun and Phalanx Close In Weapon System (CIWS), respectively.

Table 2. Comparison of U.S. Coast Guard Deepwater Vessel Characteristics

Characteristics	NSC	OPC
Number of cutters	8	25
Length overall	127.4 m (418')	estimate 106.7 m (350')
Maximum beam	16.46 m (54')	estimate 15.54 m (51')
Navigational draft	6.4 m (21')	estimate 6.4 m (21')
Displacement	4368.3 t (4300 LT)	est. 3047.6 t (3000 LT)
Sprint speed	28 kts	26.5 kts
Sprint speed range	2,600 nm	1,550 nm
Sprint speed endurance	3.91 days (94 hrs)	2.5 days (60 hrs)
Economical speed	8 kts	9 kts
Economical speed range	12,000 nm	9,000 nm
Endurance	60 days	45 days
Propulsion plant	2 Diesels, 1 Gas Turbine	4 Main Diesel Engines
Bow thruster	Yes	Yes
Gun for weapon system	57mm Gun	57mm Gun
Gunfire control	Mk-160/Mk 46/SPQ-9B	Mk-160/Mk 46/SPQ-9B
Operating days away from port	230	230
Mission days/year	200-220	200-220
Berthing capacity limit	148	106
Number of helicopter hangars	2	2

Table 3. Independent Variable Ranges

Independent Variables	Variable Ranges
Power plant type	1 or 2
Midship coefficient	0.75-0.99
Block coefficient	0.45-0.85
Length	82.3-143.3 m (270'-470')
Maximum speed	19-31 knots
Range @ cruising speed	8000-14000 nm
Number of helicopter hangars	1 or 2
Weapons system type	1, 2, or 3

Fuzzy Utility Functions

Each ship has four primary missions. The OPC and NSC each perform the National Defense, drug interdiction and living marine resources (LMR) missions. The OPC also conducts alien migration interdiction operations (AMIO) while the NSC performs general defense operations (GDO) (U.S. Coast Guard Memorandum 1995, U.S. Department of Transportation Memorandum 1996). For each mission, four ship attributes were selected to describe each ship's ability to perform these missions. The four attributes were maximum speed, number of helicopter hangars, weapon systems and endurance range. It is important to note that the fuzzy utility functions found in ensuing sections do not necessarily reflect the opinions of the U.S. Coast Guard. They were established by the authors for academic purposes only.

National Defense

The Coast Guard is "a military service and a branch of the armed forces of the United States at all times" as established in the United States Code (USC) (14 USC 1). It is required to "maintain a state of readiness to function as a specialized service in the Navy in time of war" (14 USC 2) and to operate as a service to the Navy when directed by the President (14 USC 3). It is also authorized to assist the Department of Defense in performance of any activity that the Coast Guard is qualified (14 USC 141, 145). In 1994, the Coast Guard's defense missions were more clearly defined to include Maritime Interception Operations and Deployed Port Operations, Security and Defense. The National Defense mission is accomplished using surveillance, detection, interception and sustained presence. These abilities are accomplished with a combination of maximum speed, aerial asset capabilities, weapon systems and range (U.S. Coast Guard Memorandum 1995, U.S. Department of Transportation Memorandum 1996). Figure 4 shows the assigned fuzzy utility for each of the ship attributes for the National Defense missions.

The utilities shown in Fig. 4 show the relative importance of each attribute to the accomplishment of the National Defense mission for each vessel. From these graphs, it can be seen that achieving the required maximum speed for each ship is important although some variance is accepted. At least one helicopter hangar is absolutely necessary to perform the mission. Weapon System 1 fulfills the OPC's role in the mission, but only satisfies a portion of the NSC's role. In order to maintain a sustained presence for the National Defense mission both ships must meet their minimum required ranges.

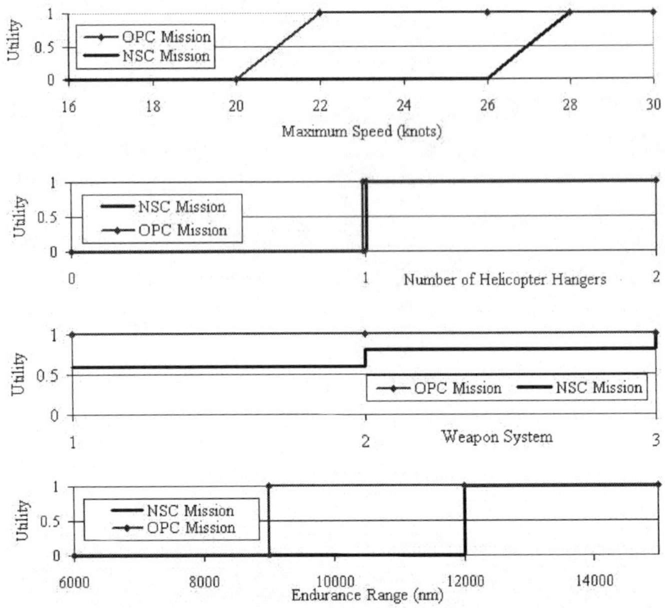

Figure 4. Maximum Speed, Helicopter Hangar, Weapon System, and Endurance Range Utilities for the National Defense Mission

Drug Interdiction

The Coast Guard is the lead agency for maritime drug interdiction. Drug interdiction operations rely heavily on the ability to detect, intercept and board vessels for compliance with U.S. and International law. These boardings are essential in deterring and interdicting drug shipments at sea. By maintaining a presence on the high seas, smugglers are required to develop new, more costly methods in order to continue the illegal transport of drugs (U.S. Coast Guard Memorandum 1995, U.S. Department of Transportation Memorandum 1996). Figure 5 shows the fuzzy utility functions for each of the four ship attributes being used to assess each ship's performance in the Drug Interdiction mission.

The utilities shown in Figure 5 shows the importance of each ship attribute in accomplishing the Drug Interdiction Mission. For this mission, the maximum speed of the vessel is not as important as for the National Defense mission as shown by the more gradual slope for the NSC. The number of helicopter hangars is more important for this mission. In order to conduct a wide range of surveillance on the high seas, the fast aerial assets are used extensively. Weapons are generally not as important to this mission and only act as a means of intimidation if close contact is made with the smugglers. Because a majority of the Drug Interdiction mission occurs in the Caribbean, land is always nearby. Therefore, ships have the ability to refuel more often which makes the need for a long range less important.

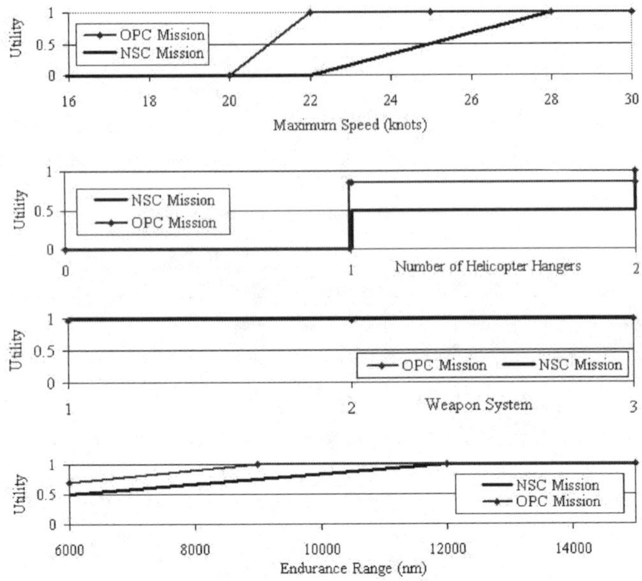

Figure 5. Maximum Speed, Helicopter Hangar, Weapon System, and Endurance Range Utilities for the Drug Interdiction Mission

Living Marine Resources

The U.S. Coast Guard provides law enforcement support to the conservation and management of our Nation's living marine resources. These resources are vital to the fragile ecosystem as well as the $67 billion that they annually contribute to the U.S. economy through commercial and recreational fisheries. In order to protect these resources, the Coast Guard must maintain a presence in the U.S. Exclusive Economic Zone (EEZ) and its borders. In order to deter illegal or unauthorized activities harmful to this maritime resource, the Coast Guard must detect and intercept vessels for boarding and inspections (U.S. Coast Guard Memorandum 1995, U.S. Department of Transportation Memorandum 1996). Figure 6 shows the need for maximum speed, aerial capabilities, weapon systems and range in the enforcement of fisheries laws.

The fuzzy utility values for the Living Marine Resource mission are very much the same as for the Drug Interdiction mission since the missions are very similar. The only difference is in the importance of having a second helicopter hangar. This attribute is not quite as important for the Living Marine Resource mission as the utility shows.

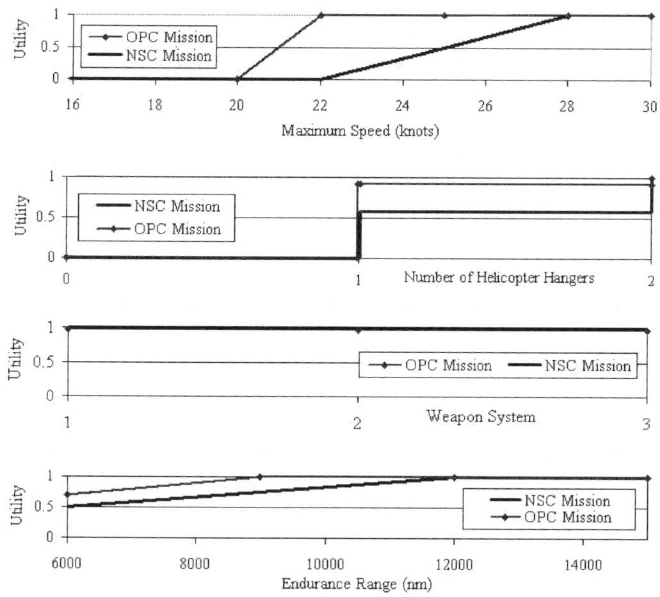

Figure 6. Maximum Speed, Helicopter Hangar, Weapon System, and Endurance Range Utilities for the LMR Mission

Alien Migration Interdiction Operations

The OPC is also tasked with the Alien Migration Interdiction operations (AMIO) mission. This mission is conducted in order to enforce U.S. migration laws. The need for this mission has grown significantly in the past 20 years. Basically, it is designed to deter the illegal flow of migrants into the U.S. In order to accomplish this mission, the Coast Guard must maintain a presence in areas where migration is likely. In addition to maintaining a presence, the ability to detect vessels carrying migrants and intercept them is crucial to the success of this mission (U.S. Coast Guard Memorandum 1995, U.S. Department of Transportation Memorandum 1996). Figure 7 shows the utilities associated with OPC attributes for the Alien Migration Interdiction mission.

General Defense Operations

The general defense operations (GDO) mission is a broad category. In addition to the National Defense mission described earlier, possible general defense missions might include surveillance, forward presence, amphibious ready group escort, sealift protection, sea lines of communication control, noncombatant evacuation, naval special forces warfare, combat operations, anti-terrorism and disaster relief. These operations are normally in support of the Navy (U.S. Coast Guard Memorandum 1995, U.S. Department of Transportation Memorandum 1996). Figure 7 shows the utilities associated with NSC attributes for the General Defense mission.

MULTICRITERION OPTIMIZATION RESULTS WITHOUT COMMONALITY

The purpose of this section is to analyze the multicriterion optimization problem of designing vessels to meet both the NSC and OPC missions in a single vessel class design. This initial investigation provided an opportunity to test the synthesis model and the optimization algorithm, develop an understanding of the tradeoffs in vessels designed for these two missions, and revealed candidates for commonality between classes designed for each of the individual missions.

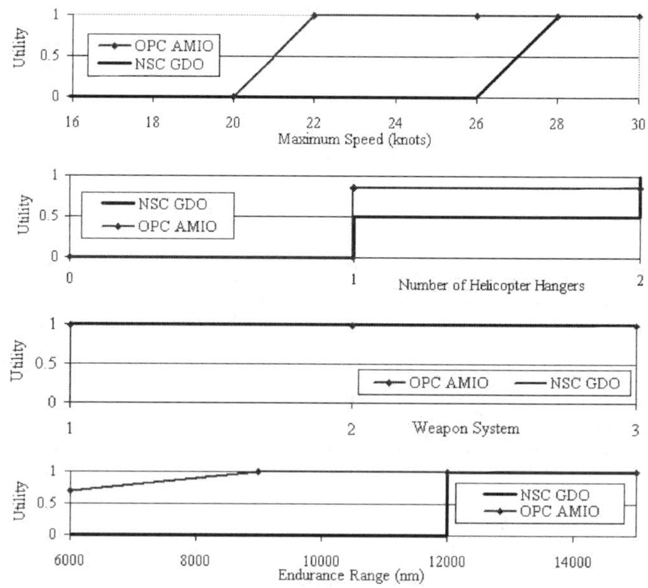

Figure 7. Maximum Speed, Helicopter Hangar, Weapon System, and Endurance Range Utilities for the AMIO/GDO Missions

The optimization was run with the following minimum settings: archive size-50, population size-150, and number of offspring per generation-100. The maximum number of generations was set at 200 and a termination condition of <1% new solutions in the archive was used.

Population History

A population history was generated for this baseline solution. At generation 108, the termination condition was met and the program stopped. Figure 8 shows the results of the baseline run in object function space. The populations of solutions at generations 0, 54, and 108 are shown. Population 0 shows the results of the random selection of input variables. The ships are spread throughout the objective function space. This illustrates the effectiveness of the random input generation function used in the optimization process. This initial population generally does not have high values in either objective function. Many of the solutions have zero NSC effectiveness.

Population 54, midway through the optimization, shows a progression of the designs toward higher objective function values. The initial formation of a Pareto front can be seen as more nondominated solutions have been created. The crossover and mutation operations appear to be effective in searching the design space for possible solutions. Mutation rates are increasing and the mutation magnitudes are decreasing. This allows for a more localized search of the design space and the start of a more clearly defined Pareto front. Population 108, the final set of solutions, shows a very distinct formation of a Pareto front. The Pareto front is densely populated as the search for new solutions has become more localized through the use of the genetic and diversity operators. The solutions cover a broad range of inputs which is a strong indication that the diversity operator has successfully served its purpose.

Overall, the optimization run demonstrated the behaviors that were expected. The run took approximately 4 minutes to complete on a 1.73 GHz PC running a compiled C++ code.

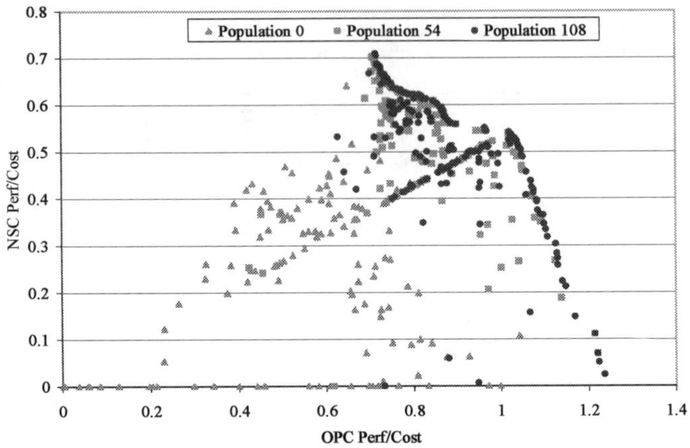

Figure 8. Population History of a Baseline Optimization Run

One interesting trend that appears in Fig. 8 is the formation of a diagonal line roughly perpendicular to the middle of the Pareto front. Upon examination, this line is revealed to be the transition from designs with one helicopter hangar to designs with two hangars. All of the designs on that line and to its right have one hangar. Designs to the left of the line have two hangars.

Baseline Run Characteristics

By only plotting the nondominated solutions from the baseline run, some more interesting characteristics can be seen. Figure 9 shows the Pareto front of the baseline optimization run. Point A shows the best NSC mission design and point D shows the best OPC mission design. The solutions found between point A and D make up compromise solutions. Two interesting areas appear on the Pareto front. Region B shows a gap in the Pareto front. This gap consistently appears in all runs. This gap is caused by the change from two helicopters to one and the resulting reduction in beam possible with the removal of one of the parallel hangar bays. This transition will be described in more detail in later sections. A second knuckle in the Pareto front can be seen at point C. This knuckle can be subtle like in Fig. 9 or more distinct as found in other runs. The cause of this knuckle is in large part due to ship service generator selection. The synthesis model uses data for real generators to outfit the ships. The knuckle seen at point C is the transition from one generator model to another. The weight and size of the generators at this transition change by a relatively large amount. This translates to a change in cost and eventually leads to the knee in the Pareto front. By populating the ship service generator database with a greater variety of generators, this knuckle could possibly be eliminated. However, the existence of this knuckle shows the methodology's sensitivity to the discrete data supplied.

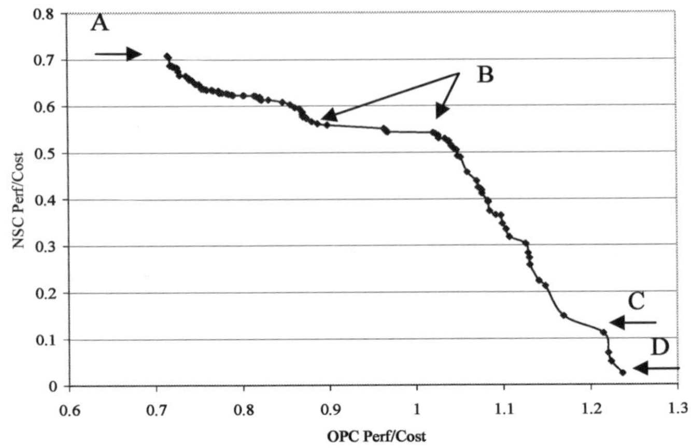

Figure 9. Baseline Optimization Run Pareto Front

Table 4 shows the design characteristics of the ship designs at points A, B, C, and D in Fig. 9. The following abbreviations are used:

W = Weapons system number
H = Number of helicopter hangars
C = Cruise engine item number
G = Diesel generator item number.

The cruise engine and ship service diesel generator numbers designate which item they are in the respective database. The engine sizes increases with higher values. In general, the larger, faster, more capable ships are at the best for the NSC mission left end of the Pareto front and the smaller, slower, less capable ships are at the best for the OPC mission right end of the front, as expected.

Numerous runs were performed to ensure that the optimization algorithm produced repeatable results. Due to the stochastic nature of the evolutionary optimization algorithm, individual runs are not identical to each other. However, the trends displayed on the Pareto front are very repeatable as seen in Fig. 10. Here four individual runs were performed and the plots of the final Pareto fronts are displayed on the same graph. The upper left hand portion of the graph shows a very strong similarity among the four runs all the way down to approximately 1.1 on the OPC performance/cost axis. At this point, the data points become a bit more sparce and the different runs show some minor differences.

Table 4. Design Characteristics for Selected Ships on Baseline Run Pareto Front

Point	L m (ft)	B m (ft)	V_{max} kts	KW_{max} (SHP)	V_{cruise} kts	KW_{cruise} (SHP)	Range nm	W	H	C	G	OPC Perf	NSC Perf	Cost $mil	OPC/ Cost	NSC/ Cost
A	122.2 (401)	16.46 (54)	28.0	17041 (22852)	18.0	3613 (4845)	12035	3	2	9	3	100.0	99.1	139.9	0.715	0.709
B (left)	107.3 (352)	16.46 (54)	26.3	12118 (16251)	18.0	3207 (4301)	9076	1	2	9	3	100.0	62.2	111.4	0.898	0.559
B (right)	103.6 (340)	12.80 (42)	25.4	8362 (11213)	18.0	2505 (3359)	9082	1	1	7	1	89.7	47.6	87.9	1.020	0.542
C (left)	92.7 (304)	12.50 (41)	22.8	6094 (8172)	18.0	2639 (3539)	9097	1	1	7	1	89.7	11.4	76.7	1.169	0.148
C (right)	96.0 (315)	12.50 (41)	22.6	5689 (7629)	18.0	2574 (3452)	9162	1	1	7	0	89.7	8.2	73.8	1.215	0.111
D	98.5 (323)	12.80 (42)	22.1	4925 (6605)	18.0	2411 (3233)	9142	1	1	7	0	89.7	1.8	72.5	1.236	0.024

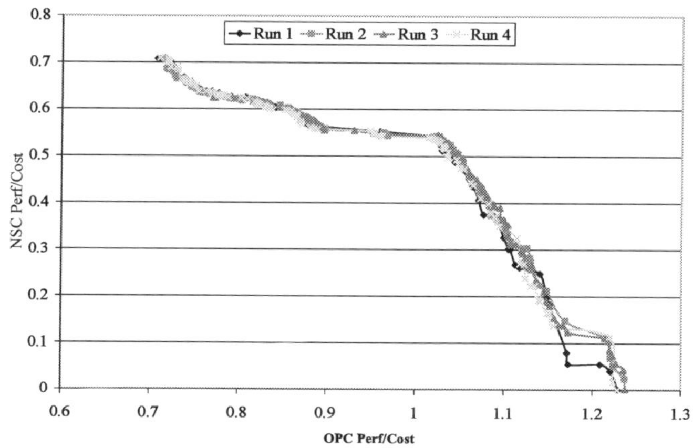

Figure 10. Repeatability Study using Baseline Optimization Conditions

There is a tendency in Fig. 10 for more designs to appear in the upper left portion of the Pareto front. The reason for this is in the nature of the optimization. Basically, a ship that is over designed will meet its mission requirements with a penalty for excessive cost. Ships that are good at performing the NSC mission will also serve the OPC mission very well. On the other hand, ships that serve the OPC mission efficiently do not generally perform well in the NSC mission. As a result, the ships at the upper left portion of the curve tend to do better in the tournament selection than the lower right ships. The diversity operator helps even this out some, but not without some imbalance in the results. Overall, there is a strong correlation between the four individual runs.

In order to more fully understand the results, a series of plots were created that show the effects of the independent variables on the shape and regions of the Pareto front. Figures 11 through 15 show the trends of the independent variables length, maximum speed, endurance range, weapons systems, and number of helicopters in the optimization results.

The results of these plots are intertwined through the overall ship designs. For example, the gap in the middle of the Pareto front is the transition from one hangar designs to two hangar designs. When a second hangar is added, the required beam necessary to accommodate the two hangars is 16.46 m (54'). Ships with one hangar tend to have a beam of about 12.80 m (42'). This large increase in beam requires the ships to increase in length in order to meet the speed requirements. In addition to the second hangar, the gap also shows an increase in maximum speed from one side to the other. Again, the desire to increase speed necessitates some increase in length. The gap represents an increase in length of about 3 m (10') as seen on the length study curve. The range requirement is higher for the NSC than it is for the OPC, therefore, the range increases from right to left on the Pareto front. The selection of weapon systems shows an interesting trend. Only ships that are near the optimal design for the NSC have the most advanced weapons system. Near that end point, the weapon system becomes the controlling factor in the overall utility and, therefore, the selection of the more advanced system becomes optimum.

The makeup of the set of solutions for the baseline condition is very broad. About half of the solutions have one hangar. Weapons system 3 was selected on only 12% of the ships with weapons system 1 on the rest. The lengths ranged from 89.6 m to 125.6 m (294'-412'), speeds from 22-28 knots and endurance ranges from 8000-12000 nm.

The distribution of length along the Pareto front generally increases from right to left. However, the length does not increase monotonically as can be seen in Fig. 11. The performance values and cost of the solutions are made up of many characteristics. Some of these characteristics complement each other nicely while others do not. This results in a fine balance of factors each having their own affect on the performance values and cost of the solutions.

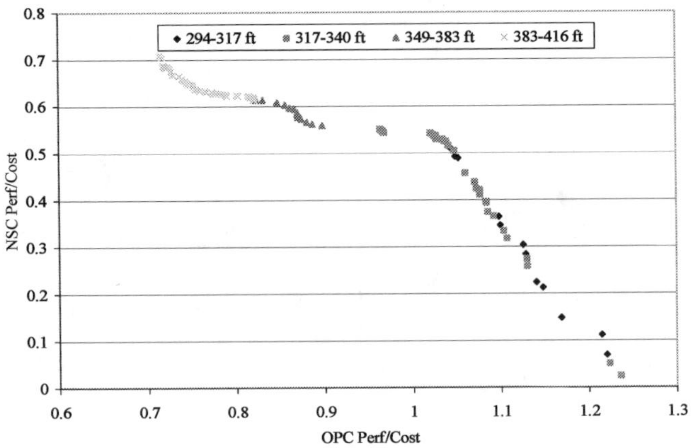

Figure 11. Detailed Study of Baseline Optimization Results for Ship Length

Figure 12 shows that the distribution of maximum speed along the Pareto front is monotonic except for the solutions marked with an A. These ships have slightly smaller speeds than the solution to the immediate right. More explanation for this will be found in the discussion of Fig. 13 which follows.

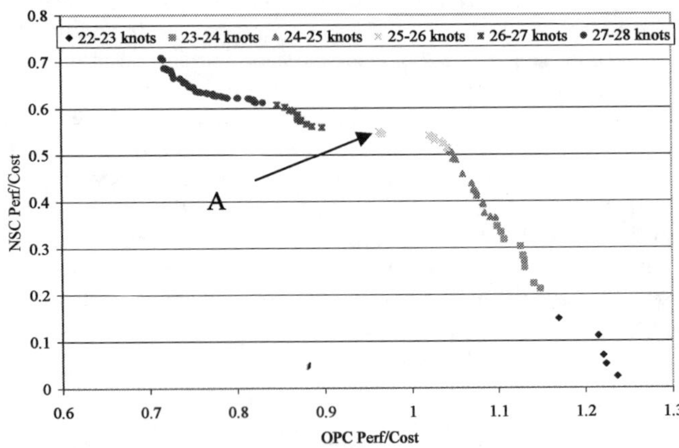

Figure 12. Detailed Study of Baseline Optimization Results for Maximum Ship Speed

Although not monotonic, the range generally increases from right to left along the Pareto front. Figure 13 shows a small cluster of solutions that do not follow the expected trend. The ships labeled with an A are the only solutions along the Pareto front with ranges less than 9000 nm. As mentioned previously, they also have a lower maximum speed than the ship to their immediate right. As a result, they will have lower OPC performance and NSC performance values than the ship to their right. However, since they have smaller ranges and maximum speeds, they will cost slightly less which results in higher performance over cost values than the ship to their right. The solutions that make up cluster A are not influenced solely by their small ranges. Their performances are also driven by maximum speed and number of hangars. Again, the fine balance of characteristics allowed these distinctly different solutions to occur.

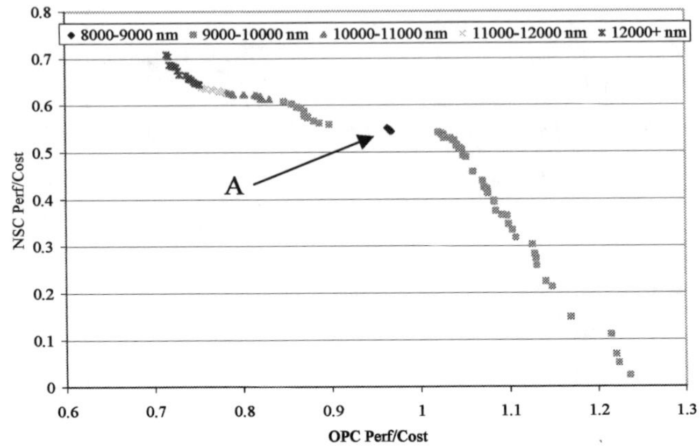

Figure 13. Detailed Study of Baseline Optimization Results for Ship's Endurance Range

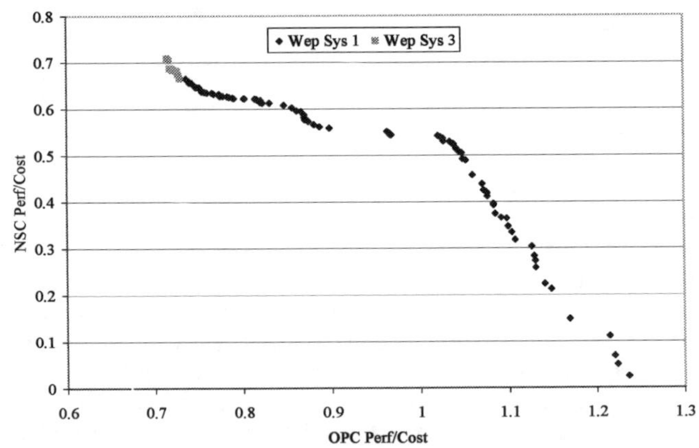

Figure 14. Detailed Study of Baseline Optimization Results for Ship's Weapon System

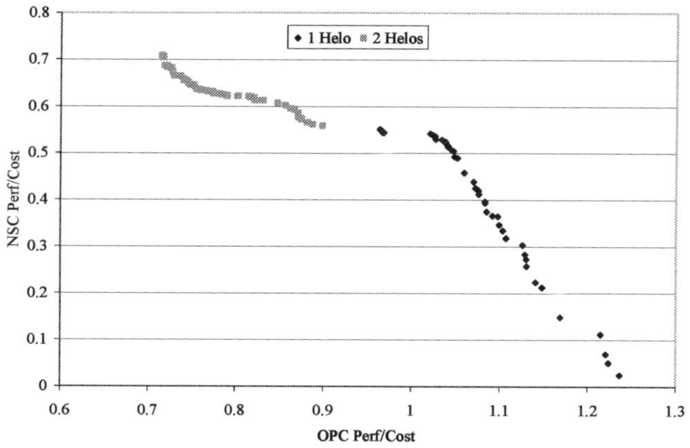

Figure 15. Detailed Study of Baseline Optimization Results for Number of Helicopter Hangars

Commonality Options

Further evaluation of the baseline optimization results showed some important trends relative to possible commonality choices. Figure 16 shows a breakdown as to which areas of the Pareto front had certain components as common. The different areas show the different combinations of weapon systems, ship service generators and cruise engines. All baseline Pareto front ships were made up of one of the four possible combinations of these three components.

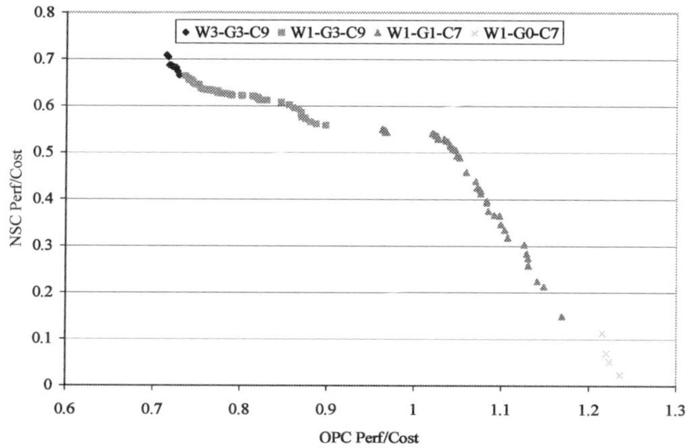

Figure 16. Natural Commonality within Pareto Front Solutions

The baseline results revealed several key factors to guide the determination of what components should be considered for common use. For components such as ship service generators and cruise engines only a few different choices were optimum from their respective databases. In addition, the superstructure volume produced by the synthesis model closely correlated with the number of hangars. Once the number of hangars was determined, the superstructures had little volume variation. Thus, the superstructures could be easily made common as either the smaller (one hangar) size or the larger (two hangar) size. A similar trend was noticed with hull depth and beam ranges. Solutions tended to have two sets of depth and beam depending on the number of hangars present. Thus, midship section hull blocks became another logical choice to make common.

MULTICRITERION OPTIMIZATION WITH COMMONALITY

Problem Formulation with Commonality

The problem formulation with commonality is similar to the formulation seen previously. The only difference is that a third fleet commonality savings objective, also to be maximized, is added:

$$\text{maximize } \mathbf{F(x)} = [f_1(\mathbf{x}_1, \mathbf{x}_C), f_2(\mathbf{x}_2, \mathbf{x}_C), f_3(\mathbf{x})]^T$$
$$\mathbf{x} = [\mathbf{x}_1, \mathbf{x}_2, \mathbf{x}_C]^T = [x_1, x_2, \ldots, x_n]^T \quad (14)$$

The objective functions are:

$f_1(\mathbf{x}_1, \mathbf{x}_C)$ = OPC Mission Ship Effectiveness /Average Ship Cost
$f_2(\mathbf{x}_2, \mathbf{x}_C)$ = NSC Mission Ship Effectiveness /Average Ship Cost
$f_3(\mathbf{x}_1, \mathbf{x}_2, \mathbf{x}_C)$ = Net Fleet Savings from Commonality

where f_1 and f_2 are subject, within the synthesis model, to the ship design constraints for their respective designs. The specific formulation of f_1 and f_2 was discussed previously. The ship design variables in $\mathbf{x}_1$ and $\mathbf{x}_2$ were also discussed earlier and remain the same. The commonality variable vector or commonality string $\mathbf{x}_C$ will be introduced below.

Basic Optimization Process

In order to add commonality, the basic optimization process seen in Fig. 3 needed to be extended. A few elements were added to create families of designs with a particular commonality. These elements include a commonality string, parallel optimization runs, the archiving of the endpoints, the fleet savings calculation and a dominance sort which includes all three design objectives. The complete optimization process can be seen in Fig. 17. The specific descriptions of each of the process elements will be discussed in detail in following subsections.

Commonality String. The optimization uses a string of n possible common design components. Each of these n possible common design components will have multiple options for commonality. The commonality string designates which design components to make common along with which component option to use for each optimization run. An example of a commonality string is

$$\mathbf{x}_C = [0\ 2\ 0\ 1\ 3]^T \quad (15)$$

There are five design components that are considered for use in common in this commonality string. These five elements can be whatever the designer feels would be good candidates to make common among a family of designs.

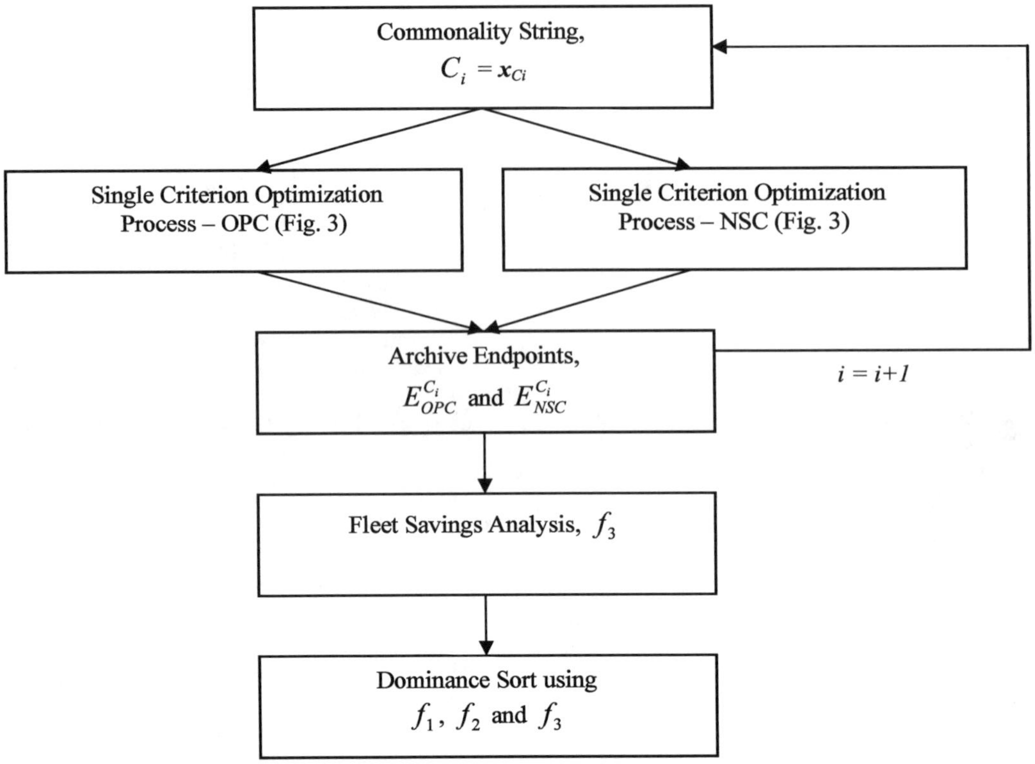

Figure 17. Basic Optimization Process with Commonality

In the example commonality string shown in eq. 15, the 0 indicates that the component will not be common while a non-zero number indicates that commonality will be used for that component. If a design component is not required to be common, the optimization is free to use whichever element it chooses to optimize the design. So in eq. 15, components two, four and five will all be common for the designated family of designs. Each of the components may have any number of options for commonality. The commonality string designates the second option for component two and that components four and five use options one and three, respectively.

If each of the n possible common components has the same number of commonality options m, the number of possible commonality strings will be $(m+1)^n$, which accounts for the possibility that a component may also be considered not common. If the components have varying numbers of options for commonality m_i, the number of possible commonality strings becomes $(m_1+1)(m_2+1)...(m_n+1)$.

For this study, the number of common components used was five. The five common components are weapon systems, ship service generators, cruise engines, superstructure and midship section hull blocks. The selection of these components was designed to illustrate the broad range of possible commonality. Each of these elements affects the designs in different ways. For each possible common component, the number of different options to choose from was determined from the optimal characteristics revealed by the results of the two-criterion optimization runs without commonality.

There are three possible weapon systems that can be used in each design. Using weapon systems for commonality is relatively simple because it is an independent variable. By simply designating which one is to be used, the synthesis model designs a family of ship variants using only the designated weapon system. The difference from one weapon system to another is primarily through its weight and cost. The weight of the weapon system will obviously have some impact on the designs, but it does not affect the design too much. The weapons systems options are listed sequence in Table 5.

Table 5. Common Weapon System Characteristics

Weapons	Weapons weight LT	Ammunition weight LT	Combat Systems wt. LT
46mm gun	11.22	10.1	73
57mm gun	14.04	11.4	75
57mm & CIWS	23.48	23.48	84

The ship service generators are a little more complex in their impact in the overall design of the ships. Since the choice of ship service generators is a dependent variable, its implementation is different from that for the weapon systems. If no generator is designated as common, the number of generators is four and the ship synthesis selects the generator. However, if the generators are designated as common, the ship synthesis selects the number of generators required to meet electrical load requirements of the ships. The effect of the generators goes beyond just weight considerations. In addition to accounting for the weight of the generators, the synthesis must also allocate space for the generators within the hull. The two objective baseline runs showed that only three of the twelve possible generators in the database were ever used in the Pareto front designs. These generators are listed in sequence in Table 6.

Table 6. Common Ship Service Generator Characteristics

kW Rating	Engine RPM	SPC (FP) in lbm/hp-hr	Total wt of DG LT	Length of DG set in m (feet)
250	1800	0.39	5.07	1.83 (6.0)
304	1200	0.35	8.00	2.41 (7.9)
444	1800	0.38	6.89	2.32 (7.6)

The cruise engines showed another aspect of the use of commonality. The choice of cruise engines limited the number of possible designs that could meet the cruise speed requirement. Similar to the ship service generators, the choice of cruise engines is a dependent variable and cannot simply be designated without further consideration. The ship synthesis was permitted to run as if no cruise engine commonality was specified. Once it had a ship designed, it checked to see if the engine used was the desired common engine. If it were, the design was kept. If not, the design was discarded. This process allowed for the iterative ship design process to take place while ensuring that the cruise engines satisfied the cruise speed requirements. During the iterative process, the synthesis model changes engines as needed to ensure the correct engine is being used for each design. If the common cruise engine is forced into the design, the iterative process is disrupted and the synthesis model may not work as intended. The two objective baseline cases consistently used only two types of cruise engines on the Pareto front. Their characteristics are shown in sequence in Table 7.

Table 7. Common Cruise Engine Characteristics

kW Rating	SPC (FP) lbm/hp-hr	Weight LT	Length m	Width m	Height m	Volume m^3
1440	0.34	3.95	2.90	1.30	1.43	5.33
1850	0.34	7.6	4.00	1.52	1.89	11.47

To this point, all the commonality components have dealt with specific shipboard equipment. In order to show another aspect of commonality in design, construction savings was considered. Two areas of commonality in ship construction, superstructure and the midship section hull blocks, were used to show the effectiveness of this optimization methodology.

As noted above, the baseline runs showed that all the ships on the Pareto front basically had one of two narrow ranges of superstructure volume. The superstructure is primarily a function of the number of helicopter hangars on the ship. In addition to the number of hangars, the beam of the ship is also related to the design of the superstructure. These characteristics along with the volume of the superstructure make up a commonality component. By using the superstructure as a common component, both independent variables and dependent variables are being designated. The number of helicopter hangars is an independent variable while the beam and volume of the superstructure are dependent variables calculated in the iterative ship synthesis model. The synthesis was allowed to calculate the beam and superstructure volume as if no commonality were being used. Once calculated, the values were overwritten to the necessary values for the designated commonality and the process was continued. The characteristics associated with the two superstructure choices are shown in sequence in Table 8.

Table 8. Common Superstructure Characteristics

Designation	Volume m^3	Beam m (feet)	Number of Helicopter Hangars
Small	1699	12.19 (40)	1
Large	3200	16.46 (54)	2

The final component that was considered to be common was the midship section region of the ships. This would enable the use of a common hull block(s) near amidships in both designs. Again, the size of the midship section was largely dependent on the number of helicopter hangars on the ships. Other midship section related characteristics included the midship coefficient (C_m), depth of the hull (D) and the beam of the ship. Again, this commonality component consists of both independent (number of hangars and C_m) and dependent variables (D and B). As above, the depth and beam of the ship were determined in the iterative process and overwritten when necessary to satisfy the commonality requirement. The characteristics used for the common midship section hull blocks are listed in sequence in Table 9.

Table 9. Common Midship Section Hull Block Characteristics

Designation	C_m	Depth m (feet)	Beam m (feet)	Number of Helicopter Hangars
Small	0.99	7.01 (23)	12.19 (40)	1
Large	0.99	8.23 (27)	16.46 (54)	2

Given the five possible commonality components and the choices associated for each, the total number of possible commonality strings was 432. However, some of these had conflicting requirements. For example, a large superstructure implies two helicopter hangars and a 16.46 m (54') beam while a small midship section implies one hangar and a 12.19 m (40') beam. Obviously both of these requirements cannot be satisfied at the same time. So options with this combination were eliminated from consideration. Similarly, a small superstructure and large midship section was also eliminated as infeasible. Other combinations of commonality were eliminated because they were counterintuitive. For example, designs using the smaller cruise engine could not adequately power ships with the large midship sections or large superstructures. As a result, the optimization program would spend excessive time trying to populate itself with ships of this nature. Instead of looking for ships that are known to be infeasible, the optimization eliminated these inconsistent combinations of commonality. As a result, there were only 288 possible commonality strings remaining.

Optimization Process. The optimization process is similar to the one seen in Fig. 3. A few changes were made, however, in order to make this process more efficient in solving the three criterion optimization problem. The most obvious change is that the optimization is run for the OPC and NSC mission designs in parallel rather than at the same time. Since finding only the endpoints of the two objective Pareto front is the goal for this problem, it is possible to optimize for each objective at higher precision when independent of the other. The dominance sorting and tournament selection processes are performed using only the respective objective for each mission ship. This single objective sorting allows for an efficient search for the best OPC ship and the best NSC ship for each commonality string.

Archive Endpoints. Each individual optimization produces the two objective Pareto front endpoints for each commonality string. Since this optimization can be run for any number of ship classes, the number of endpoints will be equal to the number of ship classes that are being considered. The endpoints are the ship designs that have the highest value of performance over cost for each of the required missions.

Fleet Savings. In each optimization in which commonality was applied, a fleet savings from the commonality was calculated. By using common engines and/or weapon systems, savings can be achieved in a variety of ways. For example, savings can be realized in crew training, spare parts, generation of manuals, and in engineering integration of components. If a fleet of ships all have these common components, training of crew members can be simplified. If a crew member were to transfer from one class of ship to another he/she would not have to be retrained on the engine or weapon system resulting in a savings of time and money. Savings could be realized in training facilities and staff.

Depending on the location of the home ports of the ships within the fleet, commonality can lead to a significant savings in spare parts. If ships of two classes are located near each other, the need for two sets of spare parts is eliminated. This results in a savings in purchasing the spare parts as well as storing the spare parts. Shore based maintenance may also be a source of commonality savings in that they will only have to service one type of engine or weapon system.

When an engine or weapon system is installed on a ship there is also a nonrecurring cost associated with that installation. If commonality is used in a fleet of ships, this cost will only occur once and can be spread out over the entire fleet of ships. If no commonality is used, the cost may occur for each class of ships and be spread out over smaller numbers of ships.

If a fleet of ships is able to use the same superstructure or midship section hull blocks, savings can be found in the construction learning curve as well as the design of those areas of the ships. As shipyards construct sections of a ship, there are lessons learned that helps them become more efficient in their work. This efficiency will save time and money in the construction process.

The fleet savings modeled in this case study was based on either the savings as a result of bulk purchasing or the savings associated with a construction learning curve. The savings model was limited to these two types of savings. The details of these models are presented in the lead author's dissertation (Corl 2007). The following sections will briefly explain the simplified fleet savings models for each commonality option. As a designer learns more about the ships and how they will be manned and operated and where they will be located, other types of savings can be added to the fleet savings model.

Weapon System Savings. The savings associated with the commonality of the weapon systems was limited here to the bulk purchase for the entire fleet. In calculating the fleet savings from the use of a common weapon system, the cost of the common fleet of ships was compared to the cost of a fleet of ships that was designed with no commonality. The optimization was run using very small parameter ranges that focused on each of the endpoints of the baseline curve. In doing so, it ensured that the ships designed were best suited for their specific missions. This resulted in baseline ships for both the National Security Cutter (NSC) and the Offshore Patrol Cutter (OPC) missions. The commonality savings incorporates a percent savings associated with bulk purchasing. The assumed savings for using weapon system 1 or 2 common was linear from zero for one ship to 10% at 33 ships. The assumed savings for using weapon system 3 was linear from zero for one ship to 20% for 33 ships.

Ship Service Generator Savings. The savings associated with using common ship service generators was performed in much the same way as the weapon system. A cost schedule was created based on the number of units purchased and the savings was the difference between the ships with no commonality and those with common generators. Because the number of generators is not constant for all ships, however, the savings had to include number of generators purchased. The assumed savings associated with using common ship service generators was linear from zero to 10% at 132 generators on all 33 ships.

Cruise Engine Savings. Again, a cost schedule was created for the savings associated with using common cruise engines. The cost savings associated with using common cruise engines was assumed linear from zero for two engines to 15 % for 66 engines for all 33 ships.

Superstructure and Midship Section Savings. The savings assumed for the common superstructures and midship section hull blocks was limited to construction labor costs. By applying a learning curve to the labor cost of construction, savings can be calculated by summing the savings for the NSC Fleet and the OPC Fleet. For a common superstructure, the savings was based upon the assumed fraction of the weight group 100 cost that is superstructure and the assumed learning curve rate. The midship section hull blocks savings was calculated in the same way assuming that 20% of the group 100 cost can be made common.

Another form of savings may also occur. In some instances, a pair of designs may have a common component despite not having it required to be common. This can occur with the weapon systems, ship service generators and cruise engines, but is unlikely with the superstructures and midship sections. When such an occurrence exists, the savings value for that component is calculated as described above.

Total Fleet Savings. The total fleet savings that results from the use of commonality in the designs was calculated by summing up each of the applicable savings associated with each of the common components.

Dominance Sorting. At this point in the three-criterion optimization, the endpoint archive has two designs for each commonality string. One design represents the best NSC mission design and the other represents and best OPC mission design for the designated commonality. The fleet savings for each pair of ship classes has been calculated relative to the pair of ship classes that were designed with no commonality. In order to determine which of these designs are the best, another dominance sort is performed. In the two-criterion part of the optimization process, dominance sorting was used to determine which designs were on the Pareto front and to help select potential parents during each generation. The dominance sort was performed each generation until the optimization was complete. In that dominance sort, ships were compared using the first two objective criteria of the optimization.

$f_1(\mathbf{x}_1) =$ OPC Mission Ship Effectiveness /Average Ship Cost
$f_2(\mathbf{x}_2) =$ NSC Mission Ship Effectiveness /Average Ship Cost

In the three-criterion case, each commonality string was optimized individually to produce the two Pareto front endpoints corresponding to these two objective functions. To this point in the overall process, the individual commonality string results have not been compared to each other. With a value calculated for the associated fleet savings, the dominance sorting can next be made using all three objective criteria.

$f_1(\mathbf{x}_1, \mathbf{x}_C) =$ OPC Mission Ship Effectiveness /Average Ship Cost
$f_2(\mathbf{x}_2, \mathbf{x}_C) =$ NSC Mission Ship Effectiveness /Average Ship Cost
$f_3(\mathbf{x}_1, \mathbf{x}_2, \mathbf{x}_C) =$ Net Fleet Savings from Commonality

As before, the ships are compared to each other to determine whether they are dominated or not. The ships are compared to each other in pairs. Each pair will have an NSC mission design, an OPC mission design and a fleet savings, which is the same for both ships in the pair. Dominance is determined by comparing the NSC mission designs using f_1, the OPC mission design using f_2, and each pairs' f_3. The solutions that make up the nondominated set of solutions are the final three-criterion discrete Pareto front anticipated in Fig. 2.

FLEET OPTIMIZATION RESULTS WITH COMMONALITY

Case Study

Similar to the previous case study, the optimization was run with the following optimization algorithm minimum settings: archive size-50, population size-150, and number of offspring per generation 100. The maximum number of generations was set at 200 and the termination condition was not used in order to ensure that the Pareto front endpoints of each individual run progress the same amount.

Optimization Results

Figure 18 shows the endpoints for each of the 288 feasible commonality strings. The results group themselves in three sections. The uppermost section consists of 128 NSC designs. The middle section consists of 160 NSC designs. The remaining section near the bottom makes up the 288 OPC designs. The reason that the NSC designs are split into two sections is due to the nature of the commonality that is being forced into the designs.

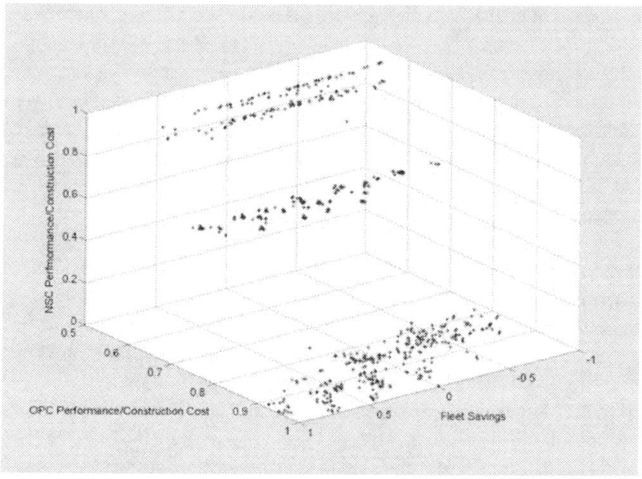

Figure 18. Optimization Run Endpoints prior to Dominance Sorting

The 160 NSC designs in the middle of Fig, 18 have certain defining characteristics. First, 144 of those designs have a small superstructure, a small midship section, or both designated as common. The remaining 16 designs have no commonality designated for the superstructure or the midship section, but have the smaller cruise engines designated as common. These three characteristics are all indicative of smaller vessels. All 160 designs have only one helicopter hangar, which is associated with the small superstructure and midship section. In order to meet the cruise speed requirement, the use of the smaller common cruise engines tends to need smaller vessels. Because of the way the fuzzy utility functions have been set in this example, the single hangar tends to be the controlling feature. The weapons systems and ship service generators selected for these designs do not influence this tendency toward the middle of the graph. If the fuzzy utility functions were modified, different results might be achieved.

One assumption that has typically been made in the platform and commonality literature to date is that commonality is inherently good. It is accepted that in using common components there will be a loss in performance. Considerable work has been done to measure how much performance will be lost in applying common components to designs. However, the loss in performance is assumed to be outweighed by the cost savings associated with using common components. This anticipated cost savings is the driving force behind the use of commonality in design.

The results in Fig. 18 show that, for this design problem, not all commonality is good. Note that a total of 129 of the 288 commonality combinations produced a negative fleet savings. This negative fleet savings is a result of over designing the OPC. The OPC that was designed without any commonality had the least expensive weapon system, ship service generators and cruise engines. These components enabled the OPC to meet its performance requirements. If more expensive options for each of these components are forced into the design through commonality, the OPC will still meet its performance requirements, but at a higher cost. Even though bulk purchase savings will occur, the cost of the more expensive components will be more expensive overall. The result is that if the more expensive weapon systems, generators or cruise engines are made common, the OPC will have a negative savings. Using a common weapon system or common cruise engines, the OPC had negative savings in half of the commonality strings (144 occurrences each). Each of those commonality strings used a larger weapon system or generator than the OPC needed to meet its performance requirements. Similarly when the larger cruise engines were made common, the OPC had a negative savings as well (126 occurrences). Making the superstructure common never resulted in a negative savings for either ship. Common midship section hull blocks resulted in a negative savings for 94 of the 288 pairs of ships. This occurred each time the larger midship section was common (64 times) and about a third of the time when the smaller midship section was common (30 times). The reason that these pairs resulted in a negative savings is due to the definition of the common midship section sizes. The larger midship section was sized so that all baseline optimization solutions were smaller than it. This was done to ensure that it would not limit the size of the ships in the optimization. The smaller common midship section was made slightly larger than the smallest OPC designs in the baseline run. Again, this was done in order to not force the OPC designs into being smaller in size and possibly making many designs infeasible. All this translates to more labor for the slightly oversized midship sections.

It is important to realize that just because the savings may be negative for one or more of these components does not necessarily mean that the overall fleet savings for that pair of ships will be negative. For the weapon systems, ship service generators and cruise engines only the OPC designs had negative savings. The NSC designs for each of those designs had a positive savings value. For a common weapon system 2, the fleet weapon system savings was always positive despite the negative savings for the OPC design. In addition, if positive savings were achieved in other common components, the overall savings could still be positive. Because of the relative cost of each of the components that are considered for commonality, some are more influential than others. Table 10 shows the relative importance of each component in its potential to create savings.

Table 10. Relative Influence of Components on Savings

Positive Savings		Negative Savings	
Component	Ratio	Component	Ratio
Generators	1.000	Generators	1.000
Cruise Engines	0.806	Cruise Engines	0.934
Superstructure	0.512	Midship Section	0.230
Midship Section	0.120	Weapon System	0.031
Weapon System	0.014	Superstructure	0.000

The components with the greatest potential for large commonality savings are ship service generators, cruise engines and superstructure. These large savings are possible when the smallest generators and cruise engines are designated as common. The cruise engines and generators can also have large negative savings. Using the largest generator or cruise engine as common will result in the greatest negative savings values. A common superstructure, which always results in a positive savings, has the largest savings when the smaller option is used. The midship section and weapon systems have a much smaller impact on savings, either positive or negative. Again, smaller options tend to result in positive savings and larger options tend to result in negative options due to the resulting over design of the OPC.

Pareto Front

When the three-criterion dominance sorting is performed, the discrete Pareto front or Pareto surface can be plotted. Figure 19 shows the Pareto front for the optimization run which consists of 20 pairs of ships.

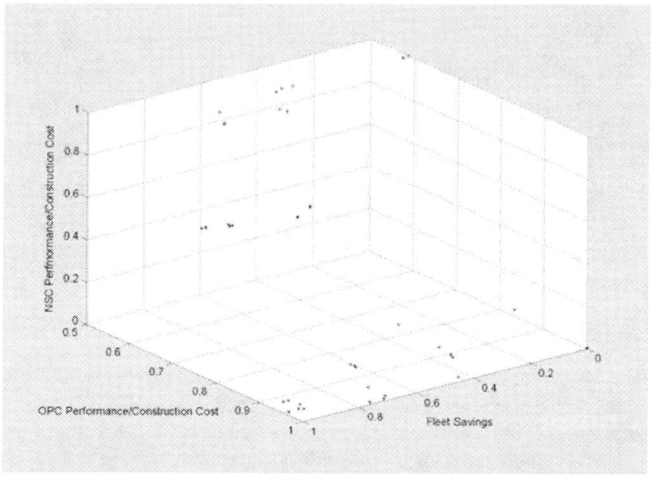

Figure 19. Pareto Front

If the plot in Fig. 19 is rotated, it shows that the solutions group themselves into four bands. The four bands appear at nondimensionalized Fleet Savings values of 0-0.01, 0.39-0.45, 0.60-0.76 and 0.93-1.00. Figure 20 shows the plot from such a perspective.

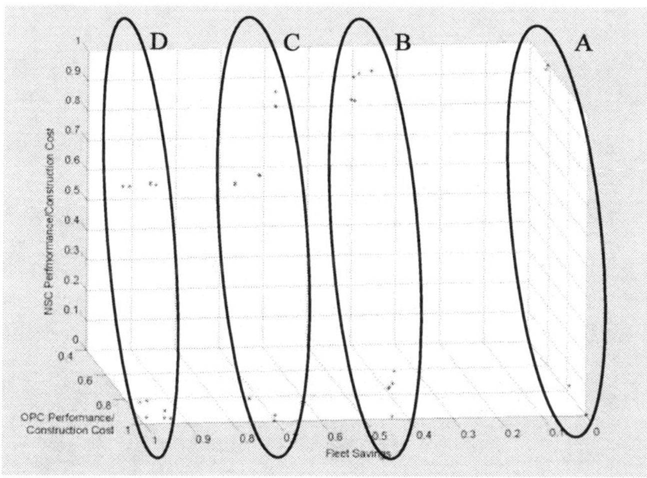

Figure 20. Pareto Front – Alternate View

The band labeled A consists of two pairs of solutions. One of the pairs is the baseline case, which has no commonality and therefore no fleet savings. The other solution has weapon system 3, small generators, large cruise engines and the large midship section all designated as common. The common weapon system, cruise engines and midship section all yield negative savings, however, the small generators yield a large savings which slightly outweighs the negatives resulting in a small fleet savings. Table 11 shows the commonality make-up of group A (N indicates no commonality).

Table 11. Commonality Summary for Group A

	Weapon System	Generator	Cruise Engine	Super-structure	Midship Section
1	N	N	N	N	N
2	3	1	2	N	Large

The B group in Fig. 20 consists of five pairs of solutions. The small common generators appear in all five pairs of solutions. Additionally, the weapon systems and midship sections may be common which result in relatively small impacts on savings compared to the common generators. Table 12 shows a commonality summary of the ships in group B.

Table 12. Commonality Summary for Group B

	Weapon System	Generator	Cruise Engine	Super-structure	Midship Section
1	3	1	N	N	Large
2	3	1	N	N	N
3	N	1	N	N	N
4	2	1	N	N	N
5	1	1	N	N	N

Group C consists of seven pairs of ships. Table 13 shows the commonality breakdown of group C. Every pair of ships in group C has at least two common components, most have three. The small common generators appear in all seven pairs. One trend that becomes evident for the higher savings values in group C is the use of smaller options of commonality. As discussed previously, the OPC tends to have negative savings when larger components are used. In some cases these negative values can be outweighed by other common components. This can be seen in the half of group C with lower fleet savings.

Table 13. Commonality Summary for Group C

	Weapon System	Generator	Cruise Engine	Super-structure	Midship Section
1	N	1	N	Large	N
2	1	1	N	Large	N
3	2	1	N	Large	N
4	N	1	1	N	N
5	1	1	1	N	N
6	N	1	N	N	Small
7	1	1	1	N	Small

The final group, D, has six pairs of ships. In this group fleet savings is maximized by selecting the smaller component options. Large impact savings is obtained by routinely using common generators, cruise engines and superstructures. Table 14 shows the commonality components for the ships in group D.

Table 14. Commonality Summary for Group D

	Weapon System	Generator	Cruise Engine	Super-structure	Midship Section
1	N	1	N	Small	N
2	1	1	1	Small	N
3	1	1	N	Small	N
4	N	1	1	Small	N
5	1	1	N	Small	Small
6	N	1	1	Small	Small

Similarity in Designs

Despite having different commonality strings, two ships may have virtually the same naval architectural characteristics. This occurs when one ship has no commonality designated for one or more components and the other ship uses commonality for those components. The ships without the commonality may optimize to similar characteristics. Take for example the commonality strings 111NN and N1NNN. Both strings have no commonality designated for the superstructure or the midship section and both use common ship service generators. The commonality designations for the weapon system and cruise engines are different for the two ships. One designates no commonality while the other designates option 1 for both weapon system and cruise engines. If the weapon system and cruise engines for the second ship are both optimized to be choice 1 then the two ships will be virtually the same ship.

In order to make this comparison of ships, an *n*-dimensional design independent variable distance formula was used to determine how different the ships actually were.

$$Similarity = 1 - \sqrt{\frac{(x_i^1 - x_j^1)^2}{x_{max}^1} + \ldots + \frac{(x_i^n - x_j^n)^2}{x_{max}^n}} \qquad (16)$$

The formula calculates how similar two ships are to each other; identical ships will have a similarity of 1.0 It would not be appropriate to compare all ships to each other. If two ships do not have the same commonality designations for superstructure and midship section, they should not be compared. By designating either of these components as common, the ship will artificially use the associated values. The ship synthesis will have been overridden to ensure commonality. Because of this alteration of the design, a ship that does not have its superstructure or midship section designated as common cannot be similar to one that does.

The Pareto set of solutions consisted of 20 pairs of ships. Comparisons were possible for seven combinations of superstructure and midship section commonality. Of these seven combinations, only six are found in the Pareto set of solutions. Table 15 shows the commonality combinations and which ship numbers coincide with each.

Table 15. Possible Commonality Combinations for Similarity

Group	Super-structure	Midship Section	Ship Numbers
1	N	N	3-7-9-12-14-19-20
2	N	Small	2-16
3	N	Large	10-11
4	Small	N	5-13-17
5	Small	Small	1-4-15
6	Large	N	6-8-18
7	Large	Large	

The ships in each group of the same commonality combination were compared using the similarity equation (eq. 16). A final analysis of the similar design for both the OPC and the NSC revealed that there are twelve distinctly different pairs of designs on the Pareto front. Based on the groupings listed in Table 16, the best OPC and NSC designs from each group were examined and matched with their corresponding ship based on common components. There were three ships that did not have a corresponding ship that was considered the best in its group. As a result, OPC ship 7, OPC ship 18 and NSC ship 10 were included to ensure that there was an even number of OPC and NSC solutions.

Table 16. Pareto Front Commonality Strings with Corresponding OPC and NSC Ship Numbers

CW	CG	CC	CS	CM	OPC Ship	NSC Ship
1	1	1	N	N	14	3
1	1	N	N	N	7	7
2	1	N	N	N	9	9
3	1	N	N	N	12	12
1	1	1	N	Small	16	2
3	1	2	N	Large	10	10
3	1	N	N	Large	11	11
1	1	1	Small	N	13	17
1	1	1	Small	Small	15	15
1	1	N	Large	N	6	6
2	1	N	Large	N	8	8
N	1	N	Large	N	18	18

Analysis of Final Discrete Pareto Surface

Figure 21 shows the baseline ships and the twelve pairs of ships determined to be distinctly different from a naval architectural viewpoint. Table 16 shows the commonality strings and the corresponding OPC and NSC solutions that make up the Pareto front. In some instances, ships that were not originally paired together have been matched with another ship. In these cases, one or more components were not designated as common. Through the course of the optimization, certain components may have naturally become common. One example of this occurring is seen in Table 16. OPC ship 14 and NSC ship 3 have been paired together. During the optimization run, OPC ship 14 only had ship service generators and cruise engines designated as common. However, the synthesis designated weapon system 1 as being the best choice for this particular ship. NSC ship 3 had weapon system one designated as common along with generators and cruise engines. Since OPC ship 14 was determined to be a better design that OPC ship 3, it was used in its place. As a result, these two ships have the same components common and can be paired together.

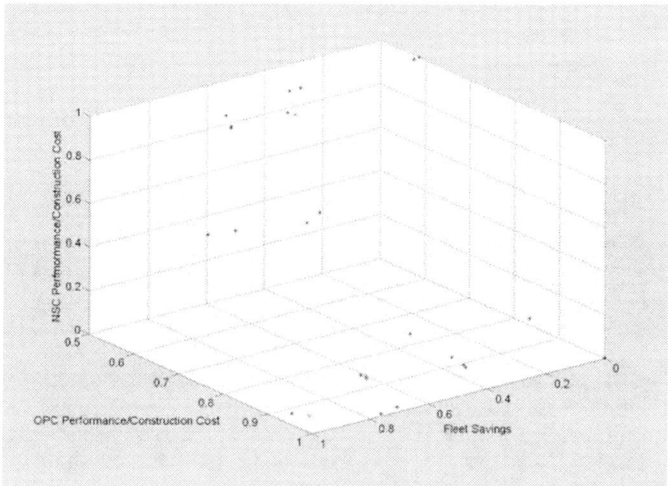

Figure 21. Pareto Front Showing Only Distinctly Different Ships

The results seen in Table 16 show some interesting trends. First, each pair of ships remaining on the discrete Pareto front has the small generator designated as common. Generally speaking, the use of the small generators has little to no impact on performance. It may add weight, and therefore cost, to NSC ships that need more than four generators to meet the electrical load requirements. However, the performance will not suffer and the net fleet savings will benefit greatly from this choice.

There is no clear cut best choice for a common weapon system. Half of the solutions on the Pareto front have weapon system 1, which allows for good savings with some loss in performance for the NSC. Weapon system 2 has a slight positive affect on savings and a loss of performance for the NSC. Weapon system 2 does not benefit the OPC's performance while increasing its cost. Weapon system 3 has a negative fleet savings while completely satisfying the requirements of the NSC. Again, the OPC suffers with the use of weapon system 3 because of increased cost with no performance gain.

Seven of the twelve pairs do no designate common cruise engines. Even though the ships will not benefit from the cost savings, this can be good. Each ship is able to optimize its performance when able to use the engine that is best suited for its requirements. One third of the designs have the small cruise engines as common. This maximizes cruise engine savings. However, the performance of the NSC tends to suffer from the use of the small cruise engines. As mentioned previously, the small cruise engines tend to require a smaller ship in order to meet the cruise speed requirements. By forcing the NSC to be smaller its performance declines. Its range has to be smaller and it can only have one helicopter hangar. The OPC is not affected in this manner. The use of the larger cruise engines is seen in one pair of designs on the Pareto front. In most cases, negative fleet savings occur when the large cruise engines are used. The performances tend to stay the same, but the OPC may have excess horsepower at cruise speed which will negatively affect its performance over cost.

The study has shown that common superstructures never result in a negative fleet savings. However, only five of the twelve solutions realize a savings from the use of a common superstructure. Generally speaking, the smaller common superstructure does not hinder the performance or cost of the OPC. The NSC is again limited in size and this will cause a decline in performance. The larger superstructure adds unnecessary costs to the OPC without a comparable increase in performance.

Common midship section hull blocks have little effect on savings, but can influence performance and cost. Similar to the superstructure, the small midship section hinders the performance of the NSC. The larger midship section benefits the NSC while at the same time adding cost and possibly hurting the performance of the OPC.

Analysis of the results illustrates how finely balanced the three objectives can be. What tends to benefit the fleet savings the most generally hurts the performance of the NSC. At the same time, what maximizes the NSC performance tends to not produce large savings by increasing the cost of the OPC. In order to maximize all three objective functions, a multi-objective balance in common components must be made.

Perhaps the most attractive pair of designs from this study are the two designs indicated as OPC_{18} and NSC_{18} on the frontispiece. This pair of designs is commonality 1 in Group C in Table 13 and OPC ship 18 and NSC ship 18 in Table 16. This pair of designs has the smaller ship service diesel generators and the large superstructure in common. This pair has the highest fleet savings from commonality possible before the NSC designs take the large loss in performance/cost and, thus, this might be the most likely choice for a design team. This commonality achieves 60.7% of the maximum fleet savings considered, but the performance of the NSC is only compromised to 97.0% of its maximum, while the OPC performance remains at 100% of its maximum. The characteristics of OPC ship 18 and NSC ship 18 are shown in Table 17 along with OPC ship 15 and NSC ship 15, also indicated on the frontispiece, which had the highest net fleet savings on the Pareto front. It is worth noting that because of the strong similarity of OPC ship 15 and NSC ship 15, consideration should be made to build a single ship to perform both ship missions. A single ship design would achieve even more fleet savings.

Table 17. Design Characteristics for Selected Ships on Pareto Front

Point	L m (ft)	B m (ft)	V_{max} kts	KW_{max} (SHP)	V_{cruise} kts	KW_{cruise} (SHP)	Range nm	W	H	C	G	OPC Perf	NSC Perf	Cost $mil	Fleet Savings $mil
OPC_{18}	107.6 (353)	16.46 (54)	22.0	5757 (7720)	18.0	2895 (3882)	9158	1	2	8	0	100.0	0.314	89.8	45.5
NSC_{18}	121.6 (399)	16.46 (54)	27.9	16687 (22377)	18.0	3585 (4807)	12074	3	2	9	3	100.0	97.0	141.7	45.5
OPC_{15}	91.4 (300)	12.19 (40)	22.2	5333 (7152)	18.0	2537 (3402)	9046	1	1	7	0	89.7	2.946	72.9	75.0
NSC_{15}	92.4 (303)	12.19 (40)	25.5	9642 (12930)	18.0	2767 (3710)	9019	1	1	7	0	89.7	47.6	91.1	75.0

CONCLUSIONS

The methodology contained in this research offers a valuable tool for use in making optimum commonality decisions. It provides a logical procedure for the use of commonality in design while taking into consideration performance loss, cost and savings. In much of the commonality literature, there is a basic assumption that commonality in design hinders the performance of products and this is accepted because of the savings associated with using common parts. Prior to this research, the amount of savings was never really quantified. Instead, it was assumed that the use of commonality always results in savings. This research has shown that positive savings is not always realized. If poor commonality decisions are made in design, products could both cost more and perform less.

The mission performance model relied on the use of fuzzy utility values. Performance was determined using the four design characteristics for each mission area and applying the corresponding fuzzy utility value to each. A designer could easily modify this model to include more design characteristics or even more mission areas. The fuzzy utilities could also be replaced with another tool for awarding value to a given design characteristic.

Commonality decisions were limited here to five components. Each of these components was integrated into the design in a slightly different manner to show the versatility of the optimization model. In this research there were a finite number of commonality options from which to choose. As a result, an exhaustive search was used to determine which commonality choices were the best. If more choices are available, another genetic algorithm could be used to more efficiently search for the Pareto optimal commonality combinations.

Bulk purchasing and construction learning curves were used here to determine the savings associated with the use of commonality. The savings model was kept relatively simple. Other forms of savings could be realized as well. These could include training of personnel, technical design costs, administrative savings, facility costs and spare parts. The type of savings and the number of different factors to consider varies with each product being designed. A designer may choose to make the savings model very elaborate when information of these other forms of savings is available or it may be kept simple as seen in this research.

Using the logical methodology presented here will enable a designer to present a much more complete analysis of commonality decisions in design. Designers can expand the optimization model in many ways to adapt it to their particular needs. Regardless of how crude or elaborate it may become, the process shown should be able to stay very much the same.

ACKNOWLEDGEMENTS

This research was possible with academic leave and related support of the U.S. Coast Guard for the lead author and support from the Office of Naval Research through N00014-03-0983 for all authors. The authors would like to thank the U.S. Coast Guard for the use of their ship synthesis model and emphasize that this research is by no means intended to be a comparison to any U.S. Coast Guard design work for either the OPC or the NSC. The mission requirements for the current Deepwater designs were used for academic purposes only.

REFERENCES

ASSET "Advanced Surface Ship Evaluation Tool." Naval Surface Warfare Center, Carderock Division, Version 5.2.0, 2005.

BOSWORTH, M. and J. HOUGH. "Improvements in Ship Affordability." *Transactions* of the Society of Naval Architects and Marine Engineers, 101 (1993): 705-726.

BROWN, A. and J. SALCEDO. "Multi-Objective Optimization in Naval Ship Design." *Naval Engineers Journal*, 115:4 (2003): 49-61.

CECERE, M.L., J. ABBOTT,, M. L. BOSWORTH and T. J. VALSI. "Commonality-Based Naval Ship Design, Production, and Support." *Journal of Ship Production*, 11:1 (1995): 1-14.

CORL, M. J., "Methodology for Optimizing Commonality Decisions in Multiple Classes of Ships." Ph.D. Dissertation, Naval Architecture and Marine Engineering, University of Michigan, June 2007.

DEB, K., *Multi-Objective Optimization using Evolutionary Algorithms*. New York: John Wiley & Sons, 2001.

FELLINI, R., M. KOKKOLARAS, N. MICHELENA., P. Y. PAPALAMBROS, K. SAITOU., A. PEREZ-DUARTE and P. FEYNES. "A Sensitivity-Based Commonality Strategy for Family Products of Mild Variation with Application to Automotive Body Structures." *Structural and Multidisciplinary Optimization*, 27:1-2 (2004): 89-96.

FELLINI, R., M. KOKKOLARAS, P. Y. PAPALAMBROS, and A. PEREZ-DUARTE. "Platform Selection under Performance Loss Constraints in Optimal Design of Product Families." *Journal of Mechanical Design*, 127:4 (2005): 524-535.

FELLINI, R., M. KOKKOLARAS and P. Y. PAPALAMBROS. "Quantitative Platform Selection in Optimal Design of Product Families, with Application to Automotive Engine Design." *Journal of Engineering Design*, 17:5 (2006): 429-446.

FUJITA, K. and H. YOSHIDA. "Product Variety Optimization: Simultaneously Designing Module Combination and Module Attributes." *Concurrent Engineering: Research and Applications*,, 12:2 (2004): 105-118.

GOLDBERG, D.E., *Genetic Algorithms for Search, Optimization, and Machine Learning*. Reading, MA: Addison-Wesley 1989.

GONZALEZ-ZUGASTI, J.P. and OTTO, K.N. "Modular Platform-Based Product Design," *Proceedings* of the 2000 ASME Design Engineering Technical Conference, Baltimore MD, 2000, paper no. DAC-14238.

GONZALEZ-ZUGASTI, J.P., K. N. OTTO and J. D. BAKER. "A Method for Architecting Product Platforms." *Research in Engineering Design*, 12 (2000): 61-72.

KOSKO, B. *Neural Networks and Fuzzy Systems: A Dynamical Systems Approach to Machine Learning*. Englewood Cliffs, NJ: Prentice-Hall, 1992.

LI, J. and M. G. PARSONS. "An Improved Method for Shipbuilding Market Modeling and Forecasting." *Transactions* of the Society of Naval Architects and Marine Engineers, 106: (1998): 157-183.

MESSAC, A., M. P. MARTINEZ, and T. W. SIMPSON. "Effective Product Family Design using Physical Programming." *Eng. Optimization*, 34 (2002): 245-261.

MICHALEWICZ, Z. *Genetic Algorithms + Data Structures = Evolutionary Programs*. Berlin: Springer-Verlag, 1996.

Naval Surface Warfare Center Carderock Division. "User's Guide USCG Performance Based Cost Model," 1998.

NAYAK, R.,W. CHEN and T. W. SIMPSON. "Variation-Based Methodology for Product Family Design." *Journal of Engineering Optimization*, 34:1 (2002): 65-81.

OSYCZKA, A. *Evolutionary Algorithms for Single and Multicriteria Design Optimization*. Berlin: Physica-Verlag, 2002.

PARSONS, M. G. "Parametric Design," Ch. 11 in *Ship Design and Construction*, Vol. I, T. LAMB (ed.), New York: SNAME, 2003.

SIMPSON, T.W. "Product Platform Design and Customization: Status and Promise," *Artificial Intelligence for Engineering Design, Analysis, and Manufacturing*. 18 (2004): 3-20.

SIMPSON, T.W., J. R. A. MAIER and F. MISTREE. "Product Platform Design: Method and Application." *Research in Engineering Design* 13 (2001): 2-22.

SIMPSON, T.W.,Z. SIDDIQUE and J. JIAO, eds. *Product Platform and Product Family Design: Methods and Application.*, Berlin: Springer, 2005.

U.S. Coast Guard Internal Documents. Various.

U.S. Coast Guard Memorandum. "Mission Analysis Report (MAR) N-001-95 Deepwater Missions." 1995.

U.S. Coast Guard Website (2006) http://www.uscg.mil/deepwater/system/cuttercomparison .

U.S. Department of Transportation Memorandum. "TSARC Decision Memorandum for the Deepwater Capability Replacement Project." (1996).

ZALEK, S. F., M. G. PARSONS and P. Y. PAPALAMBROS. "Multi-Criteria Design Optimization of Monohull Vessels for Propulsion and Seakeeping." *Proceedings* of the 9th Int. Marine Design Conference, 2 (2006a): 533-557.

ZALEK, S. F., M. G. PARSONS and R. F. BECK. "Evolutionary Multicriterion Optimization for Propulsion and Seakeeping." *Proceedings* of the 26th Symposium on Naval Hydrodynamics, 2 (2006b): 69-85.

ZALEK, S. F. "Multicriterion Evolutionary Optimization of Ship Hull Forms for Propulsion and Seakeeping." Ph.D. Dissertation, Naval Architecture and Marine Engineering, University of Michigan, January 2007.

ZITZLER, E., M. LAUMANNS and S. BLEULER. "A Tutorial on Evolutionary Multi-objective Optimization." http://citeseer.ist.psu.edu/zitzler03tutorial.html , 2003.

Application of Cross Co-Spectral Moments to Von Mises Stress

Bruce L. Hutchison, P.E. (LFL) **and Benjamin B. Ackers, P.E.** (M)

Recent advances in the analysis of joint seakeeping processes are applied to the prediction of Von Mises stress. The methods of cross co-spectral moments are applied to the estimation of the probability distribution of Von Mises stress in each stationary joint environmental and operating condition. The important class of Von Mises stress problems subject to conditions with steady stress components (i.e., still water stresses) is considered, and the probability distribution is developed through application of the method of statistical principal axes for joint processes. An explicit method for estimating the lifetime probability distribution of Von Mises stress is described. From this lifetime distribution, the expected extreme value can be determined. This expected extreme value of Von Mises stress is suitable and appropriate for use as a dynamic loads parameter (DLP) in a global dynamic loads analysis (DLA) similar to that promoted by the American Bureau of Shipping. The method of cross co-spectral moments applied to the prediction of Von Mises stress is validated by comparison with Fourier-Stieltjes time-domain simulations. As an aid to understanding, the ellipsoidal distributions of Von Mises stress and probability are discussed, and the statistical principal axes are determined.

KEYWORDS: Von Mises; DLA; structure; strength; statistical principal axis; joint processes; cross co-spectral moments.

"In a broad sense, the laws of nature are Gaussian."
- Manley St. Denis and Willard J. Pierson, Jr. (1953)

INTRODUCTION AND MOTIVATION

The three pillars of structural design and analysis are concern for strength, stability and repeated loads. Concern for strength may be regarded as concern for a first-passage failure, i.e., violating a strength criterion such as yield or ultimate in a single event of load application. Concern for stability is concern for buckling of structural elements. And concern for repeated loads is concern for fatigue. For ships and offshore structures, all of these structural concerns must be evaluated when subject to loads and resulting stress states arising from the application of a Gaussian sea. The loads and stress states are then joint Gaussian processes.

The concern of this paper is that of the first pillar described above, strength. Specifically, we consider a structural element's elastic limit when subjected to slowly-varying multi-axial loads. Maximum-distortion-energy theory assumes that ductile yield failure occurs when the portion of strain energy producing change of shape of the element exceeds the limiting value determined from a tensile test of a specimen subjected to a load resulting in only one nonzero principal stress. This theory has been shown to agree well with available test data determining the yield stress of ductile material.

As a scalar quantity, the total strain energy in an element can be calculated by summing the strain energy due to each component strain. Maximum-distortion-energy failure can be evaluated with an expression based on an algebraic sum of the multi-axial stress components. The scalar value produced by this expression, known as Henky-Von Mises (Von Mises) stress, can be directly compared to a material's yield stress to evaluate failure.

Because of its accuracy, Von Mises stress evaluation has been widely adopted in allowable stress design of homogeneous, ductile metal structures. Marine classification societies have generally adopted Von Mises stress evaluation as their sole yield failure criterion for mild steel and aluminum structures assessed with 3-D finite element analysis.

Beginning from 1992, the American Bureau of Shipping (ABS) has pioneered and promoted global dynamic loads analysis [13]. In that reference, they introduced the equivalent regular wave approach to global dynamic loads analysis (DLA)[1]. Consistent within the body of linear theory, Hutchison et al [8, 9] presented an equivalent irregular wave for dynamic loads analysis (an alternative to the ABS equivalent regular wave method [13]) and a determination of extreme lifetime responses and lifetime distributions of load cycles, both subject to a spatially and temporally varying wave climatology and subject to seamanship.

A key concept of these DLA methods is that of dynamic load parameters (DLPs). Typical candidates for dynamic load parameters include global bending (e.g., longitudinal, lateral, torsional), acceleration and roll angle. Whatever processes are chosen as the DLPs, the lifetime extreme is estimated for each, and then equivalent regular or irregular waves are determined that realize the lifetime extreme DLP. These equivalent regular or irregular waves then define the instantaneous pressure distribution on the hull and concurrent 6-DOF accelerations. Together the pressure distribution and accelerations define a global loading case suitable for analysis using finite element methods.

As presented by Liu et al [13] and by ABS [2], the equivalent regular wave method has the virtue of being unique. The

[1] Other class societies have adopted similar quasi-static global analysis procedures. Efforts within the context of IACS to develop common structural rules for both tankers and bulk carriers have been calibrated using such procedures. For a summary of class society common rule development, see the report of ISSC Committee II.1, Quasi-static Response [5].

advantage of the equivalent irregular wave is that it maintains a faithful representation of all of the components of the irregular sea at all frequency and length scales. This is believed to provide a more realistic representation of secondary and tertiary structural loading processes. The price to be paid for this more realistic representation is, first of all, greater computational effort to determine the equivalent irregular wave case, but also uniqueness is forfeit. While this loss of uniqueness is troubling, it should not lightly be disregarded, as it reflects the true nature of such loadings. The character of the conditional probability distributions of critical structural response processes, given that a higher level DLP (e.g., global bending) is satisfied, is a worthy topic for future research, but will not be further addressed in this paper.

These global dynamic load analyses are the ultimate motivator behind the present paper. It is obvious that the true interest of such dynamic load analyses is the stress state and structural stability of the elements comprising the structure. The higher level DLPs (e.g., global bending, acceleration, roll angle, etc.) are mere artifices that it is hoped are closely correlated with extreme stresses and/or buckling. The purpose of this paper is to show the way to regard stress, and in particular Von Mises stress, directly as a DLP. The problem of buckling is left to a future paper. The problem of fatigue was addressed in [11].

The authors regard their proposal that Von Mises stress be used directly as a dynamic load parameter not as a challenge to dynamic loads analysis as currently practiced, but rather as an enhancement to that practice, intended to extend and make more useful and accurate the results of DLA.

NOMENCLATURE

Constants

g	acceleration of gravity
i	imaginary constant, $i=\sqrt{-1}$

Variables

H_S	significant wave height
k	wave number, $k=\omega^2/g$
L	vessel life-time (years)
$m_{yy}^{(n)}$	n^{th} statistical moment of the y-process
N	number of cycles
p	probability density
P	cumulative probability
S	wave power spectrum
S_{yy}	power spectrum for the y-process
T	exposure time (seconds)
T_L	life-time exposure (seconds)
T_2	average zero-crossing period
T_4	average peak-to-peak period
T_P	peak (aka 'modal') period
V	vessel speed
$\phi_x, \phi_y, \phi_{xy}$	phase angles of the frequency response operators for processes σ_x, σ_y and τ_{xy} respectively
$\varphi_x, \varphi_y, \varphi_z$	direction cosines
γ	fraction of time spent at sea
$\Gamma(\tau)$	autocovariance function (aka covariance when τ=0)
$\gamma_{xy}(\tau)$	xy component of autocovariance
χ	relative direction of wave propagation
ε	spectral breadth parameter
ζ	wave amplitude
Λ	vessel load condition parameter
σ_{VM}	Von Mises stress
σ_x, σ_y	x and y components of bi-axial normal stress
$\bar{\sigma}_x, \bar{\sigma}_y, \bar{\tau}_{xy}$	stress frequency response operators
$\hat{\sigma}_x, \hat{\sigma}_y, \hat{\tau}_{xy}$	unit stress vectors
$\sigma_{x_0}, \sigma_{y_0}, \tau_0$	time invariant (aka mean or still water) stresses
$\tilde{\sigma}_x, \tilde{\sigma}_y, \tilde{\tau}$	time-dependent stresses
σ_1, σ_2	principal stresses
τ, τ_{xy}	shear stress, or τ may be a time difference, depending on context
ω	circular frequency
ω_e	frequency of encounter $=\omega - V\,k\cos(\chi)$

Operators and notation

$\overline{}$	a bar $\overline{}$ over a complex-valued variable denotes the complex conjugate, also time average operator
$\vec{x}^T$	transpose of vector $\vec{x}$
$\lvert\;\rvert$	modulus (amplitude) of a complex value or the absolute value of a scalar or determinate of matrix
$\Phi(\xi)$	standardized normal (aka Gaussian) cumulative probability distribution function
$p(x\vert y)$	conditional probability density for x given y
$P(x\vert y)$	conditional cumulative probability density for x given y

Background

A series of recent technical papers has explored joint seakeeping processes for structural responses [9] and statistical principal axes for joint processes [10]. Mansour and Hutchison exploit the power of cross co-spectral moments [7, 14].

In 2005, cross co-spectral moments were applied to the determination of statistical principal axes for structural fatigue [11]. The present paper extends the promise of these methods to bi-axial Von Mises stresses and briefly, in Appendix C, points the way forward to tri-axial Von Mises stresses.

Objective

Our concerns are stress processes in any specific finite element for a vessel, with specified, but not necessarily zero, still water stresses, operating in an irregular sea. These stress processes (e.g., $\sigma_x(t)$, $\sigma_y(t)$ and $\tau_{xy}(t)$) are functions of time, and the Von Mises stress is a scalar valued function of these basis stress processes. Hence, Von Mises stress is also a function of time.

Within the context of linear theory, there are advantages to beginning with frequency-domain representations of these basis stress processes and evaluating frequency-domain integrals to determine statistical parameters of probability distributions. As the stress processes have important harmonic character, we are particularly interested in the probability distributions that describe the set of local maxima (in time) of each of these processes. Of greatest interest is the probability distribution of extreme values of these sets of local maxima, which can be determined from the probability distributions for the set of local maxima using order statistics.

A virtue of this familiar analysis strategy, frequency domain followed by probability domain, is that one avoids what is known as the sampling problem, and arrives at statistical statements with population status. The methods for performing such analyses in the frequency and probability domains are well known for each basis process taken individually, but the problem of Von Mises stress involves a quadratic function of the basis stress processes considered jointly.

Coordinates and Space

The joint Gaussian probability density of the basis stresses (σ_x , σ_y and τ_{xy}), and that of the ,Von Mises stress, represent strictly positive scalar valued fields. The distribution of the extreme values of Von Mises stress arises from the interaction of these two scalar fields. It will be shown that both fields are distributed such that surfaces of constant scalar measure are ellipsoids. The scalar measure of Von Mises stress increases as the concentric ellipsoids grow larger, and the scalar measure of probability density decreases as the concentric Gaussian ellipsoids increase in size. As demonstrated elsewhere in this paper, the extreme values of Von Mises stress tend to concentrate about a statistical principal axis that emanates from the centroid of the Gaussian ellipsoidal field and follows a locus along what may be regarded as the points of tangency between the Von Mises and Gaussian ellipsoidal fields.

Space as it is generally understood does not figure prominently in this topic. Space is implicit in the location (in vessel coordinates) of the structural element under consideration. In order to provide a substantive notion of this location we can denote the location of the nth structural element as $X_nY_nZ_n$, where XYZ is a vessel fixed coordinate system, but we will have little use for this in the developments that follow. Likewise each structural element may have its own local xy or xyz coordinate system that establishes the orientation (not necessarily parallel to vessel coordinates) of the structural element in which the normal stresses, σ_x and σ_y , and the shear stress, τ_{xy} , are measured. In general, for ship structures, we are concerned with plate elements and plane stress, so local coordinates xy and basis stresses σ_x , σ_y and τ are sufficient. Tri-axial stress is briefly considered in Appendix C, which requires local xyz coordinates and basis stresses σ_x , σ_y , σ_z , τ_{xy} , τ_{xz} and τ_{yz} .

As such notions of ordinary space are for the most part implicit and don't enter directly into the topic at hand, they shall in general be ignored. There is, however, another sense of space that is useful to the topic under consideration. It is a more abstract space, that is, a space in which the orthogonal unit vectors defining the space are stresses, $\hat{\sigma}_x$, $\hat{\sigma}_y$ and $\hat{\tau}_{xy}$. This is the space in which Von Mises stress can be visualized as an ellipsoidal field, and the probability density of σ_x , σ_y and τ can be visualized as a Gaussian ellipsoidal field. This abstract stress 'space' will be adopted, and where there is hopefully no confusion, a more simplified xyz notation will occasionally be employed to represent σ_x , σ_y and τ_{xy} . In addition to simplicity of notation, it is hoped that allowing xyz to stand for coordinates in abstract stress space will also promote a geometric grasp of the behavior of the two scalar fields (i.e., Von Mises and probability density) in stress space.

In the case of tri-axial stress (see Appendix C), additional stress coordinates are required, and abstract stress space becomes six-dimensional. The scalar valued Von Mises field is distributed according to a hyper-ellipsoid, and similarly the probability density is also hyper-ellipsoidal. This, of course, further confounds our ability to provide graphic displays, but does not preclude our ability to carry out the required analyses.

The Three Domains of Analysis

As observed by many authors addressing topics of ocean waves and the responses of ships and offshore structures to those waves, there are three domains of analysis: 1) time domain, 2) frequency domain and 3) probability domain. All three are important, and there exist rich connections between the three which we exploit to our advantage. The time domain is the domain of ordinary experience, and that is where we will start. The frequency domain is a vital bridge between the time and probability domains, and we naval architects and ocean engineers have relied on that bridge ever since St. Denis and Pierson's landmark paper "On the Motions of a Ship in Confused Seas." [17] That bridge makes possible useful statistical statements without the need for sampling in the time domain.

"From these stochastic models, one can derive all kinds of interesting conclusions about the sea, some of which will be true." –attributed to T. Francis Ogilvie

Beginnings in the Time Domain

As noted above, we begin in the time-domain world of everyday experience and observation. There we can directly observe the components of the stress tensor, for example, $\sigma_x(t)$, $\sigma_y(t)$ and $\tau_{xy}(t)$ for a plane stress process. For its pedagogic value, the exposition that follows proceeds from these time-domain records, first to the autocovariance (a time-domain statistical function), which is then transformed to spectra and cross-spectra in the frequency domain. These are then integrated to arrive at the cross co-spectral moments. The matrix of zero-order cross co-spectral moments is equivalent to the covariance matrix that can also be developed directly from the time-domain records.

We will be primarily concerned with these cross co-spectral moments. In actual practice, we often begin with wave spectra and response operators in the frequency domain, and these provide us a path to these cross co-spectral moments that does not require any direct access to time-domain information. However, it is the ability of these frequency- and probability-domain methods to generate statistics that have meaning in the time domain that makes them valuable.

Autocovariance

The autocovariance function of zero-mean[2] time-domain processes $x_i(t)$ and $x_j(t)$ is:

$$\gamma_{ij}(\tau) = \lim_{T \to \infty} \frac{1}{2T} \int_{-T}^{T} x_i(t)\, x_j(t+\tau)\, dt \qquad (1)$$

Strictly speaking, equation 1 is true in the limit as $T \to \infty$, but as a practical matter, co-variances of zero-mean Gaussian processes are routinely evaluated merely with T large compared to the inverse of the lowest frequency of the spectra (next topic below) of $x_i(t)$ and $x_j(t)$.

The matrix of autocovariance functions of zero-mean time-domain processes x(t), y(t) and z(t) is:

$$[\Gamma(\tau)] = \begin{bmatrix} \gamma_{xx}(\tau) & \gamma_{xy}(\tau) & \gamma_{xz}(\tau) \\ \gamma_{yx}(\tau) & \gamma_{yy}(\tau) & \gamma_{yz}(\tau) \\ \gamma_{zx}(\tau) & \gamma_{zy}(\tau) & \gamma_{zz}(\tau) \end{bmatrix} \qquad (2)$$

When $\tau = 0$, the autocovariance function is also known as the covariance matrix, and the values on the diagonal are the variances of the respective processes.

[2] As our concern in this paper extends to processes with non-zero mean (e.g., still water stresses) it will simply be observed here that such processes can be separated into the sum of a constant process corresponding to the mean and a non-zero Gaussian process.

Onward into the Frequency Domain

A classical path leading from the time domain to the frequency domain, and the path hereafter presented, is to apply the Fourier transform to the autocovariance function in order to obtain the frequency (self) spectra and cross-spectra. Frequency integrals of these spectra and cross-spectra then yield the cross co-spectral moments that are of particular utility to our main topic.

While we began our exposition in the time domain, in many instances, it is not necessary to begin there in practice, as we often have access to frequency-domain response operators and incident wave spectra. It is common practice to obtain the frequency response operators either from ship motion programs, from 3-D radiation-diffraction programs, or from model tests. The wave spectra may be analytical spectra, or in some cases obtained either from measurements or hindcast.

Spectra and cross spectra

The spectra and cross spectra are given as the Fourier transform[3] of the autocovariance:

$$S_{ij}(f) = \Im[\gamma_{ij}(\tau)] = \int_{-\infty}^{\infty} \gamma_{ij}(\tau)\, e^{-2\pi i f\tau} d\tau \qquad (3)$$

and

$$S_{ij}(\omega) = \frac{S_{ij}(f)}{2\pi} \qquad (4)$$

Cross co-spectral moments

The cross co-spectral moments are defined as frequency-domain integrals of the respective spectra and cross spectra:

$$m_{ij}^{(n)} = \Re \int_0^{\infty} \omega_e^n S_{ij}(\omega)\, d\omega \qquad (5)$$

where $\omega_e = \omega - kV\cos(\chi)$ is the frequency of encounter given speed, V, relative heading, χ, and wave number $k = \omega^2 / g$.

The matrix of cross co-spectral moments is Hermitian, because all of the off-diagonal elements occur as complex-conjugate pairs, i.e., $m_{jk} = \overline{m}_{kj}$. For this reason, the operator for the real part of a complex value, $\Re$, is included in equation 5 given above. However, this is not truly necessary, as these terms invariably occur in pairs in the quadratic expansion, and hence

[3] There are numerous conventions for Fourier forward and inverse transforms. Consult FourierTransform at http://mathworld.wolfram.com. The following results in a symmetric transform pair in the radian frequency $\omega = 2\pi f$:

$g(\omega) = \Im[f(\tau)] = \frac{1}{\sqrt{2\pi}} \int_{-\infty}^{\infty} f(\tau)\, e^{-i\omega\tau} d\tau$ and

$f(\tau) = \Im^{-1}[g(\omega)] = \frac{1}{\sqrt{2\pi}} \int_{-\infty}^{\infty} g(\omega)\, e^{i\omega\tau} d\omega$

their imaginary parts invariably cancel.[4] Similar observations apply to the cross power spectral densities themselves (i.e., $S_{jk}(\omega) = \overline{S}_{kj}(\omega)$).

Higher-order frequency moments (e.g., $m^{(2)}$ and $m^{(4)}$) are of use in determining the periods, T_2, of the zero-crossing, and T_4, of the local peak and trough processes, and the spectral breadth parameter, ε, that modifies the generalized Rayleigh probability distribution, all of which will be defined in a later section.

Zero-order moments and covariance

An important identity is that between the covariance matrix (i.e., autocovariance function evaluated at τ=0, see equation 2) and the zero-order cross co-spectral moments:

$$[\Gamma(0)] = \begin{bmatrix} \gamma_{xx}(0) & \gamma_{xy}(0) & \gamma_{xz}(0) \\ \gamma_{yx}(0) & \gamma_{yy}(0) & \gamma_{yz}(0) \\ \gamma_{zx}(0) & \gamma_{zy}(0) & \gamma_{zz}(0) \end{bmatrix} = \begin{bmatrix} m_{xx}^{(0)} & m_{xy}^{(0)} & m_{xz}^{(0)} \\ m_{yx}^{(0)} & m_{yy}^{(0)} & m_{yz}^{(0)} \\ m_{zx}^{(0)} & m_{zy}^{(0)} & m_{zz}^{(0)} \end{bmatrix} \quad (6)$$

We have now established the basic theory and notation necessary to our exposition of bi-axial Von Mises stress.

Results obtained from higher-order moments

It is convenient to now introduce some results that are obtained from higher-order cross co-spectral moments, as these will be of use in sections that follow.

For the familiar single-DOF Gaussian random variable, we identify the average zero-upcrossing period, T_2, and the average period of the local peak process, T_4 :

$$T_2 = 2\pi \sqrt{\frac{m_{\sigma\sigma}^{(0)}}{m_{\sigma\sigma}^{(2)}}} \quad (7)$$

$$T_4 = 2\pi \sqrt{\frac{m_{\sigma\sigma}^{(2)}}{m_{\sigma\sigma}^{(4)}}} \quad (8)$$

Here the subscript σ is intended to denote any critical process of interest, though in the context of this paper it will primarily refer to Von Mises stress (to be introduced in the next section). That is to say, in general, where σ is not itself subscripted (as in σ_x or σ_y) then $m_{\sigma\sigma}^{(0)} = m_{VM}^{(0)}$ is intended.

The following spectral bandwidth parameter, originally defined by Cartwright and Longuet-Higgins [4], is of use in the generalized Rayleigh probability distribution that will be discussed in a subsequent section:

$$\varepsilon = \sqrt{1 - \frac{[m_{\sigma\sigma}^{(2)}]^2}{m_{\sigma\sigma}^{(0)}\, m_{\sigma\sigma}^{(4)}}} = \sqrt{1 - \left(\frac{T_4}{T_2}\right)^2} \quad (9)$$

Bi-Axial Zero-Mean Von Mises Stress

For the class of problems where the normal stresses, σ_x and σ_y, and the shear stress[5], $\tau = \tau_{xy}$, are each zero-mean Gaussian processes, the equation of Von Mises stress in general coordinates (not principal stress coordinates), is:

$$\sigma_{VM}^2(t) = \sigma_x^2(t) + \sigma_y^2(t) - \sigma_x(t)\sigma_y(t) + 3\tau^2(t) \quad (10)$$

Applying the time average operator to equation 10, the co-variance of the Von Mises stress process is obtained:

$$\gamma_{VM}(0) = \gamma_{\sigma_x\sigma_x}(0) + \gamma_{\sigma_y\sigma_y}(0) - \gamma_{\sigma_x\sigma_y}(0) + 3\gamma_{\tau\tau}(0) \quad (11)$$

where the zero argument of the autocovariance function components refers to $\tau = 0$ in equation 2, and signifies that $\gamma_{\sigma_x\sigma_x}(0)$, $\gamma_{\sigma_y\sigma_y}(0)$, $\gamma_{\tau\tau}(0)$ and $\gamma_{\sigma_x\sigma_y}(0)$ are components of the covariance matrix.

Due to the equivalence shown by equation 6, equation 11 may be equivalently stated in terms of the cross co-spectral moments:

$$m_{VM}^{(0)} = m_{\sigma_x\sigma_x}^{(0)} + m_{\sigma_y\sigma_y}^{(0)} - m_{\sigma_x\sigma_y}^{(0)} + 3\, m_{\tau\tau}^{(0)} \quad (12)$$

and by extension (without proof):

$$m_{VM}^{(n)} = m_{\sigma_x\sigma_x}^{(n)} + m_{\sigma_y\sigma_y}^{(n)} - m_{\sigma_x\sigma_y}^{(n)} + 3\, m_{\tau\tau}^{(n)} \quad (13)$$

Provided (as is the usual case) that we have stress spectra for use in equation 5, then we can determine the cross co-spectral moments and proceed directly with equations 12 and 13.

As will subsequently be demonstrated, provided that σ_x, σ_y and τ are zero-mean Gaussian processes, the local peaks of the σ_{VM}-process are distributed according to the generalized Rayleigh distribution, as described by Cartwright and Longuet-Higgins [4], Ochi [15] and other authors. The class of problems where any or all of σ_x, σ_y or τ are not zero-mean Gaussian processes will be addressed in later sections of this paper.

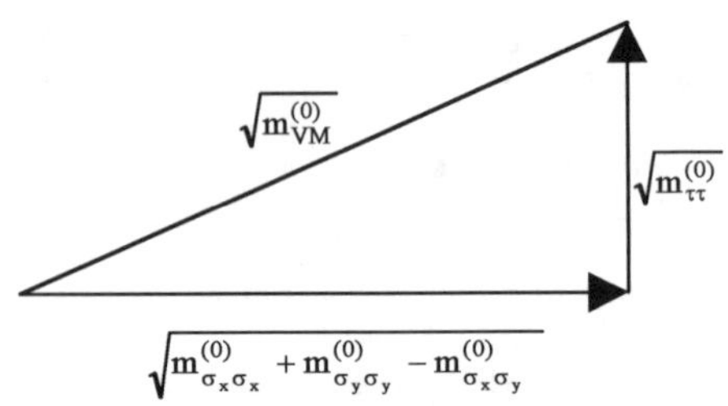

Figure 1. Illustrating the orthogonal Pythagorean relationship between the normal and shear stress components of Von Mises stress

[4] As the real parts of these complex conjugate pairs are equal, explicit notation such as m_{jk} and m_{kj}, can be omitted because

$m_{jk} = \Re\int_0^\infty S_{jk}(\omega)\,d\omega = \Re\int_0^\infty S_{kj}(\omega)\,d\omega = m_{kj}$. Thus, m_{jk} can be used to stand for either m_{jk} or m_{kj}.

[5] Hereafter the subscript notation on shear stress will be omitted wherever such omission is unlikely to result in misunderstanding.

As illustrated in Figure 1, it should be noted that $\sqrt{m^{(0)}_{\sigma_x\sigma_x} + m^{(0)}_{\sigma_y\sigma_y} - m^{(0)}_{\sigma_x\sigma_y}}$ and $\sqrt{m^{(0)}_{\tau\tau}}$ are orthogonal (Pythagorean) components of $\sqrt{m^{(0)}_{VM}}$.

Onward into the Probability Domain

As stated at the beginning of this paper, our objective is to develop the probability distribution for extreme values of Von Mises stress in a stationary Gaussian loads environment. In this section, we introduce the familiar generalized Rayleigh probability distribution on which we will rely.

Generalized Rayleigh probability

The generalized Rayleigh distribution was ably explored and presented by Ochi [15]. Selected results from Ochi's paper that are important to the current exposition are presented here for the convenience of the reader, and to frame the exposition consistently in the notation of this paper.

The generalized Rayleigh probability function for a nondimensional independent variable, $\xi = x/\sqrt{m^{(0)}_{\sigma\sigma}}$, is parametrically dependent on the spectral bandwidth parameter, ε, as given in equation 9 above.

Using the notational convention for conditional probability to indicate the parametric dependence on ε, the generalized Rayleigh probability density function is:

$$p(\xi \mid \varepsilon) = \frac{2}{1+\sqrt{1-\varepsilon^2}}\left[\frac{\varepsilon}{\sqrt{2\pi}}\exp\left\{-\frac{\xi^2}{2\varepsilon^2}\right\} + \sqrt{1-\varepsilon^2}\,\xi\exp\left\{-\frac{\xi^2}{2}\right\}\left\{1-\Phi\left(-\frac{\xi}{\varepsilon}\sqrt{1-\varepsilon^2}\right)\right\}\right] \quad (14)$$

for $0 \le \xi < \infty$ and $0 < \varepsilon \le 1.0$, where $\Phi(\xi)$ is the cumulative distribution function of the standardized normal (aka Gaussian) probability distribution.

For the special case when ε=0, the nondimensional Rayleigh probability density reduces to the familiar and classical narrow band case:

$$p(\xi \mid \varepsilon = 0) = \xi\, e^{-\frac{1}{2}\xi^2} \quad (15)$$

And again using the notational convention for conditional probability to indicate the parametric dependence on ε, the generalized Rayleigh cumulative probability function is:

$$P(\xi \mid \varepsilon) = \frac{2}{1+\sqrt{1-\varepsilon^2}}\left[-\frac{1}{2}(1-\sqrt{1-\varepsilon^2}) + \Phi\left(\frac{\xi}{\varepsilon}\right) - \sqrt{1-\varepsilon^2}\exp\left\{-\frac{\xi^2}{2}\right\}\left\{1-\Phi\left(-\frac{\xi}{\varepsilon}\sqrt{1-\varepsilon^2}\right)\right\}\right] \quad (16)$$

for $0 \le \xi < \infty$ and $0 < \varepsilon \le 1.0$.

For the special case when ε=0, the nondimensional Rayleigh cumulative probability distribution reduces to the familiar and classical narrow band case:

$$P(\xi \le \delta \mid \varepsilon = 0) = 1 - e^{-\frac{1}{2}\delta^2} \quad (17)$$

And the cumulative probability for exceeding any particular threshold, δ, is:

$$P(\xi \ge \delta \mid \varepsilon = 0) = e^{-\frac{1}{2}\delta^2} \quad (18)$$

In terms of dimensional variables, as used in the main body of this paper, the generalized Rayleigh distributions assume the following form:

$$P_R(\sigma \le \delta \mid m^{(0)}_{\sigma\sigma}, \varepsilon) = \frac{2}{1+\sqrt{1-\varepsilon^2}}\left[-\frac{1}{2}(1-\sqrt{1-\varepsilon^2}) + \Phi\left(\frac{\delta}{\varepsilon\sqrt{m^{(0)}_{xx}}}\right) - \sqrt{1-\varepsilon^2}\exp\left\{-\frac{\delta^2}{2\,m^{(0)}_{xx}}\right\}\left\{1-\Phi\left(-\frac{\delta}{\sqrt{m^{(0)}_{xx}}}\frac{\sqrt{1-\varepsilon^2}}{\varepsilon}\right)\right\}\right] \quad (19)$$

And:

$$P_R(\sigma \ge \delta \mid m^{(0)}_{\sigma\sigma}, \varepsilon) = 1 - P_R(\sigma \le \delta \mid m^{(0)}_{\sigma\sigma}, \varepsilon) \quad (20)$$

Component Stress Operators

To achieve the ultimate goal of estimating the lifetime extreme of σ_{VM}, it is first necessary to estimate $P_R(\sigma_{VM} \le \delta \mid m^{(0)}_{VM}, \varepsilon)$. The values of $m^{(n)}_{VM}$ are obtained from equation 13, and the value of ε from equation 9. Both of these equations require the cross co-spectral moments of normal stresses, σ_x and σ_y and of shear, τ, which are obtained by using complex stress operators in the frequency domain. Complex stress operators are determined in the frequency domain for the component normal and shear stress processes according to each element's local x and y axes. These may be determined using a global finite element structural analysis model (FEA). Within the context of linear theory, the process for determining the component stress operators is consistent with the established dynamic load analysis method that is well defined in ABS and Hutchison et al [2, 8]. However, the procedure herein advocated differs from the more conventional DLA, in that complex-valued stress operators are determined for each element of interest, and for each regular wave frequency, relative heading, speed and vessel load condition. Thus, the advocated procedure places a significant stress analysis step before the determination of the dynamic load parameters (DLPs). This is because it is the goal of our procedure to base some of our DLPs on Von Mises stresses.

The procedure that we advocate involves running a 3-D finite element model through a matrix of regular wave load cases. A sufficient matrix might resemble the following:

- Sixteen to 32 regular wave frequencies, where the density of the frequency mesh is determined by the complexity of the structural response vs. frequency.

- Two wave phase angles for each wave frequency, representing the real and imaginary cases.
- Twelve to 24 wave headings for heading increments of 15° or 30°.
- One to four combinations of (hydro-) static load condition and forward speed combinations, depending on the operating profile of the structure and the sensitivity of the structural response to load case.

As will subsequently be shown, when we consider the class of problems with still water stress, the (hydro-) static load conditions only need concern displacement and trim, as still water stress can be efficiently introduced at a later stage of the analysis.

Depending on the number of frequencies and load cases run, this could result in between 384 and 6144 FEA load cases. Obviously, these parameters should be selected carefully to achieve sufficient fidelity with acceptable computation time.

For each structural element, relative heading, speed and load condition,[6] the complex-valued stress operators in the frequency domain are as follows:

$$\hat{\sigma}_x(\omega) = \frac{\sigma_x(\omega)}{\varsigma(\omega)} = \left|\frac{\sigma_x(\omega)}{\varsigma(\omega)}\right| e^{i\phi_x(\omega)} \tag{21a}$$

$$\hat{\sigma}_y(\omega) = \frac{\sigma_y(\omega)}{\varsigma(\omega)} = \left|\frac{\sigma_y(\omega)}{\varsigma(\omega)}\right| e^{i\phi_y(\omega)} \tag{21b}$$

$$\hat{\tau}_{xy}(\omega) = \frac{\tau_{xy}(\omega)}{\varsigma(\omega)} = \left|\frac{\tau_{xy}(\omega)}{\varsigma(\omega)}\right| e^{i\phi_{xy}(\omega)} \tag{21c}$$

And the stress spectra that enter into equation 5 are given in the usual way as:

$$S_{\sigma_x\sigma_x}(\omega) = \hat{\sigma}_x(\omega)\,\overline{\hat{\sigma}_x(\omega)}\,S_{\varsigma\varsigma}(\omega) \tag{22a}$$

$$S_{\sigma_y\sigma_y}(\omega) = \hat{\sigma}_y(\omega)\,\overline{\hat{\sigma}_y(\omega)}\,S_{\varsigma\varsigma}(\omega) \tag{22b}$$

$$S_{\tau\tau}(\omega) = \hat{\tau}_{xy}(\omega)\,\overline{\hat{\tau}_{xy}(\omega)}\,S_{\varsigma\varsigma}(\omega) \tag{22c}$$

$$S_{\sigma_x\sigma_y}(\omega) = \hat{\sigma}_x(\omega)\,\overline{\hat{\sigma}_y(\omega)}\,S_{\varsigma\varsigma}(\omega) \tag{22d}$$

$$S_{\sigma_y\sigma_x}(\omega) = \hat{\sigma}_y(\omega)\,\overline{\hat{\sigma}_x(\omega)}\,S_{\varsigma\varsigma}(\omega) \tag{22e}$$

$$S_{\sigma_x\tau}(\omega) = \hat{\sigma}_x(\omega)\,\overline{\hat{\tau}_{xy}(\omega)}\,S_{\varsigma\varsigma}(\omega) \tag{22f}$$

$$S_{\tau\sigma_x}(\omega) = \hat{\tau}_{xy}(\omega)\,\overline{\hat{\sigma}_x(\omega)}\,S_{\varsigma\varsigma}(\omega) \tag{22g}$$

$$S_{\sigma_y\tau}(\omega) = \hat{\sigma}_y(\omega)\,\overline{\hat{\tau}_{xy}(\omega)}\,S_{\varsigma\varsigma}(\omega) \tag{22h}$$

$$S_{\tau\sigma_y}(\omega) = \hat{\tau}_{xy}(\omega)\,\overline{\hat{\sigma}_y(\omega)}\,S_{\varsigma\varsigma}(\omega) \tag{22i}$$

---

[6] A practice of strict and explicit notation would suggest that $\hat{\sigma}_x$, $\hat{\sigma}_y$ and $\hat{\tau}_{xy}$ (and the spectral densities in equation set 22) all indicate their parametric dependencies on structural element, relative heading, speed and load. However, such explicit notation is extremely cumbersome and omitted here.

where $S_{\varsigma\varsigma}(\omega)$ is the incident wave spectral density and the bar, $\overline{}$, over the stress operators denotes the complex conjugate.

It follows from equations 5 and 12 that:

$$S_{VM}(\omega) = S_{\sigma_x\sigma_x}(\omega) + S_{\sigma_y\sigma_y}(\omega) - S_{\sigma_x\sigma_y}(\omega) + 3\,S_{\tau\tau}(\omega) \tag{23}$$

Since it is possible to determine the power spectral density for Von Mises stress, it follows that a response amplitude operator for Von Mises stress can also be defined:

$$\begin{aligned}
\hat{\sigma}_{VM}(\omega) &= \sqrt{\frac{S_{VM}(\omega)}{S_{\varsigma\varsigma}(\omega)}} \\
&= \Big\{ \hat{\sigma}_x(\omega)\overline{\hat{\sigma}_x(\omega)} + \hat{\sigma}_y(\omega)\overline{\hat{\sigma}_y(\omega)} \\
&\qquad - \Re[\hat{\sigma}_x(\omega)\overline{\hat{\sigma}_y(\omega)}] + 3\,\hat{\tau}(\omega)\overline{\hat{\tau}(\omega)} \Big\}^{1/2} \\
&= \Big\{ |\hat{\sigma}_x(\omega)|^2 + |\hat{\sigma}_y(\omega)|^2 \\
&\qquad - |\hat{\sigma}_x(\omega)|\,|\hat{\sigma}_y(\omega)| \cos[\phi_x(\omega) - \phi_y(\omega)] \\
&\qquad + 3\,|\hat{\tau}(\omega)|^2 \Big\}^{1/2}
\end{aligned} \tag{24}$$

This is an useful development if the ABS equivalent regular wave approach to DLA is to be used, as that approach requires determination of the frequency, relative heading and speed associated with the peak operator, i.e., finding (ω, χ, V) such that $\hat{\sigma}_{VM}(\omega, \chi, V)$ is maximum.

Since no phase information is retained, equation 24 is a response amplitude operator, but not a frequency response operator. The Von Mises stress response amplitude operator of equation 24 is valid in the mean squared sense. The normal stress and shear stress terms making up the Von Mises stress RAO participate as orthogonal Pythagorean components in the sense illustrated in Figure 1.

In subsequent sections of this paper, results are presented that lead to unaccountably large spectral breadth parameters. Speculatively, the internal dimensionality of the Von Mises stress RAO may contribute to that large spectral breadth.

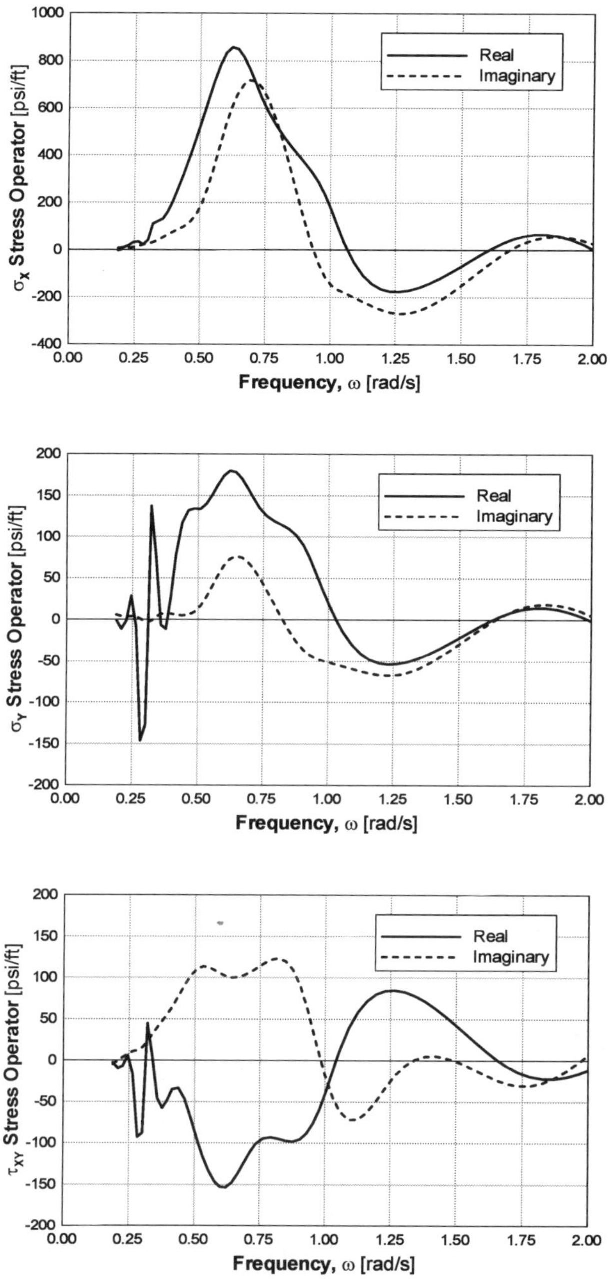

Figure 2. Example stress operators $\widehat{\sigma}_x$, $\widehat{\sigma}_y$ and $\widehat{\tau}_{xy}$ at 90 degree relative heading angle

Figure 2 presents an example of the stress operators $\widehat{\sigma}_x(\omega)$, $\widehat{\sigma}_y(\omega)$ and $\widehat{\tau}_{xy}(\omega)$ described by equations 21 a-c. The example shown is for a normal-stress-dominated structural element for a subject vessel at 90° relative heading.

Structural elements studied

Three different structural elements were included in this study. One element was normal-stress-dominated, the second was neither normal- nor shear-stress-dominated and the third was shear-stress-dominated.

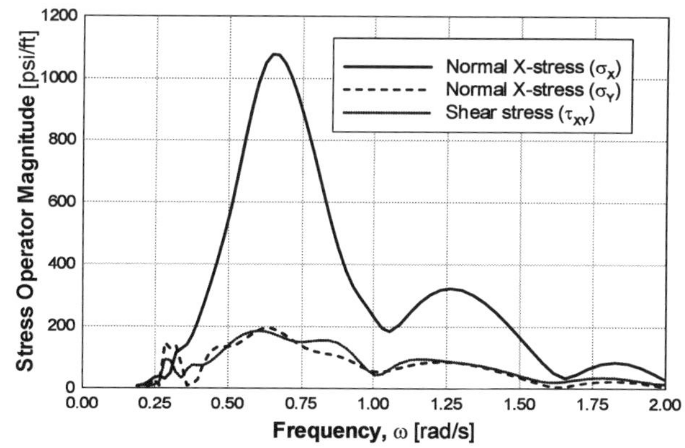

Figure 3. Example normal-stress-dominated element at 90° relative heading angle

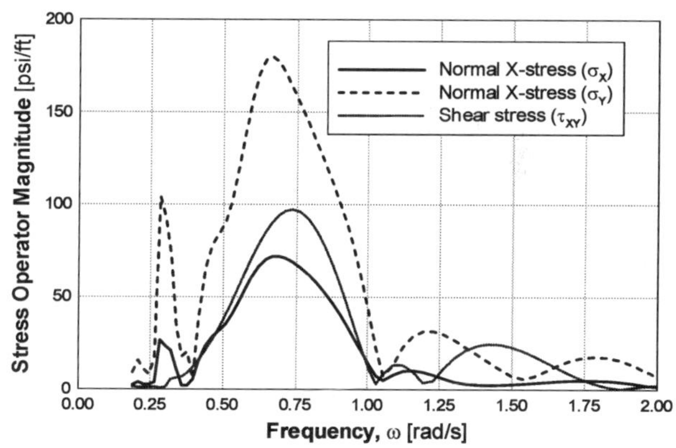

Figure 4. Example structural element that is neither normal-stress- nor shear-stress-dominated at 111° relative heading angle

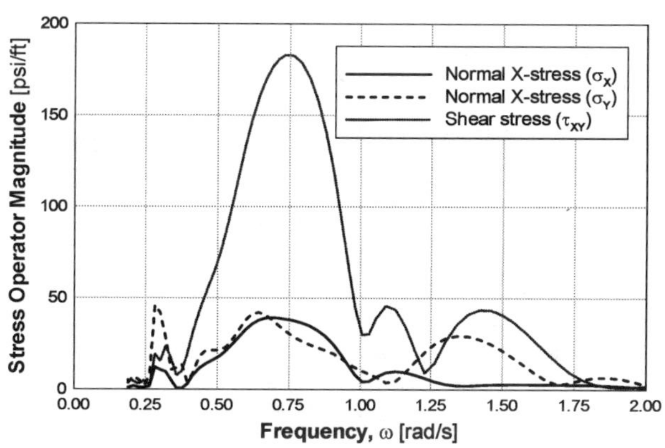

Figure 5. Example shear-stress-dominated element at 60° relative heading angle

Figures 3-5 respectively present stress components for these structural elements. Plots for each element are presented at headings selected to maximize Von Mises stress and to best represent the stress component dominating the response. .

Table 1. Characteristics of selected example structural elements with H_S=32.2 feet (≈10 m) and T_P=12 s

Structural Element stress dominance	Relative Heading Angle [deg]	H_S=32.2 feet T_P=12 s	
		σ_{VM} [psi]	$\dfrac{m^{(0)}_{\sigma_x\sigma_x} + m^{(0)}_{\sigma_y\sigma_y} - m^{(0)}_{\sigma_x\sigma_y}}{m^{(0)}_{VM}}$
Normal stress	90°	5,815	88.0%
No dominance	111°	1,292	47.5%
Shear stress	60°	1,897	5.8%

The ordinate in these figures is the magnitude of the stress operator. The magnitudes of all three stress components ($\hat{\sigma}_x$, $\hat{\sigma}_y$ and $\hat{\tau}_{xy}$) are shown in each figure as an aid to comparing magnitudes and dominance. The normal-stress-dominated element in Figure 3 is the same as that for which real and imaginary parts are presented in Figure 2.

The characteristics of the selected example structural elements are summarized in Table 1 for a sea state with H_S=32.2 feet and T_P=12 s. The last column in Table 1 shows percentage contribution of the normal stresses to the variance of the Von Mises stress process. Each element may be seen to be a good example of its dominance type.

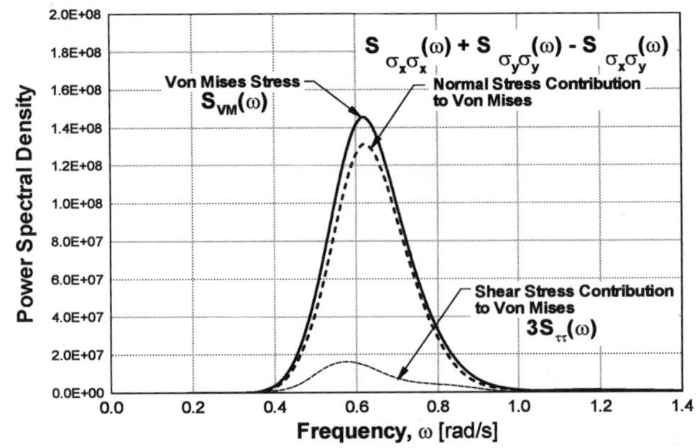

Figure 6. Example power spectral density spectrum of Von Mises stress

Figure 6 shows the power spectral density of Von Mises stress for the normal-stress-dominated element. Also shown are the power spectra of the shear stress and the normal stress components that contribute as orthogonal Pythagorean components to the Von Mises spectrum (see also Figure 1). The Von Mises spectrum of the other example structural elements is similarly well behaved, and exhibits well defined modes in the sea state with H_S=32.2 feet and T_P=12 s.

It should be noted that the frequency shown along the abscissa of these figures is the wave frequency, and not either the frequency of encounter or the frequency associated with the Von Mises stress response. Due to the quadratic character of Von Mises stress, there is a frequency doubling that will be further elucidated in a following section concerning the period structure of the Von Mises response processes. Thus, the frequency of any Von Mises response process will be twice the wave encounter frequency.

Fourier-Stieltjes Simulations

Validation of the method of cross co-spectral moments applied to zero-mean Von Mises stress is accomplished through comparison with data generated using Fourier-Stieltjes simulations of Von Mises stress processes.

The Fourier-Stieltjes simulation method is set forth by the following equations:

$$\sigma_x(t) = \Re\int_0^\infty \hat{\sigma}_x(\omega)\, e^{i\{\omega t+\varepsilon(\omega)\}} \sqrt{2\,S_{\varsigma\varsigma}(\omega)\, d\omega}$$

$$= \Re\int_0^\infty e^{i\{\omega t+\varepsilon(\omega)\}} \sqrt{2\,S_{\sigma_x\sigma_x}(\omega)\, d\omega} \qquad (25a)$$

$$\sigma_y(t) = \Re\int_0^\infty \hat{\sigma}_y(\omega)\, e^{i\{\omega t+\varepsilon(\omega)\}} \sqrt{2\,S_{\varsigma\varsigma}(\omega)\, d\omega}$$

$$= \Re\int_0^\infty e^{i\{\omega t+\varepsilon(\omega)\}} \sqrt{2\,S_{\sigma_y\sigma_y}(\omega)\, d\omega} \qquad (25b)$$

$$\tau_{xy}(t) = \Re\int_0^\infty \hat{\tau}_{xy}(\omega)\, e^{i\{\omega t+\varepsilon(\omega)\}} \sqrt{2\,S_{\varsigma\varsigma}(\omega)\, d\omega}$$

$$= \Re\int_0^\infty e^{i\{\omega t+\varepsilon(\omega)\}} \sqrt{2\,S_{\tau\tau}(\omega)\, d\omega} \qquad (25c)$$

where $\varepsilon(\omega)$ is a random phase selected at each frequency from a uniform distribution on the interval $-\pi$ to π.

Von Mises stress is then given at each instant by equation 10.

In order to generate high-fidelity time-domain realizations, and in particular to obtain good representation of extremes, the following measures were taken:

- Real and imaginary parts of the stress operators (equations 21 a-c) were interpolated to 256 random frequencies.
- Multiple ensemble realizations were generated, each 1200 seconds (20 minutes) in length, and each with unique random frequency mesh and associated random phase.

The total number of ensemble realizations for each case was variable depending on the objective of the case study.

Criterion for peaks in time domain

The Fourier-Stieltjes time-domain simulations were carried out with a time step of one-twelfth of a second. At this sampling rate, each twenty-minute ensemble results in an array of 14,401 values for each simulated variable (i.e., η, σ_x, σ_y, τ and σ_{VM}). For bona fide zero-mean Gaussian variables, a common and robust strategy for determining peaks (and troughs) is to find the greatest (and least) values between successive zero up-crossings. The zero up-crossing approach to counting peaks leads to the average period T_2 (see equation 7). However, this strategy is not suitable to the task of determining the peaks of a strictly positive function such as Von Mises stress.

To overcome these difficulties, a very simple criterion was applied to determine the peaks of each Von Mises stress time series. The essence of the criterion is that any value that was greater than both its predecessor and its successor was regarded as a peak. Additional logic was applied to detect the rare case of successive ordinates equal valued at a local peak. This definition of a local 'peak' resembles the practice used to count local peaks leading to average period T_4, (see equation 8).

Questions of what constitutes a local peak and how they are counted are of interest, both because the total number of sampled 'peaks' affects the determination of the sample cumulative probability of Von Mises (against which we will be comparing), and also because it affects the sampled average period for the Von Mises stress process.

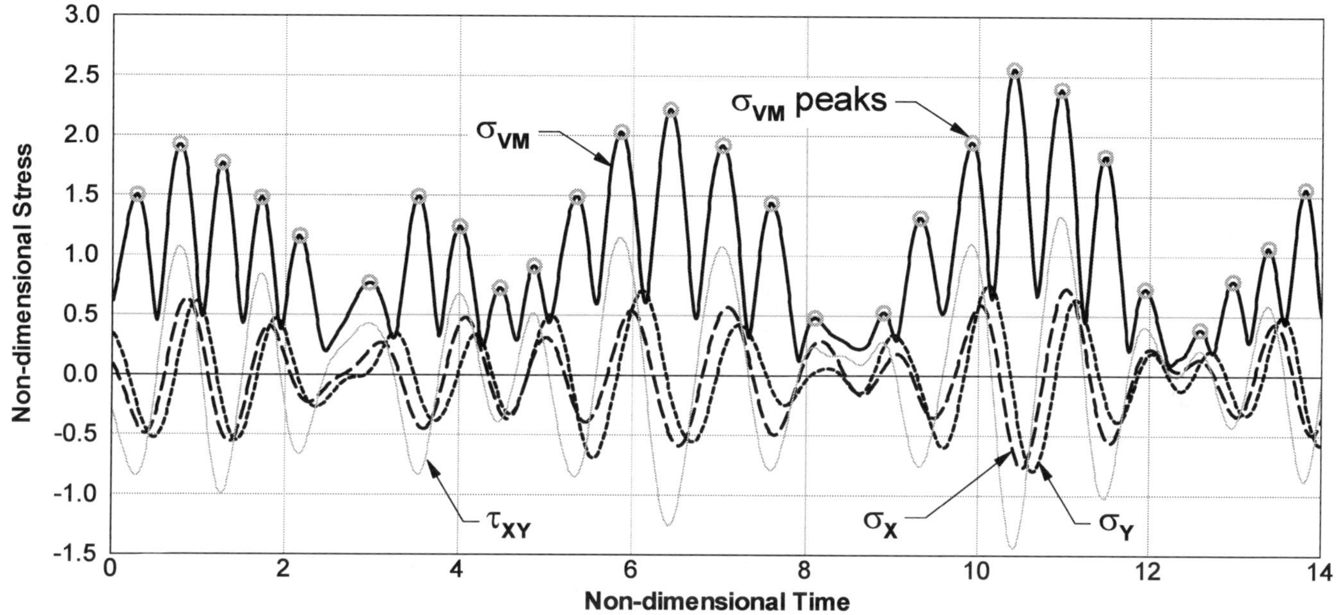

Figure 7. Example time series for $\sigma_{VM}(t)$, $\sigma_x(t)$, $\sigma_y(t)$ and $\tau_{xy}(t)$

Figure 7 provides an example time series for the structural element with no dominance. Stresses are all nondimensionalized by $\sqrt{m_{VM}^{(0)}}$, and time is nondimensionalized by $\frac{1}{2}(T_2 + T_4)$, the average of T_2 and T_4 without any consideration of frequency doubling[7] (period halving) associated with the quadratic Von Mises process.

The frequency doubling (period halving) is quite apparent in Figure 7, as there are twenty-seven peaks of the nondimensional Von Mises stress process in the fourteen nondimensional time units, whereas without frequency doubling (period halving), we would expect the number of peaks to approximately equal the integer measure of nondimensional time.

The other characteristics worthy of note are that all three of the underlying basis stresses, $\sigma_x(t)$, $\sigma_y(t)$ and $\tau_{xy}(t)$, are zero-mean processes and the Von Mises stress process is strictly positive.

Comparison with Fourier-Stieltjes

Comparison of the probability distribution given by equation 17 using the variance of the Von Mises stress process obtained from equation 12 is provided in Figures 8-10. The cumulative probability on the ordinal axis is transformed according to $\log_{10}\{1/[1-P(\sigma_{VM})]\}$ in order to provided focus and resolution on the comparison at extreme values of the Von Mises stress process.

---

[7] Recall that $\cos^2(\omega t) = \frac{1}{2}\{1+\cos(2\omega t)\}$:

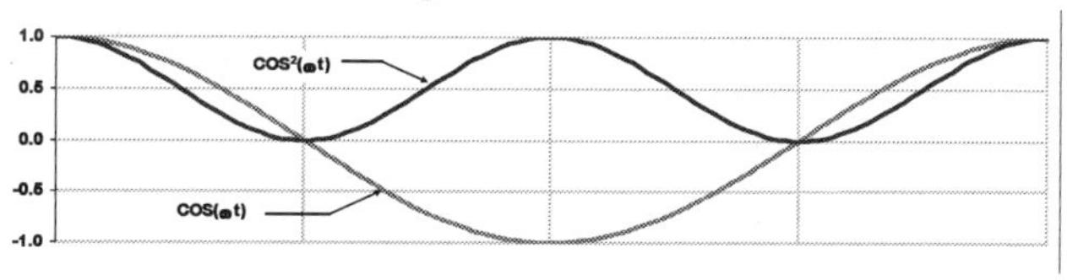

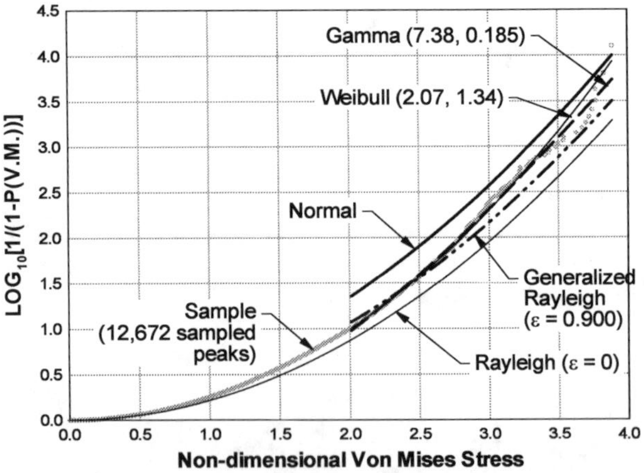

Figure 8. Normal-stress-dominated element; comparison of Fourier-Stieltjes time-domain simulation and theoretical Rayleigh cumulative with emphasis on extreme values. Heading=90 deg, H_s=32.2 ft, T_p=12 sec

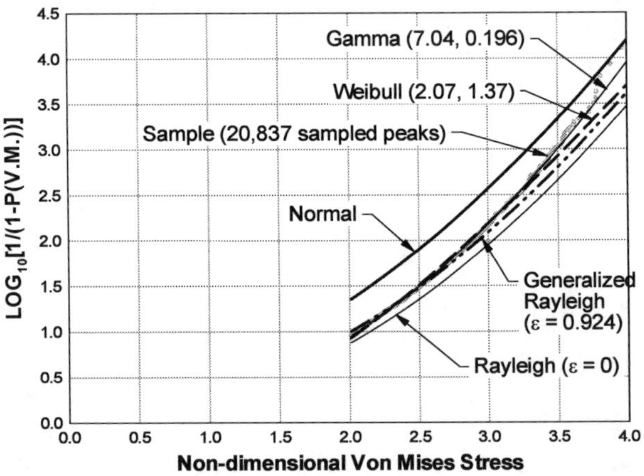

Figure 9. Structural element with no dominance; comparison of Fourier-Stieltjes time-domain simulation and theoretical Rayleigh cumulative with emphasis on extreme values. Heading=111 deg, H_s=32.2 ft, T_p=12 sec

Also shown in Figures 8 and 9 are normal, gamma, Weibull and generalized Rayleigh distributions (equation 16). The parameters of the gamma and Weibull distributions were fit using least squares regression. The spectral breadth parameter of the generalized Rayleigh was also fit using least squares regression, and the fit values of ε are significantly higher than those obtained using the higher-order frequency moments of the Von Mises process in equation 9.

Narrow-band Rayleigh probability is a special case of both the gamma and Weibull distributions. Gamma and Weibull may therefore be regarded as alternative forms of generalized Rayleigh, different from that advanced by Cartwright and Longuet-Higgins [4]. The challenge is to identify a general theory for the extreme values of Von Mises stress where all the parameters are calculable in the frequency domain.

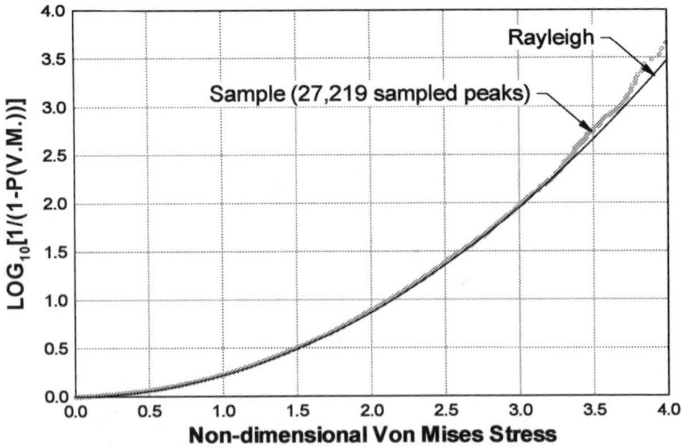

Figure 10. Shear-stress-dominated element; comparison of Fourier-Stieltjes time-domain simulation and theoretical Rayleigh cumulative with emphasis on extreme values. Heading=60 deg, H_s=32.2 ft, T_p=12 sec

Figure 10 presents an interesting and revealing contrast to Figures 8 and 9. The normal-stress-dominated element in Figure 8 and the element with no dominance shown in Figure 9 both exhibit clear departures from narrow-band Rayleigh probability, even though the spectral bandwidth parameter ε, calculated from the higher-order frequency moments, is modest (ε=0.367 in the case of the normal-stress-dominated element and ε=0.215 for the element with no stress dominance). The shear-stress-dominated element in Figure 10 adheres very closely to the narrow-band Rayleigh distribution. If one examines equation 12, it may be seen that shear enters through a single term, $3\,m_{\tau\tau}^{(0)}$. Hence, a shear-stress-dominated process is essentially one-dimensional. Taken together, these observations suggest that the departures observed in Figures 8 and 9 may be a manifestation of increased dimensionality (degrees-of-freedom).

Period structure of Von Mises stress process

Referring to equation 10, the Von Mises stress process is seen to depend on the underlying stress processes, $\sigma_x(t)$, $\sigma_y(t)$ and $\tau_{xy}(t)$. $\vec{x} = (\sigma_x, \sigma_y, \tau)$ is a random variable with a joint Gaussian distribution:

$$p(\sigma_x, \sigma_y, \tau) = p(x, y, z) = p(\vec{x}) = \frac{\exp(-\frac{1}{2}\vec{x}^T[\Gamma]^{-1}\vec{x})}{(2\pi)^{(3/2)}\sqrt{|\Gamma|}} \qquad (26)$$

Therefore the Von Mises stress σ_{VM} is a function of a Gaussian random variable, $\vec{x}$. As in the case of the more familiar single-DOF Gaussian random variable, this relationship can be exploited to determine the amount of time (on average) that σ_{VM} exceeds some threshold value δ, but, in itself, this joint Gaussian distribution is insufficient to estimate the mean time between exceedances of that threshold.

Here it is helpful to observe that the joint Gaussian distribution of $\vec{x}$ applies to a continuous function in time, whereas the process of interest consists of discrete events (the passage above the threshold, regardless of duration) that are counted on the integers. We track these discrete events by tracking the local maxima (peaks) in time. If we can then determine the average time between these peaks (period) we are able to estimate the average exposure time before any particular threshold is exceeded. For the familiar single-DOF Gaussian random variable we identify the average zero-upcrossing period, T_2, and the average period of the local peak process, T_4 .

As a quadratic process, Von Mises stress is by its very definition a strictly positive function, and the concept of zero-upcrossing therefore loses force. Furthermore, it should be observed that the squaring of a (generalized) harmonic process leads to frequency doubling (or halving of apparent periods).

Values of T_2 and T_4 were calculated for the Von Mises stress process using equations 7 and 8, respectively, as well as the frequency moments, $m_{VM}^{(0)}$, $m_{VM}^{(2)}$ and $m_{VM}^{(4)}$, as determined from equations 12 and 13. These are compared in Figure 11 with the apparent period of the local peaks of the Von Mises stress process sampled from the Fourier-Stieltjes time-domain simulations.

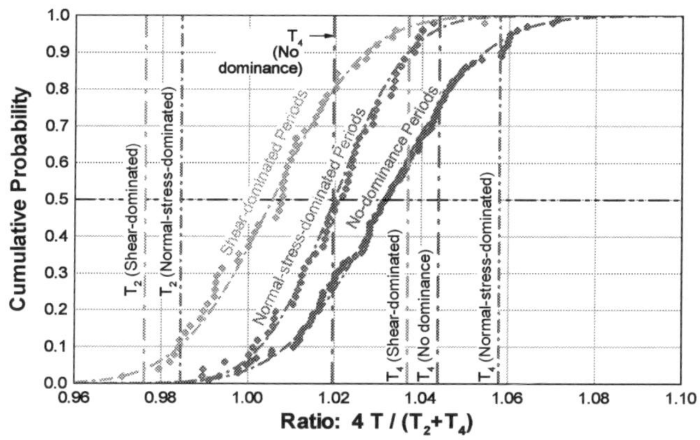

Figure 11. Comparison of Von Mises period determined from sample time-domain simulations with the average of $[T_2]/2$ and $[T_4]/2$ as determined from frequency-domain spectral moments

It may be observed from Figure 11 that the median apparent period of the Von Mises stress process is well approximated by the average:

$$T_{VM} \cong \frac{\frac{1}{2}T_2 + \frac{1}{2}T_4}{2} = \tfrac{1}{4}[T_2 + T_4] \tag{27}$$

The period estimator given by equation 27 is slightly conservative, in that the actual medians of the sampled periods are longer by one-half to about three percent. This is a conservative bias, because using equation 27 will result in a slight over-estimate of the number of Von Mises stress peaks expected during any duration of exposure.

Order statistics of Von Mises stress process

In general, the interest in Von Mises stress is in extreme values, either the short-term extreme or the long-term extreme. For use as a DLP in a dynamic loads analysis (DLA) procedure, the interest will be focused primarily on the long-term extreme, which will be addressed in a later section of this paper. However, there may well be situations, for example where a design sea state exposure has been specified, where short-term methods may be appropriate.

As ably presented by Ochi [15] and other authors, the order statistical distribution of the extreme value in N peaks is:

$$p_N(x) = N\,[P(x)]^{N-1}\,p(x) \tag{28}$$

where: $p(x)$ is the probability density distribution for peaks of x

$P(x)$ is the cumulative probability distribution for peaks of x

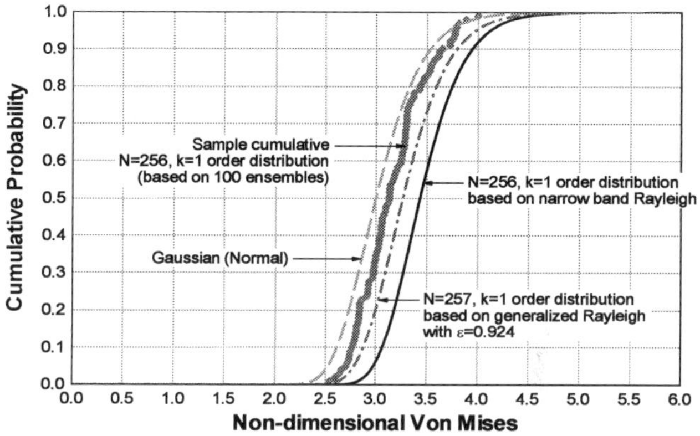

Figure 12. Comparison of sample short-term extreme value statistics with theoretical order distribution. Heading=111 deg, H_s=32.2 ft, T_p=12 sec.

Figure 12 is an example based on the structural element with no dominance. It compares the sample cumulative probability distribution of the extreme values from each of one hundred 20-minute time-domain realizations, with an average of 257 Von Mises peaks in each realization. Also shown are the corresponding narrow-band Rayleigh, generalized Rayleigh, and Gaussian (normal) distributions.

The narrow-band Rayleigh distribution is predicted using equation 28, where p(x) and P(x) are narrow-band Rayleigh. The generalized Rayleigh distribution is a predicted using equation 28, where p(x) and P(x) are generalized Rayleigh (equations 14 and 16 respectively) with ε=0.924. This is the same spectral breadth parameter that was determined using least-squares regression to provide a good fit to the sampled distribution as shown in Figure 9.

The theoretical cumulative-order distributions shown in Figure 12 were all generated through numerical quadrature of the probability density distributions given by equation 28.

Each distribution is clearly congruent with the sample. However, the narrow-band Rayleigh order distribution overpredicts the sample by an average of 10.0%. The generalized Rayleigh order distribution overpredicts the sample by an average of 4.6%. And, the Gaussian (normal) order distribution under predicts the sample by an average of 4.5%.

Von Mises with Nonzero Mean

The class of problems where the normal stresses, σ_x and σ_y, and the shear stress, τ, are each Gaussian processes but some or all possess nonzero means is far more common and important to the problem of ship structures where there are still water (i.e., mean) loads. The preceding exposition of the exceptional class of problems where stresses σ_x, σ_y and τ are each zero-mean Gaussian processes, while useful in its own right, has primarily been provided for its pedagogic value and as a stepwise introduction to the more complete but complex class of problems with nonzero mean σ_x, σ_y and τ.

At any instant in time, the Von Mises stress state may be regarded as the arising from equation 10 (here repeated):

$$\sigma_{VM}^2(t) = \sigma_x^2(t) + \sigma_y^2(t) - \sigma_x(t)\sigma_y(t) + 3\tau^2(t) \tag{29}$$

where:

$$\sigma_x(t) = \sigma_{x_0} + \tilde{\sigma}_x(t)$$

$$\sigma_y(t) = \sigma_{y_0} + \tilde{\sigma}_y(t)$$

$$\tau(t) = \tau_0 + \tilde{\tau}(t)$$

Here σ_{x_0}, σ_{y_0} and τ_0 are the still water (or time invariant) values of the basis normal and shear process respectively. $\tilde{\sigma}_x(t)$, $\tilde{\sigma}_y(t)$ and $\tilde{\tau}(t)$ are the time-dependent basis stress processes that are regarded as zero-mean Gaussian processes with covariance matrix $[\,\Gamma(0)\,]$ (refer to equation 6).

Hence:

$$\begin{aligned}\sigma_{VM}^2(t) &= [\sigma_{x_0} + \tilde{\sigma}_x(t)]^2 + [\sigma_{y_0} + \tilde{\sigma}_y(t)]^2 \\ &\quad - [\sigma_{x_0} + \tilde{\sigma}_x(t)][\sigma_{y_0} + \tilde{\sigma}_y(t)] + 3[\tau_0 + \tilde{\tau}(t)]^2 \\ &= \{\sigma_{x_0}^2 + \sigma_{y_0}^2 - \sigma_{x_0}\sigma_{y_0} + 3\tau_0^2\} \\ &\quad + \{\tilde{\sigma}_x^2(t) + \tilde{\sigma}_y^2(t) - \tilde{\sigma}_x(t)\tilde{\sigma}_y(t) + 3\tilde{\tau}^2(t)\} \\ &\quad + 2\sigma_{x_0}\tilde{\sigma}_x(t) + 2\sigma_{y_0}\tilde{\sigma}_y(t) - \sigma_{x_0}\tilde{\sigma}_y(t) - \sigma_{y_0}\tilde{\sigma}_x(t) \\ &\quad + 6\tau_0\tilde{\tau}(t)\end{aligned} \tag{30}$$

Our objective is to determine the probability distribution for extreme values of Von Mises stress. As will be shown below, cross terms, such as $\sigma_{x_0}\tilde{\sigma}_x(t)$, $\sigma_{y_0}\tilde{\sigma}_y(t)$, $\sigma_{x_0}\tilde{\sigma}_y(t)$, $\sigma_{y_0}\tilde{\sigma}_x(t)$ and $\tau_0\,\tilde{\tau}(t)$, play an important role in those extreme values. Further results concerning the time average of equation 30 are presented in Appendix B, but the exposition here will continue by considering the various terms of equation 30 under the condition that each Von Mises stress point corresponds to times at which Von Mises stress is locally maximum.

Figure 13 is for a case of the shear-stress-dominated element at 120° relative heading with H_S=32.2 feet, T_P=12 sec. In this example, each component of the mean (still water) stress is set equal to the root mean square (RMS) value of the respective time-dependent stress component processes (74.3,421.1 and 930.8).

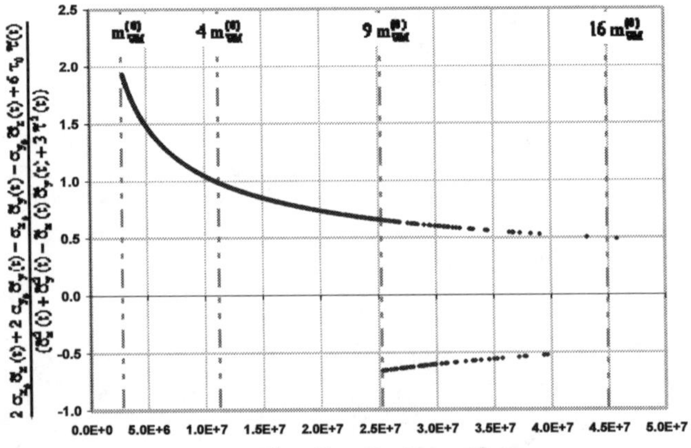

Figure 13. Example showing ratio of the sum of the cross terms of equation 30 to $\{\tilde{\sigma}_x^2(t) + \tilde{\sigma}_y^2(t) - \tilde{\sigma}_x(t)\,\tilde{\sigma}_y(t) + 3\,\tilde{\tau}^2(t)\}$.

Figure 13 shows the ratio of the cross terms to the term $\{\tilde{\sigma}_x^2(t) + \tilde{\sigma}_y^2(t) - \tilde{\sigma}_x(t)\,\tilde{\sigma}_y(t) + 3\,\tilde{\tau}^2(t)\}$, whose time average, $m_{VM}^{(0)}$, as obtained from equation 12, figures so prominently in the class of problems with zero means of σ_x, σ_y and τ. For reference, four vertical lines are shown in Figure 13, corresponding (from left to right) our values of $\sqrt{m_{VM}^{(0)}}$, $2\sqrt{m_{VM}^{(0)}}$, $3\sqrt{m_{VM}^{(0)}}$ and $4\sqrt{m_{VM}^{(0)}}$.

At large values of $\{\tilde{\sigma}_x^2(t) + \tilde{\sigma}_y^2(t) - \tilde{\sigma}_x(t)\,\tilde{\sigma}_y(t) + 3\,\tilde{\tau}^2(t)\}$, the upper (and dominant) branch of the loci in Figure 13 may be seen to be asymptotic to a cross term ratio of 0.5. The lower branch, whose origin and significance will be more apparent below, may be seen to be the symmetric reflection below the axis of the dominant branch and hence asymptotic to -0.5.

Statistical principal axis for Von Mises stress

The method of statistical principal axes for joint processes described in [9-11] is recommended as an approach to the estimation of the contribution of cross-terms such as $\sigma_{x_0}\tilde{\sigma}_x(t)$, $\sigma_{y_0}\tilde{\sigma}_y(t)$, $\sigma_{x_0}\tilde{\sigma}_y(t)$, $\sigma_{y_0}\tilde{\sigma}_x(t)$ and $\tau_0\,\tilde{\tau}(t)$ to Von Mises stress in the case where some or all of σ_x, σ_y and/or τ have nonzero means.

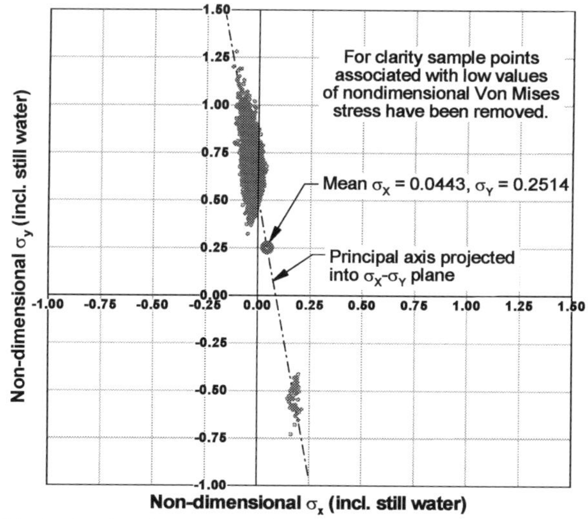

A) Projection in σ_x σ_y -plane

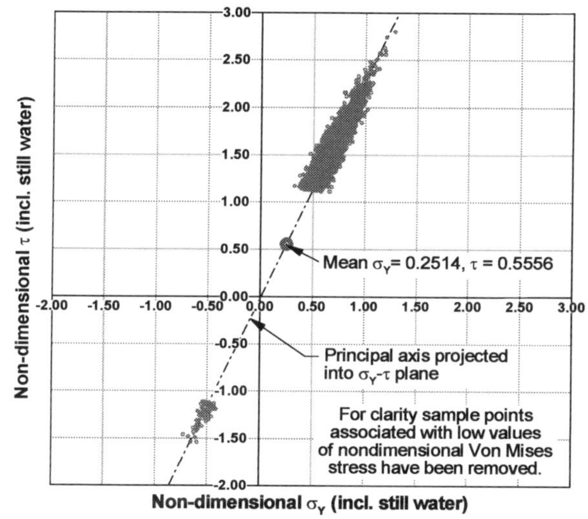

B) Projection in σ_y τ-plane

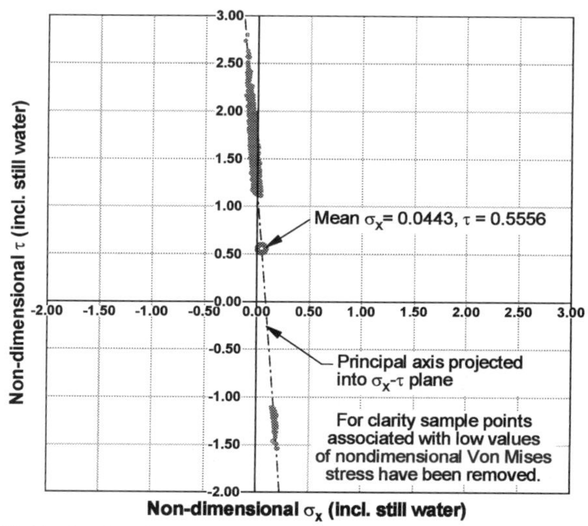

C) Projection in σ_x τ-plane

Figure 14. Orthogonal projections of 3-D correlogram of Von Mises peaks showing the statistical principal axis

Figure 14 applies to the example case of the shear-stress-dominated element at 120° relative heading with H_S=32.2 feet, T_P=12 sec and mean (still water) stresses (74.3, 421.1 and 930.8).

The direction cosines of the statistical principal axis are (-0.0672, 0.3992, 0.9144). The three plots taken together constitute a set of three orthogonal projections. Sample extreme peaks of the Von Mises stress process are shown projected in each view, along with the projection of the statistical principal axis. The salient observation is that these sample extreme peaks cluster closely about the statistical principal axis.

Determination of the orientation of the statistical principal axis for Von Mises stress

Consistent with the approach to statistical principal axes described in [10] and [11], the statistical principal axis for Von Mises stress is here defined as a line in σ_x σ_y τ-space directed from the mean (i.e., still water) stresses (σ_{x_0}, σ_{y_0}, τ_0) through the point on a surface of constant Von Mises stress at which the probability density is maximized.

As previously noted, surfaces of constant Von Mises stress are ellipsoids in σ_x σ_y τ-space (also referred to hereafter as xyz-space). Dropping the explicit notation for time dependency, the equation of bi-axial Von Mises stress given in general coordinates (not principal stress coordinates) as equation 10, may be written as:

$$\sigma_{VM}^2 = x^2 + y^2 - xy + 3z^2 \tag{31}$$

where $x = \sigma_x$, $y = \sigma_y$, $z = \tau$, which describes an ellipsoid oriented along y=x.

The Gaussian (normal) joint probability density at any point in xyz-space can be determined from the covariance and the zero-mean vector:

$$\vec{\xi} = (x - x_0)\hat{i} + (y - y_0)\hat{j} + (z - z_0)\hat{k} \tag{32}$$

where: $x_0 = \sigma_{x_0}$, $y_0 = \sigma_{y_0}$, $z_0 = \tau_0$ are the mean or still water values, and $\hat{i}, \hat{j}, \hat{k}$ are unit vectors in the x, y and z directions respectively.

$$p(\vec{\xi}) = \frac{\exp(-\frac{1}{2}\vec{\xi}^T[\Gamma]^{-1}\vec{\xi})}{(2\pi)^{(3/2)}\sqrt{|\Gamma|}} \tag{33}$$

As previously noted, surfaces of constant joint probability density also describe ellipsoids in xyz-space. These surfaces correspond to any solution of $\vec{\xi}^T[\Gamma]^{-1}\vec{\xi} = \text{constant}$. These surfaces are ellipsoids centered at x_0, y_0, z_0 with semi-axes oriented in xyz-space along the eigenvectors of $[\Gamma]$, as further described in Appendix A. In general, the orientations of the semi-axes of the Gaussian ellipsoidal field do not align with that of the Von Mises stress ellipsoidal field. In any event, nonzero-mean x_0, y_0, or z_0 destroys any possibility of symmetrical relationship between the Gaussian and Von Mises fields.

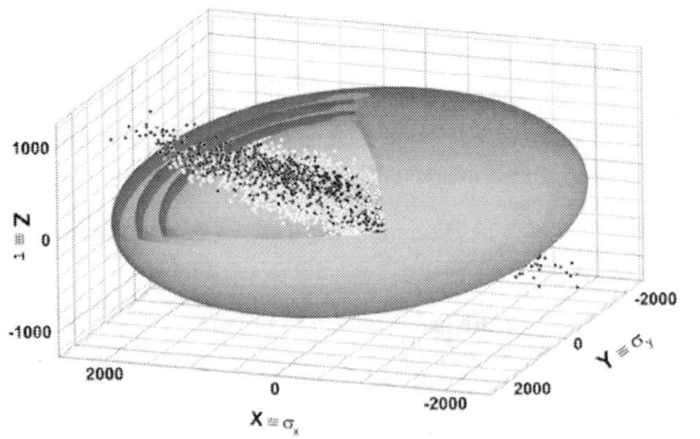

Figure 15. A 3-D depiction of the ellipsoidal character of plane Von Mises stress in xyz-space

Figure 15 shows three congruent ellipsoidal surfaces, each associated with a constant value of Von Mises stress. Note that these surfaces are shown cut away in one octant of the ellipsoids to show the underlying data. Also shown in Figure 15 is the ellipsoidal Gaussian 'cloud' of Von Mises stress sampled at short uniform time intervals (light points), and a more compact ellipsoid of points sampled at local peaks (dark points) of the Von Mises stress process.

Figure 16 shows a Von Mises ellipsoid whose surface has been 'painted' according to the Gaussian probability density. Sample data may be seen to be projecting through the surface of the Von Mises ellipsoid through the region of greatest Gaussian probability density.

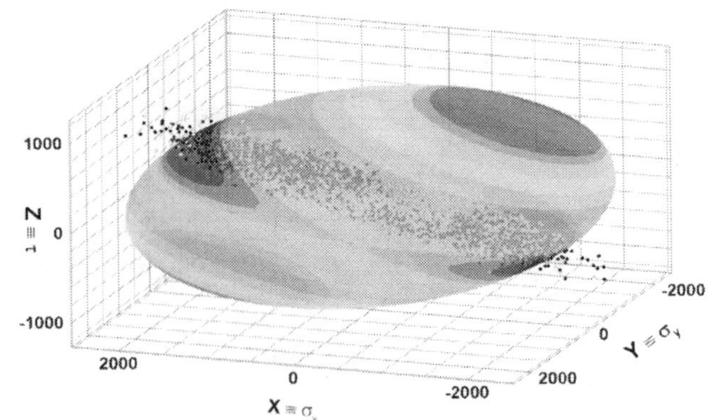

Figure 16. A surface of the Von Mises ellipsoid 'painted' according to the log of the Gaussian probability density on that surface

The statistical principal axis for the Von Mises stress corresponds to a line through the origin (σ_{x_0}, σ_{y_0}, τ_0) and the point of greatest joint normal probability density on a surface of constant Von Mises stress, $\text{MAX}\{p(x,y,z \mid \sigma_{VM}^2 = c^2)\}$. This can be determined through a search procedure.

A suggested search procedure is to make use of either cylindrical or spherical coordinates. For a specified value of $\sigma_{VM}^2 = c^2$, the two-dimensional search over the surface of the ellipsoid of constant Von Mises stress takes place along $-\sqrt{\frac{c^2}{3}} \le \tau = z \le \sqrt{\frac{c^2}{3}}$ (in cylindrical coordinates), or equivalently, along $-\frac{\pi}{2} \le \phi \le \frac{\pi}{2}$ in spherical coordinates, where $\tau = z = \sqrt{\frac{c^2}{3}}\sin(\phi)$. At each value of $\tau = z$ the second degree-of-freedom of the search may be expressed in terms of $0 \le \theta \le 2\pi$, where:

$$x = [\cos(\theta) - \sqrt{1/3}\sin(\theta)]\sqrt{c^2 - 3z^2} \tag{34a}$$

$$y = [\cos(\theta) + \sqrt{1/3}\sin(\theta)]\sqrt{c^2 - 3z^2} \tag{34b}$$

The spherical coordinates may be regarded as a cylindrical map projection of the surface of the ellipsoid onto the rectangular plane similar to the familiar Mercator projection as illustrated in Figure 17. Pursuing this analogy, the angle ϕ corresponds to latitude and the angle θ corresponds to longitude.

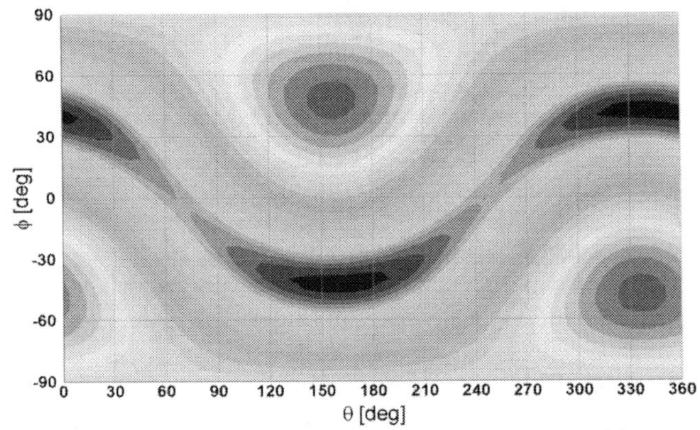

Figure 17. Cylindrical map projection of the surface of the Von Mises ellipsoid depicted in Figure 16

The probability density varies smoothly on the surface of the Von Mises ellipsoid and the $\phi\theta$-location that corresponds to the maximum of $p(\vec{\xi})$ in equation 33 yields the xyz coordinates of the point on that ellipsoidal surface through which the statistical principal axis passes. The search for that $\phi\theta$-location can be accomplished through repeated application of rectangular grid searches with successively finer mesh or through some directed search method such as Newton's. An optimal search strategy probably employs a coarse rectangular grid search to locate a well behaved region in which the desired point of maximum is known to exist, followed by application of Newton's or some equivalent directed search method to locate the point of maximum with the desired accuracy.

It is worthy of note that there is only one Von Mises ellipsoidal field in xyz-space, and that field is universal. It is the joint Gaussian ellipsoidal fields that exhibit unlimited variation according to both the mean values x_0, y_0, z_0 and cross co-spectral moments (covariance) $[\Gamma]$.

Conditioned on the chosen value of c used in the search, let the coordinates corresponding to the point of maximum $p(\vec{\xi})$ be $\breve{x}, \breve{y}, \breve{z}$. Once the point of maximum is known, the direction cosines of the statistical principal axis can be determined as:

$$\breve{r} = \sqrt{(\breve{x}-x_0)^2 + (\breve{y}-y_0)^2 + (\breve{z}-z_0)^2} \tag{35}$$

$$\cos(\varphi_x) = \frac{(\breve{x}-x_0)}{\breve{r}} \tag{36a}$$

$$\cos(\varphi_y) = \frac{(\breve{y}-y_0)}{\breve{r}} \tag{36b}$$

$$\cos(\varphi_z) = \frac{(\breve{z}-z_0)}{\breve{r}} \tag{36c}$$

Hereafter we can drop the $\breve{x}, \breve{y}, \breve{z}$ and $\breve{r}$ notation, which we reserve for the specific point maximizing $p(\vec{\xi})$ in our search, and use x, y, z and r for any point on the statistical principal axis. In terms of r, the parametric equations of the coordinates along the principal axis are then:

$$x = x_0 + r\cos(\varphi_x) = \sigma_{x_0} + \tilde{\sigma}_x(t) = \sigma_x(t) \tag{37a}$$

$$y = y_0 + r\cos(\varphi_y) = \sigma_{y_0} + \tilde{\sigma}_y(t) = \sigma_y(t) \tag{37b}$$

$$z = z_0 + r\cos(\varphi_z) = \tau_0 + \tilde{\tau}(t) = \tau(t) \tag{37c}$$

Determination of the cumulative probability of Von Mises along statistical principal axis

Applying equation 29 along the statistical principal axis, we can now determine the cumulative probability of Von Mises stress for the important class of problems where the mean (i.e., still water) stresses are nonzero. However, here we encounter a complication. The statistical principal axis extends from (x_0, y_0, z_0) in two directions: one (positive r-branch) towards the point $\breve{x}, \breve{y}, \breve{z}$, and the other (negative r-branch) extending in the opposite direction. These two branchesboth contribute local peaks to the Von Mises process. The accumulation of generalized Rayleigh probability along both branches contributes to the cumulative probability associated with the local peaks of the Von Mises stress process.

The development of these two branches and the need to evaluate the contributions towards the cumulative probability along both branches arises because the mean stresses break symmetry. For that extraordinary class of problems first considered, where mean stresses are all zero, symmetry between Von Mises stress and Gaussian joint probability exists along all axes through the origin (not just the statistical principal axes), and that symmetry simplifies the problem.

From equation 29, the Von Mises stress along the statistical principal axis is given by:

$$\sigma_{VM}^2(r) = \sigma_x^2(r) + \sigma_y^2(r) - \sigma_x(r)\sigma_y(r) + 3\tau^2(r) \tag{38}$$

$$= [x_0 + r\cos(\varphi_x)]^2 + [y_0 + r\cos(\varphi_y)]^2$$
$$- [x_0 + r\cos(\varphi_x)][y_0 + r\cos(\varphi_y)] + 3[z_0 + r\cos(\varphi_z)]^2$$

Equation 38 can be expanded to reveal the cross terms.

$$\begin{aligned}\sigma_{VM}^2(r) = &\{x_0^2 + y_0^2 - x_0 y_0 + 3z_0^2\}\\ &+ r^2\{\cos^2(\varphi_x) + \cos^2(\varphi_y) - \cos(\varphi_x)\cos(\varphi_y) + 3\cos^2(\varphi_z)\}\\ &+ r\{2[x_0\cos(\varphi_x) + y_0\cos(\varphi_y) + 3z_0\cos(\varphi_z)]\\ &- [x_0\cos(\varphi_y) + y_0\cos(\varphi_x)]\}\end{aligned} \tag{39}$$

Equations 38 and 39 are quadratic in r, and $\sigma_{VM}^2(r)$ is a parabola with shifted vertex, as illustrated in Figure 18. The difficulty presented by the two branches and broken symmetry is that $\sigma_{VM}^2(+r) \neq \sigma_{VM}^2(-r)$.

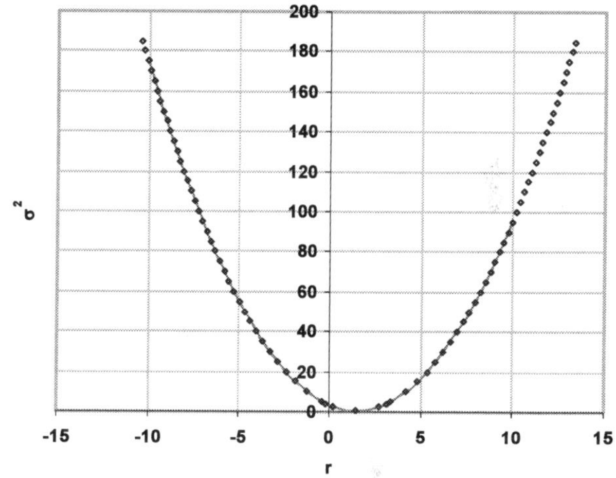

Figure 18. Parabolic distribution of $\sigma_{VM}^2(r)$

Let:

$$A = \{\cos^2(\varphi_x) + \cos^2(\varphi_y) - \cos(\varphi_x)\cos(\varphi_y) + 3\cos^2(\varphi_z)\}$$

$$B = \{2[x_0\cos(\varphi_x) + y_0\cos(\varphi_y) + 3z_0\cos(\varphi_z)] - [x_0\cos(\varphi_y) + y_0\cos(\varphi_x)]\}$$

and

$$C = \{x_0^2 + y_0^2 - x_0 y_0 + 3z_0^2\}$$

Then the r-coordinate of the parabolic focus is:

$$r_0 = \frac{-B}{2A} \tag{40}$$

and the r-coordinates on the two branches at any specified Von Mises stress, σ_{VM}, are given by:

$$r_{1,2}(\sigma_{VM}) = \frac{-B \pm \sqrt{B^2 - 4A(C - \sigma_{VM}^2)}}{2A} \tag{41}$$

The r-coordinate measures the distance along the statistical principal axis from a local r-coordinate origin centered at the mean (x_0, y_0, z_0). The metric for the Rayleigh distribution of local peaks along the r-axis is $\sqrt{m_{VM}^{(0)}}$. The squared dimensional measure of time-dependent distance along the r-axis is:

$$[\xi_{1,2}(\sigma_{VM})]^2 = [r_{1,2}(\sigma_{VM})]^2\{\cos^2(\varphi_x)+\cos^2(\varphi_y) - \cos(\varphi_x)\cos(\varphi_y)+3\cos^2(\varphi_z)\} \tag{42}$$

and $\xi_{1,2}(\sigma_{VM})$ is simply the square root of equation 42.

The cumulative probability of Von Mises stress is then:

$$P(\sigma_{VM}<\delta\,|\,m_{VM}^{(0)},\varepsilon)=\tfrac{1}{2}\{P[\xi_1(\sigma_{VM})\,|\,m_{VM}^{(0)},\varepsilon] + P[\xi_2(\sigma_{VM})\,|\,m_{VM}^{(0)},\varepsilon]\} \tag{43}$$

where $P(\xi_{1,2}\,|\,m_{VM}^{(0)},\varepsilon)$ is evaluated as generalized Rayleigh cumulative probability according to equation 19, and the one-half factor represents the fact that local Von Mises peaks associated with each of the two branches are equally likely.

Comparison of nonzero-mean cases with Fourier-Stieltjes

Figure 19 shows the sample probability distribution for the case of the element with no dominance. The abscissa is the non-dimensional Von Mises stress normalized by $m_{VM}^{(0)}$ as determined using equation 12. Observe that the nondimensional Von Mises now extends to what would be regarded as extraordinary values in the absence of mean stresses. The theoretical cumulative probability distribution obtained using equation 43 may be observed to be an excellent predictor of the sample distribution for extreme values.

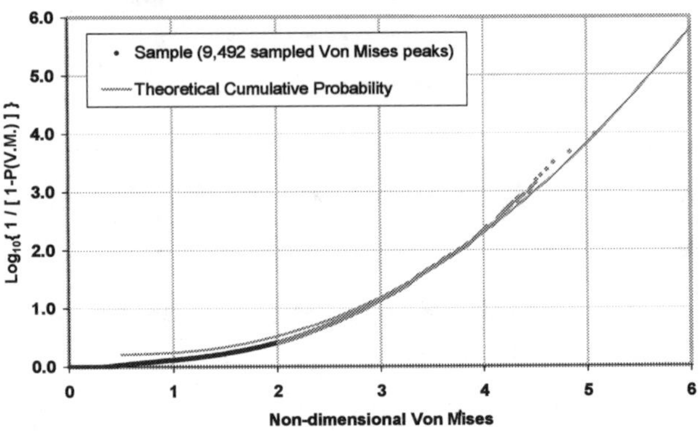

Figure 19. Element with no dominance with mean (still water) stresses equal to RMS of the time-dependent stress components. Heading=111 deg, H_s=32.2 ft, T_p=12 sec

A convenience and advantage of the methods here described for evaluating the probability of Von Mises stress in the presence of nonzero-mean stresses, is that the time-dependent covariance need only be computed once for each case, and then it may be used to evaluate the principal axis and probability distribution for any number of mean stress cases. Figure 20 shows probability distributions for the example element with no dominance, for a systematic series of mean stress cases. For comparison, the dotted curves to the right show the distributions approximated by narrow-band Rayleigh with shifted origin.

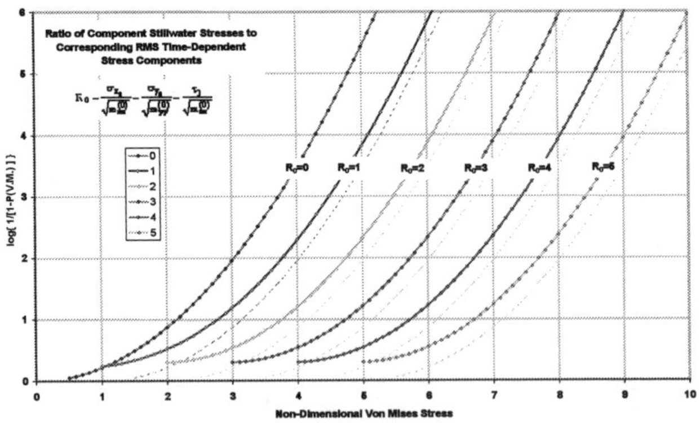

Figure 20. Variation in σ_{VM} with mean (still water) stresses

Von Mises in Short-Crested Seas

By virtue of the way equation 5 was presented, all of the foregoing applies in long-crested seas, but there is no real restriction, and all of these methods are readily extended to short-crested seas by reformulating equation 5 as follows:

$$m_{ij}^{(n)} = \Re\int_{-\pi}^{\pi}\int_{0}^{\infty}\omega_e^n S_{ij}(\omega,\chi)\,d\omega\,d\chi \tag{44}$$

where ω_e is the frequency of encounter defined previously, χ is the relative heading and $S_{ij}(\omega,\chi)$ is the directional cross-spectrum defined after the fashion of equation set 22. Once the cross co-spectral moments are determined for the short-crested sea situation, the remainder of the analysis proceeds exactly as described above.

Long-Term Statistics

The long-term (e.g., lifetime) extreme value of Von Mises stress is a desirable dynamic load parameter (DLP) for use in dynamic loads analysis (DLA). The long-term method extends over each wave height, spectrum, speed, heading and load condition combination $\vec{u}=(H_s,T_P,V,\chi,\Lambda)$ encountered during the vessel lifetime.

Distribution of stress cycles

The short-term statistics of Von Mises stress process may be determined using the methods described in preceding sections for each encountered stationary environmental and operating condition.

Long-term extreme

The long-term extreme is defined as that value which is expected to be exceeded only once in the lifetime of the vessel. The expected service life of the vessel, the fraction of total annual hours actually spent at sea, wave climatology and seamanship exercised by the vessel's master, all influence the long-term extreme.

Climatology

The encountered wave climatology is first determined as the joint probability of significant wave height and peak spectral period for the vessel operating area, i.e., $P(H_S,T_P)$ where H_S is the significant wave height and T_p is the peak spectral period.

The available data are usually given for discrete ranges of H_S and T_P in matrix form. Separate annual and seasonal distributions are typically available. The method to determine $P(H_S, T_P)$ when the vessel operates in different areas during different seasons is straightforward [6].

For unrestricted worldwide service, class societies encourage the use of climatologies such as that of ABS [1] (attributed to H. Walden), BMT [3] or IACS [12], expressed as a discrete joint probability table of H_S and T_P.

Conditional probability of heading and speed

The conditional probability of heading and speed, given H_S and T_P, is an operationally defined distribution that embodies estimates of the seamanship judgment of the vessel's master, and any standing operating orders (e.g., operational restrictions) that may have been determined for the vessel. Thus the conditional probability $P(V,\chi \mid H_S, T_P, \Lambda)$ is not a system property like $P(H_S, T_P)$, but an operational characteristic subject to variability due to human factors. Different operators will respond differently to the same stimulus (H_S, T_P), and the same operator will respond differently to the same stimulus (H_S, T_P), depending on the mission context. Here response refers to the choice of relative speed and heading, (V,χ). Note that the load condition, Λ, participates as conditional information considered by the operator when determining the appropriate operational response.

In the absence of seamanship functions, it is accepted practice to assume a uniform distribution of relative heading (χ), a single speed (V) and a single load case (Λ) for generation of the lifetime conditional probability distribution $P(V,\chi \mid H_S, T_P, \Lambda)$.

Load cycles

The total lifetime exposure (T_L) is readily calculated if the service life and the fraction of time actually spent at sea are known.

$$T_L = 31{,}536{,}000\ L\ \gamma \tag{45}$$

where T_L is the total duration of exposure in seconds, L is the service life in years and γ is the fraction of time actually spent at sea. The duration of exposure to a particular combination of H_S, T_P, V, χ and Λ is then:

$$T(\vec{u}) = P(\Lambda)\, P(H_S, T_p)\, P(V,\chi \mid H_S, T_p,\Lambda)\, T_L \tag{46}$$

where $P(\Lambda)$ is the probability for the vessel load condition. In reality, load condition is a continuous variable (e.g., fuel is being consumed on a continuous basis). Furthermore, load condition is a multi-dimensional (i.e., vector) state, as for instance, there may be different still water bending moments associated with the same displacement, depending on the longitudinal distribution of deadweight. However, in practice, the load conditions for most vessels can be adequately characterized by a few discrete load cases, i.e., $\mathrm{Set}[\Lambda] \cong \mathrm{Set}[\Lambda_1, \Lambda_2, \ldots, \Lambda_k]$.

The number of Von Mises stress cycles of a loading process in this duration is determined from the zero-crossing period computed for the process:

$$N(\vec{u}) = \frac{T(\vec{u})}{T_{VM}(\vec{u})} \tag{47}$$

Long-term extreme

The number of Von Mises stress amplitude cycles that exceed a specified load amplitude δ are:

$$N(\sigma_{VM} \geq \delta) = \sum_{\vec{u}} N(\vec{u})\, P_R(\sigma_{VM} \geq \delta \mid m_{VM}^{(0)}(\vec{u}), \varepsilon_\sigma(\vec{u})) \tag{48}$$

where $P_R(\sigma_{VM} \geq \delta \mid m_{VM}^{(0)}(\vec{u}), \varepsilon(\vec{u}))$ is the chosen form of generalized Rayleigh distribution, with spectral moments, $m_{VM}^{(n)}$, as given by either equation 5 or 44 (depending on whether the seas are long- or short-crested). The long-term, life-time, extreme Von Mises stress amplitude corresponds to δ when $N(\sigma_{VM} \geq \delta) \equiv 1.0$.

CONCLUSION

Equation 12 has been asserted by others, notably by Pitoiset et al. [16], which stands as the initial inspiration for our work. However, to our knowledge, the present paper presents the first elaboration of the derivation of equation 12, and, based on extensive time-domain simulations, the first demonstration validating that Von Mises stress peaks are Rayleigh distributed with variance parameter $m_{VM}^{(0)}$. Moreover, to our knowledge, the present paper includes the following unique contributions:

- The first elaboration of the derivation of equation 12 both from the time and cross co-spectral moment domains.
- The extension to higher moments and generalized Rayleigh probability.
- The first to explicitly extend consideration to address nonzero-mean values of the underlying Gaussian process normal stresses (i.e., σ_x and σ_y) and shear stress τ.
- We have focused attention on extreme values of Von Mises stress, including order statistical distributions that we have validated through ensemble realizations in the time domain.
- The procedure we recommend appears to adhere to the conservative engineering principle of either predicting precisely or over-predicting the actual Von Mises stress.
- Extension to long term (i.e., life time), including directional spreading (short-crestedness) and seamanship.

The generalized Rayleigh distribution was shown to provide good fit to comparable sample distributions from the time domain, but often the best fit was achieved with ε-values considerably larger than that determined directly from the higher-order spectral moments. Additional research is in order to better understand the basis for the choice of spectral breadth parameters.

The authors regard Von Mises stress as a desirable candidate for use as a dynamic load parameter (DLP) in dynamic loads analysis (DLA). As Von Mises stress is one of the structural responses of great interest during a dynamic loads analysis, it is our belief that a higher fidelity and confidence is achieved by using Von Mises stress as a DLP in lieu of, or in addition to, the use of indirectly related proxy processes, such as bending moment, roll or acceleration.

ACKNOWLEDGMENTS

The authors wish to acknowledge the generous support of The Glosten Associates, Inc., for the development of this paper.

REFERENCES

1. AMERICAN BUREAU OF SHIPPING. *Guidance Notes on Spectral Based Fatigue Analysis for Vessels*, January 2004.
2. AMERICAN BUREAU OF SHIPPING, *Guide for 'Safehull-Dynamic Loading Approach' for Vessels*, December 2006.
3. BMT Ltd 'Global Wave Statistics', BMT, London 1986
4. CARTWRIGHT, D.E. and M.S. LONGUET-HIGGINS. "The Statistical Distribution of the Maxima of a Random Function." *Proceedings*, Royal Society, A237. (1956): 212-232
5. FRIEZE, P.A. and R.A. SHENOI, ed. Proceedings of the 16th International Ship and Offshore Structures Congress (ISSC), Vol. 1, report of Committee II.1, Quasi-static Response, University of Southampton (2006).
6. HUTCHISON, B.L. "Risk and Operability Analysis in the Marine Environment." SNAME *Trans*, 89 (1981): 127-154.
7. HUTCHISON, BRUCE L. "A Note on the Application of Response Cross Spectra." *Journal of Ship Research*, 26(2) (1982): 94-96.
8. HUTCHISON, B.L., T. MATHAI, J.M. MORGAN and S. ETCHEMENDY, "Hydrodynamic Loads on SWATH Hulls." RINA conference on High Speed Craft Motions & Maneuverability, (20 February 1998).
9. HUTCHISON, BRUCE L. "Joint Seakeeping Response Processes for Determining Structural Loads." SNAME *Trans*, 110 (2002): 189-214.
10. HUTCHISON, BRUCE L., "Principal Angles for Seakeeping Response Processes." *Marine Technology*, (October 2004): 183-199.
11. HUTCHISON, B.L., B.B. ACKERS and T.S. LEACH, "Principal Axes for Structural Fatigue," SNAME *Trans*, 113 (2005).
12. IACS Recommendation No.34, Standard Wave Data, (Nov. 2001).
13. LIU, D., J. SPENCER, T. ITOH, S. KAWACHI and K. SHIGEMATSU. "Dynamic Load Approach in Tanker Design." SNAME *Trans*, 100 (1992): 143-172.
14. MANSOUR, A. "Combining Extreme Environmental Loads for Reliability Based Design." *Proceedings* of the Extreme Loads Response Symposium, sponsored by SSC and SNAME, Arlington, VA. (1981).
15. OCHI, MICHEL K. "On Prediction of Extreme Values." *Journal of Ship Research*, 17(1), (1973): 29-37.
16. PITOISET, X., A. PREUMONT and A. KERNILIS, "Tools for a Multiaxial Fatigue Analysis of Structures Submitted to Random Vibrations," Proceedings of European Conference on Spacecraft Structural Materials and Mechanical Testing, Braunschweig, Germany 46 November 1998 (ESA SP428, February 1999)
17. ST. DENIS, MANLEY and WILLARD J. PIERSON, JR., "On the Motions of Ships in Confused Seas," *Trans.*, SNAME, Vol. 81, 1953, pp. 280-357

APPENDIX A
PRINCIPAL AXES OF JOINT GAUSSIAN

As the three basis stresses, σ_x, σ_y, τ (aka xyz) are linear responses to a Gaussian wave environment. In the chosen arbitrary coordinate system, these basis stresses are associated with covariance matrix:

$$[\Gamma] = \begin{bmatrix} \gamma_{xx} & \gamma_{xy} & \gamma_{xz} \\ \gamma_{yx} & \gamma_{yy} & \gamma_{yz} \\ \gamma_{zx} & \gamma_{zy} & \gamma_{zz} \end{bmatrix} = \begin{bmatrix} m^{(0)}_{xx} & m^{(0)}_{xy} & m^{(0)}_{xz} \\ m^{(0)}_{yx} & m^{(0)}_{yy} & m^{(0)}_{yz} \\ m^{(0)}_{zx} & m^{(0)}_{zy} & m^{(0)}_{zz} \end{bmatrix} \quad \text{(A1)}$$

These basis stresses have joint Gaussian (normal) probability density according to:

$$p(\sigma_x, \sigma_y, \tau) = p(x, y, z) = p(\vec{x}) = \frac{\exp(-\frac{1}{2}\vec{x}^T [\Gamma]^{-1} \vec{x})}{(2\pi)^{(3/2)} \sqrt{|\Gamma|}} \quad \text{(A2)}$$

As previously observed, $\vec{x}^T [\Gamma]^{-1} \vec{x} \equiv \text{constant}$ describes a surface of constant probability density.

The eigenvalues of the covariance matrix are solutions to the following equation:

$$|[\Gamma] - \lambda [I]| = 0 \quad \text{(A3)}$$

This equation yields three eigenvalues, λ_1, λ_2, and λ_3,

and three eigenvectors, $\vec{r}_1, \vec{r}_2$, and $\vec{r}_3$.

The matrix of eigenvectors is the operator that diagonalizes the covariance matrix with the eigenvalues assuming the diagonal values:

$$[R] = \begin{bmatrix} r1_x & r2_x & r3_x \\ r1_y & r2_y & r3_y \\ r1_z & r2_z & r3_z \end{bmatrix} \quad \text{(A4)}$$

$$[R]^T [\Gamma][R] = \begin{bmatrix} \lambda_1 & 0 & 0 \\ 0 & \lambda_2 & 0 \\ 0 & 0 & \lambda_3 \end{bmatrix} \quad \text{(A5)}$$

And [R] is the coordinate transformation (rotation) that, when applied to the surface geometry of any ellipsoid with axes $c\sqrt{\lambda_1}$, $c\sqrt{\lambda_2}$ and $c\sqrt{\lambda_3}$ (where c is an arbitrary constant), results in an ellipsoid rotated in xyz-space whose surface has constant joint normal probability density according to the original covariance matrix $[\Gamma]$, as expressed by equation A3. Furthermore, the argument of the joint normal probability density, $\vec{x}^T [\Gamma]^{-1} \vec{x} = c^2$, is of the quadratic form, where c is the arbitrary constant multiplying the square roots of the eigenvalues used as the axes of the ellipsoid prior to rotation.

Though elegant and potentially useful, it must be observed that the principal axes for the joint Gaussian distribution of σ_x, σ_y, τ will rarely align with the invariant orientation of the principal axes of the surface of constant Von Mises stress.

APPENDIX B
AVERAGE VON MISES STRESS

While the primary objective of this paper is to investigate extreme values of Von Mises stress, there is an interesting and perhaps useful result concerning average Von Mises stress for the case where the normal stresses, σ_x and σ_y, and the shear stress, τ, are each Gaussian processes, but possess nonzero means (i.e., still water stresses). Beginning from equation 34 in the main body of this paper (here repeated):

$$\begin{aligned} \sigma^2_{VM}(t) &= [\sigma_{x_0} + \tilde{\sigma}_x(t)]^2 + [\sigma_{y_0} + \tilde{\sigma}_y(t)]^2 \\ &\quad - [\sigma_{x_0} + \tilde{\sigma}_x(t)][\sigma_{y_0} + \tilde{\sigma}_y(t)] + 3[\tau_0 + \tilde{\tau}(t)]^2 \\ &= \{\sigma^2_{x_0} + \sigma^2_{y_0} - \sigma_{x_0}\sigma_{y_0} + 3\tau_0^2\} \\ &\quad + \{\tilde{\sigma}^2_x(t) + \tilde{\sigma}^2_y(t) - \tilde{\sigma}_x(t)\tilde{\sigma}_y(t) + 3\tilde{\tau}^2(t)\} \\ &\quad + 2\sigma_{x_0}\tilde{\sigma}_x(t) + 2\sigma_{y_0}\tilde{\sigma}_y(t) - \sigma_{x_0}\tilde{\sigma}_y(t) - \sigma_{y_0}\tilde{\sigma}_x(t) \\ &\quad + 6\tau_0\tilde{\tau}(t) \end{aligned} \quad \text{(B1)}$$

Applying the time average operator all of the zero-mean terms such as $\sigma_{x_0}\tilde{\sigma}_x(t)$, $\sigma_{y_0}\tilde{\sigma}_y(t)$, $\sigma_{x_0}\tilde{\sigma}_y(t)$, $\sigma_{y_0}\tilde{\sigma}_x(t)$ and $\tau_0\ \tilde{\tau}(t)$, all go to zero, leaving:

$$\overline{\sigma^2_{VM}(t)} = \{\sigma^2_{x_0} + \sigma^2_{y_0} - \sigma_{x_0}\sigma_{y_0} + 3\tau_0^2\} + \{\overline{\tilde{\sigma}^2_x(t)} + \overline{\tilde{\sigma}^2_y(t)} - \overline{\tilde{\sigma}_x(t)\tilde{\sigma}_y(t)} + 3\overline{\tilde{\tau}^2(t)}\}$$

or:

$$\overline{\sigma^2_{VM}(t)} = \{\sigma^2_{x_0} + \sigma^2_{y_0} - \sigma_{x_0}\sigma_{y_0} + 3\tau_0^2\} + \{\tilde{m}^{(0)}_{\sigma_x\sigma_x} + \tilde{m}^{(0)}_{\sigma_y\sigma_y} - \tilde{m}^{(0)}_{\sigma_x\sigma_y} + 3\tilde{m}^{(0)}_{\tau\tau}\} \quad \text{(B2)}$$

From equation B2 it can be seen that the time-averaged mean square of the Von Mises stress is comprised of two Von Mises-like terms, one the time-independent (i.e., constant) terms and one the mean square of the time-dependent terms.

Observe that this algebraic relationship means that the two terms are orthogonal Pythagorean components of $\overline{\sigma^2_{VM}(t)}$:

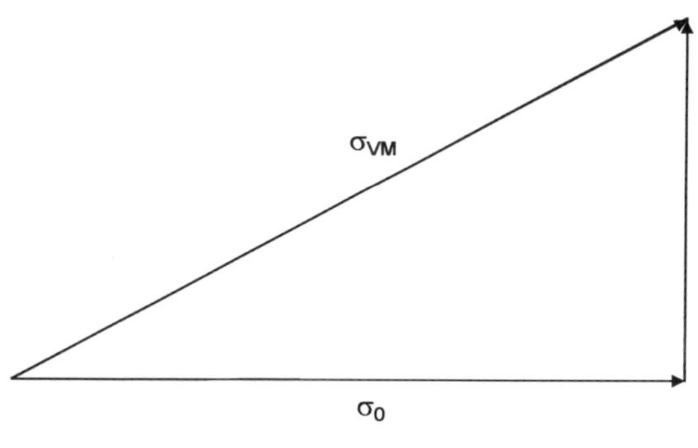

Figure B-1. Illustrating the orthogonal Pythagorean relationship between mean stress and the time average of variable stress

APPENDIX C
TRI-AXIAL VON MISES STRESS

As given in numerous texts and elsewhere, the equation for tri-axial Von Mises stress is:

$$\sigma_{VM}^2 = \tfrac{1}{2}\{(\sigma_x - \sigma_y)^2 + (\sigma_y - \sigma_z)^2 + (\sigma_z - \sigma_x)^2 + 6(\tau_{xy}^2 + \tau_{yz}^2 + \tau_{zx}^2)\}$$

$$= \tfrac{1}{2}\{(\sigma_x^2 - 2\sigma_x\sigma_y + \sigma_y^2) + (\sigma_y^2 - 2\sigma_y\sigma_z + \sigma_z^2) + (\sigma_z^2 - 2\sigma_z\sigma_x + \sigma_x^2) + 6(\tau_{xy}^2 + \tau_{yz}^2 + \tau_{zx}^2)\}$$

$$= (\sigma_x^2 + \sigma_y^2 + \sigma_z^2) - (\sigma_x\sigma_y + \sigma_y\sigma_z + \sigma_z\sigma_x) + 3(\tau_{xy}^2 + \tau_{yz}^2 + \tau_{zx}^2) \qquad \text{(C1)}$$

this last being of similar form to that used in the present paper to address bi-axial Von Mises stress, excepting that the tri-axial case introduces the additional terms σ_z^2, $\sigma_y\sigma_z$, $\sigma_z\sigma_x$, $3\tau_{yz}^2$ and $3\tau_{zx}^2$.

APPENDIX D
VON MISES STRESS IN PRINCIPAL STRESS COORDINATES

In principal stress coordinates, there are no shear stresses, so tri-axial Von Mises stress assumes the following simple algebraic forms:

$$\sigma_{VM}^2 = \frac{(\sigma_1 - \sigma_2)^2 + (\sigma_2 - \sigma_3)^2 + (\sigma_3 - \sigma_1)^2}{2}$$

$$= \sigma_1^2 + \sigma_2^2 + \sigma_3^2 - \sigma_1\sigma_2 - \sigma_2\sigma_3 - \sigma_3\sigma_1 \qquad \text{(D1)}$$

The bi-axial case in principal stress coordinates is obtained simply by setting $\sigma_3 = 0$.

However, in a Gaussian loads environment, the orientation of the principal stress axes is constantly changing with time, so despite the apparent simplicity of Von Mises in principal stress coordinates, it is more convenient to use the equations given in general (arbitrarily oriented) coordinates.

APPENDIX E
BI-AXIAL VON MISES STRESS IN MATRIX NOTATION

The exposition of this paper has favored a direct component algebra notation and presentation, but matrix notation may afford advantages when implementing these procedures in actual practice. A few observations are here offered regarding matrix notation for bi-axial Von Mises stress.

Define the matrix, [VM] and the stress vector $\vec{\sigma}$:

$$[VM] = \begin{bmatrix} 1 & -\tfrac{1}{2} & 0 \\ -\tfrac{1}{2} & 1 & 0 \\ 0 & 0 & 3 \end{bmatrix} \qquad \text{(E1)}$$

$$\vec{\sigma} = \begin{Bmatrix} \sigma_x \\ \sigma_y \\ \tau \end{Bmatrix} \qquad \text{(E2)}$$

Bi-axial Von Mises stress in arbitrary coordinates has previously been defined as:

$$\sigma_{VM}^2 = \sigma_x^2 + \sigma_y^2 - \sigma_x\sigma_y + 3\tau^2 \qquad \text{(E3)}$$

which may be expressed in matrix notation using the following quadratic form:

$$\sigma_{VM}^2 = \vec{\sigma}^T\,[VM]\,\vec{\sigma} \qquad \text{(E4)}$$

The eigenvalues of [VM] are $(\lambda_1, \lambda_2, \lambda_3) = (\tfrac{1}{2}, \tfrac{3}{2}, 3)$ and the eigenvectors are:

$$[R] = \begin{bmatrix} \frac{1}{\sqrt{2}} & -\frac{1}{\sqrt{2}} & 0 \\ \frac{1}{\sqrt{2}} & \frac{1}{\sqrt{2}} & 0 \\ 0 & 0 & 1 \end{bmatrix}$$

This means that the Von Mises ellipsoid is oriented with the major axis oriented at 45° in the $\sigma_x\sigma_y$-plane and the minor axis directed in the 135° direction, as shown in Figure E-1. The third principal axis of the Von Mises ellipsoid is coincident with the τ-axis and normal to the $\sigma_x\sigma_y$-plane. The unit length of each of these principal axes is given by the square root of the reciprocal of the respective eigenvalue, or :

$$\left(\sqrt{\tfrac{1}{\lambda_1}}, \sqrt{\tfrac{1}{\lambda_2}}, \sqrt{\tfrac{1}{\lambda_3}}\right) = \left(\sqrt{2}, \sqrt{\tfrac{2}{3}}, \sqrt{\tfrac{1}{3}}\right).$$

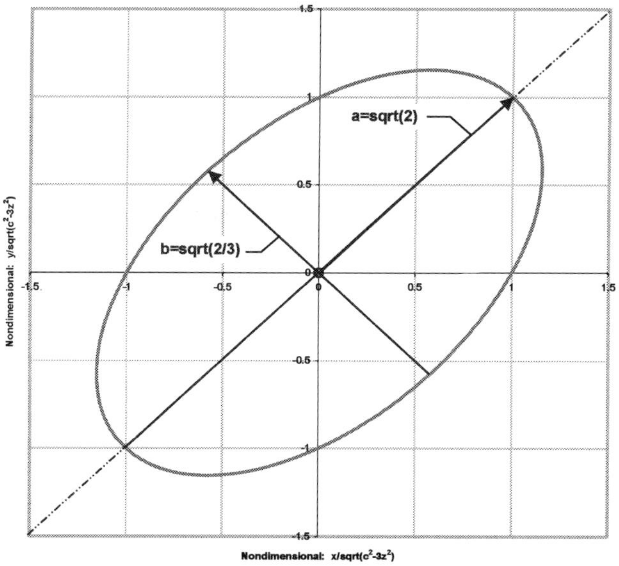

Figure E-1. Nondimensional ellipse of constant $\sigma_{VM}^2 - 3\tau^2$

Let us define two stress vectors, one time-dependent and a mean (aka still water) stress vector that is time-independent:

$$\vec{\sigma} = \begin{Bmatrix} \tilde{\sigma}_x(t) \\ \tilde{\sigma}_y(t) \\ \tilde{\tau}(t) \end{Bmatrix} \quad \vec{\sigma}_0 = \begin{Bmatrix} \sigma_{x_0} \\ \sigma_{y_0} \\ \tau_0 \end{Bmatrix} \tag{E5}$$

Then equation 32 (here repeated):

$$\begin{aligned} \sigma_{VM}^2(t) &= [\sigma_{x_0} + \tilde{\sigma}_x(t)]^2 + [\sigma_{y_0} + \tilde{\sigma}_y(t)]^2 \\ &\quad - [\sigma_{x_0} + \tilde{\sigma}_x(t)][\sigma_{y_0} + \tilde{\sigma}_y(t)] + 3[\tau_0 + \tilde{\tau}(t)]^2 \\ &= \{\sigma_{x_0}^2 + \sigma_{y_0}^2 - \sigma_{x_0}\sigma_{y_0} + 3\tau_0^2\} \\ &\quad + \{\tilde{\sigma}_x^2(t) + \tilde{\sigma}_y^2(t) - \tilde{\sigma}_x(t)\,\tilde{\sigma}_y(t) + 3\,\tilde{\tau}^2(t)\} \\ &\quad + 2\,\sigma_{x_0}\tilde{\sigma}_x(t) + 2\,\sigma_{y_0}\tilde{\sigma}_y(t) - \sigma_{x_0}\tilde{\sigma}_y(t) - \sigma_{y_0}\tilde{\sigma}_x(t) \\ &\quad + 6\,\tau_0\,\tilde{\tau}(t) \end{aligned} \tag{E6}$$

can be written in matrix notation as:

$$\sigma_{VM}^2 = \vec{\sigma}_0^T\,[VM]\,\vec{\sigma}_0 + \vec{\sigma}^T\,[VM]\,\vec{\sigma} + 2\,\vec{\sigma}_0^T\,[VM]\,\vec{\sigma} \tag{E7}$$

Discussion

Yung-Sup Shin, Member

The authors should be congratulated for their excellent work in offering a new method to determine the critical responses of the structure by using cross co-spectral moments to von Mises stress. ABS has developed the Dynamic Loading Approach (DLA) and applied it successfully to analyze the hull structure based on first principles and direct calculations of wave load and structure responses. DLA first started off with a determination of dynamic load parameters (DLP) such as wave-induced bending moments, acceleration, and roll motion, and life time maximum values as design target values, and then corresponding critical wave condition and structural loads of wave pressures and acceleration-induced inertial forces that are applied to structural FEA.

This new approach by the authors is a conceptual response-based method rather than a load or load-effect-based method used in DLA. This response-based method is more direct with fewer assumptions on the equivalent wave systems and dynamic load parameters. The critical response of the hull structure owing to the loads in design sea conditions is a natural outcome of the analysis and more accurate and can be used for the direct evaluation of hull structure. In the past spectral analysis and extreme value approach have been successfully applied to calculate linear response sprectra based on RAOs of shipmotions, vertical superposition bending moments, and normal stresses, etc. and wave spectrum. However, to calculate von Mises stress, which is not a linear function of wave and requires special treatment to get response spectrum and corresponding short-term and long-term life time maximum value. The author developed a new approach of handling this difficulty and demonstrated its practical application to determine the critical response, which is important in the evaluation and acceptance of the hull structure strength. This method offers a complete picture of the hull responses at any location in the hull in terms of more realistic and accurate extreme values subject to wave excitation. A possible drawback of this method could be more computation-intensive because it is calculating extremes of local response for stress magnitude and spatial distribution as compared to DLA, which is based on finite numbers of critical loads.

Finally I would like to ask author to clarify a couple of points. This method seems to be applicable for the strength assessment of hull structure. Would the method also be applicable for fatigue analysis, which depends on the maximum amplitude but depends more on the range of the stress variation and also the number of cycles?

I would like to congratulate the authors for presenting a new method that is definitely a step forward and useful for more accurate strength assessment of ship structure.

Robert Sielski, Member

The authors are to be congratulated on an excellent exposition of stress analysis of ships operating in a seaway. In an age when many technical papers consist of a summarization of the analysis of private data with a proprietary computer program, it is refreshing to see the methodology for analysis completely laid out. I believe that this paper will stand as a reference document for many years.

However, I take exception to the statement by the authors that the von Mises stress has been widely adopted because of its accuracy. It has been used widely only because of its convenience. In many finite element analysis programs, the von Mises stresses are an output option for describing the state of stress in an element, and in a multi-directional stress field they represent a single number to typify that field. However, the von Mises stresses are not linked to any failure mode, either collapse or fracture and are not suited for limit state analysis, which is becoming increasingly important in ship structural design. For analysis of collapse, the stresses aligned with the orientation of structural elements are necessary. For analysis of fracture, principal stresses may be of value as is strain energy density, if a single number is needed to evaluate a multi-directional stress field. For fatigue analysis, stresses normal to the direction of crack propagation are needed, possibly augmented by a hot-spot finite element stress analysis. Could the authors please comment on how their procedure could be modified to suit a limit-state analysis?

The authors have made extensive use of the ellipsoids of constant von Mises stress in the sx, sy, t space. However, these element directions are sometimes arbitrary, selected on the basis of the structural geometry for element modeling, and not related to the significant axes of the stress field. How would the ellipsoidal representation change if the axes of the principal stresses were used?

Again, I congratulate the authors on an excellent paper, and look forward to their further development of the methodology.

Armin Troesch, Member, and **Laura Alford**, Member

The authors are to be congratulated on presenting a well written and insightful paper on extreme responses. By examining the maxima of von Mises stresses (Eq (10)), new relationships involving products of zero-mean, normally distributed, random shear and normal stresses have been found.

This written comment concerns the curve fits using various distributions as shown in Figures 8, 9, and 12 and their applicability to stresses resulting from both wave bending and whipping or springing. The dynamic structural response of a vessel in extreme seas is likely to contain both high-frequency (whipping) responses and low-frequency (quasi-static wave induced) responses leading to bimodal spectra. Maxima of the combined low- and high-frequency responses will likely be greater than maxima of the individual responses by themselves. Have the authors considered how their conclusions might change if the processes were represented by bimodal spectra?

Authors' Closure

We humbly acknowledge and thank each of the discussers for their kind congratulations. We are aware that the marine industry is very busy and that there are ever increasing demands on the limited time of everyone so engaged. By commenting, our discussers have demonstrated their dedication to professional ideals exemplified through the exchange and debate of technical information. The authors wish to thank the discussers for their time and effort spent developing their comments. Such thoughtful discussion greatly enhances the archival value of the paper.

Dr. Shin correctly observes that what we describe could be regarded as a response-based method rather than a global load-based method. However, we approached this topic presuming that von Mises stress would be used as a dynamic load parameter (DLP) in essentially the same way that bending moments, accelerations and roll motion are used in the ABS Dynamic Loading Approach. Consistent with linear, stationary, Gaussian sea states, and linear response theory, the methods that we present allow the estimation of the long-term extreme von Mises stress. Then, unique equivalent regular wave, or sample (non-unique) equivalent irregular wave conditions, can be determined that reproduce the life time extreme von Mises stress. In this equivalent wave condition the instantaneous pressures over the wetted hull surface and the rigid body accelerations corresponding to dynamic equilibrium are determined. This completely describes a load case for global structural FEA. In this sense the methods we present are philosophically consistent with the ABS Dynamic Loading Approach as set forth in the ABS guides and as presented by Liu et al. [13].

As suggested by Dr. Shin, an alternative approach would also be possible in which the life time extreme von Mises stress of FEA structural elements is estimated directly using the methods we present, without proceeding to loading the finite element model with an equivalent wave. This would truly be a response-based method.

There is, however, a crucial question that lies somewhat deeper. The life time maximum DLP can be estimated using linear methods but current ABS practice is to require that the equivalent wave be applied using nonlinear hydrostatic pressures over the instantaneous wetted surface. Thus, the final equivalent wave load condition is intrinsically nonlinear and not altogether consistent with the genesis of the life time extreme DLP. This transition from linear theory DLP to nonlinear equivalent wave load condition typically leads to some imbalances that need to be adjusted before the global FEA can be completed. Thus, the DLP, whether it be bending moment, acceleration or roll, as suggested by ABS, or von Mises stress as we suggest, is merely a guide to a nonlinear load case for FEA. If these nonlinear features of the final load case for FEA are to be retained, then a direct response-based method using the methods we present must be regarded as inferior to a method that treats von Mises stress as a DLP.

Finally, Dr. Shin asks whether the methods of this paper are appropriate for fatigue analysis. We would suggest that the quite similar principal stress-based methods of Hutchison et al. [11] are more appropriate for fatigue. Principal stresses are thought to be more appropriate for fatigue, as they have an orientation in the element. In [11] we set forth methods for determining that orientation and also for determining the stress cycle spectrum suitable for use in spectral fatigue analysis.

Dr. Sielski raises some familiar questions regarding von Mises stress and expresses his preference for principal stresses. He suggests that the decision to use von Mises stress may be guided less by concern for accuracy than by convenience. We are aware that the legitimacy of von Mises stress is debated in some structural circles and agree that the choice of von Mises stress may indeed often be guided by convenience. Our own perspective is practical and reflects our experience with the classification societies. It has been our experience that, not only do the class societies accept von Mises stress, but they often encourage its use.

He asks how our procedure could be modified to suit limit-state analysis. To the extent that his question concerns the choice of characteristic stress, we would respond that, if one prefers principal stresses, those have been addressed to an equivalent level in Hutchison et al. [11], albeit with a focus on application to fatigue analysis. But the principal stress methods of [11] could, with equal ease, be applied to strength. However, the more serious question is whether the methods we describe should be extended to limit-state analysis at all. The reason for our caution is that our methods are built on the familiar foundations of linear response theory and second-order stationary, Gaussian random processes. This includes the assumption of a linear relationship between strain and stress. Thus, it may depend on what sort of limit state one wishes to address. If it is a yield or buckling inception limit state, then these methods may be appropriate. But, certainly, they are not appropriate for direct analysis (in the sense suggested by Dr. Shin, for example), if the ultimate limit state is to be addressed. However, as a guide to those cases that should be globally modeled in an equivalent (regular or irregular) wave, as in the ABS Dynamic Loading Approach, we do believe that our procedure may represent an improvement. As observed in our response to Dr. Shin, selected nonlinearities, including (if appropriate) post-yield and post-buckling behavior, can be addressed during the global FEA in response to the selected equivalent wave(s).

As observed by Dr. Sielski, it is indeed true that element x- and y-coordinates, and consequently σ_x, σ_y and τ space, are arbitrary, based solely on considerations of structural geometry for element modeling, and not on the significant axes of the stress field. We believe that is, in fact, one of the intended major points of our paper. We exploit the ellipsoidal character of not one but two different interacting ellipsoids. One is the ellipsoid of constant von Mises stress and the other is the Gaussian stress ellipsoid. In general these two ellipsoids are not aligned.

The first is invariant under rotation of element coordinates. This means that the von Mises ellipsoid is oriented with the major axis oriented at 45° in the $\sigma_x\sigma_y$ -plane and the minor axis directed in the 135° direction, as shown in Figure E-1. This relationship does not vary with rotation of coordinates, because σ_x, σ_y and τ take their meaning from the coordinates. The orientation of the major axis of the second ellipsoid, the Gaussian stress ellipsoid, is a major focus of our paper. Appendix A of this paper discusses the determination of the principal axes of the Gaussian stress ellipsoid and the description of that ellipsoid in principal axes is particularly simple and elegant. However, the principal axes of the Gaussian stress ellipsoid rarely align with the invariant orientation of the principal axes of the von Mises stress ellipsoid. Furthermore, the orientation of the principal axes of the Gaussian stress ellipsoid depend on sea state and vessel operational parameters. Hence, even if the Gaussian stress and von Mises stress ellipsoids were to align for a particular set of sea state and vessel operational parameters, they would be misaligned for many others such parameter sets.

As developed in Appendix D of this paper, von Mises stress in principal stress coordinates likewise assumes a simple and elegant form. However, one of the major results of Hutchison et al. [11] is that there are multiple methods of determining the statistical principal axis about which principal stresses can be calculated.

Professor Troesch and Dr. Alford raise the question of whipping, springing and bimodal spectra. We believe that our methods are compatible with springing, but perhaps not for whipping. We suggest this because springing response in stationary sea and operating conditions may be regarded as having a second-order stationary spectrum, whereas whipping is intrinsically a transient response. We don't believe that the bimodal character of combined rigid body seakeeping and springing responses necessarily poses any extraordinary obstacle to the methods we have presented, especially if one uses generalized Rayleigh probability. There does remain a question, that we recommend for further research, regarding the proper a priori determination of the spectral breadth parameter to be used in the generalized Rayleigh distribution, but this question exists whether or not springing is included. There may be methods based on wavelet analysis that could be appropriate for the question of whipping, but our current suggestion is that whipping is best addressed in the time domain.

In closing, we would like to take note of Dr. Sielski's favorable opening observation that we have presented our analysis methodology completely. We believe that it is important to proper and professional exchange that methods be presented completely and agree that all too often in the modern era this is not done. SNAME is one of an elite few forums where authors are afforded the space necessary to complete exposition, and we are most thankful for that noble tradition.

A Comparison of Three Types of Evacuation System

António Simões Ré (M)
Canada, **Brian Veitch** (M)

ABSTRACT

The capabilities of three marine evacuation systems have been investigated using systematic series of model experiments in a large test facility. Tests were done with a conventional davit launched twin-falls lifeboat, a similar system with the addition of a flexible boom, and a free-fall lifeboat. The performance of each system was evaluated as a function of weather conditions, ranging from calm conditions to severe storms. In addition, the effects of the configuration of the evacuation station were examined. Results of this large experimental campaign are presented and practical applications to emergency preparedness planning and evacuation system design are discussed in the context of goal-based regulations. This work aims to address a long-standing knowledge gap by providing objective, empirical data that can be used by designers, regulators and others in their decisions concerning safety.

INTRODUCTION

In the event of a marine evacuation of a ship or an offshore petroleum installation, all personnel on board must have access to an evacuation system, be able to embark and launch safely, clear the ship or installation, and survive until rescued. Further, personnel should have a reasonable expectation of avoiding harm arising from credible hazard scenarios, including those occurring in environmental conditions that can reasonably be expected to prevail during operations. As a statement of the goals of a marine evacuation system, the foregoing is realistic and reflects common sense. For it to be useful as part of a goal-based regulation framework, it must be translated into terms of expected standards of performance that are amenable to objective evaluation and that ensure the objectives are achievable.

The main aim of this paper is to provide some objective benchmarks, upon which reasonable estimations can be made of the expected performance of three marine evacuation systems: a conventional davit launched twin-falls lifeboat, a similar system with the addition of a flexible boom, and a free-fall lifeboat. The capabilities of these three evacuation systems were investigated in a large test facility using model scale experiments, an approach well suited to the circumstances [1]. The performance of each system was evaluated as a function of weather conditions, ranging from calm conditions to storms. In addition to environmental conditions, the tests examined the effects of the evacuation systems' configuration. The investigation did not incorporate issues related to equipment reliability, and although human performance effects were investigated (in terms of injury and motion sickness criteria), these are outside the scope of the present paper and will be presented elsewhere.

Results from the experiments are presented and discussed in the context of performance goals as might be used in a goal-based regulatory regime. This work aims to address a long-standing knowledge gap by providing objective, empirical data that can be used by designers, regulators and others in their decisions concerning safety.

REGULATORY REGIMES

In broad terms, performance goals will generally reflect society's values and norms, and should specifically reflect the requirements of the law. Embodied in regulations, these are matters of public policy whose application is mediated in some manner by a regulatory body, which may also have contributed to defining the goals in the first place. In practical terms, a performance goal is the objective or purpose of a piece of equipment, procedure, system, or other element of a particular installation or ship. Performance goals are set by the regulator. A good example is the regulations in [2].

A performance standard is the operator's specification of a solution to achieving a given goal. It is a verifiable statement of the performance required of the equipment, procedure, or system. Performance standards should be cast in terms of a relevant measure or measures, such as reliability, functionality, availability, time or distance. They should manifestly contribute to the overall goal of reducing the risk of harm. Each standard should provide a basis for monitoring and maintaining the requisite performance of the equipment, procedure, or system throughout its life cycle, and should account for the specific circumstances particular to the installation (or ship) and its operation.

The operator has the responsibility to achieve the goal, the opportunity to establish the performance standards by which to achieve it, and the continuing responsibility to ensure that the equipment, procedure, system or similar is monitored and maintained fit for purpose. Arising from this added responsibility is the notion that goal-based regulations promote a culture of safety rather than a compliance culture.

The regulator accepts the proposed performance standards or not, and holds the operator to the stated standards. Rather than using inspections as the key mechanism to ensure operators are in compliance with regulations, as is the general case under a specification type regulatory framework, regulators in a goal-based framework rely more heavily on audits of the operators' safety plan.

This difference – audit rather than inspection – has given rise to some criticism of goal-based regulations as entailing self-regulation by industry, and too much focus on the management of safety rather than the matter of safety. Subsequently, this has led to opposition to change from specification style regulations to goal-based regulations. In practice, the regulator is the ultimate authority under both types of regulatory system, although in practice, their activities and skill sets are likely to be quite different under the different regimes.

There are other arguments against the move away from specification regulations, including that they capture a wealth of historical knowledge and experience, are relatively easy to use by designers and operators, and are relatively easy to check by regulators and their designated inspectors.

Indeed, existing specification regulations do incorporate valuable experience, including that from accidents, although the context is sometimes lost once the specification type regulation is constituted. Routine application of regulations without clear understanding of their context then provides some unspecified level of safety that is still accepted by the regulator. This situation obtains even when the value of the specification standards derives from experience with installations that differ significantly from a given specific situation at issue.

That said, goal-based regulations can, and perhaps should, incorporate existing specification standards, at least during a transition period from one regulatory system to the other, as the specification standards provide a clear example of what has been accepted under one regime and may still be acceptable to the regulator under a goal-based system. Likewise, existing codes of practice, industry guidelines, and other accepted norms are available to operators as the basis of performance standards under a goal-based system. It is in the absence of such accepted means that the operator has an additional responsibility to propose a new performance standard and demonstrate its efficacy in achieving the performance goals.

This can be an onerous requirement and is likely to involve more uncertainty for the operator in terms of meeting the obligations of the law, but can also stimulate innovation, which is often cited as an important benefit of goal-based regulations. Indeed, the response to the opportunity for innovation is where much of the potential advantage of goal-based regulations lies.

Innovations must find their way into practice if the advantages are to be realized, though, and by this measure, a framework in which regulations are set out as high level goals, rather than detailed specification standards, should facilitate the relatively rapid adoption of evolving best practice and improved technology, because delays associated with changing legislated requirements can be avoided. When innovations with demonstrated benefits in terms of reducing the risk of harm are available at costs that are not grossly disproportionate to the benefits, the operator should adopt them; otherwise, the regulator should insist they do. Such provisions (e.g. best available technology) can be incorporated in both specification and goal-based regulations to help ensure the adoption of efficacious innovations, while simultaneously discouraging regulations becoming fossilized solutions.

It is sometimes argued that specification regulations are fair in the sense that they apply equally to all operators so that no commercial advantage can be sought through variance from the specified rules. There is also the view that as safety is often considered to be a cost, it will therefore generally be eroded over time unless specific regulations are applied and enforced. Goal-based regulations have also been criticized as relying too heavily on risk management, particularly on quantitative risk assessment and its attendant uncertainties.

PERFORMANCE MEASURES

The debate over the pros and cons of the different regulatory regimes will continue [e.g.3]. In the meantime, more information that bridges the knowledge gap concerning the performance capabilities of marine evacuation systems can be incorporated into the design of evacuation equipment, procedures, training and operational planning, including planning for emergency response. Where performance capabilities are known to deteriorate with weather and reach performance limits, decisions by designers, operators, and regulators should recognize the residual risks associated with maritime operations under conditions that approach and exceed the limits of the evacuation systems.

As stated in the notional performance goal at the start of the introduction, the success of a marine evacuation is dependent on several factors, including the safety equipment itself, the people who have to use it, the nature of the hazard that initiates an emergency response, the prevailing environmental conditions, and the interaction of all these.

Credible hazard scenarios, such as arising from a loss of structural integrity and progressive flooding, or due to a fire, form the basis of emergency response planning at the design stage, from which factors like response times and vulnerability to harm and impairment can be derived. Safety equipment, and in particular the evacuation system, has to be accessible to personnel and be maintained in functional condition. Most importantly, it has to be fit for purpose: getting people off the ship or installation and away from the emerging hazard so that personnel on board can survive until rescued.

Human factors are also critical to the success of an evacuation response, which has implications for procedural planning. Appropriate and effective training for such scenarios is also critically important, particularly when they can reasonably be expected to include a major hazard (e.g. fire and explosion, smoke, structural failure, loss of stability), difficult environmental conditions (e.g. wind and waves, sea ice and icing, poor visibility), or a combination of both. Environmental conditions are a major factor in the overall performance of evacuation systems and it is on this specific point that the work reported in this paper focuses.

Performance in our experiments was evaluated in terms of a variety of technical performance measures, such as the accuracy of the launch, the degree to which the lifeboat was setback due to its encounter with oncoming waves, the path taken by the lifeboat between launching and clearing to some distance away from the platform, and accelerations and motions relating to injury criteria and motion sickness, amongst others. Setback due to the boat's initial encounter with an oncoming wave is illustrated in Figure 1, along with progressive setback due to subsequent wave encounters. This was found to be the most important performance measure. The progression of passing waves is illustrated in the four wave profiles shown in the figure. At the top, a lifeboat is shown to have splashed down on the up-slope of the incoming wave. Despite its heading (to the right), the lifeboat is unable to make way and is pushed back by the passing wave, as shown in the second profile, until it crests the wave. The distance it is pushed back by the first wave encounter is the setback. After cresting the wave, the lifeboat begins to make way, as shown in the third profile. Progressive setback is illustrated in the fourth profile, which is additional setback due to subsequent wave encounters. Note that the dot shown on the consecutive profiles represents the original point on the wave where the boat was launched.

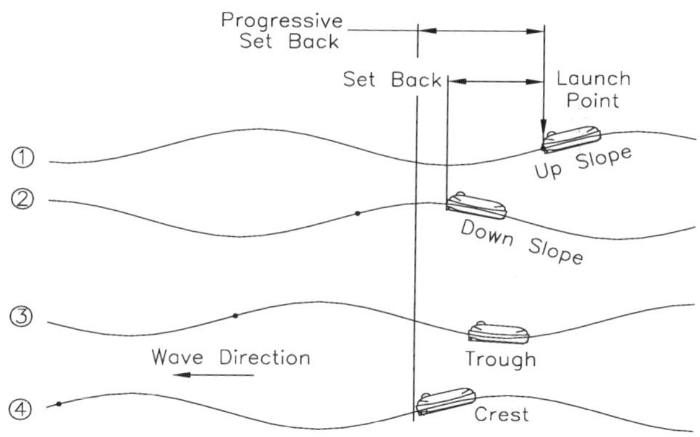

Figure 1. Setback and progressive setback.

The evacuation area can be divided into zones to help define the various performance measures; the zones are illustrated in Figure 2 for a davit launched lifeboat and Figure 3 for a free-fall boat. With reference to Figure 2, an exclusion zone extends out from the installation far enough to accommodate launching in the range of weather conditions and damaged conditions for which evacuation is a planned contingency. The lifeboat should not enter this zone under any conditions as to do so would put it at risk of colliding with the ship or installation. In practice, the exclusion zone boundary should encompass all collision hazards, whether the hull of a ship or legs of a semi-submersible, and so will be particular to each installation.

Tangent to the exclusion zone is a launch zone, shown nominally in Figure 2 as circular. The size of the launch zone depends on how much area is required to accommodate a safe launch, bring the lifeboat under control and initiate clearing towards the rescue zone. At its centre is the nominal launch target, which for the conventional davit launched system illustrated in Figure 2 is on the sea surface directly below the aftermost part of the lifeboat.

The decision on the size of the launch zone can be

based on the weather conditions that are chosen by the operator to be the upper limit for a planned evacuation. For example, the weather conditions in which the lifeboat becomes un-seaworthy is *de facto* an upper weather limit and evacuation in conditions exceeding this limit should not be entertained in the emergency response plan. Other, less extreme, upper weather limits might be implemented, as illustrated in Figure 2 where the location of the launch target moves progressively closer to the installation as the weather limit decreases from the most severe conditions denoted by *C*, to the relatively moderate conditions denoted by *B*, and on to the relatively benign conditions denoted by *A*.

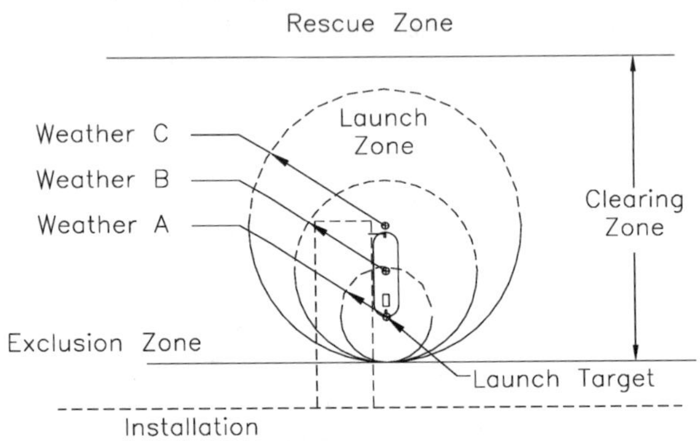

Figure 2. Evacuation area zones (and 3 design weather limits) for a conventional davit launched lifeboat.

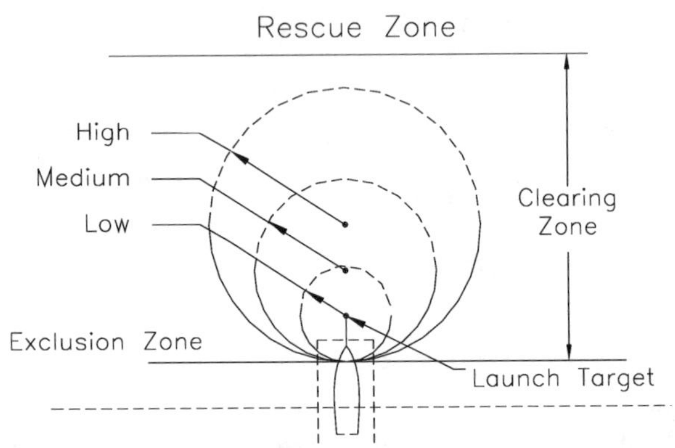

Figure 3. Evacuation area zones (and 3 target safety levels) for a free-fall lifeboat.

A target level of safety in terms of successful launches can also be used as a second criterion for setting the launch zone boundary. This is based on using quantitative data, such as the experimental results presented in this paper, to encompass a target proportion of launches in given environmental conditions. The higher the target level of safety is for a given weather condition, the bigger the size of the launch zone required to encompass the higher proportion of launches. Figure 3 shows this for three notional levels: relatively low, medium, and relatively high. For the circular launch zones used in the illustration, the size of the zone determines the position of the launch target, similar to the boundary set according to weather limits. These two criteria provide a straight forward and rational means of arranging an evacuation station using appropriate benchmark data.

Figure 3 illustrates the concept as applied to a free-fall lifeboat, where the target launch point extends out from the boat in the stowed position (rather than immediately below it) as this is where a free-fall boat would be expected to splash down.

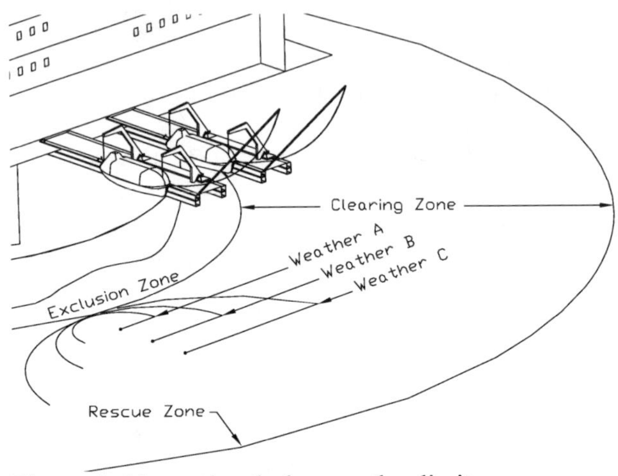

Figure 4. Alternative design weather limits.

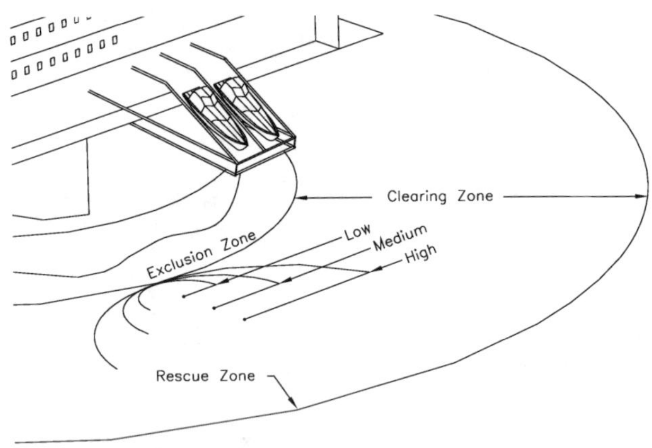

Figure 5. Alternative design safety levels.

Beyond the launch zone in Figures 2 and 3 is a rescue zone, where response vessels can execute rescue procedures without risking exposure to harm from the emerging and possibly escalating hazard (e.g. radiant heat from fire, overpressure from explosions, toxic smoke or gas from fire or blowout). Between the exclusion zone and rescue zone is a clearing zone. Figures 4 and 5 illustrate the concepts of design weather limits and design safety levels again, in the first case for a conventional lifeboat arrangement, and in the second for a free-fall station. Both criteria could be applied together. Additional performance criteria have been presented in [2 and 4].

EXPERIMENTAL APPROACH

Setup

The experimental campaign was carried out in three phases in the Offshore Engineering Basin (OEB) at the National Research Council's Institute for Ocean Technology. The OEB has a 65m×26m working area, bounded on two adjacent sides by wave makers and on the opposite sides by wave absorbers. The water depth during all the tests was 2.8m. A bank of fans was used to provide wind. A four-legged truss structure was fixed to the floor of the OEB and served as the platform for the evacuation stations. In each phase, a 1:13 scale model of a fully loaded lifeboat was used.

The free-fall lifeboat model and evacuation station are shown in Figure 6. It was a generic style model of an 11.25m long lifeboat with a capacity of about 50 persons. It was fitted with a four-bladed propeller, an active steering nozzle, an electric motor and shaft, rechargeable batteries, a wireless video camera and a radio transmitter. The model was configured to operate at a power level corresponding to a calm water speed of 6 knots at full scale. The instrumentation used to collect data during the test series included three accelerometers to record longitudinal, lateral and vertical accelerations, three rate gyros to monitor roll, pitch and yaw, a motor controller, and a Qualysis optical tracking system. The Qualysis reflectors were mounted inside the model's clear plastic canopy. A 16-channel 16-bit resolution data acquisition system was set up to sample 12 channels at a rate of 320Hz. Additional details concerning the model, remote control, and data acquisition system are given elsewhere [5-6].

The free-fall lifeboat launching station was modular, including the key component consisting of the lifeboat mounted on a launch ramp. The ramp's length and angle were adjustable. The base of the ramp was mounted on an orientation template with pre-drilled holes corresponding to different orientations of the launch ramp with respect to the wind and waves. In turn, the orientation template was mounted on a cantilever beam that extended the lifeboat station out over the water. A steel frame connected the cantilever beam to an adjustable table that was used to change the launch height, although all results presented in this paper were made from a constant height of 20m. The modular design facilitated rapid setup changes during testing.

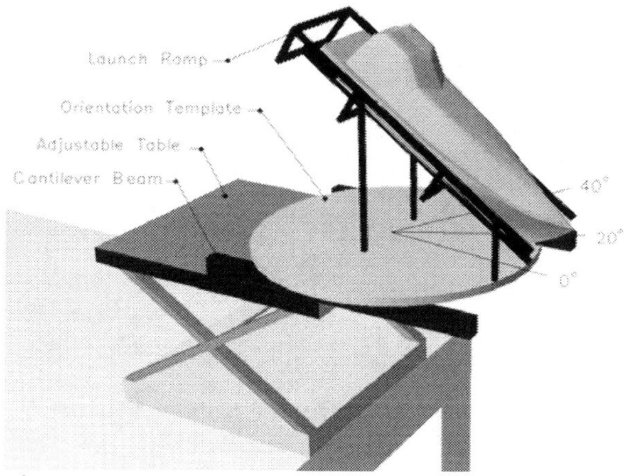

Figure 6. Free-fall lifeboat model and evacuation station.

The model of the conventional lifeboat, also a generic style model, was of an 80 person totally enclosed motor propelled survival craft (TEMPSC). This was the model used in the first two experimental phases as the conventional davit launched evacuation system and the same system with the addition of a flexible boom. The model and evacuation station arrangement are shown in Figures 7 and 8 for two cases: first when the model is orientated perpendicular to the platform (typical of an arrangement on a petroleum installation [2]), and then for the case of a parallel arrangement (typical of ship evacuation stations). In both arrangements, a flexible boom is shown, although it was not used in all tests.

The TEMPSC model was fabricated of glass reinforced plastic and powered for a full scale speed of 6 knots in calm water using an electric motor and shaft, a four bladed propeller, a steering nozzle and two rechargeable batteries. It was outfitted with a simulated hydrostatic release circuit with interlocking mechanical release servos, a radio transmitter, a wireless video camera, and a water detection light emitting diode (LED). The LED was used to signal that the hydrostatic interlock had been released so that the hooks could be released. The model's location was tracked using a Qualysis optical tracking system. The model setup is described elsewhere [7-11].

Like the free-fall station setup, the TEMPSC lifeboat station was modular. The davits, winches and lifeboat were all mounted onto a wooden deck, which was in turn fitted to a cantilever beam that allowed the lifeboat to be extended out over the water. The other end of the beam was attached to an adjustable table that was used to set the launch height and inclination of the lifeboat station in order to simulate damaged conditions. The same setup was used for the conventional twin-falls davit launched boat with and without a flexible boom. Most launches of the davit lifeboat were made from a height of 25m.

Additional applications of model testing to investigate evacuation system performance can be found in [12-13], for example.

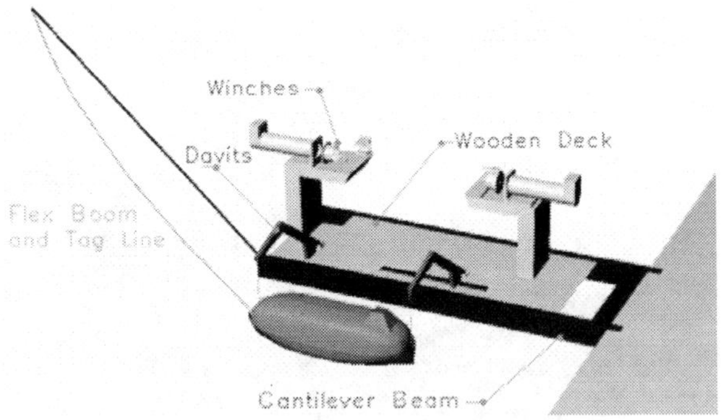

Figure 7. Davit launched lifeboat model and evacuation station arranged perpendicular to the platform.

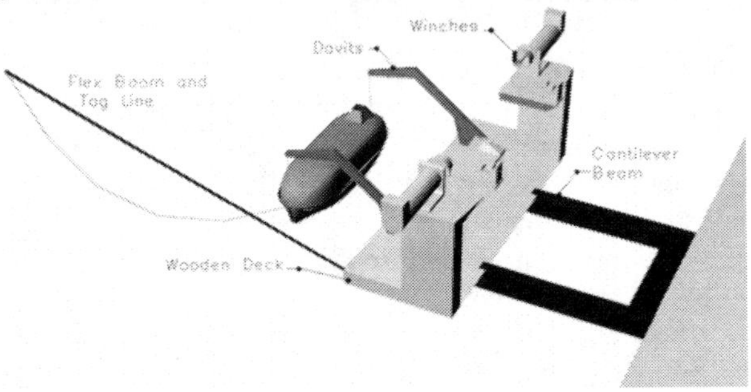

Figure 8. Davit launched lifeboat model and evacuation station arranged parallel to the platform.

Figures 9, 10 and 11 show the arrangements used for the free-fall experiments, the TEMPSC system oriented perpendicular to the platform, and the TEMPSC system oriented parallel to the platform. In the latter two cases, experiments were done with and without a flexible boom.

In the Figures 9 to 11, regular waves propagated from right to left. Each environmental condition was modeled using an average wind and a regular wave of a specific height and period that, when combined, constituted an *equivalent* Beaufort condition. Conditions ranged from calm to approximately Beaufort 8 (equivalent) and included several nominally discrete weather conditions in between. Tests with the TEMPSC systems were done in six conditions, as shown in Table 1. The same conditions were used for the free-fall system, except for the moderate breeze environment, which was not modeled. Significant wave heights from the Beaufort scale were used as the target regular wave heights in the experiments.

Table 1. Nominal environmental conditions.

(Beaufort) description	Mean wind speed	Average wave height	Significant wave height
	[$m \cdot s^{-1}$]	[m]	[m]
(0) calm water	0	0	0
(4) moderate breeze	5.6–8.2	0.42-0.88	0.67-1.40
(5) fresh breeze	8.7–10.8	1.16-1.52	1.86-2.44
(6) strong breeze	11.3–13.9	1.95-2.93	3.04-4.57
(7) moderate gale	14.4–17.0	3.35-4.88	5.49-7.92
(8) fresh gale	17.5–20.6	5.79-8.53	9.14-13.72

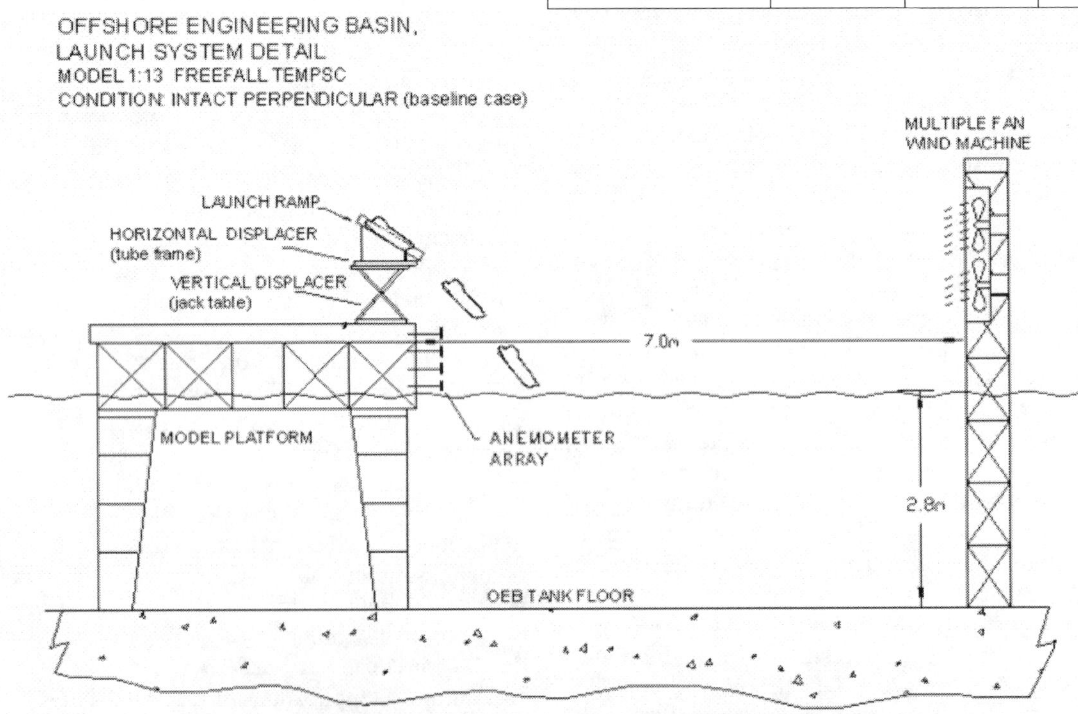

Figure 9. Test arrangement: free-fall lifeboat launching into head seas.

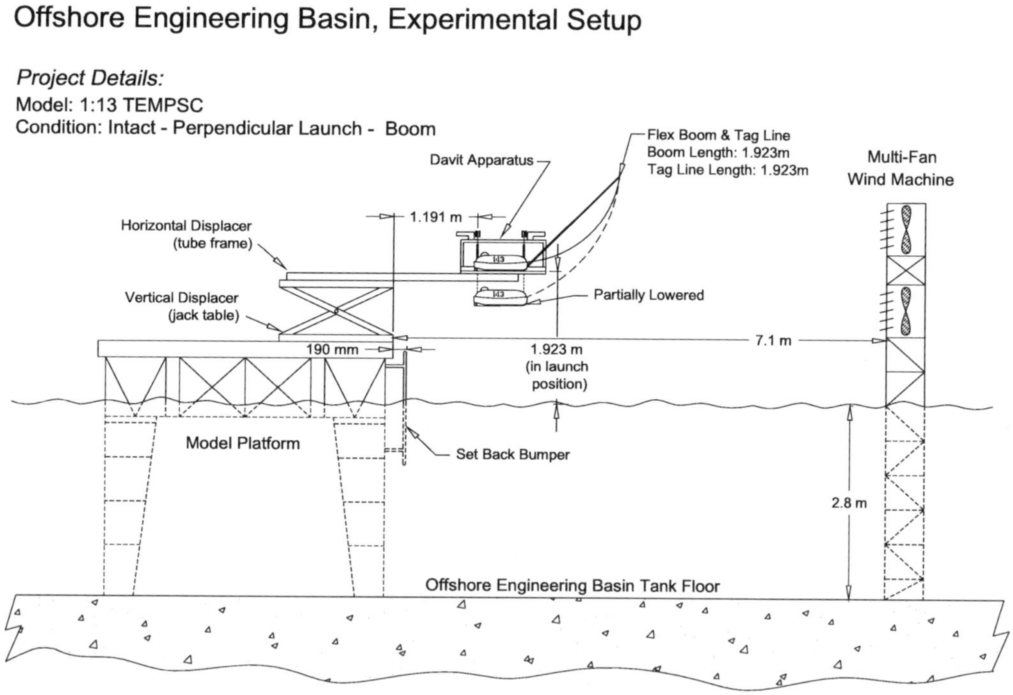

Figure 10. Test arrangement: TEMPSC lifeboat orientated perpendicular to the platform. The flexible boom is also shown.

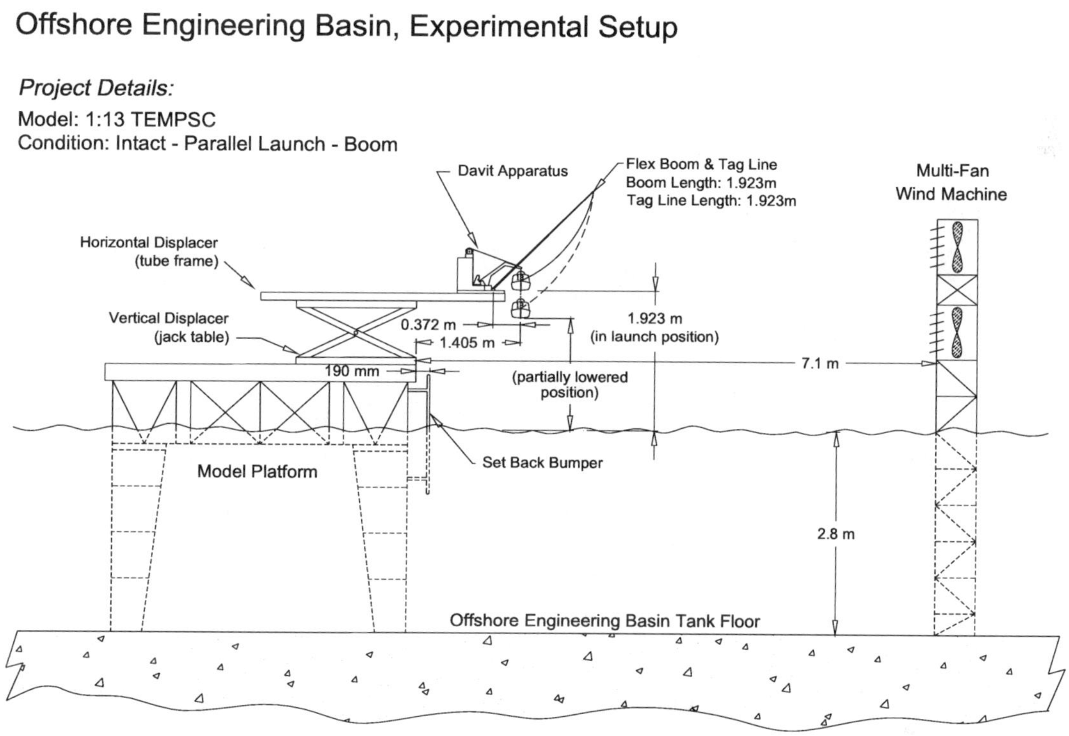

Figure 11. Test arrangement: TEMPSC lifeboat orientated parallel to the platform. The flexible boom is also shown.

RESULTS

Scope

Results are presented below for all three systems so that comparisons can be made concerning their performance. For all three systems, the main investigation concerned performance over a range of weather conditions, as indicated in Table 1. As measures of performance, the focus here is on setback and the directness of the paths taken by the lifeboats during the sail-away (or clearing) phase of the evacuation process.

Results are also presented here of important configuration variations, namely, perpendicular and parallel launch arrangements for the davit launched systems. Free-fall lifeboats would normally be arranged perpendicular to the platform, which the experiments also modeled, or at the stern of a ship. Most of the free-fall test launches were made into head seas, but some launches into oblique seas (up to 40° off the bow) were also made.

The experiments also investigated the effects of wave steepness on performance and results showing these effects are also presented. More results were acquired than can be presented here, but the interested reader can expect to see additional results in subsequent publications.

To begin, examples of the launch process are shown for each system to highlight some important attributes of each. Results of multiple launches are then presented by tracing out their paths from launch to clearing to the rescue zone. These plots show a footprint of the sail-away process for the various evacuation systems and configurations. Multiple launch results are then presented in terms of the extent to which the different lifeboats were setback after launch. The effects of the wave steepness and the splash-down position with respect to wave phase are then examined for the sail-away paths and the setback. Throughout, comparisons are made between the three systems and configurations.

Examples of launching: free-fall

An example of a successful free-fall launch is shown in Figure 12 for moderate gale weather conditions with approximately 7m high (full scale) waves with a nominal steepness of 15. Two views are illustrated. The top one shows a plan view of the launch site, including an outline of the free-fall lifeboat on its ramp prior to launch. Some notional boundaries used in the measurement of performance are also indicated in this view, including the target drop point for the lifeboat and a nominal exclusion zone boundary around the installation. An *x-y* coordinate system is superimposed on the plan view. The corresponding profile view is shown in the bottom part of the figure, with a superimposed *x-z* coordinate system. In both views, the path of a single point in the lifeboat's reference frame is traced out. The point corresponds to the intersection of the flat stern and keel line, which is the origin of the boat's coordinate system.

In this case, the *x-z* view shows that the lifeboat followed a smooth trajectory as it left the ramp and entered the water on the down-slope of a wave. The outline of the boat is superimposed at several points to illustrate its orientation during the launch process. The path dropped smoothly below the mean water level, resurfaced at the trough and then crested a wave just beyond the rescue zone.

The plan view shows that the free-fall boat hit the target launch point accurately and then moved away from the platform at a fairly straight heading during the entire launch and sail-away process, which was completed rapidly. There was only minor lateral displacement en route to the rescue zone, and little pitch and roll motion.

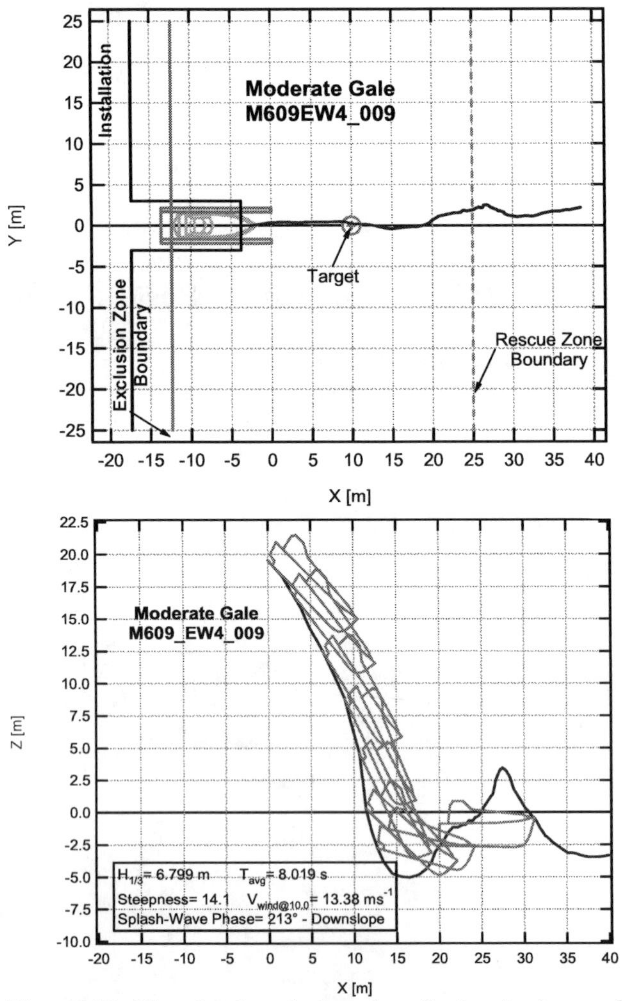

Figure 12. Plan (top) and elevation (bottom) views of path taken by free-fall lifeboat in moderate gale conditions.

For simplicity, the path traced out by only a single point on the lifeboat is shown in the figure. The paths shown in the two views do not correspond exactly due to the means used to measure it. Two video cameras were used to record the position. The more reliable of the two was the overhead camera, which is the plan view in Figure 12 and in similar figures following.

Figure 13 shows another example of a successful free-fall launch into moderate gale conditions with the same nominal wave steepness (15) as in the previous example in Figure 12. In this case, the lifeboat entered the water at a wave crest and briefly submerged before emerging on the down-slope. It then sailed into the trough and crested a second wave after having reached the rescue zone. The launch was smooth and quick, with very little pitch and roll motion, and a fairly steady heading.

from the successful launches illustrated in Figures 12 and 13, largely because the boat entered the water in a wave trough. It resurfaced on the up-slope of the following wave at which point the stern section was out of the water, but the boat maintained some forward way. The incoming wave stopped the boat's forward progress. From the profile view, it is clear that the boat was setback significantly during this encounter. After the initial setback, it made some headway after the first crest passed, but was setback by about 4m during its second wave encounter and then crested another wave before making the rescue boundary. The plan view shows that the lifeboat moved to port after it surfaced and was then brought under control and headed away from the installation.

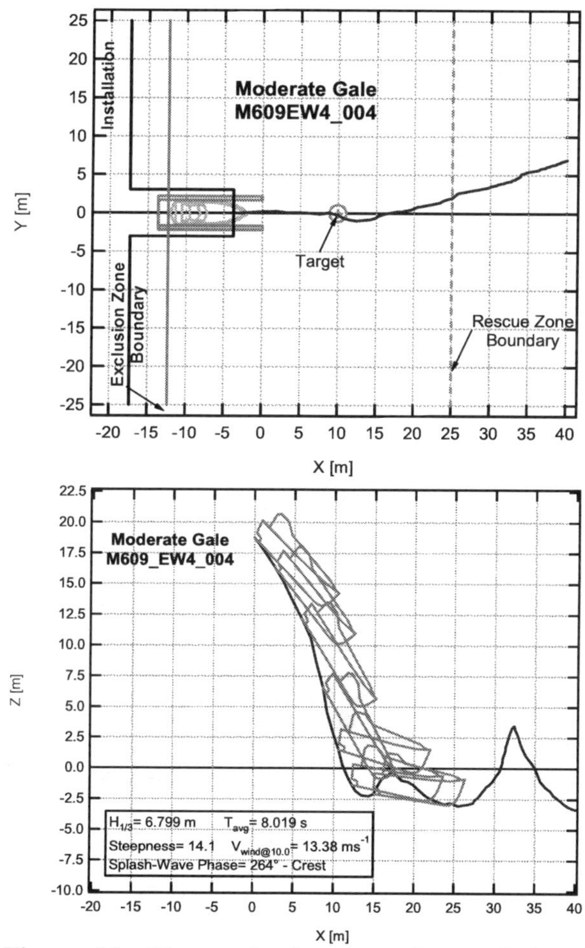

Figure 13. Plan and elevation views of free-fall lifeboat's path during launch in moderate gale conditions.

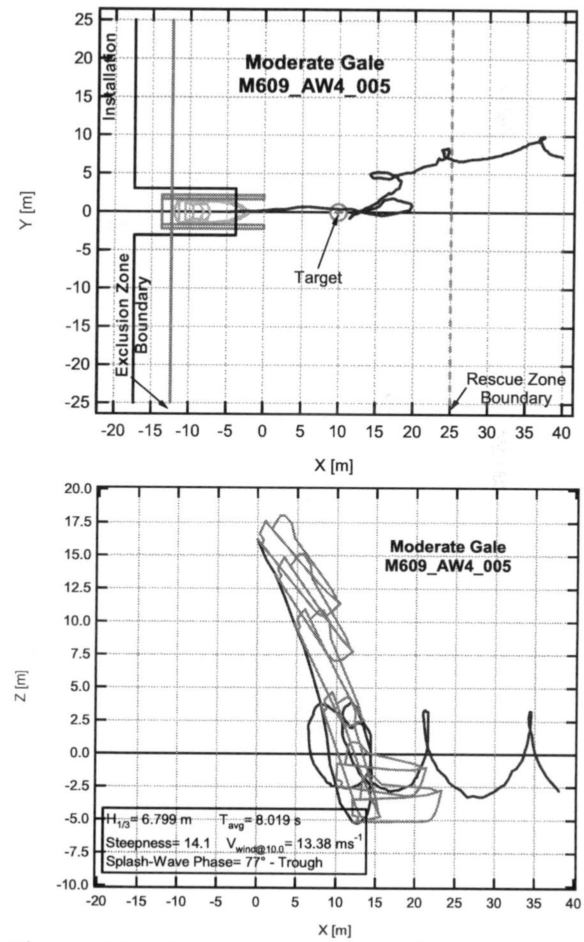

Figure 14. Plan and elevation views of free-fall lifeboat's path during launch in moderate gale conditions.

Another example of a free-fall launch into the same nominal conditions (moderate gale, wave steepness 15) is presented in Figure 14. This case is very different

The final example of a free-fall launch is shown in Figure 15, again for the same weather conditions. This time, the lifeboat entered the water on the up-slope of an incoming wave and surfaced at the crest. The consequences are dramatic. Upon surfacing, it rolled approximately 90° onto its port side, was setback significantly, and deviated more than 40° from its intended heading. It was setback further by the next wave, which resulted in a collision with the platform. It recovered its heading after the collision and made way to the rescue zone, although with difficulty at each subsequent wave encounter. Compared to the cases where the boat landed on a down-slope and a crest, the performance when landing on the up-slope was very poor. The plan view shows that the lifeboat moved quite erratically after its launch, ending up well to the port of the evacuation station.

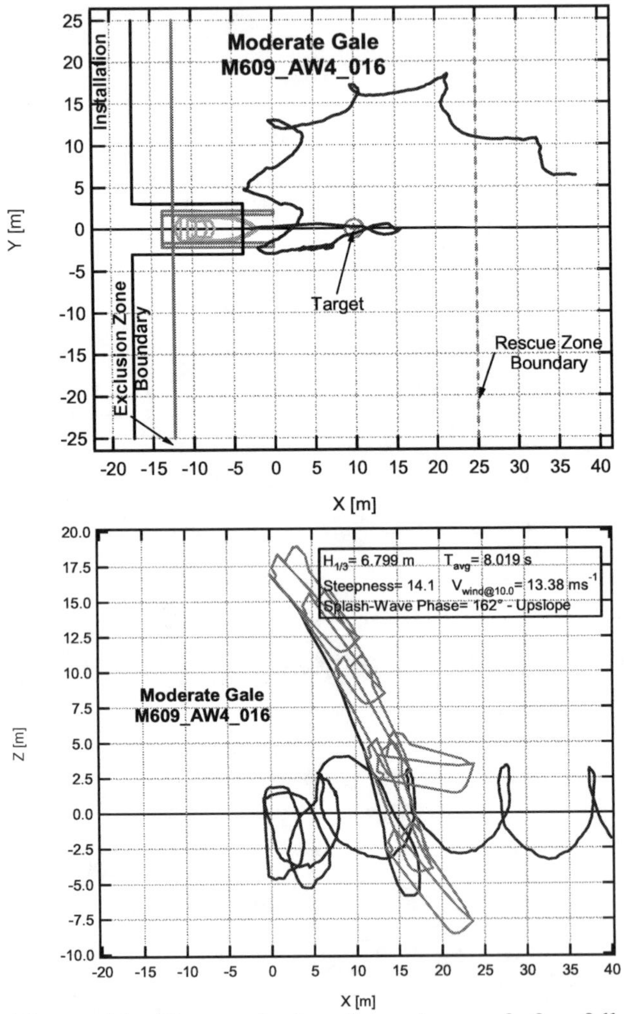

Figure 15. Plan and elevation views of free-fall lifeboat's path during launch in moderate gale conditions.

These four examples demonstrate clearly the significance of the launch point on the wave in terms of the performance of the evacuation.

Mathematical models of free-fall launches have been developed by [14-26] and used to examine the launch parameters of free-fall evacuation stations and accelerations in terms of (human) response criteria. The focus of much of these previous studies has been on ensuring that the launch arrangement (e.g. ramp angle, length, height) will result in successful launches and that occupants will avoid injury.

Using a two-dimensional numerical model of the free-fall launch process, [22-23] provided a useful classification of launch outcomes that might be expected for a given launch arrangement, all for calm water. Basically, the launches were categorized as successful or unsuccessful based on the motions of the vessel upon launch. These are described briefly here and then used to categorize the examples shown above.

Successful launches were described in [22] as one of two sorts: type Ia and type Ib. The best sort is type Ia, which was characterized by a smooth entry into the water followed by surfacing with forward headway in a straight direction, away from the launch ramp. Despite the fact that the categories in [22] were based on simulated launches in calm water, we have interpreted the examples presented in Figures 12 and 13 as being of the Ia sort. Type Ib launches were described in [22] as being similar to type Ia, but with the difference that the lifeboat has more heaving motion after it surfaces and moves away from the launch site. Still, this type is also in the successful category. None of the examples above were interpreted as type Ib, but such cases were certainly observed in the course of the experimental campaign.

In a type IIa launch, the lifeboat makes headway during the splash-down, but surfaces with negative velocity. That is, it moves backwards in the direction of the launch. In a type IIb launch, the lifeboat reverses direction during submergence and surfaces with negative velocity, that is, in the direction it entered the water. The examples shown in Figures 14 and 15 are, in our interpretation, best categorized as types IIa and IIb, respectively. Both sorts of type II launches are unsuccessful.

It bears repeating that these categories were based on calm water launches with a two-dimensional numerical model, but are still useful for categorizing in broad terms the types of launches observed during the model test campaign. Examples of all the launch classes in [22] were observed in the test program, with the added complications of waves and three-dimensional effects. Three-dimensional effects include substantial deviations in headway, and the motions (roll, yaw etc.) of the vessel during and after splash-down. The effects of wave height, steepness, and phase are described in more detail following, particularly in terms of setback.

Examples of launching: TEMPSC (perpendicular)

Figure 16 shows an example of a launch using a conventional TEMPSC (without the flexible boom) arranged perpendicular to the platform. The path taken by the lifeboat during this launch (again in moderate gale conditions) is shown as in the previous examples, where the irregular line in both views is the path of a reference point at the stern of the lifeboat as tracked by the optical tracking system. The origin of the axis system is the target drop point. The boat splashed down on the up-slope of an incoming wave and was setback more than 13m towards the platform, where it collided while on the wave crest before moving away again, cresting three waves en route to the rescue zone.

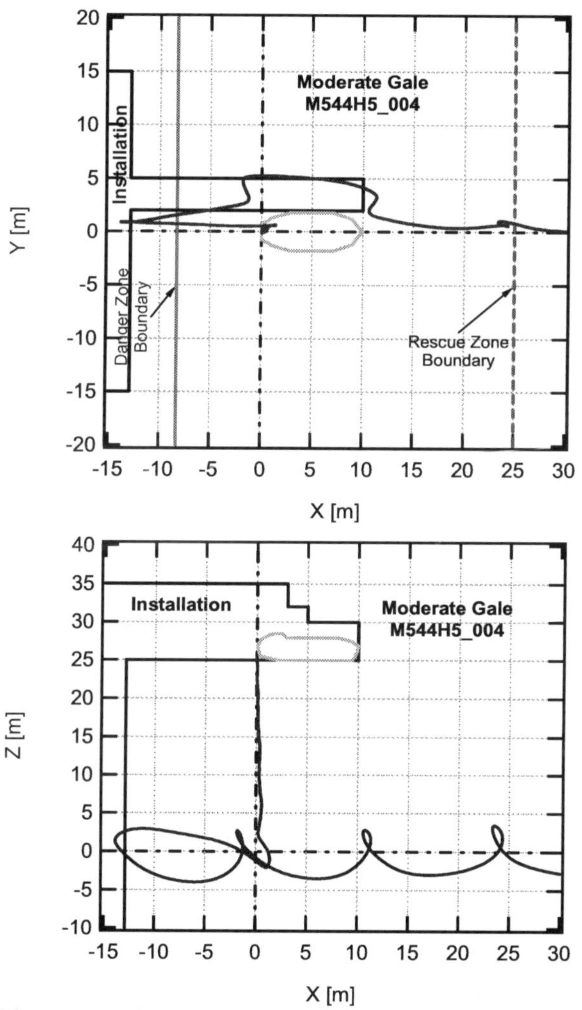

Figure 16. Plan and elevation views of TEMPSC's (w/o boom) path during launch in moderate gale conditions (perpendicular orientation).

A similar example is presented in Figure 17, this time for a launch with a flexible boom. The elevation shows that the flexible boom was already effective during lowering: the path is not vertically downward, but rather slants away from the platform and is about 2m farther out from the target drop point at splash-down. Immediately after launching on an up-slope, the lifeboat was pushed back toward the platform during its first wave encounter, but its setback was mitigated by the action of the boom. It then proceeded ahead and crested two more waves en route to the rescue zone. The tag line attached to the lifeboat's bow and the tip of the flexible boom was under tension and both pulled the boat away from the platform and kept its bow pointed out, which helped the vessel maintain direction. This guiding force was active until the boat passed under the tip of the boom and the tag line released, which occurred in this case at about 25m (full scale) out from the target drop point.

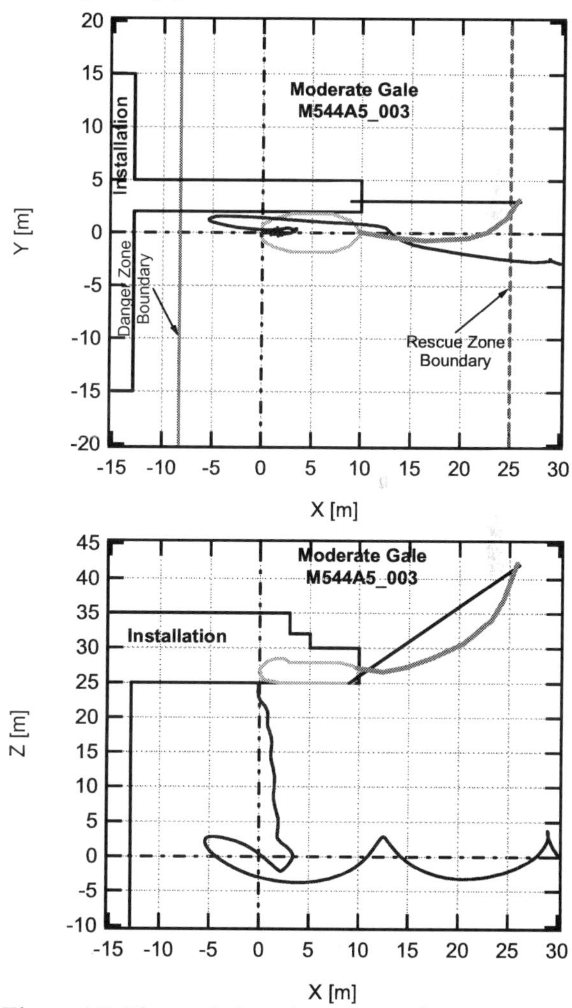

Figure 17. Plan and elevation views of TEMPSC's (with boom) path during launch in moderate gale conditions (perpendicular orientation).

This exemplifies the general finding that the flexible boom was effective in mitigating setback and reducing the likelihood of collisions with the platform.

Examples of launching: TEMPSC (parallel)

Figure 18 shows another example of a launch in moderate gale conditions, this time using a conventional TEMPSC (without the flexible boom) arranged parallel to the platform. The path taken by the lifeboat during this launch included initial setback upon launch into beam seas and further setback during the following two wave encounters with some headway made in between waves. After being progressively setback during the first few wave encounters, the lifeboat finally turned into the head seas and made its way to the rescue zone.

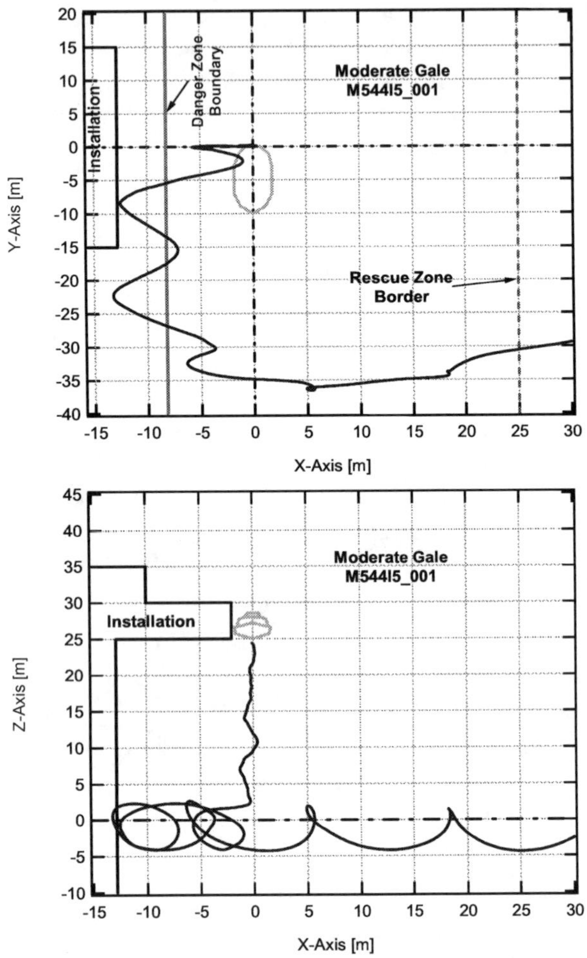

Figure 18. Plan and elevation views of TEMPSC's (w/o boom) path during launch in moderate gale conditions (parallel orientation).

Figure 19 shows an example similar to Figure 18, but this time for a launch with a flexible boom. The example shows that the effect of the flexible boom is somewhat different for the parallel orientation compared to the perpendicular arrangement. From the plan view, it is apparent that the tag line between the boat and the boom turned the bow out away from the platform already during lowering. This continued after splash-down, which made the transition to a heading 90° away from the platform faster than was the case without the boom. In both cases, the launch occurred on a similar point on the wave profile and the initial setback was comparable. The boom appeared to have a positive effect by orienting the bow of the lifeboat in the Figure 19 case into the head seas and thus avoiding the beam seas, thereby mitigating the progressive setback evident in Figure 18.

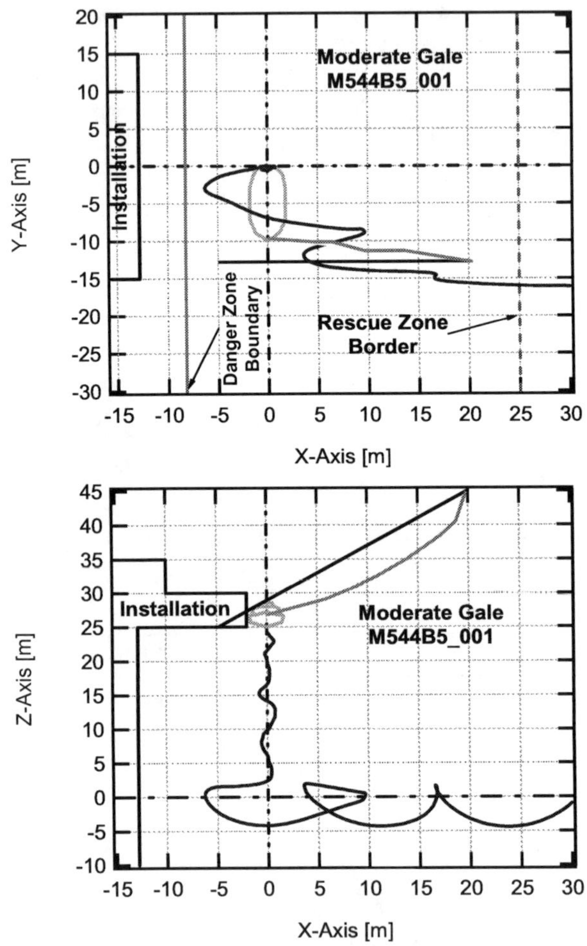

Figure 19. Plan and elevation views of TEMPSC's (with boom) path during launch in moderate gale conditions (parallel orientation).

Comparison of sail-away paths: footprints

The examples shown in Figures 12 to 19 are just a few of the several hundred test launches done with the various evacuation systems. The samples shown so far were all for the same nominal weather condition (moderate gale and wave steepness 1:15), although the test conditions included a range of weather conditions (Table 1) and a systematic variation in wave steepness.

Rather than look at more individual examples, Figures 20 to 24 show sail-away paths (plan views) for numerous launches for the five evacuation station arrangements: free-fall, TEMPSC perpendicular to the platform both without and with a flexible boom, and TEMPSC arranged parallel to the platform both without and with a flexible boom. While it is not possible to discern individual launch paths in these figures, nor the effects of weather conditions, they give an overall impression of the performance of the systems and configurations.

Figure 20 shows results for the free-fall boat. The plot is for a range of weather conditions from calm to fresh gale, all in nominal wave steepness of 1:10. The results indicate that for some launches at least, the free-fall boat experienced significant setback (and progressive setback) and had difficulty gaining control and making way to the rescue zone.

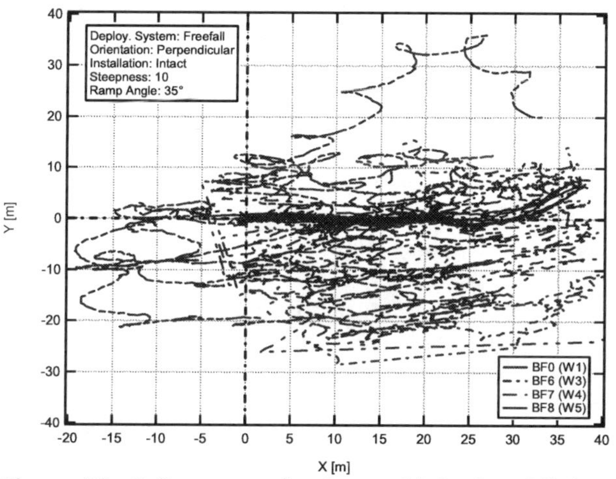

Figure 20. Sail-away paths for multiple free-fall boat launches in a range of weather conditions (1:10 wave steepness).

Figures 21 and 22 show results for the TEMPSC arranged perpendicular to the platform and Figures 23 and 24 are for the TEMPSC in the parallel arrangement. No boom was used in the launches in Figures 21 and 23. The flexible boom was used in the tests presented in Figures 22 and 24. All four plots are for results over a range of weather conditions (from calm water up to fresh gale), all in nominal wave steepness of 1:15.

Several conclusions can be drawn from comparisons of the plots. First, the sail-away paths were much more direct and in control for the TEMPSC in the perpendicular arrangement compared to both the parallel arrangement and free-fall boat. With respect to the TEMPSC options, the effect of the flexible boom was generally to reduce setback (and progressive setback) and increase control of the boat in terms of heading to the rescue zone.

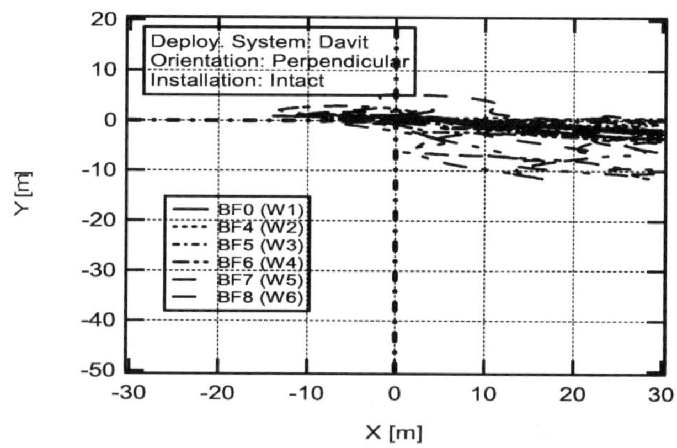

Figure 21. Sail-away paths for multiple TEMPSC (w/o boom) launches (perpendicular arrangement).

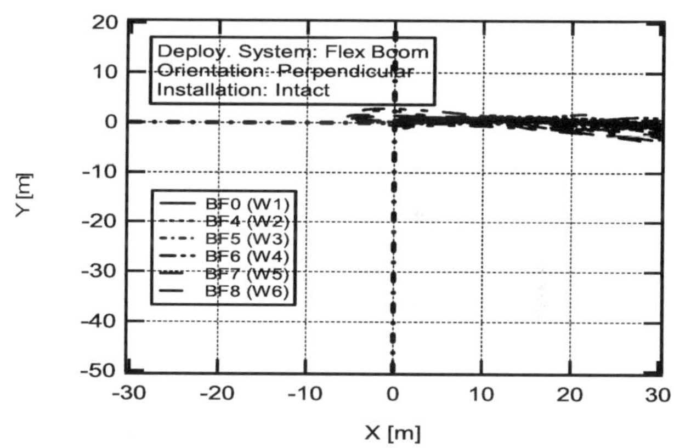

Figure 22. Sail-away paths for multiple TEMPSC (with boom) launches (perpendicular arrangement).

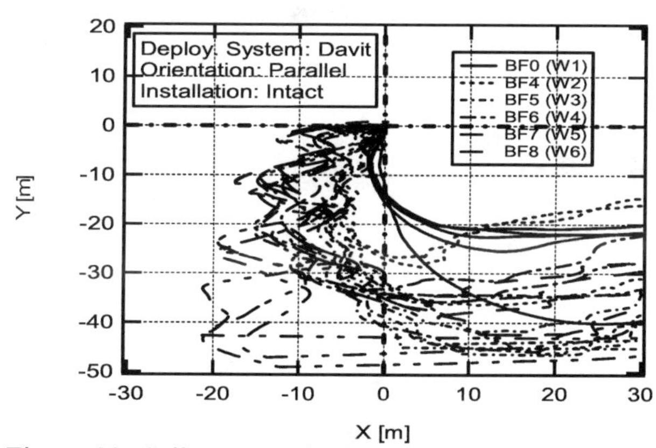

Figure 23. Sail-away paths for multiple TEMPSC (w/o boom) launches (parallel arrangement).

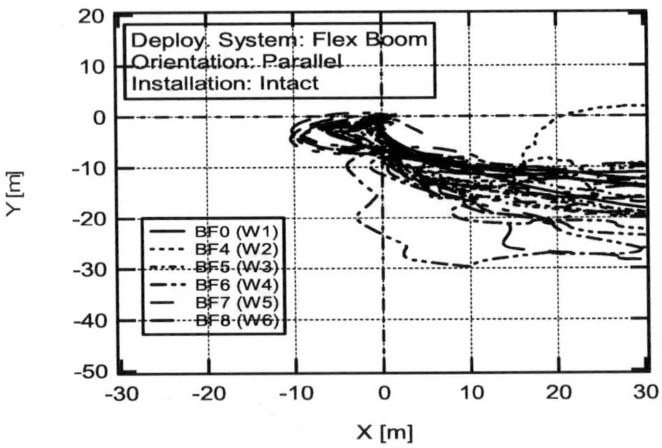

Figure 24. Sail-away paths for multiple TEMPSC (with boom) launches (parallel arrangement).

The effect of weather conditions on the sail-away phase of the evacuations can be seen in Figures 25, 26, and 27 for the free-fall system, the TEMPSC arranged perpendicular to the platform, and the TEMPSC arranged parallel to the platform, respectively. The path length from the lifeboat's splash-down to where the rescue zone was crossed is plotted against wave height.

For each case, there is considerable variability in the results for any given weather condition. Further, it is apparent too that the worst case in each weather category tends to increase with deteriorating weather conditions.

With reference to Figure 25 for the free-fall evacuation system, the baseline path length was established by the calm water launches to be about 14m to 18m, as indicated in the figure. Similar path lengths were observed for some launches in all other weather conditions too, including launches in Beaufort 8 (equivalent) conditions. However, the worst cases for the Beaufort 6, 7 and 8 weathers were about 85m, 100m and 107m respectively.

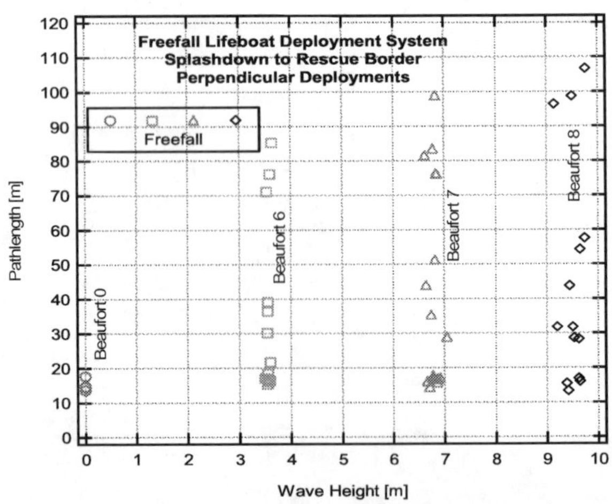

Figure 25. Sail-away path length (to rescue zone) versus wave height for free-fall evacuation tests.

The same can be observed in Figure 26 where the path length is plotted against wave height for the perpendicular TEMPSC arrangement. In general, the path lengths are longer for the free-fall launches than the TEMPSC when launched into head seas. Launches using the davits alone are indicated by solid symbols; launches with the boom are indicated by open symbols. The effectiveness of the boom in reducing the path lengths at the upper end of each weather condition is obvious for the conditions above Beaufort 5. At Beaufort 8, the maximum path length experienced without the boom was about twice that with the boom.

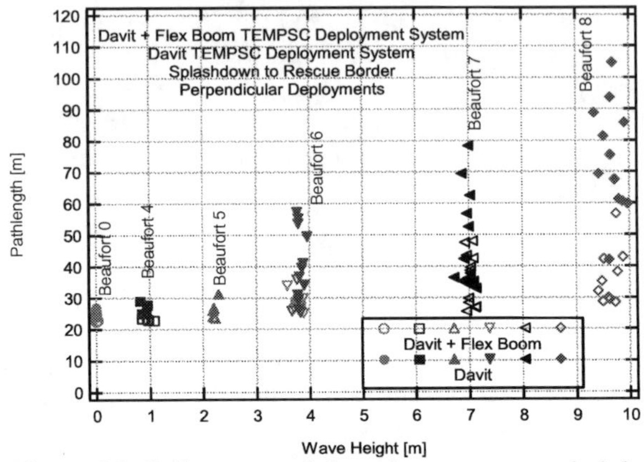

Figure 26. Sail-away path length versus wave height for TEMPSC (perpendicular arrangement).

Figure 27 displays generally similar results as Figure 26 in the sense that the worst cases for a given weather condition tend to worsen as wave height increases, and that the boom was effective in reducing the worst of the path lengths. In general, the path lengths observed in launches with the TEMPSC arranged parallel to the platform were worse that the perpendicular arrangement, although this is largely a function of the fact that one was launched into head seas and the other into beam seas.

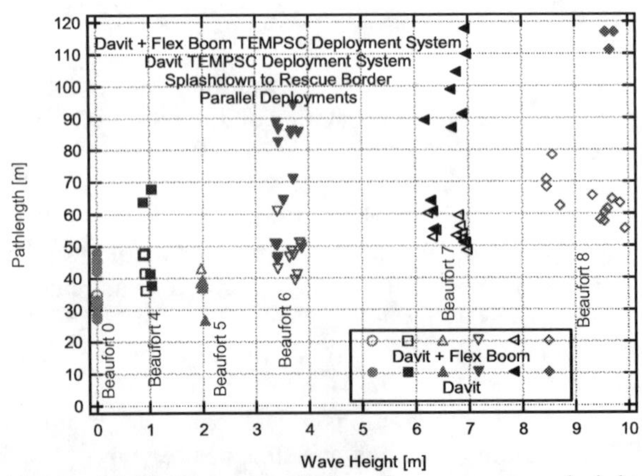

Figure 27. Sail-away path length versus wave height for TEMPSC (parallel arrangement).

Comparison of setback

Setback was found to be a key performance measure and is the focus of the results presented in Figures 28 to 34. Figure 28 gives results for free-fall launches over the full range of weather conditions, all at a nominal wave steepness of 15. The plot shows the launch target at the origin and the open symbols represent the location farthest from the target to which the lifeboat was setback after launch. In all the cases shown, the setback was primarily in the direction of the waves. The maximum setback increased as the weather conditions deteriorated. Progressive setback is not shown.

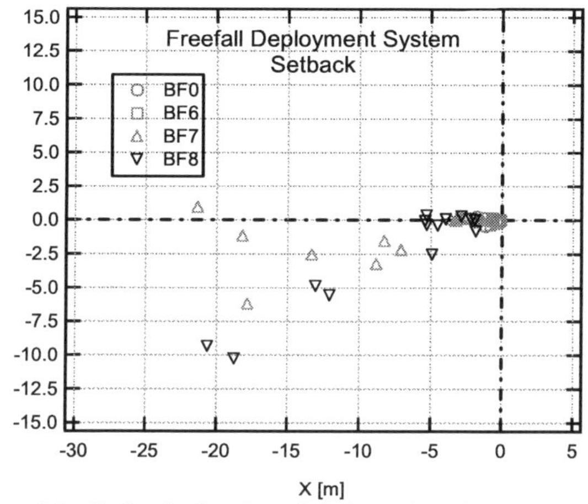

Figure 28. Setback for free-fall launches in a range of weather conditions with wave steepness 1:15.

The maximum setback measured in each test launch is presented in Figure 29 for the free-fall launches in 1:15 wave steepness conditions, for a range of weather conditions, represented in the plot by measured wave height. There was no setback in calm conditions. For any other given weather condition, the results in the plot show there was a high degree of variability, which is similar to the results seen above for the path length measure. In Beaufort 7 conditions, for example, several launches had almost no setback, but the maximum setback measured in that series of tests at that nominal weather condition was 22m. This is a significant performance measure, particularly as it relates to the potential for collisions between the lifeboat and platform, and therefore on the design of the evacuation arrangement. The line in the plot indicates roughly that the upper measured value of setback increased approximately linearly with (~2.8 ×) wave height.

Similar plots of setback versus wave height are shown in Figure 30 for the TEMPSC arranged perpendicular to the platform and in Figure 31 for the TEMPSC arranged parallel to the platform. In both of these plots, results are shown for the lifeboat with and without the flexible boom.

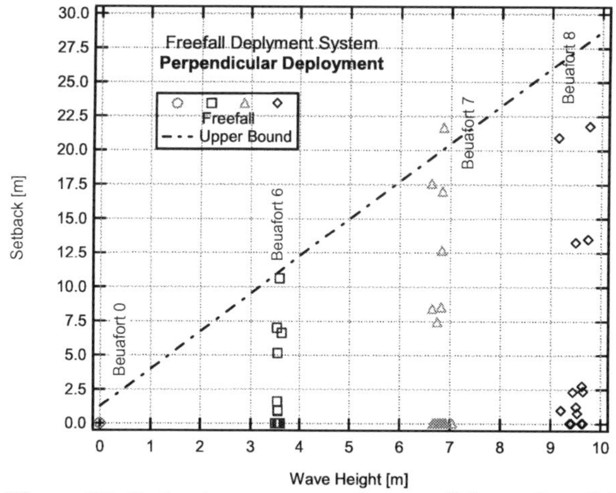

Figure 29. Setback versus weather conditions: free-fall.

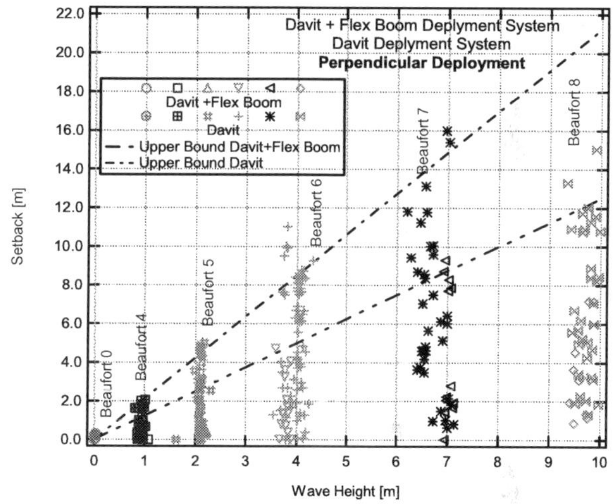

Figure 30. Setback versus weather conditions: TEMPSC in perpendicular arrangement.

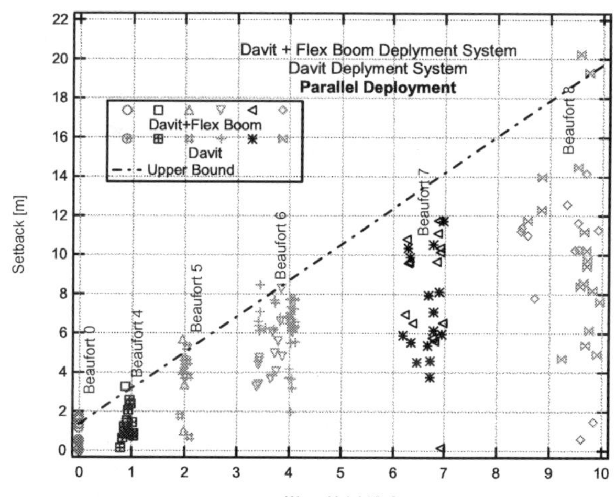

Figure 31. Setback versus weather conditions: TEMPSC in parallel arrangement.

Figure 30 shows similar trends in performance in terms of setback as are evident in Figure 29, including that the upper limit increases approximately linearly with wave height. One line is shown for the TEMPSC with the flexible boom (~1.3 × wave height) and the other is for launches without the boom (~2.1 × wave height). These results indicate that the effect of the boom when used with a TEMPSC arranged perpendicular to the platform is to reduce the maximum setback. Again, this is significant in terms of evacuation system design, for example in terms of ensuring suitable clearance between the platform and the launch target.

The boom had less effect on setback when used in conjunction with the parallel launch arrangement, as indicated in Figure 31. A single upper bound line is used for both (~2.0 × wave height) in Figure 31. The boom launches were generally just marginally better in terms of setback, although, as seen above, the boom system did improve performance in terms of path length when used in this arrangement, reflecting its utility in limiting progressive setback and improving control.

A comparison of the upper bound for each system can be made using these three figures. Interestingly, the maximum setback for the conventional TEMPSC is much the same regardless of orientation to the platform. Further, the influence of the boom on maximum setback is significant for the perpendicular arrangement but not for the parallel arrangement. Compared to the TEMPSC configurations, the performance of the free-fall boat was poorer. It bears repeating at this point that all the test launches were made into collinear wind and waves that were perpendicular to the launch platform, such that launches of the free-fall boat and TEMPSC in the perpendicular arrangements were into heads seas, and launches of the TEMPSC in the parallel arrangement were into beam seas.

Effects of wave phase

Figures 32 to 34 show setback plotted against the position on the wave of the lifeboat's splash-down point. With reference to Figure 32, the trend is clear: setback is relatively small for launches made near the crest and down-slope of the wave compared to the relatively large setback measured in cases where the launch was in the wave trough or up-slope. Results of many tests are shown in the figure, including tests done in different weather conditions.

Similar results are shown in Figures 33 and 34 for the TEMPSC evacuation station arranged perpendicular and parallel, respectively, to the platform. These exhibited similar trends as the free-fall lifeboat. One difference between the TEMPSC and the free-fall is that the relatively slower lowering speed of the conventional TEMPSC resulted in shadowing of the down-slope, making launching there unlikely.

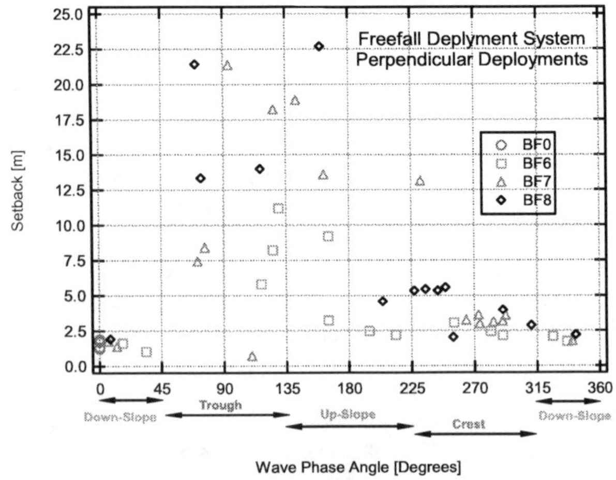

Figure 32. Setback versus wave phase: free-fall.

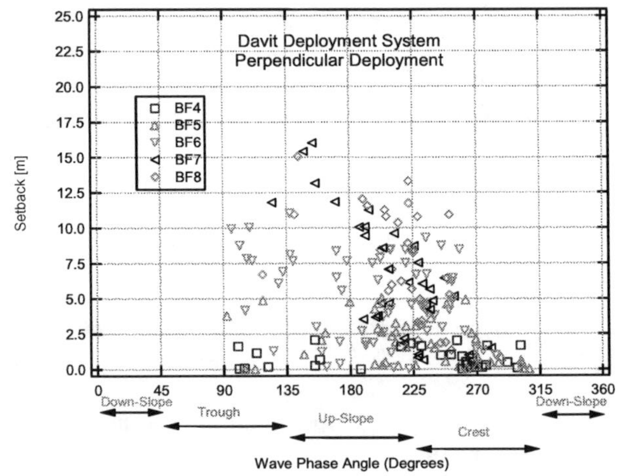

Figure 33. Setback versus wave phase: perpendicular TEMPSC.

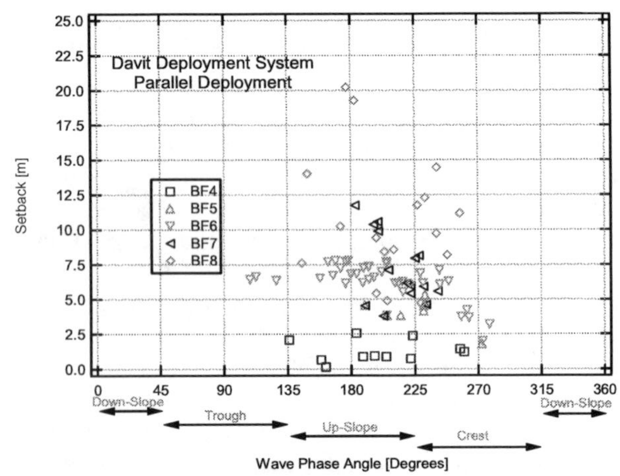

Figure 34. Setback versus wave phase: parallel TEMPSC.

These results show the importance of launch conditions in relation to wave phase on the overall performance of the launch and a corresponding opportunity to improve performance by coordinating the launch with the wave environment [27]. A comparison of the figures indicates that the worst of the free-fall cases were marginally worse than the other two arrangements.

Effects of wave steepness

Wave steepness was varied in the experiments so that its effects might be discerned. A plot of path length versus wave steepness is presented in Figure 35 for the free-fall tests. Each point represents the total length of the path taken by the vessel as it moved from splash-down to the rescue zone. For each of the nominal wave steepness values (10, 15 and 20) the results are highly scattered, reflecting again the effect of launch conditions in terms of wave phase. As wave steepness increased, the tendency was for the path length to increase as well. This is important as path length in this context is a proxy measure of the control and maneuverability of the lifeboat.

Figure 36 shows the same sort of results for the TEMPSC in a perpendicular arrangement, without the flexible boom. The trend that performance worsened with increasing wave steepness is repeated here. Note that a comparison of the results of the free-fall and TEMPSC in the perpendicular arrangement shows that the TEMPSC performance is better, even without the flexible boom. As noted above (see Figure 26), the boom improves the performance in terms of the path length measure considerably. This result reflects the relatively erratic behavior experienced by the free-fall lifeboat when launched into unfavorable parts of the wave.

Launches of the TEMPSC arranged parallel to the platform (and without the flexible boom) are shown in Figure 37. A direct comparison of these results with the other two plots (Figures 35 and 36) is not reasonable as the boat in these cases (Figure 37) had to turn through 90° after launch in order to head to the rescue zone. Still, the role of steepness is evident again: steeper waves degrade performance.

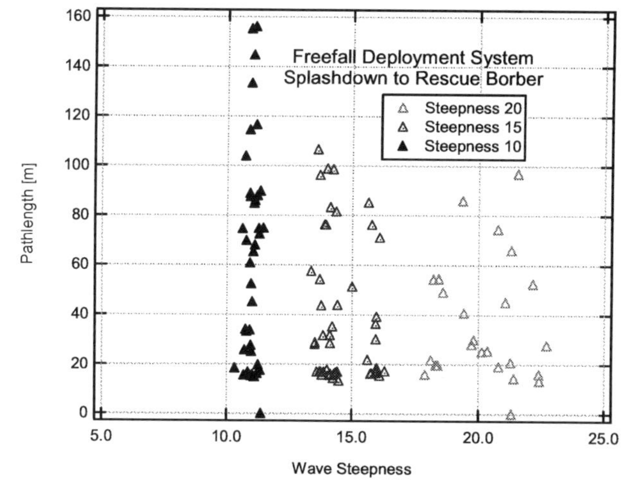

Figure 35. Steepness effects: free-fall.

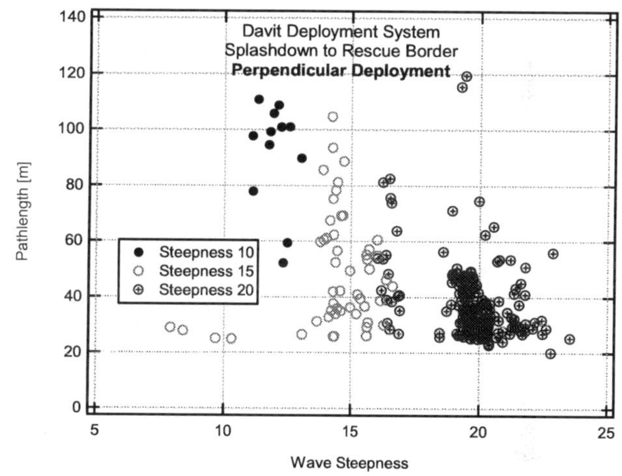

Figure 36. Steepness effects: perpendicular TEMPSC.

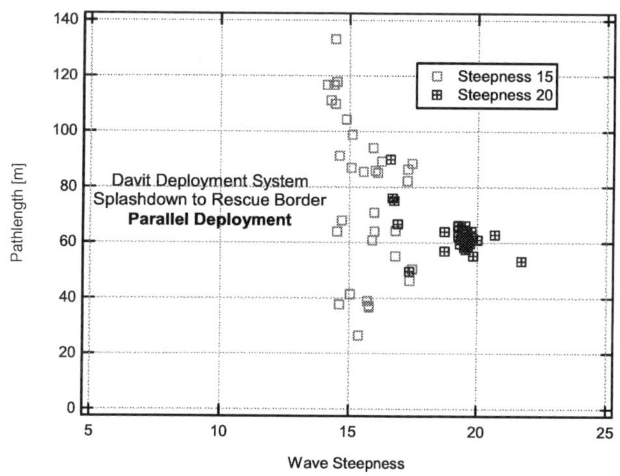

Figure 37. Steepness effects: parallel TEMPSC.

SUMMARY

Three types of evacuation system were evaluated in an extensive experimental campaign using model tests in a large test facility. The aim of the work was to contribute to the knowledge concerning evacuation system performance, which can be used in the decision-making processes of operators, regulators and designers. The scope of the work included the technical performance of the evacuation systems, but excluded most human factors (e.g. mistakes and psychological considerations) and mechanical reliability (e.g. design faults and mechanical failures).

The main performance measures considered in the paper were setback at launch and the length of the path the boat took to clear to the rescue zone after launch. The former is an indicator of the evacuation system's control at launch as well as the lifeboat's likelihood of collision with the platform or ship. The latter is a proxy for the control and maneuverability of the lifeboat after launch. All of the test launches presented in the paper were made from a fixed platform that was largely transparent to waves.

Several examples of launches with the different systems were presented, including a series of examples of the free-fall system launched into head seas for the same nominal weather condition, but with different launching conditions in terms of where the lifeboat landed with respect to the wave phase. These examples typified and highlighted the wide variation in performance observed in the free-fall launches, a result that was attributed largely to the wave phase effect.

Examples were also presented of the conventional TEMPSC evacuation system arranged perpendicular to the platform (typical of an offshore petroleum installation) and the same system with the addition of a flexible boom and tagline in the arrangement. The examples illustrated a general observation from similar tests: that the flexible boom was effective in mitigating setback and reducing the likelihood of collisions with the platform, which is an important consideration when launching into head seas as in the tests.

Examples of test launches with the TEMPSC arranged parallel to the installation, again with and without the boom, were also shown. It is important to recognize that launches in such conditions were into beam seas, rather than head seas as was the case for the free-fall launches and the perpendicular arrangement. The examples illustrated that the boom had the effect of turning the bow of the lifeboat away from the platform and into the head seas (where it had to go under the circumstances), thereby avoiding the persistent beam seas and mitigating the progressive setback that was more evident in the tests without the boom.

One of the key performance measures used to evaluate the evacuation systems was setback. For all of the systems, setback was a dominant feature and found to be dependent on the wave phase at splash-down. For any given weather, launches resulted in a wide range of setback, from a minimum baseline value corresponding to the calm water case, up to some maximum. The maximum setback for the range of weather conditions was found for all systems to be approximately linear with wave height. A comparison of the different systems revealed that the maximum setback (as a function of wave height) was larger for the free-fall lifeboat than any of the TEMPSC configurations. The maximum setback for a given weather condition was similar for both of the TEMPSC arrangements (parallel and perpendicular) without the boom. Further, setback was significantly reduced by the boom in launches of the conventional TEMPSC in the perpendicular arrangement, but was not particularly influenced in the parallel arrangement. However, the boom did reduce the progressive setback in the parallel launches, indicating its utility in terms of improving control, at least in the launch scenarios in which the tests were done.

Test results were also presented in terms of the paths taken by the lifeboats after launch, as they moved to the rescue zone. These results took the form of 'footprints' consisting of the trajectory of the lifeboat over multiple tests, as well as plots of integrated path lengths. For all the systems, path length varied widely for any given weather condition, much like the setback measure discussed above. That is, it varied from some baseline value comparable to the measurement in calm conditions, up to a maximum that was sometimes many times longer than the baseline value. Also the longest path length for the given weather conditions tended to increase with deteriorating weather.

The free-fall test results showed that the free-fall lifeboat experienced significant setback and progressive setback in some launches (type II), reflecting the difficulty in gaining control of the boat and making way to the rescue zone. Interestingly, while some of the launches (type I) of the free-fall system in waves were excellent – even in heavy weather, the performance in others indicated more difficulty in terms of control and maneuverability than was experienced with the conventional TEMPSC lifeboats. Again, this is due to the sensitivity of performance on the entry point on the wave, to which the free-fall lifeboat seemed to be more sensitive than the conventional boats. This was also reflected in the plots of path length, which were longer for the free-fall boat than for the conventional TEMPSC when launched in the perpendicular arrangement (both of which were into heads seas).

Sail-away paths for the conventional TEMPSC in the perpendicular arrangement were much more direct and in control compared to the free-fall system (launched into head seas) and the TEMPSC in the parallel arrangement (launched into beam seas). As noted elsewhere, the boom was effective in the perpendicular arrangement at reducing setback, progressive setback and path lengths.

For launches using the parallel arrangement, the boom was helpful in reducing path length due to its utility in reducing progressive setback, which was otherwise a major feature of the parallel launches (without the boom) because they were into beam seas.

Wave steepness was also found to have an effect on performance. In particular, path length, the proxy for control and maneuverability during sail-away, was found to increase with increasing steepness for all the evacuation systems.

DISCUSSION

In general, the results presented provide some performance benchmarks for the different evacuation system configurations. The scope of the work should be borne in mind when application of the results to a given design scenario is contemplated. Our experiments are concerned with only the evacuation stage of an emergency response that must include both escape before evacuation and rescue after it. The experiments modeled the launching and clearing process during evacuation, and were assessed using performance measures such as setback.

The performance of the evacuation systems in the tests did not include mechanical issues, such as the reliability of the equipment, or launch failures due to design and operational faults. Nor were issues such as availability addressed, which are relevant to the overall emergency planning exercise as lifeboats may be damaged by the emergency initiating event, or may be undergoing maintenance when needed, for example. Similarly, human factors were not included in this work, except in terms of the relationship between vessel motions, injury criteria and motion sickness. The importance of various design and operational issues is reinforced in any number of accident reports and other studies that complement the present one [e.g. 28].

It is also worth highlighting that the experiments reported here all involved launches into oncoming weather, rather than launches on the lee side (or into oblique or following seas), and so represent one of many weather orientation conditions, albeit a relatively difficult one. Also the platform from which the launches were made was fixed and largely transparent to the waves, so the otherwise complicating motions of the platform were avoided, as were the effects of wave reflection and radiation. This should be borne in mind when considering the present results in terms of ship launches, where the conventional arrangement of TEMPSC is parallel to the ship's side and of free-fall is at the stern. The latter in particular would be worth investigating further, specifically in following seas.

Indeed, launching a conventional lifeboat (in the conventional parallel arrangement) from a ship that is oriented into the weather might be preferred to the other arrangements used in the experiments as in such a case, the boat would be launched into head seas, rather than beam seas. In fact, we have investigated this situation for an FPSO [7] and discussed the implications for passenger ships elsewhere [29]. Interestingly, in the test launches from the FPSO, there were no collisions observed between the lifeboat and vessel in the over 150 launches that covered a wide range of weather conditions. This was despite the relatively small clearance between the lifeboat and the vessel during the launch process. An investigation of lifeboat arrangement and deployment for this same sort of scenario, but based on a numerical approach that included waves and ship motions, as well as lifeboat motions during lowering, has been presented in [30].

Space limitations prevent us from including additional results that we have measured from the experimental campaign. Of particular interest are results of launches of all the systems from a damaged platform (with a combined list and trim), launches of the free-fall lifeboat into oblique seas, and measurements of lifeboat accelerations and motions that are relevant for human performance evaluation, including in terms of injury criteria and motion sickness. We hope to publish these additional results in due course.

CONCLUSIONS

The experimental campaign yielded extensive, objective benchmark measures of the performance of several evacuation systems and system configurations. None of the various alternatives can be described categorically as better than the others. Indeed, it is apparent that the type of evacuation system that best suits a particular set of circumstances is likely to be different for different circumstances. With this in mind, the experimental results presented here should be useful to designers, operators and regulators in making decisions concerning evacuation systems that are fit for their specific purpose.

ACKNOWLEDGMENTS

The authors acknowledge with gratitude the financial support provided by the following organizations: Transport Canada, Natural Resources Canada, Canadian Association of Petroleum Producers, National Research Council, Petroleum Research Atlantic Canada, Natural Sciences and Engineering Research Council, National Search and Rescue Secretariat, and the Terra Nova Project partners led by Petro-Canada. We also acknowledge the role of peers who have offered constructive criticism of this work, colleagues at the Canada-Newfoundland & Labrador Offshore Petroleum Board who have provided advice over a period of years, and the technical staff who have executed a lot of model tests.

REFERENCES

[1] Royal Society Study Group. 1992. Risk: analysis, perception and management. *The Royal Society*, London.

[2] Health and Safety Executive. 1995. Offshore installations regulations: Prevention of fire and explosion, and emergency response on offshore installations. *HSE S1 1995/743*, London.

[3] Hoppe, H. 2005. Goal-based standards – a new approach to the international regulation of ship construction. *Journal of Maritime Affairs, WMU*, 4(2): 11 p.

[4] Kingswood, A. 2000. Evacuation standards development and quantification. *Proc. Evacuation, Escape and Rescue*, Aberdeen, 18 p.

[5] Simões Re, A. Veitch, B. and Kennedy, E. 2005. Systematic experimental evaluation of free-fall evacuation in a range of weather conditions - Phase III. *TR-2005-12*, Institute for Ocean Technology, St. John's.

[6] Simões Ré, A. and Veitch, B. 2007. Experimental investigation of free-fall lifeboat performance. *Proc. ISOPE 2007*, Lisbon, 8 p.

[7] Simões Ré, A. and Veitch, B. 2001. Experimental evaluation of lifeboat evacuation performance. *Trans. SNAME*, Vol.109, 18 p.

[8] Simões Ré, A., Veitch, B. and Pelley, D. 2002. Systematic investigation of lifeboat evacuation performance. *Trans. SNAME*, Vol. 110, 20 p.

[9] Woolgar, R., Simões Ré, A., Veitch, B. and Pelley, D. 2002. Safe evacuation from offshore petroleum installations. *Proc. OTC*, Houston, 10 p.

[10] Simões Ré, A., Veitch, B. and Pelley, D. 2003. Evacuation performance. *Proc. OTC*, Houston, 9 p.

[11] Simões Ré, A. and Veitch, B. 2004. Evacuation performance of davit launched lifeboats. *Proc. OMAE2004-51528*, Vancouver, 8 p.

[12] Rutgersson, O. and Tsychkova, E. 1999. Safety management of the mustering and evacuation of damaged passenger ships. *Proc. Learning from Marine Incidents, RINA*, 13 p.

[13] Campbell, I., Claughton, A. and Kingşwood, A. 1983. Development of lifeboat launching systems for offshore rigs and platform. *Proc. International Conference on Marine Survival Craft, RINA*, 11 p.

[14] Tasaki, R. and Ogawa, A. 1990. Numerical simulation and its application on the falling motion of free-fall lifeboat. *Journal SNAJ*, 167: 147-158.

[15] Nelson, J.K., Fallon, D.J., Verhoef, J. and Hirsch, T.J. 1991. Effects of mass distribution on free-fall lifeboat behavior. *Proc. OMAE*, I-B: 587-593.

[16] Boef, W.J.C. 1992a. Launch and impact of free-fall lifeboats, Part I: Impact theory. *Ocean Engineering*, 19(2): 119-138.

[17] Boef, W.J.C. 1992b. Launch and impact of free-fall lifeboats, Part II: Implementation and applications. *Ocean Engineering*, 19(2): 139-159.

[18] Frazer-Nash Consultancy Ltd. 1993. Feasibility of computer simulation of the launch of free-fall lifeboats. *OTH 92-391*, Health and Safety Executive, London.

[19] Arai, M., Khondoker, M.R.H. and Inoue, Y. 1995. Water entry simulation of free-fall lifeboat: analysis of motion and acceleration. *Journal SNAJ*, 178: 193-201.

[20] Arai, M., Khondoker, M.R.H. and Inoue, Y. 1996. Water entry simulation of free-fall lifeboat: effects of accelerations on the occupants. *Journal SNAJ*, 179: 205-211.

[21] Nelson, J.K., Waugh, P.J. and Schwickhardt, A.J. 1996. Injury criteria of the IMO and the hybrid III dummy as indicators of injury potential in free-fall lifeboats. *Ocean Engineering*, 23(5): 385-401.

[22] Khondoker, M.R.H. 1998. Effects of launching parameters on the performance of a free-fall lifeboat. *Naval Engineers Journal*, 110(4): 67-73.

[23] Khondoker, M.R.H. 1999a. Evaluation of optimum conditions of launching of free-fall lifeboat. *ISP*, 46(447): 305-318.

[24] Khondoker, M.R.H. 1999b. Numerical investigation of the behavior of a hook launching free-fall lifeboat during water entry. *Proc. OMAE-99-OFT4243*, St. John's, 17 p.

[25] Khondoker, M. and Arai, M. 2000. A comparative study of the behaviour of free-fall lifeboat launching from a skid and from a hook. *Proc. IMechE Part C Journal of Mechancal Engineering Science*, 214(2): 359-370.

[26] Wisniewski, Z., Carlomagno, G.M. and Brebbia, C.A. 1999. Computer-aided prediction of the launching behavior of a free-fall lifeboat. *Computational Methods and Experimental Measurements*, 7 p.

[27] Simões Ré, A., Finch, T., Veitch, B., Janes, G., Sullivan, M. and Kennedy, E. 2003. Controlled lifeboat deployer. *US Patent No. US 2003/0221605*.

[28] Peachey, J. and Pollard, S. 2006. Development of lifeboat design. *MCA Research Project 555*, Burness Corlett-Three Quays Limited for The (UK) Maritime and Coastguard Agency.

[29] Simões Ré, A., Lam, E., Igloliorte, G. and Veitch, B. 2003. Quantitative assessment of escape and evacuation. *Proc. Passenger Ship Safety Conference, RINA*, London, 7 p.

[30] Ekman, P. 2006. A numerical study of the influence of launch parameters on lifeboat deployment from a ship in beam seas. *Marine Technology*, 43(2): 91-99.

Discussion

Walter Spring, Visitor

The paper presents important results related to the safety of launching survival craft in high sea states, specifically the definition and determination of setback once a craft is launched. The dangers of launching into a head sea were understood, but not quantified until the research presented in this paper. As such the results presented in the paper should be incorporated into the planning of EER programs for offshore and shipping facilities.

Can you describe the effort being made by yourselves and regulatory bodies to ensure these results are used by the offshore and shipping industries to assist in the design of specific EER programs? I imagine that presentations to varied audiences, such as this one, will be a part of this effort. Also on the same track, are you aware of any recent offshore or ship installations that incorporated your results into their EER programs?

Finally, the work presented here is several years old. Have you performed additional research in the area of safety of launching survival craft into other environmental conditions?

Authors' Closure

We agree with the discusser's view that our results should be incorporated into emergency response planning for shipping and offshore petroleum industry applications. To this end, our main efforts have focused on publishing the results in venues like the SNAME Maritime Technology Conference, as well as other technical conferences and journals. We have also been involved in formulating new regulations and regulatory guidelines, where we believe our work can add support to the goal-oriented design and regulation approach. We have also been able to draw attention to the outcomes of some of this work at IMO.

In response to the question concerning where the results have found practical application, we know of several cases where our work has been used as a basis for decisions concerning evacuation systems. These include fixed and floating offshore installations and ships.

The discusser's final question was about our additional work on evacuation craft. We have done model experiments of lifeboats in ice and have recently started a multiyear field program focused on evacuation technology and human factors in ice covered water. Further, we and our colleagues have investigated performance of life rafts, both at model and full scales. Some of this work has also involved human subjects, which we believe to be an important element of our research program on offshore and maritime safety.

References

Simões Ré, A. and Veitch, B. 2007 Lifeboat operational requirements in cold environments. *Proceedings*, RINA Conference Design and Construction of Vessels Operating in Low Temperature Environments, London.

Effects of Geometry Variations on the Performance of Podded Propulsors

Mohammed F. Islam, (SM), **Brian Veitch**, (M), **Susan Molloy**, (SM), **Neil Bose**, (FL), and **Pengfei Liu**, (M)

This paper presents results and analyses of an experimental study into the effects of geometric parameters on the propulsive characteristics of puller and pusher podded propulsors in straight course open water conditions. Five geometric parameters were chosen for the current study and a design of experiment technique was used to design a series of 16 pods that combined the parameters. Tests on the 16 different pod-strut-propeller combinations in puller and pusher configurations were completed using a custom designed podded propeller test rig. The dynamometry consisted of a six-component global dynamometer and a three-component pod dynamometer. The test rig was used to measure the thrust and torque of the propellers, and forces and moments on the whole unit in the three orthogonal directions. The design of experiment analysis technique was then used to identify the most significant geometric parameters and interaction of parameters that affect propeller thrust, torque and efficiency as well as unit thrust and efficiency in both the puller and pusher configurations. An uncertainty analysis of the measurements is also presented.

KEY WORDS: Podded propulsors; puller and pusher propulsors; pod geometry variation; propulsive performance; design of experiments;

INTRODUCTION

Podded propulsors have become attractive to the cruise, ferry and other shipping sectors. The geometry variations of the pod that encases the motor and shaft of a podded propulsor have been guided primarily by the size of available motors. In their study on podded propulsor optimization, Goubault and Pérrée (2004) concluded that the pod motor parameters are not as influential on the hydrodynamic performance as the geometric parameters. This emphasizes the need for further research on pod-strut shape to better understand the effect of geometry on podded propulsors' hydrodynamic performance. Karafiath and Lyons (1998) presented a study on the effect of pod geometry on the performance of podded propulsors. Pod length and strut position were varied using four pods to study their effects on pod drag and pod-propeller interactions.

As motor design becomes more flexible, the relationship between various geometric parameters and performance becomes an important design consideration. The determination of the geometric parameters needs to be supported by detailed investigations into their individual and their combined (interaction) effects on the hydrodynamic performance of the propulsor. There are a number of geometric parameters that can be used to optimize the design of a pod and five were chosen for the current study, specifically, pod diameter, pod length and taper length, strut distance from the propeller plane, and propeller hub taper angle. Our study focuses on the effects of geometry variations on the hydrodynamic performance of both puller and pusher podded propulsors. It is important to study the performance of pusher and puller propulsors separately because of the necessary different flow conditions involved.

This research program on podded propellers is being undertaken jointly by the Ocean Engineering Research Centre (OERC) at Memorial University of Newfoundland (MUN) and the National Research Council's Institute for Ocean Technology (IOT), with the support of Oceanic Consulting Corporation, and Thordon Bearings Ltd. The program combines parallel developments in numerical prediction methods and experimental evaluation. Amongst the hydrodynamic issues that have been identified are questions regarding the effects of hub taper angle (Islam, 2004; Islam *et al.*, 2004; Islam *et al.*, 2005; Taylor *et al.*, 2005 and Taylor, 2005), pod-strut configuration (Islam, 2004 and Taylor, 2005), pod-strut interactions (He *et al.*, 2005a and He *et al.*, 2005b), gap pressure (MacNeill *et al.*, 2004), and pod-strut geometry (Molloy *et al.*, 2005) on podded propeller performance.

THE GEOMETRIC SERIES AND EXPERIMENTAL APPARATUS

Using existing commercial pods as a baseline, five geometric parameters were selected that allowed variation in the primary dimensions of the pod-strut-propeller hub (Molloy *et al.*, 2005). The pod length, diameter and taper length, as well as strut distance from the propeller plane, and propeller hub taper angle were chosen as key defining parameters of the propulsor (Fig. 1). Values of these parameters were chosen so that there was one set of parameters in the series that is higher than the average commercial dimensions and one set lower. The high and low values of the parameters (Table 1) were then combined to give a series of 16 pods based on a fractional factorial design of experiment technique (Molloy, 2003). Two propellers with hub taper angles of -15° and -20° (denoted as Pull-15° and Pull-20°, respectively) were used with the pod series in puller configuration, and two propellers with hub taper angles of 15° and 20° (denoted as Push+15° and Push+20°, respectively) were used with the pod series in pusher configuration. These four propellers had the same blade sectional geometry but different hub taper angles. Details of the geometry of the propellers can be found in (Liu, 2006).

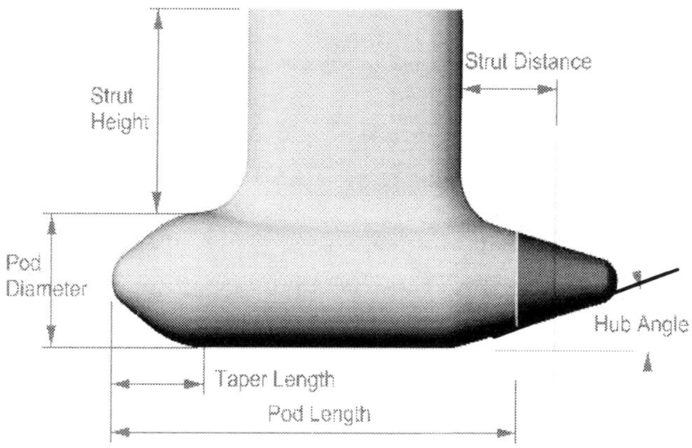

Fig. 1. Geometric parameters used to define pod-strut geometry.

Table 1. High and low values of the geometric parameters of the pod models.

External Dimensions of Model Pod	Avg. Values mm	Low Values mm	High Values mm
Propeller Diameter,	270	270	270
Pod Diameter, D_{Pod}	139	128	166
Pod Length, L_{Pod}	430	430	524
Strut Distance, S_{Dist}	100	75	133
Taper Length, TL	110	69	150
Hub Angle, H_{Angle}	15° & 20°	15°	20°

For the research work, 16 combinations of dimensions were designed and manufactured as listed in Table 2. The pod combinations were selected to include one combination with all dimensions low and one with all dimensions high, denoted as Pod 1 and Pod 16. The list of symbols of the parameters used in the paper is shown in Table 2. The physical models of the 16 pods tested in the geometric series are shown in Fig. 2.

Table 2. Combinations of dimensions of the 16 pods (used in fractional factorial design). Here, A is the ratio of pod diameter to propeller diameter, D_{Pod}/D_{Prop}, B isthe ratio of pod length to propeller diameter, L_{Pod}/D_{Prop}, C is the ratio of pod taper length to propeller diameter, TL/D_{Prop}, D is the ratio of strut distance to propeller diameter, S_{Dist}/D_{Prop}, and E is the propeller hub taper angle, H_{Angle}.

	Factor A	**Factor B**	**Factor C**	**Factor D**	**Factor E**
Pod No.	D_{Pod}/D_{Prop}	L_{Pod}/D_{Prop}	TL/D_{Prop}	S_{Dist}/D_{Prop}	H_{Angle}
	Ratios are based on propeller diameter, D_{Prop} of 270 mm				
Pod 1	0.474	1.593	0.256	0.370	15
Pod 2	0.474	1.593	0.256	0.489	20
Pod 3	0.474	1.593	0.556	0.489	15
Pod 4	0.474	1.593	0.556	0.370	20
Pod 5	0.474	1.941	0.256	0.370	20
Pod 6	0.474	1.941	0.256	0.489	15
Pod 7	0.474	1.941	0.556	0.370	15
Pod 8	0.474	1.941	0.556	0.489	20
Pod 9	0.615	1.593	0.256	0.370	15
Pod 10	0.615	1.593	0.256	0.489	20
Pod 11	0.615	1.593	0.556	0.370	20
Pod 12	0.615	1.593	0.556	0.489	15
Pod 13	0.615	1.941	0.256	0.370	20
Pod 14	0.615	1.941	0.256	0.489	15
Pod 15	0.615	1.941	0.556	0.370	15
Pod 16	0.615	1.941	0.556	0.489	20

The pod series tests used a fractional factorial design technique, a method of experimentation used to examine the effects of single parameters and interactions between parameters for multifactor experiments. In factorial experiment designs, a factor is a major independent variable. In this method, the significance of individual factors is ranked in an ascending order based on the estimate of their effects on the overall result of the experiments. An experimenter can go through the process of logical elimination of insignificant factors and rerun the tests using only the most influential factors, thus making the experimentation simpler. Some factors or interaction factors are aliased with each other and cannot be differentiated (Montgomery, 2005). The details of the factorial effect aliases in our test series design can be found in (Islam, 2004). The factorial analysis was used to determine which geometric parameters of a pod-strut-propeller have the most significant effect on the measured performance values (of propeller and unit thrust, propeller torque, propeller and unit efficiency) in an open water propulsion unit test in puller and pusher configurations.

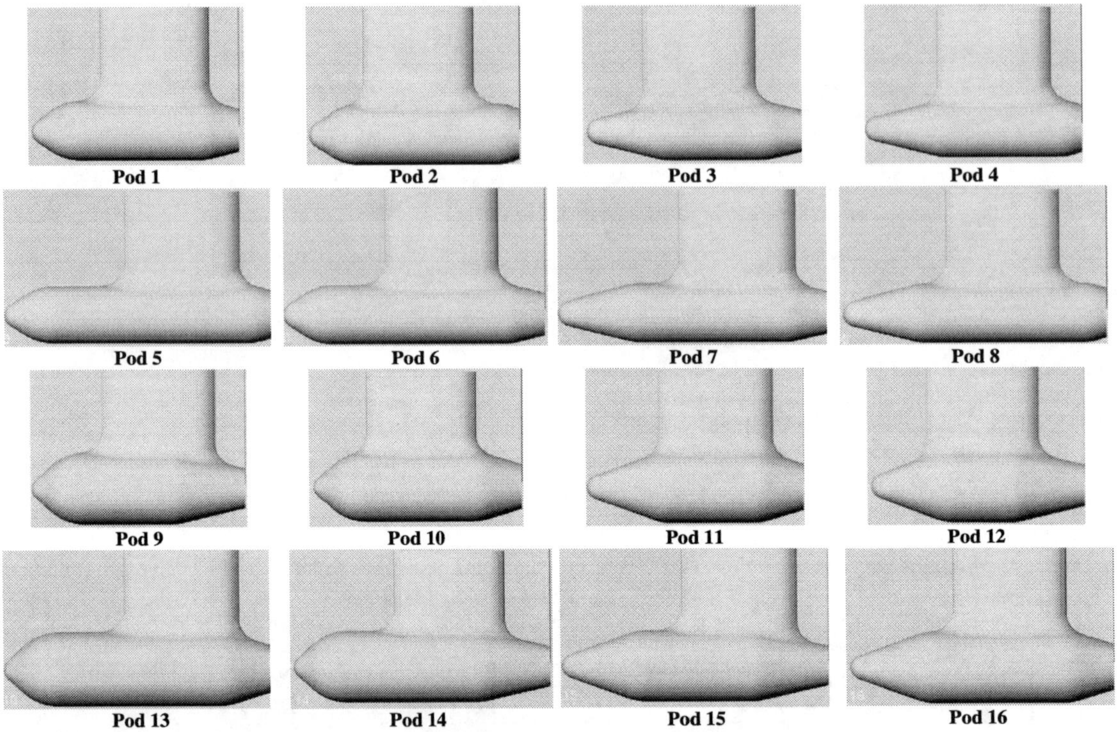

Fig. 2. Geometric models of the pod series used in the experiments.

The open water tests of the 16 pods were performed in accordance with the ITTC recommended procedure (2002a) and the description provided in Mewis (2001). A custom-designed dynamometer system was used to measure propeller thrust, torque and unit thrust of the pod series in both puller and pusher configurations. A motor was fitted above the propeller boat to drive the propeller via a belt system. The center of the propeller shaft was $1.5D_{\text{Prop}}$ below the water surface. The part of the shaft above the strut went through the boat. The propeller boat stayed 3 to 5 mm above the water surface to avoid waves caused by the strut piercing the surface. Water temperature, carriage speed, V_A, and the rotational speed of the propeller, n, were measured. Details of the experimental apparatus can be found in MacNeill *et al.* (2004).

RESULTS AND ANALYSIS

The dynamometer system measured propeller and pod forces and moments as follows: propeller thrust at the propeller end (T_{Prop}), propeller thrust at the pod end (T_{Pod}), propeller torque (Q), unit longitudinal force (T_{Unit} or F_X), unit transverse force (F_Y) and unit vertical force (F_Z). Two thrust load cells were used to measure propeller thrust at the two ends of the propeller shaft (T_{Prop} and T_{Pod}). The difference between the two measurements was insignificant taking into account the uncertainty in the system. In the current paper, only the propeller thrust at the propeller end (T_{Prop}) is presented. The experimental results were analyzed in terms of propeller thrust coefficient, $K_{T\text{Prop}}$, propeller torque coefficient, $10K_Q$, propulsive efficiency, η_{Prop}, unit thrust coefficient, $K_{T\text{Unit}}$, and unit efficiency, η_{Unit}, all versus propeller advance coefficient, J. $K_{T\text{Prop}}$, K_Q, η_{Prop}, $K_{T\text{Unit}}$, η_{Unit} and J are defined in equations 1-6, respectively. Additional experiments were conducted to study *Reynolds Number* effects on the puller and pusher propulsors' performance (Islam, 2006a). The study showed that the *Reynolds Number* effects became negligible at rotational speeds of 11 (equivalent to Re=6.50×10^5) and above, as both the propeller thrust and unit thrust stabilized at that rotational speed for all advance coefficient values. In all of the tests for pod geometry effects, a constant shaft rotational speed of 11 rps was used. Equations 1 to 6 were used to present the measurements.

$$K_{T\text{Prop}} = T_{\text{Prop}} / \left(\rho n^2 D^4\right) \quad (1)$$

$$K_Q = Q / \left(\rho n^2 D^5\right) \quad (2)$$

$$\eta_{\text{Prop}} = \left[J / (2\pi)\right] \times \left(K_{T\text{Prop}} / K_Q\right) \quad (3)$$

$$K_{T\text{Unit}} = T_{\text{unit}} / \left(\rho n^2 D^4\right) \quad (4)$$

$$\eta_{\text{Unit}} = \left[J / (2\pi)\right] \times \left(K_{T\text{Unit}} / K_Q\right) \quad (5)$$

$$J = V_A / (nD) \quad (6)$$

EXPERIMENTAL RESULTS

Pod Series in Puller Configuration

The propeller thrust coefficient, $K_{T\mathrm{Prop}}$, propeller torque coefficient, $10K_Q$ and propulsive efficiency, η_{Prop} values for each of the 16 pods in the puller pod series experiments (in the range of J=0.0~1.20) are presented in Figs. 3, 4 and 5, respectively. The tests were conducted at 17 different advance coefficients with repeated tests at 4 advance coefficients. Tables A-1 and A-2 present the $K_{T\mathrm{Prop}}$ and $10K_Q$ values, respectively, for the pods in puller configuration.

With reference to the Figures, for the puller propulsors in the series, the $K_{T\mathrm{Prop}}$ values at J = 0 of the 16 pods ranged from 0.464-0.495, an approximately 7% spread based on the lowest $K_{T\mathrm{Prop}}$ given by pod 15. At J = 0.8, the $K_{T\mathrm{Prop}}$ values covered 0.152-0.179 (approximately 17% spread based on the lowest $K_{T\mathrm{Prop}}$). At J=0.8, the highest $K_{T\mathrm{Prop}}$ was given by pod 16 and the lowest $K_{T\mathrm{Prop}}$ was given by pod 3. The torque coefficient ($10K_Q$) values of the different pods ranged from 0.679-0.692 (approximately 2% spread based on the lowest $10K_Q$ given by pod 3) at J = 0 and 0.275-0.312 (approximately 14% spread based on the lowest $10K_Q$ given by pod 4) at J = 0.8. The η_{Prop} values of the different pods ranged from 0.647-0.753 at J = 0.8 and 0.625-0.824 at J = 1.0. The trends showed that for puller configuration propulsors, there was significant variation in $K_{T\mathrm{Prop}}$, $10K_Q$ and η_{Prop} values with the change of the geometric parameters. At J=0, the thrust coefficients of the pods 1, 3, 6, 7, 12, 14, 15 were lower than those of pod 9 and the thrust coefficients of the remaining pods were higher than those of pod 9. At J=0.8, the propulsive efficiencies of the pods 1, 3, 6, 7, 12, 14, 15 were lower than those of pod 9 and the propulsive efficiencies of the remaining pods were higher than those of pod 9. This indicated that the efficiency of the propeller attached to pod 9 was approximately the average of all the other pods. Among the pods, pod 3 had the lowest efficiency (η_{Prop} =0.647) and pod 16 had the highest efficiency (η_{Prop} =0.753) at the design advance coefficient of J=0.8.

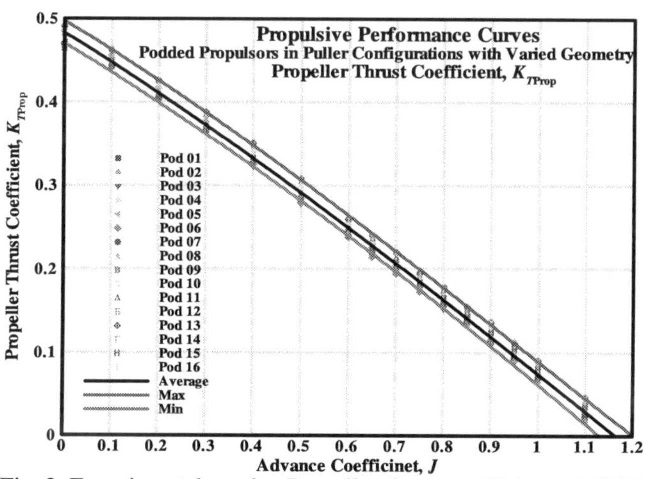

Fig. 3. Experimental results: Propeller thrust coefficient of all 16 puller pods in the series.

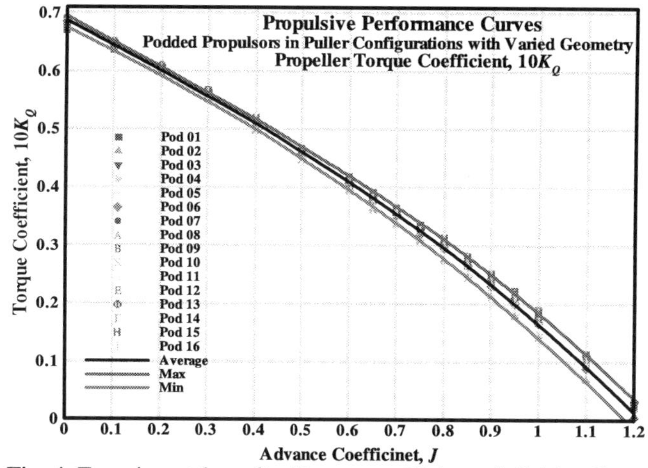

Fig. 4. Experimental results: Torque coefficient of all 16 puller pods in the series.

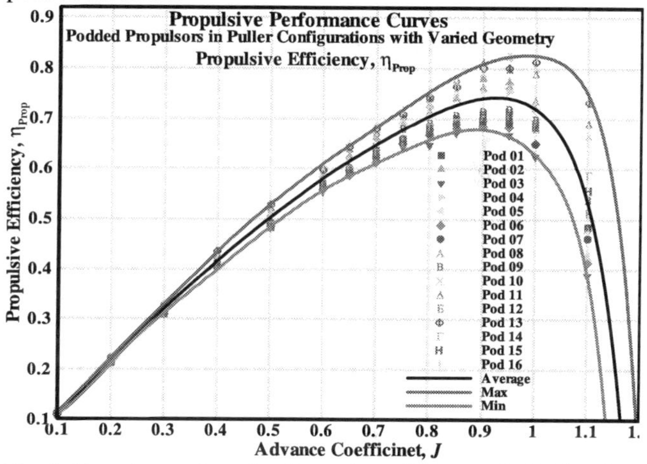

Fig. 5. Experimental results: Propeller efficiency of all 16 puller pods in the series.

The unit thrust coefficient, $K_{T\mathrm{Unit}}$ and unit efficiency, η_{Unit} values for each of the 16 pods in the pod series experiments (in the range of J=0.0~1.20) are presented in Figs. 6 and 7. The unit thrust coefficient values at J = 0 for the different pods ranged from 0.458 to 0.484, an approximately 6% spread based on the lowest $K_{T\mathrm{Unit}}$ given by pod 7. At J = 0.8, the $K_{T\mathrm{Unit}}$ values ranged from 0.13 and 0.153, an approximately 18% spread based on the lowest $K_{T\mathrm{Unit}}$ given by pod 4. The η_{Unit} values of the different pods ranged from 0.564 to 0.645 at J = 0.8 and from 0.366 to 0.565 at J = 1.0. The trends showed that there is significant variation in $K_{T\mathrm{Unit}}$ and η_{Unit} values with the change of the geometric parameters. At J=0, the thrust coefficients of the pods 1, 3, 6, 7, 12, 14, 15 were lower than those of pod 9 and the thrust coefficients of the remaining pods were higher than those of pod 9. Among these pods, pod 9 had the lowest unit efficiency (η_{Unit} =0.565) and pod 5 had the highest unit efficiency (η_{Unit} =0.645) at the design advance coefficient of J=0.8. Table A-3 presents the $K_{T\mathrm{Unit}}$ values for the pods in puller configuration.

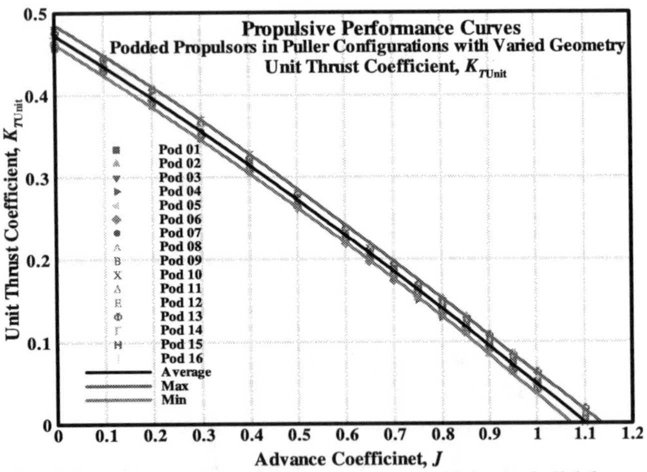

Fig. 6. Experimental results: Unit thrust coefficient of all 16 puller pods in the series.

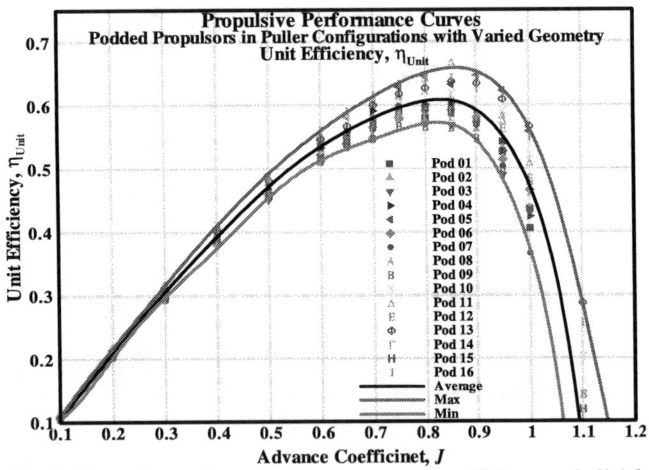

Fig. 7. Experimental results: Unit propulsive efficiency of all 16 puller pods in the series.

Pod Series in Pusher Configuration

The propeller thrust coefficient, K_{TProp}, propeller torque coefficient, $10K_Q$ and propulsive efficiency, η_{Prop} values for each of the 16 pods in the pusher pod series experiments (in the range of J=0.0~1.20) are presented in Figs. 8, 9 and 10, respectively. The tests were conducted at 17 different advance coefficients. Tests were repeated at a minimum of 4 advance coefficients. Table A-4 and A-5 present the K_{TProp} and $10K_Q$ values for the pods in pusher configurations, respectively.

For the pusher propulsors in the series, the K_{TProp} values at $J = 0$ of the 16 pods ranged from 0.4564 to 0.4715, an approximately 4% spread based on the lowest K_{TProp} given by pod 7. At $J = 0.8$, the K_{TProp} values were in a range of 0.1469-0.1724 (approximately 17% spread based on the lowest K_{TProp}). At J=0.8, the highest K_{TProp} was given by pod 4 and the lowest K_{TProp} was given by pod 13. The torque coefficient ($10K_Q$) values of the different pods ranged from 0.6532 to 0.6852 (approximately 5% spread based on the lowest $10K_Q$ given by pod 12) at $J = 0$, and from 0.275 to 0.312 (approximately 14% spread based on the lowest $10K_Q$ given by pod 1) at $J = 0.8$. The η_{Prop} values of the different pods ranged from 0.666-0.712 at $J = 0.8$ and 0.580-0.702 at $J = 1.0$. The trends showed that for the pusher configuration propulsors, there was significant variation in K_{TProp}, $10K_Q$ and η_{Prop} values with changes of the geometric parameters. At J=0, the thrust coefficients of the pods 2, 7, 8, 12, 13, 14 were lower than those of pod 9 and the thrust coefficients of the remaining pods were higher than those of pod 9. At J=0.8, the propulsive efficiencies of the pods 5, 7, 9, 13, 14, 15, 16 were lower than those of pod 8 and the propulsive efficiencies of the remaining pods were higher than those of pod 8. This indicated that the efficiency of the propeller attached with pod 8 was approximately the average of all the other pods. Among the pods, pod 13 had the lowest efficiency (η_{Prop} =0.666) and pod 12 had the highest efficiency (η_{Prop} =0.712) at the design advance coefficient of J=0.8.

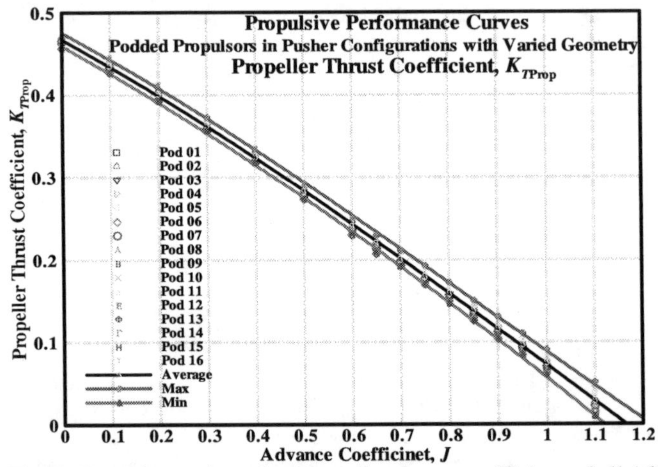

Fig. 8. Experimental results: Propeller thrust coefficient of all 16 pusher pods in the series.

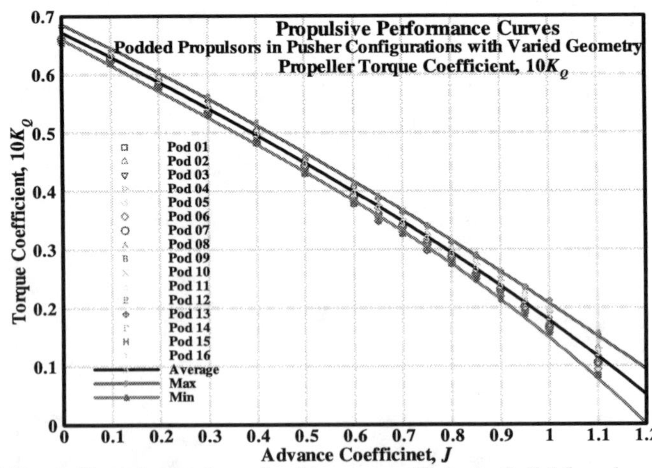

Fig. 9. Experimental results: Torque coefficient of all 16 pusher pods in the series.

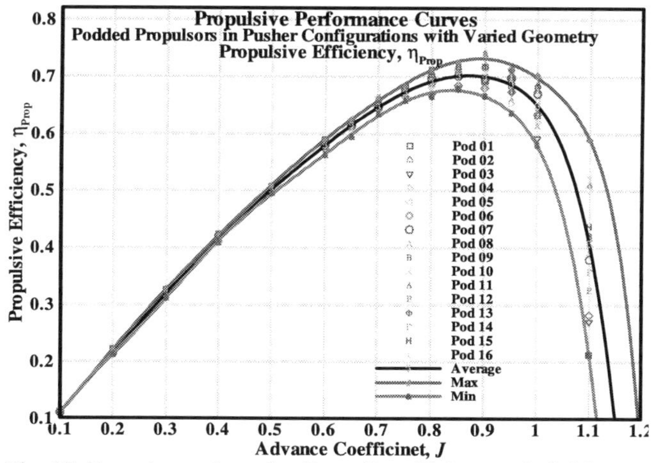

Fig. 10. Experimental results: Propeller efficiency of all 16 pusher pods in the series.

The K_{TUnit} and η_{Unit} values for each of the 16 pods in the pusher pod series experiments (in the range of J=0.0~1.20) are presented in Figs. 11 and 12, respectively. The unit thrust coefficient (K_{TUnit}) values at J = 0 of the different pods ranged from 0.440-0.462, an approximately 5% spread based on the lowest K_{TUnit} given by pod 1. At J = 0.8, the K_{TUnit} values cover a ranged of 0.112-0.143 (approximately 18% spread based on the lowest K_{TUnit} given by pod 12). The η_{Unit} values of the different pods ranged from 0.514-0.634 at J = 0.8 and 0.232-0.415 at J = 1.0. The trends showed that there was significant variation in K_{TUnit} and η_{Unit} values with the change of the geometric parameters. At J=0, the unit thrust coefficients of the pods 1, 3, 4, 5, 6, 8, 16 were lower than those of pod 9 and the thrust coefficients of the remaining pods were higher than those of pod 9. Among the pods, pod 4 had the lowest unit efficiency (η_{Unit} =0.514) and pod 11 had the highest unit efficiency (η_{Unit} =0.634) at the design advance coefficient of J=0.8. Table A-6 presents the K_{TUnit} values for the pods in pusher configuration.

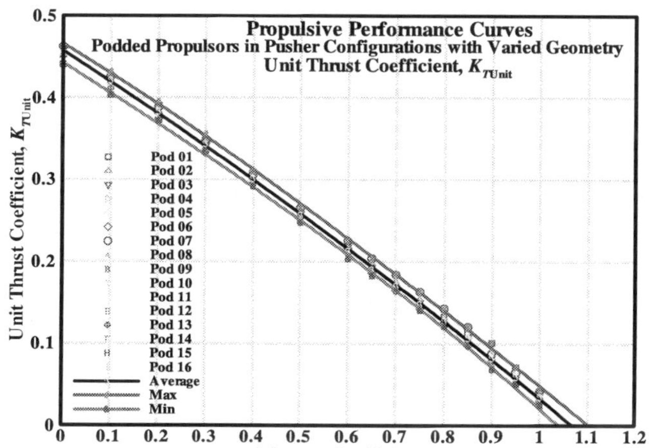

Fig. 11. Experimental results: Unit thrust coefficient of all 16 pusher pods in the series.

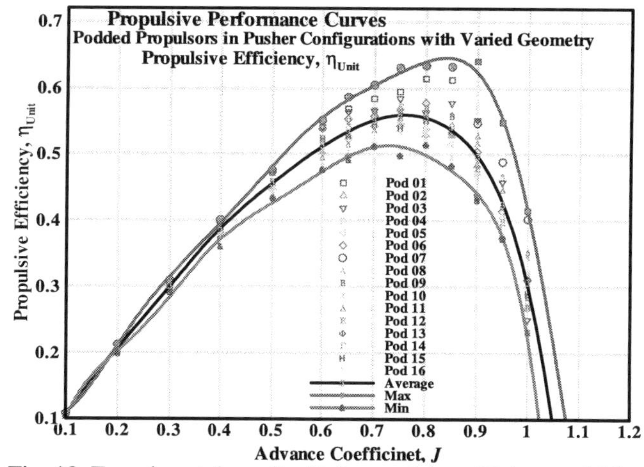

Fig. 12. Experimental results: Unit propulsive efficiency of 16 pusher pods in the series.

UNCERTAINTY ANALYSIS

While the details of uncertainty analysis are beyond the scope of this paper, a brief discussion of the levels of uncertainty in the above results is given below. To assess the uncertainty in each set of experiments and to identify the major factors influencing these results, a thorough uncertainty analysis was conducted (Islam, 2006b). The techniques used were based on adaptations of uncertainty analysis techniques outlined in ITTC recommended Procedure (2002b); Bose and Luznik (1996); Coleman and Steele (1989) and Hess *et al.* (2000).

The overall uncertainty in the non-dimensional performance coefficients of the podded propulsors required proper identification of all the variables contained within the data reduction expressions (equations 1-6). The experimental approaches used to obtain the data for each of the variables in the expressions were influenced by a variety of elemental sources of error. These elemental sources were estimated, and combined using the root-sum-square (RSS) method to give the bias and precision limits for each of the variables. The bias errors consisted of many elemental sources of error, which depended on the approaches followed to measure the variables. However, for the precision error estimates of most variables, only one source of error (repeatability) was considered significant. In order to calculate the uncertainty due to calibration of the six-component dynamometer measurement, it was required to determine how the uncertainties in the calibration data propagates into each element of the interaction matrix and into the measured forces and moments (Hess *et al.*, 2000).

The error estimates used in the determination of the bias and precision errors in this study were considered to be 95% coverage estimates. The bias uncertainty and the precision uncertainty were combined using the root-sum-square (RSS) method to provide estimates of overall uncertainty levels in these variables. The overall uncertainty was thus considered to be a 95% coverage estimate.

The final step in the methodology of uncertainty analysis was to determine how uncertainties in each of the variables propagate through the data reduction equations. Using the approaches described in Bose and Luznik (1996) and Coleman and Steele (1989), the uncertainty expressions for each set of experiments were developed as shown in equation 7 to 10, where U denotes the uncertainties in the corresponding coefficients denoted by the subscripts. It is to be noted that, in deriving the expressions, the cross-correlated bias limits (Coleman and Steele, 1989) were ignored. Strictly they should have been included, but they would have, in the current case, reduced the total uncertainty.

$$\left(\frac{U_{K_{TProp}}}{K_{TProp}}\right)^2 = \left(\frac{U_{TProp}}{T_{Prop}}\right)^2 + \left(\frac{U_\rho}{\rho}\right)^2 + 4\left(\frac{U_n}{n}\right)^2 + 16\left(\frac{U_D}{D}\right)^2 \quad (7)$$

$$\left(\frac{U_{K_Q}}{K_Q}\right)^2 = \left(\frac{U_Q}{Q}\right)^2 + \left(\frac{U_\rho}{\rho}\right)^2 + 4\left(\frac{U_n}{n}\right)^2 + 25\left(\frac{U_D}{D}\right)^2 \quad (8)$$

$$\left(\frac{U_{K_{TUnit}}}{K_{TUnit}}\right)^2 = \left(\frac{U_{TUnit}}{T_{Unit}}\right)^2 + \left(\frac{U_\rho}{\rho}\right)^2 + 4\left(\frac{U_n}{n}\right)^2 + 16\left(\frac{U_D}{D}\right)^2 \quad (9)$$

$$\left(\frac{U_J}{J}\right)^2 = \left(\frac{U_{V_A}}{V_A}\right)^2 + \left(\frac{U_n}{n}\right)^2 + \left(\frac{U_D}{D}\right)^2 \quad (10)$$

In the expressions for the podded propulsors' tests, it should be noted that for both thrust and torque coefficient uncertainties, the tare thrust and frictional torque were imbedded in the corresponding measurements. Since the tare thrust and frictional torque were part of the same data stream as the thrust and torque readings, they were not treated as an independent contributor of error to the corresponding coefficients, but rather as a bias error on the static-zero value of the thrust and the torque measurements. The resulting error estimates for the podded propulsor tests are given in Table 3 below.

Table 3. Uncertainty in the measurements of the pod series data.

J	U_J (+/-)	U_J/J (+/-)	U_{KTProp} (+/-)	U_{KTProp}/K_{TProp} (+/-)	U_{KQ} (+/-)	U_{KQ}/K_Q (+/-)	U_{KTUnit} (+/-)	U_{KTUnit}/K_{TUnit} (+/-)
0.00	-	-	5.78E-03	1.21	7.61E-03	1.11	4.75E-03	1.01
0.10	5.20E-03	5.20	5.59E-03	1.24	7.57E-03	1.17	4.39E-03	1.00
0.20	5.27E-03	2.63	5.28E-03	1.28	7.25E-03	1.20	4.05E-03	1.02
0.30	5.37E-03	1.79	5.02E-03	1.34	6.68E-03	1.19	3.74E-03	1.05
0.40	5.52E-03	1.38	4.74E-03	1.42	6.27E-03	1.23	3.40E-03	1.08
0.50	5.70E-03	1.14	4.62E-03	1.60	5.90E-03	1.28	3.07E-03	1.13
0.60	5.90E-03	0.98	4.22E-03	1.71	5.92E-03	1.45	2.66E-03	1.17
0.70	6.15E-03	0.88	4.02E-03	1.96	5.18E-03	1.47	2.44E-03	1.33
0.80	6.40E-03	0.80	3.33E-03	2.03	5.30E-03	1.79	2.22E-03	1.71
0.90	6.71E-03	0.75	2.89E-03	2.40	4.66E-03	2.00	2.21E-03	2.30
1.00	7.02E-03	0.70	2.43E-03	3.21	4.94E-03	2.94	1.43E-03	2.86
1.10	7.32E-03	0.67	1.43E-03	4.85	4.15E-03	4.38	9.99E-04	2.80
1.20	7.67E-03	0.64	-2.26E-03	12.58	3.82E-03	45.91	-1.34E-03	2.70

Applying the uncertainty limits to the performance curves of pod 16 in the form of error bars yields a plot as shown in Fig. 13. From the figure, it is observed that the curves fitted to the data lie inside the error bars. Therefore, the fitted curves provide a good representation of the trends indicated by the results.

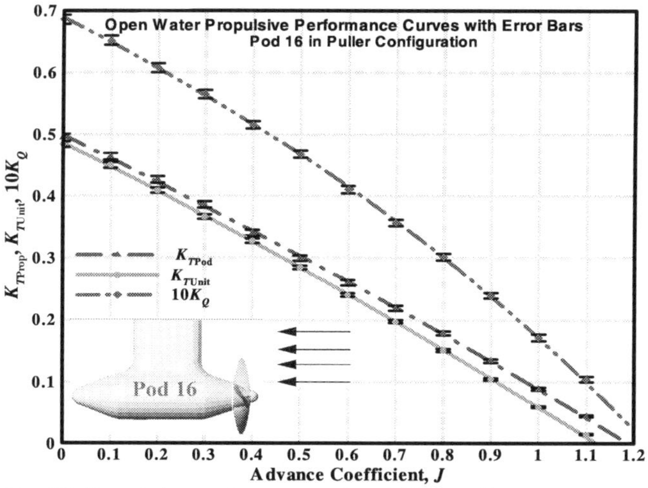

Fig. 13. Propulsive performance of pod 16 in puller configuration with error bars.

DOE ANALYSIS

The design of experiment data analysis was completed using the commercial software, Design Expert® (2005). The software allows the user to choose a factorial design that meets specific research configuration requirements; in this case the design includes one pod with all-low factors and one pod with all-high factors. The analysis of the data using a fractional factorial design technique resulted in the identification of the most significant factors and interactions of factors that affect the propulsive performance of the podded propulsors both in puller and pusher configurations.

DOE Analysis of Puller Propulsors

The Analysis of Variance Approach (ANOVA, see Montgomery, 2005) incorporating with a 95% confidence interval was used to examine the geometric parameters of the series that have the most significant impact on the performance of the podded propulsors. A separate analysis was completed for each advance coefficient. Table 4 lists all the factors and interactions of factors that have significant influence on the performance coefficients: propeller thrust coefficient, K_{TProp}, propeller torque coefficient, K_Q, propeller efficiency, η_{Prop}, unit thrust coefficient, K_{TUnit} and unit efficiency, η_{Unit} for the puller propulsors in the pod series. The factors are designated as shown in Table 2. Table 4 shows that the significant factors that come up repeatedly over the range of J values are A (the ratio of pod diameter to propeller diameter, D_{Pod}/D_{Prop}), D (the ratio of strut distance to propeller diameter, S_{Dist}/D_{Prop}), E (hub taper angle, H_{Angle}), AE (interaction of D_{Pod}/D_{Prop} and H_{Angle}) and BD (interaction of L_{Pod}/D_{Prop} and S_{Dist}/D_{Prop}). The factor C (the ratio of pod taper length to propeller diameter, TL/D_{Prop}) did not show any significant influence on the performance coefficients at any advance coefficient.

The factors H_{Angle} and D_{Pod}/D_{Prop} had significant impact on propeller thrust coefficient (K_{TProp}), unit thrust coefficient (K_{TUnit}) and torque coefficient (K_Q) for almost all values of advance coefficients, but as the value of advance coefficient increased from 0, interaction of D_{Pod}/D_{Prop} and H_{Angle} became as significant as the single factors D_{Pod}/D_{Prop} and H_{Angle} (in the range of J=0.3~1.0). The factor B (L_{Pod}/D_{Prop}) appeared to have significant effect in the form of interaction with D (S_{Dist}/D_{Prop}) on K_{TUnit} at high values of J (in the range of J=0.75~0.95). The factor D (S_{Dist}/D_{Prop}) appeared to have significant effect on K_{TProp} and K_Q at moderate advance coefficients, as shown in Table 4. The propeller and unit efficiencies, η_{Prop} and η_{Unit} were mostly affected by H_{Angle} and D_{Pod}/D_{Prop}; H_{Angle} being the most influential one at all advance coefficients. D_{Pod}/D_{Prop} played an important role when the advance coefficient was higher than 0.6.

Table 4. Fractional factorial design results: List of significant factors and interaction of factors for puller propulsors. Here, A is the ratio of pod diameter to propeller diameter, D_{Pod}/D_{Prop}, B isthe ratio of pod length to propeller diameter, L_{Pod}/D_{Prop}, C is the ratio of pod taper length to propeller diameter, TL/D_{Prop}, D is the ratio of strut distance to propeller diameter, S_{Dist}/D_{Prop}, and E is the propeller hub taper angle, H_{Angle}.

Significant factor and interaction of factors										
J	K_{TProp}		K_{TUnit}		K_Q		η_{Prop}		η_{Unit}	
0.00	A, E		E		A, E		-	-	-	-
0.10	A, E		E		A, E		E		E	
0.20	A, E		E		A, E		E		E	
0.30	A, E	AE	A, E	AE	A, E	AE	E		E	
0.40	A, E	AE	A, E	AE	A, E	AE	E		E	
0.50	A, E	AE	A, E	AE	A, E	AE	E		E	
0.60	A, D, E	AE	A, E	AE	A, D, E	AE	A, E		E	
0.70	A, D, E	AE	A, E	AE	A, D, E	AE	A, E		E	
0.80	A, D, E	AE	A, E	AE, BD	A, D, E	AE	A, E		E	
0.90	A, D, E	AE	A	AE, BD	A, D, E	AE	A, E		A, E	
1.00	A, E		A	BD	A, E		A, E		A, E	
1.10	A, E		A	BD	A, E		A, E	AE	A, E	

As shown in Fig. 14, the parameter D_{Pod}/D_{Prop} had a significant effect on K_{TProp} at J=0.8. For a fixed propeller diameter, as the pod diameter increased the K_{TProp} increased. In the puller configuration, the larger pod diameter created larger blockage to the flow behind the propeller. The flow blockage reduced the local inflow velocity, thus the propeller operated at a lower effective advance coefficient, which resulted in increases in the *KT*Prop. The analysis also indicated that the factor A (D_{Pod}/D_{Prop}) was involved in an interaction. This means that while the information in Fig. 14 is valid, there might be indirect impact on K_{TProp} due to the interaction of D_{Pod}/D_{Prop} and H_{Angle}. The parameter A (D_{Pod}/D_{Prop}) had an effect on K_{TUnit}, K_Q, η_{Prop} and η_{Unit} similar to that of K_{TProp} for the advance coefficient values shown in Table 4.

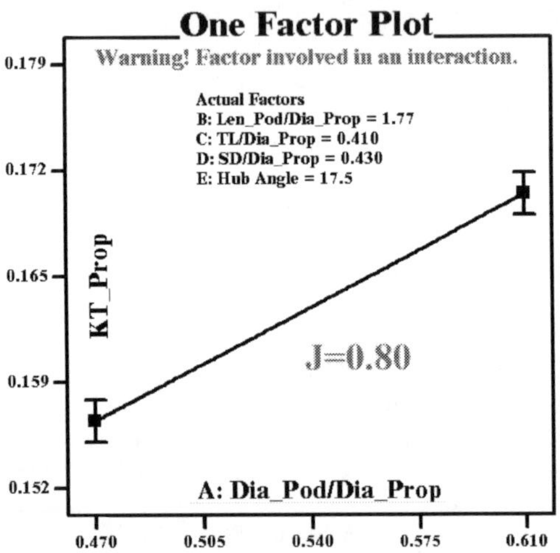

Fig. 14. DOE Analysis: The effect of D_{Pod}/D_{Prop} (A) on propeller thrust at J=0.8 for puller propulsors.

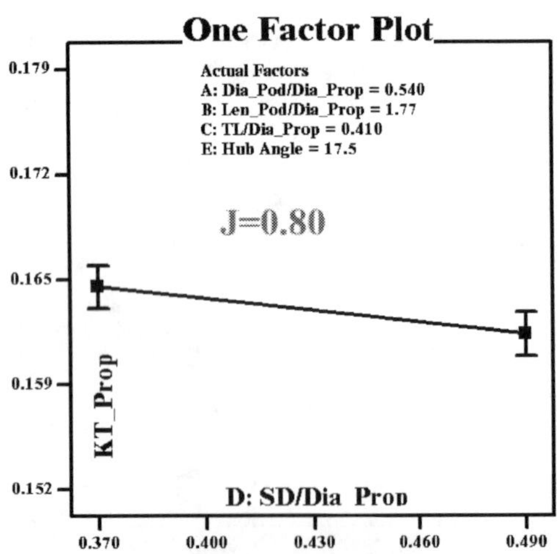

Fig. 15. DOE Analysis: The effect of significant factors, S_{Dist}/D_{Prop} (D) on propeller thrust at J=0.8 for puller propulsors.

Fig. 15 shows the effect of the strut distance, D (S_{Dist}/D_{Prop}) on K_{TProp}; it shows that as the strut distance increased the K_{TProp} tended to decrease slightly because of the blockage effect of the strut. A similar effect was found for K_Q. S_{Dist}/D_{Prop} did not show any effect on *KT*Unit as an individual factor but had influence on *KT*Unit as an interaction effect as described later (Fig. 21). Fig. 16 shows the effect of hub angle, E (H_{Angle}) on K_{TProp}. As the hub angle increased, the K_{TProp} also increased. The increasing effect of the hub angle on K_{TProp} was also found in a previous study using the same instrumentation (Islam *et al.*, 2006). Hub angle had a similar effect on K_{TUnit}, η_{Prop} and η_{Unit}. Figs. 17 and 18 show that the parameter hub angle, H_{Angle}, had a significant but opposite effect on K_Q at J=0 and J=0.8, respectively. As the hub angle increased, the K_Q increased at J=0 but decreased at J=0.8.

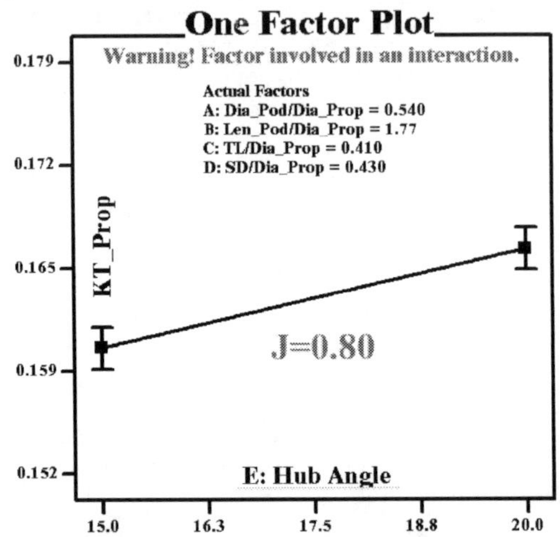

Fig. 16. DOE Analysis: The effect of significant factor, H_{Angle} (E) on propeller thrust at J=0.8 for puller propulsors.

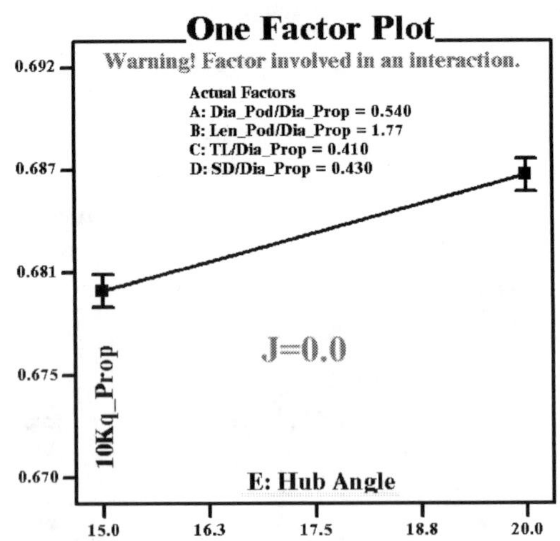

Fig. 17. DOE Analysis: The effect of significant factor, H_{Angle} (E) on propeller torque at J=0 for puller propulsors.

Fig. 19 shows the interaction effect AE (D_{Pod}/D_{Prop} and H_{Angle}) on K_{TProp}; it shows that the influence of the factor D_{Pod}/D_{Prop} was more obvious at a high hub angle i.e. increasing the ratio increased K_{TProp} at a faster rate at high hub angle. When the factor D_{Pod}/D_{Prop} was low, there was little change in K_{TProp} (within 2% based on the lower K_{TProp}) due to change in hub angle. However, when the ratio was high, there was a highly significant effect of hub angle on K_{TProp} (approximately 8% based on the lower K_{TProp}). This indicated that for a fat pod with respect to the propeller (higher value of D_{Pod}/D_{Prop}), the hub angle had more effect on K_{TProp} (as the hub angle increases, the K_{TProp} increases) than a slender pod with a low value of the factor D_{Pod}/D_{Prop}. A similar interaction effect was also found on K_{TUnit} and K_Q.

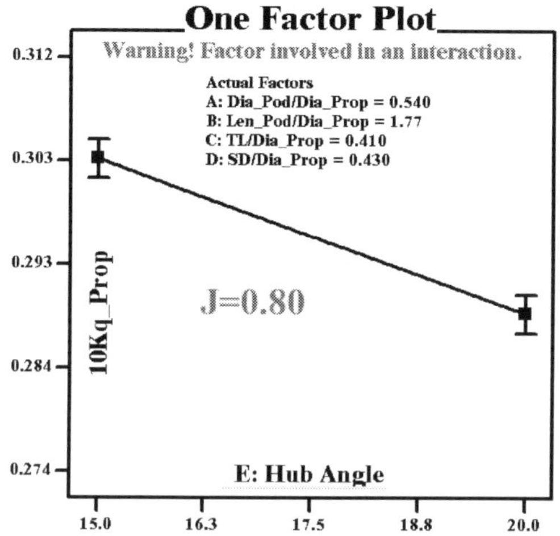

Fig. 18. DOE Analysis: The effect of significant factor, H_{Angle} (E) on propeller torque at J=0.8 for puller propulsors.

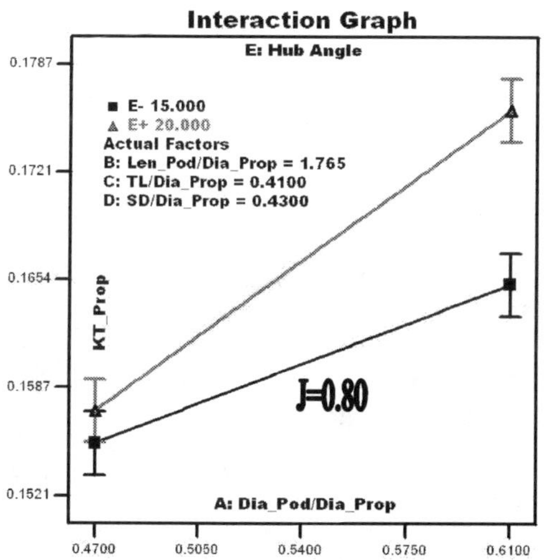

Fig. 19. DOE Analysis: The interaction effect of significant factors, D_{Pod}/ D_{Prop} (A) and H_{Angle} (E) on propeller thrust at J=0.8 for puller propulsors.

Fig. 20 shows the interaction effect BD (L_{Pod}/D_{Prop} and S_{Dist}/D_{Prop}) on K_{TUnit} at J=0.8; it shows that the effect of the factor L_{Pod}/D_{Prop} was opposite at the high and low values of S_{Dist}/D_{Prop}. At the low L_{Pod}/D_{Prop} value, the increase of S_{Dist}/D_{Prop} increased the K_{TUnit}, whereas at high L_{Pod}/D_{Prop} value, the increase of S_{Dist}/D_{Prop} decreased the K_{TUnit}. In other words, for the low S_{Dist}/D_{Prop} case, increasing L_{Pod}/D_{Prop} increased the K_{TUnit}, but for the high S_{Dist}/D_{Prop} case, increasing L_{Pod}/D_{Prop} decreased the K_{TUnit}.

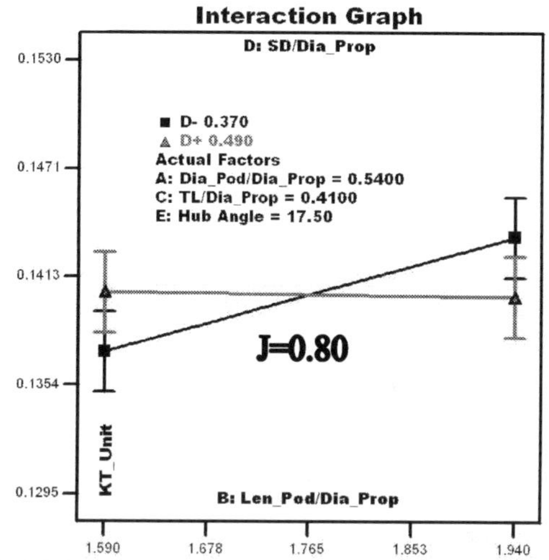

Fig. 20. DOE Analysis: The interaction effect of significant factors, L_{Pod}/D_{Prop} (B) and S_{Dist}/D_{Prop} (D) on unit thrust at J=0.8 for puller propulsors.

DOE Analysis of Pusher Propulsors

Table 5 lists all the factors and interactions of factors that have significant influence on the performance coefficients K_{TProp}, K_Q, η_{Prop}, K_{TUnit} and η_{Unit} for the pusher propulsors in the pod series. Table 5 shows that the significant factors that came up repeatedly over the range of J values were A (D_{Pod}/D_{Prop}), B (L_{Pod}/D_{Prop}), C (TL), D (S_{Dist}/D_{Prop}), E (H_{Angle}), AB (interaction of D_{Pod}/D_{Prop} and L_{Pod}/D_{Prop}) and BC (interaction of L_{Pod}/D_{Prop} and TL).

The factor H_{Angle} had significant impact on K_{TProp}, K_Q, η_{Prop} and η_{Unit} for all values of advance coefficient, whereas the factor A (D_{Pod}/D_{Prop}) had significant impact at moderate and high values of advance coefficient (in the range of J=0.4~1.1). The factor B (L_{Pod}/D_{Prop}) had significant influence in the form of interaction with D (S_{Dist}/D_{Prop}) on K_{TUnit} at moderate values of J (in the range of J=0.4~0.7) and as an individual factor at high values of J (in the range of J=0.9~1.1). The factor D (S_{Dist}/D_{Prop}) appeared to have significant effect on K_{TProp} and η_{Prop} at advance coefficients of 0.8 and higher. The factor C (TL/D_{Prop}) had a noticeable impact on K_{TUnit} at low and moderate values of advance coefficient as shown in Table 5.

Table 5. Fractional factorial design results: List of significant factors and interaction of factors for pusher propulsors. Here, A is the ratio of pod diameter to propeller diameter, D_{Pod}/D_{Prop}, B isthe ratio of pod length to propeller diameter, L_{Pod}/D_{Prop}, C is the ratio of pod taper length to propeller diameter, TL/D_{Prop}, D is the ratio of strut distance to propeller diameter, S_{Dist}/D_{Prop}, and E is the propeller hub taper angle, H_{Angle}.

Significant factor and interaction of factors										
J	K_{TProp}		K_{TUnit}		K_Q		η_{Prop}		η_{Unit}	
0.00	E	BC	C	AB	E		-	-	-	-
0.10	E	BC	C	AB	E		E		E	AB
0.20	E	BC	C	AB	E		E		E	AB
0.30	E		C	AB	E		E		A, E	
0.40	A, E		A, C	BD	E		A, E		A, E	
0.50	A. E		A, C	BD	A, E		A, E		A, E	
0.60	A, E		A, C	BD	A, E		A, E		A, E	
0.70	A, E		A, C	BD	A, E		A, E		A, E	
0.80	A, D, E		A, C		A, E		A, D, E		A, E	
0.90	A, D, E		B, C		A, E		A, D, E		A, E	
1.00	A, D, E		B		A, E		A, D, E		A, D, E	
1.10	A, E		B		A, E		A, D, E		A, D, E	

Fig. 21 shows the effect of the factor D_{Pod}/D_{Prop} on K_{TProp} at J=0.8. For a fixed propeller diameter, as the pod diameter increased the K_{TProp} increased. The larger pod diameter created larger blockage to the flow in front of the propeller. The blockage in the flow reduced the local flow velocity, thus the propeller operated at lower effective advance coefficient, which resulted in an increase in KTProp. The parameter, D_{Pod}/D_{Prop} had similar effect on K_{TProp}, K_Q, η_{Prop} and η_{Unit} for the other advance coefficient values shown in Table 5. Fig. 22 shows the effect of the factor D_{Pod}/D_{Prop} on K_{TUnit} at J=0.8. For a fixed propeller diameter, as the pod diameter increased the K_{TProp} decreased. A larger pod diameter means higher drag on the pod, along with a larger blockage effect, which resulted in lower unit thrust.

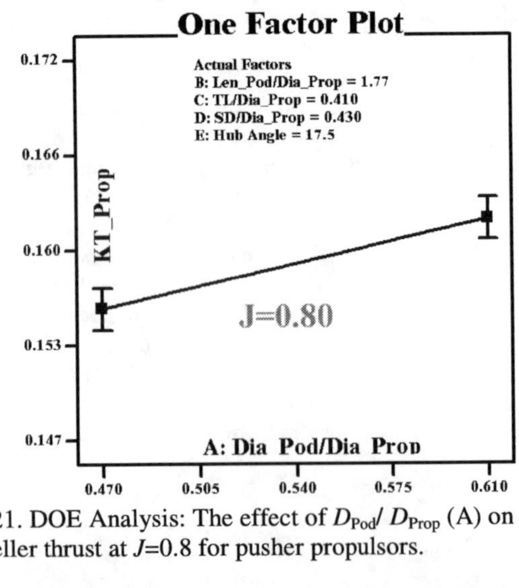

Fig. 21. DOE Analysis: The effect of $D_{Pod}/\ D_{Prop}$ (A) on propeller thrust at J=0.8 for pusher propulsors.

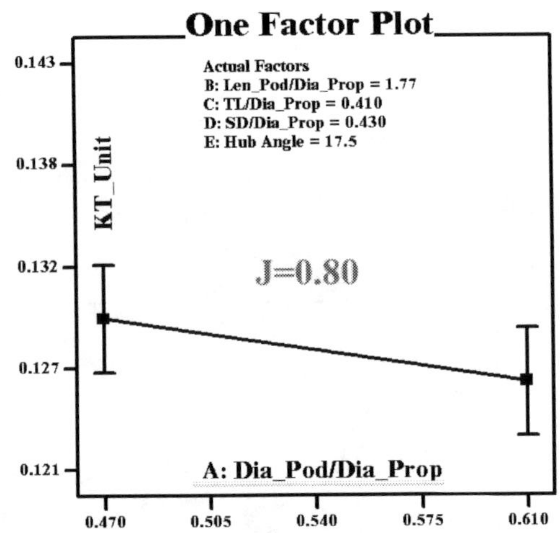

Fig. 22. DOE Analysis: The effect of $D_{Pod}/\ D_{Prop}$ (A) on unit thrust at J=0.8 for pusher propulsors.

Fig. 23 shows the effect of the factor taper length, C (TL/D_{Prop}) on K_{TUnit} at J=0.8. For a fixed propeller diameter, as the pod taper length increased the K_{TUnit} increased. Taper length showed a similar effect on unit thrust at other advance coefficients but it did not have significant effect on K_{TProp}, K_Q, η_{Prop} and η_{Unit} at any of the advance coefficients.

Fig. 24 shows the effect of the strut distance, D (S_{Dist}/D_{Prop}) on propeller thrust; it shows that at J=0.8, as the strut distance increased, the K_{TProp} tended to decrease slightly because of blockage effect of the strut. The factor, S_{Dist}/D_{Prop} showed similar effect on η_{Prop} and η_{Unit} at high advance coefficients but did not show any effect on K_{TUnit} and K_Q at any advance coefficients.

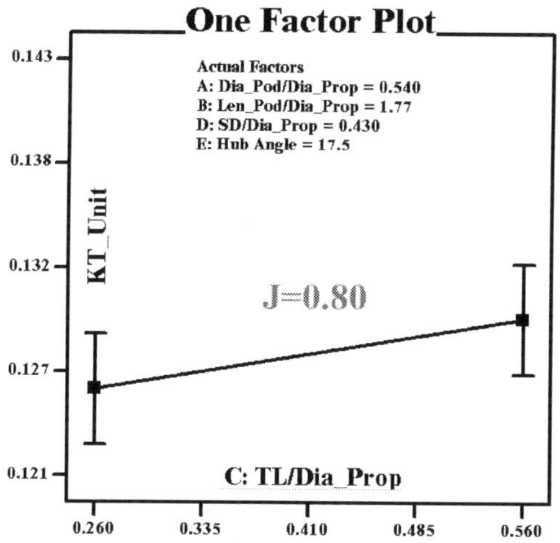

Fig. 23. DOE Analysis: The effect of TL/D_{Prop} (C) on unit thrust at J=0.8 for pusher propulsors.

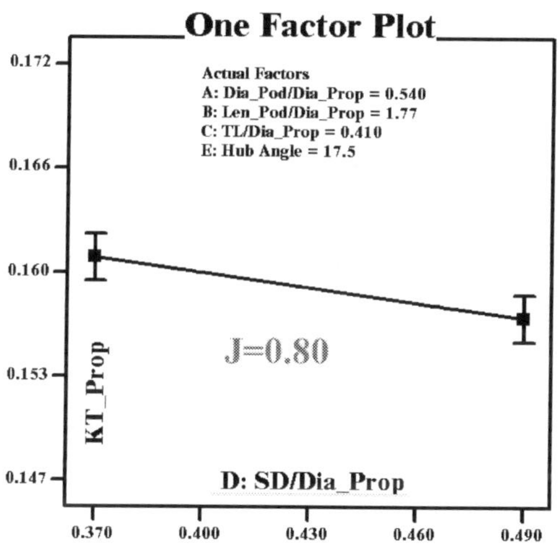

Fig. 24. DOE Analysis: The effect of S_{Dist}/D_{Prop} (D) on propeller thrust at J=0.8 for pusher propulsors.

Fig. 25 shows the effect of hub angle, H_{Angle} on K_{TProp} at J=0.8. As the hub angle increased, the K_{TProp} also increased. The parameter, H_{Angle} had a similar effect on K_{TProp}, K_Q, η_{Prop} and η_{Unit} for the other advance coefficient values as shown in Table 5 but did not have any effect on K_{TUnit} at any advance coefficients. Fig. 26 shows the effect of pod length, L_{Pod}/D_{Prop} on K_{TUnit} at J=0.9. As the pod length increased, the K_{TUnit} decreased meaning longer pods had lower unit thrust. The parameter L_{Pod}/D_{Prop} showed a similar effect on only unit thrust at higher advance coefficients (J greater than 0.9).

The interaction of factors B and C, i.e. BC, (L_{Pod}/D_{Prop} and TL/D_{Prop}) had noticeable effect on K_{TProp} when J was equal to 0.3 or less. Fig. 27 shows that for constant propeller diameter, at J=0 and at lower taper length, increasing pod length resulted in lower propeller thrust, whereas at higher taper length, increasing pod length resulted in higher propeller thrust.

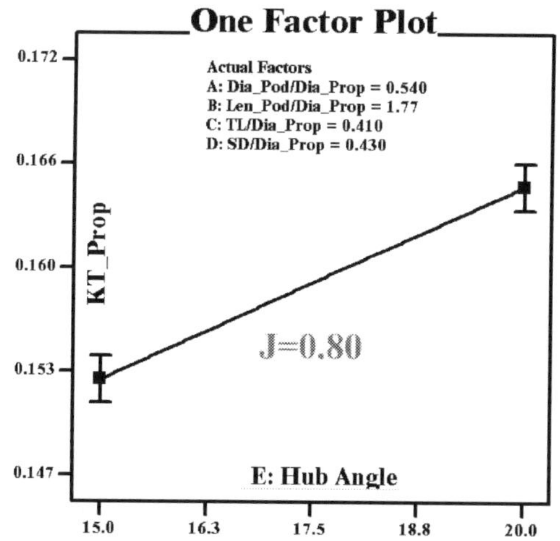

Fig. 25. DOE Analysis: The effect of H_{Angle} (E) on propeller thrust at J=0.8 for pusher propulsors.

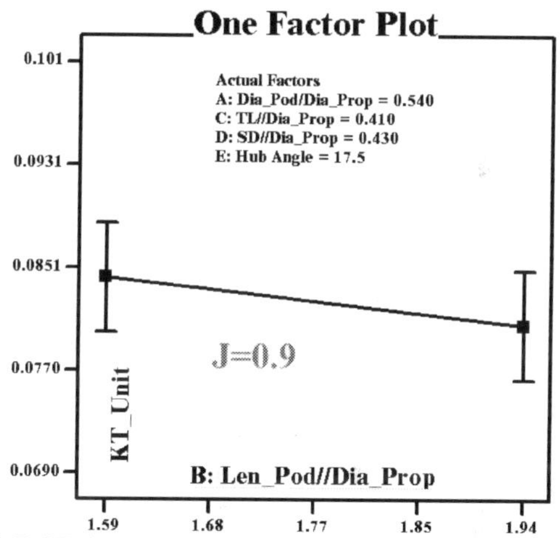

Fig. 26. DOE Analysis: The effect of L_{Pod}/D_{Prop} (B) on unit thrust at J=0.9 for pusher propulsors.

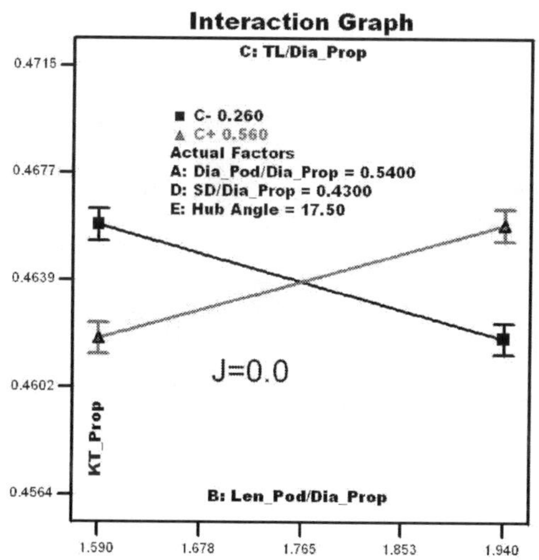

Fig. 27. DOE Analysis: The interaction effect of L_{Pod}/D_{Prop} (B) and TL/D_{Prop} (C) on propeller thrust at J=0 for pusher propulsors.

Fig. 28 shows the interaction effect of AB (D_{Pod}/D_{Prop} and L_{Pod}/D_{Prop}) on unit thrust at J=0. It shows that for constant propeller diameter, at lower pod diameter, increasing pod length resulted in an increase in unit thrust, whereas at higher pod diameter, increasing pod length produced a lower unit thrust. A similar effect was found in the range of J=0 to 0.4.

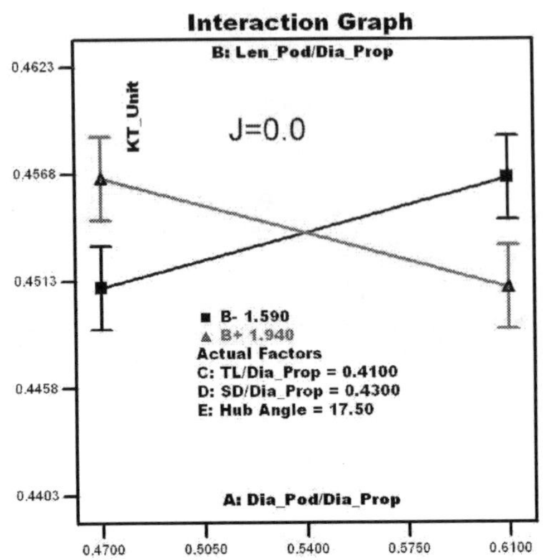

Fig. 28. DOE Analysis: The interaction effect of D_{Pod}/D_{Prop} (A) and L_{Pod}/D_{Prop} (B) on unit thrust at J=0 for pusher propulsors.

CONCLUDING REMARKS

A series of 16 pods were designed using a fractional factorial design technique to study the effects of five geometric parameters (pod diameter, pod length, pod taper length, strut distance and propeller hub angle) of podded propulsors in pusher and puller configurations. The experimental data on the pod series was acquired using a custom-designed pod testing system at the OERC towing tank at Memorial University. The findings of the analysis can be summarized at follows:

For a fixed propeller diameter, as the pod diameter increased, the propeller thrust, torque and efficiency increased for both puller and pusher propulsors; this might be attributed to the blockage effect of the pod. However, the increase in pod diameter resulted in a decrease in unit thrust for the pusher propulsors.

Pod length did not show an obvious effect on the performance coefficients of the puller propulsors, but it showed some effect on unit thrust coefficient of the pusher propulsors at higher advance coefficient regions. At advance coefficients of 0.9 or higher, for a fixed propeller diameter, as the pod length increased the propulsor unit thrust coefficient decreased, meaning that longer pods had lower unit thrust in the pusher configuration.

For the puller propulsors, as the hub taper angle increased, both propeller and propulsor unit thrust coefficients and efficiencies increased for all ranges of advance coefficients; this was more pronounced at low advance coefficients. However, hub angle had opposite effects on torque coefficients at low and high advance coefficients. The torque increased with increasing taper angle at advance coefficients of 0.7 or lower but decreased at higher advance coefficients. For the pusher propulsors, as the hub taper angle increased, propeller and unit thrusts and propeller torque coefficients and efficiencies increased for all ranges of advance coefficients

The ratio of strut distance to propeller diameter had moderate effects on propeller thrust and torque coefficients for puller propulsors at moderate advance coefficients but only on propeller thrust for pusher propulsors. As the distance of the strut leading edge from the propeller plane increased, the propeller thrust and torque coefficient decreased.

Taper length of the pod aft end, the end away from the propeller, did not have a significant influence on performance of the puller propulsors within the range tested. However, it had significant effect on unit thrust of the pusher propulsors at all advance coefficients. For a fixed propeller diameter, as the pod taper length increased, the unit thrust also increased due to less pod drag on pods with high taper length.

The interaction of the factors pod diameter and hub angle had significant effect on both propeller and unit thrust and torque coefficients at moderate advance coefficient for the puller propulsors. For a fat pod, the influence of increased hub taper angle, which causes an increase in propeller and unit thrust coefficients, was more pronounced. The interaction of the factors showed little or no effect for the pusher propulsors.

For the pusher propulsors, the interaction of the factors pod length and pod taper length had noticeable effect on propeller thrust for low advance coefficients. When the ratio of the pod taper length to propeller diameter was low, increasing pod length resulted in lower propeller thrust, whereas at higher taper length increasing pod length resulted in higher propeller thrust.

For the pusher propulsors, the interaction effect of pod diameter and pod length was significant on unit thrust coefficient at low advance coefficients. The analysis showed that for a slender pod (low value of pod diameter to propeller diameter), increasing pod length resulted in an increase in unit thrust, whereas the opposite is true for the pod with a high value of the ratio of pod diameter to propeller diameter.

For the puller propulsors, the interaction effect pod length and strut distance was significant on unit thrust coefficient at moderate advance coefficients. The analysis showed that the impact of pod length was opposite at high and low values of strut distance. At the low pod length value, the increase of strut distance increased the unit thrust, whereas at high pod length value, the increase of strut distance decreased the unit thrust.

The measurement showed that there were significant variations in the propeller thrust, torque, unit thrust and propeller and unit efficiencies values due to the variations of the geometric parameters of the pods. The uncertainty analysis of the measurements showed that the level of uncertainty was within acceptable limits. The variations of the performance coefficients due to the geometry variations were outside of the errors, which implied that the results were significant.

ACKNOWLEDGEMENTS

The authors thank the Natural Sciences and Engineering Research Council (NSERC) Canada, the National Research Council (NRC), Oceanic Consulting Corp., Thordon Bearings Inc., and Memorial University for their financial and other support. Thanks are also extended to Andrew MacNeill of Oceanic Consulting Corporation and Memorial University Technical Services Staff.

REFERENCES

BOSE, N. and LUZNIK, L. "1996 Uncertainty Analysis in Propeller Open Water Tests." Intl Shipbuilding Progress, Vol. 43, no. 435, pp. 237-246.

COLEMAN, H W. and STEELE, W. G. "1989 Experimentation and Uncertainty Analysis for Engineers." Wiley Interscience, 1989.

DESIGN EXPERT, Version 6, Stat Ease, www.statease.com, 2005.

GOUBAULT, P. and PÉRRÉE, J. "2004 Parametric investigations designed to help focused pod technology development." Proceedings of the 1st International Conference on Technological Advances in Podded Propulsion, Newcastle University, pp 27-28, UK, April 2004.

HE, M., VEITCH B., BOSE N., BRUCE C. and LIU P. "2005a Numerical Simulations of Propeller Wake Impacting on a Strut." Proceedings of the CFD2005, St John's, NL Canada, 8p, August 2005.

HE, M., VEITCH B., BOSE N. and LIU P. "2005b An Investigation on Wake/Strut Interaction of a Tractor-Type Podded Propulsor." Proceedings of the 7th CMHSC, Halifax, NS Canada, 8p, September 2005.

HESS, D.E., NIGON, R.T., BEDEL, J.W. "2000 Dynamometer Calibration and Usage." Research and Development Report No. NSWCCD-50-TR-2000/040, Hydromechanics Directorate, Carderrock Division, Naval Surface Warfare Center, West Bethesda, Maryland, 31p.

ISLAM, M. F. "2004 Numerical investigation on effects of hub taper angle and Pod-strut geometry on propulsive performance of pusher Propeller configurations." Master of Engineering thesis, Memorial University of Newfoundland, Canada, 136 p.

ISLAM, M. F., TAYLOR, R., QUINTON J., VEITCH, B., BOSE, N., COLBOURNE, B. and LIU, P. "2004 Numerical investigation of propulsive characteristics of podded propeller." Proceedings of the 1st International Conference on Technological Advances in Podded Propulsion, Newcastle University, UK, pp. 513-525, April 2004.

ISLAM, M., VEITCH, B., BOSE, N. and LIU, P. "2005 Cavitation Characteristics of Pushing and Pulling Podded Propellers With Different Hub Taper Angles." Proceedings. of the 7th CMHSC, Halifax, NS Canada, 7p, September 2005 .

ISLAM, M. F. "2006 Experimental Results: Open Water Performance of Puller Podded Propulsors with Varied Geometry." Ocean Engineering Research Center (OERC) Report No. OERC-2006-02, St. John's, NL, Canada, 109p, 2006.

ISLAM, M.. F. "2006a Calibration and Usages of NSERC-NRC Pod Dynamometer System." Ocean Engineering Research Center (OERC) Report No. OERC-2006-04, St. John's, NL, Canada, 86p, 2006.

ISLAM, M. F. "2006b Uncertainty analysis of NSERC-NRC pod dynamometer system." Ocean Engineering Research Center (OERC) Report No. OERC-2006-05, St. John's, NL, Canada, 96p, 2006.

ISLAM, M. "2004 A proposal for Doctoral research in the Faculty of Engineering and Applied Science." Engineering and Applied Science, Memorial University of Newfoundland, 85p.

ISLAM, M. F., VEITCH, B., BOSE, N. and LIU, P. "2006 Hydrodynamic characteristics of pod propeller units of highly tapered hub." Proceedings, Propellers/Shafting, Society of Naval Architects and Marine Engineers, Virginia Beach, USA, 12p, September 2006.

ITTC – Recommended Procedures, "2002a Propulsion, Performance - Podded Propeller Tests and Extrapolation" 7.5-02-03-01.3, Revision 00, 9p.

ITTC – Recommended Procedures, Propulsion, "2002b Propulsor Uncertainty Analysis, Example for Open Water Test." 7.5-02-03-02.2, Revision 00, 2002.

KARAFIATH, G. and LYONS, D. "1998 Hydrodynamic performance with Pod Propulsion - U.S. Navy experience." American Towing Tank Conference, Iowa City, 23p, 1998.

LIU, P. "2006 The Design of a Podded Propeller Base Model Geometry and Prediction of Its hydrodynamics." Technical Report no. TR-2006-16, Institute for Ocean Technology, National Research Council Canada, 16 p,

MACNEILL, A., TAYLOR, R., MOLLOY, S., BOSE, N., VEITCH, B., RANDELL, T. and LIU, P. "2004 Design of model pod test unit, Proceedings of the 1st International Conference on Technological Advances in Podded Propulsion." Newcastle University, UK, pp. 447-458, April 2004.

MEWIS, F. "2001 The efficiency of pod propulsion." HADMAR 2001, Bulgaria, October, 7p, 2001.

MOLLOY, S., ISLAM, M. F., HE. M., VEITCH, B., BOSE, N., WANG J., AKINTURK, A. and LIU, P., "2005 Use of factorial design in podded propulsors geometric series." Proceedings of the 7th CMHSC, Halifax, NS Canada, 8 p, September 2005.

MOLLOY S. "2003 A proposal for Doctoral research in the Faculty of Engineering and Applied Science." Engineering and Applied Science, Memorial University of Newfoundland, 86p.

MONTGOMERY, D.C. "2005 Design and analysis of experiments." sixth Edition, Wiley & Sons, USA, 189p, 2005.

TAYLOR, R. VEITCH, B. and BOSE, N. "2005 The Influence of Hub Taper Angle on Podded Propeller Performance: 'Propeller Only' Tests vs. 'Podded Propeller Unit' Tests." Proceedings of the 7th CMHSC, Halifax, NS Canada, 8p, September 2005.

TAYLOR R. "2005 Experimental Investigation of the Influence of Hub Taper Angle on the Performance of Push and Pull Configuration Podded Propellers." Master's of Engineering Thesis, Memorial University of Newfoundland, Canada, 120p, 2005.

APPENDIX

Table A-1. Puller Configuration: Propeller Thrust Coefficient, $K_{T\text{prop}}$

J	Pod 01	Pod 02	Pod 03	Pod 04	Pod 05	Pod 06	Pod 07	Pod 08	Pod 09	Pod 10	Pod 11	Pod 12	Pod 13	Pod 14	Pod 15	Pod 16
0.00	0.4688	0.4865	0.4693	0.4850	0.4833	0.4691	0.4689	0.4826	0.4763	0.4900	0.4924	0.4721	0.4903	0.4719	0.4642	0.4950
0.10	0.4425	0.4542	0.4410	0.4549	0.4556	0.4430	0.4430	0.4544	0.4490	0.4615	0.4638	0.4436	0.4617	0.4448	0.4424	0.4640
0.20	0.4048	0.4179	0.4038	0.4174	0.4179	0.4044	0.4041	0.4157	0.4092	0.4200	0.4260	0.4052	0.4260	0.4054	0.4055	0.4267
0.30	0.3663	0.3782	0.3636	0.3731	0.3777	0.3640	0.3666	0.3741	0.3723	0.3857	0.3860	0.3688	0.3885	0.3689	0.3675	0.3867
0.40	0.3308	0.3327	0.3230	0.3290	0.3363	0.3246	0.3281	0.3330	0.3311	0.3440	0.3481	0.3269	0.3507	0.3274	0.3321	0.3410
0.50	0.2825	0.2858	0.2807	0.2874	0.2906	0.2788	0.2822	0.2902	0.2881	0.3013	0.3059	0.2839	0.3083	0.2828	0.2829	0.2999
0.60	0.2403	0.2432	0.2386	0.2439	0.2448	0.2378	0.2400	0.2422	0.2483	0.2565	0.2601	0.2420	0.2601	0.2435	0.2466	0.2604
0.65	0.2179	0.2200	0.2162	0.2189	0.2266	0.2146	0.2192	0.2194	0.2268	0.2341	0.2374	0.2230	0.2408	0.2265	0.2227	0.2403
0.70	0.2022	0.2004	0.1945	0.2023	0.2045	0.1941	0.1994	0.2000	0.2067	0.2153	0.2149	0.2007	0.2201	0.2071	0.2041	0.2193
0.75	0.1773	0.1776	0.1765	0.1792	0.1795	0.1733	0.1796	0.1787	0.1899	0.1907	0.1966	0.1817	0.1954	0.1831	0.1834	0.1966
0.80	0.1587	0.1579	0.1521	0.1553	0.1616	0.1549	0.1581	0.1551	0.1683	0.1729	0.1760	0.1631	0.1786	0.1657	0.1649	0.1787
0.85	0.1376	0.1365	0.1334	0.1395	0.1379	0.1339	0.1379	0.1348	0.1472	0.1481	0.1529	0.1411	0.1548	0.1422	0.1429	0.1549
0.90	0.1130	0.1166	0.1129	0.1136	0.1139	0.1102	0.1139	0.1134	0.1230	0.1271	0.1341	0.1187	0.1368	0.1231	0.1243	0.1344
0.95	0.0946	0.0922	0.0883	0.0916	0.0933	0.0909	0.0942	0.0916	0.1057	0.1090	0.1097	0.0999	0.1115	0.1025	0.1046	0.1138
1.00	0.0720	0.0662	0.0652	0.0671	0.0644	0.0684	0.0697	0.0666	0.0833	0.0871	0.0845	0.0762	0.0906	0.0816	0.0801	0.0890
1.10	0.0262	0.0168	0.0205	0.0170	0.0186	0.0207	0.0242	0.0197	0.0332	0.0379	0.0409	0.0310	0.0468	0.0384	0.0356	0.0447
1.20	-0.0299	-0.0233	-0.0371	-0.0228	-0.0201	-0.0291	-0.0294	-0.0209	-0.0120	-0.0070	-0.0049	-0.0163	-0.0015	-0.0101	-0.0145	-0.0084

Table A-2. Puller Configuration: Propeller Torque Coefficient, $10K_Q$

J	Pod 01	Pod 02	Pod 03	Pod 04	Pod 05	Pod 06	Pod 07	Pod 08	Pod 09	Pod 10	Pod 11	Pod 12	Pod 13	Pod 14	Pod 15	Pod 16
0.00	0.6698	0.6828	0.6798	0.6806	0.6848	0.6856	0.6829	0.6815	0.6808	0.6891	0.6921	0.6830	0.6916	0.6825	0.6795	0.6854
0.10	0.6367	0.6444	0.6462	0.6480	0.6443	0.6471	0.6446	0.6440	0.6492	0.6497	0.6511	0.6433	0.6511	0.6454	0.6453	0.6515
0.20	0.5972	0.5991	0.6037	0.6049	0.6008	0.6054	0.6025	0.6002	0.6082	0.6019	0.6078	0.6038	0.6092	0.6057	0.6067	0.6077
0.30	0.5564	0.5519	0.5601	0.5571	0.5559	0.5625	0.5621	0.5520	0.5669	0.5638	0.5625	0.5635	0.5667	0.5648	0.5656	0.5653
0.40	0.5066	0.4986	0.5099	0.4991	0.4982	0.5123	0.5126	0.4986	0.5120	0.5062	0.5122	0.5136	0.5124	0.5102	0.5192	0.5155
0.50	0.4587	0.4460	0.4633	0.4510	0.4445	0.4568	0.4591	0.4491	0.4662	0.4602	0.4655	0.4656	0.4643	0.4623	0.4659	0.4677
0.60	0.4018	0.3954	0.4112	0.3949	0.3988	0.4076	0.4088	0.3955	0.4184	0.4103	0.4110	0.4079	0.4152	0.4123	0.4177	0.4109
0.65	0.3772	0.3645	0.3789	0.3656	0.3679	0.3787	0.3796	0.3629	0.3898	0.3772	0.3835	0.3844	0.3873	0.3909	0.3897	0.3832
0.70	0.3534	0.3374	0.3506	0.3368	0.3391	0.3533	0.3548	0.3387	0.3630	0.3525	0.3561	0.3538	0.3605	0.3643	0.3645	0.3572
0.75	0.3221	0.3089	0.3272	0.3107	0.3095	0.3186	0.3249	0.3091	0.3336	0.3261	0.3325	0.3288	0.3288	0.3341	0.3346	0.3268
0.80	0.2955	0.2767	0.2993	0.2745	0.2802	0.2960	0.3000	0.2760	0.3101	0.2974	0.3024	0.3038	0.3063	0.3093	0.3121	0.3021
0.85	0.2656	0.2495	0.2689	0.2484	0.2515	0.2661	0.2705	0.2462	0.2816	0.2655	0.2653	0.2766	0.2741	0.2774	0.2798	0.2698
0.90	0.2330	0.2142	0.2342	0.2146	0.2145	0.2306	0.2344	0.2129	0.2464	0.2299	0.2363	0.2417	0.2442	0.2494	0.2508	0.2396
0.95	0.2032	0.1802	0.2002	0.1835	0.1852	0.2013	0.2059	0.1804	0.2221	0.1997	0.2074	0.2123	0.2104	0.2190	0.2213	0.2089
1.00	0.1678	0.1461	0.1660	0.1481	0.1468	0.1672	0.1708	0.1438	0.1895	0.1727	0.1701	0.1783	0.1772	0.1885	0.1845	0.1719
1.10	0.0947	0.0699	0.0923	0.0708	0.0761	0.0881	0.0916	0.0715	0.1078	0.0994	0.1036	0.1060	0.1118	0.1142	0.1119	0.1046
1.20	-0.0053	-0.0208	-0.0100	-0.0159	-0.0171	0.0042	0.0057	-0.0155	0.0317	0.0213	0.0244	0.0230	0.0322	0.0310	0.0258	0.0183

Table A-3. Puller Configuration: Propulsor (Unit) Thrust Coefficient, $K_{T\text{unit}}$

J	Pod 01	Pod 02	Pod 03	Pod 04	Pod 05	Pod 06	Pod 07	Pod 08	Pod 09	Pod 10	Pod 11	Pod 12	Pod 13	Pod 14	Pod 15	Pod 16
0.00	0.4620	0.4836	0.4632	0.4780	0.4707	0.4622	0.4576	0.4685	0.4659	0.4792	0.4788	0.4607	0.4760	0.4611	0.4577	0.4840
0.10	0.4307	0.4450	0.4289	0.4439	0.4431	0.4353	0.4322	0.4319	0.4351	0.4450	0.4479	0.4342	0.4453	0.4300	0.4305	0.4499
0.20	0.3841	0.4085	0.3926	0.4040	0.4060	0.3934	0.3932	0.3945	0.3892	0.4067	0.4102	0.3895	0.4075	0.3880	0.3924	0.4095
0.30	0.3407	0.3647	0.3497	0.3525	0.3646	0.3495	0.3560	0.3537	0.3487	0.3693	0.3709	0.3516	0.3666	0.3467	0.3528	0.3670
0.40	0.3032	0.3208	0.3092	0.3135	0.3188	0.3063	0.3151	0.3158	0.3094	0.3273	0.3301	0.3099	0.3236	0.3058	0.3163	0.3281
0.50	0.2645	0.2674	0.2624	0.2699	0.2742	0.2634	0.2696	0.2715	0.2660	0.2798	0.2833	0.2685	0.2801	0.2637	0.2677	0.2849
0.60	0.2212	0.2248	0.2233	0.2201	0.2274	0.2183	0.2295	0.2261	0.2234	0.2373	0.2365	0.2229	0.2376	0.2204	0.2262	0.2406
0.65	0.1866	0.1969	0.2003	0.1972	0.2067	0.1972	0.2035	0.2050	0.2010	0.2125	0.2171	0.2040	0.2122	0.2019	0.2028	0.2182
0.70	0.1735	0.1761	0.1767	0.1786	0.1822	0.1735	0.1833	0.1825	0.1779	0.1904	0.1956	0.1833	0.1943	0.1832	0.1867	0.1974
0.75	0.1522	0.1530	0.1585	0.1506	0.1630	0.1548	0.1629	0.1595	0.1581	0.1668	0.1711	0.1596	0.1701	0.1618	0.1658	0.1732
0.80	0.1351	0.1307	0.1385	0.1295	0.1419	0.1330	0.1386	0.1350	0.1375	0.1475	0.1530	0.1415	0.1507	0.1410	0.1460	0.1511
0.85	0.1126	0.1115	0.1130	0.1107	0.1176	0.1164	0.1178	0.1120	0.1173	0.1212	0.1311	0.1188	0.1291	0.1216	0.1236	0.1282
0.90	0.0884	0.0839	0.0931	0.0884	0.0969	0.0934	0.0931	0.0870	0.0941	0.1017	0.1072	0.0980	0.1081	0.0997	0.1001	0.1046
0.95	0.0762	0.0633	0.0647	0.0639	0.0763	0.0684	0.0685	0.0696	0.0726	0.0780	0.0791	0.0790	0.0847	0.0814	0.0769	0.0857
1.00	0.0466	0.0401	0.0454	0.0395	0.0514	0.0489	0.0393	0.0433	0.0509	0.0566	0.0544	0.0542	0.0630	0.0547	0.0505	0.0594
1.10	-0.0013	-0.0104	-0.0077	-0.0018	0.0014	-0.0051	-0.0023	0.0012	0.0038	0.0115	0.0084	0.0086	0.0183	0.0072	0.0076	0.0153
1.20	-0.0612	-0.0500	-0.0719	-0.0457	-0.0413	-0.0572	-0.0619	-0.0503	-0.0525	-0.0435	-0.0501	-0.0400	-0.0309	-0.0466	-0.0458	-0.0409

Table A-4. Pusher Configuration: Propeller Thrust Coefficient, $K_{T\text{prop}}$

J	Pod 01	Pod 02	Pod 03	Pod 04	Pod 05	Pod 06	Pod 07	Pod 08	Pod 09	Pod 10	Pod 11	Pod 12	Pod 13	Pod 14	Pod 15	Pod 16
0.00	0.4655	0.4681	0.4598	0.4638	0.4643	0.4564	0.4641	0.4715	0.4605	0.4712	0.4667	0.4576	0.4672	0.4601	0.4595	0.4660
0.10	0.4354	0.4408	0.4302	0.4356	0.4392	0.4267	0.4328	0.4429	0.4327	0.4445	0.4393	0.4322	0.4371	0.4335	0.4317	0.4390
0.20	0.4038	0.4019	0.3978	0.3975	0.4059	0.3924	0.3936	0.4116	0.3968	0.4119	0.4095	0.3960	0.4054	0.3949	0.3962	0.4029
0.30	0.3618	0.3667	0.3566	0.3616	0.3654	0.3571	0.3574	0.3733	0.3665	0.3692	0.3685	0.3609	0.3731	0.3586	0.3579	0.3632
0.40	0.3210	0.3283	0.3213	0.3264	0.3288	0.3213	0.3186	0.3271	0.3203	0.3347	0.3343	0.3200	0.3320	0.3239	0.3208	0.3334
0.50	0.2745	0.2838	0.2740	0.2855	0.2867	0.2728	0.2778	0.2871	0.2767	0.2914	0.2934	0.2741	0.2855	0.2842	0.2770	0.2877
0.60	0.2329	0.2415	0.2358	0.2416	0.2435	0.2297	0.2371	0.2438	0.2378	0.2481	0.2500	0.2357	0.2465	0.2393	0.2332	0.2478
0.65	0.2101	0.2223	0.2108	0.2216	0.2261	0.2073	0.2136	0.2198	0.2162	0.2302	0.2291	0.2135	0.2295	0.2168	0.2159	0.2276
0.70	0.1921	0.2012	0.1921	0.2006	0.2042	0.1923	0.1971	0.2018	0.1974	0.2103	0.2060	0.1960	0.2112	0.1956	0.1974	0.2102
0.75	0.1710	0.1801	0.1708	0.1786	0.1830	0.1695	0.1755	0.1833	0.1782	0.1909	0.1860	0.1740	0.1926	0.1741	0.1766	0.1892
0.80	0.1511	0.1587	0.1469	0.1600	0.1634	0.1497	0.1571	0.1600	0.1561	0.1695	0.1670	0.1549	0.1724	0.1523	0.1571	0.1660
0.85	0.1317	0.1405	0.1269	0.1413	0.1433	0.1261	0.1334	0.1431	0.1359	0.1487	0.1464	0.1330	0.1508	0.1339	0.1369	0.1481
0.90	0.1133	0.1211	0.1033	0.1171	0.1191	0.1039	0.1123	0.1191	0.1122	0.1275	0.1274	0.1088	0.1312	0.1108	0.1120	0.1244
0.95	0.0892	0.0979	0.0881	0.0973	0.1009	0.0845	0.0932	0.1003	0.0943	0.1083	0.1054	0.0887	0.1108	0.0924	0.0958	0.1089
1.00	0.0644	0.0773	0.0602	0.0754	0.0776	0.0626	0.0700	0.0773	0.0730	0.0887	0.0870	0.0682	0.0911	0.0701	0.0730	0.0883
1.10	0.0102	0.0284	0.0160	0.0299	0.0352	0.0151	0.0229	0.0304	0.0275	0.0429	0.0428	0.0228	0.0518	0.0226	0.0285	0.0440
1.20	-0.0412	-0.0225	-0.0351	-0.0192	-0.0163	-0.0296	-0.0246	-0.0205	-0.0186	-0.0019	-0.0086	-0.0106	0.0017	-0.0199	-0.0176	-0.0020

Table A-5. Pusher Configuration: Propeller Torque Coefficient, $10K_Q$

J	Pod 01	Pod 02	Pod 03	Pod 04	Pod 05	Pod 06	Pod 07	Pod 08	Pod 09	Pod 10	Pod 11	Pod 12	Pod 13	Pod 14	Pod 15	Pod 16
0.00	0.6571	0.6552	0.6579	0.6818	0.6852	0.6588	0.6609	0.6832	0.6585	0.6790	0.6834	0.6578	0.6828	0.6656	0.6532	0.6740
0.10	0.6188	0.6349	0.6268	0.6311	0.6314	0.6220	0.6240	0.6421	0.6324	0.6421	0.6343	0.6243	0.6322	0.6373	0.6269	0.6316
0.20	0.5789	0.5925	0.5871	0.5853	0.5956	0.5812	0.5808	0.6048	0.5891	0.6054	0.5956	0.5834	0.5964	0.5903	0.5872	0.5981
0.30	0.5311	0.5380	0.5424	0.5420	0.5445	0.5361	0.5344	0.5619	0.5509	0.5568	0.5470	0.5383	0.5550	0.5494	0.5407	0.5528
0.40	0.4828	0.4978	0.4920	0.5071	0.4996	0.4981	0.4836	0.5072	0.4927	0.5168	0.5064	0.4941	0.5044	0.4931	0.4890	0.5077
0.50	0.4304	0.4487	0.4360	0.4570	0.4507	0.4342	0.4441	0.4612	0.4386	0.4581	0.4619	0.4319	0.4543	0.4494	0.4377	0.4611
0.60	0.3775	0.4014	0.3847	0.4013	0.3965	0.3832	0.3906	0.4139	0.3967	0.4089	0.4098	0.3837	0.4078	0.3903	0.3838	0.4071
0.65	0.3501	0.3734	0.3559	0.3740	0.3738	0.3490	0.3600	0.3823	0.3658	0.3911	0.3844	0.3573	0.3879	0.3649	0.3603	0.3882
0.70	0.3267	0.3468	0.3324	0.3512	0.3428	0.3292	0.3397	0.3537	0.3403	0.3656	0.3556	0.3329	0.3636	0.3356	0.3363	0.3650
0.75	0.2989	0.3188	0.3075	0.3232	0.3227	0.2988	0.3093	0.3319	0.3161	0.3380	0.3274	0.3073	0.3406	0.3067	0.3109	0.3355
0.80	0.2750	0.2949	0.2806	0.2984	0.3000	0.2765	0.2869	0.3027	0.2860	0.3094	0.2990	0.2856	0.3122	0.2806	0.2808	0.3099
0.85	0.2509	0.2647	0.2524	0.2731	0.2722	0.2487	0.2572	0.2851	0.2593	0.2850	0.2730	0.2627	0.2887	0.2566	0.2580	0.2908
0.90	0.2261	0.2408	0.2154	0.2451	0.2384	0.2189	0.2307	0.2504	0.2318	0.2577	0.2463	0.2338	0.2620	0.2285	0.2293	0.2626
0.95	0.1968	0.2106	0.1926	0.2247	0.2138	0.1896	0.2017	0.2292	0.2108	0.2300	0.2252	0.2106	0.2353	0.2042	0.2073	0.2349
1.00	0.1621	0.1752	0.1619	0.1933	0.1872	0.1574	0.1661	0.2002	0.1823	0.2055	0.2008	0.1869	0.2120	0.1736	0.1716	0.2084
1.10	0.0833	0.1189	0.1034	0.1295	0.1215	0.0935	0.1054	0.1352	0.1142	0.1441	0.1462	0.1224	0.1539	0.1108	0.1139	0.1566
1.20	-0.0097	0.0400	0.0149	0.0406	0.0478	0.0145	0.0213	0.0344	0.0379	0.0819	0.0731	0.0505	0.0887	0.0404	0.0391	0.0786

Table A-6. Pusher Configuration: Propulsor (Unit) Thrust Coefficient, $K_{T\text{unit}}$

J	Pod 01	Pod 02	Pod 03	Pod 04	Pod 05	Pod 06	Pod 07	Pod 08	Pod 09	Pod 10	Pod 11	Pod 12	Pod 13	Pod 14	Pod 15	Pod 16
0.00	0.4403	0.4505	0.4583	0.4524	0.4502	0.4607	0.4617	0.4623	0.4582	0.4448	0.4537	0.4610	0.4497	0.4475	0.4560	0.4517
0.10	0.4031	0.4186	0.4238	0.4211	0.4213	0.4236	0.4260	0.4319	0.4262	0.4184	0.4176	0.4282	0.4114	0.4189	0.4203	0.4190
0.20	0.3725	0.3823	0.3912	0.3783	0.3859	0.3874	0.3883	0.3951	0.3856	0.3842	0.3801	0.3852	0.3737	0.3855	0.3829	0.3754
0.30	0.3347	0.3424	0.3494	0.3437	0.3396	0.3484	0.3463	0.3562	0.3444	0.3398	0.3381	0.3486	0.3422	0.3437	0.3424	0.3432
0.40	0.2953	0.3009	0.3058	0.3020	0.2987	0.3107	0.3050	0.3066	0.3004	0.2922	0.2989	0.3019	0.2964	0.3045	0.2999	0.3011
0.50	0.2553	0.2570	0.2592	0.2564	0.2521	0.2610	0.2651	0.2660	0.2605	0.2505	0.2541	0.2567	0.2478	0.2555	0.2520	0.2544
0.60	0.2134	0.2121	0.2181	0.2169	0.2090	0.2163	0.2255	0.2188	0.2140	0.2060	0.2127	0.2108	0.2036	0.2101	0.2101	0.2097
0.65	0.1924	0.1902	0.1940	0.1901	0.1887	0.1866	0.2041	0.2014	0.1875	0.1859	0.1915	0.1897	0.1864	0.1831	0.1836	0.1893
0.70	0.1712	0.1696	0.1687	0.1715	0.1702	0.1648	0.1844	0.1783	0.1682	0.1687	0.1714	0.1692	0.1672	0.1660	0.1644	0.1739
0.75	0.1489	0.1456	0.1505	0.1479	0.1449	0.1420	0.1635	0.1568	0.1469	0.1441	0.1493	0.1452	0.1420	0.1470	0.1407	0.1533
0.80	0.1328	0.1228	0.1241	0.1262	0.1257	0.1256	0.1429	0.1337	0.1250	0.1255	0.1297	0.1277	0.1261	0.1221	0.1214	0.1290
0.85	0.1136	0.1034	0.1077	0.1077	0.1039	0.0979	0.1203	0.1182	0.1069	0.1044	0.1102	0.1029	0.1031	0.1009	0.1022	0.1076
0.90	0.1011	0.0814	0.0830	0.0806	0.0778	0.0772	0.0881	0.0904	0.0808	0.0779	0.0909	0.0783	0.0804	0.0792	0.0690	0.0821
0.95	0.0714	0.0524	0.0584	0.0617	0.0583	0.0520	0.0652	0.0711	0.0588	0.0623	0.0666	0.0556	0.0582	0.0526	0.0511	0.0614
1.00	0.0423	0.0297	0.0255	0.0336	0.0377	0.0307	0.0421	0.0434	0.0327	0.0409	0.0446	0.0316	0.0416	0.0253	0.0312	0.0344
1.10	-0.0085	-0.0240	-0.0264	-0.0186	-0.0206	-0.0333	-0.0082	-0.0045	-0.0098	-0.0073	-0.0091	-0.0051	-0.0014	-0.0177	-0.0228	-0.0088
1.20	-0.0827	-0.0760	-0.0875	-0.0726	-0.0803	-0.0916	-0.0753	-0.0514	-0.0506	-0.0583	-0.0662	-0.0439	-0.0430	-0.0763	-0.0742	-0.0746

Discussion

Ayhan Akinturk, Member

It was my pleasure to read this paper on the development of a series for podded propulsors. It has been a couple of decades since these novel propulsors have hit the market and they have gained in popularity ever since. This is the first comprehensive work on this subject, i.e. development of a pod series, published in open literature. The authors should be congratulated for such a scholarly work and the contributions to the understanding of the performance of such propulsors. It should be of particular interest to the people who design podded propulsors.

One of the main attraction points of podded propulsors is the capability they provide during maneuvering. Unlike conventional screws they can direct the thrust in any direction 360 degrees encompass. Hence, their performance at other azimuthing (steering) angles can be important for all ships but in particular, tug boats. Additional parameters of interest would be the steering moment and the unit side force. Would the authors comment on this? Are they planning to publish such data in the future?

With regard to the second paragraph in the conclusions, it is not clear from the paper whether increased pod diameter also increased the unit thrust or not in puller configuration. For push configuration a decrease associated with increased drag is reported. Would the authors clarify this point?

If the unit thrust is assumed as (propeller thrust minus the drag on the pod-strut system), I think this is a fair assumption for in particular, zero degree steering angle; were there any propulsor resistance tests done and if so, would the authors comment on the accuracy of this assumption.

In closing, I'd like to thank the authors again for such an important work.

William B. Morgan, Honorary Member, Fellow

This paper presents a prodigious amount of experimental data on the effect of pod shapes on the performance of podded propulsors. The experiments are very professionally performed and the uncertainty of the data is well documented. The procedures used adhere to the ITTC recommendations. Also, the design of experiment data analysis (DOE) is very useful in determining the effectiveness of various configurations. The paper should be very useful to those designing podded propulsors. This discusser thoroughly appreciates the complete documentation of the experimental data.

It would have been useful to include a complete nomenclature, as some of terms used are not standard, such as factors "A", etc. Also, including the propeller characteristics would have been useful as the referenced document was not readily available to this discusser.

The significant information in the paper appears to me to be the effect of the pod shape on the unit thrust. It is hard to judge whether the efficiency data presented are even relevant for assessing the pod shape. Presumably in an actual ship design, one would optimize the design of the propeller for the thrust required to propel the ship and for the actual wake inflow to the propeller. I would appreciate the authors' comments on my observation.

Gabor Karafiath, Member, **Scott Black,** Member, **Jonathan Slutsky,** Member, and, **Jennifer Gardner,** Member

We would first like to compliment the authors on a very interesting paper. The results of this work have been greatly anticipated as portions of it have appeared in other venues, and the systematic examination of a series of podded propulsors is a very valuable contribution to the literature.

A surprising aspect of the results is the strong influence of hub taper angle, particularly the apparent superiority of the 20 degree configuration. Inclusion of the open water propeller performance curves would provide a very valuable context for the pod designer in interpreting the data presented here. In one of your previous papers (Liu 2006), you have shown how a conventional cylindrical hub blade section on a tapered hub leads the root portions of the blade sections to operate at off-design conditions. Since the propellers are not designed for the different hub angles, do you feel that the performance differences should be reported as the primary design parameter? If hub angle effect were not included, what would be the next most critical design parameter?

Based on the changes in the zero-crossing of the propeller thrust curves on the pods, it is obvious that the propellers are operating in inflows with different wake fractions. How much of the efficiency change at a given advance coefficient is due to changes in wake fraction? Since modern propellers are designed for a specific wake fraction, how might the propellers that are designed for those different wake fractions affect the results of this study?

The design space of the pod series covers a commendably large range of pod sizes with correspondingly large differences in wetted surface and, presumably, residuary resistance. Have you considered the apparent differences in unit level performance in terms of these elementary resistance considerations? Along the same lines, the paper addresses Reynolds Number effects on the propulsor, but not on the rest of the apparatus. What ranges of Reynolds Numbers were tested if either the pod length or strut chord are used in the calculation? Was turbulence stimulation used during the test?

The baseline pod design is stated to be derived from existing commercial pod designs. What do the results of the experiment suggest about the relative performance of the parent designs?

Are current commercial best practices with attention to propeller inflow characteristics likely to produce efficient pod designs? By referencing strut position to the propeller reference plane, you create a number of configurations in which a significant portion of the pod length extends beyond the strut on the side opposite the propeller. Is this consistent with the structural and cooling arrangements requirements of existing commercial pod designs?

These different proportions of the pod and strut position relative to the propeller produce different average and unsteady inflows into the propeller, particularly in the pusher configuration. Is it reasonable to base your results for propeller geometries that were derived from the NSWCCD 4119 propeller, which was designed for a uniform spanwise inflow and a cylindrical hub geometry?

It appears that the strut cross section must be fore and aft symmetric in order for the pods to be run in both puller and pusher configurations. Is that shape representative of current design practice? If not, is the strut wake disturbance into the propeller sufficiently representative of commercial pods to have confidence in the trends reported in this paper?

Thank you for your attention to these questions. We look forward to using these valuable results in our ongoing podded propulsion research and design.

Andrew Macneil, Visitor

Congratulations on a paper well done. This work clearly shows some of the performance effects of geometry changes as they relate to the design of podded propulsor shapes. A number of important implications stem from this experimental work. One is that depending on whether the propulsion mode is push or pull, the pod housing geometry influences the performance characteristics differently. Another is that with an improvement in the understanding of the effect that shape has on hydrodynamic performance of pods, motor design can become more a function of the pod shape that is best suited for the vessel application and propulsion characteristics desired.

To that end it might be beneficial to include a few brief statements on the state-of-the art on current pod energy densities and which motor geometry factors have the largest influence on power output. The experimental work presented here shows that both pod diameter and length have a significant effect on performance, and these parameters are intuitive to think about in terms of electric motors.

Of the geometric parameters chosen, four were related to the pod shape itself. The fifth parameter was the distance from the leading edge of the strut to the propeller plane. The shape of the strut remained unchanged. In terms of this experimental work, the pods were designed as they were to allow the shells to be interchanged easily. The struts were made as a symmetric form, so as not to introduce any effects of the strut being used as a pusher in one set of tests, and a puller in the next set. This also had the benefit of reducing the machining time and kept the number of pieces to a minimum. However, the instrumentation design makes it possible to allow one to change geometric aspects of the strut as well. It would be interesting to design a series of experiments that alters the strut geometry. Some of the strut geometric parameters that could be altered are:

- strut foil shape
- strut length
- strut thickness
- strut asymmetry or twist angles

Overall it is good to see that the instrumentation in these experiments yielded such interesting results.

Author's Closure

The authors appreciate the time taken by the discussers in writing their discussions and the contributions they have made to the paper. A separate reply is given below to each discusser.

We thank Dr. Morgan for his kind remarks concerning the execution of the experimental work and its presentation. In the spirit of completeness, we have incorporated a fuller description of the propeller characteristics here in the discussion (Table 6), along with a clarification of the terms used in the experimental design.

Table 6. Geometric characteristics of the propellers used in the current experimental study.

Model Propeller Name: PP00+00C0	
Diameter (m)	0.27
No. of blade	4
Design advance coefficient, *J*	0.8
Hub-Diameter (*H/D*) ratio	0.26 (based on regular straight hub)
Angular speed (rps)	15
Section thickness form	NACA 66 (DTMB Modified)
Section meanline	NACA = 0.8
Blade planform shape	Blade planform shape was based on David Taylor Model Basin model P4119
Expanded area ratio, *EAR*	0.60
Pitch distribution	Constant, *P/D*=1.0
Skew distribution	Zero
Rake distribution	Zero

The factors A, B, C, D, and E are shorthand notations used in the design of the experiments to denote, respectively, the ratio of pod diameter to propeller diameter, the ratio of pod length to propeller diameter, the ratio of pod taper length to propeller diameter, the ratio of strut distance to propeller diameter, and the propeller hub taper angle. These can be understood more clearly by referring to Figure 1 and Table 2 on page 2 of the paper.

In response to Dr. Morgan's observation concerning the utility of efficiency in the present context, we agree that in practice, one would adapt the propeller design to the ship wake to ensure adequate thrust to achieve the design speed. We presented the results of our experiments in a format that is standard for open water propellers: KT-KQ-η curves.

Dr. Akinturk raised several points. His first question concerns performance of podded propulsors at steering angles, rather than straight ahead conditions. We agree with his statement that these non-zero azimuth conditions are important from a performance point of view. In recognition of this, we have done a series of experiments with podded propellers at static azimuth angles in uniform, oblique flow. Following these experiments, we did another seri5es of tests with pods moving dynamically through a wide range of azimuth angles. Both series of experiments have been completed and we will publish the results in due course.

In response to Dr. Akinturk's second question, it was found for the puller propeller that the pod diameter had a similar effect on unit thrust as on propeller shaft thrust: increases in pod diameter increased unit and shaft thrust. We attributed this effect to the blockage effect of the pod on the inflow conditions to the propeller, as described on page 9 of the paper.

As for his final question, we did not do experiments to measure the resistance only of the pod-strut assembly.

Karafiath et al. asked about the increased propeller thrust of the pods with 20° hub taper angles compared to those with 15° taper angles. We observed increases in the propeller thrust for both the puller and pusher configurations over a range of advance coefficients when the taper angle was increased. These tests included interaction effects, for example between the hub angle and the pod diameter. In an earlier experiment where we examined only the hub taper angle (Islam 2004 and Taylor 2005) and did not change the pod or strut shapes, we found broadly similar results, although there was a difference. Specifically, while increasing the hub angle in pusher configuration increased the propeller thrust for low and high advance coefficients, the effect in puller configuration showed a decrease in propeller thrust at high advance coefficients.

It is likely that the higher hub angles leads to a lower overall inflow speed at the propeller due to the potential flow effect around the hub (and pod), whether in pusher or puller configuration. If this is the case, the influence would be greatest near the root sections and would diminish towards to the tips.

Another way to explain the higher thrust of the propellers with higher hub angle is to look at the blade geometry at the roots. For the pulling propeller with a high hub angle of 20° compared to that of 15°, the intersection between the blade root section and the hub surface makes the blade root section have a smaller angle of attack (this will reduce the thrust) and to have a substantially higher leading edge area (this will increase the thrust), the combined effect at this hub angle may have given an increase in thrust. For the pushing propeller with a high hub angle of 20°, the angle of attack at the blade root section is increased substantially and the leading area is decreased; this combined effect at this hub angle range, may have resulted in an increased thrust as well (refer to Figures 29 to 31 and Table 7).

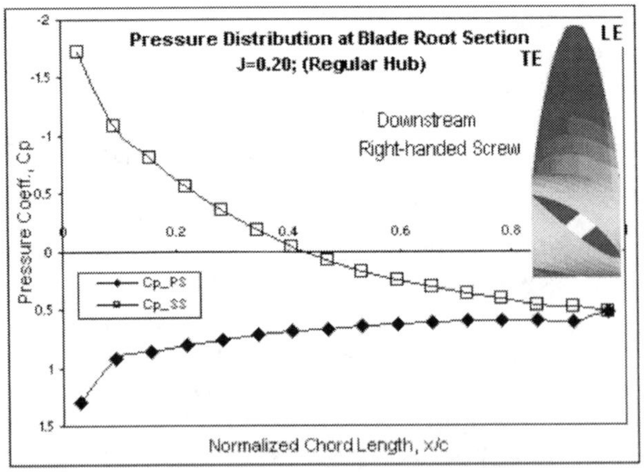

Fig. 29. Numerical results showing the pressure distribution at the blade root section of the propeller with straight hub at J=0.20.

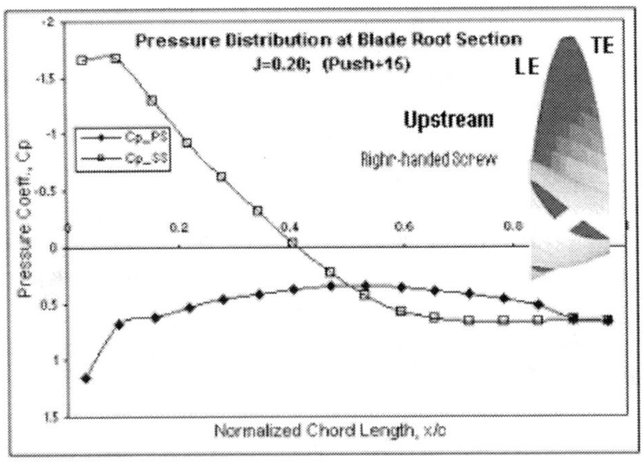

Fig. 30. Numerical results showing the pressure distribution at the blade root section of the propeller Push+15 at J=0.20.

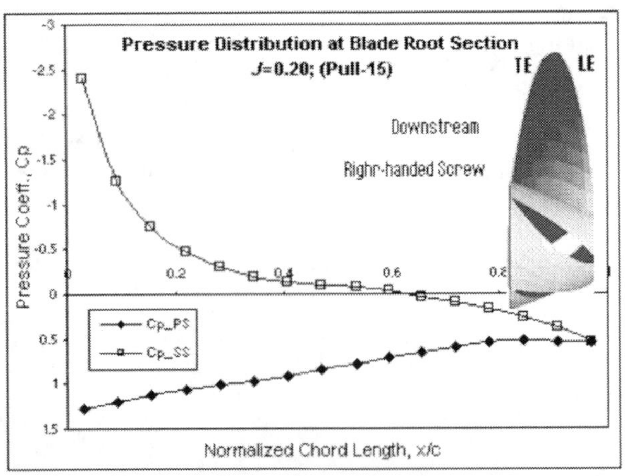

Fig. 31. Numerical results showing the pressure distribution at the blade root section of the propeller Pull-15 at J=0.20.

Table 7. The blade sections at the root of the propellers with straight hub, -15° taper angle, -20° taper angle, 15° taper angle, and 20° taper angle hubs. The inflow direction is from right to left.

Straight Hub	15° Pull	20° Pull	15° Push	20° Push
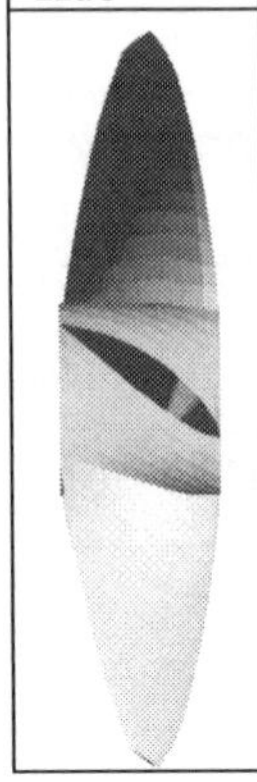	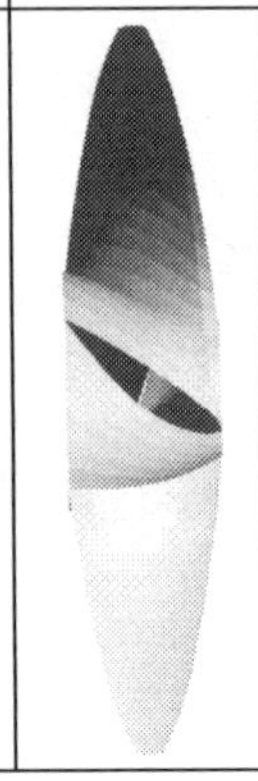	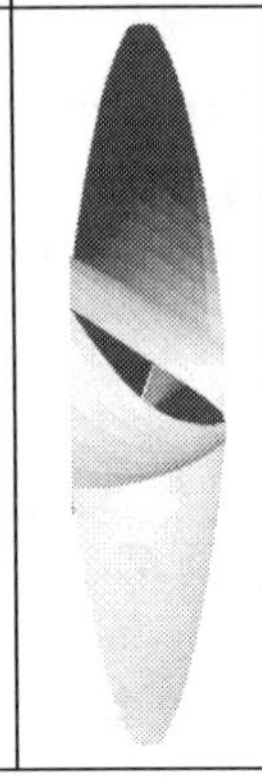	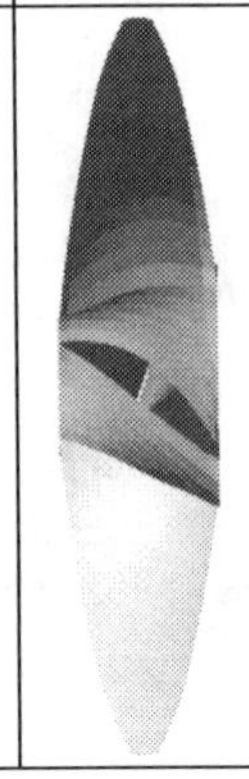	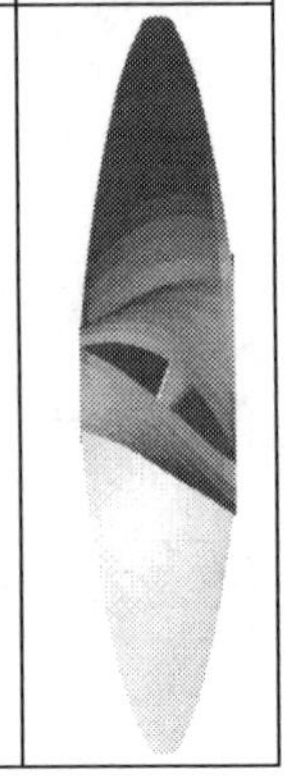

In reality, without further study, we don't know exactly why this increased thrust occurs at higher taper angle. It would be possible to check the hypotheses by measuring the inflow conditions just forward of the propeller, using PIV for example.

The authors agree with the comment that inclusion of the open water propeller performance curves would provide very valuable context for the pod designer in interpreting the data presented in the paper. The open water propeller performance curves of the four propellers are provided in Figure 32.

In response to Karafiath et al.'s question regarding the inclusion of hub taper angle as a primary design parameter, we would like to mention that we looked at the effect of hub angle within a factorial design, which means we were looking at the effect of the parameters to determine which have the most influence on the overall performance coefficients. The performance differences we presented are representative of the significance of the effect of the hub taper angle. The hub angle should be a primary design parameter in the early stages of designs and further analysis of the propeller performance at a given hub angle should be included in a design analysis. However if hub angle effect were not included, based on the findings of the study, the next most critical design parameter would have been pod diameter.

Karafiath et al. asked about the effects of wake fraction in designing a podded propulsor. To design a good podded propulsor, the inflow wake has to be considered. However the aim of the current work was to compare performance by making as few changes to the geometry as possible. Hence the propeller design was not changed between configurations; the emphasis was on changes on pod geometry. We have not investigated the wake yet as part of this study, thus we cannot estimate the efficiency change at this point, but may include it in a future study.

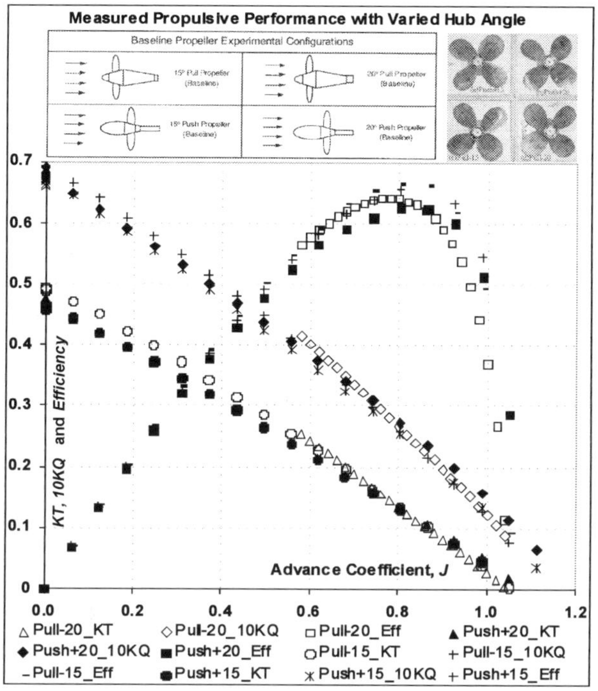

Fig. 32. Open water performance curves of the propellers: Pull-15, Pull-20, Push+15 and Push+20.

For this study, redesigning the propeller for each geometric variation would have involved changing too many parameters to run a valid factorial design. However, a propeller study that could be run factorially with one pod design would be a valuable future study.

In the current tests with the pod series, the Reynolds Number based on pod length ranged approximately from 5.5E+05 to 1.5E+06 and the Reynolds Number based on strut chord length ranged approximately from 2.6E+05 to 8.0E+05. We did not use any turbulence stimulation.

In response to Karafiath et al.'s comment regarding current pod design practices, the authors would like to mention that at this point, the range of pod designs seem to be dependent on the dimensions of the motor and required apparatus. None of the pod shapes we looked at has come out "on top" as a very efficient design. The aim was not to model and compare actual designs in use, but to cover the range in geometries in current practice.

Regarding Karafiath et al.'s comments on the extent and applicability of pod shapes, we would like to argue that we did not make this decision based on commercial pod designs. It was intended to be an extreme design based on the extreme dimensions used by the commercial pods. The method was used to determine how the interaction of these parameters affected the overall performance, so we don't know if it was consistent with commercial pods but most important it doesn't matter. This is a factorial design and is supposed to include extreme designs. We are looking at the effects of the interactions, not the quantitative values; although the quantitative values are important.

Regarding Karafiath et al.'s comments on the selection of the DTMB P4119 propeller as the reference propeller, our reason is that, in examining the hub angle effect, the tests for podded propellers were performed without pod and strut and under open water conditions; we therefore selected DTMB P4119 as the reference propeller. Another reason to choose P4119 propeller as the reference propeller is that P4119 propeller geometry has been one of the most panel method friendly propellers. This would give us the highest confidence on the accuracy and reliability of the results produced by the in-house panel method code PROPELLA.

Karafiath et al. asked about the shape of the strut and its applicability. We chose a symmetric strut design so that we could use some of the same pod pieces in push and pull mode and reduce the number of factors used in the factorial design. The shape of the strut is not representative of current design practice. Most of the commercial struts are asymmetric (wake adapted strut) and usually an extension is used in the aft side and around the base for strength purposes. The strut is usually a streamlined shape and the strut wake only extends to a small portion of the propeller (in pusher configuration). Thus, it is a fair approximation that the strut wake disturbance into the propeller in the current study is sufficiently representative of commercial pods to have confidence in the trends reported in this paper.

In response to Mr. MacNeill's comments regarding the motor geometry factors, it is found that the most influential motor parameter is the motor diameter (Goubault and Pérrée, 2004). An optimized motor usually has a short-fat geometry but the current trend in the pod design is to host a motor with the diameter as low as possible without sacrificing the motor feasibility, its proper cooling and hence its reliability. A proper interaction between the naval architect and pod manufacturer is required to ensure that the pod housing has sufficient diameter to house an optimized motor (Goubault and Pérrée, 2004). Another important reference to this end is the work by Kalsi (2006). In the work the author showed that smaller diameter pods can be achieved by employing high-temperature superconductor (HTS) synchronous propulsion motors. HTS motors have been under development since the late 1990s. Having up to three-times higher torque (and power) density than alternative technologies, HTS machines are significantly more compact and lighter in weight. Compact and lighter HTS motors reduce pod structural requirements and simplify interfaces. Their inherent characteristics such as quietness and absence of vibrations make them most suitable for Naval and cruise ship applications.

Mr. MacNeill made a valuable suggestion regarding further study on the effect of the strut shape on pod performance. The suggested strut parameters can be studied numerically using our in-house panel method code, PROPELLA, and the outcome from that work can be verified experimentally using our existing pod instrumentation. To this end, our colleague Moqin He performed an experimental and numerical study to investigate the wake impingement effect on strut for puller podded propulsors (He et al., 2005a and He et al., 2005b) where he recommended a future study to design an optimized wake adapted strut.

Overall it needs to be understood that the focus of the current study was to perform a factorial design and analysis to find out the most influential pod parameters on its performance. A factorial design shows which factors/parameters have the most significant effect and should be investigated further.

References

KALSI, S. S. 2006 HTS Ship Propulsion Motors for Podded Applications, *Proceedings*, Second International Conference on Technological Advances in Podded Propulsion, Brest, France, October

Computational Predictions vs. Model Testing for a High Speed Vessel with Lifting Bodies

John P. Hackett[1] (LM), Jessica Calix St. Pierre[1] (AM), Christopher Bigler[1] (AM), Todd J. Peltzer[2] (M), Frans Quadvlieg[3] (M), Frans van Walree[3] (M)

ABSTRACT

The need for higher performing naval ships is spurring interest in non-traditional hull forms. As these tend to be more complex than conventional hull designs, the use of computational techniques to analyze hydrodynamic performance in early stages of design is a key to shortening the design spiral. This was the approach used for the 2,000 t Composite High Speed Vessel project, which incorporated lifting body technology to achieve improvements in resistance and seakeeping. Multiple computational studies, using a variety of software tools, were completed in the hydrodynamic design process. These were followed by an extensive model testing program to validate and extend the computational results. The purpose of this paper is to compare the results obtained from the computational analyses with the model test data.

KEY WORDS: Hydrodynamics; CFD; validation; CHSV; lifting bodies; hydrofoils; predictions.

INTRODUCTION

Current transformational naval philosophy of operations is demanding increased performance from naval vessels, requiring vessels to travel at higher speeds in a seaway. To meet these needs, naval architects have been pursuing non-traditional hull forms. This prompted the Office of Naval Research (ONR) to study patented lifting body technology (Loui et al. 2006) through the Composite High Speed Vessel (CHSV) program, which incorporated lifting body technology into a monohull design to achieve improvements in resistance and seakeeping. Lifting bodies are large underwater appendages with significant volume and foil-shaped, cambered cross sections designed to generate lift at speed, resulting in improved resistance and seakeeping.

Increased interest in non-traditional hull forms is posing challenges for many computational tools. Complex geometries require the naval engineering community to modify or write new software packages to accurately predict vessel performance. Due to the complexity of the geometry of lifting bodies, the CHSV program provided an opportunity to expand and extend some computational tools to handle this novel geometry and to further test tools that currently have this capability. An extensive hydrodynamic scale model testing program was performed for the project providing the opportunity for software validation. This paper concentrates on comparing the computational work performed with the model test results.

BACKGROUND

Project Methodology

The principal goal of the hydrodynamic portion of the CHSV program was to quantitatively assess the performance benefits provided by lifting body technology. This was accomplished by comparing a composite monohull design incorporating lifting body technology, the CHSV, to an equivalent, well-designed monohull, the High-Speed Vessel or HSV. Both the CHSV and HSV vessel designs began with the same baseline or parent hull form. The HSV was refined and improved to encompass all of the qualities of a well-designed monohull, and the CHSV was modified to incorporate lifting body technology. The baseline hull form used for both designs is shown in Figure 1, and the principal characteristics are provided in Table 1. For a discussion of the HSV hull design, see Hackett et al. (2006).

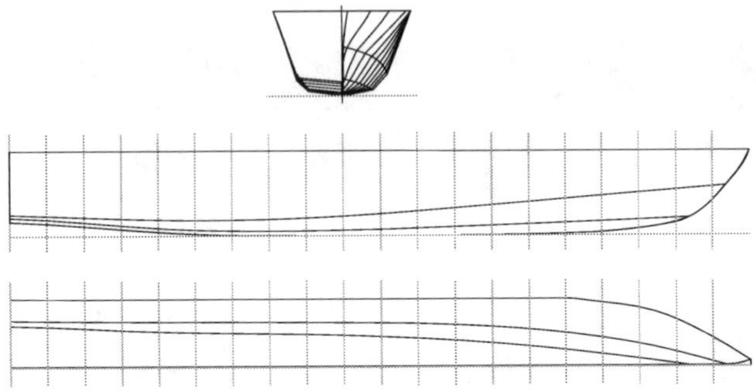

Figure 1: CHSV and HSV Baseline Hull Form

Table 1: CHSV and HSV Baseline Hull Form Characteristics

Parameter	Value
Length Overall (LOA)	89.15 m
Length on Waterline (LWL)	83.61 m
Beam Overall (BOA)	16.15 m
Beam on Waterline (BWL)	12.03 m
Draft, Design	3.61 m
Displacement, Design	2,000 t

The governing project philosophy was to perform computational studies during the design process, followed by model testing to confirm performance predictions. The goals of the model tests were to demonstrate the feasibility of the design, to observe aspects that may have been neglected by the computational studies, to generate data that could not be calculated, and to validate numerical predictions. The model test results were compared with the predictions to determine if the analytical tools used provided adequate results for a complex-shaped hull/appendage vessel such as the CHSV. Due to the extensive nature of the model test program, a substantial amount of data now exists for the validation and extension of numerical programs.

Lifting Body Technology

The general approach to designing lifting body ships blends three complementary technologies:

- Innovative hull forms,
- Lifting bodies, and
- Active control systems that provide either flight control, ride enhancement, or both.

The goal of this approach is to produce more efficient ships and boats with improved motions across the entire operational speed range, from zero to maximum, with simultaneous improvements in powering efficiency. The methods used to achieve these goals include extensive analytical and computer modeling, model testing, and significant at-sea testing using scale demonstration craft. The following definitions will facilitate the remainder of the discussion.

Lifting bodies. Lifting bodies are underwater appendages with cambered foil-shaped cross sections, that generate dynamic lift at speed. They are characterized by substantial volume, large planform areas, and low lift coefficients. Figure 2 shows computed surface pressure on three examples of lifting bodies with struts, designated the "G-Body," the "H-Body," and the blended wing body, or Blended Wing Body (BWB).

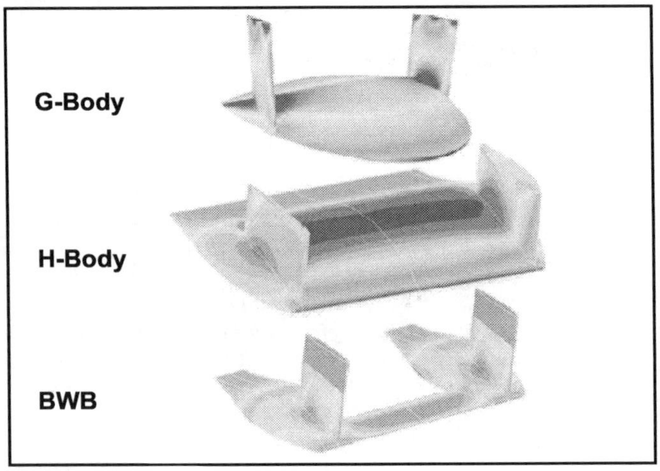

Figure 2: Examples of Lifting Bodies

Figure 3: 19.8 m *Midfoil* Hybrid Lifting Body Ship

Hybrid lifting body ships. These are ships that combine a parent hull form (monohull or multi-hull) with lifting bodies. Figure 3 shows the 19.8 m, 50.8 t technology demonstrator *Midfoil*, which is an example of a hybrid lifting body ship. It combines a "G" type lifting body with a multi-hull. The CHSV is considered a hybrid lifting body ship.

Variable immersion. A key aspect of hybrid lifting body ships is the ability to reduce the wetted surface area of the parent hull through a combination of buoyant and dynamic lift from the lifting bodies and control surfaces. This is referred to as variable immersion.

Amas. For CHSV, these are paired demi-hulls located outboard of the center hull to provide additional stability.

Sponson. This is the above the water structure, connecting the center hull to the amas or demi-hulls.

The potential benefits offered by innovative hull forms combined with lifting bodies and active control systems as integral components of a hybrid lifting body ship include improved seakindliness and seakeeping, increased maximum speed, and increased carrying capacity of payloads (Hackett et al. 2006; Peltzer 2005).

The three examples of patented lifting bodies (Gornstein et. al 2001, Loui et. al 2006) shown in Figure 2 were the available candidates for the CHSV project. All three were proven designs, with extensive analytical and computational underpinnings and at-sea experience at large scale. A brief description of each follows.

G-Body. Developed for *Midfoil*, this lifting body is characterized by a doubly parabolic planform. Its cambered foil section provides sufficient lift to offset the sinkage generated at speed, a phenomenon experienced by all bluff bodies operating near the free surface.

H-Body. This lifting body design uses the same hydrodynamic section as the G-Body, but distributes it across a constant chord length to produce a two-dimensional cross section with highly curved side caps.

BWB. This lifting body design was created in response to a need for a cavitation-free high-speed capability (up to 50 knots) that would provide a specified lift. The result was a configuration of two transversely separated H-Bodies (referred to as lifting bodies for convenience hereafter), joined by a high aspect ratio hydrofoil (termed the cross-foil), at the leading edge. Extensive experience with this lifting body configuration was accumulated through building and testing a scale version of the lifting body (2.23 t displacement). A larger scale version (16.3 t displacement lifting body) was built and installed on the 30.5 m, 101.6 t hybrid lifting body technology demonstrator HDV-100 (Figure 4).

Figure 4: 30.5 m HDV-100 Technology Demonstrator

CHSV Design Goals

Project requirements dictated a comparison between the CHSV and HSV hull forms. The common condition for comparison between the two vessels was the displacement, which was determined to be 2,000 t to produce a militarily meaningful platform while providing an achievable, though challenging, composite structural design. The hydrodynamic design constraints are provided in Table 2 with the project seakeeping requirements outline in Table 3.

Table 2: CHSV Design Constraints

Design Constraint	Value
Threshold Design Speed	40 knots
Design Speed Objective	50 knots
Seakeeping	CHSV capable of meeting requirements in 1 sea state higher than HSV
Displacement	Nominally 2,000 t

Table 3: CHSV Seakeeping Requirements

Motion	Requirements
Roll	8° SSA
Pitch	2.5° SSA
Vertical Acceleration	Less than 0.2g SSA

Selection of the lifting body will be discussed later in the context of the overall CHSV hydrodynamic design.

Composite manufacturing considerations dictated a hard chine, developable surface hull form for both the CHSV and HSV.

Waterjet propulsors were selected for the design because of their high-speed performance and their availability in the power range required.

COMPUTATIONAL TOOLS

Multiple computational tools were used by the project team to independently predict the full-scale performance of the CHSV in order to refine and improve the design and to provide predictions of model-scale and full-scale performance to compare with model test results.

The principal computational tools used by Navatek, Ltd. for analysis of the CHSV design were USAERO and CFX.

USAERO. This is a commercially-available three-dimensional time-domain panel code with a coupled integral boundary layer designed to analyze transient aerodynamic and hydrodynamic characteristics of complex configurations in arbitrary motion, including nonlinear ship free surface simulations. This tool was used for a variety of purposes during the design process including lift and resistance predictions of the lifting bodies and of the entire CHSV configuration, and to make resistance predictions of the model both prior to and after the model testing program.

CFX. This is a commercially-available three-dimensional time-domain Reynolds Averaged Navier-Stokes (RANS) code. Some of the analysis capabilities of CFX include steady-state and transient flows; laminar and turbulent flows; subsonic, transonic and supersonic flows; buoyancy; transport of non-reacting scalar components; multiphase flows (including cavitation); flows in multiple frames of reference; and particle tracking. This tool was used for a variety of purposes, principally for analysis of flow interactions between the hull, lifting bodies, and struts; it was also used to make resistance predictions of the model after the model testing program.

These complementary CFD programs provide a broad range of capabilities with respect to predicting the hydrodynamic performance of hybrid lifting body ships. An extensive, and continuing, series of validation tests of these tools has provided confidence in their use (Peltzer 2005).

The computational tools used by the Maritime Research Institute the Netherlands (MARIN) in their analysis of the CHSV included the following:

HYDRES. This is a semi-empirical method for hydrofoil craft design and analysis for steady flow conditions. It combines a lifting line method for lifting surfaces with a database for hull force components based on the Series 65 model test data. This tool is used to calculate the powering requirements and steady state equilibrium positions of the CHSV in calm water.

PANSHIP. This is a time-domain simulation method used to investigate the performance of ships. It is based on a panel method that accounts for linearized free surface effects through the transient free surface Green's function. The method is described by Van Walree (1999, 2002). This tool is used to calculate the performance of the CHSV in waves. The most 'interesting' wave conditions for model testing and settings for the ride control system are selected with this tool.

FREDYN. This is a semi non-linear time-domain simulation tool capable of computing the motion of a ship in a seaway in six degrees of freedom. Additionally, it is capable of studying the performance of ships in severe sea states, up to the point of capsizing and extreme motions because FREDYN predicts the non-linear wave excitation for the actual submerged hull portions. Details of this program are discussed by De Kat (1998). This tool is used for a final evaluation of the maneuvering and course keeping properties of CHSV at all operational speeds. The mathematical model of CHSV in FREDYN is custom-made based on the captive tests.

Maneuvering Simulation. This is a time-domain maneuvering simulation code based on hydrodynamic coefficients obtained from PANSHIP or PMM test results. This tool is used in the pre-design stage of the project to evaluate whether the selected steering gear would be adequate.

CHSV HYDRODYNAMIC DESIGN

The fundamental concept for the CHSV was to create a hybrid lifting body ship capable of meeting the project design goals using a high-performance monohull as the parent hull and an appropriate lifting body system. The main steps in the design process, some of which were performed in parallel, were as follows:

- Select lifting body ship configuration
- Design lifting body
- Adapt parent hull
- Size and place lifting bodies
- Refine design to improve performance

These steps will be discussed in the following sections.

Lifting Body Design

Of the three candidate lifting bodies discussed earlier, the BWB (Figure 2) was selected as the baseline lifting body for the CHSV based on two principal factors:

Cavitation-free operation at high speed. The BWB was designed to operate free of cavitation at speeds up to 50 knots. This was confirmed using CFD and also using model testing in a cavitation tunnel.

Compatibility with monohull geometry. Of the three candidates, the BWB was best suited for integration with a monohull. Both the G-Body and H-Body would cause a navigational draft penalty when placed below the keel of a monohull. The BWB, by virtue of the separation between its lifting bodies and the relatively thin cross foil, could be "tucked up" closer to the hull than either the G-Body or the H-Body.

The design of the BWB assemblies went through a series of refinements as the overall CHSV design evolved. The starting point for the design was the result of a previous numerical optimization study to improve hydrodynamic performance of the BWB. Included in the resulting changes to the geometry is a taper in the planform of the cross-foil from the centerline to the intersection with the lifting bodies, and also a slight twist along the span of the cross-foils that results in a spanwise variation in the angle of attack. The cross-foil is also at a slight positive angle of attack relative to the lifting bodies. Further modifications for the CHSV application included changes to the section shape of the cross-foils and several changes in the cross-foil dimensions, some of which were in response to structural requirements. The final cross-foil design, employing a proprietary section with a parabolic leading edge, was 28 cm thick with a maximum chord length of 3.4 m, yielding a thickness-to-chord ratio of approximately 8%. The final BWB geometry is shown in Figure 5. Although camber on the lifting bodies is not distinguishable by the eye in the drawing, it is present in the geometry. All changes in BWB geometry were evaluated using CFD to verify satisfactory cavitation performance and lifting capability.

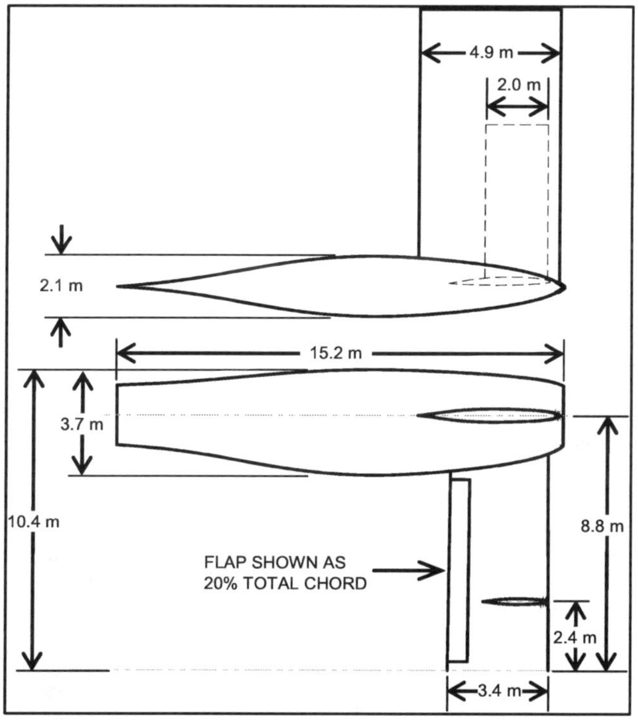

Figure 5: Final BWB Geometry

Lifting Body Ship Configuration

In developing a lifting body arrangement for the CHSV, the overall design objective was to provide benefits to ship operations through increased lift-to-drag ratio, and improved motions in a seaway at all speeds. Other design considerations included waterjet inlet ventilation, lifting body support structure, ship static stability at speed, and enhanced performance in the 30-40 knot speed range.

The starting point in selecting the lifting body ship configuration for the CHSV was a previous concept design (Figure 6) that had similar design objectives (Hackett et al. 2006). This configuration used a deep-vee hull and two BWB assemblies, one located forward of amidships and one located near the stern.

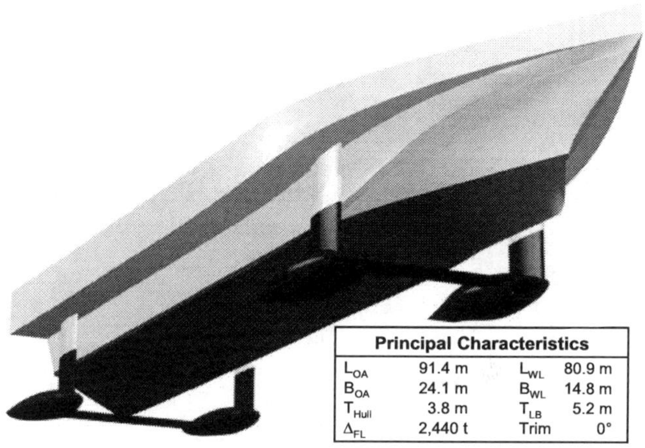

Figure 6: Notional CHSV Lifting Body Configuration

Propulsion for the CHSV included twin LM-2500+ gas turbine engines, each driving two outboard KaMeWa 200SII waterjets, and one MTU 20V diesel driving a KaMeWa 160SII waterjet on centerline waterjet.

Parent Hull Adaptation

The baseline hull form chosen for the CHSV was a twin-chine hull with developable surfaces and a transom depth sufficient to accommodate the selected waterjets (Figure 1). Three variants of the baseline hull geometry were developed in the course of the design process. Refinements resulted from concerns about possible waterjet ventilation due to the presence of the aft lifting body, which affects transom immersion. For the first variant, rocker was removed from the maximum section aft, resulting in a flat keel line. The section shape was held constant from the maximum section to the transom to develop a no rocker variant. While this addressed the waterjet ventilation issue, the BWBs would need to be deeper in the water resulting in a significantly larger navigational draft. This led to the second and third variants. A compromise was established between the initial full rocker hull and the no rocker variant resulting in a hull form where partial rocker was restored in the aft portion of the hull. Additionally, a finer entry was given to the bow.

Figure 7 shows the profile views of the original hull form at the top, the first variant in the middle, and the final variant at the bottom. This design modification was based on analyses performed using USAERO. Table 1 gives the principal characteristics of the baseline parent hull; Table 4 gives those of the final model tested hull.

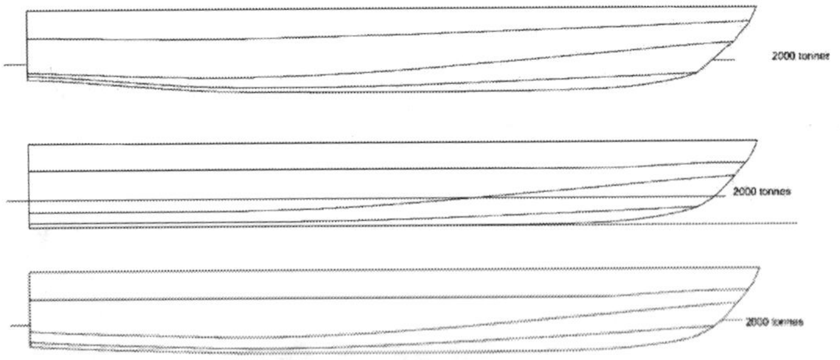

Figure 7: CHSV Hull Form Variants (Original (top), First Variant (middle), Final Variant (bottom))

Table 4: CHSV Final Principal Characteristics Inclusive of BWBs

Parameter	Value
Length Overall (LOA)	90.60 m
Length on Waterline (LWL)	74.18 m
Beam Overall (BOA)	20.87 m
Beam on Waterline (BWL)	19.12 m
Draft, Design	3.68 m
Displacement, Design	1,995 t

Lifting Body Sizing and Placement

Initial lifting body sizing was accomplished by targeting a total lift at maximum design speed such that roughly half the mass of the ship would be supported by the lifting bodies. This was adjusted in response to the hull geometry changes noted in the previous section, with the end result a static displacement for each BWB assembly of 147.25 t.

To determine the longitudinal placement of the BWB assemblies, the center of dynamic lift of the parent monohull alone at various operating conditions (speed, trim) was determined, and then the resulting trimming moment was calculated. This data, along with the known lift characteristics of the BWB, was used to compute the required moment arm to balance the trimming moment. Using this procedure, several BWB configurations were selected for further analysis. Figure 8 depicts the BWB longitudinal positions studied and Table 5 shows the complete list of BWB configurations studied. The final configuration is shown in heavy lines in Figure 8.

Using USAERO, the lift and drag for each selected configuration was computed at the design speed of 40 knots with the parent hull fixed at zero trim (but free to heave to balance the displacement), and the forward BWB assembly fixed at 2.0 degrees angle of attack with respect to the hull. To simplify the analysis, the intermediate struts were not included at this stage. The aft BWB assembly was allowed to rotate freely to balance the moment about the longitudinal center of gravity (preserving the zero trim condition of the parent hull). The goal was to find the arrangement with the highest lift-to-drag ratio. The resulting positions of the two BWB assemblies were 2.8 m and 48.5 m forward of the transom, measured to the leading edge of each BWB. See Hackett et al. (2006) for a more detailed discussion of the BWB placement optimization.

Table 5: BWB Configurations Studied

	Long'l Location		Angle of Attack		
Version	AFT	FWD	AFT	FWD	Lift/Drag
1	5.1 m	54.6 m	N/A	2.0°	Note 1
2	5.1 m	53.0 m	N/A	2.0°	Note 1
3	5.1 m	51.5 m	N/A	2.0°	Note 1
4	5.1 m	50.0 m	N/A	2.0°	Note 1
5	5.1 m	48.5 m	3.66°	2.0°	12.29
7	2.8 m	54.6 m	3.22°	2.0°	13.36
8	2.8 m	51.5 m	2.52°	2.0°	13.92
9	2.8 m	46.9 m	0.96°	2.0°	15.30
10	2.8 m	48.5 m	0.38°	2.0°	16.09
11	3.1 m	48.5 m	0.53°	2.0°	15.86
10a	2.8 m	48.5 m	0.68°	2.0°	16.37

Note 1: N/A, aft body could not balance moments

Design Refinement

A number of refinements were accomplished during the CHSV hydrodynamic design spiral; two specific refinements will be discussed here: BWB strut alignment, and ama sizing and placement.

As noted earlier, intermediate struts were not included in the analysis of lifting body placement since their intersection with the hull changed as the BWBs were moved longitudinally and also since the aft BWB was usually free to rotate to change the angle of attack.

These factors made inclusion of the intermediate struts in the USAERO model impractical. Consequently, the flow interactions between the hull and the BWBs were not explicitly studied in the early design stages.

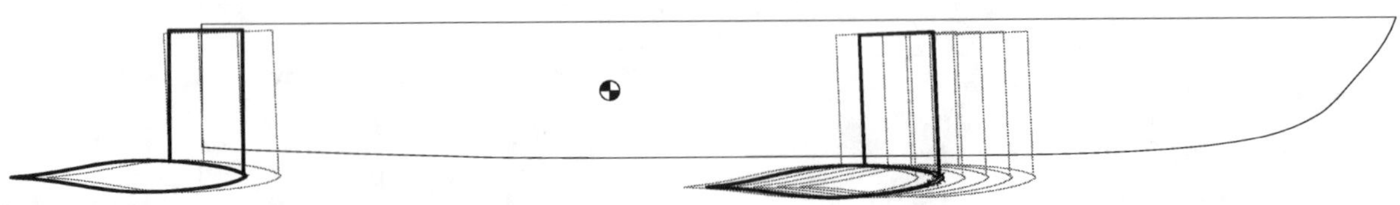

Figure 8: BWB Longitudinal Positions Studied

In preparation for the model testing program, however, analysis of these flow interactions was deemed essential to ensure the best hydrodynamic arrangement.

Based on previous work, it was known that the horizontal hydrodynamic forces acting on the vertical struts due to hull-induced cross flow could be significant. The induced drag associated with these side loads was of particular concern since it would degrade the overall hydrodynamic performance. Better alignment of the struts to the flow was expected to significantly reduce this induced drag. This could be accomplished by giving the struts a variable span-wise twist, since the cross flow is a function of depth, or the struts could be rotated uniformly to align with the average flow. The latter approach was chosen in part to simplify model construction.

The following process, using computational methods, was followed to determine the optimum strut alignment configuration:

1. Examine the flow field around the struts at design speed and find a mean apparent angle of attack for each strut.
2. Using this angle of attack as an initial estimate, rotate each strut about the leading edge in an attempt to better align the strut with the flow.
3. Analyze the resulting geometry using CFX and determine the new side loads.
4. Assuming a linear relationship between lift (side load, F) and strut angle of attack (α), calculate the slope of α vs. F using the data from Step (3). Calculate a new strut angle of attack to reduce the side loads.
5. Repeat Steps (3) and (4) as necessary to minimize the side loads for all struts.

Several iterations of strut alignments were analyzed at full scale using CFX. Figure 9 plots the resulting predicted side loads as a function of speed for the inner struts for the baseline and final configurations. Similar results were obtained for the main struts (Hackett et al. 2006).

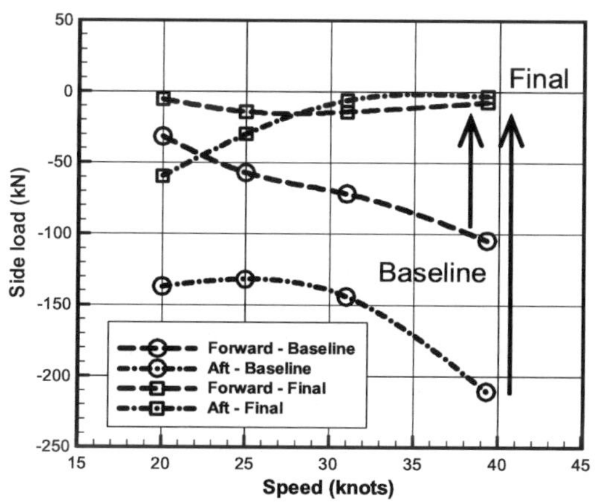

Figure 9: Inner Strut Side Loads, Baseline and Final Configurations

Figure 10 shows an example of the computed streamlines around the outboard struts, and Figure 11 shows an example of the pressure distribution on the lifting bodies, cross foils, and struts. The hull is omitted from this image for clarity, but it was included in the analysis. The improvement in the final configuration is quite dramatic: the side loads have been essentially eliminated.

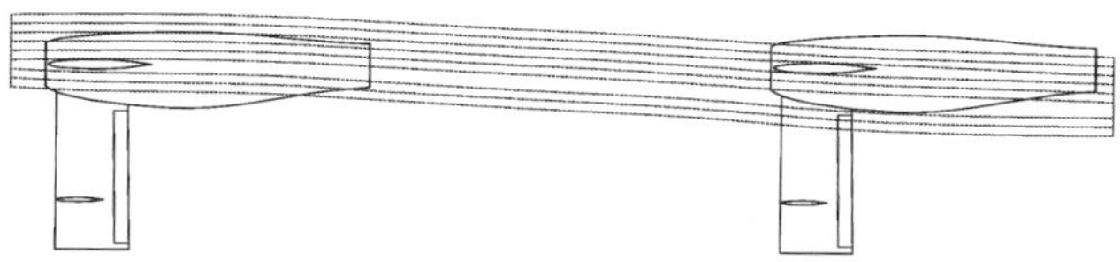

Figure 10: Example of Computed Streamlines Around Main Struts

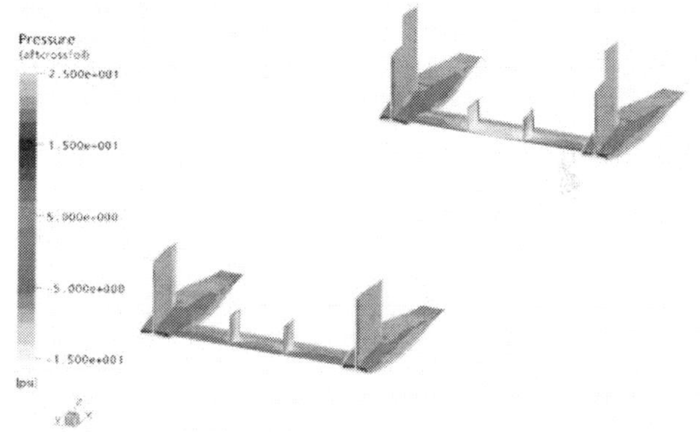

Figure 11: Example of Computed Pressure Distribution on Lifting Bodies, Cross-Foils, and Struts

The sizing and placement of the four amas, which provide additional waterplane area for static stability as well as structural attachment points for the main BWB struts, were also refined during the course of the CHSV hydrodynamic design spiral.

Of particular concern was the vertical placement of the forward amas, for which CFD analysis showed there would be significant wetting at all speeds analyzed (25 to 50 knots). Figure 12 illustrates the results of the analysis at 31 knots. The upper image gives a profile view showing the shape of the dynamic waterline along the hull at this speed with the outline of the amas superimposed; the amas themselves were not included in this particular series of USAERO runs. The lower image gives a section view through the forward ama showing similar information. Based on this data, it was recommended that the forward ama be raised by 1.6 m so that it would remain unwetted at speeds above 25 knots. Subsequent USAERO analysis of the new ama location indicated this would succeed. However, during model testing, the forward amas were removed because of their interaction with the spray from the struts.

MODEL TESTING PROGRAM

An extensive model test program was performed on the CHSV. The complexity of the CHSV geometry and the extensive nature of the test program provided many challenges. One of the major challenges was the scaling of the model. The only way to perform this task without scale effects present is to satisfy Froude and Reynolds number similarity simultaneously, which is impossible with a scale model. This is generally not an issue for typical monohulls because enough model testing has been performed and compared to full-scale vessels trials to permit adequate correlation allowances to be established accounting for these effects. However, this issue is more of a challenge for hydrofoil and lifting body vessels. The Reynolds number of a hydrofoil model is smaller than at full scale causing the laminar to turbulent transition point to be different at model scale than at full scale and also causing the lift developed at model scale to vary from the actual lift developed at full scale. While it is possible to use artificial boundary layer tripping methods to address this issue, the transition point from laminar to turbulent flow on a lifting body does not typically occur at the leading edge. Depending on the lifting body shape, the transition point could be located 20 to 50 percent of the chord length aft of the leading edge. Additionally, the low Reynold's number of the lifting body at model scale favors the re-establishment of laminar flow aft of where the artificial boundary layer tripping method has been established. This means that scale effects would be present during testing requiring that the magnitude of the effects be calculated and accounted for during the testing and the extrapolation of results. Additionally, a significant amount of measured data was necessary to validate numerical results and to develop a strong understanding of the performance differences between the CHSV and HSV. This required extensive instrumentation on the model, some of which demanded special considerations in the installation and in the data extrapolation due to the CHSV geometric complexity. The complete configuration of the final CHSV design is shown in Figure 13.

Model Construction

One model was constructed for all model testing performed on the CHSV. This required a careful balance and considerable engineering. The same model was used to study resistance, powering, cavitation, and free sailing seakeeping and maneuvering. During selected seakeeping tests, structural loads were measured. The model and its instrumentation package were designed to measure slamming loads, BWB lift, and midship bending.

Scaling the CHSV model was a challenge for three main reasons. First and typical of naval architectural hydrodynamic model tests, the CHSV was scaled in accordance with Froude's law with adjustments made to account for viscous effects. In order to reduce scale effects, the model needed to be constructed as large as possible. Second, the model needed to be scaled for both the resistance and the seakeeping testing simultaneously. For resistance, the major concerns are depth and wall effects. For seakeeping, the wavemaking capability in the basin, the maximum speed in the basin, the maximum sea state to be tested, and the vessel's mass properties usually require a relatively small light model. The third challenge was the scaling of the propulsion system. The CHSV waterjets needed to be scaled so that the required thrust would be developed in order to meet the necessary speed. Based on these considerations a 1:18 scale factor was selected.

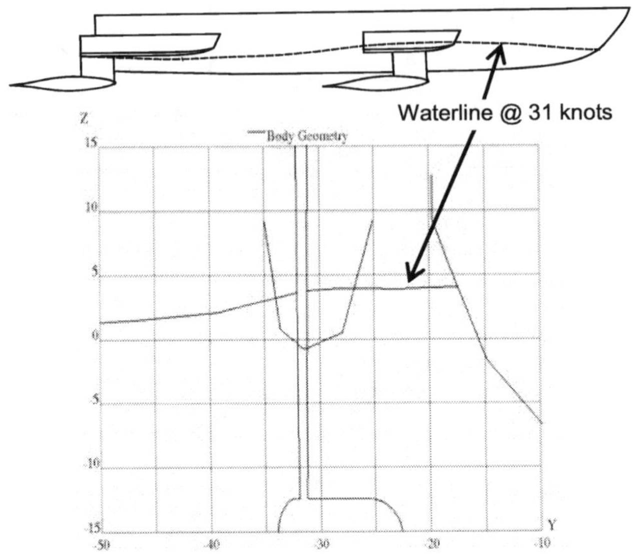

Figure 12: Forward Ama Vertical Placement at 31 knots

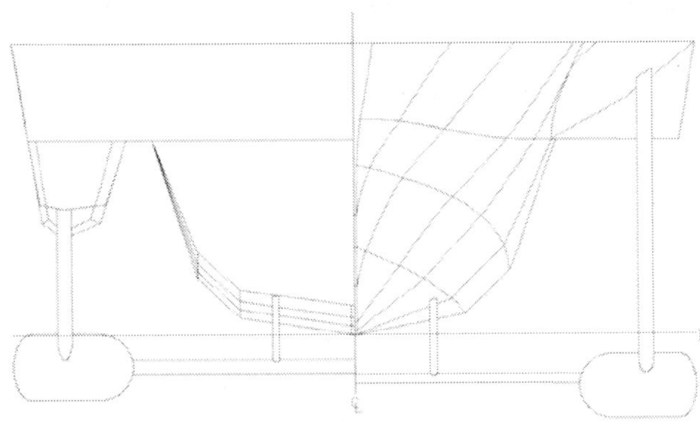

Figure 13: CHSV Final Design Configuration

Since the CHSV was to be a large vessel constructed of composite materials with a unique hull form and appendages, accurate data on the structural loads was desired and thought to be imperative. To develop the required results, a segmented model with a "back bone" beam incorporated in the model connecting the two halves was necessary. The beam was designed to have the appropriate stiffness and to have the correct 2-node bending mode. The bending moment was determined by measuring the deformation of the beam. Additionally, accelerations were measured in the fore and aft halves of the model. The inertial forces imposed on each half of the model were determined from the measured accelerations and the weight of the model. These inertial forces and moments were subtracted from the measured values to obtain the pure hydrodynamic structural forces and moments.

In order to guarantee model stability at speed, it was necessary to install an active ride control system on the model. Two controllable flaps were incorporated on the cross-foils of both BWB systems, and vertical controllable flaps (rudders) were installed on the aft outboard struts. The three waterjets were equipped with steerable and reversing buckets to increase the maneuverability of the model. Additionally, both BWBs were mounted on force transducers to determine the loads acting on each BWB system. Figure 14 shows the completed CHSV model.

Figure 14: CHSV Model

Seakeeping

For the seakeeping testing, the CHSV model was instrumented to measure wave heights along the hull, accelerations at specific points, and structural loads. In addition, CHSV motions were measured in all six-degrees of freedom. Relative wave probes were located at three positions along the length of the model to measure the wave heights. Pressure transducers were installed at the bow and at the forward sponson to measure slamming loads. An accelerometer was placed inside the model to provide acceleration values at specific locations. Six-component load cells were installed on the BWB systems between the outboard struts, providing loads acting on both the forward and aft BWB systems. The ride control system and the steering buckets on the waterjets were active throughout the seakeeping testing to simulate the actual motions of the CHSV. Motions and track were measured by an optical KRYPTON tracking system.

An initial set of seakeeping tests preceded the full seakeeping program. These tests were designed to validate seakeeping predictions and to determine whether or not there would be an issue with waterjet inlet ventilation. This was a concern because the BWB systems develop a significant amount of lift reducing the draft of the vessel (the "variable immersion" concept discussed earlier). When combined with the motions of the vessel in a seaway and heel associated with the CHSV banking while maneuvering it was conceivable that air could enter the waterjet inlets. During these tests, underwater cameras were positioned to observe any ventilation of the inlets, and the thrust produced by the waterjets was measured to determine the effect of such ventilation on the propulsion system. Figure 15 shows sample underwater photographs of these observations. The top photograph shows initial ventilation observations, and the bottom photograph shows ventilation observations once the forward amas were removed. Removing the forward amas from the model significantly reduced the observed ventilation.

A full CHSV seakeeping test program was performed including tests in regular seas and in irregular seas. All tests were performed with the model in rigid and segmented modes. Table 6 outlines the tests performed.

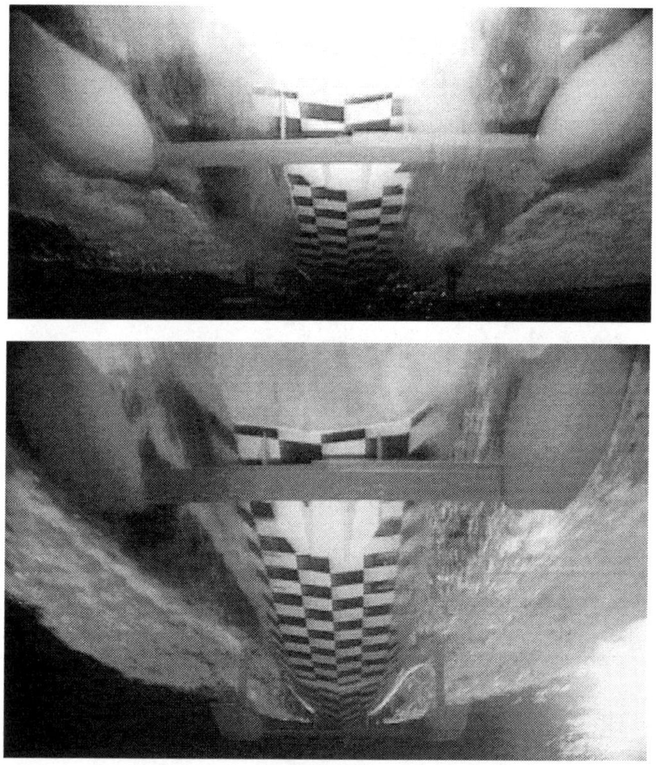

Figure 15: Underwater Observations of Ventilation – With Forward Amas (top), Without Forward Amas (bottom)

Some additional captive seakeeping tests were performed in order to measure foil-to-foil interaction and to measure wave excitation forces.

Maneuvering

The maneuvering testing for the CHSV consisted of both free sailing and an expanded Planar Motion Mechanism (PMM) testing program. For free sailing, the model was instrumented as in the seakeeping testing and outfitted with reversing buckets on the waterjets for crash stop testing. The free sailing tests consisted of zig-zag tests, turning circle/pull out tests, turn-on-the-spot tests, and crash stop tests. Tests were performed at several speeds. The results of these tests were used not only to quantify the steering behavior of the vessel, but also to generate validation material for the calculations. This validation material was used to tune the FREDYN model.

Table 6: Seakeeping Test Program

Irregular Waves – Pierson Moskowitz Spectrum	
Headings	Head, bow quartering, beam, and stern quartering seas
Sea States	4 (Hs = 1.88m, T = 8.8s), 5 (Hs = 3.25m, T = 9.7s), and 6 (Hs = 5m, T = 12.4s)
Speeds	20, 30, and 40 knots
Regular Waves	
Headings	Head, bow quartering, and stern quartering seas
Wave height	2.0 meters
Speeds	20, 30, and 40 knots

PMM tests, or captive tests, were carried out for several reasons. These tests enabled measurement of the total forces and moments acting on the vessel for a number of equilibrium positions. These forces and moments could be used to directly compare to the forces and moments calculated using CFD and other computations. A number of tests were carried out in the captive condition:

- PMM tests, consisting of pure drift, pure yaw, and combined drift and yaw tests. Tests are carried out at several trim and heel angles.
- Foil interaction tests
- Waterjet steering force tests
- Measurement of excitation forces by the ride control system
- Effect of heeling and trimming
- Wave excitation tests
- Resistance

During these tests, the forces on the forward and aft BWBs were measured separately so that foil-to-foil interactions and foil-to-hull interactions could be measured. Knowledge of these contributions is of particular importance in understanding the correspondence between numerical predictions and the model-scale measurements.

Based on the captive experiments, a theoretical/empirical model of the vessel was created. This mathematical model was implemented in a subroutine of FREDYN. The FREDYN model was then used to further study the motion behavior of the vessel. With this simulation model, a variety of maneuvers in calm water and in waves were studied. Additionally, the impact of various changes to the ride control system was studied and quantified. Figure 16 is taken from the visualization of one of these maneuvers.

The maneuvering model that was implemented in FREDYN was unique in the following ways. During the PMM model tests, the non-linear coefficients were obtained along with the linear coefficients allowing the model to be fully non-linear. Furthermore, it was important that the mathematical model be free to move in six degrees of freedom because for high speed craft the directional stability is largely influenced by the trim of the vessel. The mathematical model should therefore be run for a range of trim angles. The results presented in Figures 43 and 44 are just a small collection of input that is necessary for the mathematical model in FREDYN.

Figure 16: Simulation of maneuvers with CHSV

Calm Water Testing

Most of the calm water testing was conducted in the Deep Water Basin (252 m x 10.5 m x 5.5 m). Calm water testing began with paint streak tests to verify the flow along the hull, BWB systems, and additional appendages. These tests were performed in the design displacement condition at speeds of 20 and 40 knots. Figure 17 shows the results of the paint streak test at 20 knots.

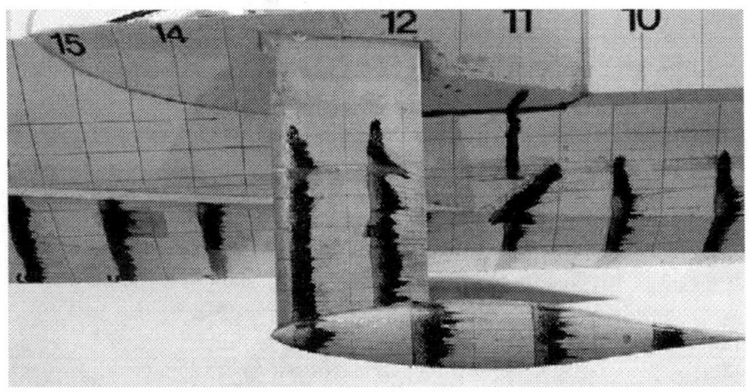

Figure 17: Paint Streak Test Results at 20 knots

Foil angle optimization tests were performed to determine the cross-foil control flap angle that produced the least amount of drag. In order to perform these tests, the model was connected to the carriage by an air-lubricated rod, which provided the friction correction force, applied along the axis of the waterjet stream and connecting to the model aft of the CHSV's longitudinal center of gravity. Eight flap angles were investigated in the appended resistance test at the design displacement. Four flap angles were tested in a short set of self-propulsion tests at design displacement. From these tests, the optimum flap angles were chosen for the forward and aft BWB systems.

Bare hull resistance, appended resistance, and self-propulsion tests were performed on the CHSV. In addition to the standard instrumentation suite, three wave probes were mounted in the towing tank to take longitudinal wave cuts at distances of 1, 2, and 3 meters, at model scale, off the centerline course of the model. Table 7 outlines the resistance testing that was performed and Figure 18 shows a photograph of the model during testing.

Table 7: Calm Water Testing Program

Test	Condition	Speed Range
Bare hull resistance	3 displacements	5.5 to 50 knots at 1.5 knot increments
Appended hull resistance	5 displacements	5.5 to 50 knots at 1.5 knot increments
Self-propulsion	Design displacement	5.5 to 50 knots at 1.5 knot increments
Self-propulsion	4 additional displacements	5.5 to 50 knots at 3.0 knot increments
Cavitation observations	Design displacement	40 knots

Figure 18: Resistance Test at 40 knots (full scale)

At the conclusion of the standard appended resistance tests, a special resistance test/wake survey was performed measuring the three-dimensional wake behind the vessel and lifting bodies. This was performed by mounting a 5-hole pitot tube in the slip stream behind the centerline hull and the lifting bodies. Figure 19 shows this setup and Figure 20 provides an example of the transverse wake results.

Figure 19: Five-Hole Pitot Tube Setup

Additionally, resistance testing was performed in the depressurized towing tank (240 m x 18 m x 8 m). The purpose of this testing was to analyze any scale effects on the resistance measurements due to the basin size. The depressurized tank is deeper and wider than the deep water basin. In addition, the depressurized towing tank was used to make observations for any cavitation on the BWB system, particularly on the cross-foils. The depressurized towing tank allowed the rather significant wave train of the CHSV to influence the flow and pressure fields around and associated with the BWBs, while running at a reduced ambient tank pressure, making a more realistic condition for cavitation observations. There was no cavitation observed in the depressurized towing tank.

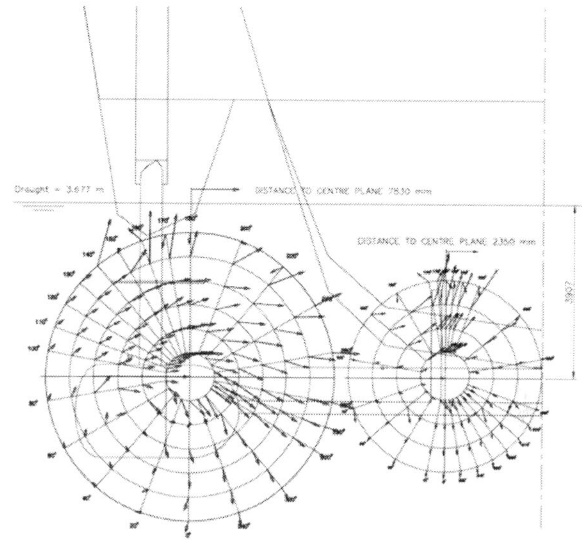

Figure 20: Example of Wake survey Results

RESISTANCE COMPARISON

Resistance predictions of the full-scale CHSV were performed using computational methods as an integral part of the hydrodynamic design process. Figure 21 shows the results of USAERO analyses of the full-scale design in calm water at four specific speeds from 25 to 50 knots. The plot shows both resistance (solid line) and lift-to-drag ratio (dashed line) as a function of speed. Note that these results include no margin or allowance for roughness (frictional effects), air resistance, etc.

Prior to conducting the model test program, USAERO analyses were also performed at model scale, with the model tank cross-sectional geometry held constant along the length of the tank. The intent was to "model the model test" in an effort to avoid scaling issues. Figure 22 shows the results of these model scale predictions for three specific speeds from 30 to 50 knots (full-scale equivalent); as in Figure 21, both resistance (solid line) and lift-to-drag ratio (dashed line) are plotted as a function of speed. The model scale speed in m/s is shown on the lower *x*-axis and the equivalent full-scale speed in knots is shown on the upper *x*-axis. Again, these results include no margin or allowance for roughness, air resistance, etc.

Some resistance predictions were also completed using PANSHIP. PANSHIP was originally developed to evaluate the seakeeping characteristics of ships with lifting surfaces. However, it is often necessary to obtain an initial assessment of resistance and maneuvering/course keeping characteristics early in the design. For the present case PANSHIP was extended with relatively simple viscous force contributions and a ride control/autopilot system, so that resistance and maneuvering/course keeping could be studied in addition to seakeeping.

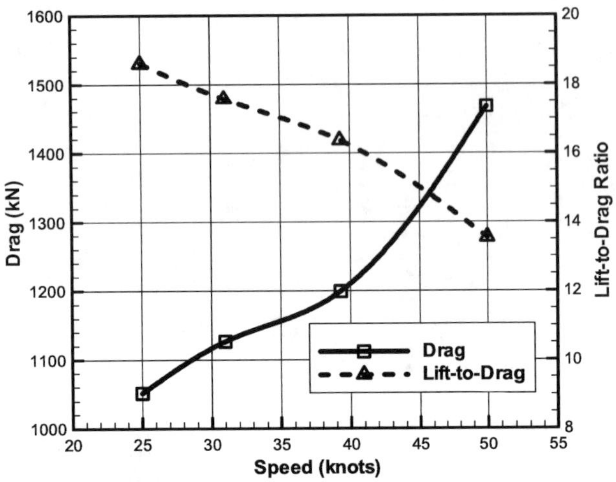

Figure 21: USAERO Predictions of Full-Scale Drag and Lift-to-Drag Ratio

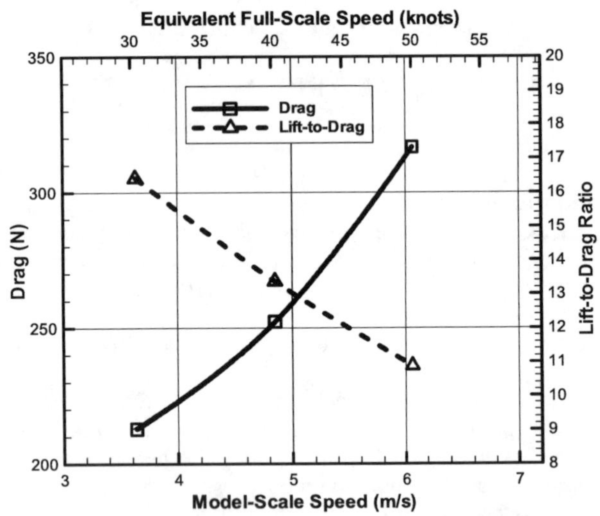

Figure 22: USAERO Predictions of Model-Scale Drag and Lift-to-Drag Ratio

PANSHIP simulations were performed to obtain the rise of CG, trim, and resistance for the baseline flap angle setting. Figures 23 through 25 provide a comparison between the experimental and PANSHIP results.

The trim and rise of CG are reasonably well predicted. The absolute values may differ, but the correct trends with speed are noted. The resistance coefficient is over-predicted at low speed, probably since PANSHIP assumes a fully ventilated transom. At higher speeds, the resistance prediction seems to be adequate for a preliminary design stage estimate. The presented resistance coefficient is the total resistance divided by ½ ρV^2 and the wetted area.

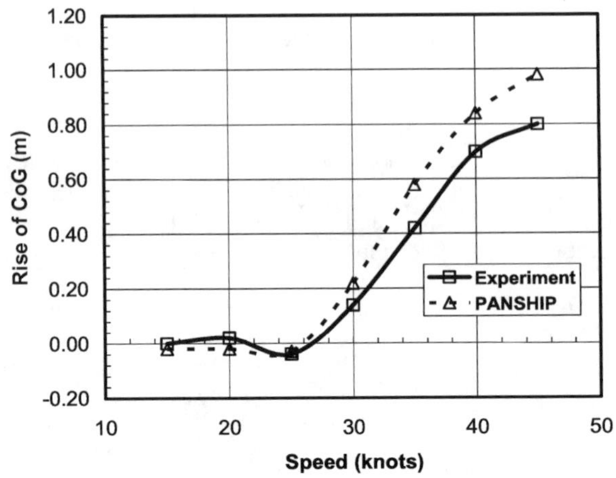

Figure 23: Rise of center of gravity, self-propelled

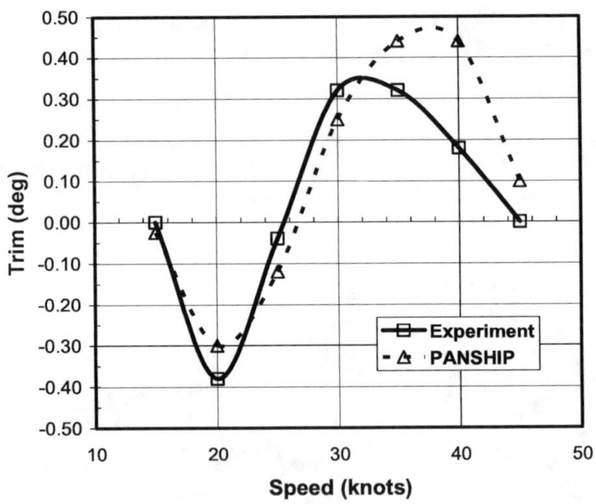

Figure 24: Trim angle versus speed

COMPARISON OF FORCES BASED ON CAPTIVE RESISTANCE TESTS

In order to provide a more rigorous basis for comparison between CFD predictions and model test results, a series of captive resistance tests was performed with the model constrained by a six degree of freedom force block mounted rigidly to the hull. The static draft for the initial condition at each speed was set based on estimates of the mean dynamic draft for that speed obtained from CFD predictions. Trim was fixed at zero degrees for all runs in this series. Table 8 gives the conditions for the three test runs. Figure 26 shows a photograph from the captive resistance tests.

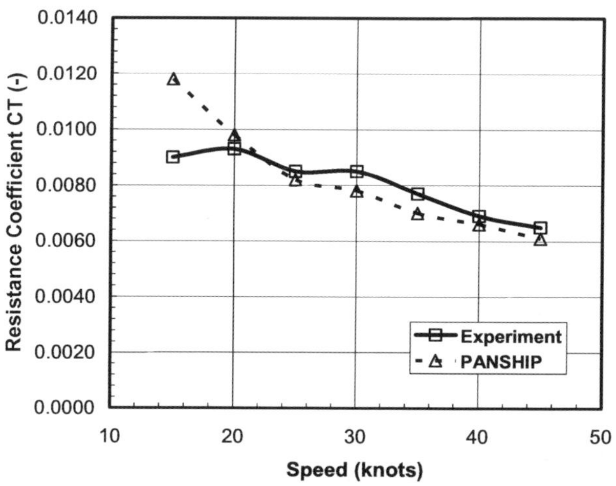

Figure 25: Resistance coefficient versus speed

Table 8: Captive Model Resistance Test Conditions

Speed (Full Scale) [kts]	Speed (Model Scale) [m/s]	Static Draft (Model Scale) [m]
31.0	3.76	0.193
40.0	4.85	0.163
50.5	6.12	0.118

Figure 26: Photograph of Captive Resistance Test at 40 knots

Given these conditions, an additional series of CFD analyses was performed to provide steady-state resistance predictions. The expectation was that these predictions should more closely match the model test data since resistance is sensitive to both draft and trim. With the captive model tests, both draft and trim are known explicitly, as described above. The CFD analyses were performed using both USAERO and CFX.

Figure 27 shows an example of the output from the USAERO analysis of the captive resistance tests. The predicted non-dimensional pressure coefficient ("CP") on the wetted hull and on the lower surface of the two BWB assemblies for the 31 knot case (top image) and for the 40 knot case (lower image) are shown for the test conditions listed in Table 8. The color scale at the right showing the magnitude of the pressure is the same for both images.

An example of an output from the CFX analysis of the captive resistance test is shown in Figure 28. This image shows a visualization of the disturbed free surface at a speed of 50.5 knots as seen from below the model.

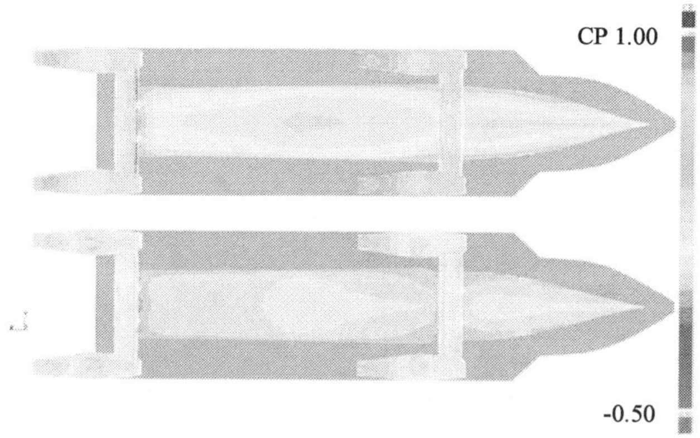

Figure 27: USAERO pressure predictions for captive resistance tests at 31 knots (top) and 40 knots (bottom)

Figure 28: Example of CFX output showing the disturbed free surface at 50.5 knots

Table 9: Captive Model Resistance Test Data vs. CFD Predictions

	Resistance (N)		
Speed (Full Scale) (kts)	Model Test	CFX	USAERO
31.0	276.8	264.0	260.6
40.0	349.3	355.3	329.8
50.5	450.6	433.8	411.0

The results of these analyses are presented in Table 9 and in Figure 29 with the model test data. Both the USAERO and the CFX data are computed based on fully turbulent flow. The agreement between the CFD predictions and the model test data is good. On average, the CFX predictions are about 2% lower than the model test results, ranging from about 5% lower to about 2% higher, while the USAERO predictions are on average about 7% lower than the model test results, ranging from about 6% lower to about 9% lower.

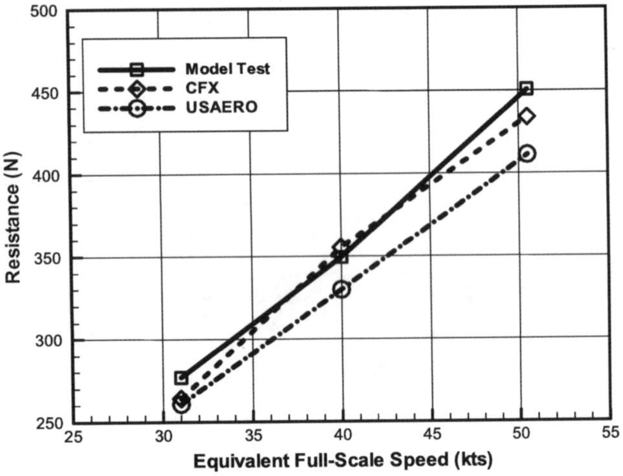

Figure 29: Captive Model Resistance Test Data vs. CFD Predictions

COMPARISON OF FORCES BASED ON PMM TESTS

Predictions determined from PANSHIP simulations were also compared to the model test results. Figure 30 shows an example of the PANSHIP output. This figure highlights the suction areas on the hull above the lifting body systems. This corresponds well with results from the captive experiments. Figure 31 shows the CHSV model during the captive tests. The amount of instrumentation on the model can be observed.

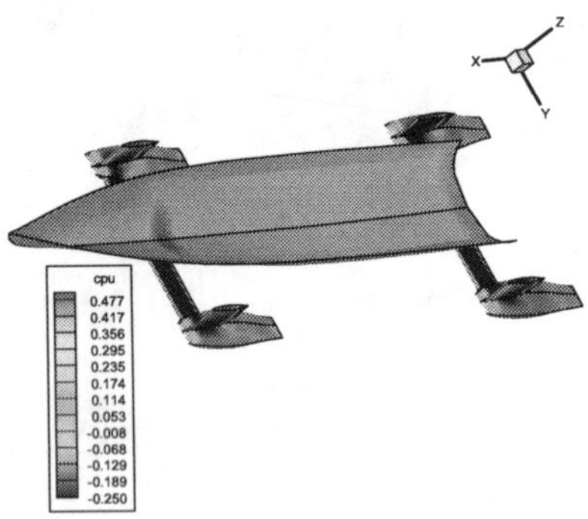

Figure 30: Pressure on hull and BWBs calculated with PANSHIP

SEAKEEPING COMPARISON

Figure 32 shows the CHSV model during the seakeeping tests. Samples of the comparisons made are included in Figures 33 through 37. These figures show the correspondence between PANSHIP predictions and model tests of the response amplitude in regular waves for five motions (surge is not presented here because it is not of importance for the determination of the operability). All simulations and model tests were performed with an active ride control system.

Figure 31: CHSV during captive tests

Figure 32: CHSV in Seakeeping Tests (Sea State 6, Beam waves at 30 knots)

These plots demonstrate that the response in regular waves is predicted quite well. Individual responses vary, but the trends are well predicted. It should be kept in mind that the roll and pitch responses are low, making it difficult to obtain a high accuracy in predictions. When considering the complexity of the calculation model, these results are acceptable.

MANEUVERING COMPARISON

Early stage maneuvering predictions were performed using PANSHIP calculations for the initial CHSV configuration at a number of drift angles and rotation rates. This procedure closely simulates a rotation arm or PMM test program. From these simulations, a mathematical model was derived. Using this mathematical model, maneuvering simulations were performed, resulting in data such as turning circle diameters, overshoot angles, and behavior due to steering action. It was calculated that the vessel would have a considerable amount of directional stability at 40 knots. Therefore, it was decided to add more steering capacity to the vessel. Trailing edge rudders were incorporated in the aft outboard struts, along with ventilation plates to prevent rudder ventilation. Figure 38 shows this configuration. Captive and free sailing model tests were carried out in this configuration. Figure 39 shows the CHSV model during a 25-knot turning circle test.

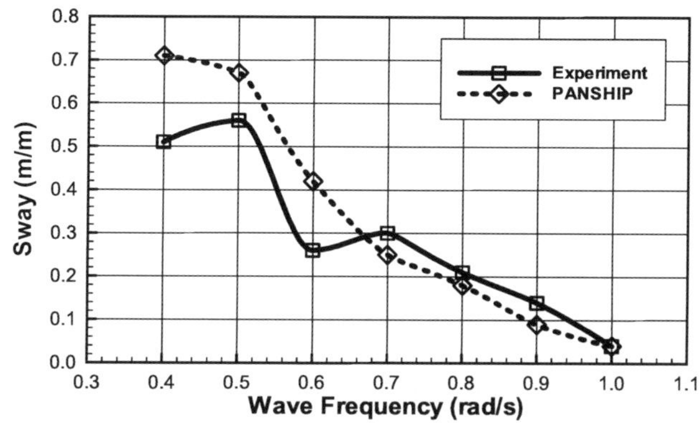

Figure 33: Sway Response in Bow Quartering Waves

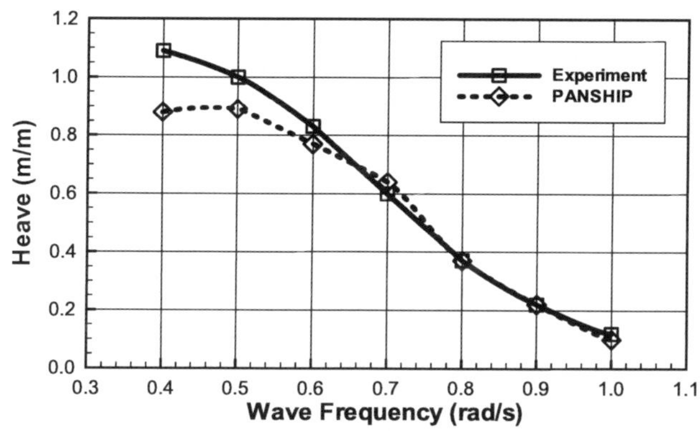

Figure 34: Heave Response in Bow Quartering Waves

Figure 40 and Figure 41 show the comparison between several predictions of the turning rate at constant rudder angle and experimental results. The following predictions were made:

- Free sailing model tests
- Initial predictions based on a simple model. This simple model is based on linear coefficients derived from PANSHIP calculations and empirical non-linear coefficients.
- FREDYN calculations, which incorporates the mathematical model, is based on the captive model tests
- PANSHIP simulations. Because PANSHIP was extended to have the capability of predicting maneuvering after project completion, these simulations were performed after the end of the test program

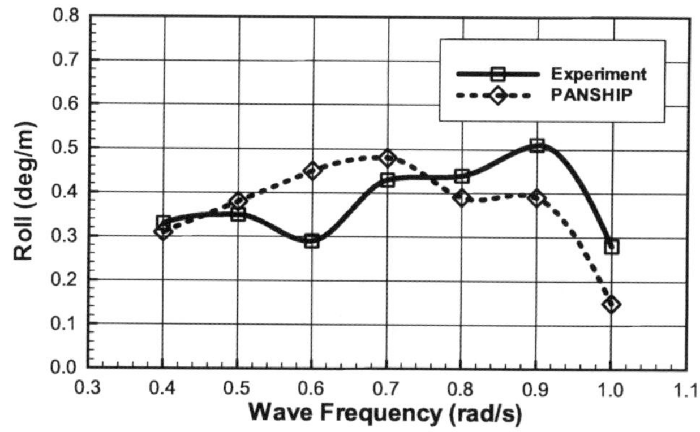

Figure 35: Roll Response in Bow Quartering Waves

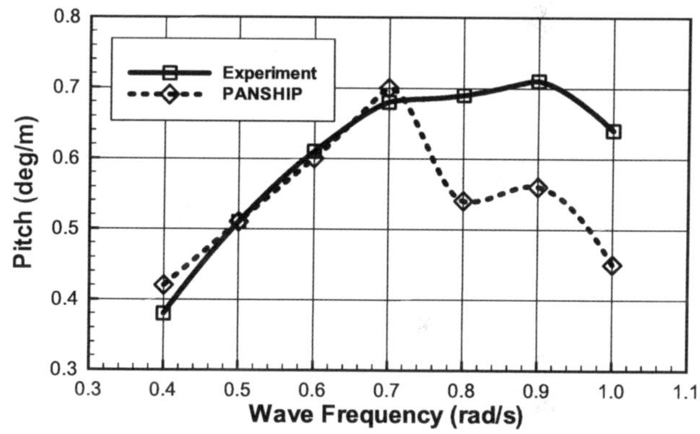

Figure 36: Pitch Response in Bow Quartering Waves

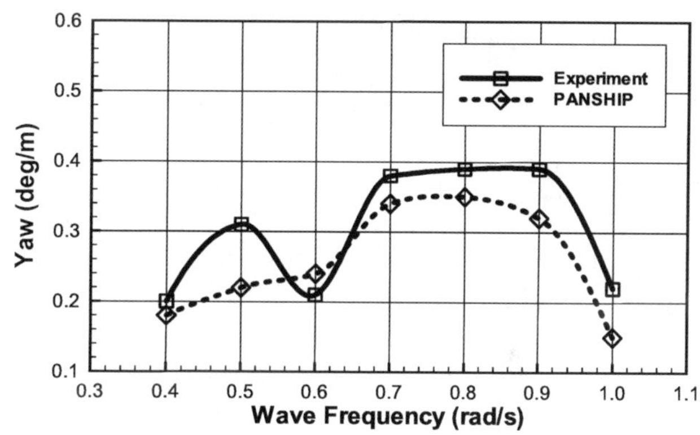

Figure 37: Yaw Response in Bow Quartering Waves

The maneuvers included in these figures were banked turns for constant rudder angles (waterjet buckets and strut rudders) at 10 and 40 knots. The non-dimensional turning rate γ is defined from the yaw rate r, forward speed U, and length L as

$$\gamma = \frac{rL}{U}$$

Figure 38: Outboard Strut Trailing Edge Rudders (top) Ventilation Plates (bottom)

Figure 39: Model during 25 knots Turning Circle Test

PANSHIP predicts a low turning rate at large rudder angles in comparison to the model test results, see Figure 40. However, for use in early design stage, this is a valuable tool. The heel angles during the banked turn are a bit over-predicted, but quite close to the dedicated FREDYN results. The FREDYN results and the model test results are on top of each other, as they should be, because the FREDYN simulations were based on the model tests. The initial calculation varies significantly from model test results at 10 knots. The turning ability is over-estimated. For the speed of 40 knots, the initial calculations seem to underestimate. However, the initial calculations were performed without the trailing edge outboard strut rudders.

During these maneuvers, the ride control system attempted to achieve a commanded heel angle φ defined by

$$\varphi = \arctan\left(\frac{-Ur}{g}\right)$$

Figure 42 shows a comparison of predicted vs. measured heel angle in degrees as a function of rudder angle.

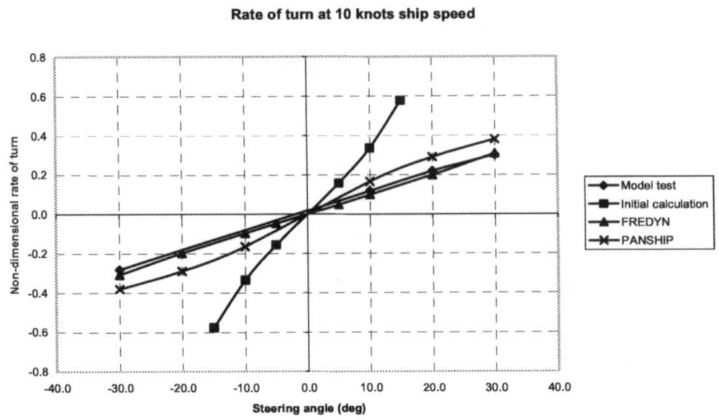

Figure 40: Non-dimensional turning rate at 10 knots

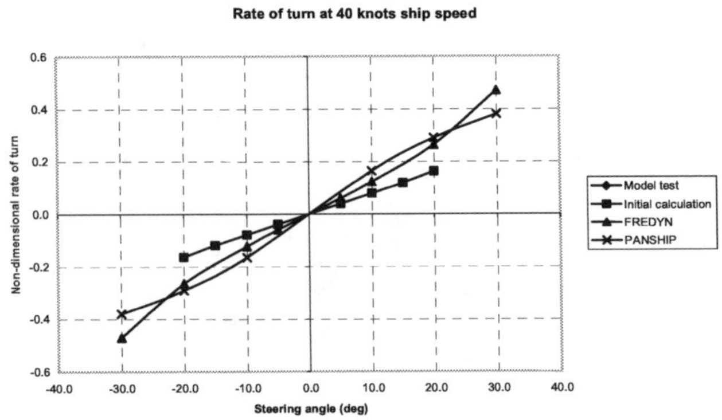

Figure 41: Non-dimensional turning rate at 40 knots

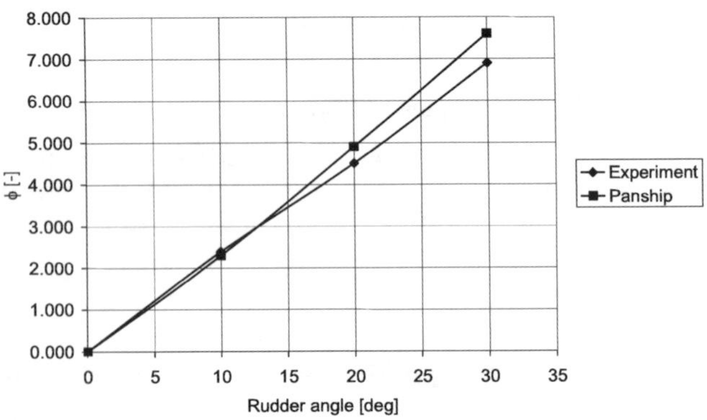

Figure 42: Heel angle during banked turn at 40 knots

Examining the details of the forces used in the calculations more closely, it is found that the waterjet steering forces (which are the excitation forces) were not well predicted initially. The hydrodynamic forces on the ship itself during maneuvering are dominated by the four main vertical struts that generate significant lift forces in the horizontal plane, due to their angle of attack to the flow when the ship is, for example, drifting. Figure 42 presents a sample of the results of the captive tests. It demonstrates that during lateral motions of the ship, the horizontal lift forces, especially due to the forward struts, are dominant. The results in Figure 43 show that 50% of the lateral forces are generated by the forward BWB system, 20% by the aft BWB system, and 30% by the hull. The forward pair of struts therefore delivers about 2.5 times more horizontal lift than the aft pair of struts. The conclusion to be drawn is that a significant amount of interaction exists between the different components of the CHSV: the aft BWB system is in the shadow of both the hull and the forward BWB system. The magnitude of the horizontal lift generated by the main hull due to transverse motion is about as expected.

Figure 44 illustrates the point of application of the individual forces: the forces of the forward and aft BWB systems have their longitudinal point of application at the location of the BWB system. The hull has its point of application in the forward part of the ship at 20% of the length from the bow.

Analysis of the captive model tests led to these insights and has enabled extension of the CFD codes, thereby improving predictions significantly. The updates that took place providing these results are described by van Walree et al. (2006). One of the aspects that also proved to be very important was a correct description of the steering forces due to the three waterjets installed in CHSV. A correct description of the waterjet steering forces was one of the major factors influencing the difference between the initial maneuvering calculations and the final maneuvering calculations.

CONCLUSIONS

This paper has described the extensive use of several computational tools during the course of the CHSV project both in its hydrodynamic design and in support of the model testing program. The results of the model testing provided an excellent opportunity to make direct comparisons between the numerical and the experimental results.

Captive resistance (model test condition) predictions made using USAERO and CFX showed good correlation with the experimental data. The CFX predictions ranged from approximately 5% lower to 2% higher than the model test results over the speed range, while the USAERO predictions ranged from about 6% to 9% lower than the model test results over the speed range.

The comparisons between PANSHIP calculations and model tests with respect to resistance, seakeeping, and maneuvering show, although not perfect, very acceptable results. It can be concluded that for the case of the CHSV (with its relatively large blended wing body systems), using PANSHIP's inviscid method in combination with simple viscosity corrections seems to be successful. It should be mentioned that the addition of these viscosity corrections was necessary to get this good match; the captive experiments provided the required data that made this possible. The accuracy of PANSHIP for an early assessment in a pre-model test phase has been increased by this effort.

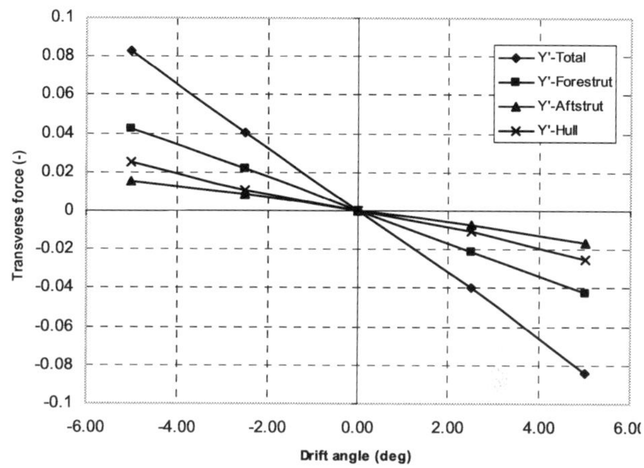

Figure 43: Measured side force on CHSV while drifting

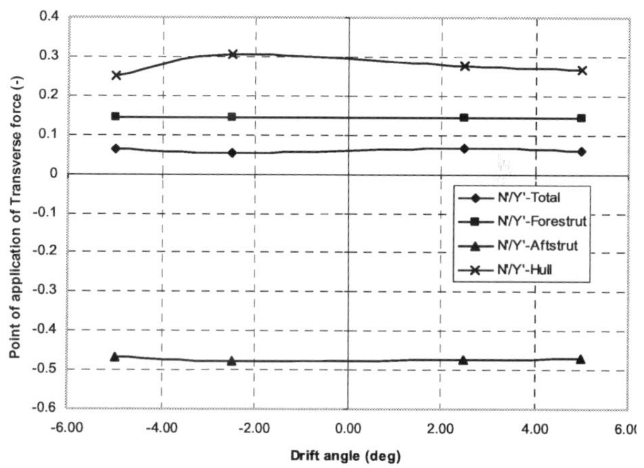

Figure 44: Measured point of application of side force

ACKNOWLEDGEMENTS

The work described in this paper was made possible with support from the Office of Naval Research. The authors also wish to express their appreciation to Northrop Grumman Ship Systems, Navatek, Ltd., and MARIN for their assistance and willingness to make this information available for publication.

Additionally, the authors wish to thank other key MARIN researchers: Dr. Cornell Thill, Ir. Jaap van Heerd, and Ir. Marc Levadou.

REFERENCES

DE KAT, J. O., and W.L. THOMAS, "Extreme Rolling, Broaching and Capsizing – Model Tests and Simulations of a Steered Ship in Waves," *22nd Symposium on Naval Hydrodynamics*, August 1998.

GORNSTEIN, R., S. LOUI, G. SHIMOZONO, and S. LAWREY, "Low Drag Submerged Displacement Hull," *U.S. Patent No. 6,263,819 B1*, July 24, 2001.

HACKETT, J.P, J. CALIX ST. PIERRE, M. LEVADOU, and T.J. PELTZER, "Composite High Speed Vessel: Study of lifting body technology," *ASNE Ships and Ship Systems Technology Symposium*, November 2006.

LOUI, S., G. SHIMOZONO, and T. KEIPPER, "Low Drag Submerged Asymmetric Displacement Lifting Body," *U.S. Patent No. 7,004,093 B2*, February 28, 2006.

PELTZER, T.J., "Application and validation of computational tools in the design of lifting body ships," *ASNE high Speed High Performance Ships and Craft Workshop 2005*, July 2005.

WALREE, F. VAN., "Computational Methods for Hydrofoil Craft in Steady and Unsteady Flow," *Ph.D. Thesis, Delft University of Technology, Department of Naval Architecture and Marine Engineering*, March 1999.

WALREE, F. VAN., J.P. HACKETT, J. CALIX ST. PIERRE, and C. BIGLER, "The use of a time domain panel code for hydrodynamic design support for complex ships" *RINA conference on high speed craft, ACVs, WIGs and hydrofoils.* London, 31 October - 1 November 2006.

Probabilistic assessment of ship stability

Author Name(s): Nikos Themelis (Visitor) **& Kostas J. Spyrou** (Member)

ABSTRACT

Whilst the task of developing a sound probabilistic stability assessment method is a formidable one given the current depth of knowledge in stochastic dynamical systems, a practical solution is proposed here exploiting the groupiness characteristic of high waves. Rather than attacking directly the complete problem, an effort is made to define a path where, the rigour of the deterministic approach in eliciting the nature of instability is combined with suitable analysis of the probabilistic seaway. Critical wave encounters that could generate instability are identified on the basis of deterministic analysis of ship dynamics. A rational procedure is put forward for calculating the probability of such wave encounters. The probabilities of different capsize modes are distinguished. The method is not biased towards any specific type of mathematical model of ship motions and is easily integrated within a risk assessment framework.

KEY WORDS: ship, stability, dynamics, assessment, criteria, performance, probabilistic, wave group.

INTRODUCTION

After Moseley (1851) and Froude (1861), the dynamics of ship rolling enjoy constant popularity as a theme, challenging the research community till our days. Few could doubt that fundamental understanding is prerequisite for establishing stability criteria that have a rational basis. It is no surprise therefore that, over the years several research efforts have shown a tendency to cluster around a treatment of nonlinear ship dynamics under deterministic excitation; because this promised to hold the key towards the elucidation of the root mechanisms of instability. Nevertheless, the seaway bears probabilistic wind and wave loads on a ship and moreover, several operational uncertainties influence behavior. Efforts to integrate these important elements are not novel. On the one hand, the nature of the seaway is well reflected by a number of parametric models that abound in the literature of sea wave statistics. However, the capturing of the probabilistic properties of the response is deterred by the nonlinearity of mathematical models of extreme ship dynamics. In plain terms, knowing the probability density functions (pdf) of wind and waves in the vicinity of a ship does not suffice for deducing the exact pdf of her responses especially if the latter are large. Thus, to characterise the response, a number of simplifications will be obliged on the way. The sort of simplifications defines in essence the research philosophy on the subject.

Those mathematically inclined often prefer to focus narrowly on an abstract version of the problem, assuming a simple structure for the mathematical model and certain idealising properties for the random excitation. These enable a rigorous treatment, permitting sometimes analytical solution as regards the pdf of the response. However, generalisations towards more realistic conditions are restricted. At the other end, and in the urge for practicality, those of engineering background often adopt a "brute force" simulation approach, paying little attention to the intricacies of dynamical phenomena that possibly underlie the occurrence of instability. The general feeling is that still, there is no reliable probabilistic method for ship stability assessment with the entailed depth and breadth for general engineering application.

A method that shows promise to fill this gap is presented in the current paper, interfacing the deterministic analyses of ship dynamics with the wealth of probabilistic wind/wave models and statistics. The idea was surfaced in Spyrou and Themelis (2005) and some further elaboration is found in Themelis and Spyrou (2006). Here is given fully the method of probability calculation, together with complete applications. From the outset however, one should acknowledge the fact that a perfect, non-compromising solution is not viable given the state-of-the-art in stochastic dynamical systems. A number of observations have provided the inspiration for this development:

- Subject to satisfying well-known relationships between the range of frequencies that appear in the excitation and the system's natural frequencies, near regularity of excitation is conducive to large amplitude responses.
- As noted by Draper (1971), higher waves tend to appear in groups.
- For certain instabilities to appear, some regularity in the excitation is required and the amplitude is built-up gradually (beam-sea resonance, parametric roll, cumulative broaching). Other instabilities are the outcome of the encounter of a single critical wave that is capable to disturb the ship from her safe bounded motion (capsize in breaking waves, pure-loss of stability, direct broaching). A final case is the cumulative effect of random waves (accumulation of water on deck for intact or damaged ships).

Accounting for all the above facts, the core idea behind the proposed approach is spelt out as follows: the probability of occurrence of some instability event could be assumed as equal to the probability of encountering the critical (or "worse") wave groups that generate this instability. Subsequently, the task could be disassembled into two parts:

- a mainly deterministic one, for producing the specification of critical wave groups as represented by their height, period and run-length, obtained entirely from ship motion dynamics; and
- a purely probabilistic, focussing on statistical wave models, in order to calculate the probability to encounter these critical wave groups.

The paper is structured as follows: after a short literature review on the application of probabilistic methods for ship rolling, the proposed novel assessment methodology is presented as a number of steps. These are followed by an exposition of the basic theory related to wave group probability calculation. The feasibility of the methodology is demonstrated by means of two applications.

BRIEF REVIEW OF GENERAL PROBABILISTIC METHODS FOR SHIP ROLLING

St Denis and Pierson (1953) were the first to apply stochastic process theory in linear seakeeping. They exploited the Wiener – Khinchin theorem that, for a linear system, supplies the relationship between the spectra of excitation and response. As no general method has been developed yet for treating rigorously nonlinear systems driven by stochastic processes, several concepts were proposed for dealing with it in an approximate way.

In an effort to use results from linear theory, linearization techniques have been developed, where the idea is that the original nonlinear system is substituted with a linear equivalent. One method is statistical linearization; see for example Vassilopoulos (1971). Here, the stochastic excitation is assumed ergodic, while nonlinear roll restoring is substituted with an equivalent linear, under the requirement that the mean square of the error incurred is minimal. The linear response is, like the excitation, Gaussian with mean value zero and its variance should be minimized in order to obtain the "equivalent" linear coefficient. Another method based on linearization techniques is the energy statistical linearization of Gerasimov (1979) cited in Belenky & Sevastianov (2003). This method assumes equivalence of statistical characteristics in terms of work/energy balance. Specifically, the concept of energy equivalent cycle is proposed where a linear undamped and unforced oscillation with frequency equal to the averaged frequency of the original system and same roll velocity variance is sought. A drawback of these methods is that the output process is assumed Gaussian, which is not representative for a nonlinear system even though the input due to the wave surface elevation could be adequately assumed as a Gaussian process.

Analytical calculation of the probability distribution of the response is often attempted by assuming a stochastic "Markov process". This means that the current value of the process depends only on the value in a previous moment of time and not on the previous history of the process. A Markov process can be fully characterized by the conditional distribution at two consecutive moments of time; and the function of the conditional distribution $f(t_1,x_1;t_2,x_2)$ can be considered as a solution of two partial differential equations (the so-called Fokker – Planck – Kholmogorov equations), referring to a previous and to the current moment of time. Stationary solutions of the Fokker – Planck – Kholmogorov (FPK) equations for the case of nonlinear rolling under the assumption of Markov process were presented for example in Haddara (1974) and Roberts (1982). However, a disadvantage of this approach is that the excitation has to be modeled as an ideal white noise process in order to obtain analytically the stationary solution. Sea wave spectra have usually a distinct peak and limited bandwidth. Appropriate forming filters should then be used to describe the wave input. Even so, to derive analytic formulas in manageable form, the procedure can be applied successfully for one degree of freedom (Belenky et al. 1998). Problems arise also if nonlinear damping is assumed (Haddara and Zhang 1994).

In order to overcome these difficulties while keeping the assumption of a Markov process, numerous treatments have been presented. Roberts (1982) considered the energy envelope of the response, using the assumption that the energy of the envelope is a slowly varying parameter of time. Applying then a stochastic averaging technique, the two dimensional FPK equations are reduced to one for the energy envelope, implying that the process of the energy envelope is modeled approximately as one dimensional Markov process. This equation can be solved for any type of nonlinearity while the input process is not necessarily assumed as white noise. A stationary joint distribution for the roll displacement and roll velocity was obtained considering a related phase process. Haddara and Nassar (1986) combined also the averaging technique with the FPK approach to formulate the partial differential equation describing the conditional probability distribution for roll amplitude and employing also the Galerkin technique in order to derive the related distribution.

Other analytical methods dealing with the problem are non - Gaussian closure techniques which can be incorporated in order to develop expressions for the probability density functions for nonlinear systems (Ibrahim 1985). Another analytical method is Hermitte's which is based on diffusion process theory and it can be utilized for deriving the non - Gaussian response statistics of nonlinear systems driven by Gaussian or non Gaussian excitation (Ness et al 1989). In particular for parametrically excited ship rolling, a detailed review on probabilistic methods can be found in Spyrou (2005).

Recently, Naito et al. (2006) proposed a non linear relationship of a certain form between the short-term significant wave height and the standard deviation of ship response, where two parameters could be identified from measurements; or on the basis of the severest outcome. Thereafter long-term response could be obtained with linear superposition of short-term exceedence probabilities. Assessments by this philosophy where a limited number of short-term sea states are used for obtaining a long-term assessment are popular also in neighboring fields (see for example Baarholm and Jensen 2004). The log normal can be a satisfactory model for the probability distribution of the significant wave height, while the three-parameter Weibull is believed to be suitable for extreme significant heights (Ochi 1998). For the joint distribution, scatter diagrams that present information of the combined values of significant wave height and average zero-crossing period are sometimes available. The joint log normal distribution is a popular fit, although it seems to diverge for severe sea states (Ochi 1978). The joint distribution could be obtained from the product of the marginal distribution of significant wave height and the conditional distribution of mean period for a specific significant wave height. Mathiesen and Bitner – Gregersen (1990) used the three dimensional Weibull for the marginal and the log normal for the conditional.

From another standpoint, Arnold et al (2004) discussed how some concepts from the theory of dynamical systems like attractors and bifurcations are generalised in the case when a system is perturbed by random noise. Then, the statistical behaviour of the random dynamical system is described by invariant measures, which in this case are probability measures. Furthermore, Lyapunov exponents, random attractors and their domain of attraction as well as stochastic bifurcation theory are utilized to describe the random system and possible qualitative changes of behaviour. In nonlinear dynamical systems' theory is rooted also the work of Hsieh et al. (1994) who have used the Melnikov function for a randomly excited system in order to assess the tendency of a ship for capsize (Simiu 2002). Melnikov's function expresses the distance between the stable and unstable invariant manifolds (Wiggins, 1988; Falzarano et al., 1992). Their intersection generates chaos and potentially capsize. In this approach, the time-fluctuating part of the Melnikov function is associated with the stochastic excitation. It was assumed that this part represents a stationary and ergodic process and has a Gaussian distribution, so with further elaborations, a probabilistic measure of capsize tendency was proposed.

On the other hand, numerical simulation techniques based e.g. on the Monte-Carlo method turn to be nowadays more popular. A profound advantage of these techniques is that there are fewer constraining assumptions about the type of the mathematical model. Generally, Monte-Carlo simulations are a kind of stochastic simulations that use a random generator. For irregular rolling a generator of irregular waves is needed and then the rolling equation is solved numerically. However a large number of realizations are needed while the results depend strongly on satisfactory representation of the random excitation (Belenky et al. 2006). The analytical piece-wise linear method of Belenky (1994), including the fluctuating part of wind's excitation, was used by Paroka and Umeda (2006) for comparisons with numerical predictions and for testing the effectiveness of Monte-Carlo simulations. A Poisson process was assumed for capsize.

Repetitive numerical simulations are contemporarily a practical choice for estimating the probability of capsize. For example, McTaggart and DeKat (2000) used a well-known simulation code and the random wave phase approach for generating wave realizations in order to estimate the probability of capsize. They adopted the Gumbel distribution in order to fit the maximum roll angles from each realization. However, the number of realizations can affect the Gumbel fit especially when hourly probabilities are greater than 10^{-3}. They examined different scenarios of speed, heading and weather and obtained practical polar diagrams of capsize probability. Whilst embodying a "black-box" approach and being short of accounting for factors like the initial conditions, polar diagrams are a popular means within the naval architecture community for expressing capsize tendency. New approaches like the time-to-capsize concept of Ayyub et al (2006) may enhance the potential of this assessment philosophy.

Nearer to the spirit of the current paper, Tikka and Paulling (1990) contemplated the analysis of parametric rolling in terms of the encounter of a high run of waves and referred also to the combination of ship's speed and heading that could lean toward such an encounter. DeKat (1994) referred to the use of joint distributions of wave length and steepness for predicting their encounter. Similarly, Myrhaug et al (1999) investigated synchronous rolling by using joint distributions of successive wave periods. Boukhanovsky and Degtyarev (1996) studied statistical issues related with the group structure of waves and their correlation with the group structure of parametric roll response in longitudinal seas. They claimed that a sequence of short groups with similar periods is more dangerous than one long wave group, repeating also that swell seas are more propitious to the development of wave groups than wind waves.

DESCRIPTION OF METHODOLOGY

The methodology is arranged as a number of successive steps that are summarized in the following:

Type of assessment
Assuming as known, at least in general terms, the specification of the ship under investigation, different perspectives are possible concerning the character of her assessment, determined by the length of time of the ship's exposure to the environment. In the current context, as "short-term" will be labeled an assessment where stability is gauged during a single voyage. Therefore a route, or alternative routes, need to be specified beforehand and the assessment will be driven by the "few hours" forecast of the representative environmental parameters, in the vicinity of the prescribed route(s). A note of caution is needed here: whilst short-term wave data are an essential input of this assessment, one should not confuse the, say, one hour validity of short-term wave data with the reference duration of a short-term stability assessment under the current methodology which could last for several hours as "fresh" environmental data continually feed the model. Such a short-term assessment could support decision-making concerning ship departure control and weather routing (Spyrou et al. 2004).

On the other hand, "long-term" assessments might be undertaken for a variety of reasons. Probabilities of occurrence of instability on a seasonal or annual basis, either for specific reference routes or for wider areas of operation, can be determined with obvious utility at corporal and national administration levels. Moreover, by projecting the annual statistics to the ship's life-span, a long-term assessment could be tied to ensuring a satisfactory stability safety level by design.

Type of service
Especially for a long-term assessment, the anticipated service profile needs to be specified. Restricted service refers to operation on specific routes (e.g. the standard route of a Ro/Ro ferry in the Mediterranean Sea). Unrestricted service sets no narrow limits for the navigational area, e.g. the North Atlantic, and thus it might lead to a different assessment result.

Portfolio of stability criteria
In theory, instabilities of a different nature could arise during a voyage. Furthermore, different ship types often show different tendencies for unstable behaviour. Acknowledging these, the criteria of the assessment will be linked to the occurrence of the known types of instability. A portfolio of stability criteria reflecting the main instability phenomena are thus incorporated into the assessment. Suitable norms (see next paragraph) take care of the level of acceptable performance per criterion. A recommended generic set of stability criteria should include resonance instabilities (beam-sea resonance and parametric rolling in longitudinal seas); pure – loss of stability on a wave crest in following seas; for smaller vessels, instability due to breaking waves from the side and "water-on-deck"; and lastly broaching (including also the cumulative type).

Norms of unsafe behaviour
For the criteria to become meaningful, norms representing quantified levels of unsafe behaviour should be "attached" to each one of them. One could envisage the setting of quantitative "warning" and "failure" levels per criterion and ship type, on the basis of threshold angular and linear displacements and accelerations, referring respectively to the safety of the ship and her cargo. A warning level should play a cautionary role and exceedence should be permitted with controlled probability. To the contrary, failure levels should not be exceeded for the acceptable level of risk.

Specification of critical wave groups
This step deals with the specification of critical wave groups, represented by the wave height, period and group run-length. A wave group is a sequence of waves with heights exceeding some preset level and having nearly equal periods (Masson and Chandler, 1993; Ochi, 1998). In Fig. 1 is shown schematically a wave group comprised of three successive waves (i.e. the run length is equal to three) with periods lying within a predefined time interval and corresponding wave heights above a threshold height that had been determined in advance as critical. Individual waves are measured here using the standard zero – upcrossing method, with their height taken as the maximum vertical excursion of the surface elevation between two zero – upcrossings; and their period, as the time interval separating these two events.

In this step of the assessment, advantage is taken of the strengths of deterministic analysis. With a specific instability mode in mind (criterion) one could subject the ship to a regular wave group and observe the outcome in terms of exceedence of the threshold values (norms). Numerical simulation tools can be employed for this, based on detailed mathematical models of large-amplitude ship motions. Alternatively, analytical techniques that capture the key systems dynamics may be utilized; for example, global stability analysis methods (Melnikov's method), or analytical expressions of the growth of roll amplitude per encounter wave cycle. Different methods have their own strengths and weakness. However the conceptual disengagement of ship dynamics from the probabilistic seaway allows one to reach an informed decision about what constitutes the critical wave environment for a ship, on the basis of more than one tools. Therefore, results from different simulation codes or combinations of simulation with rigorous stability analysis techniques could be used.

Let us discuss further the procedure for identifying critical wave groups for resonance-based instabilities like beam-seas resonance and parametric rolling in longitudinal seas. Notably, critical wave groups are specified for the entire range of wave periods that could be met in a realistic seaway. Firstly, the range of wave periods is discretised in, say, ℓ, equal parts, with step δT. As initial condition, the unbiased upright state is assumed with zero roll velocity. There is some theoretical background from ship dynamics in support of adopting this initial condition (Rainey and Thompson 1991). Nonetheless, the effect of alternative initial conditions is a matter that we are currently investigating.

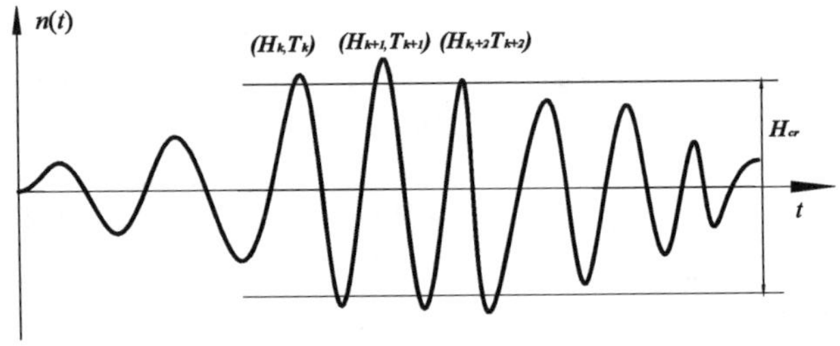

Fig. 1: Wave group

Then, the set of wave groups that satisfy the critical norm is determined utilizing simulations or direct analytical methods as mentioned earlier. More specifically, the critical surface of Fig. 2 is determined according to the following procedure:

- Consider a wave period T_i in the defined range.
- Take n waves in the group.
- Calculate the corresponding critical wave height $H_{cr}(T_i,n)$.
- Continue for run length $n+1$.
- Take the next wave period $T_{i+1}=T_i+\delta T$.

The procedure can be repeated for other loading conditions on the basis of the loading profile of the ship. Moreover, rather than a single speed U, a range of speeds may be deemed necessary to be examined, especially for a longitudinal sea where speed affects seriously the frequency of encounter. For each discrete value of speed, the wavelengths λ, and thus wave periods T, that could give rise to parametric resonance are determined. For example, to attain exact principal parametric resonance, the frequency of encounter can be identified from the condition $\omega_e=2\omega_0$ and thereafter the wave length from $\lambda=\frac{2\pi}{\omega_e}(c+U\cos\psi)$. Symbols are used with their customary meaning: ω_0, c, ψ are respectively, ship's natural roll frequency, wave celerity and heading angle relatively to the wave direction of propagation (0^0 for head-seas).

For non-directly excited modes, some (probabilistic) initial heel angle should be assumed; otherwise, even within the instability region, no growth of roll could be expected. For all parameters that come into this calculation, some distribution of their values should be assumed.

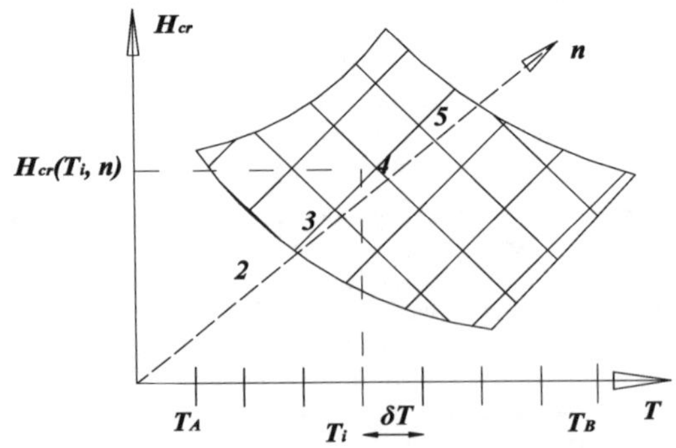

Fig. 2: Critical surface of wave group characteristics.

Calculation of probability of critical wave groups

The procedure "assumes" mapping of the weather data (representing either a forecast or statistical information) on a grid of "weather nodes". For a restricted ship service profile they should be placed, with sufficient density, close enough to the route; otherwise, for unrestricted service they should be scattered in the wider navigational area that is under consideration. Each node "influences" an area that surrounds it. It will be assumed that inside an area of this kind, the weather conditions are uniform, represented by the weather at the node. Then, the voyage is split into a number of segments whose number depends on the distance between nodes in the "weather grid". This notion is illustrated in Fig. 3. For a short-term assessment, stationarity of wave data can reasonably be assumed if a sufficiently dense grid has been used.

Following the placement of weather nodes, the time that the ship spends in the vicinity of each node as a percentage of the duration of the voyage in beam, following and head seas is calculated from the ship route and the distribution of wave direction. Speed reduction due to adverse weather could be taken into account in the above calculations.

Generally, for the calculation of critical wave groups, appropriate conditional probability density functions $f(\boldsymbol{\tau}/h)$ (reduced to a joint $f(\tau,h)$ for the case of single wave encounter) should be utilized. Such pdfs are presented in the following section. As is easily realised, it is more practical to base the analysis on frequency spectra.

Probabilities to encounter wave groups with the following characteristics need to be determined:

- n successive waves in group (maximum considered run length k),
- periods within the group lying in an interval $\left[T_i, T_j\right]$,
- wave heights exceeding the critical height: $H_n \geq H_{cr}$.

Then the associated probability can be calculated from the formula:

$$P\left[T_1,T_2,...,T_k/H_n>H_{cr}\right]=\underbrace{\int_{\tau_i}^{\tau_j}\int_{\tau_i}^{\tau_j}...\int_{\tau_i}^{\tau_j}}_{k} f\left(\boldsymbol{\tau}/h_n>h_{cr}\right)d\tau,\ n=1,...,k \quad (1)$$

where $\boldsymbol{\tau}=\left[\tau_1,\tau_2,...,\tau_k\right]^T$. $\tau_i=T_i/T_m$, $h_n=H_n/H_{rms}$ are respectively dimensionless wave period and height. T_m and H_{rms} stand for mean spectral period and root–mean–square wave height respectively. For phenomena where the period of encounter is different from the wave period, the above calculations should be based of course on the encountered spectrum.

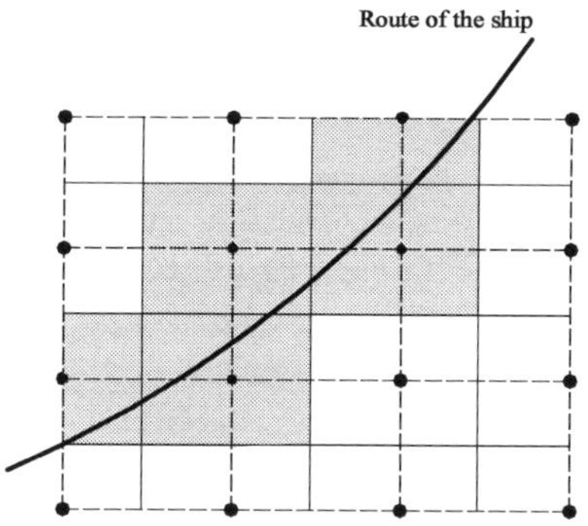

Fig. 3: Weather nodes with their areas of influence.

To illustrate the calculation procedure, let us assume a scenario of beam-sea resonance where $\ell \times k$ critical wave groups have been determined, with ℓ corresponding to the number of the discretised segments of the wave period range and k to the number of considered wave group run lengths. The probability P_S for some arbitrary weather node S is expressed as:

$$P_S = \sum_{i=1}^{\ell} \sum_{n=2}^{k} P_{i,n} \qquad (2)$$

In the above we considered that for resonance to be realized a sequence of waves is essential; therefore we assumed as lowest run length value equal to 2. Probabilities for each segment are summed up. Events of wave-group-encounter in a range $\left[T_i, T_j\right]$ with $n = 2,3,...,k$ successive waves are treated as independent but not mutually exclusive. So, the internal sum follows the well-known rule:

$$P_i = \sum_{n=2}^{k} P\left[A_n\right] - \sum_{n=2}^{k-1} \sum_{m=n+1}^{k} P\left[A_n A_m\right] + \sum_{n=2}^{k-2} \sum_{m=n+1}^{k-1} \sum_{q=m+1}^{k} P\left[A_n A_m A_q\right] - ...$$
$$+(-1)^k P\left[A_n A_m A_q ... A_k\right], \quad i \in \left[1,2,...,\ell\right] \qquad (3)$$

where $A_1 A_2 = A_1 \cap A_2$ and A_j refers to the encounter of a critical wave group that has a certain specification. As these events could be assumed as independent, $P\left[A_i \cap A_j\right] = P\left[A_i\right] P\left[A_j\right]$. Also, the probabilities from each discretised part ℓ are mutually exclusive; thus the total probability for some weather node S is obtained by summing up according to eq. (2).

Arguably, it is more meaningful to transform these probabilities (which by definition refer to the number of encountered waves irrespectively of their periods) into a representation that conveys the percentage of time that the ship is expected to be in peril, with respect to the entire time of exposure. We shall label this as "critical time ratio". The calculation is based (approximately) on the next simple transformation:

$$\bar{t}_S = \sum_{i=1}^{\ell} P_i \frac{T_i}{T_m} \qquad (4)$$

If along the considered route m weather nodes have been placed, then the total probability for beam-sea resonance and respectively the total critical time will be:

$$P_{beam} = \sum_{S=1}^{m} P_S \tau_{S\ beam} \qquad (5)$$

$$\bar{t}_{beam} = \sum_{S=1}^{m} \bar{t}_S \tau_{S\ beam} \qquad (6)$$

Then, as already discussed:

$$\tau_{S\ beam} = \frac{\text{time spent in beam seas in the } S \text{ rectangle}}{\text{whole duration of the voyage}} \qquad (7)$$

The calculation is further elucidated by considering a scenario of parametric rolling where additional parameters should be involved. The probability in the area of influence of some weather node S is:

$$P_S = \sum_{a} \sum_{n} \sum_{\varphi_0} \sum_{U} P_{a,n,\varphi_0,U} \qquad (8)$$

In this case the speed determines the wave periods where parametric resonance could occur and hence it affects the probability calculation. Also, initial roll angles φ_0 should be taken into account. The parameter a takes care of frequency tuning, but only through the wave frequency. The critical time ratio for the S weather node is then determined by combining equations (4) and (8).

For other stability criteria, such as pure loss of stability and direct broaching, the probability of encountering a single critical wave should be calculated, assuming the wave period to take values in such a time interval $\left[T_i, T_j\right]$ so that the wave length is around the ship length and the wave height exceeds the critical.

If the character of the assessment is long-term, scatter diagrams or fitted joint distributions of significant wave height and peak period $P_L(H_S, T_P)$ will be necessary. Then, to obtain (per node) the long-term probability from the short-term data, the following summation should be performed (Ochi 1998):

$$\bar{P}_S = \frac{\sum_{H_S} \sum_{T_P} N_S \, P_S(H_S, T_P) \, P_L(H_S, T_P)}{\sum_{H_S} \sum_{T_P} N_S \, P_L(H_S, T_P)} \qquad (9)$$

N_S is the average number of waves, a quantity that depends on the duration of each short-term case. The dominator of eq. (9) accounts for the total number of waves encountered during the assumed long–term period (e.g. one year).

PROBABILITY OF WAVE GROUPS – SPECTRAL METHODS

The so-called spectral methods will be employed in order to derive the relevant group statistics and probabilities, as all data that is necessary for the calculations can be extracted from the frequency wave spectrum. The general assumptions introduced at this stage are the following (Ochi 1998):

- Waves are considered to be a stationary ergodic random process.
- Waves are a Gaussian random process; thus wave profiles are distributed following the normal probability distribution with zero mean and a variance representing the sea severity.

- The spectrum is concentrated at a particular frequency, i.e. the wave spectral density function is 'narrow banded'.

Next, two very well-known spectral approaches are briefly presented as an introduction to wave group theory and properties. These approaches are: the envelope theory of Longuet – Higgins (1957, 1984) and the Kimura theory combined with its modification by Battjes and van Vledder (1984). Both approaches target the estimation of a standard measure of wave groupiness, the mean group length $\overline{j}$, obtained by averaging the number of consecutive waves exceeding a threshold height H_{cr} over a record of sea surface elevations. The mean group length is related through the narrow spectral bandwidth approximation to some characteristics of the wave spectrum: the spectral bandwidth parameter ν for the first approach; and the spectral correlation coefficient γ_s for the second.

It is worth to mention here that the group analysis according to these two approaches is restricted to linear wave theory and therefore, wave group formation as a result of nonlinear instability is not considered. Usually, in linear theory, wave groups are assumed to be associated with wave components concentrated around the peak frequency and thus, in these approaches, only unimodal spectra are appropriate for wave group analysis.

Envelope approach

In this linear approach, sea surface elevation is considered as narrow band Gaussian noise, and the group properties are derived from the envelope function of surface elevation. Longuet Higgins (1957, 1984) developed an estimate of the average number of waves in a high run, $\overline{\mathcal{H}}$ in terms of the spectral band parameter. Sea surface elevation $n(t)$ can be represented as the sum of sinusoidal components with angular frequency ω_n:

$$n(t)=\mathrm{Re}\left(\sum_n c_n e^{i(\omega_n t+\varepsilon_n)}\right) \tag{10}$$

where the random phases ε_n are normally distributed over the range $[0,2\pi]$. The amplitude is estimated from the frequency spectrum $S(\omega)$ according to: $c_n=\sqrt{2S(\omega)\delta\omega}$. Selecting a carrier wave frequency $\overline{\omega}$, the mean frequency $\overline{\omega}=m_1/m_0$ may be a representative midband value. Then $n(t)$ may be written as:

$$n(t)=\mathrm{Re}\left(e^{i\overline{\omega}t}\sum_n c_n e^{[(\omega_n-\overline{\omega})t+\varepsilon_n]}\right)=\mathrm{Re}\left(R(t)e^{i\overline{\omega}t}\right) \tag{11}$$

A wave envelope function can be specified as:

$$R(t)=a(t)e^{ix(t)} \tag{12}$$

$a(t)$ is the amplitude of the envelope function and $x(t)$ is its phase. The variation of the wave envelope function $R(t)$ with time is slow compared to the carrier wave $e^{i\overline{\omega}t}$ and the wave crests and troughs tightly follow the surface elevation $n(t)$. Moreover, a Rayleigh and a Gaussian pdf are assumed for the wave envelope amplitude $a(t)$ and its time derivative respectively. If these distributions are known, the average number of waves in a group of the envelope function $\overline{\mathcal{H}}$ can be derived by dividing the average length of episodes for which the wave envelope exceeds the critical level $H_{cr}/2$, by the mean zero – upcrossing period. According to Longuet Higgins (1957, 1984):

$$\overline{\mathcal{H}}=\sqrt{\frac{2m_0}{\pi}}\frac{\sqrt{1+\nu^2}}{\nu}\frac{1}{H_{cr}} \tag{13}$$

The spectral bandwidth parameter ν, is defined as:

$$\nu=\sqrt{\frac{m_2 m_0}{m_1^2}-1} \tag{14}$$

The spectral moment will be: $m_j=\int_0^{\infty}\omega^j S(\omega)d\omega$. However, the groupiness parameter $\overline{\mathcal{H}}$ derived by the continuous envelope theory is defined in a different way compared to the parameter $\overline{j}$ that is based on discrete counting. Longuet Higgins (1984) provided a discrete counting correction expression where:

$$\overline{j}\approx\overline{\mathcal{H}}+0.5 \tag{15}$$

The expression assumes that successive upcrossings of the critical level by the envelope are uncorrelated.

Modified Kimura approach

In this approach proposed by Kimura (1980), the sequence of wave heights is treated as a Markov chain, allowing for non-zero correlation between consecutive waves. The mean group length $\overline{j}$ depends on the correlation coefficient γ_h of successive wave heights.

In fact, the process of successive waves can be modeled by a first order autoregressive model (Wist et al 2004). Generally, for an autoregressive representation the value of variable Y at time t depends on the values of its own past, plus a random variable. For an autoregressive process of order r:

$$Y(t)=n_1Y(t-1)+....+n_rY(t-r)+\varepsilon(t) \tag{16}$$

Where $n_1,...,n_r$ are weights that can be related with correlation coefficients and $\varepsilon(t)$ is a zero mean Gaussian white noise process. Furthermore, a first-order autoregression process has the Markov chain property, meaning that the value of $Y(t)$ is completely determined by the knowledge of $Y(t-1)$. The Markov chain property is expressed by:

$$P\left(Y(t+1)=y_{t+1}/Y(0)=y_0,...,Y(t)=y_t\right)=P\left(Y(t+1)=y_{t+1}/Y(t)=y_t\right) \tag{17}$$

Kimura proposed the bivariate Rayleigh distribution as the joint pdf of successive wave heights H_1, H_2:

$$f_{HH}(H_1,H_2)=\frac{4H_1H_2}{(1-\kappa^2)H_{rms}^4}e^{-\frac{(H_1^2+H_2^2)}{(1-\kappa^2)H_{rms}^2}}I_0\left(-\frac{2\kappa H_1H_2}{(1-\kappa^2)H_{rms}^2}\right) \tag{18}$$

where κ is a correlation parameter, H_{rms} the root–mean-square wave height and I_0 the modified Bessel function of the zeroth order. The correlation parameter κ is a function of correlation coefficient between successive wave heights γ_h, according to Kimura (1980):

$$\gamma_h=\frac{1}{\sigma^2(H)}\frac{1}{N-1}\sum_{i=1}^{N-1}(H_i-H_{rms})(H_{i+1}-H_{rms}) \tag{19}$$

where $\sigma(H)$ is the standard deviation of a large number N of wave heights. The correlation parameter κ is associated with the correlation coefficient of successive wave heights γ_h. According to Kimura (1980):

$$\gamma_h = \frac{E(\kappa) - (1-\kappa^2)\frac{K(\kappa)}{2} - \frac{\pi}{4}}{1 - \frac{\pi}{4}} \approx \frac{\pi}{16 - 4\pi}\left(\kappa^2 + \frac{\kappa^4}{16} + \frac{\kappa^6}{64}\right) \quad (20)$$

E, K are complete elliptic integrals of the first and second kind respectively. The approximation [second part of eq. (20)] is according to Stansell et al (2002). Kimura (1980) calculated the probability of a sequence of high waves using the conditional probability that a wave height exceeds the threshold level H_{cr}, given that the previous wave also exceeds H_{cr}. The conditional probability is computed from the joint pdf of eq. (18):

$$p_{22} = P\left[H_{i+1} \geq H_{cr} / H_i \geq H_{cr}\right] = \frac{\int_{H_{cr}}^{\infty}\int_{H_{cr}}^{\infty} f(H_1, H_2)dH_1 dH_2}{\int_{H_{cr}}^{\infty}\int_{0}^{\infty} f(H_1, H_2)dH_1 dH_2} \quad (21)$$

The probability that a wave group has a run length j is:

$$p(j) = (1 - p_{22})p_{22}^{j-1} \quad (22)$$

As shown by Kimura (1980) the obtained mean group length is:

$$\bar{j} = \frac{1}{1 - p_{22}} \quad (23)$$

A disadvantage of Kimura's analysis is that the group properties are not connected with the properties of an energy spectrum but depend on the parameter γ_h computed from the series of discrete wave heights. In order to overcome this drawback, Battjes & Van Vledder (1984), modified Kimura's theory, introducing a new correlation parameter κ_s, determined by the frequency spectrum:

$$\kappa_s^2 = \frac{1}{m_0}\left|\int_0^{\infty} S(\omega)e^{i\omega t}d\omega\right|, \quad t = T_{m2} \quad (24)$$

$T_{m2} = 2\pi\sqrt{m_0/m_2}$ is the average period between zero – upcrossings. Using this spectral parameter κ_s, a new correlation coefficient γ_s can be produced in a similar way according to eq. (20). In fact, parameter γ_s represents the correlation coefficient between points of the wave envelope function $\alpha(t)$, separated by a constant time interval T_{m2}. Therefore, the correlation coefficient between discrete waves is replaced by the correlation coefficient between points of the wave envelope. These two parameters are identical under the following assumptions.

- The amplitude of the discrete waves obeys the envelope function.
- The waves are separated by a time interval approximately constant and equal to T_{m2}.
- The wave heights can be approximated by twice the value of the envelope function.

These assumptions are strictly satisfied in the limit of very narrow spectra. In a step towards confining the repercussions of the above assumptions, Stansell et al (2002) proposed to calculate the correlation parameter κ_s not only at the average zero upcrossing period T_{m2} but also at $T_{m2}/2$ and $3T_{m2}/2$. Thereafter a new averaged correlation coefficient can be produced according to the following expression:

$$\bar{\gamma}_h = \frac{\gamma_h(T_{m2}/2) + 2\gamma_h(T_{m2}) + \gamma_h(3T_{m2}/2)}{2 + 2\gamma_h(T_{m2}/2)} \quad (25)$$

Probability distribution functions

For the calculation of the probability to encounter a single critical wave or a critical wave group (dependending on the examined stability criterion) joint pdf of wave height and period $f(\tau, h)$ and conditional multivariate pdf of wave periods and heights $f(\tau / h)$ should be sought from the literature.

A variety of bivariate distributions of wave height and period have been proposed, see for example Longuet – Higgins (1975, 1983), Cavanié et al. (1976); Tayfun (1993). The bivariate pdf according to Longuet – Higgins (1975) is:

$$f(h,\tau) = \frac{\pi}{2}\frac{(1+v^2)}{v + \frac{v}{\sqrt{1+v^2}}}\left(\frac{h}{\tau}\right)^2 \exp\left(-\frac{\pi h^2}{4}\left(1 + \frac{\left(1 - \frac{\sqrt{1+v^2}}{\tau}\right)}{v^2}\right)\right) \quad (26)$$

Where: $h = \frac{H}{H_{rms}}$, $\tau = \frac{T}{T_m}$ are respectively the dimensionless wave height and period; $T_m = 2\pi\frac{m_0}{m_1}$ is the mean spectral period. Tayfun (1993) proposed a joint pdf of wave height and period that is appropriate for large wave heights. This joint pdf comes from the product of the marginal pdf f_h of large wave heights and the conditional pdf $f_{\tau/h}$:

$$f_{h\tau} = f_h f_{t/h} \quad (27)$$

$$f_h = \frac{2h}{[2\kappa(1+\kappa)]^{1/2}}\left(1 + \frac{1-\kappa^2}{32\kappa h^2}\right)\exp\left(-\frac{2h^2}{(1+\kappa)}\right) \quad (28)$$

The correlation parameter κ can be calculated from the spectrum according to eq. (24). Regarding $f_{\tau/h}$, Tayfun (1993) noticed that for large h, the observed $f_{\tau/h}$ presents a characteristic symmetry about a mean value $\mu_{t/h}$, that can be larger than $\mu_{t/h} = 1$ by as much as 25 – 30% that depends on spectrum's bandwidth v. One possible reason of this effect may be ascribed to nonlinear effects, modifying wave phases in a manner that their distribution is no longer uniform over $[0, 2\pi]$.

It is justified to approximate therefore $f_{\tau/h}$ by a Gaussian form, modifying the conditional mean and standard deviation as follows:

$$f_{\tau/h} = C_1 \exp\left(-\frac{1}{2}\left(\frac{\tau - \mu_{\tau/h}}{\sigma_{\tau/h}}\right)^2\right) \quad (29)$$

where $h > 1$, $0 < \tau < 2$ and C_1 is a parameter that satisfies the condition:

$$\int_0^2 f_{\tau/h} d\tau = 1 \quad (30)$$

In eq. (29) the conditional mean and standard deviation are respectively:

$$\mu_{\tau/h} = 1 + v^2(1+v^2)^{-3/2} \quad (31)$$

$$\sigma_{\tau/h} = 2v/[\sqrt{8}\,h(1+v^2)] \quad (32)$$

According to eq. (30) the parameter C_1 approaches the value $C_1 \approx 1/\left[\sqrt{2\pi}\sigma_{\tau/h}\right]$. The explicit form of the joint pdf of large wave height and associated period is:

$$f_{h\tau}(h,\tau) = C_T h\left(1+\frac{1-\kappa^2}{32\kappa h^2}\right)\exp\left(-\frac{1}{2}\left[\frac{4h^2}{1+\kappa}+\left(\frac{\tau-\mu_{\tau/h}}{\sigma_{\tau/h}}\right)^2\right]\right) \quad (33)$$

where $h > 1$, $0 < \tau < 2$ and C_T a normalizing factor such that:

$$C_T = \frac{2\sqrt{2}}{2\sqrt{4\pi\kappa(1+\kappa)}\sigma_{\tau/h}} \quad (34)$$

Tayfun (1993) approximated the conditional distribution of successive wave periods given the wave height on the basis of the Gaussian distribution for one wave period. Wist et al. (2004) noted that, at least for up to three wave periods, the multivariate Gaussian distribution is a satisfactory model of the conditional distribution. Their conditional pdf of p successive wave periods $\mathbf{T} = \left[T_1,...,T_p\right]^T$, given that each wave height in the group exceeds the threshold H_{cr}, is given below by eq. (35). It is worth to mention that a general formula for variable threshold $H_{cr,i}$ per wave in the group could also be extracted. This is of interest since it could lead to a more realistic representation of a wave group and for this reason it is currently under investigation.

$$f_{\mathbf{T/H}}(\tau/h_i > h_{cr}) = \frac{1}{(2\pi)^{p/2}\left|\Sigma_{\tau/h_{cr}}\right|^{1/2}} e^{\left(-\frac{1}{2}\left(\tau-\mu_{\tau/h_{cr}}\right)^T \Sigma^{-1}_{\tau/h_{cr}}\left(\tau-\mu_{\tau/h_{cr}}\right)\right)} \quad (35)$$

$\Sigma_{\tau/h_{cr}}$ is the matrix of covariance given as:

$$\Sigma_{\tau/h_{cr}} = \begin{bmatrix} \sigma^2_{\tau/h_{cr}} & \mathrm{Cov}\left[T_1,T_2/H_{cr}\right] & \mathrm{Cov}\left[T_1,T_p/H_{cr}\right] \\ & \cdots & \\ \mathrm{Cov}\left[T_1,T_p/H_{cr}\right] & \cdots & \sigma^2_{\tau/h_{cr}} \end{bmatrix} \quad (36)$$

And $\mathrm{Cov}\left[T_i,T_j/H_{cr}\right] = \rho_{ij}\sigma^2_{\tau/h_{cr}}$. The mean values $\mu_{\tau/h_{cr}}$ and the standard deviations $\sigma_{\tau/hcr}$ are calculated from eq. (31) and (32). The correlation coefficients ρ_{ij} are determined assuming a Markov chain:

$$\rho_{1j} = \rho_{12}^{j-1},\ j \geq 2 \quad (37)$$

The correlation ρ_{12} of two successive wave heights is calculated from eq. (25) (γ_h is identical to ρ_{12}).

APPLICATIONS

To demonstrate the viability and the various facets of the developed methodology, two applications have been undertaken. The first concerns a Ro/Ro ferry on a specific route in the Mediterranean Sea. Probabilities and corresponding critical time ratios for three modes of instability have been calculated.

The basic particulars of the ferry are collected in Table 1. In Fig. 4 is shown her general arrangement. To reduce the volume of calculations, one draft corresponding to full-load departure has been considered. The route Barcelona – Piraeus was selected and in Fig. 5 it is schematically shown on a Google-Earth (2005) map of the Mediterranean Sea. The length of the trip was calculated as 1209.30 nautical miles. Given the service speed and ignoring any voluntary or involuntary speed reduction, it should be covered theoretically in 53.75 hours.

The objective was to assess the ship in the "long-term" concerning her winter operability. Mediterranean wave statistics have been accounted through placement of "weather nodes" along the selected route, a concept explained in earlier Sections. Each node lies at the centre of an assumed square "influence" area (Fig. 6). To ensure sufficient resolution, a grid with density $1^0 \times 1^0$ has been built surrounding the route. In total, 23 nodes were cast. Tagged (per node) wave statistics were contributed by Athanassoulis et al (2006). In particular, the following wave parameters were required:

- Univariate distributions of significant wave height H_S, peak period T_P and mean wave direction Θ_M.
- Bivariate distributions of significant wave height and peak period.

The obtained mean weather values for winter season along the route are shown in Figs. 7 to 9 (preliminary analysis confirmed that, in this region, the most extreme values of wave parameters arise in winter).

Time of exposure to weather (per node)

The time spent by the ship within the influence area of some arbitrary node depends on the route's length that is inside the associated rectangle (obtained with a simple geometrical calculation) and the speed sustained by the ship in that part of the journey. Judging from the values of the most probable significant wave heights and periods, the assumption of constant speed is not completely unrealistic. In Table 2 are summarized the segmented travel times associated with the 23 nodal points. Thereafter, the time of exposure to beam, head and following seas can be deduced taking into account the heading of the ship (according to the defined route) and the distribution of mean direction of the local wave field around each node. The time of exposure to specific wave scenarios, scaled with reference to the duration of the whole voyage, is shown in Fig. 10.

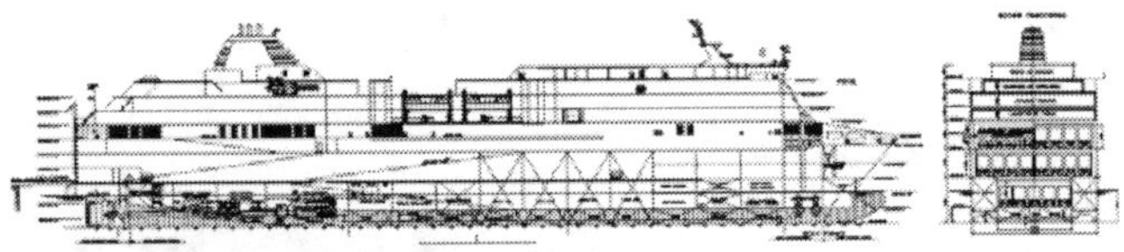

Fig. 4: General arrangement.

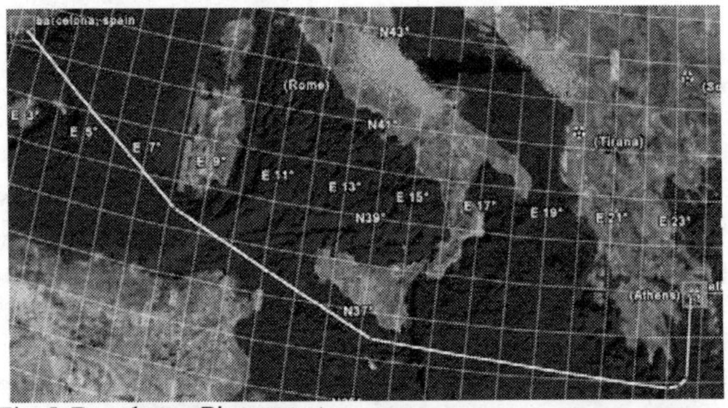

Fig. 5: Barcelona – Piraeus route.

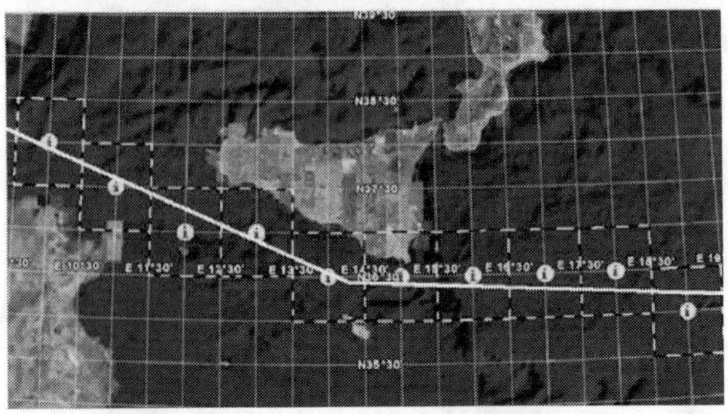

Fig. 6: Middle part of route, nodes and areas of influence are shown.

Table 1. Basic particulars of RO/RO ferry.

L_{BP} (length)	157 m	GM (metacentric height, corrected)	2.08 m
B (beam)	26.2 m	T_0 (natural roll period)	15.26 s
D (depth to upper deck)	9.2 m	b_{BK}, l_{BK} (breadth, length of bilge keels)	0.26 m, 60.9 m
T_d (mean draught)	6.2 m	KG (vertical position of the center of gravity above keel)	12.72 m
C_b (block coeff.)	0.626	Trailers	99
V_S (speed)	22.5 kn	Cars	166

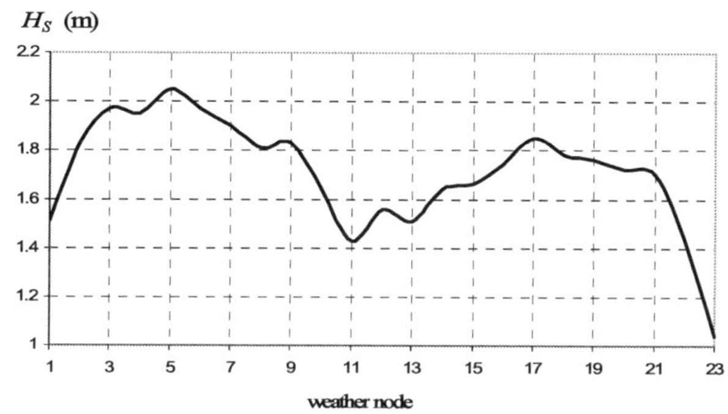

Fig. 7: Variation of mean value of H_S along the route for winter.

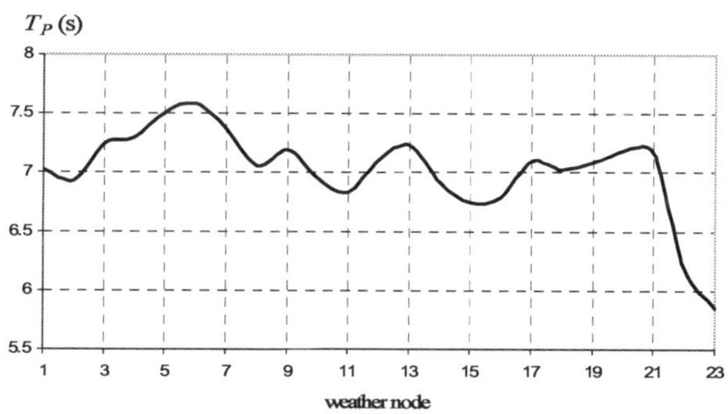

Fig. 8: Variation of mean value of T_P, as above.

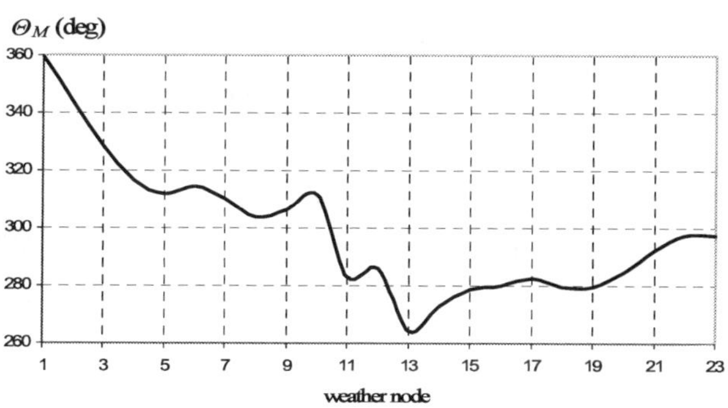

Fig. 9: Variation of mean Θ_M (0^0 waves coming from North, 90^0 East).

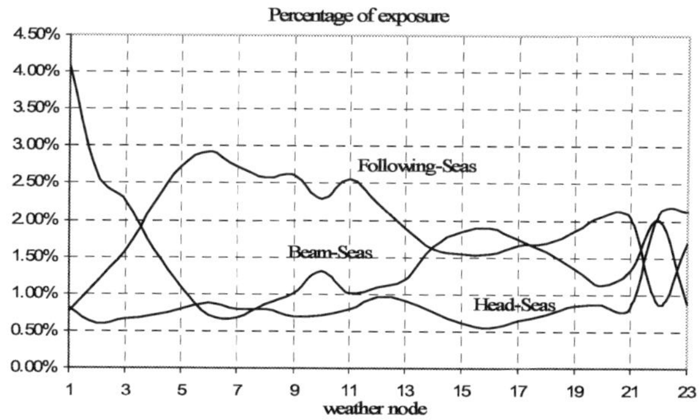

Fig 10: Exposure to beam, head and following seas, per node.

Table 2. Time spent in each grid sub-area.

	Distance (nm)	Time (hr)	%
node 1 (41^{0}N, 03^{0}E)	68.56	3.05	5.67
node 2 (40.5^{0}N,04^{0}E)	53.28	2.37	4.41
node 3 (40^{0}N,05^{0}E)	54.43	2.42	4.50
node 4 (39.5 ^{0}N,06^{0}E)	54.64	2.43	4.52
node 5 (39^{0}N,07^{0}E)	55.71	2.48	4.61
node 6 (38.5^{0}N,08^{0}E)	54.62	2.43	4.52
node 7 (38^{0}N,09^{0}E)	50.78	2.26	4.20
node 8 (38^{0}N,10^{0}E)	51.21	2.28	4.23
node 9 (37.5^{0}N,11^{0}E)	52.00	2.31	4.30
node 10 (37^{0}N,12^{0}E)	52.01	2.31	4.30
node 11 (37^{0}N,13^{0}E)	52.76	2.34	4.36
node 12 (36.5^{0}N,14E)	51.81	2.30	4.28
node 13 (36.5^{0}N,15^{0}E)	48.37	2.15	4.00
node 14 (36.5^{0}N,16^{0}E)	48.50	2.16	4.01
node 15 (36.5^{0}N,17^{0}E)	48.57	2.16	4.02
node 16 (36.5^{0}N,18^{0}E)	48.31	2.15	3.99
node 17 (36^{0}N,19^{0}E)	48.81	2.17	4.04
node 18 (36^{0}N,20^{0}E)	48.48	2.15	4.01
node 19 (36^{0}N,21^{0}E)	48.87	2.17	4.04
node 20 (36^{0}N,22^{0}E)	48.95	2.18	4.05
node 21 (36^{0}N,23^{0}E)	51.00	2.27	4.22
node 22 (36.5^{0}N,24^{0}E)	60.84	2.70	5.03
node 23 (37.5^{0}N,24^{0}E)	56.79	2.52	4.70
Total	1209.30	53.75	100.0

Norms of unsafe response for ship and cargo

At this stage failure norms need to be set, addressing the safety of the ship and her cargo. These norms are expressed respectively through a critical roll angle for the ship and a critical acceleration for the cargo as explained next:

To determine a critical roll angle as threshold of imminent "capsize", the principle of the weather criterion was adopted. This angle should thus be the minor of: the angle of vanishing stability (determined at $\varphi_c = 63^0$) ; the flooding angle ($\varphi_f = 35^0$); and the prescriptive limit of $\varphi_a = 50^0$. Subsequently 35^0 has been set as the critical roll angle.

Regarding the cargo, the acceleration that imperils the lashing of the remotest trailer has been targeted as the critical response. The tendency of the trailer for transverse sliding and tipping was checked for three different lashing arrangements (Fig.11). Background calculations have been reported in detail earlier (Themelis and Spyrou 2006). Ship and trailer parameters that are essential for the current analysis are collected in Table 3. The most critical transverse acceleration was found to correspond to transverse sliding for a vertical securing angle of 60^0 (Fig. 12). The specific value of this critical acceleration was calculated as $a_y = 6.04$ m/s^2.

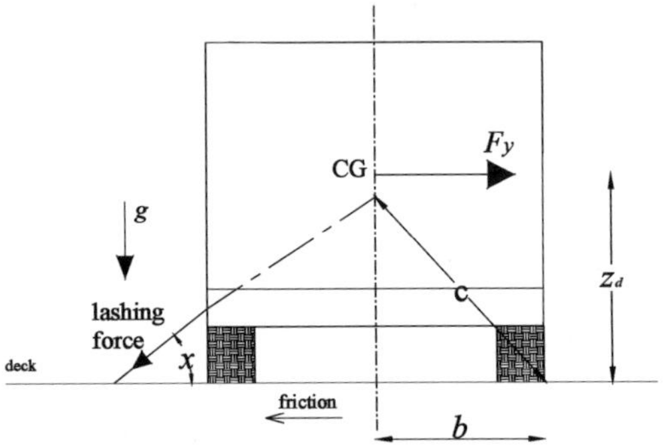

Fig. 11: Lashing arrangement (transverse view) and key forces.

Table 3. Trailer and lashings.

Trailer mass:	$m_t = 49$ t
Most arduous position of trailer:	$x = 57.2$ m (from mid-ship section) y=10.2 m $z_{bl} = 14.84$ m (from base-line)
Centre of gravity above deck:	$z_d = 2.36$ m
Lever-arm for tipping:	$b = 1.261$ m
Frictional coefficient:	$\mu = 0.3$ (rubber against steel)
Lashing arrangement:	3 chains with maximum securing load 100 kN on each side, symmetrical. Vertical securing angle: $x = 30°/45°/60°$

Beam-seas resonance

Numerical simulation based on regular excitation has been employed, using a mathematical model of coupled roll, heave and sway motions that is briefly described in Appendix A. In Figs. 13 and 14 are summarized the obtained critical wave heights, parameterized with respect to wave period and run length, that generate exceedence of norm. They have been identified by taking into account ship response (Fig 13) and trailer shift (Fig. 14). The realistic range of periods of ocean waves has been spanned. It should be noted however that, in the low range of periods (about $T_w < 8.5$ s), it was not feasible to determine critical heights because the required steepness exceeded the wave breaking limit of Airy waves $(H/\lambda \simeq 1/7)$.

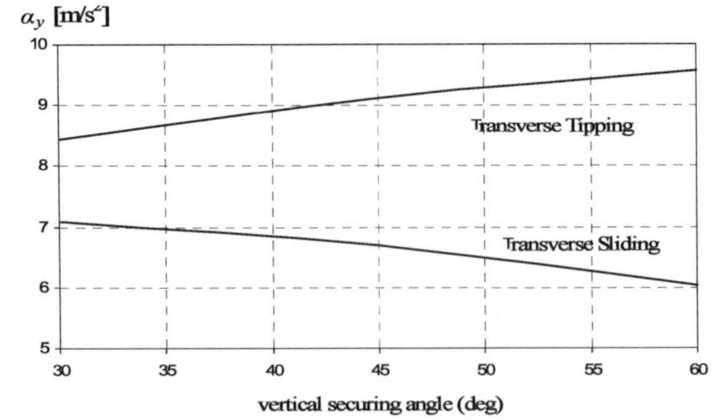

Fig. 12: Critical transverse accelerations (with respect to sliding and tipping) for a range of angles of lashing arrangement.

Parametric rolling of critical amplitude

For the assumed speed of $V_S = 22.5$ kn the ship could be liable to head-sea parametric rolling at the principal region of instability when the wavelength receives values like those shown in Fig. 15. The diagram was obtained by setting the frequency parameter a defined below, to a value around 1.0 so that, one of the conditions of principal resonance is satisfied.

$$a = \frac{4\omega_0^2}{\omega_e^2} \tag{38}$$

It is admitted that, at a high speed in head-seas, the heave and pitch effects should be important and a coupled model should be more appropriate. Here it is not intended to elaborate on this aspect further as the emphasis was set on the workings of the probabilistic method.

A convenient analytical formula that expresses the transient part (growth) of parametric rolling has been used. More details are found in Appendix B. The formula supplies the critical magnitude of parametric excitation h_{cr} that is sufficient for realising q-fold increase from an initial roll disturbance, within a set number, say p, of roll cycles. Although here a range of a values is examined (from 0.7 to 1.2), it is noted that at exact resonance it receives the following simple form:

$$h - \frac{4k}{\omega_0} = \frac{0.693 + \ln q}{1.571p} \tag{39}$$

By $2k$ is meant the dimensional linear damping divided by the moment of inertia (mass plus hydrodynamic). For the current Ro/Ro ship the calculated damping value was $k = 0.01039$ s^{-1}, according to the already discussed numerical model of beam-sea rolling.

The parametric excitation h was linked exclusively to GM fluctuation:

$$h = \frac{GM_{trough} - GM_{crest}}{2GM_{mean}} \tag{40}$$

The q-fold growth is defined as the ratio of the critical roll angle (35^0 or the amplitude of the roll oscillation during which the critical acceleration $a_y = 6.04$ m/s^2 is realized) to an uncertain initial roll disturbance. To simplify matters, this initial roll angle disturbance (combined with zero initial roll velocity) was assumed uniformly distributed in the range $\varphi_0 = 0^0$ - 6^0 with 3^0 discretisation step. In Fig. 16 is shown the obtained critical parametric amplitude for the sub-range φ_0=3^0 - 6^0.

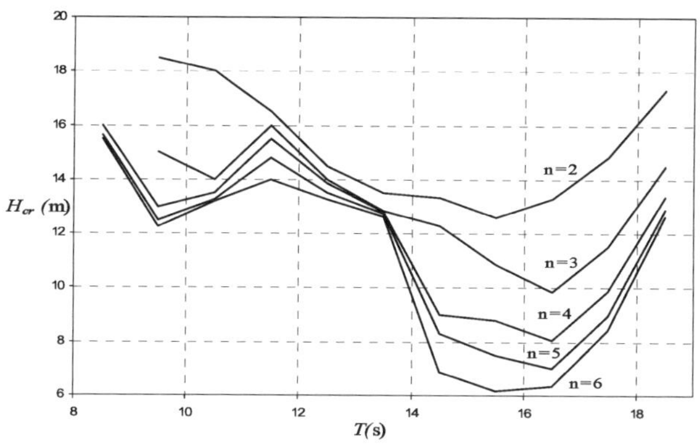

Fig. 13: Wave groups producing critical ship inclination.

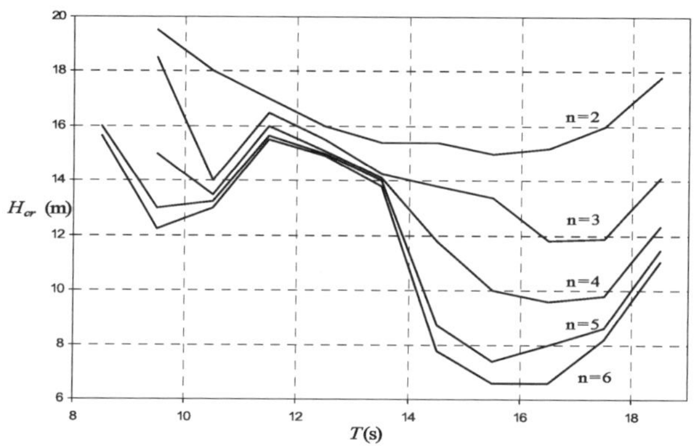

Fig. 14: Wave groups producing critical cargo acceleration.

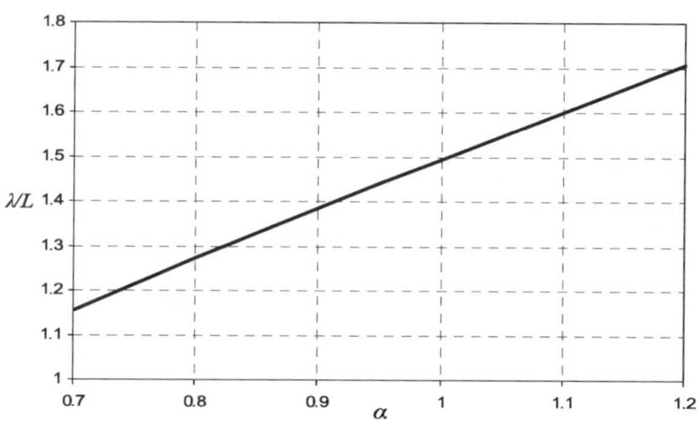

Fig. 15: Wavelengths conducive to principal parametric rolling.

To realize p parametric roll cycles, a wave group with run length equal to $2p$ should be encountered; i.e. for, say, 4 parametric roll cycles, approximately 8 successive critical waves will be needed. Should this be combined with a requirement of high waves in the group, such an encounter would constitute a very low probability event.

In the next step, the identified critical parametric excitations need to be converted into critical wave heights taking into account the ship form. Characteristic variations of submerged volume as the ship passes from a crest and from a trough, for one of the critical combinations of wave steepness and length ratio, is shown in Fig. 17. The obtained critical wave heights according to run length are shown in Fig. 18. The sensitivity of righting-arm variation to wave height can be assessed from the two characteristic examples shown in Fig. 19.

It is remarked that, some truly excessive wave heights appear in Fig. 18. However, their true relevance will emerge from the probability calculation and there is no need to be excluded at this stage.

Critical waves for pure – loss of stability

Pure-loss of stability in astern seas is based on the single encounter of a critical wave and in this respect this phenomenon can be treated separately from parametric rolling although both accrue from the fluctuating righting-arm. The procedure for the identification of the critical waves that could rouse pure-loss of stability is based on a procedure described briefly in Appendix C. The key idea exploited is that the critical fluctuation of *GZ* can be identified on the basis of the following condition: the time of experiencing negative restoring in the vicinity of a crest should be at least equal with the time that is necessary for developing capsizal inclination, assuming an initial roll disturbance. Unlike a resonance phenomenon, this growth is relatively slow, almost insensitive to damping, and the condition on the ship norm should prevail over the condition on the cargo. The critical wave height for several values of the length ratio λ/L was determined according to the above procedure, taking into account the detailed form of the *GZ* curve. As expected, the required wave height proves to be extremely high and thus very unlikely to be encountered in practice (Fig. 20).

Probability of "instability" per mode

The identification of critical wave groups per stability criterion (with reference to the corresponding limit-state parameter and norm) should be followed by the calculation of the probability of encountering these (or "worse") wave groups as well as the associated critical time ratio of ship exposure. The assumption of JONSWAP spectrum underlies the ensuing calculation of probabilities (Hasselmann 1973). As is well-known, its spectral density function is expressed as:

$$S(\omega) = a_w g^2 \omega^{-5} \exp\left(-\frac{5}{4}\left(\frac{\omega}{\omega_p}\right)^{-4}\right) \gamma^{\exp\left(-0.5\left(\frac{\omega-\omega_p}{\sigma\omega_p}\right)^2\right)} \tag{41}$$

where

$$a_w = \frac{5}{16}\left(\frac{H_S^2 \omega_p^4}{g^2}\right) A_\gamma \tag{42}$$

$$A_\gamma \cong 1 - 0.287 \ln(\gamma) \tag{43}$$

$$\sigma = \begin{cases} 0.07 \text{ if } \omega \le \omega_p \\ 0.09 \text{ if } \omega > \omega_p \end{cases} \tag{44}$$

a_w is the generalized Philips' constant, A_γ a normalizing factor, γ the peakness parameter, σ the spectral width parameter and ω_p the angular spectral peak frequency. For the peakness parameter γ the following formulas can be used (DNV 2002).

$$\gamma = \begin{cases} 5 \text{ for } \frac{T_P}{\sqrt{H_S}} \le 3.6 \\ e^{5.75-1.15\frac{T_P}{\sqrt{H_S}}} \text{ for } 3.6 \le \frac{T_P}{\sqrt{H_S}} \le 5 \\ 1 \text{ for } 5 \le \frac{T_P}{\sqrt{H_S}} \end{cases} \tag{45}$$

The encounter spectrum is obtained by the next transformation:

$$S(\omega_e,\psi) = \frac{S(\omega)}{\left|1 - \dfrac{2\omega U_S \cos\psi}{g}\right|} \tag{46}$$

For each weather node placed along the route, mean "winter" values of H_S and T_P have been used.

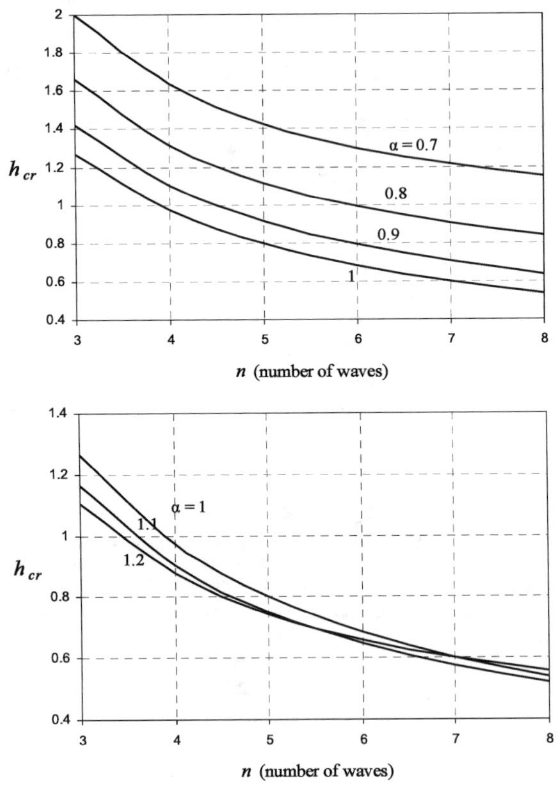

Fig. 16: Critical parametric roll amplitudes assuming an initial roll angle in the range φ_0=3^0 - 6^0.

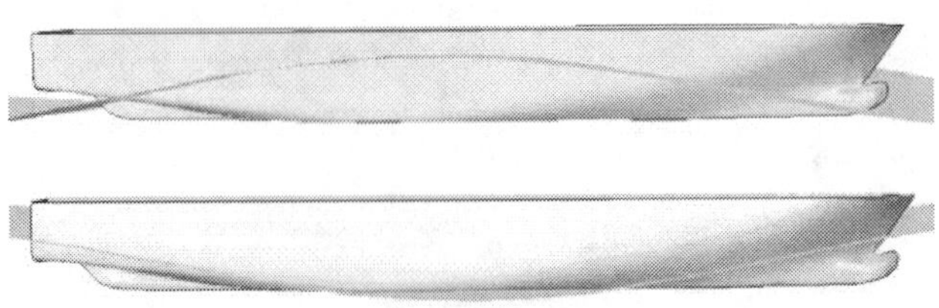

Fig. 17: Variation of the wetted part of the hull in a wave ($H/\lambda = 0.065$, $\lambda/L = 1.493$).

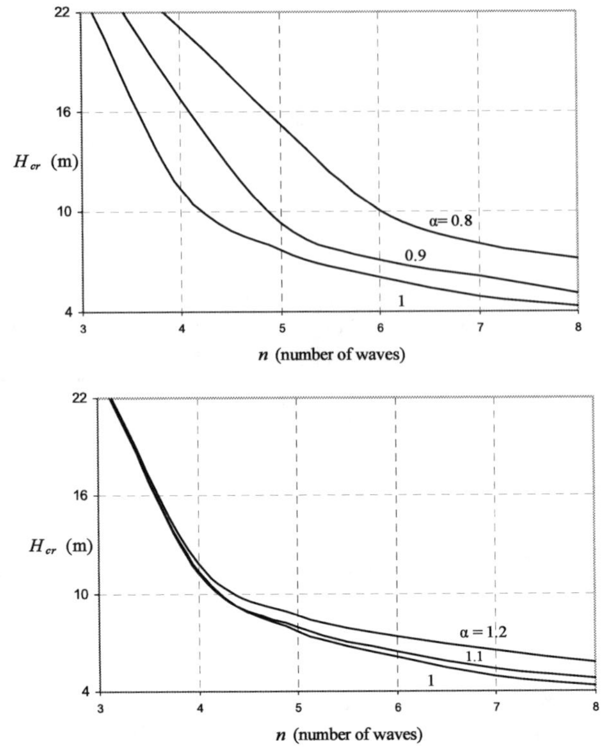

Fig. 18: Corresponding wave height for reaching the critical roll angle.

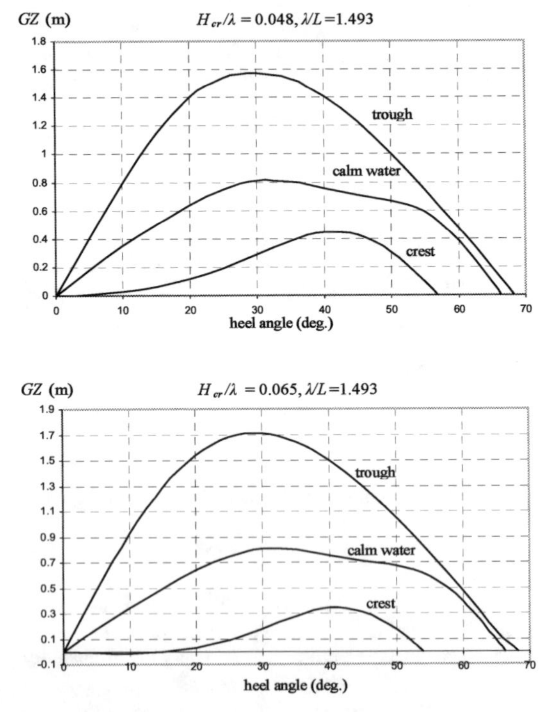

Fig. 19: Variation of the *GZ* curve for $a = 1$ and two different wave heights.

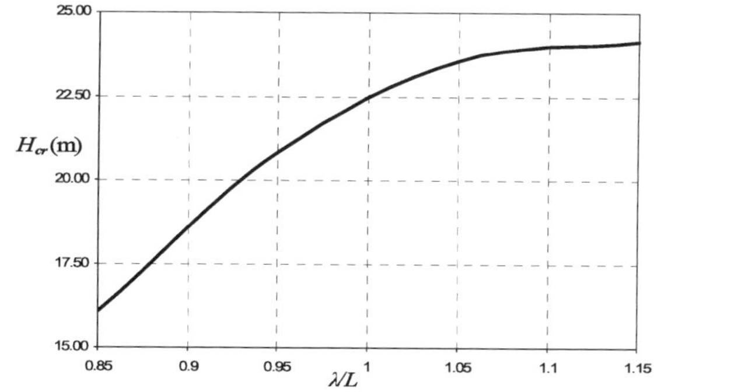

Fig. 20: Extreme wave heights are required for realizing pure-loss of stability of the Ro/Ro ferry.

Resonance in beam-seas

The probability calculations described in the earlier sections lead us to propose a new type of ship stability performance diagram, illustrating the variation of the critical time ratio at various locations along the desired route. Such a diagram is shown in Fig 21, targeting specifically the occurrence of beam-sea resonance and with reference to the two specified norms of ship and cargo safety. The applied time scaling was local, in accordance to the time spent in the vicinity of the nearest node; i.e. within the associated rectangle.

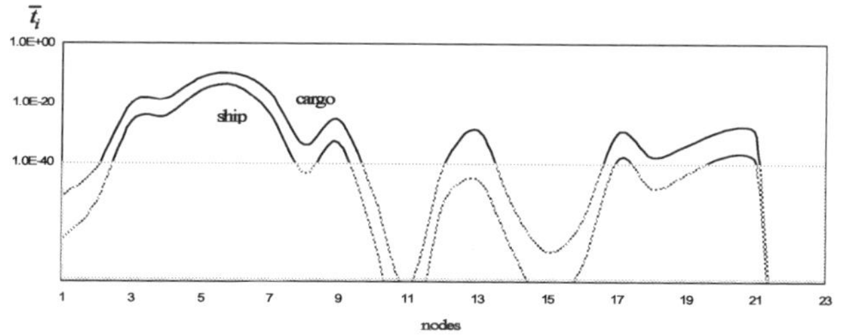

Fig. 21: Variation of critical time ratio during voyage, for extreme rolling in beam-seas (ship and cargo norms).

Head-seas parametric rolling

For a more meaningful calculation of probabilities, some fluctuation of speed ± 1 kn around the service speed $\left(V_S = 22.5 \text{ kn}\right)$ has been assumed, given that this affects the frequency of encounter which, in a random seaway, is a probabilistic quantity. The obtained diagram of critical time ratio for head-seas parametric rolling can be seen in Fig.22. As expected, for most part of the journey the ratio of critical time is negligibly small.

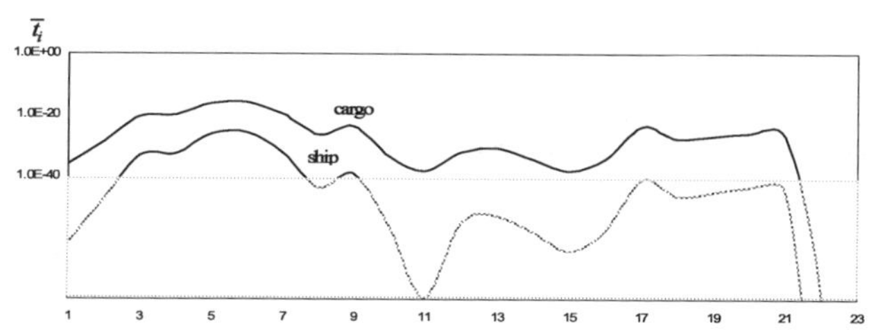

Fig. 22: Critical time ratio curves for head-seas parametric rolling (ship and cargo norms).

Pure-loss of stability

In Fig. 23 is shown the probability of occurrence of a "pure-loss" event.

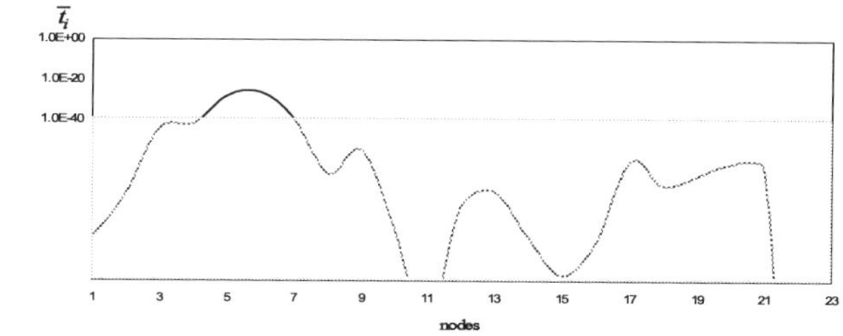

Fig. 23: Critical time ratio for pure-loss of stability (ship norm).

Collective probability of instability (all considered modes)

The percentage of time spent in beam, head and following seas is shown in Fig. 24. The three critical time ratio curves are overlaid in Fig. 25. The probability of exhibiting instability of anyone of the considered types may thus be determined. The last figure is very illuminating also about the ship's propensity to instability of some selected type for individual stages of the journey. Critical times for ship and for cargo (beam-sea resonance) per node can be compared in Fig. 26. Finally, the tendency for instability concerning the entire voyage can be characterized from the probability values shown in Table 4.

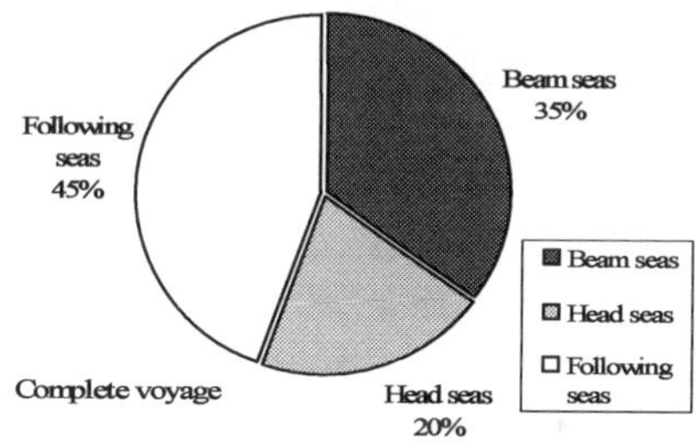

Fig. 24: Percentage of time in beam, head and following seas.

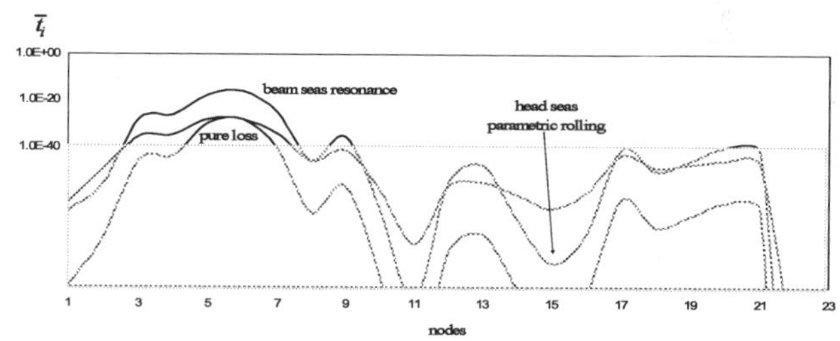

Fig. 25: Comparison of tendencies for different types of instability during journey (ship norm).

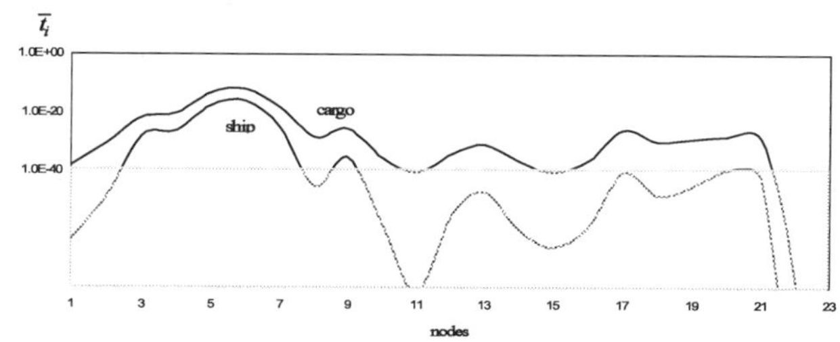

Fig. 26: Critical time ratio diagrams for ship and for cargo.

Table 4. Summed probability and associated critical time ratio for entire journey.

	P_i	$\bar{\xi}_i$
Ship: ($\varphi > 35^0$)	6.58E-17	9.55E-17
Cargo: ($a_y > 6.04$ m/s^2)	7.84E-13	1.14E-12

Probability values are expressed as number of critical waves over the total number of encountered waves. Thus, to determine the probability of a single loss given a certain period of exposure, a suitable transformation is entailed. Let for example one assume 25 years of exposure (winter operability) and let service time represent 60% of the actual time. A mean wave period of 6.5 s along the route leads to 0.6 x 25 (years) x 3 (months) x 30 (days) x 24 (hr) x 3600 (s) / 6.5 (s) = 1.8 x 10^7 encountered waves approximately. Then the life-time probability of a single cargo shift event (this is the worst case according to Table 4) is about 1.43 x 10^{-5}.

Parametric rolling of post-panamax containership

In this second application the examined ship is a fictitious post-panamax containership (Fig. 27). Her particulars are collected in Table 5. The objective of this study is to test the capability of the method to reflect realistically the severity of the environment in the extracted figure of the probability of instability. Taking as basis "North Atlantic" conditions of navigation, parametric studies will be performed for a range of significant wave heights H_s and peak periods T_P. These will be assumed to represent uniformly the wave environment. No specific route will be defined. The ship is tested in the "short-term" for parametric rolling. As before, the method of calculation is the one described in Appendix B.

Investigations of this kind are believed to be relevant, for example, to a weather routing decision support system. Specific H_s and T_P values could have been obtained by means of weather forecasting. Here, for simplicity, their values are taken as certain (probability equal to 1.0).

Fig. 27: Hull form of the examined containership.

A roll angle of 15^0 has been specified as the value whose exceedence should be avoided. The critical parametric amplitude h_{cr} that could provoke such roll motion within a limited number of wave cycles has been calculated for a range of initial angular disturbances ($\varphi_0 = 0^0$ - 6^0). A 2^0 discretisation step and 50%, 30% and 20% probabilities for the first, second and third range respectively have been assumed. Furthermore, rather than a single speed, a range of speeds $V_S \in [0,8]$ kn has been considered, due to its effect on the encounter frequency which is a probabilistic quantity. To reduce the number of calculations, we took two equal sub-ranges; with 60% probability of occurrence for the first sub-range and 40% for the second.

The linear roll damping coefficient k has been determined as $k = 0.0129\ s^{-1}$. In Fig. 28 are presented the calculated critical wave amplitudes H_{cr} for the third sub-range of initial roll angles and the first sub-range of speeds. The intensity of variation of the containership's restoring in waves can be assessed from Fig. 29. The calculation of probabilities relied again on the JONSWAP spectrum.

Table 5. Ship particulars.

L_{BP} (length)	288.87 m	Δ (displacement)	113,956 t
B (beam)	42.80 m	To (natural roll period)	30.26 s
D (depth, upper deck)	24.40 m	KG (vertical position of the center of gravity above keel)	18.83 m
T_d (mean draught)	14.00 m	GM (metacentric height)	1.08 m

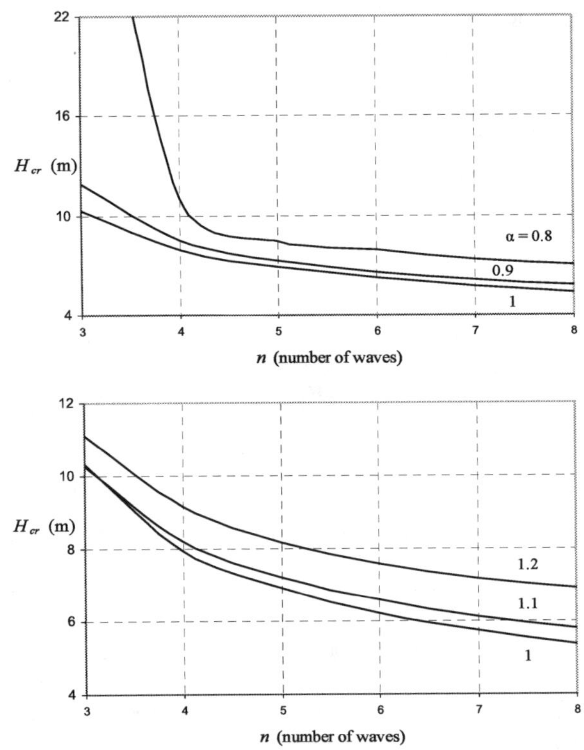

Fig. 28: Required wave height for reaching the critical roll angle (initial roll angle in range $\varphi_0 = 4^0$ - 6^0 and speed in range $V_S = 0-4$ kn).

The principal results of this investigation are summarized in Figs. 30 and 31. As the peak period of the spectrum is shifted to higher values, the probability of occurrence of parametric rolling that exceeds the prescribed amplitude is increased. Nearer to the limit of realistic wave periods this trend is weakened. In general, realistic conditions are encountered where the probability of parametric rolling exceeds the level of 10^{-5}. As for the dependence on significant wave height, once a certain threshold level has been exceeded (about 5 m), the probability increases slowly with the wave height. This is explained as follows: as soon as the containership is caught in the parametric rolling regime, she exhibits a tendency for large amplitude rolling (30^0 and above). In consequence, given such a realization, the prescribed roll amplitude is easily surpassed even for the lower (but above critical) wave heights.

Lastly, let us estimate the probability of exhibiting instability for a fixed time of exposure to a severe sea state characterized by $H_S = 8$ m, $T_P = 16$ s, assuming that our containership encounters the waves always from the bow. The ensuing probability of critical waves is 5.1 x 10^{-5}. Assuming a mean ship speed of 4 kn, then the mean encounter wave period is 12.66 s and the number of waves encountered by the ship in one hour should be 3600 (s) / 12.66 (s) = 2.84 x 10^2. Hence the probability of instability for one hour of exposure should be 1.45% which is a rather high value.

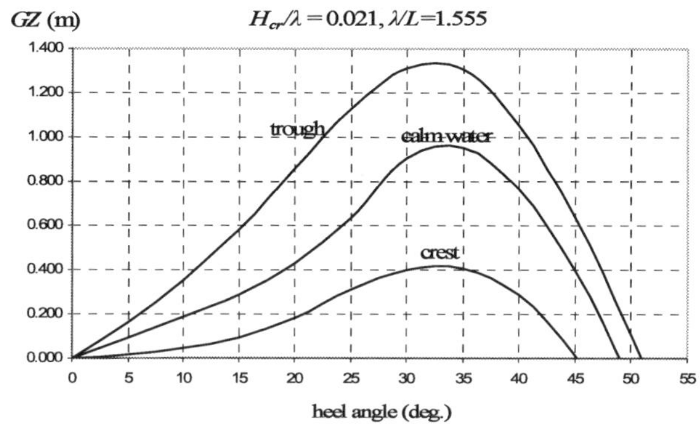

Fig. 29: Variation of the *GZ* curve on a characteristic steep wave.

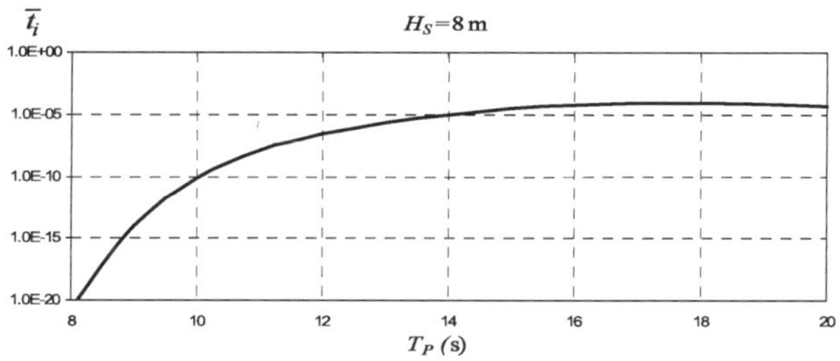

Fig. 30: Critical time ratio as functions of T_p

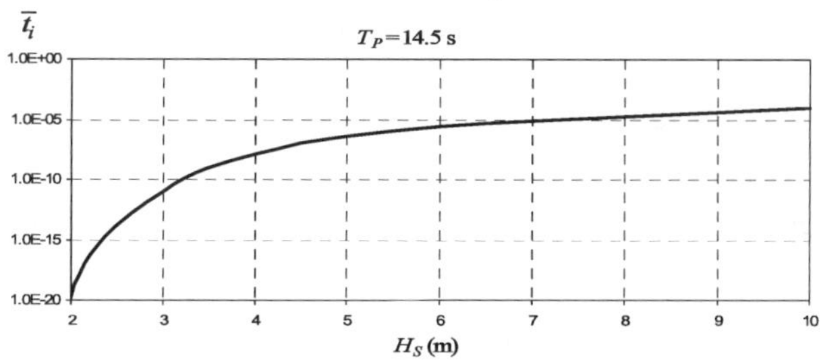

Fig. 31: Critical time ratio as functions of H_s

CONCLUSIONS

A novel method for the probabilistic assessment of ship stability has been developed that can be easily integrated within a risk assessment framework. By exploiting the groupiness characteristic of high waves, the method combines the strengths of deterministic analyses of ship dynamics with state-of-the-art probabilistic analysis of the seaway. The calculation of joint pdfs of successive wave heights and periods is the most laborious part of the work, which however can be easily automated since this is not tied directly to the ship dynamics calculations. It is believed that all established ship capsize modes can be handled by the current method. To facilitate understanding, the method was laid out as a number of specific steps.

Practicality was proven through application to a Ro-Ro ferry that was set to operate on an invented long route across the Mediterranean Sea. Also, a second study was undertaken, targeting the probability of parametric rolling of a containership operating in "North-Atlantic like" conditions. Mathematical models of different level of detail have been used. In general, the proposed method does not depend exclusively on a specific mathematical model and predictions of instability based on different models or techniques can be combined for reaching a more informed decision. The probability figures obtained from these two application studies appear to be quite logical.

ACKNOWLEDGMENTS

This research has been partially supported by the integrated project SAFEDOR funded by the European Commission (relevant sub-project: "SP 2.3: Probabilistic assessment of intact stability; alternative method 2"). In addition, the first author wishes to thank the Greek Scholarship Foundation for supporting his PhD research. Finally, the contribution of Mr. Stavros Niotis, a recent graduate of NTUA, to the calculations presented in the paper is acknowledged.

REFERENCES

ARNOLD, L., CHUESHOV, I. AND OCHS, G. "Stability and capsizing of ships in random sea – a survey." *Nonlinear Dynamics*, 36 (2004): 135 – 179.

ATHANASSOULIS, G.A., BELIBASSAKIS, K.A. AND GEROSTATHIS, T.P." Description of wave data along ship routes in the Mediterranean Sea." *NTUA Internal Report, SAFEDOR S.P.2.3.6,* Athens, 2006.

AYYUB, B.M., KAMINSKIY, M., ALMAN, P.R., ENGLE, A., CAMPBELL, B.L. AND THOMAS, W.L. III. "Assessing the probability of the dynamic capsizing of the vessels." *Journal of Ship Research*, 50: 4 (2006): 289-310.

BAARHOLM, G.S. AND JENSEN, J.J. "Influence of whipping on long-term vertical bending moment." *Journal of Ship Research*, 28:4 (2004): 261-272.

BATTJES, J.A AND VAN VLEDDER, G.PH. "Verification of Kimura's theory for wave group statistics." *Proceedings*, 10th International Coastal Engineering Conference, (1984): 642- 648.

BELENKY, V.L " Piece – wise linear method for the probabilistic stability assessment for ship in a seaway." *Proceedings*, 5th International Conference on Stability of Ships and Ocean Vehicles, STAB 1994, Melbourne, (1994): 13-30.

BELENKY, V.L., DEGTYAREV, A. B. AND BOUKHANOVSKY, A. V. "Probabilistic quantities of nonlinear stochastic rolling." *Ocean Engineering*, 25 (1998): 1 – 25.

BELENKY, V., SEVASTIANOV, N. *Stability of Ships: Risk of Capsizing*, Vol. 2, Elsevier, Ocean Engineering Book Series, Kidlington, UK, 2003, ISBN 0080443540.

BELENKY, V., YU, HAN-CHANG AND WEEMS, K. "Numerical procedures and practical experience of assessment of parametric roll of container carriers." *Proceedings*, 9th International Conference on Stability of Ships and Ocean Vehicles, STAB 2006, Rio de Janeiro, (2006): 119-130.

BOUKHANOVSKY, A.V. AND DEGTYAREV, A.B. "Nonlinear stochastic ship motion stability in different wave regime." *Proceedings*, CFR'96: 3rd International Conference in Commemoration of the 300th Anniversary of creating Russian fleet by Peter the Great, St. Petersburg, 2 (1996): 296 – 306.

CAVANIÉ, A., ARHAN, M. AND EZRTATY, R. "A statistical relationship between individual heights and periods of storm waves." *Proceedings*, Conference on Behaviour of Offshore Structures, Trondheim, (1976): 354-360.

CUMMINS W.E. "The impulse response function and ship motions." *Schiffstechnik*, 9 (1962): 101–109.

DET NORSKE VERITAS. "DNV MaxWave." Report No. 2001 – 1491.

DRAPER, L. "Severe wave conditions at sea."*Journal of the Institute of Navigation*, 24 (1971): 273- 277.

FALZARANO, J., SHAW S. AND TROESCH, A. "Application of global methods for analyzing dynamical systems to ship rolling motion and capsizing." *International Journal of Bifurcation and Chaos*, 2 (1992): 101-115.

FROUDE, W. "On the rolling of ships." *Transactions of the Institute of Naval Architects*, 2 (1861).

GERASIMOV, A.V. "*Energy statistical theory of nonlinear irregular ship motion.*" Sudostroenie Publishing, Leningrad (in Russian).

GOOGLE EARTH, "*http://earth.google.com*" (2005).

HADDARA, M.R." A modified approach for the application of Fokker - Planck equation to the nonlinear ship motions in random waves." *International Shipbuilding Progress*, 21 (1974): 283 – 288.

HADDARA, M.R. AND ZHANG, Y. "On the joint probability density function of nonlinear rolling motion." *Journal of Sound and Vibration*, 169 (1994): 562- 569.

HADDARA, M.R. AND NASSAR, M.A. "A stochastic model for the analysis of rolling motion in a realistic seaway." *International Shipbuilding Progress*, 33 (1986): 144- 150.

HASSELMANN, K. "Measurements of wind-wave growth and swell decay during the Joint North Sea Wave Project (JONSWAP)." *Ergänzungsheft zur Deutschen Hydrographischen Zeitschrift*, Reihe (A8) 12 (1973).

HAYASHI, C. *Nonlinear Oscillations in Physical Systems*, Princeton University Press, Princeton, New Jersey, 1986, ISBN 0691083835.

HSIEH, S.R., SHAW, S.W. AND TROESCH, A.W. "A nonlinear probabilistic method for predicting vessel capsizing in random beam seas." *Proceedings A*, The Royal Society, London, 446 (1994): 195 – 211.

IBRAHIM, M.R. *Parametric Random Vibration*, Research Studies Press, Letchworth, England, 1985, ISBN 0471908304.

IMO, "Code of safe practice for cargo stowage and securing and amendments thereto". *Resolution A.714 (with 1994/95 amendments, Annex 1-13)*, 1991.

DeKat, J.O. "Irregular waves and their influence on extreme ship motions." *Report No. 208883-OP1*, ONR'94 (1994).

KIMURA, A. "Statistical properties of random wave groups." *Proceedings*, 17th International Conference on Coastal Conference, Sydney, Australia, (1980): 2955 – 73.

LONGUET-HIGGINS, M.S. "The statistical analysis of a random moving surface." *Philosophical Transactions A*, The Royal Society, London, 249 (1957): 321 – 387.

LONGUET-HIGGINS, M.S. "On the joint distribution of the periods and amplitudes of sea waves." *Journal of Geophysical Research*, 80 (1975): 6778–89.

LONGUET-HIGGINS, M.S. "On the joint distribution of wave periods and amplitudes in a random wave field." *Proceedings A*, The Royal Society, London, 389 (1983): 241–258.

LONGUET-HIGGINS, M.S. "Statistical properties of wave groups in random sea state." *Philosophical Transactions A*, Royal Society, London, 312 (1984): 219 – 250.

MASSON, D. AND CHANDLER, P. "Wave groups: a closer look at spectral methods." *Coastal Engineering*, 20 (1993): 249-275.

MATHIESEN, J. AND BITNER – GREGERSEN, E. "Joint distribution for significant wave height and zero crossing period." *Applied Ocean Research*, 12 (1990): 93-103.

MCTAGGART, K. AND DEKAT, J.O. "Capsize risk of intact frigates in irregular seas." *SNAME Transactions*, 108 (2000): 47 – 177.

MOSELEY, H. "On the dynamical stability and on the oscillation of floating bodies." *Philosophical Transactions A*, The Royal Society, London, 207 (1850): 609–643.

MYRHAUG, D., DAHLE, E.A., RUE, H. AND SLAATTELID, O.H. "Statistics of successive wave periods with application to rolling of ships." *International Shipbuilding Progress*, 47 (1999): 253-266.

NAITO, S. MINURA, M. HAMANAKA, S. AND YAMAMOTO, T. "Long-term prediction method based on ship operation criteria." *International Shipbuilding Progress*, 53 (2006): 229-252.

NESS, O.B., MATHISEN, J., MCHENRY, G. AND WINTERSTEIN, S. "Nonlinear analysis of ship rolling in random beam waves." *Proceedings*, STAR Symposium, New Orleans.

OCHI, M. "On long term statistics for ocean and costal waves. *Proceedings*, 16th Conference on Coastal Engineering, Hamburg, (1978): 59-75.

OCHI, M. *Ocean Waves: The Stochastic Approach*, Cambridge University Press, Cambridge, England, 1998, ISBN 052156378X.

PAROKA, D. AND UMEDA, N. "Capsizing probability prediction for a large passenger Ship in irregular beam wind and waves: comparison of analytical and numerical methods." *Journal of Ship Research*, 50 (2006): 371 – 377.

RAINEY, R.C.T AND THOMPSON, J.M.T. "Transient capsize diagram – a new method of quantifying stability in waves." *Journal of Ship Research*, 41 (1991): 58 – 62.

ROBERTS, J.B. "A stochastic theory for nonlinear ship motion in irregular seas." *Journal of Ship Research*, 26 (1982): 229 – 245.

SIMIU E. *Chaotic Transitions in Deterministic and Stochastic Dynamical Systems – Applications of Melnikov Processes in Engineering, Physics and Neuroscience.* Princeton Series in Applied Mathematics, Princeton University Press, New Jersey, 2002, ISBN 0691050945.

SPYROU, K.J. "Pure – loss of stability revisited: analytical and numerical design aids." *NTUA Internal Report*, Athens, 2000.

SPYROU, K.J. "Design criteria for parametric rolling." *Oceanic Engineering International*, 9 (2005): 11-27.

SPYROU, K.J. AND THEMELIS, N. "Probabilistic assessment of intact stability." *Proceedings*, 8th International Ship Stability Workshop, Istanbul, (2005).

SPYROU, K.J., POLITIS, K., LOUKAKIS, T. AND GRIGOROPOULOS, G. "Towards a risk – based system for the departure control of passengers ships in rough weather in Greece."*Proceedings*, 2nd International Maritime Conference on Design for Safety, Sakai, Japan, (2004): 255 – 261.

ST. DENIS, M. AND PIERSON, W. "On the motion of ship in confused ship." *SNAME Transactions*, 61 (1953): 280-357.

STANSELL, P., WOLFRAM, J. AND LINFOOT, B. "Statistics of wave groups measured in the northern North Sea; comparisons between time series and spectral predictions." *Applied Ocean Research*, 24 (2002): 91–106.

TAYFUN, M.A. "Joint distributions of large wave heights and associated periods." *Journal of Waterway, Port, Coastal and Ocean Engineering*, 119 (1993): 261–273.

THEMELIS, N. AND SPYROU, K.J. "A coupled heave-sway-roll model for the analysis of large amplitude ship rolling and capsize of ships in beam seas on the basis of a nonlinear dynamics approach." *Proceedings*, 16th International Conference on Hydrodynamics in Ship Design, Gdansk, Poland, (2005).

THEMELIS, N. AND SPYROU, K. "Probabilistic Assessment of Resonant Instability." *Proceedings*, 9th International Conference on Stability of Ships and Ocean Vehicles, STAB 2006, Rio de Janeiro, (2006): 37-48.

TIKKA, K.K AND PAULLING, J.R. "Prediction of critical wave conditions for extreme vessel response in random seas." *Proceedings*, 4th International Conference on the Stability of Ships and Ocean Vehicles, STAB '90, Naples, (1990).

VASSILOPOULOS, L. "Ship rolling at zero speed in random seas with nonlinear damping and restoration." *Journal of Ship Research*, 15 (1971): 289 – 294.

WIGGINS, S. *Global Bifurcations and Chaos: Analytical Methods*, Springer-Verlag, New York, 1988, ISBN 0387967753.

WIST, H. MYRHAUG D. AND RUE H. "Statistical properties of successive wave heights and successive wave periods." *Applied Ocean Research*, 26 (2004): 114–136

APPENDIXES

APPENDIX A: Mathematical model of coupled roll in beam seas

The presented investigation of ship rolling in beam seas was based on the mathematical model that is outlined below. For more details see Themelis and Spyrou (2005). From basic kinematics, the equations of motion in heave, sway and roll are written as follows:

$$m(\dot{v} - \dot{\varphi}w) = \sum F_y \tag{A1}$$

$$m(\dot{w} + \dot{\varphi}v) = \sum F_z \tag{A2}$$

$$I_G \ddot{\varphi} = \sum M_G \tag{A3}$$

where v, w are the sway and heave velocity of the ship's centre of gravity and $\dot{\varphi}$ is the roll angular velocity, m and I_G are, mass and mass moment of inertia around the longitudinal axis x of the ship that passes from the centre of gravity. The well-known transformation between the inertial and body fixed coordinate systems is applied.

The two forces and the moment (about the center of gravity) that appear at the right-hand-side of eq. (A.1)-(A.3) can be decomposed as follows:

$$\sum F = F_{Hs} + F_W{}^{FK} + F_W^D + F_R + F_V \quad \text{(A4)}$$

where F_{Hs} is hydrostatic, F_W^{FK} is Froude – Krylov, F_W^D is diffraction, F_R is radiation and F_V is the viscous force.

The hydrostatic and Froude – Krylov (hydrodynamic) forces are estimated by the integration of the incident wave pressure (static and dynamic respectively) over the wetted surface of the ship. For regular waves, the incident wave potential is calculated from:

$$\Phi_I = \frac{Ag}{\omega} e^{kZ^*} \sin(kY - \omega t) \quad \text{(A5)}$$

where Z* is the vertical distance from the water surface.

The hydrostatic and Froude - Krylov forces are repetitively

$$F_{HSi}(t) = -\rho g \iint_{S(t)} Z^* \vec{n}_i ds, \text{ for } i=2, 3, 4 \quad \text{(A6)}$$

$$F_{Wi}^{FK}(t) = -\iint_{S(t)} \rho \frac{\partial \Phi_I}{\partial t} \vec{n}_i ds, \text{ for } i=2, 3, 4 \quad \text{(A7)}$$

where i = 2, 3, 4 correspond respectively to sway, heave and roll motion, $\vec{n}_i$ are projections of the unit vector $\vec{n}$ (vertical towards the body) to the i direction (however $\vec{n}_4 = \vec{r} \times \vec{n}$, where $\vec{r}$ determines the examined position on the wetted surface). Also, $S(t)$ is the instantaneous wetted surface. As usual, ρ is seawater density and g is acceleration of gravity. It should be noted that the integration is performed over the instantaneous wetted surface and pressures are calculated from the exact wave elevation. As a matter of fact, at least some of the nonlinear part of the wave excitation is taken into account, which is important for more accurate simulation of large motions.

Currently in the mathematical model are calculated also: the radiation force, the sway drag force, roll damping; and cross coupling forces between sway, heave and roll. The drag force due to bilge keel is calculated according to the following expressions (see also Fig. A.1):

$$F_{BY} = \frac{1}{2}\rho(\dot{Y}_G - r_A \dot{\varphi}\cos(\theta) - u_2) * \left|\dot{Y}_G - r_A \dot{\varphi}\cos(\theta) - u_2\right| C_D A_{BK} \quad \text{(A8)}$$

$$F_{BZ} = \frac{1}{2}\rho(\dot{Z}_G - r_A \dot{\varphi}\sin(\theta) - u_3) * \left|\dot{Z}_G - r_A \dot{\varphi}\sin(\theta) - u_3\right| C_D A_{BK} \quad \text{(A9)}$$

$$F_B = \sqrt{F_{YB}^2 + F_{ZB}^2} \quad \text{(A10)}$$

$$M_B = -r_A\left(F_{BY}\sin\theta_1 + F_{BZ}\cos\theta_1\right) \quad \text{(A11)}$$

C_D is the drag coefficient and A_{BK} is the total bilge keel area. Other symbols are explained in Fig. A1. The method takes into account the local relative velocities along the hull, using the sway, heave and roll velocities ($\dot{Y}_G, \dot{Z}_G, \dot{\varphi}$), the wave particle velocities u_2 and u_3 as well as the detailed geometry of the hull.

The numerical model is programmed completely in a Mathematica environment. The code creates panels over the hull whereon the static and dynamic pressures are calculated at successive time steps, as well as the angle between the horizontal plane and the normal vector of the panel.

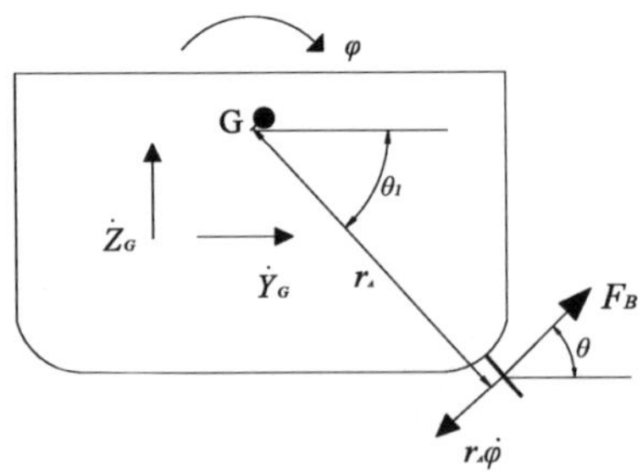

Fig.A1: Bilge keel damping forces.

APPENDIX B: Growth of parametric roll in principal region.

The following analysis is based on Spyrou (2005). We make the gross assumption that a rudimentary linear Mathieu-type model can describe adequately parametric rolling. Let us take the undamped system first: for the first region of instability, the unstable motion should build up according to the following approximate general solution (Hayashi 1986):

$$\varphi(t) = c_1 e^{\mu\omega_0 t}\sin(\omega_0 t - \sigma) + c_2 e^{-\mu\omega_0 t}\sin(\omega_0 t + \sigma) \quad \text{(B1)}$$

The parameters μ, σ are functions of frequency and parametric excitation (a, h) and they can be determined by the next relationships (only first-order terms are kept):

$$\cos 2\sigma \approx \frac{2(a-1)}{ah} \quad (-\frac{\pi}{2} \le \sigma \le 0) \quad \text{(B2)}$$

$$\mu \approx -\frac{\sqrt{a^2h^2 - 4(a-1)^2}}{4} \quad \text{(B3)}$$

For exact resonance $(a=1)$, parameters μ, σ obtain the specific values $\mu = -h/4$ and $\sigma = -\pi/4$. Subsequently the corresponding growth per roll cycle is captured by the following expression:

$$\frac{\varphi(T_0)}{\varphi(0)} = \frac{\frac{\sqrt{2}}{2}\left(c_1 e^{\frac{\pi h}{2}} - c_2 e^{-\frac{\pi h}{2}}\right)}{\frac{\sqrt{2}}{2}(c_1 - c_2)} \quad \text{(B4)}$$

Assuming the initial conditions $\varphi(0) = \varphi_0$ and $\dot{\varphi}(0) = 0$, the growth of roll amplitude after p roll cycles should be according to eq. (B4):

$$\frac{\varphi(pT_0)}{\varphi_0} = \frac{e^{\frac{p\pi h}{2}} + e^{-\frac{p\pi h}{2}}}{2} \quad \text{(B5)}$$

Introducing linear damping the roll model becomes:

$$\ddot{\varphi} + 2k\dot{\varphi} + \omega_0^2\left[1 - h\cos(\omega_e t)\right]\varphi = 0 \quad \text{(B6)}$$

Setting $\varphi = we^{-kt}$ the above is transformed to an equivalent equation with no explicit damping term:

$$\ddot{w} + \left(\omega_0^2 - k^2\right)\left(1 - \frac{\omega_0^2}{\omega_0^2 - k^2} h\cos\omega_e t\right)w = 0 \quad \text{(B7)}$$

It can be shown by combining (B6) with (B7) that the growth of roll amplitude after p roll cycles for the condition of exact resonance (but this time taking into account roll damping), is:

$$\frac{\varphi(pT_0)}{\varphi_0} = q = e^{-\frac{2p\pi k}{\omega_0}} \left(\frac{e^{\frac{p\pi h}{2}} + e^{-\frac{p\pi h}{2}}}{2} \right) \qquad \text{(B8)}$$

Given that the experienced growth is dominated by the positive eigenvalue of the characteristic equation, a q-fold increase in roll amplitude (from some initial angular disturbance φ_0) in terms of the number p of roll cycles can be expressed approximately as:

$$h - \frac{4k}{\omega_0} = \frac{0.693 + \ln q}{1.571\, p} \qquad \text{(B9)}$$

APPENDIX C: Time-to-capsize during pure-loss of stability

We have correlated the time that it takes to reach a large inclination during a pure-loss event, with the time spent by the ship in the vicinity of a wave crest with negative restoring. This work is still unpublished (Spyrou 2000). Consider firstly a generic roll equation in non-dimensional form:

$$\ddot{z} + 2k\dot{z} + cz = 0 \qquad \text{(C1)}$$

where z is the roll angle scaled by the angle of vanishing stability; and c is an equivalent linear 'restoring' for the region where the latter receives negative values. The value of c can be obtained by equalising the potential energy of the original system with that of the equivalent linear one, for the period of time per wave cycle that the ship suffers from negative restoring. For a harmonic fluctuation of *GM*, this time can be identified from the following expression:

$$t_{nr} = \frac{2\arccos\frac{1}{h}}{\omega_e}, \text{ with } h > 1 \qquad \text{(C2)}$$

Equalization of potential energies produces the linear restoring coefficient c:

$$c = \frac{2mg \int_0^{t_{nr}} \int_0^{z_{cr}} GZ(z,t)\, dz\, dt}{z_{cr}^2\, t_{nr}} = \omega_0^2 \left[1 - \frac{\sqrt{h^2 - 1}}{\arccos\frac{1}{h}} \right] \qquad \text{(C3)}$$

The last term in the above equality is the one obtained when a linear *GZ* curve has been considered for the region of negative restoring of the original system.
With some processing it is possible to derive from (C1) that the time for reaching the critical scaled roll angle z_{cr} is approximately:

$$t_{cap} \approx \frac{\ln z_{cr} - \ln\left(\frac{k + \sqrt{k^2 - c}}{2\sqrt{k^2 - c}} z(0) + \frac{1}{2\sqrt{k^2 - c}} \dot{z}(0) \right)}{-k + \sqrt{k^2 - c}} \qquad \text{(C4)}$$

The instability condition is obtained finally by requesting:

$$t_{nr} \geq t_{cap} \qquad \text{(C5)}$$

It is worth to mention that a nonlinear version of the above analysis is also possible; but unfortunately the expressions obtained lack simplicity.

Discussion

Bilal M. Ayyub, Member

This timely and interesting paper describes a proposed methodology to assess ship stability by examining the grouping characteristics of high waves based on a deterministic analysis of ship dynamics that correspond to different capsize modes and to compute the encounter probabilities of these instability-inducing wave groupings. As noted in the paper, high waves tend to appear in groupings, and some regularity in excitation produces ship instability events. The method's two components of deterministically identifying these instability-causing groupings using ship dynamic analysis and computing their encounter probabilities offer an innovative solution to this problem definition.

According to the method, wave groupings are defined by the number of waves in a group (up to some maximum value *k*), periods of the waves in the group falling within a particular range of interest, and wave heights exceeding some critical height. Critical wave groupings for a particular vessel are defined as instability-causing groupings based on subjecting the vessel to representative regular waves and examining deterministic dynamic responses that exceed some critical thresholds defining warning and failure levels. The definition and exhaustive identifications of instability-causing wave groupings is a foundational challenge for this method. How could analysts define a set of wave groupings that are mutually exclusive and collectively exhaustive, their correspondence to different instability events, and their joint occurrences? For example, would the following groupings defined differently from the groupings in the paper result in instability:

- A grouping of several waves (up to some maximum value *k*);

- Periods of the waves in the group falling within a particular range; and

- Wave heights exceeding some critical height with one or two minor down-crossings within the grouping?

It seems that defining these groupings requires an educated guess by an analyst, and they are then verified by simulation tools based on vessel responses. Can this definition process be enhanced? How can we ensure completeness in defining these groupings? Even if these groupings are fully defined, i.e., collectively exhaustive, the probabilistic part of the method should consider the joint occurrences of different groupings so that probabilities of stability loss are not overestimated by not subtracting joint occurrences of these groupings.

The paper describes the identification of wave groupings using simulation tools to produce different results depending on vessel headings, speed, loading condition, and wind within a particular sea environment. The representation of the sea environment as a stationary and ergodic process for the purpose of computing the probabilities of realizing such groupings ignores the initial vessel conditions when encountering these wave groupings, although the authors note in the well-written literature review of the paper that initial conditions are important contributing factors to the occurrence of stability-loss events. Ayyub, et al. (2006) performed sensitivity analysis of initial conditions required in the input file of a well-known simulation tool on the time to capsize considering both regular and random seas. In the sensitivity analysis, the examined parameters included the initial position of the center of gravity (COG) with respect to the origin of the earth-fixed coordinate system (Xe, Ye, Ze), heeling angle PHI0, trim angle THETA0, and initial surge velocity. Two sea wave conditions, regular and random waves, are considered. The analysis was based on a fixed random seed and on perturbation of one factor at a time over subjectively established ranges.

Based on the simulation results, the following observations were made:

- With respect to the initial position of the vessel, Xe has a significant impact on the observed time to capsize in both random and regular waves because Xe affects the phase angle with respect to the waves.

- Ye has no effect on the observed times in both wave conditions.

- Ze has little effect on the observed times in regular wave conditions, and no effect on observed times in random waves.

- The range of the heeling angle PHI0 was limited to (-10,10) degrees. In random seas, PHI0 has no effect on the observed time to capsize. In regular waves, in some cases, PHI0 has no effect on the observed times, but in other cases, PHI0 has a significant effect on the observed times to capsize.

- The effect of the trim angle THETA0 was assessed over the range of (-5,5) degrees. THETA0 has no effect on the observed times to capsize in random seas.

It also has an insignificant effect on the observed times to capsize in regular waves.

• Initial surge velocity UG0 has a significant effect on the observed times to capsize in both wave conditions.

I would like to thank you the authors for their contribution to the literature and the editors for the opportunity to provide this discussion.

Reference

Ayyub, B. M., Kaminskiy, M., Alman, P. R., Engle, A. Campbell, B. L., and Thomas, III, W. L., 2006 Assessing the probability of the dynamic capsizing of vessels, *Journal of Ship Research*, 50, 4, 289-310.

Alberto Francescutto, Member, and **Gabriele Bulian**, Visitor

The authors are to be congratulated for the comprehensive work presented on a subject which has recently been considered of paramount importance in the development of novel stability criteria.

There are, however, several points that are worth deeper discussion.

Through the introduction of the concept of critical wave height, the authors implicitly assume a kind of semi-continuous and monotonic dependence of the relevant motion characteristic (amplitude or acceleration) on the wave height. This is reasonably true in beam sea, in the absence of bifurcations and jumps of amplitude (Francescutto 1993, Contento & Francescutto 1999), but it is only partly true in longitudinal waves, due to the possible existence of an upper threshold in terms of wave amplitude for the disappearance of parametric roll (Neves & Rodriguez 2007).

The analysis of the loss of stability in waves, which is conducted along the same lines developed in Bulian and Francescutto 2007, suffers the necessity of fixing a distribution of the initial conditions of roll motion in order to have an estimation of the time to capsize. This distribution is unknown, even approximately, in longitudinal waves, since in such condition there is no support from the linear approach to ship motions when roll is of concern, and previous works have therefore envisaged the use of some semi-empirical expression for the time to capsize (Vermeer 1990, Helas 1982).

Considering the case of beam sea in detail, again the problem is that of the selection of initial conditions. In this case it could be reasonable to assume an approximate Gaussian distribution as obtained from a linear/linearized or even a simplified non-Gaussian approach as a starting point, but the framework of the proposed approach would be impaired by the impossibility of defining a deterministic critical wave height for each group length (see, e.g., Blocki 1994, Blocki 1986).

In addition to the issues related to the deterministic models for roll motion as mentioned above, there are also some concerns coming from the underlying theory of wave groups, that is fundamental in formulating the final results. The uncertainty inherent in such approximate theories propagates to the final result, passing through a series of more or less arbitrary, although surely necessary, assumptions. The question is then finally, what does the final numbers mean? Can them be related to an actual probability of exceeding some critical conditions? In principle, and from a purely probabilistic point of view, they could, but which is the level of confidence in the results? Are the results as obtained from this framework comparable with direct simulations, even in a single scenario?

The idea of considering separately a beam sea modelling and a longitudinal sea modelling is of course very valuable, since, at present, there are no sufficiently simple and fast models able to deal with all the angles of encounter. Some ships are more prone to suffer of excessive angles in beam waves, while others are more endangered in longitudinal waves, and this shall be acknowledged by any intact stability assessment. It is however our personal opinion that the difficulties and the uncertainty introduced by mixing the wave grouping theory with a deterministic nonlinear approach to ship motions (in particular roll) could be reduced by trying to pursue the way of developing simplified analytical models directly embedding a stochastic excitation (Blocki 1986, Bulian & Francescutto 2004). Of course the capabilities of stochastic models is limited by their simplifications with respect to the complex ship dynamics, but at the same time their inherent stochastic nature could be exploited for the development of a less involved framework introducing a more limited number of, in any case likely necessary, assumptions.
The opinion of the Authors on these points would be welcome.

References

Francescutto, A. 1993 Nonlinear ship rolling in the presence of narrow band excitation, *Nonlinear Dynamics of Marine Vehicles*, ASME/DSC, 51, 93-102.

Contento, G. and Francescutto, A. 1999 Bifurcations in ship rolling: experimental results and parameter identification technique, *Ocean Engineering*, 26, 1095-1123.

Neves, M. A. S. and Rodriguez, C. A. 2007 Nonlinear aspects of coupled parametric rolling in head seas, *Proceedings*, Tenth International Symposium on Practical Design of Ships and Other Floating Structures (PRADS2007), Houston, TX.

Bulian, G. and Francescutto, A. 2006 Probabilistic assessment of intact stability (Code CRIESA001/00002), (In the framework of SAFEDOR-SP2.3), Dept. DINMA, University of Trieste, reported in 2007 Review of procedures and derivation of base line capsize probabilities using comprehensive approach, SAFEDOR D 2.3.3 2007 02 07 MARIN REVIEW STUDY, rev 2, February.

Vermeer, H. 1999 Loss of stability of ships in following waves in relation to their design characteristics, *Proceedings*, STAB90, Napoli, September, 369-377.

Helas, G. 1982 Intact stability of ships in following waves, *Proceedings*, STAB82, Tokyo, Japan, 689-699.

Blocki, W. 1994 Ship's stability safety in resonance case, Proceedings, Fifth International Conference on Stability of Ships and Ocean Vehicles, STAB94, November, Melbourne, FL.

Blocki, W. 1986 *Proceedings*, Third International Conference on Stability of Ships and Ocean Vehicles (STAB86), September , Gdansk, Poland, 143-149.

Bulian, G. and Francescutto, A. 2004 A simplified modular approach for the prediction of the roll motion due to the combined action of wind and waves, *Journal of Engineering for the Maritime Environment*, 218, M3, 189-212.

Bulian, G. and Francescutto, A. 2006 Safety and operability of fishing vessels in beam and longitudinal waves, *International Journal of Small Craft Technology*, Royal Institution Naval Architects, trans., 148, Part B2.

Dag Myrhaug, Member, and **Emil Aall Dahle**, Member

The authors have presented a comprehensive and well-written paper on a practical method that enables probabilistic assessment of ship stability. The method utilizes the advantage of combining a deterministic approach dealing with instability mechanisms of a ship in a seaway and the probabilistic approach of determining the frequency of occurrence of the wave phenomena causing the instability mechanisms. Thus, a rational procedure is presented which should be a useful tool for practical purposes in, for example, risk assessment of ships. The methodology presented by the authors is novel, although the idea of combining a deterministic approach of the ship response with a probabilistic approach describing the wave phenomena leading to this response also has been utilized previously by other researchers.

Overall, the literature review is thorough and comprehensive. However, capsize in breaking waves is referred to as a critical event, but references to relevant literature are missing, for example, Dahle and Myrhaug (1996). Moreover, pure-loss of stability has also been treated by, for example, Dahle and Myrhaug (1995). These works combine the deterministic ship responses with the probabilistic description of the critical wave events.

The literature review is followed by a description of the methodology and an outline of the probabilistic calculation of a wave group which represents a critical wave event in the applications of the method, beam-seas resonance, parametric rolling of critical amplitude, and critical wave for pure-loss of stability.

For practical application in a risk analysis an acceptance criterion should be specified, for example, say 10^4 per year of exceeding a specific roll angle, or pure-loss of stability. This could lead to tactical weather routing of the ship to limit the exposure. This leads to summing up time periods for high exposure of dangerous events. In practice this could lead to very careful operations at the end of the year to be within the acceptance criterion. This calls for a good understanding of the application of the approach by governmental agencies, ship owners and ship operators.

Finally, some minor comments can be summarized as follows:

Page 1: Moseley (1851) or (1850) as in "References" on page 16.

Page 3: Myrhaug et al. (1999) should be read (2000); see "References" with the complete reference including errata.

Page 8: The statement regarding Tayfun (1993) is confusing; note that Tayfun (1993) presented a joint pdf of a wave height and period for large wave heights (as stated correctly on page 7), while Wist et al. (2004) utilized this pdf to study the conditional distribution of successive wave periods when conditioning on the wave height is larger than a given value.

Page 11: Hasselmann et al. (1973); it is a multiauthored paper, also on page 16. DNV 2002 is not given in "References," or is it the "DNV Max Wave" report, which is not referred to elsewhere?

Page 16:

1) Should read "CAVANIÉ, ------ and EZRATY, ----"
2) Where is "CUMMINS" referred to?
3) GERASIMOV, include "1979"
4) Where is "IMO" referred to?

Page 17: NESS--------, include "1989"

References

Dahle, E. Aa. and Myrhaug, D. 1995 Risk analysis applied to capsize of fishing vessels, *Marine Technology*, 32, 4, 235-257.

Dahle, E. Aa. and Myrhaug, D. 1996 Capsize risk of fishing vessels, *Schiffstechnik*, 43, 4, 164-171.

Myrhaug, D., Dahle, E. Aa., Rue, H. and Slaattelid, O.H. 2000 Statistics of successive wave periods with application to rolling of ships, *International Shipbuilding Progress*, 47, 253-266; errata: 425-426.

Naoya Umeda, Member

It is my great pleasure to find a methodology using a combination of deterministic analysis under critical situations and probabilistic analysis on the occurrence of critical situations. This methodology has been utilized by several researchers, and sometimes is known as "time-split method," etc. The main advantage of this approach, in my opinion, is that statistical output has a definite relationship with physical phenomena, which were identified by mariner experience and existing model experiments.

For parametric rolling, Blocki (1986) calculated the probability of capsizing due to parametric rolling in irregular beam seas. Here the occurrence probability of wave group length is estimated with Goda's theory and critical roll angular velocity for capsizing under the given wave group length is deterministically calculated. The current authors utilize Kimura's theory for calculating probability of wave group so that its prediction accuracy might be significantly improved because Goda's theory assumes white noise. By contrast, while Blocki assumes that pdf of roll angular velocity is related to sea state, the current authors assume that initial roll disturbance is independent of sea state. This could be a drawback of the current methodology in some cases.

For capsizing due to stability loss on a wave crest in following and quartering waves, the discusser (Umeda et al. 1990, Umeda and Yamakoshi, 1994) attempted to calculate capsizing probability as the product of the probability that a ship suffers restoring reduction on a wave group and the conditional probability that a set of ship roll angle and roll angular velocity is outside a deterministic safe basin defined by the specified reduced restoring moment. This means the statistical correlation between the restoring variation, roll angle and roll angular velocity is fully taken into account. It was confirmed that probabilities obtained by this method agree with those from the Monte Carlo simulation. In my opinion, the statistical relationship between wave events and the initial condition for deterministic analysis should be properly evaluated.

For surf-riding, the discusser (Umeda 1990) proposed a method for calculating probability of surf-riding in following waves. Here it was regarded as the product of the probability of a critical set of wave height and wave length for surf-riding, which was deterministically estimated, with some consideration of the effect of initial conditions. Recently, this approach was applied to broaching in quartering waves (Umeda et al. 2007) and obtained good agreement with the Monte Carlo simulation without statistical initial condition effect.

These results seem to indicate that the importance of initial condition could depend on types of deterministic dynamics of the subject phenomena. Therefore, the methodology should be developed to cover this statistical correlation of wave events and initial conditions for deterministic ship dynamics. And then it would be desirable to quantify the effect of such correlation on capsizing probability.

References

Blocki, W. 1986 Probability of non-capsizing of a ship as a measure of her safety, *Proceedings*, Third International Conference on Stability of Ships and Ocean Vehicles, Gdansk, Poland, II, 143-149.

Umeda, N., Yamakoshi, Y. and Tsuchiya, T. 1990 Probabilistic study on ship capsizing due to pure loss of stability in irregular quartering seas, *Proceedings*, Fourth International Conference on Stability of Ships and Ocean Vehicles, Naples, Italy, 328-335.

Umeda, N. 1990 Probabilistic study on surf-riding of a ship in irregular following seas, Proceedings, Fourth International Conference on Stability of Ships and Ocean Vehicle, Naples, Italy, 336-343.

Umeda, N. and Yamakoshi, Y. 1994 Probability of ship capsizing due to pure loss of stability in quartering Seas, *Naval Architecture and Ocean Engineering*, 30, 73-85.

Umeda, N., Shuto, M. and Maki, A. 2007 Theoretical prediction of broaching probability for a ship in irregular astern seas, *Proceedings*, Ninth International Ship Stability Workshop, Hamburg, Germany, 1.5.1-1.5.7.

Vadim Belenky, Member

The authors are to be commended for their original, comprehensive and very timely paper describing excellent research.

The subject chosen by the authors is extremely important as new types of ships appear in the world fleet. The variety of unconventional hull configurations and operational modes motivated IMO to consider the development of a new generation of intact stability criteria based on physical principles, thereby being highly adaptable for new types of vessels. The importance and timeliness of the paper are first and foremost in relation to this development.

Another important aspect of this paper is the adaptation of solutions generated by nonlinear dynamics for the real environment that are intrinsically stochastic. Problems of stability in waves involve consideration of large-amplitude motions in a case of partial stability failure or a complete transition to another stable equilibrium in the case of capsizing, which is total stability failure. Nonlinear dynamics offer a set of formal procedures allowing for analyzing and predicting these motions, but only for periodic excitation. At the same time, the available description of realistic wind and wave environment conditions are stochastic, creating a gap between the theory and practical application. This paper bridges this gap by employing a critical wave group/critical wave approach. Application of this approach can also be considered as a treatment of problem of rarity — the computational problem caused by the necessity of dealing with very small probabilities of stability failure of an intact vessel.

The paper also considers problems of stability failures both in the short term, assuming that the environment is described by stationary stochastic processes, and in the long term, by simulating the route that demonstrates how the proposed approach could be modified for different applications.

Several particular points deserve to be mentioned. Application of the envelope description of wave group is, indeed, a powerful tool as the theory envelope is well developed, especially for the narrow band stochastic process. The process is substituted by the combination of two other stochastic processes: time-dependent amplitude and phase. These two new processes are changing slowly in comparison with the original process that they represent. The amplitude process plays a "modulating" role, making sure that the variance (therefore energy) of the original process is represented properly. The phase part is characterized by the autocorrelation function normalized by unity and representing the "memory" of the process. Therefore, the role of phase in the envelope presentation is quite different from the one in spectral representation, where the phases provide just a "randomization" of the process. As a result, the application of envelope presentation requires special attention to modeling phases.

Another formidable task is evaluation of the GZ curve in waves. Change of stability in waves is the physical reason of two phenomena leading to stability failures of parametric roll and pure loss of stability. Several components make these changes: hydrostatic (change of the waterline caused by waves), change pressure in waves (Smith effect), contribution from wave-body interaction, and the influence of ship waves. Computational methods and codes for hydrostatic and Smith effect components are available (program Eureka/STABW by J.R. Paulling). Other components at this moment are included in the Nechaev method which is based on the series model test (available in English in Belenky and Sevastianov 2007). Calculation of stability in waves definitely requires attention when the approach proposed by the authors is considered for practical use.

The paper associates "pure loss of stability" with an event when the GZ curve becomes completely negative in following seas and with enough time to reach a roll angle constituting stability failure. Without a doubt, this is the classic pure loss of stability. However, if one considers quartering seas, where some heeling moment exists, a capsizing or catastrophic heel angle may happen even if there is some positive stability, but it is significantly degraded in comparison with calm waters.

These particular points should be considered as directions of future study rather that a critique. No single paper can comprehend the sheer size of the problem of probabilistic assessment of intact stability.

The authors also have to be commended for citing many reference sources, not all of which are well known, while deserving more circulation.

Reference

Belenky, V.L. and Sevastianov, N.B. 2007 Stability and safety of ships, *Risk of Capsizing*, 2nd ed. SNAME, Jersey City.

Authors' Closure

The authors are thankful to all discussers for their valuable comments and questions. Their generally positive regard of the current work is a source of encouragement for taking it further. The authors' thanks extend also to SNAME for publishing this paper.

In his thought-provoking discussion, Prof. Ayyub wonders how could one identify systematically the set of critical wave groups, so that completeness and mutual exclusivity of elements are ensured? This is currently accomplished by sequentially spanning the range of wave periods, heights and group-run-lengths to determine the boundary curve of critical wave groups for each capsize scenario. Ship control parameters like the speed and the type of wave encounter (following, beam and head seas) are participants in this process. Thereafter, the probability calculation addresses all wave group occurrences that "lie" above this boundary. Indeed, if one is not cautious, overlaps in the probability calculation might arise. A typical example is when certain wave group specifications have the potential to inflict capsize of more than one type. Such wave group subsets should be identified and accounted once. As long as a wave group is assumed to consist of waves with the same height, this represents no major burden for the calculation process. But Prof. Ayyub hints further at a more advanced wave group scenario where a combination of one or two waves slightly below the threshold, combined with one or two well above it, could still produce capsize. In fact, this points to the direction of a possible improvement of the method, which is in line with the authors' thinking. It could mean to turn the currently deterministic ship dynamics part of our method into a probabilistic one too, but without taxing too much the currently straightforward calculation process. Such a platform could take on also issues related to the effect of initial conditions. About the latter Prof. Ayyub has given some useful insights, deriving from his own work. Undoubtedly there are some interesting prospects for further research; yet, the authors believe that, even in its current form, the method "sees" the real world with satisfactory resolution.

The authors share with Dr. Belenky the spirit and essence of his comments. Indeed, bringing in advances from the nonlinear dynamics field to the probabilistic analysis of ship motion has been a driving idea for the current work. As he notably points out, the developed approach does not suffer from the problem of rarity that has been specifically identified in the past as a significant impediment of conventional probabilistic analysis of extreme ship motions. With respect to the problem of phase that is intrinsic to the envelope approach, the following clarification is necessary: two well-known approaches for the statistical analysis of wave groups have been discussed in the paper. The first is the envelope approach where the wave group is regarded as a "level crossing" phenomenon. In the second approach, described mainly by Kimura and modified later by other authors, the sequence of wave groups is assumed as a Markov chain process. In the current work the second approach has been utilized for the calculation of the probability occurrence of wave groups. Therefore, problems that might arise in the modeling of the phases do not appear. The discusser's suggestion of looking also into more complex capsize scenarios, like the one of "pure-loss" in quartering seas with low yet positive GZ, is a truly worthy one. As soon as some event has been identified to be of importance, it can be targeted by the wave group methodology for probability calculation. The only concern should be whether one has enough confidence in the available numerical simulation or analytical tools that could capture the dynamics of such events. The built-in flexibility in the use of tools is certainly an advantage over other methods that are tied to a single modeller of ship dynamics.

Prof. Francescutto and Dr. Bulian have raised some very interesting issues of detail where research at an international level is currently underway. Of course, no claim could be made that a single piece of work is able to handle rigorously all open issues of a field of scientific endeavor. Basically, here one has looked for a logical construction that could serve toward an approximate assessment of ship dynamic stability. At the same time, the authors are quite reserved about the necessity of integrating within their assessment methodology the alleged disappearance of parametric rolling at extreme heights. The nonlinear roll responses in extreme wave heights pose no fundamental obstacle to the implementation of the method. The authors have argued in favor of employing nonlinear boundaries whenever these could be effectively captured, e.g. by means of analytical methods. In other words, there is no implicit assumption of monotonic or similar relationship as implied. Admittedly, bifurcations could make the implementation more "interesting," especially if, for example, multiplicity of responses invokes complex consequences for the instability boundary. In such a case, initial conditions from different subspaces of the system's state space could lead to qualitatively different motion patterns. To account for this multiplicity of responses, the calculation of probabilities then needs to be properly weighted - something that inevitably ties in the treatment of initial conditions. The way to integrate within the current methodology the effect of initial conditions was also queried by the discussers. Because of its deterministic ship dynamics kernel, the "wave group approach" can in principle handle this. Should there be a question, this would be how to realize it cost effectively. This topic is still open and research contributions on this are very welcome. The authors feel that some fundamental work is required on this topic, and they hope to be able to publish some relevant work soon. Issues like the distribution of initial conditions in a longitudinal seaway may be academically interesting; here however one focuses on wave encounters that can destabilize the ship from some initial state that is not too far from rest.

The question about possible accumulation of errors due to successive approximations is relevant to any theory that has been hitherto applied for the assessment of ship dynamics. Mathematical models of ship motions and statistical environmental models alike, all these represent, at best, approximations as no definitive theories exist. Comparison with other simulation-based methods could, and perhaps should, be performed; yet the real value of the outcome would still be debatable as recent ITTC benchmarking exercises have xlearly demonstrated.

Finally, concerning the recommendation to revert to simplified models, the authors think that these are valuable for developing a fundamental understanding of nonlinear ship behavior. Whether the direct embedding of stochastic excitation in a model of this philosophy could achieve true benefits in a quantitative sense is something that the future will show as the field is open. However, practical methods of assessment like the current one have to rely on existing knowledge and understanding of phenomena instead of endeavoring to generate these on the go.

Prof. Myrhaug and Dr. Dahle should be thanked for indicating a few more references that the authors also consider as useful. Indeed, there has been a broad literature base behind this work, and these two discussers have made some critical contributions to it. Capsize due to breaking waves, which they mention, is a truly interesting topic, and one hopes that mathematical models with reasonable representation of the physics of the phenomenon become available. Their recommendation of defining explicitly the acceptance threshold is of course at the back of one's mind when the foundations of a risk assessment methodology are laid. However, such a derivation should not be tackled haphazardly. Societal values and current practice should be reflected upon it, but the current work has not covered these issues. The authors could not agree more that maritime decision makers should develop a good understanding of the method.

Prof. Umeda correctly remarks that the physical phenomena are clearly represented in a methodology like the current one and that clear practical benefits should accrue from this. It is true that some other researchers have in the past also attempted to utilize deterministic ship roll dynamics in a probabilistic calculation framework. The authors' modest assessment is, however, that these works did not reach similar levels of detail in the probabilistic calculation of wave group encounters. Also, their potential to address all modes of ship capsize had not been demonstrated. The work of Blocki is indeed interesting as he attempted to correlate wave groups with parametric rolling. However, there was no connection of the heights of the waves in the group with the probability of occurrence of parametric rolling for a given sea state. The initial roll velocity was treated as a random variable while the roll angle was always zero. This is a different simplification than the one adopted here. Nevertheless the effect of initial conditions is justifiably pointed out as one issue worthy of attention, and the authors have given their opinion earlier on a similar question raised by Francescutto and Bulian. With regard to the studies of probabilistic pure-loss referred to by the discusser, it is felt that there is conceptual difference if one follows Grim's effective wave idea. In any case the term wave group does not appear in the referenced earlier works. Things are different in the recent paper on broaching referenced by Prof. Umeda where the philosophy of the current approach seems to have been adopted.

In closing this discussion, the authors wish to express their gratitude to all discussers for contributing to the deeper understanding and clarification of the presented methodology.

Naval Vessel Rules: Taking Naval Ship Classification to a New Plateau

Authors: Glenn M. Ashe
Raymond R. Finney

ABSTRACT

In the wake of acquisition reform, many process reengineering changes have taken place across the Naval Ship engineering and construction enterprise since the late nineties. One key element of those changes has been wider use of commercial practices, such as ship classification to Classification Society standards, which is a process in broad use across the commercial shipbuilding industry. This paper chronicles the adaptation of the ship classification process for application on U.S. Navy combatant ships, as well as the development and evolution of the ABS Naval Vessel Rules (NVR) which have been jointly developed by the American Bureau of Shipping and the Naval Technical Authorities of the U.S. Navy. The multiple dimensions of fundamental change are described, along with areas where this initiative has served as a catalyst for companion improvements and an enabler for process efficiencies.

KEY WORDS: Ship Classification, Naval Combatant, NVR

INTRODUCTION

The commercial shipping industry has for many decades employed the process of ship classification, using the independent third-party certification services of one of the recognized Classification Societies. The U.S. Navy has gained intimate familiarity with that commercial process through the many construction programs involving ships built to commercial standards, which generally have been those ships operated by the Military Sealift Command. Fifty-five ships have been built by the U.S. Navy using that model over the period from 1990 to present, with all of those ships having been delivered in Class with the American Bureau of Shipping (ABS). In fact, about half of all U.S. Government owned vessels have been built to ABS Class. Through participation in the ABS Technical Committee structure, the U.S. Navy also has maintained a long-standing relationship with ABS on the development of rules for the design and certification of commercial ships, which includes Naval Auxiliary Ships, SEALIFT Ships and Oceanographic Ships built by NAVSEA. It is these relationships that have served as the foundation for the broader partnership described in this paper.

International Context

A number of similar relationships have also evolved between non-US navies and classification societies. In addition to ABS, the following have developed Rules to address a range of naval and naval auxiliary vessels:

Lloyd's Register
Det Norske Veritas
Germanisher Lloyd
Bureau Veritas
Registro Italiano Navale

Navies in addition to that in the United States which have applied the process of classification to some degree to either the acquisition or in-service maintenance of their vessels are numerous, and include the following:

United Kingdom	South Africa
France	Australia
Spain	New Zealand
Germany	Canada
Italy	The Philippines
Denmark	Singapore
Saudi Arabia	India
Oman	Chile
Egypt	Ecuador
Israel	Columbia

Some navies choose to deal predominantly or exclusively with the classification society located in their home country while others have chosen to deal with several. There are benefits and drawbacks to each approach. In the United States, the U.S. Government is directed in U.S. Code to use ABS as their sole agent for classification or classification-related matters.

THE DEVELOPING ENVIRONMENT

What Led the Navy to Pursue the NVR Initiative?

The Naval Sea Systems Command (NAVSEA) is assigned responsibility, under Title X, as the Technical Authority for

Navy ships. In its technical authority role, NAVSEA is responsible for the safety and performance of ships and ship systems. Therefore, NAVSEA develops or approves the contract technical package for ship procurements, including the ship specification and other major ship design engineering products. Since 1992, staffing cuts and funding reductions have challenged NAVSEA to find innovative ways to fulfill its technical authority responsibilities. Particularly problematic has been maintaining the currency of Navy shipbuilding specifications and standards for combatant ships. By the late 1990's that base of standards, embodied in the General Specifications for Ships of the United States Navy (GENSPEC), had fallen into disuse, and had not been updated for a period of several years. In 1998, the GENSPEC was formally cancelled due to safety concerns relative to the potential use of an out-of-date standard leading to a compelling incident that stemmed from preventable causes.

Meeting the Need

At that point, new combatant ship design projects were on the horizon, and a complex set of influences led the Navy to consider a stream of turning point questions:

- Where can we take some risk?
- Where can we use commercial processes?
- Where can we use commercial standards?
- Where can efficiencies be realized?

For the emerging projects, it was recognized broadly that the Navy needed to accept measured risk in its acquisition programs in the interests of: innovation and improved architectures, minimizing the number of unique R&D investments, cross-platform design reuse, application of transformational technologies, and lower total ownership cost. At that time, the Navy was also struggling with how best to implement the mandates of Acquisition Reform including use of performance-based contracting, the single-process initiative (elimination of Government-unique processes), and broader use of industry standards and processes. In addition, the Navy was looking for efficiencies and process enablers to facilitate doing more with less resources, while not sacrificing safety or the satisfactory perfomance of our ships in a military environment. Regarding a refresh of our base of combatant design standards, a number of options were considered. Through lengthy dialogue with the U.S. Shipbuilders via the National Shipbuilding Research Program (NSRP) Executive Control Board, a consensus was established that the Navy must maintain the body of military-unique design standards used on our combatant programs. On the one hand, the Navy desired to use commercial industry standards where the resulting performance could be determined to be acceptable. And on the other hand, it was desired that our retained military requirements be overhauled to inject industry influence, and where practical, apply a performance language approach. It was recognized that ABS has extensive experience in applying commercial standards, and ABS already had an industry committee structure in place that involved ship owners and shipbuilders in the process of maintaining a body of ship design and ship certification rules. The Navy, therefore, concluded that an expanded partnering relationship with ABS on design rules for combatants would be of great benefit to the Navy, and the combatant design community at large.

Setting the Course Ahead

In order to address the challenges of technical authority in an austere environment, a two-fold joint decision was made by NAVSEA and Program Executive Officer Ships (PEO Ships) in February 2003. First was to refresh and migrate the Navy's existing base of combatant ship design standards in partnership with ABS, so that those standards could be applied in an adaptation of the commercial Ship Classification process. The rules developed, and subsequently authorized for use by the ABS Naval Technical Committee, are a set of rules that are Navy controlled. The ABS Naval Vessel Rules is the product that resulted from that effort. The second companion decision was to require ABS Classification for DDG 1000 and all LCS Ships. That decision rested on the rationale that the Navy wanted to leverage the commercial processes that were already in place and well understood. The Navy had already crafted extensive working relationships with ABS, and had reason to believe that extension of those processes and working relationships for use on combatants could be a key ingredient for the success of those programs. The key difference envisioned would be that ABS would be Classing DDG 1000 and LCS using a set of military design rules jointly developed with NAVSEA.

THE END IN MIND

Developing the Naval Vessel Rules

To enable the process of Classing combatant ships, the military design and certification rules had to be written. Some early collaborative efforts between ABS and NAVSEA had taken place over the period between 2000 and 2002, but only in a small number of high interest areas like standards for materials and prime mover certification. So to facilitate the broader partnership, in February of 2003 an intensive effort was tasked for the Naval Technical Authorities (NTAs) for Hull, Mechanical and Electrical (HM&E) Systems to work directly with ABS to develop the NVR for use on the DDG 1000 and LCS programs. In October 2003, this initiative was further underscored as a long term commitment with the issuance of a formal Cooperative Agreement between ABS and NAVSEA, that was signed by the Commander, Naval Sea Systems Command and the President of ABS Americas. This extensive overhaul of the combatant design standards base yielded a first complete draft of the NVR in February of 2004. Many sections were available earlier in the Fall of 2003 and those had been circulated to the ABS Naval Technical Committee (NTC) Members for comment. At that time, the NTC membership consisted of:

- NAVSEA Deputy Commander for Ship Design, Integration and Engineering
- NAVSEA Director Future Concepts & Surface Ship Design
- NAVSEA Director Machinery Systems
- NAVSEA Director Hydrodynamics Group
- NAVSEA Director Materials & Environmental Engineering
- NAVSEA Director Structures & Survivability Group
- NAVSEA Director SUPSHIP Management Division
- American Bureau of Shipping (ABS Americas)
- Representatives from:

- US Coast Guard Headquarters (CG-45)

- Commander Fleet Forces Command (CFFC)

- Program Executive Officer Ships

- NSWC Carderock Division

- NSWC Combatant Craft Division

- Military Sealift Command

- Massachusetts Institute of Technology

- Computer Sciences Corp (CSC) Marine

- George G. Sharp

- Gibbs and Cox

- John J. McMullen Associates/Alion

- M. Rosenblatt & Son/AMSEC LLC

- Shipbuilders Council of America (VT Halter)

- Bath Iron Works

- NASSCO

- Northrop Grumman Ship Systems

- Northrop Grumman Newport News

Shortly thereafter, in March and April of 2004, NAVSEA and ABS hosted technical reading sessions on the entire rules package, which was the first coordinated attempt to obtain industry input, feedback, and buy in on the combatant rules. Industry comments and the results of the reading sessions were folded into the rules, and in May 2004 a Final Draft set of rules was provided to the ABS NTC for their consideration and authorization for release. In July 2004 the NTC authorized release of the NVR for initial use by design teams, recognizing that full formal approval of the rules would take place over time, section by section, after all necessary concurrences had been obtained. Figure 1 below shows the breakdown of major topics addressed in the NVR.

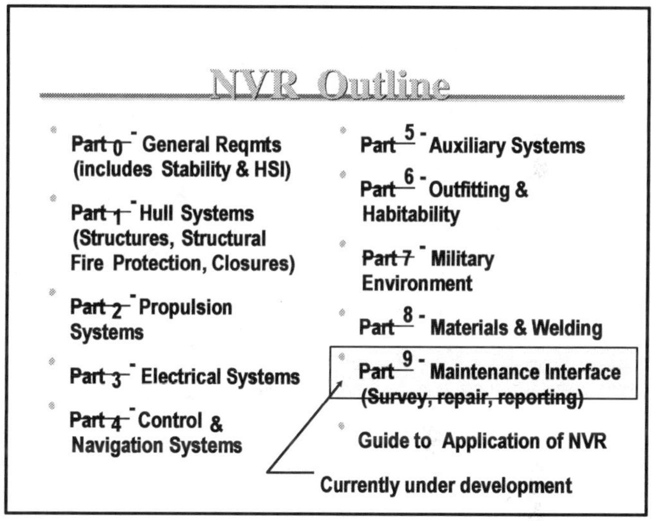

Figure 1. - Outline of ABS Naval Vessel Rules Content

In crafting the rules, NAVSEA essentially designated ABS as their certification agent for lower-risk HM&E systems. That decision was considered low risk because of the extensive past experience with ABS performing the same functions, albeit to a lesser commercial set of standards, on the many MSC-operated ships NAVSEA had built in recent years.

Benefits for the Navy

This approach was recognized as having several advantages for the U.S. Navy. It allowed the great volume of lower risk HM&E certification work to be vectored to ABS for execution during the Detail Design and Construction phase of a program (i.e., subsequent to Navy Design Approval at the end of the Contract Design Phase). Doing so allowed the Navy engineering resources to be focused in the higher risk areas, including mission systems and the military-unique aspects of a ship like shock, signatures, combat systems integration and battle force interoperability. The NVR essentially defines the perimeter within which the Navy is willing to accept an ABS certificate, with ABS ship classing activity being considered to be sufficient effort as delineated in the rules. It should be noted, however, that the Navy only wanted to engage ABS in doing those things they were already good at. The Navy understood that the ABS processes, and to some extent their talent base, would be shared with their many commercial shipping customers. Those resources would represent a shared infrastructure that could be leveraged to meet the needs of the naval combatant ship programs. For those reasons, the NVR does not contain requirements for the military-unique or warfighting aspects of a ship. It was intended that design standards for those aspects would be maintained in separate documents, and those areas of a ship would be certified by the NTAs or their designated certification agents (usually from one of the Navy Warfare Centers).

Another recognized advantage was the closed-loop nature of the ship classification process. The fact that ABS would be engaged with the Navy in developing the combatant rules, and then apply them directly in their typically rigorous and comprehensive fashion was highly valued by the Navy. That aspect was viewed as building a legacy of expertise and corporate memory at ABS that would begin during the early design phases of a program, crest with initial classing of each vessel during new construction, and then enable additional benefits when those combatant ships are maintained in ABS Class (in accordance with the same rules set) during in-service operation. The intent has been for ABS to be empowered to handle the majority of day-to-day involvement directly with the shipbuilder and/or design agent, with all of that activity being tasked and funded under the ship classification contract. That posture allows the Navy to be involved in Class related matters only by exception, when specific circumstances arise, as described later in this paper.

HOW ARE THE RULES APPLIED DURING DESIGN?

How Does the NVR Compare

There are several basic differences between the NVR and the former GENSPEC. First, is the intended range of applicability. Whereas the GENSPEC was a generic document, addressing all ship types including nuclear aircraft carriers and submarines, the NVR focuses only on non-nuclear surface combatant ships. That means the range of intended use for the NVR is limited to those ship types typically designed to combatant standards (i.e., Frigates, Destroyers, Cruisers, Amphibious Ships, Fast Combat Support Ships, and Mine Warfare platforms). By contrast, the Navy intends that future Military Sealift Command operated ships, and other like ships designed to commercial standards (i.e., Naval Auxiliary Ships, Sealift Ships, Maritime Prepositioning Ships, and Oceanographic Ships), will be designed and Classed using the ABS Steel Vessel Rules, as in the past.

Another basic difference is how the two documents were intended to be applied. While both were intended to serve as a "content model" for design requirements that would appear in a Shipbuilding Specification, the GENSPEC was never invoked directly in a shipbuilding contract. The NVR was intended to be invoked by citation in a Request for Proposals (under a competitive Industry design program) for early stage design, and also later by direct citation of the NVR in a shipbuilding specification, which is invoked by contract. Similar to past practice on Military Sealift Command operated vessels, the NVR is invoked by citing individual Chapters and Sections in the appropriate section of the shipbuilding specification. For ships built to the NVR, the companion contract requirement is for the ship to be delivered with an ABS Class certificate. It should be noted here that, as is the case for the commercial Rules, the NVR is invoked using the *Class Notations* that are set out in the first part of the rules. That means that the ship owner (in this case the Navy) elects which parts of the rules are effective for a particular design and construction program. Those designations are part of the detailed records kept by ABS for the life of each hull, to more specifically indicate exactly which pieces of the Rules set a given hull complies.

One more difference is that the NVR includes some requirements (e.g., design margins) that are primarily applicable to the early phases of design activity. Providing those requirements is necessary in a competitive design environment where industry is selecting basic design concepts, and performing the overall sizing of the ship and installed capacities (e.g., propulsion powering). Concerning content of the NVR generally, in most cases the GENSPEC was used as the starting point. Figure 2 describes the process that was applied in making the transition from the GENSPEC to the NVR.

GENSPEC to NVR – Process Applied

- Revalidated all technical requirements as valid; or
- Redefined those requirements that were deficient or out-of-date
- Went to performance language where appropriate
- Removed military-unique aspects not in the ABS lane (e.g. shock, signatures, combat systems). Replaced with hooks to other documents
- Folded in additional language to facilitate ABS execution
- Folded in language addressing early stage design considerations (e.g., design margins, service life allowances)
- Added requirements that did not exist in GENSPEC (e.g., networks, dynamic stability)
- Added State-of-the-Practice techniques (e.g., requirement for finite element modeling of structure, spectral fatigue analysis)
- Where tiered references are cited, industry standards were invoked or allowed where the resulting performance was acceptable

Figure 2. - Process Applied in Transitioning GENSPEC Content to the NVR

The NVR Must Be Tailored

The Navy considers the NVR to be a minimum set of acceptable requirements, but it was not their intent to limit innovation or truncate dialogue on better or more cost effective ways to achieve equivalent combatant-level performance. From the designers' vantage point, a reasonable way to look at the NVR is to consider those Rules an easy default. That means, if the designer and Shipbuilder deliver what the NVR requires, it is virtually guaranteed that ABS and the Navy will accept the product. It is recognized, however, that many NVR requirements may not apply to a particular design, and that many existing requirements need to be tailored at the time they are invoked. Tailoring is generally conducted to adapt the published requirements to the specifics of a design, to identify exceptions to certain rules where the designer does not wish to comply, or to identify areas where the designer has identified an alternate means of compliance that may be approved as a variation. Even within a given Rules section, the designer may propose that certain requirements be foregone as a cost savings measure, provided the technical impacts have been evaluated and the outcome is judged acceptable by the Navy. All programs have a context that may vary significantly from one program to the next. For instance, the NVR was initially prepared to meet the needs of the Littoral Combat Ship and Land Attack Destroyer Programs. There are many occasions where capital warship design requirements, like those for DDG 1000, are either not applicable, or gauged as being not appropriate for LCS ships. It is best to develop the picture of which NVR Class requirements the designer does not want to comply with as early in a program as possible. Some of the basic root-level requirements can drive overall ship size and weight, so not being upfront about not meeting them can lead to gross misunderstandings and rework on a grand scale. Such misunderstandings can also have distinct negative impacts on cost, that can lead to the need to reprice a ship program in the middle of a program. After a basic understanding has been developed (including ship sizing) during Concept Design and Preliminary Design, the place where this thought process is intended to play out is the Contract Design phase of a program. Concept Design embraces the formulation of an initial ship concept; Preliminary Design establishes a complete functional baseline for the ship; and, the Contract Design phase has as its chief objective the preparation of the shipbuilding specification. For those portions of the ship where ABS is involved, individual sections of the NVR are tailored and invoked in the ship specification. For example, a given ship specification section may make a direct cite of a particular Part and Chapter of the NVR, followed by the words "except as modified below." Several exemptions from specific requirements, or more specific requirements language may be inserted there. That tailoring may also refer to alternate reference documents, beyond those listed as required for detailed aspects in the NVR, provided the impacts have been evaluated and shown to be acceptable to the NTAs.

How Did the NVR Process Work for LCS and DDG 1000

The Industry Teams for both programs were tasked to develop the shipbuilding specifications for their respective ships as part of the Contract Design efforts. On both programs this was called the Final System Design (FSD) Phase. An integral piece of those efforts was to tailor the NVR in the course of invoking it (or other alternate Rules) in each section of the shipbuilding specification. One means undertaken to facilitate this process, and provide a clear understanding of how various elements of the ships would be certified, was for ABS to prepare a Certification Matrix. The matrix approach provided a simple line item display of specific Rules (or a range of Rules) that would be acceptable for a given Ship Work Breakdown Structure (SWBS) category. The matrices were provided to the Industry Teams to guide their efforts, along with directing ABS participation with the teams to assist as needed in understanding and tailoring the design Rules. On the LCS Program, the Industry Teams were also required to task ABS to conduct a *Preliminary Design Evaluation (followed by a Final Design Evaluation)* to give them early perspective on their level of compliance with the design Rules and identify areas that were of technical concern. On the DDG 1000 Program, similar ABS involvement with the Industry Team was achieved by their participation (along with the NTAs) in the Drawing Board Reviews process, which evaluated the planned configuration and tailored Rules for major systems on the ship.

Controlling Configuration is Key

While the Industry Teams were responsible for managing their ship configurations in consonance with the flow-down of system requirements, a companion approach was needed to configure and control how the NVR was being tailored for each ship application. A mechanism was developed to meet this need in a very trim fashion. A paper instrument was developed called a *Justification for Technical Determination (JTD),* whereby all involved parties could screen, vet and approve individual design issues. JTDs serve multiple purposes, but they are mainly written to accept and document individual variances from the published Rules. Just as any involved party can propose an alternate standard or other variance in design requirements, anyone can submit a JTD to amend or tailor the NVR for a specific application. The originator identifies a topic for consideration, includes proposed tailoring language to be included in the ship specification, provides rationale and supporting technical justification, and the JTD process routes that issue through the designer, ABS, and the affected NTAs for a decision. Of course, not all JTDs are approved, and they are not all generated at the same time. While most are generated during preparation of the shipbuilding specification, JTDs can be processed at any point where needed to document an excursion from the Rules, even during Detail Design and Construction when unanticipated conditions arise. Once JTDs are approved, ABS maintains them as part of the body of documentation for each ship. One specific purpose that JTDs serve is to form the basis of what the ABS Surveyor uses in Classing the ship, along with the NVR and ABS approved design drawings. One important aspect of configuration control for the Rules is to capture lessons learned from program-specific tailoring, in order to influence and improve the published Rules, as described later in this paper. Approximately twenty percent of approved JTDs have been immediately folded back into the published Rules, and made available for other programs. In many instances, Industry brought to the mix new approaches that were evaluated and found to be acceptable. Often these cases were areas where the Navy could widen the range of acceptable solutions generally. In other cases, alternate solutions were adopted in the NVR as being more appropriate for ships like LCS. Figure 3 below shows a recent snapshot of JTDs that have resulted thus far on the LCS and DDG 1000 Programs, and how many of these have resulted in NVR changes during the update cycle for the published Rules.

JTDs Processed

	Proposed JTDs	Approved JTDs	JTDs Transitioned to Rule Changes
• LCS 1	126	91	15
• LCS 2	204	134	22
• DDG 1000	448	259	61

Figure 3. - Number of Justification for Technical Determination Forms Processed

Who Has the Lead

With respect to certifying compliance with the NVR and other military unique design rules, the NTAs have the lead until the design and shipbuilding specification have received Navy Design Approval at the conclusion of the Contract Design phase. During Contract Design, ABS assists the Navy Ship Design Team in evaluating the tailoring of the Rules, including any proposed excursions from the NVR. To the extent that some top level Class related engineering deliverables may have been generated during Contract Design, ABS will begin reviewing them along with the Navy Design Team. During this phase, the industry teams could choose (as was done on both LCS platforms) to engage ABS to review design products prior to, or concurrent with, Navy review. Contract Design involves a continuous stream of difficult, risk-based, trades and value judgments on technical issues, many of which require Rules interpretations and approval of exceptions. The Navy will seek input from ABS on making such determinations. The Navy Design Approval is intended to freeze the design requirements that will be used to govern the Detail Design and Construction phase. Further changes to those requirements do occur, but only as individual serialized changes under the control of the Ship Program Manager's Configuration Control Board. The original intent was, as far as the Class related aspects of the ship are concerned, if the Detail Design and ship are completed in accordance with the tailored Rules, as laid out in the NVR, JTDs and ship specification, the NTAs need only be involved in extraordinary cases. At the start of Detail Design, the lead shifts to ABS to monitor the Design Agent's or Shipbuilder's execution of Class related technical requirements. Early on this takes the form of engineering plan review. ABS reviews drawings to confirm Rules compliance, and it is the ABS approved drawings that the ABS Surveyor uses to verify compliance in the ship as built. ABS reviews and approves other supporting engineering products such as calculations, finite element models of hull structure, analyses (such as the Spectral Fatigue Analysis), and process related deliverables such as the weld sequencing plan. ABS coordinates with the Shipbuilder or Design Agent to establish a delivery and turn-around schedule for ABS reviews, so that production delays are avoided. Detailed records of all ABS drawing comments are generated, and ABS manages a closed-loop process for retiring all comments before approval is provided. Some Class deliverables additionally require NTA approval in specific areas of higher risk. When such matters arise, ABS coordinates their activity through the Navy Ship Design Manager. During initial applications, there has in reality been more involvement of the NTAs (including the waterfront NTAs) in specific areas (e.g., drawing reviews). This is natural as a confidence base and comfort level with ABS expertise and the adapted class process must be established.

HOW ARE THE RULES APPLIED DURING CONSTRUCTION?

Applying the Rules During Survey

ABS Surveyors are assigned who have received specialized training in the differences between the ABS commercial Rules and the more demanding requirements in the NVR. The Surveyor becomes thoroughly familiar with the details of the design as articulated in the ABS approved drawings, and any excursions from the published Rules as documented in the approved JTDs for a particular ship. ABS establishes a survey office at the building shipyard, where early activity focuses on approval of Class-related processes (e.g., weld procedures), and review of welder qualifications with regard to NVR welder certification requirements. In addition to approving engineering plans for major items of machinery (e.g., propulsion engines), ABS will provide source inspection at the vendors' facilities to confirm compliance with the Rules. Once construction has begun, the ABS Surveyor oversees the fabrication of each hull module, the erection of the ship, and the installation of all systems covered by the NVR. While this is happening, the surveyor provides feedback to the Shipbuilder on a continuous basis regarding the level of compliance with Rule requirements. Noted deficiencies on scheduled callouts are documented by the ABS Surveyor and provided to the Shipyard Quality Assurance staff. A running detailed record of all Class-related deficiencies is maintained throughout the construction period and shared with the Navy Supervisor of Shipbuilding (SUPSHIP) Quality Assurance staff. ABS witnesses the conduct of system and equipment completion tests as part of the Shipbuilder's Integrated Test Program. The ABS Surveyor will also participate in ship's trials to witness the satisfactory performance of systems addressed in the NVR, in order to support issuance of a ABS Class Certificate once all necessary outstanding items have been resolved.

The Two Cardinal Rules

When examining ship Classification activity on combatant programs where the NVR has been invoked, two aspects are easily recognizable as being different from past practice on ship programs where the ABS commercial Rules were applied. The first is that free and open communications must exist between ABS and the Navy. Contractual relationships on past programs presented constraints that could not be tolerated on combatant programs. One prominent example is that many issues require a formal interpretation of the Rules, and where the NVR is concerned, the Navy established a cardinal rule that no formal interpretations would be issued by ABS without Navy concurrence. A second aspect that differs on NVR programs is that when design or technical construction issues arise that may, by their nature, require a determination to be made by the NTAs, ABS is required to consult the NTA. Therefore, the second cardinal rule is that when a situation comes up that is clearly beyond what the Rules allow, or Rules compliance is for any reason at dispute, the NTAs must be advised so that they can become involved, evaluate the situation themselves, and render a technical authority decision when appropriate. First contact to coordinate ABS and NTA interaction on such matters is made through the SUPSHIP Chief Engineer for the shipyard involved. In order to support the shipbuilding process and avoid disruption or delay, it was necessary to focus on ABS and the NTAs being timely in their responses.

Business Rules are Necessary

When pilot implementation began on the LCS program, it was recognized that a different set of dynamics would need to be in place for combatants. ABS reached out to SUPSHIP and the shipyards to generate coordination plans that would form the basis of a more structured working relationship. Early on, those plans received concurrence from the SUPSHIP and the Shipyards, but they have evolved over time, incorporating experience and perspective on how best to operate efficiently. As a result, a more detailed set of *Business Rules* were established in the Fall of 2006. The Business rules cover many aspects of day-to-day relationships, but one aspect that needed more clarity was a means for managing disagreements over technical matters. The Business Rules require the use of a local *Adjudication Board* to clear the air on emergent issues as quickly as possible. A meeting of that board includes representatives from ABS, the SUPSHIP and the Shipyard. Any party can call such a meeting to facilitate movement toward resolution of issues in real time. When issues are not easily resolved at the local level, additional members are added to form a *Technical Authority Adjudication Board.* The expanded participation includes the Navy Ship Design Manager, SUPSHIP Chief Engineer for the affected Shipyard, and a Program Office representative. A brief period of time is allowed for fact finding, and then the board meets to thoroughly understand each party's position and evaluate options. If a resolution is not adopted by the board, the technically acceptable options are defined by the affected NTAs and presented to the Program Manager for a decision. In the past, it has not been uncommon to arrive at differences in what ABS and the NTA determined to be technically acceptable. So, one dimension of the new relationship is that ABS and the NTAs are committed to work together to establish a single view on what the acceptable solutions are for any issue, before transmitting that view to the Program Manager. These measures are being emphasized as a means to minimize any disruption to the Shipbuilder's orderly process for construction while clearing technical issues. Operating relationships under the Business Rules are being monitored at the highest levels within NAVSEA and PEO Ships. Regular 90-day reviews of performance under the Business Rules have been established, so that any needed adjustments can be factored in as the need becomes apparent.

HOW ARE THE RULES MAINTAINED

The NVR has been placed on a regular yearly update cycle. As a result of pilot NVR implementation on LCS and DDG 1000, many good ideas have been brought to light and many lessons learned have been documented for consideration as potential Rule changes. Updates can stem from a number of needs like: simple corrections or omissions discovered as part of implementing the Rules, additional requirements to address newly discovered safety concerns, expansion of the scope of systems or aspects to be certified by ABS, a broadening of the range of acceptable design solutions, or inclusion of lower cost alternate solutions that are identified by Industry. Though there is a pattern to the update cycle, good ideas and recommended changes are welcome at any time. The cycle begins anew each year with the issuance of the prior year's change package. For example, the January 2007 edition of the NVR incorporates changes collected and vetted during the 2006 Rules update cycle. Inputs for suggested Rules changes are solicited during the period of January through June. A committee process, similar to that applied by ABS on maintaining the commercial Steel Vessel Rules, has been established for the NVR to screen and evaluate proposed changes. For the NVR, they have designated the ABS Naval Panels (aka, Industry Panels) as having the responsibility to handle that work. Once the yearly package for each Part of the NVR has been assembled, it is circulated through the Naval Panel members for initial consideration. The Panel interacts either electronically or in a series of meetings, depending on the need. The Naval Panel proceedings take place during the period of June to September. Once the change packages have reached a consensus end-state and have been vetted at that level, they are submitted for broader review by the ABS Naval Technical Committee. The NTC members receive a consolidated input from all of the Naval Panels in the September timeframe. Many of the NTC members have representatives on the Naval Panels, so the NTC review is a secondary higher level review of proposed changes. The NTC members are given about six to eight weeks to review the consolidated package of changes and submit comments. Then an annual ABS NTC Meeting is scheduled in the October or November timeframe to consider the completed package as a whole, and make a formal decision to release the next version of the NVR, pending the resolution of any issues that had not been resolved prior to the NTC Meeting. That gives ABS sufficient time to format any final Rules changes and publish the next edition is following January. With the release of the January 2007 Rules, the ABS NTC has thus far authorized the release of four versions of the NVR. Though annual changes to the NVR are not folded into ongoing programs, emergent critical requirements can be addded on a case basis where a contract change is approved by the Program Manager.

Cost sensitivity has grown to be a central consideration in affixing new design requirements. The NTC Leadership has recently applied a more purposeful focus on the cost of proposed Rules changes. They have tasked ABS to begin screening proposed changes to highlight those that are perceived as cost increases. Then conscious NTC decisions are made for each of those cases, with only those having a compelling technical rationale (e.g., a safety issue) being authorized for inclusion in the revised Rules set. A new requirement has also been established for future submissions of proposed changes to include a cost versus risk evaluation as part of the supporting technical justification. That means the days are gone where more and better design requirements can be piled on without regard to cost impacts.

HOW HAS ABS DEVELOPED IT'S WORKFORCE TO SUPPORT CLASSING COMBATANTS

ABS has established and maintains a vigorous Quality System which forms the foundation for accreditation as an ISO 9000 2000 company and a recognized class society by the International Maritime Organization and the International Association of Classification Societies. In addition, it provides the mechanism whereby these parties as well as a host of other clients, Governments and interested parties may audit ABS procedures and processes to ensure that ABS is meeting their expectations. As the Naval Vessel Rules and classification of naval combatants have been introduced into the work carried out, appropriate processes and procedures have been developed, implemented and documented.

Training

In reality, the skill sets used by the Engineers and the Surveyors to carryout their classification duties do not fundamentally change from those employed in classing any other marine platform or vessel. Their skills in naval architecture, marine engineering and materials sciences, as well as the art of surveying, are applied directly to the central issue of determining the degree of compliance of the vessel with the established Rule set. The training that they must receive is in establishing a baseline understanding of the specific content of the Naval Vessel Rules and unique processes associated with working with the Navy as both a ship owner and as the Naval Technical Authority. ABS maintains the ABS Academy, a training organization with the primary focus of ensuring the engineers and surveyors are adequately prepared to carryout their duties. There is an established curriculum for both engineers and surveyors, which includes introductory and continuing education courses focused on the skill sets they require. Failure to accomplish required training results in the engineer or surveyor being barred from carrying out assigned work until the required training is completed. Naturally, the most effective training accrues from on-the-job training received in accompaniment of seasoned and skilled individuals carrying out their duties. Training in the Naval Vessel Rules has been introduced into the curriculum of the Academy, and engineers and surveyors targeted for that work are receiving the background they need to begin work.

Qualification

Once trained, engineers and surveyors cannot be assigned plan review or specific survey tasks until they have been qualified by an authorized supervisor in a specific plan review or survey activity. When a job request is received, the first thing a supervisor does is to check the qualification database maintained on-line by ABS in order to establish which of his engineers or surveyors are qualified to carryout the specific tasking in question. This qualification is based on the process instructions and supporting check sheets that exist for each task to be accomplished.

Management Structure

Two significant structural changes have been accomplished at ABS specifically to support classification of naval combatants:

1. **The position of Assistant Chief Surveyor for Naval Programs has been created** to support the related survey activities. The first individual assigned to this position was carefully selected to have been a highly seasoned surveyor who has previously been in an equivalent position for commercial work, and was also a Vice President for Operations overseeing all survey work in an ABS division.
2. **The Naval Engineering Department was created** to ensure that a trained and experienced engineering force is maintained to accomplish the work related to Naval Vessel Rules.
3. **Senior Management** - Overseeing all of this is a fully dedicated Vice President of Government Operations, whose sole focus is on meeting the expectations of government clients as they implement the classification process for their ships.

THE USCG IS ALSO A PART OF THIS PARTNERSHIP

The United States Coast Guard also acquires and operates vessels which are inherently military in their mission and projected operating environment. Since the very beginning, when considering application of the classification process with appropriate Rules to military vessels, it has been envisioned that the Coast Guard as well as the Navy could benefit. Non-military type Coast Guard platforms such as the two most recent buoy tender platforms have been classed with ABS to the appropriate criteria – the ABS Rules for Building and Classing Steel Vessels Under 90 Meters in this case. When the Deepwater Program was conceived, the question of a set of baseline technical criteria was considered. Naval Vessel Rules did not exist. Thus, in the earliest stages of the Deepwater Program, an effort was begun to identify in a holistic manner the criteria and standards to which the Deepwater vessel assets could be designed and constructed. Because the Naval Vessel Rules were not completed at that time, these criteria were to be identified by a joint USCG/ABS team and captured in a Cutter Certification Matrix. The matrix would identify by Ship Work Breakdown Structure (SWBS) number for each system on the vessel the projected criteria to be used for design, construction, test and certification. As such, the Certification Matrix could function as the index for the Rule set to which the vessel could be certified. The vision at that time was that ABS would play a significant role in certifying the Deepwater assets acquired in accordance with the matrix. ABS worked closely with the USCG Engineering Logistics Center (ELC) to develop these matrices and from that effort a close working relationship was formed. The standards cited included some drawn from existing ABS Rules, some internal USCG criteria and US Navy military standards, and some commercial standards such as those from standards organizations like the American Society for Testing Materials (ASTM) and the

Institute of Electrical and Electronic Engineers (IEEE). In addition, certification criteria for some areas specific to individual shipyard practices were left to the shipyard to propose.

Subsequently, the ELC participated to a great extent in development of the Naval Vessel Rules. This effort culminated in their development of a USCG Addendum to the NVR, which provides alternative criteria in a number of NVR areas to be applied when Coast Guard Cutters are classed to the NVR. Based on this effort, the Coast Guard Technical Authority (CG-4) has stated that it will be Coast Guard policy to build future surface assets to the NVR. The Coast Guard Chief Engineer is a joint partner with the Chief Engineer of NAVSEA in chairing the Naval Technical Committee as it carries out its charge to keep the NVR current and correct.

Initial application of the Cutter Specific Certification Matrix to the first Deepwater platform – the National Security Cutter – was left largely to the Contractor. That is, the Coast Guard accepted a great amount of "self-certification" by allowing the Contractor to attest compliance with the specified standards. ABS was retained to certify approximately 3% of the criteria. In retrospect, this approach has not served the purpose of compliance assessment to the degree expected by the Coast Guard and the policy is under review. At this point in the Deepwater Program, it appears that the Coast Guard will re-insert ABS into the certification process for each of the various surface assets to some degree. The Fast Response Cutter will be required to be classed with ABS to the ABS High Speed Naval Craft Guide, while the Offshore Patrol Cutter will be classed to the Naval Vessel Rules. The National Security Cutters, due to the maturity of the program, will continue to be built to the Cutter Specific Certification Matrix, but the role of ABS as a primary certification agent for compliance with that matrix will be increased.

MAINTAINING NAVAL COMBATANTS IN CLASS DURING SERVICE LIFE

Once the Navy had committed to build both LCS ships and the DDG 1000 in Class with ABS, the question was immediately posed concerning whether or not the Navy would also elect to retain those ships in Class during in-service operation. That question led to extensive study and a business case analysis to support making a determination. Because LCS 1 will be the first combatant to be delivered in Class, the thought process described below will be expressed in terms of making that decision for the LCS ships.

The LCS ships have been designed and are being constructed in accordance with the ABS *Guide for Building and Classing High Speed Naval Craft*, the *ABS Naval Vessel Rules*, and other referenced ABS Rules and Guides as necessary to comply with the designated class notations (as modified with the concurrence of the Navy Program Office and approved by the Naval Technical Authority). The vessels will receive, as a minimum, the following classification notation:

> ✠A1 (E) Naval Combatant ✠AMS, SFA(30), NIBS, FAS, CONREP, VERTREP, ROTARY WING

The unique nature of LCS will necessitate that the hull forms, materials and some systems will incorporate features normally associated with "high speed naval craft." As a result, the applicable ABS Rules will include some parts of the ABS Guide for Building and Classing High Speed Naval Craft, most specifically for the hull, the main propulsion and reduction gear components. There are also areas that will be addressed by criteria other than the ABS Rules, including requirements of a statutory nature and criteria for militarily unique features.

The LCS ships will be delivered to ABS classification in accordance with the above. In order to take full advantage of the opportunities afforded by this process, it has been decided that the Navy will retain the vessels in class with ABS during their service lives. This is the approach that the Military Sealift Command, the National Oceanographic and Atmospheric Administration and the Maritime Administration take for the vessels they operate.

LCS Maintenance Philosophy and Approach

The Navy is not constrained by existing maintenance and logistics paradigms with regards to the Littoral Combat Ship platforms. Due to minimal shipboard personnel, more of the traditional maintenance workload will be transferred to shore-based support. Ship crews will be expected to carryout only basic maintenance and repairs, all other corrective maintenance (I and D level) will be dealt with by shore based teams. These teams will be part of the Maintenance Support Detachment (MSD).

The process will be as follows: When a problem is found onboard an LCS, the crew identifies and assesses the problem. If it can be easily rectified, they do so. Otherwise the MSD is brought into the process and makes a decision as to what needs to be done. If the problem can be resolved remotely, the MSD supports the crew to do so. If the problem is not critical but cannot be resolved by the crew with the MSD's support, the corrective action is deferred to either the next port visit or completion of deployment. The 3M coordinator is part of the maintenance team that is charged with resolving these issues onboard the LCS platforms, and those maintenance actions will be scheduled by the MSD from shore.

The envisioned role of ABS classification is to dovetail into the work carried out by the MSD, whereby class surveyors will assist the Navy in oversight and verification of the LCS platforms. Class will be a cost-effective approach to supplementing the Navy's maintenance programs with an outside third-party that is extremely familiar with these vessels and will bring tremendous value to its overall operation.

Overview of ABS Classification In-Service

Presently, the two LCS platforms, though distinct in their designs, are being constructed in accordance with the appropriate ABS Rules under the continuous surveillance of a team of experienced ABS surveyors who are well versed in the applicable rules and supported by a network of expertise throughout this verification process. Upon delivery, provided that all technical and survey comments are dealt with, the Lead Surveyor for each program will issue an Interim Classification Certificate, thereby attesting, with decisive confidence and responsibility, that the vessel he has lived with during its construction has indeed been built in accordance with the approved plans and applicable ABS Rules.

At this juncture, the Navy could either choose to drop classification or continue the classification relationship throughout the vessel's life-cycle. By dropping class, ABS would no longer have a role or responsibility to the Navy with regard to the condition of the vessel. Additionally, the Navy would take on the exclusive responsibility for verifying the existing condition of these vessels, as well as overseeing all repairs and modifications to ensure continued compliance with the established baseline set of technical criteria (the Rule set to which they are built). The alternate course of action which the Navy has chosen is to maintain full classification throughout the life of the LCS vessels. By choosing to do so, the Navy benefits by having an independent certifying third party, intimately familiar with the construction standards and the ship itself, complementing its own "sustainment" programs and protocols for these vessels. As noted earlier, the Navy is already challenged with the task of breaking away from traditional maintenance programs and tailoring a new approach which takes into account the reduced crew sizes and structural configurations requiring more frequent and comprehensive inspection techniques. Therefore, it was logical for the Navy to take full advantage of an already engaged and available resource, such as ABS, to assist them in verifying that the LCS class of vessels is not only well maintained, but more importantly available for deployment whenever necessary to accomplish the Navy's intended mission.

By retaining the LCS in class a series of events and available services will come into play. Periodic surveys will be carried out in order for the vessels to comply with the classification requirements. Periodicity of these surveys remains to be determined and agreed upon between ABS and the Naval Technical Authority. The scope and focus of the periodic surveys will be promulgated in Part 9 of the Naval Vessel Rules, which covers the surveys after construction. Part 9 is currently in development and will include any necessary requirements that the Navy feels must be added in order to adapt the classification system to the unique nature of the LCS platform operations. As was done with the NAVSEA Warranted Technical Authorities, the Navy team charged with developing the LCS maintenance plan will review and contribute to Part 9 in order to ensure complete compatibility. As part of the classification package, the LCS platforms will be automatically entered into the ABS SafeNet system. Thus, all classification related information will be continuously accessible, via the Internet, to the Navy through a secure link. Through the SafeNet system, the Navy will be able to coordinate their periodic surveys with the planned maintenance periods. Additionally, all class related information and vessel survey status will be available for review and subsequent action. Finally, as part of the classification relationship, the Navy has decided to implement the Hull Maintenance Module to SafeNet – a graphical interface which tracks hull condition history including gaugings, damage, repair and alterations.

ABS Survey Organizational Structure and How It Will Align With the Navy's In-Service Maintenance Structure

ABS has 200 port offices in 78 countries. It employs 1700 exclusive surveyors, engineers and support staff. It is logistically positioned to aid in the support of the Littoral Combat Ships virtually anywhere in the world. It will, however, be necessary to develop a protocol whereby survey support can be provided to the LCS without jeopardizing the security and sensitivity of its mission. This arrangement will be advantageous to the LCS when deployed in the far reaches of the world and faced with a casualty that necessitates on-site technical expertise. In the case of battle damage, an ABS surveyor can be the first responder as was the case in recent history with the USS STARK and USS SAMUEL B. ROBERTS. Additionally, from a security and decision making perspective, limiting the number of personnel involved with a casualty could also prove advantageous to the Navy.

ABS' survey organization is made up of administrative and field personnel as depicted in Figure 4.

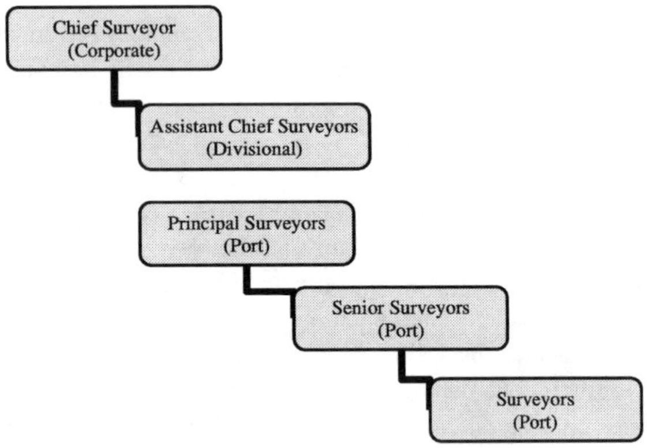

Figure 4. - ABS Survey Organization Structure

The Chief Surveyor is a corporate entity who has authority over and is responsible for all survey matters and survey personnel within ABS. He also holds the title of Vice President and Corporate Officer, and answers only to the Chairman and CEO. He attends the many IMO (International Maritime Organization) and IACS (International Association of Classification Society) meetings, as well as ABS Classification Committee meetings and other technical committee meetings relevant to the administration of the ABS Rules. Directly reporting to the Chief Surveyor are the three Division Assistant Chief Surveyors (Houston, London and Singapore) who are ultimately responsible for the day-to-day technical issues that arise with regards to survey matters within their respective divisions. In addition, a fourth Assistant Chief Surveyor for Naval Programs has been appointed to provide this same function for Government vessels that are classed. They are the catch-all for questions and concerns that filter up from both their own surveyors as well as the clients at large. The next tier consists of the Principal Surveyors who are the respective heads of their assigned port offices. The Principal Surveyor is responsible for all business within his port and any stations that fall under his port's jurisdiction. He is responsible for ensuring that all of his personnel are properly trained to carryout the surveys within his area. He is also responsible for resolving any issues that may arise between the attending surveyor and the owner's representative. Finally, he is responsible for reviewing and approving all reports prior to their release into the vessel's SafeNet records. Larger ports typically have one or two Senior Surveyors in their ranks to assist the Principal Surveyor with training of junior Surveyors, managing the job book, and carrying out day-to-day operations. Finally, we have the Surveyor whose experience level spans from brand new trainee to seasoned veteran, much like an experienced chief petty officer. Most of the Owner's dealings are directly with the attending Surveyor or, in some cases, the Senior Surveyor.

On the Navy's side, the primary contact would come from the Maintenance Support Detachment (MSD), which will provide an on shore infrastructure that will be the Navy Supply Chain's face to the fleet. The MSD will provide a single decision center for the LCS platforms. The MSD Officer will be responsible for the function and performance of the MSD. He will be the single point of contact for the LCS Supply Officer. Within this umbrella of the MSD will be the Ship's Maintenance Team (SMT) who will coordinate the sea frame corrective maintenance. They will be responsible for planning and scheduling shore-based and depot level maintenance. Maintenance will adhere to the already established SHIPMAIN process. This maintenance will be carried out through a combination of Regional Maintenance Center (RMC) personnel. The SMT will include the Commanding Officer of the LCS, as well as a designated Port Engineer. The Port Engineer will be the natural point of contact with ABS. During seaframe maintenance ashore or when an LCS is deployed away from its homeport, the SMT will be responsible for carrying out all required maintenance, repairs, and all deferred maintenance actions reported by the individual vessel's Supply Officer to the MSD. At this point the Port Engineer will determine whether the items that are on the maintenance and repair list warrant attendance by class in accordance with the Rules. Until he is comfortable with this or at any time he is not sure, immediate advice can be sought from ABS as to the necessity for an ABS attendance. The Port Engineer will ensure that ABS attendance is arranged for the required periodic surveys. He will also call in the ABS Surveyor as necessary to observe repairs requiring attendance and to examine damages and make recommendations as to extent of repairs necessary. Subsequently, the ABS Surveyor would verify that the work was in fact completed as required by visual examination, and testing, if applicable, and would complete his report attesting to that fact.

What systems, equipment and components are covered by ABS Classification

ABS classification covers all areas of the LCS which are called out in those sections of the *Guide for Building and Classing High Speed Naval Craft* and the ABS *Naval Vessel Rules* which were invoked by its build specification. A general list of covered equipment and components includes:

- Hull and Structures,
- Propulsion and Maneuvering Systems,
- Electrical Systems,
- Control and Navigation,

- Auxiliary Systems,
- Habitability and Outfit,
- Military Environment, and
- Materials and Welding.

Matching ABS Surveys with the Envisioned Maintenance Periodicity of the LCS

The traditional ABS survey interval and periodicity for commercial vessels revolve around a five-year cycle. This is due to the fact that most statutory requirements also revolve around a five-year cycle. In an attempt to keep both class and statutory surveys and inspections harmonized, the five-year cycle has been adopted as the standard. In the case of the LCS platform and the nature and uniqueness of its mission, it will be necessary to customize survey intervals so that they can be accomplished within windows of opportunity offered by the LCS deployment schedules and operational requirements. The Navy plans for the LCS platform to be deployed for a period of 2 years. It will be in port once every thirty days. Shore-based maintenance, repairs and class surveys would have to be scheduled during the in-port periods. Crew change periods could also be utilized as maintenance, repair and survey opportunities as well. The Navy will establish the drydocking periodicity. Once established, periodic surveys will be carried out in a systematic fashion in-line with the deployment, maintenance and port visit schedules. If based on a notional eight-year drydocking cycle, the proposed periodic survey intervals are as follows:

- **Annual Surveys** – carried out annually within a ± 3-month window on either side of the anniversary date. Given the estimated deployment schedules of the LCS, beginning from the anniversary date, the first, third, fifth and seventh Annual Surveys would be carried out while the LCS is deployed during its in-port or crew change period. The second, fourth and sixth Annual Surveys can be carried out while the LCS is in its homeport between deployments.

- **Intermediate Hull Surveys** – beginning after the first Special Continuous Survey, it would be carried out within a 30-month window which begins 3 months prior to the third Annual Survey and ends 3 months past the fifth Annual Survey anniversary dates. This would give the Navy a great deal of flexibility to schedule this survey.

- **Drydocking Surveys** – since the LCS platforms are built using specialized materials and reduced scantlings, it is not recommended to go more than 2 years without examining the underwater hull structure and its appendages. This could be accomplished during the biennial hull cleaning periods. However, in trying to rationalize drydocking to match the vessels needs, the LCS will undergo an extensive underwater inspection during its first drydocking period, which would fall within the 30-month window of the Intermediate Survey and could technically be carried out at the LCS' home port. An out-of-water drydocking survey would be carried out within 15 months of the Special Continuous Survey.

- **Special Continuous Surveys** - including Continuous Hull Survey and the Continuous Machinery Survey, respectively. Optimally, approximately 15% of hull and machinery items are to be dealt with for credit each year with the eighth year completing the cycle. In some cases this may not be possible and it may be necessary to cluster items in a different manner. The ultimate goal is to complete examination 100% of all necessary items within the 8-year period. It must also be noted that a percentage of non-vital items, or items agreed upon between ABS and the Navy, may be examined and recorded by the vessel's crew or Port Engineer. This would be verified for credit by an attending ABS Surveyor during the next scheduled survey. Additionally, utilizing other methods or programs to carryout these surveys, such as Risk Based Inspections or Reliability Centered Maintenance methods, in unison, may be an option. Finally, an extension may be granted by the Chief Surveyor's office under exceptional circumstances or in the case of Force Majeure.

The Mechanics of the Process

First, and most important of all, ABS has found that it is essential to establish an overall single point of contact so that, if questions arise regarding the process, the Navy has a consistent place to go to get an answer or be directed further. This individual will be assigned as the in-service LCS Program

Manager. It is desirable, to the greatest degree possible, to allow the process to proceed normally with contacts made as described below but we have found the role of the in-service Program Manager invaluable.

1) Making contact – the primary point of contact on the Navy's side is the individual LCS Port Engineer. In order to schedule a survey, the respective Port Engineer will contact the ABS in-service LCS Program Manager, whether the vessel was at its home-port or deployed overseas. The LCS Program Manager subsequently contacts the relevant local ABS office. Once the contact is made the ABS office will then generate that particular vessel's survey status from the SafeNet system. This is the same survey status that is available to the Navy and therefore both the Port Engineer and the assigned ABS Surveyor will have access to identical information. It would be during this time that the Port Engineer and the assigned Surveyor will confirm which surveys are due or coming due and which items may be available for examination. A date and time of survey is typically set at this point.

2) Developing a Survey Plan – once the ABS Port Office has been informed, a survey status has been generated and a date for attendance has been tentatively scheduled, it is the duty of the assigned ABS surveyor to maintain contact with the LCS' Port Engineer and develop a concise survey plan. This will keep everyone on the same page and will allow the LCS' crew and maintenance support personnel to prepare the vessel for survey, including scheduling external specialists that may be necessary. This also eliminates any surprises and focuses all parties to the tasks at hand.

3) Expectations on board – when the ABS Surveyor arrives at the vessel with survey plan in hand, it will be necessary to have a short meeting with the Port Engineer and other relevant parties to go over the survey plan and add or delete any items that may have arisen. If tank inspections are included as part of the task, the vessel's crew should have opened and ventilated those spaces previously with evidence proving that the spaces are safe for entry. On the machinery side, any items to be opened up should be prepared ahead of time. Items for which the crew would like credit as a result of repairs or inspections they have conducted without ABS attendance will be discussed and credited as appropriate. Also any items to be tested should be ready and preferably pre-tested prior to the ABS Surveyor's arrival.

4) Post survey activities – once all items, both in the survey plan and any that were added later, have been examined, it will be prudent to carryout a closing meeting, whereby the LCS' Port Engineer and attending ABS Surveyor can compare notes and confirm with each other what the results of the survey are and what items will be credited on the vessel's survey status. Additionally, any outstanding recommendations will be noted and explained. It is critical that the Port Engineer is aware as to what will be written in the ABS Survey Report prior to the Surveyor's departure from the vessel. In some cases, gauging reports or underwater inspection reports will not be available during the ABS Surveyor's attendance and therefore will need to be forwarded to him as soon as possible in order that he may include them in his final report.

5) Technical Authority – in matters requiring interpretation of the applicable Rules or variances from prescriptive requirements, the in-service ABS Program Manager will consult with the Naval Technical Authority through the NAVSEA In-Service Ship Design Manager. Consultation will involve all relevant parties representing the ship, the Surveyor, the MSD and the ABS Assistant Chief Surveyor for Naval Programs.

Reports and Access to Data

1) The Report – following every survey and attendance by an ABS Surveyor, the Navy will receive a report indicating what was accomplished during that attendance. In some cases the report will simply indicate that a particular survey is being commenced. In other cases the report will include entire surveys with outstanding recommendations as well as rectifications of those outstanding recommendations. The reports will also indicate all items that are considered complete and are recommended for crediting. On completion, the ABS Survey Report will be reviewed and approved by the Port Principal and the information is automatically uploaded electronically into the vessel's SafeNet status reflecting its current condition of class. Subsequently, a hard copy of the report will be forwarded to the appropriate designee within the Navy for review, distribution and action.

2) SafeNet – is a data management program that is provided to all ABS classed vessel owners. Its primary function is to manage each classed vessel's survey status. The Navy has continuous secure access to the system and can look at any or all of its vessels at the same time. SafeNet gives the Navy the ability to review their vessels survey schedules, vessel particulars, outstanding recommendations, current status with regards to items credited towards Continuous Surveys, condition of coatings within ballast tanks and overall history of surveys carried out. It is a vital tool in managing and scheduling classification surveys and is accessible by computer from the world-wide web through a secure link virtually anywhere in the world. The Navy and ABS have begun looking for opportunities to interface or coordinate SafeNet with the Navy's Current Ship Maintenance Project (CSMP).

THE NAVY RETAINS TECHNICAL AUTHORITY

It should be underscored that the Navy has in no way relinquished its Technical Authority under the NVR initiative. The Navy considers that it has final authority over the content of the NVR and any interpretations of the Rules. The Navy desires that ABS function as an independent certification agent, and that the ABS Ship Classification process be executed as close to autonomously as circumstances permit. It is to be expected, however, that numerous situations will arise on any combatant program where the NTAs must be involved to perform their Technical Authority duties, for which they alone are responsible. When situations develop that necessitate a technical determination be made by one or more of the NTAs, the deciding view on what is ultimately acceptable will be theirs. That aspect is supported by Public Law, and reinforced by SECNAV Policy.

The Navy has also recognized that ABS is learning the combatant design dimension of naval shipbuilding, and that implementation of the NVR is much more challenging than Classing to the ABS commercial Rules. As a result, the established relationship between ABS and the Navy requires monitoring of ABS Classing activity on combatant programs, so that more confidence, greater understanding and better cooperation can be built over time. This monitoring will take multiple forms, and be conducted in numerous contexts. For instance, the fidelity of engineering plan reviews performed by the recently established Naval Engineering Department at ABS Houston may be sampled randomly by any of the NTAs. And the work of the ABS Surveyors in the Shipyards may be monitored by the SUPSHIP Chief Engineers. It is believed that through these efforts the Classification process for combatants can be continually improved.

CONCLUSIONS

Developing the NVR and adapting the Ship Classification process for combatants has been a huge and challenging undertaking, stretched across many years. The authors began exploratory discussions regarding this initiative in 1996. Serious consideration of pursuing this initiative at the COMNAVSEA level began in 1998. Initial reservations from several segments of the combatant design community were gradually addressed and eventually retired. The steadfast desire has been for the members of the combatant design community to take ownership of the process and share in the development of the Rules that we all would use. Much broad-based buy-in has been achieved, and some industry leaders involved in pilot implementation have stated publicly that the ships that will be delivered under NVR programs will be better ships than would otherwise have been the case. Getting accustomed to a new process is always a challenge. But now that many of the start-up problems are behind us, it is important to remember that any process can be improved. The next thrust will be to work as a community-of-practice to apply lean methodologies so that efficiencies can be realized. Such a commitment will truly take Ship Classification to a new plateau.

REFERENCES

Cooperative Agreement for *Development, Implementation and Maintenance of Naval Vessel Rules*, signed by Commander, Naval Sea Systems Command and the President of ABS Americas on 1 October 2003.

ABS Naval Vessel Rules (Published as the *ABS Guide for Building and Classing Naval Vessels,* (Distribution Statement-D, available on a need-to-know basis), American Bureau of Shipping, Houston, Texas (most recent edition released January 2007).

U.S. Statute: U.S.C. Title 10, Chapter 503, Section 5013. (Requires The Secretary of the Navy to execute Technical Authority - Statutory requirement is delegated (for ships) to Commander, Naval Sea Systems Command.)

Discussion

Dale G. Gabel, Member

This is a very timely paper on a topic of significant relevance to both the U.S. Navy and the U.S. Coast Guard. The reason it is so timely and relevant is because it documents a fundamental swing back of the pendulum to better control by government of its purchases in the surface, non-nuclear shipbuilding arena. When GENSPECS became outdated and unsupportable in the late 1990s, the naval services lost a reference document that had helped ensure surface combatants were built to a consistent set of design and construction standards. GENSPECS constituted a rule set that served the country very well for many years — its loss was significant.

Over the past decade, lacking the consistent rule set of GENSPECS, the naval services have not had the tool necessary to ensure all of the relevant requirements are included in a ship design and construction acquisition. But the pressure to allow more industry-based solutions of technical requirements is unrelenting. The NVR will serve to swing the pendulum back to a consistent rule set for naval designers — but with much more inherent flexibility and opportunities for emerging technology insertion than was the case with GENSPECS.

In the Naval Vessel Rules, naval combatant best practices and requirements are merged with commercial industry best practices and standards to create a rule set that represents the best of both worlds. By administering NVR through ABS via a rigorous updating process, the Navy and Coast Guard take advantage of ABS's broad experience and knowledge of the commercial shipbuilding industry and its capability to rapidly identify and incorporate emerging and relevant best practices and standards into the NVR. Through ABS the government manages the NVR's content, with significant and ongoing input from the private sector and thereby provides the best environment possible for building operationally effective and cost-efficient ships.

A couple of aspects of this topic do bear watching over time. Although NVR does not apply to "military unique or warfighting" aspects of a warship design, there are elements of the lower risk hull, mechanical and electrical (HM&E) systems that must be designed and constructed differently from similar systems on a commercial vessel. Certain structural and system components must be designed for blast effects and, while a commercial vessel will, if possible, avoid heavy weather and the attendant structural loads and fatigue accumulation, U.S. Navy surface vessels and U.S. Coast Guard cutters often must enter and operate in those extreme environments. So the "bright line" between "military unique" and otherwise is not always that bright. Two problems can arise from this. The first is that the areas in which a more robust standard is needed may not always be recognized or acknowledged; the second is that in an excess of caution more robust standards will be required everywhere. Both of these tendencies must be guarded against. The first would lead to less effective vessels; the second to unnecessarily expensive vessels.

A topic of current discussion within the Coast Guard is the practicality of maintaining in-service ships in Class. Although the Navy has decided to keep at least the LCS in Class, the Coast Guard has not chosen to maintain ships built to Class standards in Class. When that decision was originally made the primary reason was that keeping the vessels in Class would be redundant to the existing maintenance program and therefore an unnecessary cost. To validate that thinking the Coast Guard is working with ABS to compare a cutter maintained under the existing program to a notional cutter that had been surveyed periodically and maintained to Class. Pending the outcome of that comparison, the Coast Guard Continuity may ultimately determine that an ABS-led condition assessment, coupled with a single authoritative data store related to a specific platform, will more readily support lifecycle sustainment efforts. If so there is a much greater likelihood that future major Coast Guard cutters, which will be built and classed to NVR, will be maintained in Class.

The authors are to be complimented on a well-written and topical paper. Use of Classification Society rules is well understood in the commercial shipbuilding world — it needs to be well understood in the naval ship design and construction world as well because it is the way associated business will be conducted in the future. A topic for a follow-on paper by these same authors could be the experience of ABS and the Navy in transitioning to NVR on the LCS class of ships. That experience was not without its challenges and could generate lessons learned that would prove valuable as this way of doing business becomes more common in the naval design and construction arena, and as NVR requirements are refined over time.

Authors' Closure

The authors would like to thank RADM Gabel for his discussion and the words of encouragement he has provided. As chief engineer of the U.S. Coast Guard, he joins with RADM Kevin McCoy USN (chief engineer of the Navy) as primary stakeholders in the success of this initiative. They, together with VADM Paul Sullivan USN and many others in their respective organizations, have contributed a great deal of guidance and labor in partnering with ABS to bring this initiative to the point at which we stand. We agree with RADM Gabel in all points and look forward to his continued oversight and guidance to ensure that we maintain course. Since the time of this presentation, progress has continued to be made and plans for maintaining the LCS vessels in class have advanced. Recent training classes to Navy waterfront implementers and the vessel Class Squadron have yielded valuable feedback in areas which require work. We look forward to being able to provide the Society with an update in the very near future.

Alternative Propulsion Methods for Surface Combatants and Amphibious Warfare Ships

James S. Webster, Visitor
Howard Fireman, Member
Dillon A. Allen, Visitor
Adrian J. Mackenna, Visitor
John C. Hootman, Associate Member

ABSTRACT

The US Navy has been studying the technical and cost impacts associated with the availability and cost of fossil fuel contrasted with nuclear energy alternatives for surface combatants and amphibious warfare ships. Over the past two years these efforts have grown in maturity to examine the tactical and strategic implications of our Navy's dependence on fossil fuels from technical, economic, and military perspectives.

This paper describes the NAVSEA evaluation of alternative propulsion methods in Small, Medium, and Large Surface Combatants performed during FY2005 in response to Chief of Naval Operations Guidance, and the efforts conducted during FY2006 directed by Public Law that added Amphibious Warfare Ships.

Ship concepts were developed and analyzed in terms of acquisition cost, life-cycle cost, and operational effectiveness. Power and propulsion technologies including nuclear power, gas turbines, diesel engines, mechanical power transmission systems, hybrid power transmission systems, integrated power transmission systems, combined power plants (e.g., diesel and gas turbine), and various propulsor systems. Operational effectiveness in warfare, mobility and vulnerability areas are evaluated for alternative power and propulsion systems and associated architectures in non-program-of-record small (~7,500 to ~12,000 metric ton) and medium (~21,000 to ~26,000-metric ton) surface combatants and in amphibious warfare (~34,000 to ~38,000-metric tons) ship concept designs.

The techniques and models used in the studies are discussed, examples provided, and results presented. These results include a discussion of viable near term alternative power and propulsion system technologies and architectures, break-even cost of oil (where life cycle costs of nuclear and fossil powered ship concepts are equivalent), and correlation of alternative architectures with operational benefits.

KEY WORDS

Ship Design, Ship Concepts, Ship Propulsion, Naval Surface Ships, Trade-off Study, Mechanical Power Transmission Systems, Hybrid Power Transmission Systems, Integrated Power Transmission Systems, Combined Power Plants, Nuclear Power, Gas Turbines, Diesel Engines, Surface Combatants, Amphibious Warfare Ships

NOMENCLATURE

AoA	Analysis of Alternatives
ASSET	Advanced Surface Ship Evaluation Tool
ASW	Antisubmarine Warfare
AWS	Amphibious Warfare Ship
$B	Dollars in Billions
BBL	Barrel of Oil (42 United States gallons)
BSP	Baseline Security Posture
CG-47	*Ticonderoga* Class Guided Missile Cruisers
CER	Cost Estimating Relationship
CH-46	Sea Knight Helicopter
CH-53	Sea Stallion Helicopter
CNO	Chief of Naval Operations
CODAG	Combined Diesel and Gas Turbine
CODLAG	Combined Diesel Electric and Gas Turbine
CODOG	Combined Diesel or Gas Turbine
COGAG	Combined Gas Turbine and Gas Turbine
CVN-21	21st Century Aircraft Carrier, now designated CVN-78, formerly CVN(X)
DC	Direct Current
DD-963	*Spruance* Class Destroyers
DDG-51	*Arleigh Burke* Class Guided Missile Destroyers
DDG -1000	*Zumwalt* Class Guided Missile Destroyers (Formerly called DD-21 and DD(X))
DESC	Defense Energy Support Center
DFM	Diesel Fuel Marine
DG	Diesel Generator
DRM	Design Reference Mission
DSL	Diesel
DoD	Department of Defense
DOE	Design of Experiments
DPS	Defense Planning Scenarios
F76	Diesel Fuel Marine
FFG	*Oliver Hazard Perry* Class Guided Missile Frigates
FP	Full Power
FPU	Forward Propulsion Unit
FY	Fiscal Year
GFE	Government Furnished Equipment
GT	Gas Turbine
GTG	Gas Turbine Generator

HR	House Resolution
Hz	Hertz
IFTP	Integrated Fight Through Power
IPS	Integrated Power System
AWS	Amphibious Warfare Ship
Kts	Knots, nautical miles per hour
kV	kilo-Volts
LCC	Life-cycle costs
LCS	Littoral Combat Ship
LHA-1	*Tarawa* Class Amphibious Assault Ships
LHA-6	LHA(R) Lead Ship
LHA-7	Formerly LHA(R) Flight 1
LHD-1	*Wasp* Class Amphibious Assault Ships
LHD-8	*USS Makin Island*
$M	Dollars in Millions
MAPS	Manpower Analysis and Prediction Systems
MCO	Major Combat Operation
MIW	Mine Warfare Ships
MOE	Measure of Performance
MOP	Measure of Effectiveness
MSC	Medium Surface Combatant
MT	Metric Ton
MW	Mega Watts
NA	Not Applicable
NAVSEA	Naval Sea Systems Command
NPV	Net Present Value
NRE	Nonrecurring Design and Engineering
NSWCCD	Naval Surface Warfare Center Carderock Division
NSWCDD	Naval Surface Warfare Center Dahlgren Division
O&S	Operating and Support
ONI	Office of Naval Intelligence
ONR	Office of Naval Research
OPNAV N81	Office of the CNO, Capability Analysis and Assessment Division
OPSITS	Operational Situations
OPTEMPO	Operating Tempo
OSD	Office of the Secretary of Defense
PDR	Preliminary Design Review
POM-08	FY 08 Program Objectives Memorandum
POE	Projected Operational Environment
RDT&E	Research, Development, Test, and Engineering
ROC	Required Operational Capability
SC	Surface Combatant
SHU	Steaming Hours Underway
SSC	Small Surface Combatant
SSN-21	*Seawolf* Class Attack Submarines
TACSITs	Tactical Situations
TBMD	Theater Ballistic Missile Defense
USN	United States Navy
V-22	Osprey Tilt-rotor Aircraft
VAMOSC	Visibility and Management of Operating and Support Costs

INTRODUCTION

One role of engineering in the US Navy is to provide quantifiable evidence supporting informed decisions between alternative solutions to complex problems. Naval surface combatant system complexity is a primary source of performance prediction uncertainty in early stage design. Innovative technologies with expected performance benefits are often represented as point solutions with little or no cost or performance validation data. System complexity aggregates the subsystem risk associated with integrating multiple unproven technologies, operating legacy technologies in different natural or warfare environments or a combination of both further reducing the attractiveness of new technology. Accordingly, Department of Defense (DoD) and the United States Navy (USN) have generally been restricted to procuring systems that are small technological excursions from existing technological baselines.

Evidence supporting decisions at tactical, strategic and national policy levels is valued if provided in a timely manner with quantifiable levels of confidence. Congressional testimony provided by senior DoD representatives during the past two years has highlighted the importance of developing a national strategy that is responsive to the increasing price and reduced availability of fossil fuel. The FY2006 Report to Congress (RTC) (Public Law, 2006) was written in response to Section 130 of the FY2006 National Defense Authorization Act (Public Law, 2006) that directed the Navy to evaluate alternative propulsion methods for surface combatants and amphibious warfare ships. Alternate propulsion studies performed during the FY2005-2006 time frame exercised recently developed modeling techniques that quantify cost, schedule, and operational effectiveness with acceptable confidence.

Energy requirements were developed for each ship type based on Design Reference Missions (DRMs). The DRMs are comprised of Tactical and Operational Situations that suggest an employment of energy management systems (energy storage, conversion, distribution, and thrust generation systems) in ship energy states linked to mobility, survivability, and warfare mission system energy demands. The mission driven peak and lifetime energy demands were used in power and propulsion system architecting and systems engineering for each ship type and its associated energy management system component suite.

The RTC study addressed power and propulsion technologies and architectures such as nuclear power, gas turbines, diesel engines, mechanical power transmission systems, hybrid power transmission systems, integrated power transmission systems, combined power plants (e.g., diesel and gas turbine), and various propulsor systems. It identified aspects of these technologies and architectures that are anticipated to mature in time to transition to ship acquisition programs within the next ten to twenty years.

The study evaluated twenty-three different ship concepts. These concepts were variants of three ship baselines: Small Surface Combatant (SSC), Medium Surface Combatant (MSC), and Amphibious Warfare Ship (AWS). Mission systems for the ship concepts are similar to current or projected system architectures. The ships represent energy demand surrogates for the ships projected to comprise the US Navy

2020 fleet. The unique methodology developed in this study relates future warfighting needs to energy demand.

Section 130 of the FY 2006 National Defense Act identifies several important and detailed matters to be addressed, including: the key assumptions used in carrying out the analysis; the methodology and techniques used in conducting the analysis; a description of current and future technology relating to surface ship propulsion; a description of each propulsion alternative and an analysis and evaluation of each such alternative from an operational and cost-effectiveness standpoint; a comparison of the life-cycle costs of each propulsion alternative; an analysis of when the nuclear propulsion alternative becomes cost effective as the price of a barrel of crude oil increases ("break even" analysis); conclusions and recommendations of the study; and the Secretary of the Navy's intended actions, if any, for implementation of the conclusions and recommendations of the study.

STUDY METHODOLOGY AND TECHNIQUES:

The US Navy study and evaluation of alternative propulsion systems for US Navy Ships has resulted in improvements in analytical techniques and modeling tools that increased resolution and reduced uncertainty in each succeeding study. The modeling development efforts culminated in the use of over 20 design/analysis tools being exercised during FY2006 in the systems architecting, systems engineering, ship synthesis, performance evaluation, cost estimation, and operational effectiveness assessments. Table 1, below, summarizes the modeling functions and modeling efforts exercised during the past two years. In the context of this table, historic implies the direct use of known designs, systems, or data; parametric implies the use of design models based upon data regression or other modeling techniques; and engineering implies the use of design based upon engineering principles and highly detailed modeling techniques.

Study Attribute	FY2005 Economic Analysis	FY2005 CNO Guidance Study	FY06 Congressional Study
Concepts:	Historic	Engineering	Engineering
Acquisition Costs:	Historic / Parametric	Eng./Parametric	Eng./Parametric
O&S Costs:	Historic	Parametric	Parametric
Fuel Costs:	Historic Cost	Engineering	Engineering
Defuel / Deac:	Historic	Historic	Engineering
Effectiveness Analysis:	N/A	N/A	Capabilities Based

Table 1: Evolution of Alternate Propulsion Studies

The FY05 Economic Analysis used historic data for amphibious warfare ships reflecting oil fired steam turbine mechanical drive propulsion plants. The FY2006 Report to Congress used a more modern combined gas turbine and diesel propulsion plant. These distinct propulsion types have significantly different efficiencies and therefore significantly different fossil fuel consumption profiles. For example, the LHD-8 diesel-electric drive with gas turbine boost propulsion plant consumes less than one-half the fuel of an LHD-1 oil fired boiler propulsion plant when operating with the same speed time profile.

The FY2006 study used a markedly higher ship service electric load for the MSC to meet mission requirements. This drove increased fossil fuel consumption rates for this ship. Consequently, the markedly increased fossil fuel usage spread the nuclear acquisition premium over a larger quantity of fossil fuel decreasing the per-barrel break-even cost. The differences in operational tempos between the ship concepts drove differences in fossil fuel consumption rates and therefore break-even costs.

Studies performed during FY2006 reflect the current state of the art in modeling capability. The study process consisted of project elements executed in sequence to provide a response to public law directed products:

- Cost versus operational effectiveness
- Break even costs of nuclear and fossil fuel plants
- Survey of current and future technology

Figure 1, below, illustrates the relationship between the various project elements and the overall process flow. The process requires iteration of the ship and energy plant designs until performance requirements are met and mission effectiveness is sufficient. Physical and cost descriptions of mature technologies are included in technology models exercised in block 6, "Propulsion Plant Architecting and Systems Engineering."

The overall study process consisted of the following pieces (numbers correspond to blocks in Figure 1):

1. Initial Capabilities: Identify ship types to be studied and baseline their warfare system performance requirements. Characterize baseline ships as well as variants of those ships with alternative propulsion systems.

2. Technology Survey (Current): Survey industry and consult with the Office of Naval Research (ONR) to identify and describe *current* technologies relating to propulsion and power systems and architectures.

3. Technology Survey (Future): Survey industry and consult with ONR to identify and describe *future* technologies relating to propulsion and power systems and architectures.

4. Technology Findings: The findings of the current and future system technology surveys were summarized.

5. Energy Requirements: Determine life-cycle energy requirements for each ship (energy demand signal) to perform missions within the context of the DoD Defense Planning Scenarios. Develop the energy demand signal from the speed range in given warfare system energy usage states for each ship over its expected lifetime. Exercise each variant in energy usage states to determine propulsion and electrical power demands. Develop an analytical model to determine lifetime energy demand based on the energy requirements and expected plant line-ups.

6. Propulsion Plant Architecting and Systems Engineering: Develop the basic architecture for each variant. Architectures will include traditional mechanical, and

electric propulsion architectures as well as "hybrid" architectures that blend integrated propulsion (electric and steam) with mechanical and electric transmissions to satisfy mobility, vulnerability and warfare system service demands. Characterize nuclear and fossil fuel power plants that meet peak, endurance, and, in the case of nuclear powered ships, the lifetime energy needs. Select the type, number, and general location of prime movers and propulsion equipment in the ship. Determine the sizes, weights, and costs associated with various propulsion plant options appropriately scaled for the surface combatants and amphibious warfare ships under study.

7. Ship Synthesis Model: Develop a total ship concept for each variant that incorporates the alternate propulsion plants and defined mission systems that is suitable for cost estimating and operational effectiveness analysis. The ship synthesis model used for this effort is the Advanced Surface Ship Evaluation Tool (ASSET) V5.3. It is a design tool which balances area, volume, weight, and basic performance characteristics of a notional ship based on parametric analysis. The result of this effort is not a full-up ship design; rather it is an engineering estimation indicative of the potentially feasible design solutions.

8. Ship Performance Assessment: Evaluate the performance of each ship baseline/variant in all energy management system areas: energy storage, energy conversion, energy distribution, energy transmission, and thrust generation. Relate the energy management system and architecture to ship speed, range, and service to warfare mission system performance areas.

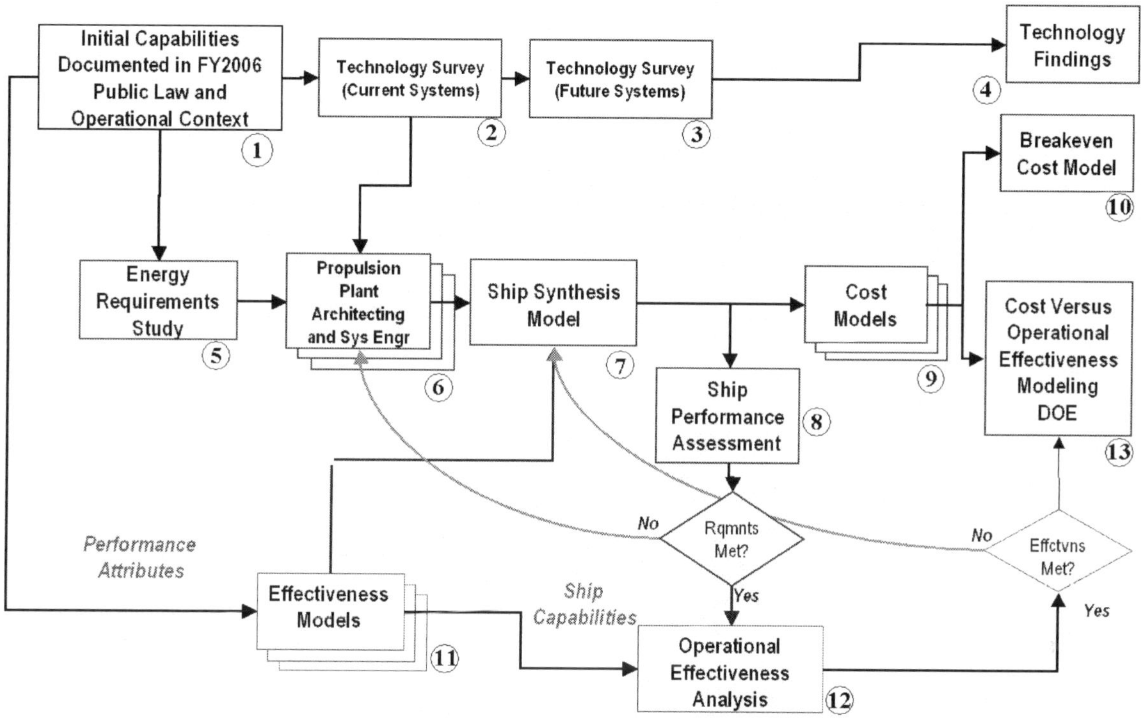

Figure 1: Alternate Propulsion Study Process Overview

9. Cost Models: Estimate fifth ship acquisition cost and Life Cycle Cost (LCC) for each ship baseline and associated variants. Acquisition costs include actual cost return data or vendor quotes for power and propulsion system material, when possible, due to the maturity of power and propulsion system equipment. The LCC estimates incorporate the following costs: inactivation, defueling (nuclear variants only), disposal, burdened fuel costs, manpower costs, and maintenance. Non-recurring costs are not specified as they are dependent on capability growth (both military and industrial) which is outside the scope of this study.

10. Breakeven Cost Model: Perform breakeven cost analyses to compare the nuclear and fossil-fueled ship concepts. Perform a correlation analysis between the breakeven cost of oil with operational tempo, operational profile, and service life.

11. Effectiveness Models: Develop analytical models to evaluate the vulnerability, operational, and mobility effectiveness of the ship variants in mission scenarios.

12. Operational Effectiveness Analysis: Evaluate each ship concept in terms of mobility, survivability, and operational effectiveness in the context of operational scenarios and include attributes such as timeliness, percent mission complete, and endurance.

13. Cost and Operational Effectiveness Analysis: Quantify the relationships between mission effectiveness and cost using a design of experiments (DOE) approach. Develop a DOE for performance versus cost and for performance versus operational effectiveness for each ship type.

ENERGY REQUIREMENTS

Ships evaluated in the RTC are non-program-of-record ship concept designs serving as energy management system surrogates bound by the capabilities of ships programmed in the Navy's 2006 30-Year Shipbuilding Plan. Warfare mission capabilities, and therefore power loads, were kept constant across all variants in each class but are reflective of the warfare capabilities of the ships envisioned in the 30-Year Shipbuilding Plan. Thus, the ships of this study are considered 'energy management system surrogates' of the ships in the 30-Year Shipbuilding Plan.

Official Defense Planning Scenarios (DPS) are used to define the quantities and capabilities of the 313 ship future fleet. The DPS is a set of scenarios commonly used throughout the DoD for analysis and planning purposes. They form a consistent analytical baseline for the evaluation of potential warfare scenarios in an uncertain future context. These scenarios were also used to develop the warfare and mobility system energy requirements and operating tempos for the ships modeled.

The energy demand signal was derived from the speed range in given warfare system energy usage states for each ship over its expected lifetime. Each variant was exercised in energy usage states to determine propulsion and electrical power demands. An analytical model determined the lifetime energy demand based on the energy requirements and expected plant line-ups.

The RTC study assumes a one-to-one relationship between nuclear and fossil fuel ships to sustain a notional force structure. The RTC study did not make any assumptions regarding the affordability of alternative propulsion ships as it relates to either a notional $13.4B per year (FY05$) shipbuilding plan, or total costs to achieve the Navy's 313 ship fleet.

The study recognized substantive historic and predicted growth in electrical power requirements. Figure 2, above, illustrates the growth in maximum margined electric loads of recent ships.

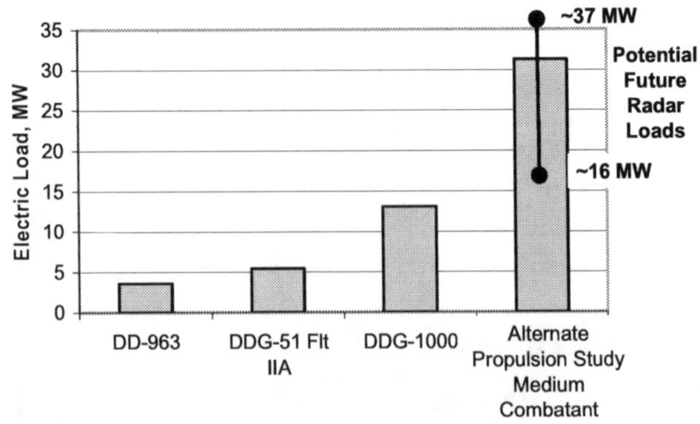

Figure 2: Historic and Predicted Growth in Maximum Margined Ship Service Loads

The ~31 MW maximum margined load shown above in Figure 2 was used in this study's MSC. This significant increase in ship service loads is primarily attributable to Theater Ballistic Missile Defense (TBMD) radar system requirements and associated margins. Electric load growth similar to that of the DDG-51 to DDG-1000 was reflected for the SSC and Amphibious Warfare Ships due to the lower radar system capability requirements.

Energy requirements were developed for each ship type based on DRM's comprised of Tactical and Operational Situations that suggest an employment of energy management systems in ship energy states, thus driving mobility, survivability, and mission system energy demands. Figure 3, below, depicts the process used to determine the energy and power requirements for the various ship concepts and therefore the "engine" and "gas tank" capacities needed.

Figure 3 shows how the results of the Energy Requirements Study defined the power conversion peak static and transient capacity and the lifetime energy requirements for each ship. Peak values and the frequency of peak values supported the architecting of the ship concept energy management system. The range of Tactical Situations suggested ship "states" for the employment of warfare, mobility, and survivability systems. Engagement level activities were decomposed in terms of times in certain speed and ship service power consumption states. These were then aggregated as shown by Figure 3.

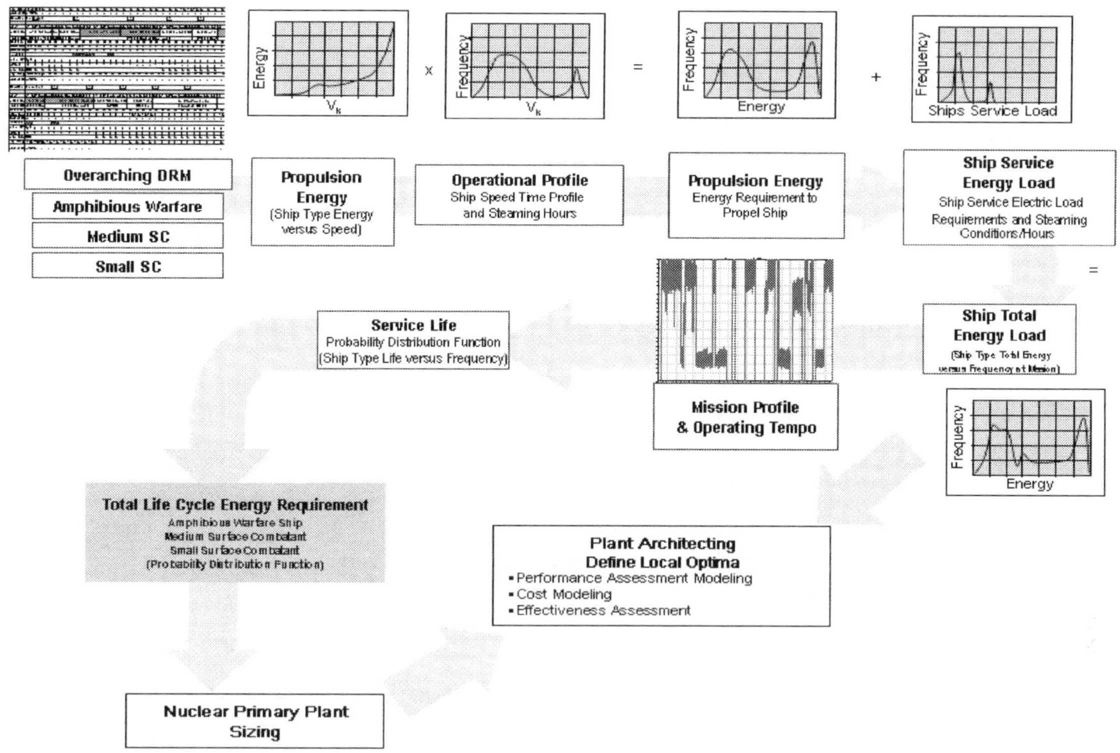

Figure 3: Energy Requirements Study Process

The energy management systems were then architected and engineered to meet those specific state conditions and the transitions between states. Based upon the probability distribution functions developed by this process, summary speed-state-time profiles were developed, an example of which is provided as Figure 4. The percent time in this table represents percent of the total year.

Speed	% Time at Cold Iron	% Time at Anchor	% Time at Condition 3 Alone	% Time at Condition 3 Networked	% Time at Condition 2 Alone	% Time at Condition 2 Networked	% Time at Super Battle
Cold Iron	52.98%						
Anchor		1.58%					
6			0.33%	15.08%	1.23%	8.17%	0.08%
10			0.00%	1.16%	1.89%	3.49%	0.16%
15			7.07%	0.00%	1.08%	0.99%	0.08%
20			0.00%	0.00%	1.46%	1.37%	0.08%
MCR			0.00%	0.00%	1.11%	0.44%	0.16%

Figure 4: Notional Speed-State-Time Profile Example

Warfare mission system loads were modeled as either being active (full power) or deactivated, (minimum connected load) in plant architecting and systems engineering, as dictated by the specific scenarios.

The DoD 2012 Baseline Security Posture (BSP) and the 2010-2014 Major Combat Operations (MCO) scenarios were the basis for the master DRM. Linkage to the BSP and MCOs permits analytical results of defense planning guidance to be compared and contrasted with the body of analysis done to support the FY2008 Program Objectives Memorandum process by the Navy Staff, Joint Staff, and Office of the Secretary of Defense (OSD). Major drivers seen in these operational scenarios include:

1. Stressing Operational and Tactical Situations from MCOs dictate the energy intensive future combat systems and the capability for high speed transits in future fleet operations.
2. The BSP and the MCO scenarios were used to model future fleet deployment and employment profiles for:
   a. Presence at Home
   b. Presence Overseas
   c. Lesser Contingencies
   d. Major Combat Operations

   These features together with overhaul and maintenance defined the "Demand Signal" for the fleet and constitute the basis for the "Ship Lifetime Power Estimate."
3. The Navy's 30-Year Shipbuilding Plan (USN 2006) is informed by analysis performed by the Navy Staff using the DoD Analytic Agenda in coordination with the Joint and OSD Staff. This plan represents the Future Naval Force of the 313 ship Navy. The Alternate Propulsion Study fleet represents a different fleet architecture with different physical characteristics but roughly equivalent capability of the Navy's 30-Year Shipbuilding Plan. Each ship is an energy demand surrogate for ships of similar type within the 313 ship Navy.

The results of the energy requirements study provided speed-time, steaming hour profiles, and ship service electrical demand signals for the three baseline ships. This information is required to set high level propulsion plant functional requirements, determine operations and support costs, as well as to determine the needed sustainment levels for fossil fuel ships and reactor lifetimes of nuclear plants. As a result of this analysis, it was determined that the reactor plants used in this study (existing non-developmental submarine and aircraft carrier designs) do provide life of ship cores; therefore, no reactor plant refuelings would be necessary.

Three operating profiles were developed from DPS for the three ship types of the study. The low operating profile is the baseline scenario that models peacetime operations that are dominated by ordinary presence, training exercises and lesser contingency operations. The medium speed-time profile considers the addition of a single Major Combat Operation in a seven-year period to the baseline low operating profile. The high profile adds two Major Combatant Operations in a seven-year period to the baseline low operating profile. An example of a notional high operating profile over a time period is shown as Figure 5.

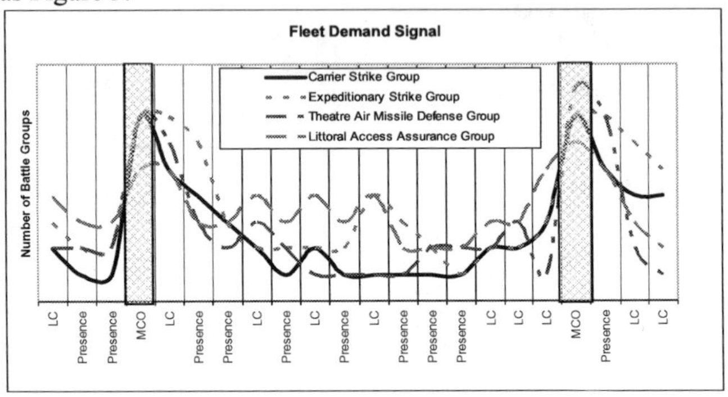

Figure 5: Notional High Operating Profile

The preceding analysis was performed without reference to historical and current fleet operating tempos and profiles. This was done intentionally so that the methodology could be validated by comparing its analytic results with data from historical and current fleet operations.

Historic operational demand signals for destroyers, cruisers and amphibious warfare ships over the seven-year period FY2000 through FY2006 were analyzed. Steaming hours underway reflect peacetime operation from FY2000 through FY2001, the FY2002 lesser contingency, and a peak in FY2003 reflecting a Major Combat Operation-like deployment. FY2004 through FY2006 reflects redistributed fleet ordinary presence in support of protracted MCO activity.

Figure 6, Figure 7, and Figure 8 present the average ship steaming hours underway by year along with the number of ships operating each year. Noteworthy is a continued increase in steaming hours underway for ships after the recovery from the MCO-like event in FY2003. Both destroyers and cruisers have a steady decrease in steaming hours not underway over the period from FY2000 to FY2006 suggesting an increased operating tempo for Small and MSC mission sets over time.

Approximately 8% to 12% of surface combatant and amphibious warfare ships' annual fuel consumption is historically consumed during the 20% to 30% of the steaming hours per year that a ship is not underway.

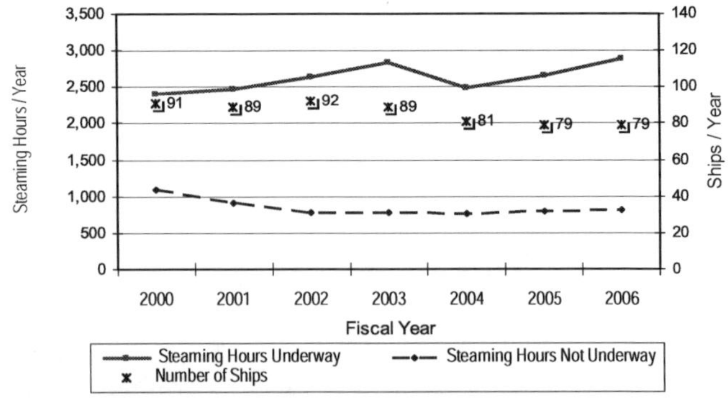

Figure 6: Destroyer and Frigate Average Historical Operational Tempo (2000-2006)

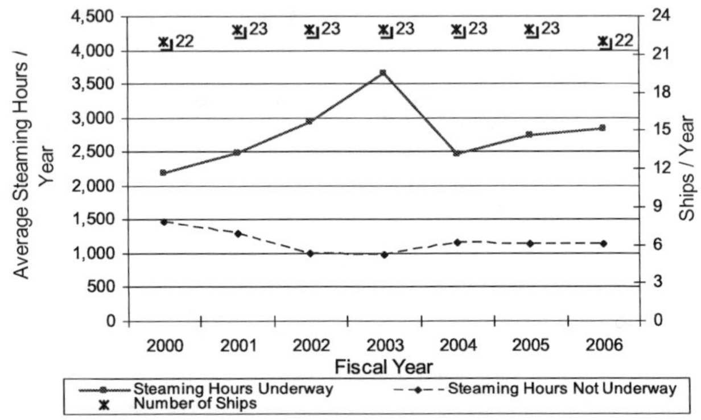

Figure 7: Cruiser Average Historical Operational Tempo (2000-2006)

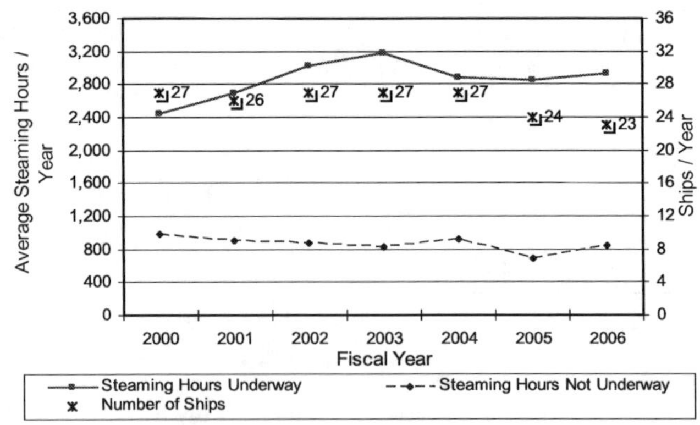

Figure 8: Amphibious Ships Average Historical Operational Tempo (2000-2006)

Figure 9, Figure 10, and Figure 11 compare the low, medium and high operating profile steaming hours underway developed by the Energy Requirements Study with historic data. They provide a basis of comparison between the DPS derived and historic operating tempos in which 70% and 80% of the annual steaming hours in which a ship is underway and consuming 80% to 90% of its lifetime energy.

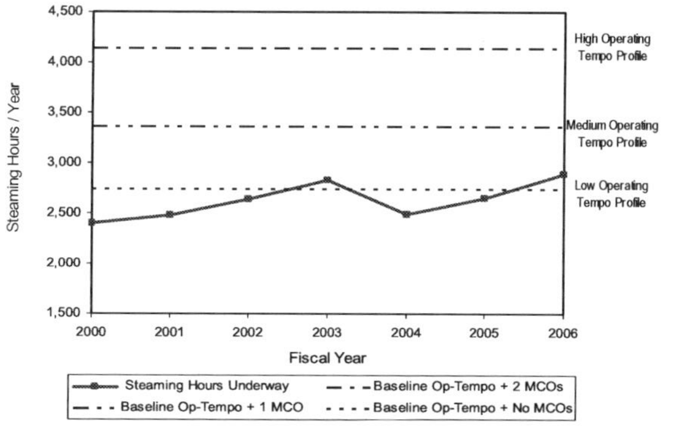

Figure 9: Comparison of DD/DDG/FFG Historic and Modeled Steaming Hours Underway

Figure 9 compares the three DPS based operating tempos of the SSC with historic data for the FFG-7, DD-963, and DDG-51 Classes. The SSC steaming hours underway in peacetime conditions (low operating profile) are higher than those of the comparison ships operating from FY2000 to FY 2002. Likewise the steaming hours underway for the single MCO operating profile (medium) are greater than comparison ships historic performance in FY2003. This reflects the lower capability of the SSC's light-destroyer mission suite (e.g., fewer Vertical Launch System cells, less capable radar suite) relative to the comparison ships requiring greater numbers and steaming hours to conduct similar missions. Since the SSC is less capable than DD/DDGs (which compose the bulk of this data), it takes more SSCs, and consequently more steaming hours, to deliver the same level of capability. The future operating tempo of the SSC also reflects the addition of an Access Assurance mission currently assigned to the ships of the Littoral Combat Ship program. The access assurance workload further increases the SSC steaming hours underway per year. This is an area where power and propulsion system architectures that are sensitive to sustainability will demonstrate discrimination through greater operational capability in theater.

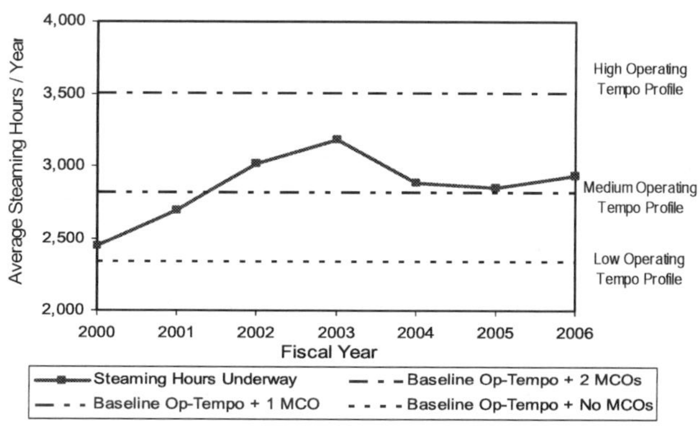

Figure 10: Comparison of CG-47 Historic and Modeled Steaming Hours Underway

Figure 10 compares the three DPS based operating tempos of the MSC with historic data for the CG-47 Class. In this case, the MSC performs the roles of existing CGs, albeit with the addition of greater capabilities in the area of Theater Ballistic Missile Defense. The historic CG 47 data compares favorably with the DPS derived optempos when the number of MCOs is correlated with historic fleet deployments. The MSC area defense and TBMD capabilities provide mission area overlaps in the air defense mission capability currently performed by cruisers. The MSC can be employed in roles and numbers similar to CGs in the 2020 timeframe assessed in this study.

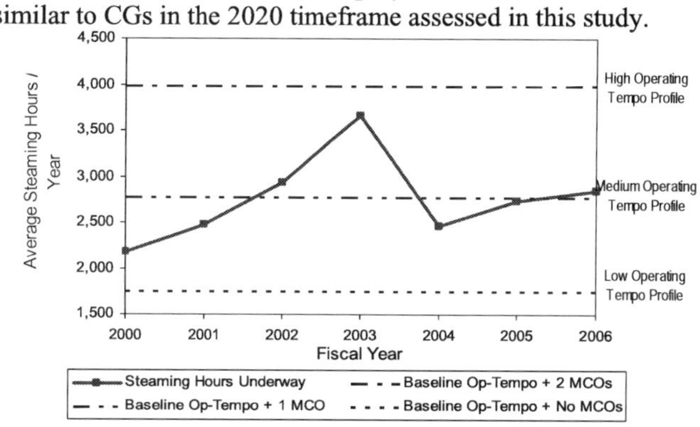

Figure 11: Comparison of Amphibious Warfare Ship Historic and Modeled Steaming Hours Underway

Figure 11 compares the three DPS based operating tempos of the AWS with historic data for the peer amphibious warfare ships of the current fleet. Again, historic ship data validates the range selected for the AWSs.

The DPS derived operational tempos profiles provide a valid model for plant architecting, lifetime energy estimation, and determination of LCC based on the strong correlation with historic return data from surface combatant and AWSs of the current fleet.

Finally, electric loads for IPS and mechanical transmission variants of surface combatants and AWSs were modeled in this study for mission specific load conditions. The electric loads for the SSC and AWS are marginally higher than the electric loads of current navy similar ship types. In contrast, the MSC's 7 to 27 MW non-margined ship service loads associated with Summer Cruise and Battle conditions, respectively, represents a significant increase relative to current navy electric loads. The higher loads are reflective of next navy warfare mission system energy demands Plant architectures were developed using the electric loads with a 20% design margin and a 20% service life allowance combined with mission consistent ship propulsion loads.

TECHNOLOGY SURVEY

Alternative propulsion systems were evaluated in a construct that decomposed the propulsion system into the systems, subsystems, and components which support ship's energy management. The holistic evaluation of a propulsion system as an energy management system supports tracing sub-system and component attributes to the total ship system life cycle cost and the impacts on operational effectiveness. Energy management systems, sub-systems and components were mapped into energy storage, energy conversion, energy distribution, and thrust generation functions. Functional attributes provide a level of discrimination that supports ship energy management system architecting.

(Webster 2007) presents the results of the USN evaluation of alternate propulsion systems that support current navy, next navy (new programs of record), and the navy after next (new technology development programs). Table 2, below, summarizes the technologies evaluated in the FY2005 and modeled in FY2006 studies. Although future prime mover technologies were investigated, the constraints of the study required the use of prime movers capable of integration into ships in the next ten years. As such, the gas turbines and diesel engines used in the ship concepts developed represent current technology and fuel consumption rates.

All currently in-service surface combatants and many amphibious warfare ships use conventional mechanical drive where prime movers such as gas turbines or diesel engines drive propellers through reduction gears. Electrical loads are provided power with a segregated electrical distribution system. In the past, nuclear powered surface ships used steam turbines with mechanical power systems.

The Navy's newest destroyer design, the DDG-1000, uses an Integrated Power System (IPS) where all prime movers produce electrical power that is shared between propulsion and ship service loads. Discriminating attributes of IPS are flexibility of arrangements, mechanical decoupling between prime movers and propulsors, and an increased level of energy conversion and transmission redundancy. Perhaps the single greatest benefit of IPS systems is the improvement in vulnerability resulting from mechanically decoupling prime movers from propulsors enabling segregation and separation of ship propulsor components.

Nuclear ship steam turbine plants are also an integrated system where propulsion power and electrical generators are driven by steam turbines with steam produced from the energy in a nuclear reactor. Similar to plants that share electric energy between mobility and warfare mission areas, the thermal energy in steam is also shared between ship service turbogenerators and propulsion steam turbines.

Finally, hybrid systems combine the attributes of the other three ship types. For example, the hybrid mechanical drive / IPS propulsion system for LHD 8 combines a gas turbine mechanical drive system with IPS diesel generator powered auxiliary propulsion motors to gain fuel efficiencies at low speeds without sacrificing transmission efficiencies at high speeds.

Table 3, on the next page, summarizes current power and propulsion architectures and technologies that have been incorporated in recently designed surface combatants and amphibious warfare ships.

The Navy and industry are investing in propulsion technologies that offer the potential to improve future naval ship designs. The goals of the current and planned technology development efforts are to improve affordability, power density, efficiency, and satisfy the energy demands of future mission systems. Table 4, below, lists power and propulsion plant architectures that should or are being considered for ships in design or construction (Next Navy) and for future ships (Navy after Next).

Navy after Next power system components and architectures were integrated into ships evaluated in the FY 2006 studies except for fuel cells. Uncertainty in fuel quality and improvements in power density preclude models of Navy after Next fuel cells into the synthesis process. Podded propulsors meeting mission speed requirements cannot withstand the weapons effects of current undersea threats. Podded propulsors have been integrated into ship concepts meeting grade B shock levels. If damaged when deployed, these pods can be retracted. The pod can be deployed after a shock event as it is isolated when in the stowed position. Ongoing technology transition programs are aimed at improving the shock resiliency of podded propulsors.

Table 2: Technologies Evaluated in FY2005 and FY2006 Studies

Ship	Energy Source	Energy Conversion	Ship Service	Transmission Types	Propulsor Types
Small Surface Combatant	Fossil Fuel	Diesel	450 Volt AC	Electric 4160 Volt AC	Fixed Pitch Propulsor
		Gas Turbine	4160 Volt AC	Electric 13,800 Volt AC	Controllable Pitch Propulsor
				Mechanical	Advanced Forward Propulsion Unit
	Nuclear	Pressurized Water Reactor			
Medium Surface Combatant	Fossil Fuel	Diesel	1000 Volt DC	Electric 4160 Volt AC	Fixed Pitch Propulsor
		Gas Turbine	450 Volt AC	Electric 13,800 Volt AC	Controllable Pitch Propulsor
			4160 Volt AC	Mechanical	Advanced Forward Propulsion Unit
	Nuclear	Pressurized Water Reactor			
Amphibious Warfare Ship	Fossil Fuel	Diesel	450 Volt AC	Electric 4160 Volt AC	Fixed Pitch Propulsor
		Gas Turbine	4160 Volt AC	Electric 13,800 Volt AC	Controllable Pitch Propulsor
				Mechanical	Advanced Forward Propulsion Unit
	Nuclear	Pressurized Water Reactor			

Table 3: Current Ship Propulsion Technology

Ship	Energy Management Systems				
	Storage	Conversion	Distribution	Transmission	Thrust Generation
DDG-51	Fossil Fuel (DFM/F76)	Gas Turbine	Segregated Electric	Mechanical COGAG	Controllable Pitch Propeller
LHD-8 / LHA-6		Diesel Cruise / Gas Turbine Boost		CODLAG	
LPD-17		Diesel		Mechanical CODAD	Controllable Pitch Propeller
LCS (LM)		Diesel Cruise / Gas Turbine Boost		Mechanical CODAG	Waterjet
LCS (GD)					
DDG-1000		Gas Turbine	Integrated Electric	Electric Motor	Fixed Pitch Propeller

CODAD: Combined Diesel and Diesel Power Plant
COGAG: Combined Gas Turbine and Gas Turbine Power Plant
CODLAG: Combined Diesel Electic and Gas Turbine Power Plant
CODAG: Combined Diesel and Gas Turbine Power Plant

Table 4: Power System Components & Architecture Considerations for the Next Navy and Navy after Next

	Next Navy (Ships in Design and Construction)	Navy after Next (Future Ship Designs)
Small Surface Combatant	LCS – Combined Gas Turbine and Diesel Mechanical Propulsion with Diesel Generator Sets	Combined Gas Turbine and Diesel Plants
		Longitudinally separated forward podded propulsion unit.
		IPS for ships with high mission system electrical loads.
		Fuel Cells
		Nuclear Power
Medium Surface Combatant	DDG-1000 – Gas Turbine-Electric Integrated Power System	IPS for ships with high mission system electrical loads and ships requiring high levels of survivability.
		Longitudinally separated forward podded propulsion unit.
		Fuel Cells
		Combined Diesel and Gas Turbine plants.
		Nuclear Power
Amphibious Warfare	LHD-8, LHA-6 – Hybrid Gas Turbine Mechanical and Diesel Electric	IPS or Hybrid Plants
		Longitudinally separated forward podded propulsion unit.
		Fuel Cells
		Nuclear Power

SHIP ENERGY MANAGEMENT SYSTEM ARCHITECTING

Ship energy management system architectures were selected to provide discrimination between energy storage, conversion, distribution, and thrust generation among peer technology alternatives in operational effectiveness areas. Energy management system attributes that can be modeled are power quality, power quality of service, system sustainability, responsiveness to transient loads, and vulnerability.

Three discreet propulsion energy management system architectures were modeled:

1. Current Navy Federated Architecture
2. Next Navy and Current Programs of Record Distributed Architecture
3. Hybrid Distributed Architecture

Current US Navy destroyer and cruiser propulsion plants are a federation of power system and propulsion plant components that are centrally controlled to meet deterministic plant architecture guidelines.

Figure 12, below, shows the two mechanically driven shaft lines powered from main machinery rooms separated by subdivisions of length defined by US Navy criteria. Ship service electric power is provided by segregated fossil fuel powered gas turbine or diesel generators. This architecture is supported by fossil fuel or nuclear powered prime movers driving mechanical or electric transmissions. Vulnerability capability is achieved by this architecture through separation and redundancy of power and propulsion plants in the two machinery spaces.

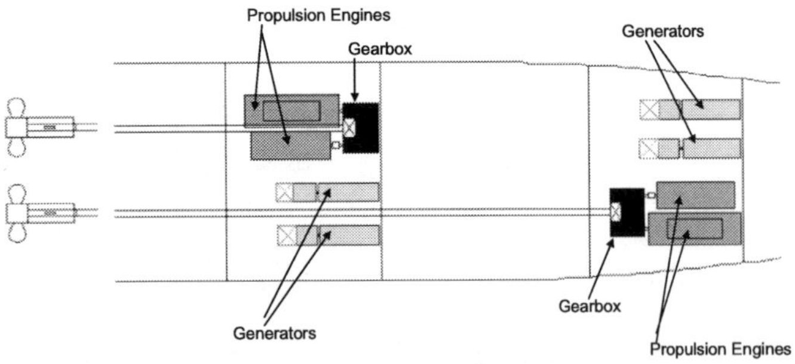

Figure 12: Federated Energy Management System Architecture (SFM-1)

The distributed architecture evaluated in FY2006 studies, as shown in Figure 13 below, provides electric energy, with prime movers distributed by zone, to meet propulsion system and ships service electric loads. The distributed architecture mechanically decouples energy conversion connectivity to the propulsor(s). The electric transmission provides for greater dispersion of power and propulsion system components than the federated architecture enabling the architecting to susceptibility and vulnerability metrics that support mission effectiveness objectives.

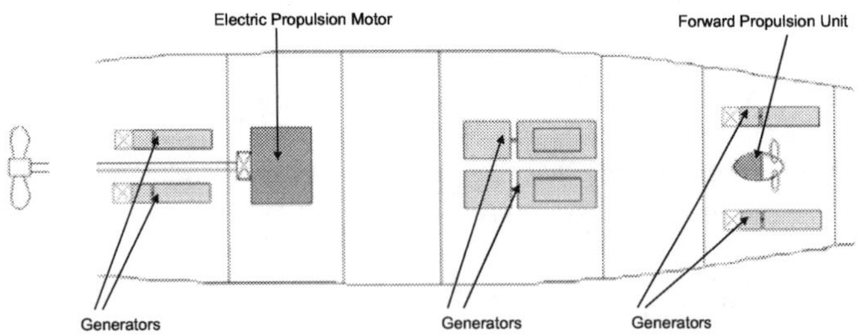

Figure 13: Distributed Energy Management System Architecture (SFI-1)

The hybrid architecture combines a zonal electric power and propulsion distribution system with a mechanical primary propulsion transmission line, as shown in Figure 14, below. This architecture captures the high efficiency of a mechanical transmission line and is able to provide propulsion redundancy with a segregated electric drive forward propulsor sized to provide warfare mission speed. Electric power generation is dispersed in zones similar to the distributed architecture. Primary propulsion energy is located in a main machinery compartment. Electric and mechanical (steam turbine, gas turbine, and/or diesel) prime mover power is combined mechanically in a reduction gear and electrically in the hybrid electric power distribution bus.

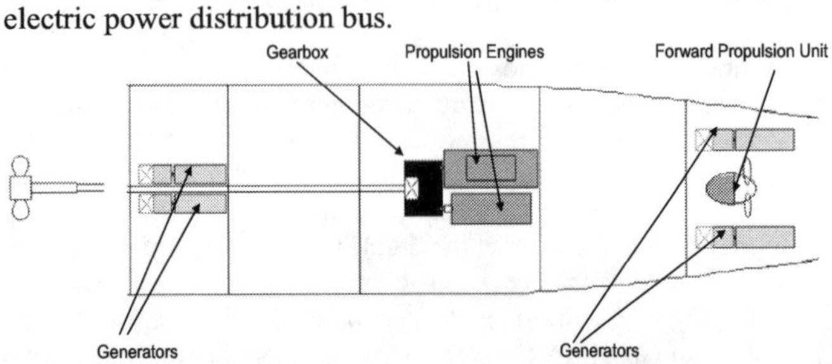

Figure 14: Hybrid Energy Management System Architecture (SFH-1)

Architectural consistency with energy management systems was imposed on each SSC, MSC, and AWS energy management system variant. The Federated architecture philosophy, for example, was applied to all mission system areas, warfare and non-warfare. This permitted a consistent treatment of vulnerability throughout the energy management system; the power generation, conversion, distribution, thrust generation loads, and non-mobility mission electric loads.

SHIP ENERGY MANAGEMENT SYSTEM SYSTEMS ENGINEERING

The design methodology used in this study differs from a traditional requirements-based design approach by embedding operational effectiveness assessments into each iteration of design synthesis. The design process architected sub-systems and total ship system capabilities into warfighting, mobility, and vulnerability effectiveness vice static performance metrics (e.g.: sustainability and surge capability vice endurance range and speed) within the ship synthesis process. The process sequentially evaluated multiple effectiveness objectives thus permitting multi-objective discrimination between propulsion system alternatives.

Energy management system Systems Engineering balances the ship and propulsion system energy demands with main and auxiliary machinery plant architectures. Energy demands are met by zonal load centers in each of the architectures. Propulsion and ship service loads are served by prime movers and generators satisfying power quality of service criteria. Load centers serve loads from dedicated ship system, chilled water – air conditioning, and firemain. Figure 15 shows an inboard profile for the distributed architecture propulsion plant for the Small Surface Combatant (SSC) described in the ship energy management system architecting section of this paper. Warfare and mobility mission energy demands, both direct and derived from dedicated distributive systems such as cooling water, air conditioning, firemain, etc., were evaluated in a common level of zonal redundancy and load distribution. The following example shows the modeling approach for the distributed, all electric drive architecture, for SSC fossil fuel, gas turbine primary and secondary generator variant.

The ship shown below, in Figure 15, employs a distributed architecture and uses an electric transmission located in Main Machinery Rooms 1 and 2 (MMR1&2). Electric loads associated with mission systems were provided with redundant service from load centers located in segregated electric zones. Figure 16 shows the alignment of energy supply with dedicated propulsion and ship service loads. Energy conversion is provided by gas turbine generators which supply power converters (PC) and load centers (LC) that are distributed within zones supporting a common level of vulnerability capability.

The chilled water and firemain systems are shown below in Figure 17 and Figure 18 respectively. Service node locations are aligned with ship system demand for primary ship mobility, warfare system and vulnerability performance. Energy demands are balanced within zones using rules-based optimizers that balance loads within load centers while providing redundant load center sources to each system

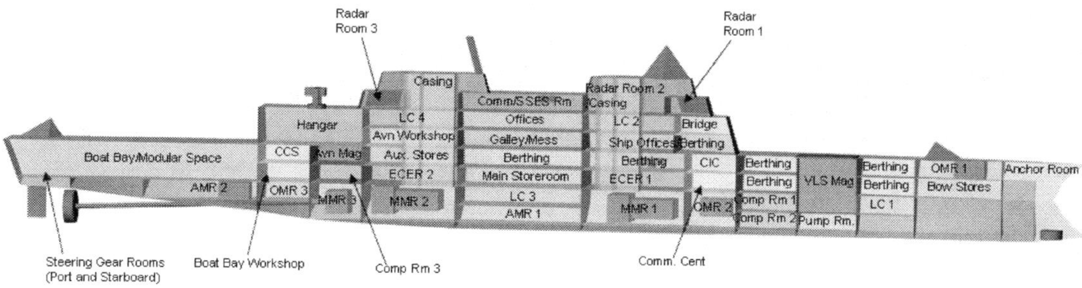

Figure 15: Inboard Profile for a Distributed Architecture Propulsion Plant

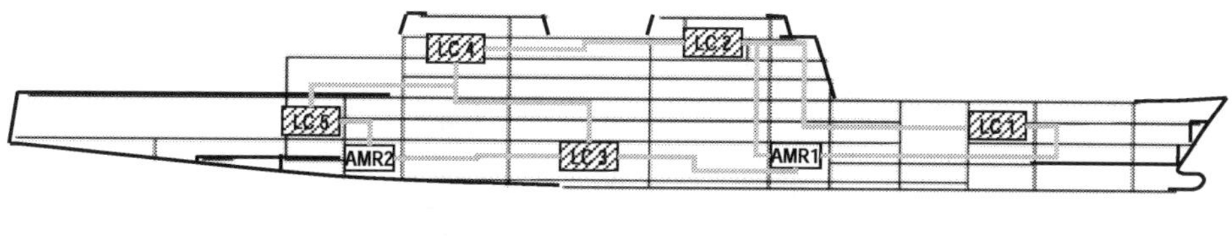

Figure 16: Distributed Architecture Energy Supply

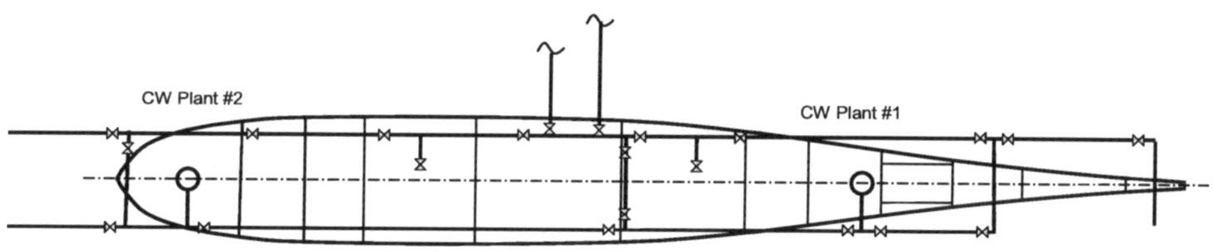

Figure 17: Distributed Architecture Chilled Water Supply

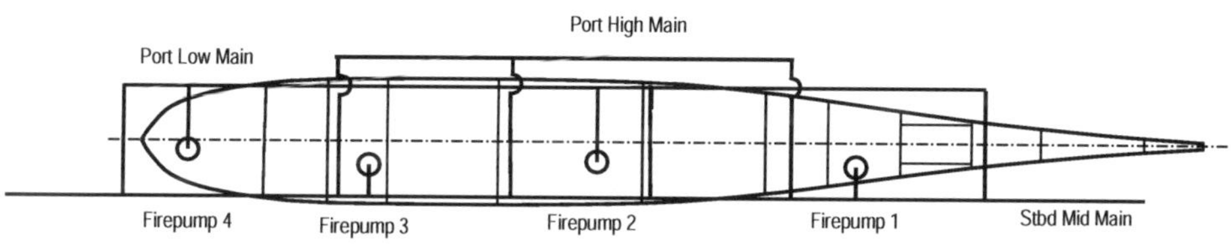

Figure 18: Distributed Architecture Firemain

demand. The distributed voltage is determined by comparison with existing current limits for power conversion and distribution system protection devices.

Integration complexity is introduced when the power generation bus is integrated with both ship propulsion and ship service busses. The same load-by-zone balance process is performed treating propulsion as another mission system load. The ship service and propulsion busses distribute current at voltages determined by rules based protection device limits of existing hardware. For instance, electric power was generated at 13.8k volts AC vice 4160 volts AC when propulsion load fault isolation had to be retained within current protective device limits.

Ship service distribution system current type, AC or DC, and voltage were evaluated for each of the plant architectures. A sensitivity study was performed to determine if cost or performance discrimination existed between the distribution current alternatives. The sensitivity study compared a 1000 volt DC versus 450 volt AC ship service distribution systems for the Distributed – Electric Drive Architecture. Both ships use a combination of gas turbine and diesel prime movers to generate power at 13.8 k-volts AC, driven by the propulsion bus loads and limitations imposed by current protection devices. The 450 volt AC system required dedicated AC to DC power conversion to satisfy warfare mission power quality needs. Equivalent performance could be provided by either system. The difference in cost between the DC distribution system and AC distribution system with dedicated power conditioning for the MSC Distributed Architecture Electric drive variant was dominated by uncertainty. Electric drive alternatives used a 450 volt AC ship service zonal distribution system with dedicated power conditioning for the warfare mission loads.

Nuclear-steam mechanical transmission variants used a 13,800 volt AC distribution system whereas the electric drive nuclear-steam variants were modeled with the 450 volt AC distribution system. Use of a common distribution bus voltage for nuclear and fossil fuel variants permitted a direct comparison between fossil fuel and nuclear fuel variants cost and operational effectiveness.

The hybrid architecture used a 450-volt power generation and ship/propulsion bus distribution system. The low power requirements of the electric-drive forward propulsion units, 3 to 5 megawatts, permitted low-voltage power distribution within current power protection technology alternatives. Higher propulsion system and warfare mission system loads increase current demands above electric power system protection device limits on the MSC that shares a common distributed, electric drive architecture with the ship shown in Figure 16.

The output from power conversion and power distribution architecting and systems engineering analyses was input into ship system synthesis where total ship system capabilities were balanced.

Ship Design Exploration Matrix							
Small Combatants							
Ship Designator	Sustained Speed (kts)	Transmission Type	Primary Propulsion Power Source	Secondary Propulsion Power Source	Separate Ship Service Power Source	Main Propulsor	Secondary Propulsor
SFM-1	30	Mechanical	Gas Turbine	Diesel Engine	Diesel Generator	2 Shafted Props	None
SFM-2				Gas Turbine	Gas Turbine Gen.		
SFH-3		Hybrid		Diesel Engine	Diesel Generator	1 Shafted Prop	FPU
SFM-5		Mechanical		Gas Turbine		2 Shafted Props	None
SFM-6	25		Diesel Engine	Diesel Engine			
SFM-7	35		Gas Turbine				
SNH-1	30	Hybrid	Nuclear Steam Turbine	None	Nuc Steam Turb Gen.	1 Shafted Prop	FPU
SNH-2	25						
SFI-1	30	Electric	Gas Turbine Gen.	Diesel Generator	N/A	2 Shafted Props	None
SFI-2				Gas Turbine Gen.			
SFI-3				Diesel Generator		1 Shafted Prop	FPU
SNI-1			Nuc Steam Turb Gen.	None	Nuc Steam Turb Gen.		
Medium Combatants							
MFM-1	30	Mechanical	Gas Turbine	Diesel Engine	Diesel Generator	2 Shafted Props	None
MFM-2				Gas Turbine	Gas Turbine Gen.		
MNH-1		Hybrid	Nuc Steam Turb	None	Nuc Steam Turb Gen.		FPU
MFI-2		Electric	Gas Turbine Gen.	Diesel Generator	N/A		None
MFI-3				Gas Turbine Gen.			
MFI-4				Diesel Generator			
MNI-1			Nuc Steam Turb Gen.	None	Nuc Steam Turb Gen.		FPU
Amphibious Ships							
AFM-1	25	Mechanical	Gas Turbine	Diesel Engine	Diesel Generator	2 Shafted Props	None
ANH-1		Hybrid	Nuc Steam Turb	None	Nuc Steam Turb Gen.		FPU
AFI-1		Electric	Gas Turbine Gen.	Diesel Generator	N/A		None
ANI-1			Nuc Steam Turb Gen.	None	Nuc Steam Turb Gen.		FPU

Table 5: Ship Design Exploration Matrix

The vulnerability analysis relates expected damage to the mobility and warfare capabilities served by zonal power and distributive system architectures and arrangements using deactivation logic. Sub-system component and architecture deficiencies that reduce mobility or warfighting capability were remedied through addition or relocation of existing equipment. Electric power generation and distribution system architectural modifications were cycled through the zonal electric design ASSET wizard and the ship was re-synthesized. The synthesized ship provided a converged solution that provides a cost comparison across common effectiveness objectives.

SHIP CONCEPTS

Each concept has a designator with three letters plus one number, such as SFM-1. The first letter represents the type of ship (S=Small, M=Medium, A=Amphibious). The second letter represents the type of fuel (F=Fossil, N=Nuclear). The third letter represents the type of energy transmission system (M=Mechanical, I=Integrated Power System, H= Hybrid Mechanical Electric Transmission System). The number provides discrimination for other changes.

The three baseline ships along with twenty variants with alternative propulsion systems are as shown in Table 5. The underlined cells indicate a change from the baseline configurations (SFM-1, MFM-1, and AFM-1):

The use of the term "small" and "medium" in the context of surface combatants in this paper is not intended to directly relate to the physical size or the capabilities of the concepts studied. The use of this nomenclature is a result of preserving consistency with previous studies and Congressional language. A basic description of the three baseline designs follows.

Small Surface Combatant

The SSC is the energy management system "surrogate" for the ships included in the CNO's 30 year ship building plan that perform destroyer and frigate functions, including the protection of shipping missions such as convoy escort and maritime interdiction. The ships were modeled to include LCS mission modularity supporting active anti-submarine warfare, small boat
defense, and anti-mine warfare in the littoral environment. The SSC is provided with a 30-knot sustained speed and a 20-knot endurance range of 5000 nautical miles.

The ship's total installed power was modeled 60-80 Megawatts (MW) including propulsion and ship service power (variation due to differences in possible power system configurations). The 24-hour average electrical power load for the SSC is approximately 2.5 MW with a maximum design load of approximately 5.7 MW. A total of 12 SSCs were designed. Figure 19, below, captures the genealogy of the variants, showing fossil fuel and nuclear power variants, as well

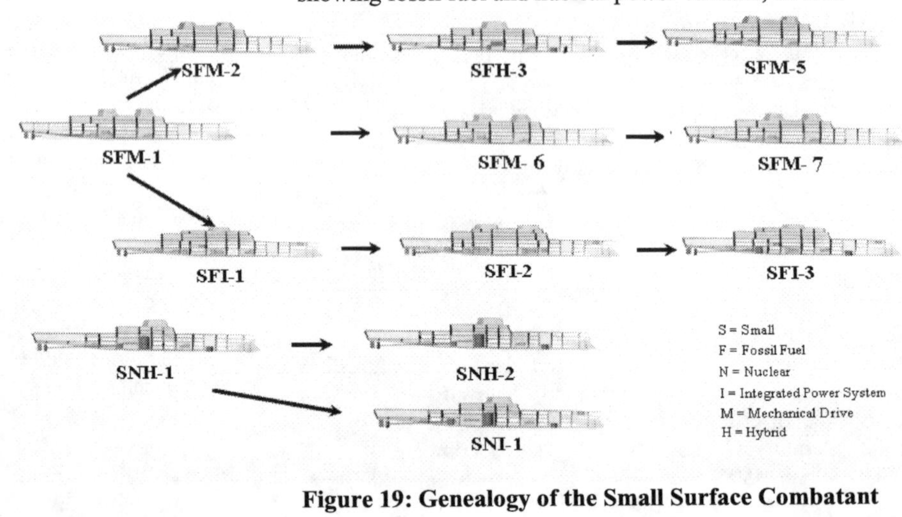

Figure 19: Genealogy of the Small Surface Combatant

as power transmission variants, using either mechanical or IPS transmission. Prime movers were also varied between gas turbines, diesel engines, and steam turbines. A detailed summary of the variants was provided earlier as Table 5. For the nuclear propulsion variants, one modified existing submarine class propulsion plant was used to meet ship power plant needs.

One of the SSC propulsion system architectures that resulted from the multi-objective design process is the hybrid electric drive propulsion system integrated into the small surface combatant. The single mechanically driven shaft with a controllable pitch propeller provides superior transmission efficiency at high speed compared to electric drive or twin shaft alternatives. Decoupling acoustic requirements from the single controllable pitch propulsor provided further increases in propulsive efficiency. The resulting propulsion sub-system efficiency coupled with the lower drag of a single shaft line with single rudder permitted lower installed power to be dedicated to the propulsion system. The loading on the single shaft propulsor meets propulsor design criteria at speeds up to the surge to theater speed when acoustic requirements are decoupled from the sub-system.

Ship effectiveness metrics are supported only if the acoustic performance can be provided by an alternate propulsion system where mission capabilities are translated from the high speed surge propulsor that has been optimized for speed. Single shaft vulnerability drove the location of the alternate propulsion unit forward of the traditional stern mounted propulsor by a distance defined by weapons effects. The capabilities transferred to the forward propulsion system include providing mission speed mobility capability within acoustic thresholds and Grade B shock performance when deployed; it can be retracted if disabled by an underwater explosion event or it can be deployed if the aft propulsor was disabled from weapons effects.

Medium Surface Combatant

The MSC is the energy management system "surrogate" for the ships included in the CNO's 30 year ship building plan that perform strike group air defense missions, carrier escort, theater missile defense, and anti-submarine warfare functions. The MSC is provided with a 30-knot sustained speed and a 20-knot endurance range of 8000 nautical miles.

The 24-hour electrical power load for MSC was estimated as 15 MW with a maximum design load of approximately 31 MW. Again, these 24-hour electric power load estimates assume the development of a "powered-down" mode for the radar, which does not currently exist. If "powered-down" technology does not come to fruition, the estimates of electric power and lifetime energy requirements for the MSC would need to be significantly increased.

A total of 7 MSC concepts were evaluated. Figure 20, below, captures the genealogy of the variants, showing fossil fuel and nuclear power variants, as well as power transmission variants, using mechanical or IPS transmission.

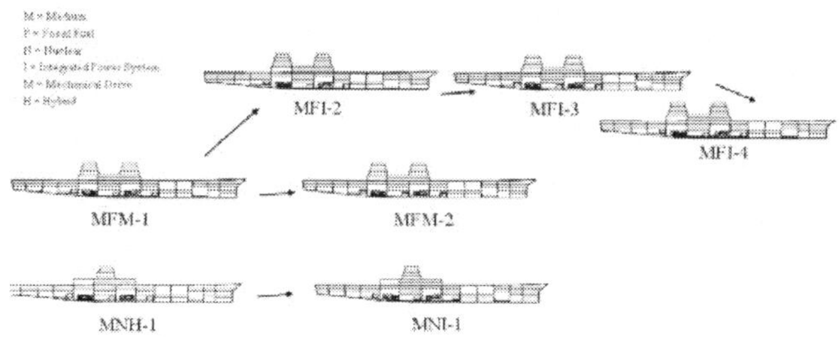

Figure 20: Genealogy of the Medium Surface Combatant

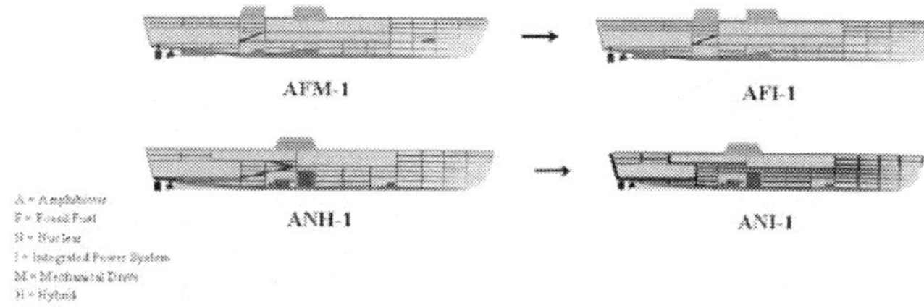

Figure 21: Genealogy of the Amphibious Warfare Ship

Two separate power distribution systems (one AC and one DC) were investigated for the medium combatant. Weight and space demands for these systems were integrated into the MSC designs. However, the uncertainty in modeling of installation cost of high-current distribution cable as well as the material costs of power conversion equipment dominated the expected differences. Accordingly, only AC distribution alternatives were modeled for the Small Surface Combatant and the Amphibious Warfare Ship as the warfare system loads were less stressing than that for the MSC where cost discrimination was not observed.

Prime movers were also varied between gas turbines, diesel engines, and steam turbines. A detailed summary of the variants was provided earlier as Table 5. For the nuclear propulsion variants, one modified next generation aircraft carrier propulsion plant was used to meet ship power plant needs.

Amphibious Warfare Ship

The AWS is a surrogate for existing amphibious assault ships. The AWS was modeled to carry one-third of a Marine Expeditionary Unit (MEU); therefore, three AWSs would form an Expeditionary Strike Group. The flight deck and hangar accommodate 17 CH-46 equivalent aircraft. The ship has the capability of operating CH-46, CH-53, V-22, and Short Takeoff and Vertical Landing Joint Strike Fighter aircraft. The AFM-1 is designed to have a sustained speed of 25 knots and a range of 10,000 nautical miles at 20 knots. The 24-hour electrical power load for AWS is approximately 7.6 MW with a maximum design load of approximately 16 MW.

A total of 4 AWS power and propulsions system surrogate variants were modeled. Figure 21, above, captures the genealogy of the variants, showing fossil fuel and nuclear power variants, as well as power transmission variants, using either mechanical or IPS transmission.

Table 6: Summary of Ship Characteristics

	Small Surface Combatants										
	SFM-1	SFM-2	SFH-3	SFM-6	SFM-7	SNH-1	SNH-2	SFI-1	SFI-2	SFI-3	SNI-1
LBP (m)	165.0	165.0	165.0	165.0	165.0	175.0	175.0	165.0	165.0	165.0	175.0
BWL(m)	18.1	18.2	18.0	18.0	18.2	19.4	19.4	18.3	18.3	18.2	19.5
Draft (m)	5.4	5.6	5.2	5.2	5.7	6.4	6.3	5.7	5.9	5.6	6.7
Full Load Displacement (MT)	7,916	8,395	7,433	7,279	8,530	11,127	10,882	8,613	8,997	8,504	11,926
Propulsion	CODAG	COGAG	CODAG	CODAD	CODAG	Nuc Mech'l	Nuc Mech'l	IPS	IPS	IPS	Nuc Electric
Propulsors	2 CPP	2 CPP	1CPP, 1 APU	2 CPP	2 CPP	1 FPP, 1 APU	1 FPP, 1 APU	2 FPP	2 FPP	1 FPP, 1 APU	1 FPP, 1 APU
Sustained Speed (kts)	30.9	30.3	31	25.2	34.8	~30	~25	30.6	30.5	31.3	~30
Range @ 20 kts (nm)	5,000	5,000	5,000	5,000	5,000	n/a	n/a	5,000	5,000	5,000	n/a
Aircraft	2 x SH-60 (VTUAV capable)										
Mission Bay Space (m2)	600 (capable of accommodating program of record interchangeable mission modules)										
C4ISR & Weapons	SPY-1F, SPQ-9B, 2 - SPG-62 directors, MF and HF SONAR, MFTA, 2 - SVTTs, 32 - Mk57 VLS, 1 - SEARAM, 2 - 57 mm guns, 2 - 25 mm guns										

	Medium Surface Combatants							Amphibious Warfare Ships			
	MFM-1	MFM-2	MNH-1	MFI-2	MFI-3	MFI-4	MNI-1	AFM-1	ANM-1	AFI-1	ANI-1
LBP (m)	220.0	220.0	220.0	220.0	220.0	220.0	220.0	229.0	229.0	229.0	229.0
BWL(m)	25.0	25.0	25.5	25.5	25.5	25.2	25.5	31.0	31.0	31.0	31.0
Draft (m)	7.8	8.0	8.8	8.2	8.3	8.3	8.9	7.3	7.9	7.5	7.9
Full Load Displacement (MT)	21,605	22,419	25,562	23,533	23,712	23,800	26,178	34,098	37,489	35,520	37,744
Propulsion	CODAG	COGAG	Nuc Mech'l	IPS	IPS	IPS	Nuc Electric	CODAG	Nuc Mech'l	IPS	Nuc Electric
Propulsors	2 CPP	2 CPP	2 CPP	2 FPP	2 FPP	2 FPP	2 FPP, 1 APU	2 CPP	2 CPP	2 FPP	2 FPP
Sustained Speed (kts)	31.2	31.4	~30	31	30.9	30.9	~30	26.2	~25	25.4	~25
Range @ 20 kts (nm)	8,000	8,000	n/a	8,000	8,000	8,000	n/a	10,000	n/a	10,000	n/a
Aircraft	2 x SH-60 (VTUAV capable)							17 x CH-46 equivalents, MV-22 capabale			
C4ISR & Weapons	Notional Next Generation Radars, 192 - Mk57 VLS, 2 - SEARAMs, 1 - AGS, 4 - 30mm CIGS, 2 - 57mm guns, MF and HF SONAR, MFTA, 2 -Mk32 SVTTs							2 - LCACs, SPY-3 MFR / VSR, 4 - RAM launchers, 4 - Mk38 25mm guns			

Prime movers were also varied between gas turbines and steam turbines. A detailed summary of the variants was provided earlier as Table 5. For the nuclear propulsion variants, one next generation aircraft carrier propulsion plant was used to meet ship power plant needs.

Summary of Ship Concepts

Table 6, above, summarizes the principal characteristics of the ship concepts developed for this study.

MANPOWER ANALYSIS

Rough order of magnitude manpower estimates for three baseline mechanical drive ships: the SSC, MSC, and AWS were developed using a total ship manpower assessment tool which analyzes total ship manpower requirements by mission, function, and functional workload for a ship configuration and applies functional workload to manpower determination rules and policy changes to arrive at a manpower estimate. This analysis required tools that incorporated the following functions:

- Workload re-distribution
- Workload driver tracking
- Navy-unique manpower determinations

Since each billet performs multiple functions, workload reduction does not necessarily translate into manpower saving. For instance, reducing one billet could involve workload re-distribution to other billets. Thus, workload driver tracking by division is very important.

Officer requirements are driven by organization and leadership positions since they exist for reasons of command authority (CO, XO, Department Heads) and special skills / knowledge (Medical, Lawyers, etc.). Officers are assigned based on the functions that need to be accomplished on the unit and for command/leadership/training functions. Chief Petty Officer estimates are based on a rate-by-rate analysis and were determined by the population size of each rate in accordance with the Distribution Spread Record Listing (formerly Paygrade Distribution Records Listing).

Manpower estimates for engineering plant manpower requirements for nuclear fueled ships are based on single plant adaptations of current and recently designed propulsion plant manning models. IPS variants conservatively use the mechanical drive manpower estimates. The resulting estimates are shown in Table 7 and Table 8. The ship's company is composed of those personnel aboard the ship dedicated to the operation of the ship. These numbers do not include detachments, embarked staff, or the like, which are not counted in the LCC estimate but would be the same across propulsion variants.

Due to continuing decreases in nuclear powered submarine force levels as well as a fifty percent reduction in nuclear trained manning requirements for the CVN-78 class, the nuclear propulsion training pipeline would be more than capable of supporting increased manning needs in both the near term and longer. The training projection for a single ship class of those evaluated in this study would represent a less than 10% impact to the nuclear training pipeline, even without the benefits of reduced manpower needs in the current fleet. A long term evaluation of the nuclear training pipeline would be required should the Navy decide to wholesale changeover to nuclear power for the entire surface ship fleet.

Table 7: Baseline Fossil Fueled Ship's Company Manpower Requirements

	SSC	MSC	AWS
Officers	20	20	64
CPOs	8	19	82
Enlisted	107	244	895
Ship's Company	135	283	1041

Table 8: Baseline Nuclear Ship's Company Manpower Requirements

	SSC	MSC	AWS
Officers	25	25	69
CPOs	13	22	83
Enlisted	138	277	921
Ship's Company	176	324	1073

LIFE-CYCLE COST RESULTS AND ANALYSIS

The cost analyses quantified the acquisition and LCC of each variant to support calculating the breakeven point of nuclear versus fossil fuel propulsion, as well as the cost deltas between various power and propulsion system architectures (e.g., diesels versus gas turbines, IPS versus mechanical transmission). The breakeven analysis is provided later in this paper.

The baseline market price of fuel used in this analysis is $74.15 per BBL of crude oil, and its burdening buildup is shown below in Figure 22.

Fully Burdened		$ 152.95
Direct (DESC)	**$**	**96.60**
Crude Oil	$	74.15
Refinement	$	13.76
Transportation	$	2.67
Facilities / Operations	$	5.93
Mark-up	$	0.09
Indirect	**$**	**56.35**
Storage & Handling	**$**	**0.05**
Navy FISC	$	-
Navy Barge	$	0.05
Storage & Handling	**$**	**52.10**
Facilities / Operations	$	14.67
Mark-up	$	37.43
Environment	**$**	**4.20**

Figure 22: Fuel Burdening

As the pricc of Crude Oil increases or decreases, the other elements of the burdened rate are assumed to remain constant with the exception of Oiler O&S / Charter costs. Fuel makes up 20% of the Oiler O&S / Charter costs; therefore, 20% is varied based on Crude Oil cost.

The acquisition cost premium of the nuclear propulsion plant options include the cost of the reactor cores that power the ship throughout its service life. This is a very important distinction when addressing LCC. The naval nuclear propulsion program is actively working to reduce the cost of propulsion plant Government Furnished Equipment (GFE) for VIRGINIA Class submarines. If successful this would provide acquisition cost savings that would be realized in surface ship nuclear propulsion plant GFE. Should these efforts realize the expected 10% reduction in up front cost savings, $32M savings for *Virginia* Class plant and $53M savings for a *Ford* Class plant would reduce the nuclear acquisition cost premium by 2 to 4% for the nuclear powered ships in this study.

Operating and support costs considered manning, maintenance, consumables, and inactivation and disposal costs. Each of the alternate propulsion systems were assumed to have a common reliability as this is a capability that can be designed in along with other RMA and supportability metrics. Maintenance and other consumable costs are based on empirical models that are sensitive to ship type and are valid for ships with SHU less than those of the ships used to develop O&S cost relationships. The operationally derived steaming hours underway per year correlated positively with recent world events for each ship size permitting the use of historic O&S cost models.

Fuel quantities were determined for each propulsion plant variant for each baseline. Lifetime fuel consumption values were modeled for two operating tempos and profiles; one that assumes one major combat operation every seven years and the more stressing scenario that assumed that two major combat operations are conducted every seven years. The operationally derived fuel quantities were input into the O&S cost model where the burdened cost of fuel was varied to determine break-even costs with nuclear fueled variants.

Nuclear variants life cycle costs include the deactivation and disposal of the nuclear plant. These costs were scaled from historic plant deactivation returns and projected costs for future plants.

O&S crewing costs assign the cost per billet to the number and type of ratings in each ship variant. Crewing costs were developed using historic data for the cost per year per billet. Thus changes in plant type are reflected in both acquisition and O&S costs with consistency. All costs included in this report are in Fiscal Year 2007 dollars. Table 9 shows the LCC Premium for nuclear propulsion compared to fossil fuel propulsion, expressed in terms of percentage.

$74.15/BBL Crude High to Medium OPTEMO	Small Surface Combatant	Medium Surface Combatant	Amphibious Warfare Ship
Nuclear LCC Premium (% of Ave Fossil Fuel Ship LCC)	17% to 37%	0% to 10%	7% to 8%

Table 9: Life-cycle Cost Premium for Nuclear versus Fossil Fuel

These values correspond to intersection points on Figure 23, Figure 24, and Figure 25 at a crude oil cost of $74.15/BBL.

OPERATIONAL EFFECTIVENESS ANALYSIS RESULTS

The Operational Effectiveness analyses quantify the impact of alternate propulsion and power architectures on the operational effectiveness of the three ship types studied. Elements of Operational Effectiveness that were modeled include:

- Warfare Area Mission Effectiveness
- Surge to Theater
- Operational Presence
- Vulnerability.

Operational context is provided from a subset of the DoD DPS used to determine the lifetime energy demand signals for the three ship types.

Warfare Area Mission Effectiveness

Warfare Area Mission Effectiveness is evaluated as the percentage of missions completed by a variant in three tactical scenarios. The ship performance parameters and capabilities modeled in this analysis were: radar cross section, ship speed, ship acceleration, and ship turning. The warfare scenarios were small boat interdiction/denial of access and missile defense. The tactical situation engagements were short duration events. The variation in electrical and propulsion plant architectures did not significantly change the outcome of the engagements at the gross level of evaluation.

Future warfare mission systems are expected to place higher transient load demands on the ship's power system than current warfare systems. Once the powering characteristics of these future warfare mission systems are known, the Navy anticipates that this method of analysis will help determine the impact of power system transients on Warfare Area Mission Effectiveness.

This method will also enable the evaluation of non-concurrent mission capability. Current plant power sizing criteria requires that the power generating plant be sized to meet the sustained maximum speed requirement (i.e. propulsion load) plus the maximum connected load (mission plus ship service loads) with design/construction and service life margins. Depending on the ship's mission, these two conditions may not occur concurrently within the operational context of the design. In that case, the power generating plant can be sized to meet the worst case propulsion load and margined mission loads that occur simultaneously. In the case of high-energy radars, allowing for non-concurrency could mean that one less prime mover is needed for a ship design compared with that required by current criteria. The capabilities modeled did not create significant discrimination in Warfare Mission Area Effectiveness; however, future evaluations may lengthen the duration of tactical situations studied to expose power and propulsion system discrimination.

Surge to Theater

For the purposes of this study, Surge to Theater was evaluated in two ways:

1. in terms of the number of refuelings and the amount of fuel required to reach a theater of operations from a homeport at maximum surge speeds of 30 knots for Small and MSCs and 25 knots for AWSs, and
2. the best speed attainable for those ships without refueling.

Both metrics are very significant to naval operations and force structure as they drive the required number of fleet oilers, the ships which must be staged to provide underway replenishment of fuel. Systems that provide high-energy storage capacity and density, high energy conversion (i.e. engine) efficiencies and high thrust generation (i.e. propulsor) efficiencies improve these metrics.

Of great significance are the numbers of propulsion architectures that provide more operational flexibility than the pure gas turbine architecture of SFM-2, which is the architecture in the fleet today. It should also be noted that this analysis assumed that ships refueled when they had burned 50% of the fuel in their tanks. The fleet is more conservative than this, which would only drive up the number of underway replenishments, and hence increase the demands on oiler infrastructure.

Nuclear powered ships are superior to all fossil fuel variants in the transit scenarios modeled as non-nuclear surface combatants required between 2-3 refuelings to support a 4,500 nautical mile surge. Other technologies providing high levels of performance relative to the mission timeliness metric are diesel prime movers and single screw propulsors.

With the current and expected increase in energy-intensive ship service electrical loads (high energy radar systems, ship self defense systems, etc.) a much more detailed analysis of a ship's energy demands is required. A methodology to do this was applied in this study as was discussed in the energy requirements section of this paper. Historically, fossil fueled ship's fuel tanks were sized based on the ship's range requirements since electrical loads were relatively small. However, this is no longer the case. Proper sizing of ship's fuel tanks requires analysis of real-time electrical loads combined with propulsion loads in an operational context. Increased fuel tankage could be pursued to improve endurance of the fossil fuel variants; however, this will result in higher acquisition and LCC than those of the ships analyzed in this study. In this version of the surge analysis, the 50% fuel burn restriction is lifted and the ships are evaluated over an approximately 4,200 nautical mile transit.

Again, the nuclear variants exhibited the best operational performance with the most efficient fossil fuel surface combatant maximum transit speeds ranging from 70 to 90% of the nuclear variants. None of the variants that used only gas turbines as their principal means of producing mobility power were able to complete an unrefuled transit. Combined diesel and gas turbine plants in either pure or hybrid IPS architectures exhibited the best performance of the fossil fueled ships.

This analysis did not consider the need for other underway replenishments for non-fuel stores (e.g., food, other consumables) which could be either connected (traditional) or vertical (via aircraft).

Operational Presence

Operational Presence is evaluated as the time a ship variant can remain on station while conducting missions in theater. As discussed earlier, the DoD DPS provided the basis for the speed time profile and ship service electric loads modeled in the Operational Presence analysis. Battle loads (Condition 1) were modeled in-theater and summer cruising loads with radars on (Condition 3) were modeled in transit to and from the Sea Base refueling point.

The nuclear powered variants are superior to fossil fuel powered variants in providing operational presence on station limited only by ship stores capacity. Fossil fuel plant variants provide between 89% and 95% of the nuclear powered plant operational presence for SSCs and between 87% and 90% of the medium nuclear powered surface combatant. Fossil fuel plant variants with diesel prime movers had a significant advantage over gas turbine variants. The best performing fossil fuel variant was SFH-3, the fossil fuel mechanical-electric drive single shaft hybrid variant. This variant best captures the system efficiencies and flexibility provided by an IPS system. Similar improvements in operational presence can be expected by employing hybrid IPS architectures. Again, increased fuel tankage could be pursued to improve fossil fuel variants operational presence at increased acquisition and LCC.

Vulnerability

Vulnerability is evaluated as the probability of losing mission capability following damage from threat weapons. This analysis purposely looked at the ability of the various options to sustain damage. Ship susceptibility (i.e., ability of ship to avoid being hit) was not modeled in this study. Therefore, any benefits of propulsion plant concepts that decrease ship susceptibility (higher sustained speed, fewer refueling, and signature impacts) were not evaluated.

The primary focus of this analysis was the SSC, because the smaller size of the platform is likely to exacerbate those stressors that drive vulnerability successes or failures. In all, 12 variants were evaluated for the SSC. The MSC was assessed using analysis by analogy. The vulnerability assessment results of the SSC are analogous because the extent of damage resulting from the threat weapons associated with DoD DPS do not discriminate between the relatively small differences in ship subdivision sizes or in the high-level compartment de-activation logic for both ships.

The results of ship vulnerability assessment studies suggested that power and propulsion systems and architectures reduce ship vulnerability through:

- Redundancy
- Zonal (vertical and longitudinal alignment between energy sources and loads) distribution systems
- Longitudinally separated distribution of propulsion systems (retractable propulsors located at the forward end of the ship)
- Flexible energy conversion systems (electric or steam integrated power systems) providing for distributed conversion architectures.

INSIGHTS ON COST AND EFFECTIVENESS ANALYSES

For the SSC, the diesel cruise engine variants were superior to the gas turbine variants in the Surge to Theater and the Operational Presence Analysis. This result, coupled with the LCC advantage of the diesel cruise engine variant suggests that ships similar to SSCs should seriously consider diesel cruise engines.

Further, the reduced fuel usage, reduced number of required refuelings, the best fossil fuel Surge to Theater and Operational Presence performance, and slightly lower LCC of the single shaft variant suggests that future SSCs also consider a single shaft propulsion plant with a longitudinally separated forward propulsion unit. The use of diesel prime movers opens the possibility of lifetime fuel savings in the range of 44% for the single shaft and 30% for the twin shafted small combatant.

Similar to the SSC results, the LCC savings of the diesel cruise engines coupled with the superiority of this configuration in the Surge to Theater and Operational Presence analyses suggest that future MSC concepts also consider diesel cruise engines. This would represent a possible lifetime fuel savings of 31-33% for the medium combatant.

Longitudinally separated propulsors as enabled by IPS or hybrid propulsion plants were the single largest discriminator among SSC peers in the vulnerability analysis. Since the LCC analysis did not significantly discriminate between IPS and mechanical drive plants, small and, by analogy, MSC designs should consider IPS and hybrid propulsion plants.

The vulnerability reduction, as well as the other benefits due to IPS comes at a LCC premium of less than 6% for the SSCs, 2% for the MSC, and less than 1% for the AWS.

For the MSC and the AWS, the superiority of the nuclear variants in the Surge to Theater analysis and Operational Presence analysis comes at a LCC premium of less than 10%, but an acquisition cost premium of 22% and 46%, respectively.

BREAK EVEN ANALYSIS FOR NUCLEAR PROPULSION ALTERNATIVES

Based on the fuel usage projections inherent in this study, the break even costs per barrel of fossil fuel at which nuclear propulsion becomes economical for the various options are:

- Small Surface Combatants: $210/BBL to $670/BBL
- Medium Surface Combatants: $70/BBL to $225/BBL
- Amphibious Warfare Ships: $210/BBL to $290/BBL

Break even regions are defined for the high (baseline plus two MCOs) and medium (baseline plus one MCO) operational tempos. Analysis correlated historic and DPS derived steaming hours underway over the period from FY2000 through FY2006. The ship type demands during this period strongly correlate with the medium operational tempo (baseline plus one MCO) excepting the addition of the FY2002 lesser contingency.

The ranges of breakeven costs for each ship concept are driven by both OpTempo and propulsion plant efficiencies. The high operating tempo includes two MCOs in a seven-year period and the medium operating tempo includes only one MCO in a seven-year period. For example, the MSC breakeven range is $70/BBL - $155BBL for the high operating tempo, while the range is $115/BBL - $225BBL for the medium operating tempo.

Figure 23, Figure 24, and Figure 25 show the breakeven analysis comparing the LCC of a nuclear variant to fossil fuel variants for increasing crude oil costs. The breakeven point, where the costs for the nuclear and fossil fuel variants are the same, depends on the ship's power requirements, operational tempo, ship mission, ship design characteristics, and service life. In this analysis, the variation in break even points is dominated by operating tempo and by the machinery configuration of the propulsion and electrical plant.

In the three following graphs, the upper and lower red fossil fuel lines bound the fossil fuel power and propulsion architecture trade space. Some architectures lie firmly in the middle of this trade space, while others may be quite close to the upper or lower bounds. These charts should not be taken as absolutes, but rather as relative indicators of the feasible trade space.

The lower line on these three graphs represents the most efficient power and propulsion architecture operating at the medium operating tempo. Similarly, the higher line represents the least efficient architecture and the high operating tempo. Changes in assumptions/projections of fuel usage for the various ship concepts will significantly impact these break even costs.

For the SSC, Figure 23, this analysis has shown that the most efficient power and propulsion system architecture at the medium operating tempo is that of a combined diesel and gas turbine plant, driving a single shaft with a longitudinally separated forward propulsion unit (dashed red line). The least efficient at the high operating tempo is the fossil fueled power and propulsion architecture of a completely gas turbine IPS ship (solid red line).

For the MSC, Figure 24, this analysis has shown that the most efficient power and propulsion system architecture at the medium operating tempo is an architecture that uses diesel engines at cruise speeds (dashed red line). The least efficient fossil fueled power and propulsion architecture at high operating tempo uses gas turbines at cruise speeds (solid red line)

For the conventional AWS, Figure 25, which has a combined gas turbine and diesel powered IPS and mechanical architectures, the range in breakeven prices is relatively tight for these ship versus other fossil fueled ship classes studied. This is due to the smaller set of designs studied and relatively small difference in propulsion transmission efficiencies at this scale. The primary driver seen here is operational tempo profile.

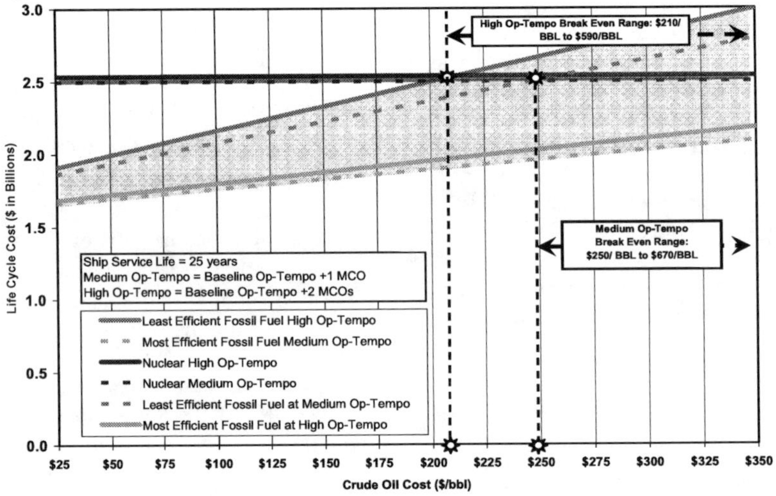

Figure 23: Life-cycle Cost vs. Crude Oil Price for the Small Combatants (5th Ship)

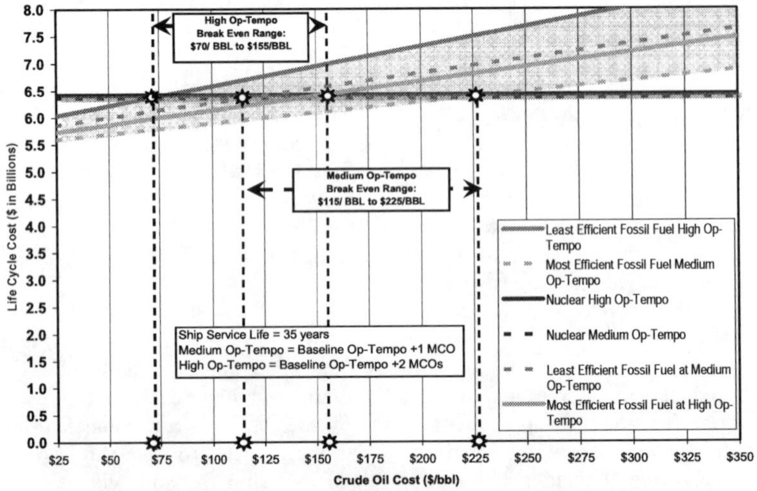

Figure 24: Life-cycle Cost vs. Crude Oil Price for the Medium Combatants (5th Ship)

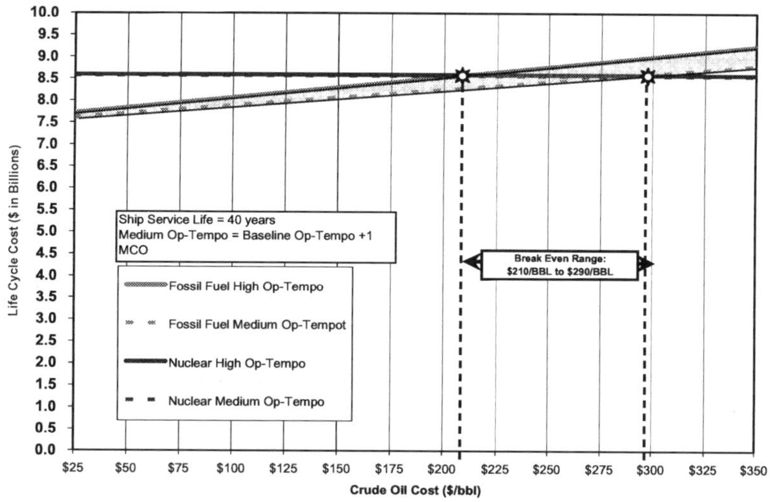

Figure 25: Life-cycle Cost vs. Crude Oil Price for the Amphibious Ships (5th Ship)

FINDINGS AND CONCLUSIONS

In response to Section 130 of the Fiscal Year 2006 National Defense Authorization Act, the US Navy evaluated alternative propulsion methods for small and medium surface combatants and amphibious warfare ships. Twenty three ship concepts were developed and analyzed in terms of acquisition cost, LCC, and operational effectiveness. The different concepts varied power and propulsion technologies including nuclear power, gas turbines, diesel engines, mechanical power transmission systems, hybrid power transmission systems, integrated power transmission systems, combined power plants, and various propulsor systems.

The major findings and conclusions of the report are:

- Mission and operating requirements drive the need for particular power and propulsion system architectures, not ship displacement. For instance, it was found that ships with constant, high demands for energy may benefit from nuclear power, whereas ships with constant low demands for energy may be more suitable with combined diesel and gas turbine plants with hybrid power transmission systems (such as a single shaft with a forward propulsion unit).

- The range of breakeven costs for each ship concept showed the greatest sensitivity to operating tempo and propulsion plant efficiencies. For example, the breakeven range for the MSC at the high operating tempo is $70/BBL - $155/BBL, while the breakeven costs increase at the medium operating tempo to $115/BBL - $225/BBL. The wide range of breakeven costs regardless of operating tempo reflects propulsion plant efficiency differences, where the highest efficiency plant has the highest breakeven cost.

- Ship vulnerability is sensitive to power and propulsion plant architecture. Vulnerability can be reduced by using components that enhance redundancy, longitudinal separation of prime movers and propulsors (e.g., forward propulsor units) and the use of flexible energy conversion (e.g., integrated propulsion systems) and distribution systems.

- Surge to theater performance is improved by propulsion and power systems that use efficient energy conversion systems and high energy densities. The most effective means to achieve this operational effectiveness advantage are:
    - Nuclear energy.
    - Propulsor systems that maximize propulsor efficiency and reduce drag by minimizing the number of shafts/screws, or in the future, incorporating podded propulsors and/or low drag hull forms.
    - Plant architectures matched to operational requirements.
    - Large fuel tankage capacities for fossil fuel ships.

- Operational Presence, measured by time on station and low speed operations, is enhanced by energy efficient plant architectures. The most effective architectures include:
    - Nuclear energy.
    - Diesel engines and generators alone or in combined power plants with boost gas turbine prime movers if high speed is also a requirement.
    - Integrated power and propulsion systems.

- Nuclear propulsion systems are technically feasible for small and medium combatants and for amphibious ships using existing reactor designs.

- Nuclear propulsion options provide operational advantages in surge to theater and time on station for all variants studied. Trends in ship weapons and sensors toward significantly higher power and energy demands will further highlight these advantages.

- Future sensors and weapons are expected to have significantly higher electrical power demands. As energy demands increase for these mission systems, the value of efficient power and propulsion systems increases.

- Based on the mission requirements assumed in this study, the LCC premiums for nuclear propulsion compared to fossil fuel propulsion, expressed in terms of percentage, for the various ship concepts are:
  - Small Surface Combatants: 17% to 37%
  - Medium Surface Combatants: 0% to 10%
  - Amphibious Warfare Ships: 7% to 8%

- Acquisition cost premiums for nuclear propulsion compared to fossil fuel propelled ships for the various ship concepts are:
  - Small Surface Combatants: ~80% (~$600M)
  - Medium Surface Combatants: ~22% ($600-$700M)
  - Amphibious Warfare Ships ~46% (~$800M)

  The nuclear propulsion ship acquisition costs include the cost of the nuclear fuel, which lasts for the entire service life of the ship. Fossil fuel powered ships acquisition costs do not include the lifetime cost of fossil fuel.

- Based on the fuel usage projections for the ships in this study, the break even costs per barrel of fossil fuel for the various options are:
  - Small Surface Combatants: $210/BBL to $670/BBL
  - Medium Surface Combatants: $70/BBL to $225/BBL
  - Amphibious Warfare Ships: $210/BBL to $290/BBL

As a result of this study, the Navy has committed to use the methods and processes developed herein in future ship analyses. These analyses will include quantifiable evaluations of ship vulnerability, sustainability, and timeliness that can be measured against acquisition and life cycle cost.

The Navy is also currently considering ship options with nuclear power and combined plant architectures (e.g., cruise diesels combined with gas turbine boost) in studies for future ships.

Additionally, the Navy is continuing to invest in Research, Development, Test, and Evaluation (RDT&E) efforts to improve affordability, power density and efficiencies of technologies for Naval ship power generation (e.g., fuel cells), power distribution, propulsion transmissions, and technologies to reduce hull drag. The Navy is also investing in RDT&E for propulsors providing improved efficiency and increased longitudinal segregation.

The design methodology used in this study differs from a traditional requirements-based design approach by embedding operational effectiveness assessments into each iteration of design synthesis. The design process architected sub-systems and total ship system capabilities into warfighting, mobility, and vulnerability effectiveness vice static performance metrics (e.g.: sustainability and surge capability vice endurance range and speed) within the ship synthesis process. The process sequentially evaluated multiple effectiveness objectives thus permitting multi-objective discrimination between propulsion system alternatives.

ACKNOWLEDGEMENTS

The Commander of the Naval Sea Systems Command (NAVSEA) directed this study with oversight from the Deputy Assistant Secretary of the Navy for Research and Development, Deputy Assistant Secretary of the Navy for Ships, and the Chief of Naval Operations' Capability Analysis and Assessment Division (OPNAV N81). Study team members included representatives from NAVSEA 017 (Cost Engineering), NAVSEA 05 (Ship Design, Integration and Engineering), NAVSEA 08 (Naval Reactors), NAVSEA 03 (Human Systems Integration), the Naval Surface Warfare Center (Carderock and Dahlgren Divisions), Office of Naval Intelligence, Naval Operational Logistics and Support Center, and the Program Executive Office Integrated Warfare Systems.

The views expressed herein are those of the authors, and not necessarily those of the U.S. Navy or Department of Defense.

REFERENCES

MULLEN, ADM M., "Winning the Fight and Bridging to the Future," CNO Guidance, 2005. (Mullen 2005)

NAVSEA Concept Study Report, CNOG 2005-068 Sea Basing: Alternate Propulsion for Submarines and Surface Combatants, WEBSTER, James S., Ser 05D/062, Mar 2006 (NOFORN). (NAVSEA 2006)

Public Law 109-360, Section 130, National Defense Authorization Act, 2006. (Public Law 2006)

USN FY 2007 Annual Long-Range Plan for the Construction of Naval Vessels Report to Congress, January 31, 2006. (USN 2006)

WEBSTER, James S., Howard Fireman, Dillon A. Allen, Adrian J. Mackenna, and John C. Hootman, "US Navy Studies on Alternative Fuel Sources and Power and Propulsion Methods for Surface Combatants and Amphibious Warfare Ships," ASNE Day 2007. (Webster 2007)

NAVSEA Concept Study Report, "Report to Congress on Alternative Propulsion Methods for Surface Combatants and Amphibious Warfare Ships, Webster, James S., Ser 05D/062, Mar 2006. (NAVSEA 2007)

Discussion

John Amy, Member

The paper is remarkable in its application of mission performance to energy requirements. This approach, an innovation of the authors', needs to be matured, generalized and applied, in greater detail, in more ship design studies so that it becomes ship design modus operandi.

The alternate propulsion study appears to attempt to answer the question of how sensitive the cost of naval ships is to fuel costs. That question cannot be answered for all ships at once. Rather, each unique set of ship requirements and cost circumstances come into play. To completely and accurately perform a "breakeven cost" comparison, the entire investment needed to realize nuclear-fueled power plants, including the "nuclear acquisition premium" *and* nonrecurring costs, must be analyzed relative to the entire investment needed to realize fossil-fueled power plants, including acquisition, nonrecurring *and* fuel costs. Placing nonrecurring costs outside the scope of the study puts the conclusions of the study in serious question. The stated congressional mandate was to conduct a "... comparison of life-cycle costs ...." Life-cycle costs definitely are taken to include nonrecurring costs. It is a certainty that, somewhere within the Naval Sea Systems Command, the total nonrecurring costs to develop USS Virginia's, DDG 1000's and CVN 78's power plants are known. The total nonrecurring costs to modify the LHD power plant to yield LHD 8's must also be known. Why were these costs assumed to be outside of the scope of the study? The paper's assertion that nonrecurring costs are outside the scope of the study does not explain why a key cost component was ignored.

Despite this serious flaw in the study's assumptions, the paper serves as excellent information for future ship acquisition efforts. The analysis of alternatives for each of these future ships must consider the results of this paper when endeavoring to develop a preferred ship design that meets the articulated set of ship requirements. It'll be those future ship acquisition programs' budgets, within the 10-20 years as stated in the paper, that must contain the funds to pay the nonrecurring costs to realize their informed power plant choice. I wonder which NRE estimates they'll use in their budget requests.

Mark Bebar, Member

The authors are to be congratulated on a very comprehensive and rigorous assessment of alternative propulsion system concepts that offer promise for surface combatants and amphibious ships in the Navy After Next. The use of a holistic evaluation of a propulsion system as an energy management system is an excellent approach because it imposes a level of engineering rigor that is lacking in most studies of design alternatives with comparable levels of technical complexity. The following detailed comments are provided.

Abstract – The definition of small and medium surface combatants is somewhat skewed, especially with reference to the world's navies. "Small" typically implies corvettes/frigates in the 1,500–6,000 ton range, and medium would typically be 6,000–15,000 ton destroyers. 21,000–26,000 tons would equate to battle cruisers (e.g., Russian KIROV class). Amphibious classes would range upwards of 40,000–50,000 tons for big deck amphibious ships and approximately 25,000 tons for LPDs/LSDs. It would be helpful if the authors could provide some context here, perhaps relating to the Affordable Future Fleet (AFF) Study.

Page 2 – Please explain what the "RTC" study scope was.

Page 9 – The authors refer to podded propulsors that can be retracted if damaged. The paper should elaborate further on this technology as it appears to be fundamentally different than that used on recent cruise ships. Are these podded propulsors rated at 3–5 MW as stated later in the paper for the forward propulsor?

Page 10 – The discussion of ship energy management system architecting refers to vulnerability and on the next page, to susceptibility and vulnerability. For completeness, it would be better to first summarize total ship survivability (TSS) attributes of susceptibility, vulnerability and recoverability and frame the discussion in terms of all three of these attributes when comparing system alternatives.

Page 11 – The authors refer to use of a segregated electric drive forward propulsor capable of providing "warfare mission speed." Given the limitations on propulsor power levels (3–5 MW as stated later) and efficiencies for such units, this statement needs further elaboration.

Page 17 – The authors refer to warfare scenarios of small boat interdiction/denial and missile defense. These would apply to SSC and MSC but not the AWS.

Page 17 – The authors refer to possible non-concurrency of worst case propulsion electrical load and use of high-energy radars, with possible reduction of one prime mover. This should be considered an optimistic assumption given the wide range of potential operational situations and associated speed-time profiles.

Page 18 – The vulnerability discussion should mention recoverability as one aspect not considered in the study and whether it should be considered in follow-on analyses.

Page 19 – Insight on cost and effectiveness states that ships similar to SSCs should seriously consider diesel cruise engines. The reason that SSCs in the U.S. Navy have not traditionally utilized diesel cruise engines (in contrast to most NATO navies) is that radiated noise (acoustic) requirements for most surface combatant ships have precluded diesel engines. This historical perspective would be helpful.

James Bennett, Member

The authors have done an admirable job of addressing a very complex and controversial issue through sound engineering methodology. This paper is the first presented in this forum to expose the integrated engineering and economic models that form the basis of making investment and force structure decisions. The authors are to be applauded for the accomplishment and the Navy should be commended for socializing methods and processes described in this paper through SNAME.

I have several comments, which fall into two areas: the first focusing on assumptions and constraints and the second on the sensitivity of cost predictions to the evolving economics of shipbuilding in the U.S. given the current state of the industry.

Assumptions and constraints have a major influence on conclusions of the analysis. The study calculates the fuel consumed by various ship classes based on power demand signal. Variation in the power demand signal has a significant influence on the conclusion of the study. A basic assumption is that all ship classes see underlying power demand growth in proportion to change from DDG 51 to DDG 1000 class. This may overstate the power demand signal because a large portion of the power growth is due to ship size difference between DDG 51 and DDG 1000 classes.

Medium surface combatant (MSC) power growth is driven by ballistic missile defense (BMD) radar requirement. The power demand of the radar is an order of magnitude greater than current generation radars. The persistence of this load over the operational life of the MSC makes this load predominate over all other loads including propulsion. Therefore variability in this load has large influence on the result of analysis and variability should be further investigated.

There are opportunities to vary radar loads considerably depending on the design choices made concerning aperture size, transmit/receive element spacing, the number of faces, efficiencies of the electronic components, array occupancy, utilization and the duty cycle relative to mission requirements. A further study should be undertaken to understand the influence of the radar technologies and design decisions on the CG(X) as a total ship system.

Fuel consumption is based on medium- and high-intensity operational profiles. The low-intensity profile is excluded. It would be useful to expand the study to include a low-intensity operational profile in balance with medium- and high-intensity operations to establish the influence of op tempo on outcome. If a higher initial cost of ships results in erosion of force structure levels, op tempo will necessarily increase or mission capability will decline. The affect of this on future force capability would be useful to understand.

Uncertainty inherent in cost calculations is greater than uncertainty in performance calculations. Accuracy of cost modeling is the most difficult aspect of achieving overall accuracy in this kind of analysis. The authors chose not to address the subject of cost modeling in order to limit scope of study effort. Variability in cost due to shifting in industrial base between conventional and nuclear propulsion industries will have major influence over the conclusions of the study. These cost differences need to be better understood and included in the analysis before this work is used to set policy for shipbuilding plans within NAVSEA. The sensitivity of the study result to variability in cost parametrics needs to be understood and quantified.

Jose Femenia, Member

The authors must be congratulated for an excellent paper documenting the very thorough analysis of propulsion plants and systems for naval vessels.

I sincerely want to thank the authors for commencing the paper with an extensive list of acronyms. Without the list I would have had a very difficult time reading and understanding the paper.

Based on the cost of oil — at the time of the study crude oil was $74.15/bbl — it is not surprising that the study concluded that the simple gas turbine plant is the least desirable prime-mover for the three classes of vessels investigated. Nor is it surprising to me to see that nuclear power rated as a good alternative. With crude currently selling for approximately $90/bbl, the argument for nuclear power for naval ships and commercial ships becomes more acceptable.

As a person whose roots are more aligned with commercial marine engineering, I often questioned the concept of using gas turbine prime-movers for most naval vessels instead of combined gas turbine/diesel plants or even combined cycle plants. I believe the authors make a very good case for the Navy's departing from the all gas turbine ship to alternate prime-mover energy converters. As the authors point out, the advantages range from economics to sustainability of forward deployed vessels.

I was also pleased to see that the authors concluded that nuclear power should be considered for the medium surface combatants and even amphibious warfare ships and small surface combatants. I understand why the authors focused on submarine class and next generation aircraft carrier propulsion plants, but I believe that this resulted in excessive prime-mover costs for the ships in question. I would recommend that the authors extend their work and partner with the commercial nuclear industry to develop two or three standard size marine nuclear reactor and primary-loop packages that would be suitable for naval vessels, commercial vessels and utilities interested in distributed power generation. I will further suggest that three size packages be designed; 200 mWt, 400 mWt and 600 mWt. By sharing the development costs with the commercial world, the development and support infrastructure costs could be better controlled.

The use of standard, off-the-shelf nuclear reactors has the potential of not only benefiting the Navy, but also extending President Eisenhower's Atoms for Peace dream into the marine industry. The commercial marine use of standard small reactors could be the nation's follow-on to the Atoms-for-Peace commissioned vessel, the N/S Savannah.

William H. Garzke, Jr., Member, **David Byers**, Member, **Matthew McCarton**, Member

Almost 80 percent of the world's reserve of oil is concentrated in the Arabian Gulf and some of the former Soviet republics that border the Caspian Sea in that area of the world. Another rich oil region is the Lake Maracaibo region of Venezuela. These regions have seen much unrest during the latter part of the twentieth century, and there is every indication that it will continue into the twenty-first century. Saudi Arabia and Iraq have very large reserves, but these two countries have been the targets of destabilizing influences by Muslim extremists. Iran, another large oil producer in that region, is hostile to the interests of the United States.

Another factor that must be considered is the status of oil reserves of the world. During the 1950s predictions were made that the world would run out of oil in fifty years. That prediction proved incorrect because new oil sources have been found in the Gulf of Mexico, Alaska, southern Russia, off the Canadian Atlantic Coast, the North Sea and recently off Brazil. Norway, for example, was oil and gas poor in 1957, but is now a significant producer. Nevertheless, oil is a limited resource and will eventually be exhausted. When that policy commences, countries dependent on oil could be the targets of those who have the resources. Unfortunately it is human nature to hoard a precious commodity in short supply and that will result in friction between the "haves" and the "have-nots." In fact, to gain a political concession, an oil-rich country might put pressure on one that does not have an oil resource by threatening to cut off their supply. The most likely areas for this type of confrontation are the Arabian Gulf and Venezuela.

The lesson of the Arab Oil Embargo of October 1973 is an important consideration in any future power plant for a naval combatant. The cessation of oil shipments after the 1973 Arab-Israeli War brought a major disruption to oil shipments to United States and some European countries. That embargo also caused an oil shortage that saw the price of oil skyrocket to an all-time high and eventually led to increased prices in operating costs of vehicular traffic, aircraft, and ships for several years. These political considerations must be part of an evaluation process when the choice of a future cruiser's propulsion plant is made.

The cruiser of the future should feature a propulsion system that is not completely dependent on fossil fuels. In considering the choice of propulsion plant for whatever cruiser design is selected, ships of that design will almost certainly be in service for 35 years. Therefore, one must study the type of alternate fuels that would be available. The most likely choices based on present-day technology are:

- Nuclear
- Fuel cells
- Hydrogen
- Super conducting electric plant

The U.S Navy built nine nuclear-powered cruisers between 1957 and 1980. Nuclear cruisers are expensive. Historically, their first cost was more than any propulsive system. However, a nuclear plant does have some advantages, and life cycle cost is one important feature. Depending on how the reactor core is used in service, the normal nuclear refueling cycle of a nuclear plant in the first nuclear-guided missile frigate, USS Bainbridge (DLGN 25, later CGN 25) was four years, but in subsequent nuclear cruisers this was extended to 10-15 years. During that time the only fossil fuel that these ships would embark was JP-5 for their aircraft complement. The first cost of the CGN 36/37 was $200 million in 1968. The cost in 2004 dollars, due to inflation, would have beem approximately $1 billion for the same ship and armament. However, that ship class was not equipped with AEGIS or the Mk 41 vertical launch system

(VLS). The first cost of a DDG 51 with gas turbines, Module A and B Mk 41 systems, and AEGIS is nearly $1 billion. Therefore, the cost of a nuclear cruiser with an AEGIS and Mk 41 system that would likely be two Module B launchers would be $1.5 billion. However, that latter cost does not include the cost of fuel for a conventionally powered warship over a period of 20 years (present core life). To make a fair comparison between a nuclear ship and a fossil fuel vessel, the cost of fossil fuel must be factored into the comparison, along with the disposition of the nuclear reactor, which is not insignificant. The cost of fuel taking into account inflation and no sudden increases in fuel costs over a 20-year period could be as much as $600 million. Also factored into this must be strategic immobility — there may not be places to refuel.

Emission standards are likely to increase over time as the quality of oil decreases during that same period. The refining process of the oil becomes more expensive with the need to remove undesired impurities. This cost is passed onto the consumer. It is unlikely that oil prices will stabilize over time.

There are other factors to consider. The refueling of a fossil fuel ship requires replenishment tankers to make the transfer of needed fuel at sea. This requires a fleet of tanker-type ships and their attendant construction, manpower, maintenance, and insurance costs.

A nuclear ship can steam at full power from one trouble spot to another without the need to refuel. When that crisis has been settled it can return to its first troubled area at the same or slower speed without the consideration of replenishing fuel. The only replenishment required might be ammunition and stores. The nuclear cruiser would still need stores and ammunition from such ships as TAOEs and TAKEs. So there will be a need for CLF ships with the exception of TAOs in this category.

The nuclear ship does have several disadvantages. It has a limited number of ports that it can enter. However, Japan has recently decided to allow one of the latest CVNs to be home-based in Yokosuka when the Kitty Hawk (CVA 63) is retired. New Zealand has not rescinded that policy. Also, the nature of its plant requires a cadre of officers and crew who require specialized training in the operation and maintenance of a nuclear propulsion plant. To help offset that requirement, the nuclear plant of the future cruiser should be the same as or a derivative of the nuclear aircraft carriers of the CVN 78 class or the SSN 774 class submarines.

Although a gas turbine propulsion plant requires less maintenance and personnel, is quiet in operation (assuming appropriate noise isolation), and gives instant response, it requires much space for their intakes and uptakes and are very light in weight. That lighter weight is countered by the amount of fuel that the gas turbine must burn. As that fuel is burned, ballast must be provided to maintain a vessel's stability. Those ballast tanks require volume, making the ship larger. Nuclear plants, by contrast, have a quick response but have a heavier plant and some ballast capacity may still be needed for trim and list purposes only.

There are many factors to weigh in the choice of a cruiser's propulsion plant. However, the fact remains that oil for a fossil fuel plant is dependent on an element that may be in short supply from questionable sources during its lifetime.

Paris Genalis, Member

This is a valuable paper not only for the results it displays but also for the methodology it develops. The Navy will be able to use the algorithm discussed in the paper to examine new data as it is developed and answer questions, that will undoubtedly arise, in a consistent and rational way. The authors have provided a great service to the Navy and the profession.

The results of breakeven costs of fuel, while interesting in themselves, also point out the high sensitivity to mission profiles. As the profiles will vary to address changing national security strategies, further refinement of the analysis will permit closing the range of breakeven costs and thus help decision makers in a more definitive way. Conversely, as NAVSEA considers commonalities among ship classes and possibly mission variability among members of each class, the breadth of mission profiles results in bounding values of breakeven costs to guide further discussion of commonality options.

The paper discusses, and more importantly, offers a process for analyzing, the growth in power requirements that we know will be imposed on future ships. Further, while the paper recognizes cost effects of the need for more (or less oilers) to support the proposed ships, a definitive analysis depends largely on the composition of the entire fleet, and many assumptions are needed to reach conclusions for the costs associated with a particular member of the fleet. In addition, the paper recognizes the need to account for disposal costs in the case of nuclear ships.

The paper addresses the merits of mission effectiveness as well as a technical analysis can. However, the intangibles of tactical flexibility, mobility and readiness are very difficult to quantify and yet may well make the difference in some decisions; they are evident in the obvious advantages of the nuclear ships discussed in the paper. This, of course, has to be balanced against the additional cost (which can be estimated, reflected in numerical values and used in an analysis). Thus, having the numbers will be helpful, but the decision may well hinge on intangible, but very real attributes of the capability offered by the increase in cost of nuclear-powered ships.

The method used to compare costs was based on constant year dollars, but not on discounted dollars. The issue of whether discounting is appropriate for a government program is an issue for the Navy and OMB to resolve. Whether the use of discounted dollars results in significantly different breakeven costs or ranking of solutions depends on the discounting rate. Resolution of this issue would enhance the usefulness of the process developed for the calculation and further strengthen the validity of the breakeven costs presented.

The costs of the fifth ship were used as a basis of comparison. This is as good a choice as any for the purposes of the study and reflects the Navy's desire to build more ships in fewer classes to increase construction and operational efficiencies. Nonetheless, it would be interesting to see the results of comparisons of the third ship, especially for expensive ships, because it may well come to pass that fewer of these may be afforded in the very tight shipbuilding budgets of the next decade.

The paper also uses construction differentials between fossil fuel and nuclear ships. It would be useful to discuss the method by which these differentials were developed. Of particular concern is the shipyard loading while nuclear ships are constructed. Nuclear-capable yards incur higher costs because of the special handling of material, safety and security issues, to name but a few of the differences with a non-nuclear-capable yard. Even assuming an arrangement like the Virginia class submarine construction, where ship sections are built by one yard and shipped to another for completion, it would be beneficial to display the cost increase due to this process. In addition, the effect of some decrease of work at a non-nuclear-capable yard (as compared with building the complete ship) would also have an effect on the overhead rates, which could affect the costs of building other classes of ships at the same yard.

The results presented in the paper are enlightening, but of more importance, the authors have developed a methodology that will enhance the Navy's ability to develop options for more rational and consistent decision making.

Michael Harrington, Member

I would like to compliment the authors on an impressive synopsis of a comprehensive and complex study. I look forward to reading the full study. The results certainly validate some established tenets of naval surface combatant ship design, while broadening the scope of study. Of particular note is the inclusion of the operational effectiveness analysis component, which offers unique insight not typically considered in propulsion plant studies and evaluations.

Recognizing the need to constrain the scope and complexity of the study and analyses, several parameters were held constant or excluded from the evaluations. Of course, this exposes areas for further companion studies and next phase studies. In particular, the following:

A validation of the tactical and operational need for both the 20 knot endurance speed and the 30 knot sustained speed requirement applied to most of the small combatants and all of the medium combatants is needed. Current fleet operations and future needs should be assessed. Excursions considering lower values for both of these conditions will likely alter the evaluation of the fossil fuel plants, particularly those with diesel cruise engines, as well as resultant ship displacements. This will become a less influential factor with increasing 24-hour average electric loads expected on future ships, so excursions on that value should be included. Recent evaluations in this area performed by Gibbs & Cox, Inc. for the Preliminary Design of the Royal Australian Navy Air Warfare Destroyer, with a CODAG propulsion plant, confirmed the beneficial effect on range and / or displacement of incremental reductions in each speed value.

General arrangement, topside arrangement and machinery arrangement studies for selected ship configurations analyzed should be pursued in the next phase. We have found that topside length requirements, driven by the competition for space and volume of the combat system elements and machinery plant requirements (e.g., intakes, uptakes, engine removals, signature influences), often drive variances from synthesis model predictions. It would be beneficial to confirm the viability of the ship concepts studied. Machinery arrangement studies and topside arrangement studies for both the IPS and nuclear plant in the small and medium Combatant groups will be invaluable in this regard. (It is conceivable that an advantage for nuclear plants would be discovered by topside arrangement comparisons.) Gibbs & Cox, Inc. has supported NAVSEA's development of IPS from the outset and can attest to the difficulty of attaining viable IPS machinery arrangements on ships in these size ranges.

With the nuclear plant options showing so positively in the findings of the study, an assessment of the capital, infrastructure and skills improvements needed at the major US shipyards to effect construction of an all-nuclear or a largely nuclear future fleet should be made. Currently, only two of the six major naval shipbuilding yards have the requisite capabilities for nuclear ship construction. An argument could be made that these costs, which would ultimately be passed on to the ship procurement cost, should be considered in the acquisition component of the life cycle cost model and breakeven analyses. Similarly, an assessment of the personnel and training needed to crew a largely nuclear future fleet should be addressed, as well as the associated cost.

Gabor Karafiath, Member

This paper deals with a very critical subject, especially at the current time of rising fuel prices and I thank all of the authors for sharing the results of their studies. For most of my career at NSWCCD I have been tasked to investigate ways of minimizing propulsion fuel through improved hydrodynamics. A major roadblock to many good ideas has been net present value (NPV) analysis which works against retrofitting fuel saving devices because fuel is saved in the out years and the device needs to be paid for up front. Would the authors please address this issue? Has NPV analysis been done, along with sensitivities with regard to assumed future interest rate, fuel cost rise, and the salaries of the 35 or more people associated with the nuclear option?

Another concern is with regard to the risk -- not necessarily the doomsday nuclear explosion risk, but rather technology obsolescence risk. The navy has recently decommissioned some of the FFG7s, DD 963s and even a CG 47 class ship mostly because of obsolete weapons systems that were to expensive to upgrade and possibly because of large manning costs. This decommissioning was well before the ships reached their hull and machinery service life. Similarly, in years past there were several nuclear-powered Navy cruisers that were laid up. Would the authors please comment on this type of obsolescence risk and incurred cost associated with both the nuclear and the fossil fuel option?

Finally, I suspect that nuclear and oil fuels are the only viable options on the table; however, have the authors investigated any other promising technologies as suggested by the paper title?

Once again, thanks for this paper dealing with a very complex subject.

Bob Keane, Member

Reading this robust technical paper on naval ship design, I am encouraged not only about NAVSEA's progress in sustaining but also advancing the Navy's capabilities to do early stage ship design. Recently, the Navy's Center for Innovation in Ship Design (CISD) identified key foundational capabilities for designing cost-effective warships. They are:

- Experienced ship designers
- Integrated design tools
- Agile processes
- Mature specs and standards
- Enterprise-wide communications.

Looking at the four authors, whom I know, I see two highly experienced ship design leaders who spent a number of years in technical specialties such as ship arrangements and hull form design. The other two co-authors have been doing ship design for about five years since they graduated from school; however, they have already had a number of different assignments and have been given significant technical responsibilities. Three of the four co-authors started doing ship design by completing a two- to three-year entry-level development program, such as NAVSEA's Engineer-In-Training (EIT) Program (no longer in existence), the Navy's Acquisition Intern Program, or the Scientist and Engineer Development Program (SEDP) at the Naval Surface Warfare Center Carderock Division. The critical factor is that they learned or are learning ship design by doing ship design and working with experienced ship designers. Having an experienced ship design workforce that has worked together before on previous ship designs is absolutely necessary to have successful ship design and acquisition outcomes.

The next foundational capability is integrated design tools. The authors mention that they used design of experiments (DoE) or multi-objective design optimization (MDO) methods. What little I know about these new, powerful tools comes from my colleagues in academia, Al Brown, NAVSEA Professor for Ship Design at Virginia Tech, and Cliff Whitcomb at the Naval Postgraduate School. I also know that as a result of the Office of Naval Research (ONR) National Naval Responsibility for Naval Engineering (NNRNE) initiative, acquisition interns on assignment at CISD (including one of the co-authors) were able to learn these new methods and experiment with applying them to the design of naval ships. A description of the DoE tools used in this design study and how they could be used in subsequent design stages would be appreciated.

How experienced ship designers use these new tools can significantly contribute to the foundational capability of agile processes, that is, processes that can be quickly adapted to your particular design and add value and eliminate waste. Hopefully, new tools permit the ship designer to improve ship design processes. These new tools have dramatically changed the early stage design processes and now enable ship designers to more rigorously explore the requirements and design solutions spaces to increase the likelihood of arriving at a global optimal versus a local sub-optimal solution. One of these improved processes mentioned by the authors is "systems architecting." This term is probably new to many of us, especially the older we are. The authors are requested to describe "systems architecting" and how it improves the ship design process.

The foundational capability that receives the least attention is mature specs and standards. Introducing any new propulsion system also introduces uncertainty that is translated into risk. One of these risks is delay in schedule and this risk needs to be quantified in terms of cost of delay. It would be helpful to get the authors' views on the costs of delay for the alternative propulsion systems considered in their paper. Furthermore, for ship designers, specs and standards include our ship design practices and criteria. It is not reasonable to expect the design practices and criteria that are appropriate for the design of a "conventional" warship with ship service power separate from ship propulsion power are also applicable to the design of future integrated electric warships. Retaining existing design practices and criteria without examining the implications of integrated electric power generation and distribution systems and electric weapons systems will inherently lead to more costly warship designs. There must be another way.

The better way is to form a multi-disciplinary team (shipbuilders, Ccmbat system integrators, design engineers and academia), similar to the Sealift Engine Room Arrangements Modeling (ERAM) team, and examine many of the Navy's existing design practices and standards; for example, formulation of speed–rangerequirements, sustained speed definition, endurance fuel calculation approach, powering and endurance fuel margins and criteria, specific fuel consumption (SFC) margins. The derivation, rationale and validity of explicit and derived design practices and criteria should be thoroughly questioned. Since future combat systems will be a far greater consumer of electric power, combat systems engineers should also be involved. This is not an issue that can be addressed in a year or two. It is a "grand challenge" that will take many years to achieve and involve our whole community. In my experienced opinion, such a "grand challenge" requires a long-term national ship design research program along the model of the National Shipbuilding Research Program (NSRP). We need to start now!

I thank the authors for a very solid technical paper that has stimulated a broad scope of discussions.

Ronald Kiss, Member

The authors have obviously tried to condense a monumental parametric design study into a concise paper for our Society. Unfortunately, simply explaining how and what they have tried to do in the design and analysis of so many alternatives has resulted in a lack of detail with regard to the results. For this reason my discussion is primarily a series of questions hoping to elicit a response that will further enhance a fine paper for final publication in our Transactions.

I have two general questions and a number of specific questions – first the general ones.

Was any account taken of the increased maintenance time for nuclear powered ships during their availabilities? I have heard it may take three surface nuclear ships to get the same number of underway steaming days as a fossil-fuel-powered ship. Historic information on the CGNs of yesteryear may shed some light on this.

Would the Navy be content with building nuclear - powered surface ships at Electric Boat and Newport News? If not, what increased costs would be associated with establishing additional qualified nuclear shipyards, and have these been factored into the analyses in this paper?

My specific comments and observations are as follows:

The last sentence in the "Energy Requirements" section states, "Plant architectures were developed using the electric loads with a 20% design margin and a 20% service life allowance...." While such margins and allowances may not affect the results of this study I am afraid that they are indicative of why the cost of naval ships has been increasing to the detriment of building the size navy needed by this country. I can anticipate at least part of the response stating that historically we have used such allowances. However, in my judgment, space or power capacity will be used if it is available. So having such margins becomes a self-fulfilling prophecy – like Field of Dreams: "Build it and they will come."

Midway through the "Ship Energy Management System Systems Engineering" section the following statement appears: "The difference in cost between the DC distribution system and the AC distribution system with dedicated power conditioning for the MSC Distributed Architecture Electric drive variant was dominated by uncertainty." Please explain what is meant?

In Table 6, why are sustained speeds for nuclear variants shown as approximate while other propulsion plant speeds are shown to the nearest tenth of a knot?

If the blue water surface ship fleet evolved toward increasing use of nuclear power, the Navy's need for oilers would diminish. Was any effort made to account for this cost savings? Conversely, an increasing number of nuclear naval ships would drive a requirement for increased numbers of nuclear-trained ship repair workers worldwide. Has this been considered?

The paper states: "The nuclear powered variants are superior to fossil-fuel-powered variants in providing operational presence on station...." I don't understand why this is so, provided a replenishment oiler can service the fossil-fueled ship while it is on station. Please elaborate on this point.

Are the consequences of major damage to a nuclear ship versus fossil fuel taken into consideration? For example, a USS Cole type attack may have had vastly different consequences had the Cole had a nuclear power plant.

It is inferred that the breakeven analysis assumes a fixed cost for nuclear fuel. If so what is that cost and how may it be expected to change in the future? And how might such cost changes be influenced by the projected major renewal in nuclear shore power stations?

Also in the "Breakeven Analysis" Section the paper states "...the upper and lower red fossil fuel lines bound the fossil fuel power and propulsion architecture trade space." Yet in Figure 23 it appears that the solid red and solid green lines bound this space. Similarly the text says "...the most efficient power...system...at the medium operating tempo is...(represented by the dashed red line)." Yet, the legend on Figure 23 labels the dashed red line as "Least Efficient Fossil Fuel at Medium Op-Tempo." Please clarify and also verify that Figure 24 is correctly linked to the text.

Also, with respect to Figures 23-25 it would be most useful if the actual life cycle costs were shown somewhere in the paper for all variants.

With respect to the seventh bullet in the conclusion, it is respectfully suggested that the second sentence should be revised to say, "If trends in ship weapons systems and sensors toward significantly higher power and energy demands continue they will further highlight these advantages."

With respect to the ninth bullet in the conclusions should not only the best fossil fuel variant be compared with the best nuclear? A reader seeing that there is a 0% difference between fossil- and nuclear-powered medium surface combatants might be led in an entirely different direction than he would be if the table showed a 10% difference.

This is a fine paper and will be a great addition to our literature, and I thank the authors for their efforts and their contributions to understanding the tradeoffs in a very complex subject. Hopefully, the additional information elicited by my questions will further enhance its value to future assessments of propulsion alternatives.

David Palmer, Member

The integration of future war fighting capabilities and requirements with overall energy demand is an extremely critical component in this type of naval design analysis. This, especially, is important with the projected increase in electric power requirements as a result of expected higher transient load demands, including theater ballistic missile defense radar system requirements. The discussion of the details related to this and the variants of the three baseline ships was exceptionally well documented and explained. Moreover, the consideration of operational effectiveness iterated throughout this analysis is agreeably a critical component of this type of design analysis and this, too, was clearly explained. It should also be noted that it was methodically presented that breakeven costs for each variant were driven by propulsion plant efficiencies and operating tempo.

I am not surprised by the conclusion that nuclear propulsion options provide advantages in terms of operational presence and surge to theater performance. I further agree that higher power demands of cutting edge ship weapons will further emphasize the advantage of nuclear power. It is further understandable that the medium speed combatants (MSC) have the most attractive breakeven numbers as a result of the mission requirements, which are driven by the higher ship service electric load. As a "powered-down" mode for the radar does not currently exist, it may be prudent to attach an addendum study based on the significantly larger 24-hour electric power load requirement for the MSC. This may show an even more attractive breakeven point for nuclear. Further, the probable reduction in nuclear acquisition cost premium by 2-4% will further enhance this breakeven point.

Integrated power systems (IPS) clearly provide flexibility in arrangement and significant improvement in vulnerability when one considers the ability to segregate and separate ship propulsion methods. The LHD 8s gas turbine mechanical and diesel electric drive system is advantageous in that "it gains fuel efficiencies at low speeds without sacrificing transmission efficiencies at high speeds." A speed versus efficiency graph would have been helpful to clarify the level of this comprehensible advantage.

This article, which describes the higher level of design which goes into the design of a naval vessel compared to traditional design criteria for a Merchant Marine vessel, will be incorporated into the United States Merchant Marine Academy's engineering curriculum. It goes a step further in analyzing alternate propulsion plant architecting and systems engineering to satisfy warfare system service demand and vulnerability which is a higher level of design that will intrigue the midshipmen.

Finally, although it is acceptable that the Navy does not procure systems that are large technological excursions from existing baselines, it is unacceptable that more resources are not put into researching alternative nuclear power technology. Specifically, gas-cooled reactors have great promise for nuclear power and propulsion in the areas of overall efficiency, safety and operational effectiveness. I hope that a follow-up study is initiated to look into "Navy after Next" gas-cooled reactors that operate on the highly efficient Brayton Cycle. I understand that standardization is extremely important in the area of nuclear power but feel that we may need to transition at some point, if needed in the future, to respond to greater technology that will enhance our Navy and the economic well being and energy security of our nation. Further, I hope that the Navy will begin a long-term evaluation of the needs associated with nuclear training. This is urged as I firmly believe that a full changeover to nuclear power for the entire medium surface combatants and amphibious warfare ships is appropriate as a result of national security related to the availability and dependence on fossil fuel, operational effectiveness superiority of nuclear power, environmental issues related to the curbing of greenhouse gases and the relatively small deltas in breakeven cost at the medium operating tempo related to the projected rise in the cost of oil as detailed so eloquently in this article.

The authors are to be congratulated for an excellent paper!

Barrick Tibbitts, Member

This is an excellent, well-written and thoughtful paper about an important and timely issue. The authors are to be congratulated for writing it, as is SNAME for publishing it. The paper describes in summary form a recent comprehensive cost study which examined a large number of power generation and propulsion plant alternatives as applied to three notional ship concepts. This is an area of technology that is often viewed as mature, but rather than being a compliment, the word "mature" connotes that little remains to be known, and therefore deserving of little or no investments in R&D. This paper presents a compelling counterargument that we need to reexamine this premise because of several dominant factors. First, the cost of fossil fuel will continue to increase while its availability will decrease. Second, new combat systems (sensors and weapons) have the potential to increase ship service loads many fold. And third, as well described in the paper, advances in technology offer significant advantages.

The Navy study, which is described in the paper was quite comprehensive, and the discussion of individual aspects in the paper is necessarily constrained by publication page limitations and only summaries are provided. There are many areas that I would like the authors to respond to or expand upon, but I will limit my comments to the subject of cost.

The paper discusses small surface combatants, amphibious warfare ships, and medium surface combatants (SSC, AWS, and MSC). Recommendations regarding SSC and AWS will likely be viewed as "Navy after Next" issues. This is because the current shipbuilding plan includes five major construction programs in these two ship categories: DDG 1000, LCS (two variants), LPD 17, and LHA 8. It will take years (possibly decades) to "build out" these classes and gaining support for R&D programs with applicability to their future replacements will regrettably not be easy.

This is not the case with the medium surface combatant (MSC). Although it, too, is not a program-of-record ship, some will view the MSC as a surrogate for CG(X) and concentrate on the conclusion that the lifecycle cost (LCC) premium for a nuclear-powered medium surface combatant is a modest 0% to 10%. This is good news indeed for those who advocate that the next class of surface combatants, CG(X), be nuclear powered. It is also appropriate to examine the realism behind these numbers.

Julio Vergara, Member

Thank you for the opportunity to comment on the paper D39. It is an interesting exercise on innovative power alternatives for surface combatants and amphibious warfare ships. This comment will focus on the nuclear variant of these ships.

Nuclear propulsion has been utilized in navies of six countries (U.S., Russia, France, U.K., China and India) in submarines for air independent stealth and in large aircraft carriers for high power demand and maximum jet fuel capacity.

Nuclear propulsion was adopted in a few cruiser types (~10,000 tons) in the dawn of the nuclear age several decades ago as an upgrade from the oil-fired steam propulsion plant. A few decades later, inexpensive air derivative gas turbine development consolidated and provided simple propulsion systems to almost all of the current surface combatant fleet in the world. Such technology development, coupled with inexpensive fuel resources with no environmental constraints, removed almost all reasons to maintain nuclear energy for those surface ships, with added costs due to larger crews and higher M&R cost. Furthermore, GT were best fitted to the task, being low capital power sources and high fuel cost for intrinsic intermittent operation. There was no reason to have a large capital nuclear power source for discontinuous operation and fractional power demand, albeit a theoretically low fuel cost.

The only remaining advantage was endurance, which also spared fuel volume for extra payload and perhaps to compensate for the reactor plant and shielding. Nevertheless, endurance became less important with the advent of longer range missiles. Today, only a few exceptions apply in Russian cruisers.

Today, a new scenario is developing with two new increasingly important issues. One is oil resource depletion which is about to reach a production peak, after which oil products will become even more expensive than the current level (ref: US$94.5/bbl WTI).

The rate of change of liquid fuel costs will not escalate forever, returning to moderate values depending on events. Two potential options may arise to moderate cost escalation, stretching the use of the relatively inexpensive gas turbines or diesel engines. One is the use of biofuels as a partial solution. The other one is coal gasification. The latter is supported on a large resource base and available Fisher-Tropsch technology.

The other important issue is climate change driven by the consumption of fossil fuels. This issue is not of second order since the consequences of global warming (temperature, sea level and precipitation patterns) may alter food supply, labor, immigration, security, etc., that may largely restructure the role of armed forces and civil defense corps. Just consider the idea of human relief support vessels instead of amphibious warfare ships.

This issue moderates the first one, because the possibility of coal gasification and synfuel production collides with the requirement of expensive GHG capture technologies in the transformation facilities. There is much to see in the energy field and a brief analysis on that would enrich the paper.

The large availability (more than two centuries) of coal at stable and moderate costs plus F-T synfuel conversion and CCS technologies may provide moderate liquid and gaseous fuels for about a century on top of low cost gas turbines.

Conventional nuclear propulsion (robust naval PWR technology with expensive, highly enriched uranium) would be too expensive without the stealth, tactical or strategic benefits in SSNs and CVNs. Furthermore, it may not be applicable yet for surface combatants with low or moderate power demand profiles.

The authors discuss the option of hybrid conversion technologies ("energy management systems" in the paper). However, from the economic side and given the operation profile, it might also be convenient to consider a hybrid power source. One of such options is to downsize the nuclear power plant to a fraction of the ship rated power, i.e., X% of the full speed power demand plus Y% of hotel load, and provide intermittent boost with a gas turbine. The GT is not expensive, would operate only occasionally, and could be refurbished at the end of the ship's life.

Precise X and Y values and architecture would depend on the energy management system and the associated technologies (number and array of shafts for mechanical/hybrid drive, suitability of pods, load configuration, electric drive, etc.).

The following figure was published in a SNAME paper in 2001. The paper proposed the use of commercial reactors for high-speed cargo vessels in a hybrid plant array, where about 80% of the rated power was nuclear and the boost was provided by a large aero-derivative GT. Such power plant architecture may fit into the amphibious warfare ship and other logistic ships. The GT may be useful for maneuvering and landing. Based on anticipated ship quantity, the reactor may be tailored to its intended use.

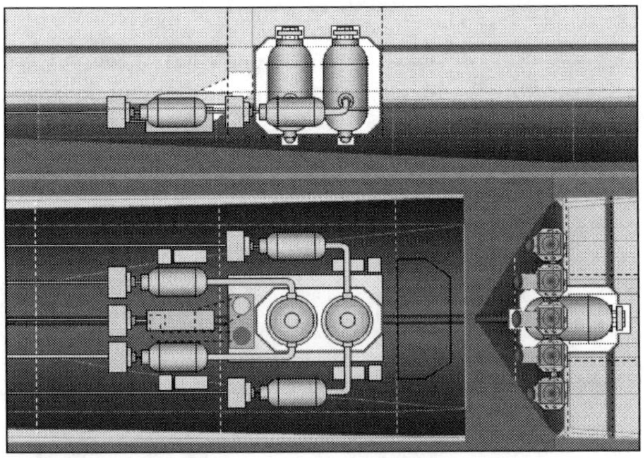

Regarding nuclear technologies, the paper freezes the nuclear reactor type to the PWR. Today, in what has been called the "nuclear renaissance," there is active concept development in several organizations within different countries, including some international collaborative projects. There is a potential new capacity of about 130 GW. A few of those projects deserve attention for nuclear propulsion. In the PWR branch, there is some development in the area of integrated PWRs, which in principle may reduce volume and may enhance safety. The latter aspect should be revised in the AWS concept if it is intended for beach landing (achieve reactor cooling under all events).

Another suitable reactor that may return to the propulsion scene is the liquid metal-cooled reactor (LMR), with several variants. This concept is interesting owing to the higher thermal efficiency and compact size (thin vessel). These concepts may be integrated as well, reducing radiation hazards.

Another reactor type is the gas-cooled reactor (GCR). In this area there has been some work in the direct and indirect Brayton cycle high temperature reactor (HTR), which uses inert helium as the coolant. The interesting aspect of this reactor is that given certain reactor parameters the reactor may be immune to severe accidents. The plant may not need such a high temperature, and it could be cooled by less expensivelycoolants.

The GCR liberates the stand-up position of the reactor. With the same figure in mind, a laid-down reactor may provide better flexibility to the designer, independence of location, a lower center of gravity, single shaft layout and other features, including potential cost reduction.

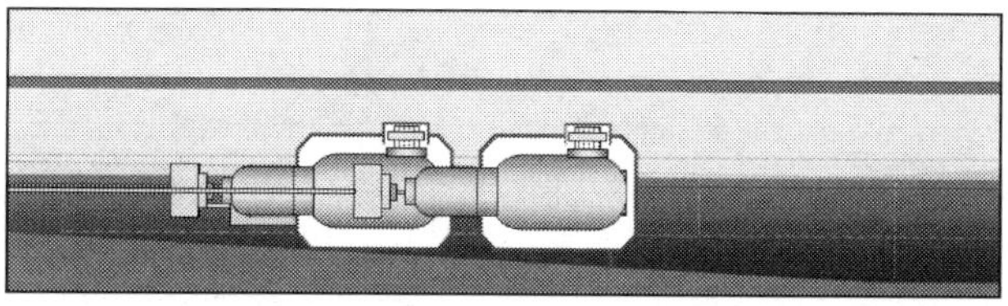

In summary, further studies are suggested in this topic, to explore more flexible nuclear propulsion solutions, along with fresh ideas to the nuclear community, required to develop systems that are less expensive, more versatile, responsible and environmentally friendly. In such a way nuclear power and other potential nuclear energy services may help to relieve the problem of climate change.

Authors' Closure

The authors would like to thank the many discussers for their insightful and valuable comments. They have helped us correct a few mistakes and clarify some of the more difficult-to-understand portions of our paper. More importantly, the combination of everyone's experience has provided a multifaceted collection of historical overview and valuable insight that is in itself a valuable educational contribution to the discussion.

We are going to use the authors' closure as an opportunity to answer many questions and delve a bit deeper into critical areas deserving of discussion. Unfortunately, owing to the nature of the paper's subject, there are certain topics that are too sensitive for this forum. The topics that we will discuss include:

• Innovative design features explored in the study
• Design tools and processes used or developed in this study
• Cost analysis, to include net present value and non recurring engineering costs

First, we would like to begin by reemphasizing that this study was not about determining a single best power and propulsion system architecture; rather, it was an opportunity to improve our design tools and knowledge base to improve our ability to support informed decisions. Ongoing science, technology, and research and development efforts being coordinated through the PEO Ships Electric Ship Office continue to build upon the foundational methodology developed in the paper.

As a review, energy requirements were developed for each ship type based on design reference missions (DRMs). The DRMs comprise tactical and operational situations that suggest an employment of energy management systems (energy storage, conversion, distribution, and thrust generation systems) in ship energy states linked to mobility, survivability, and warfare mission system energy demands. The DRM for each ship comprises speed-time probability distributions for each energy state. The mission-driven peak and lifetime energy demands were used in power and propulsion system architecting and systems engineering for each ship type and its associated energy management system component suite and were used to determine optimal plant component line-ups permitting life time energy demand to be linked to the DRM.

The study used a markedly higher ship service electric load for the medium surface combatant than any previous design studies to meet postulated ballistic missile defense mission requirements. This drove increased fossil fuel consumption rates for this ship. Consequently, the markedly increased fossil fuel usage spread the nuclear acquisition premium over a larger quantity of fossil fuel, decreasing the per-barrel breakeven cost. The range of operational tempos between the ship concepts drove differences in lifetime fossil fuel consumption rates and therefore breakeven costs.

In addition to developing breakeven costs, operational effectiveness tools were developed to facilitate more informed decision making, which turned out to be key to understanding alternatives with superior cost and operational performance.

Mr. Bennett and Dr. Genalis both address the fact that the operational profile and projected energy demand of a ship should be the key driver in the selection of propulsion plant architecture. The operational profile and energy demand need to be quantified at an early stage of the design so that it is possible to influence early design decisions. Plant selection criteria should favor options that provide the highest level of readiness and mission effectiveness with the lowest fuel consumption given the operational profile. These factors strongly influenced the ship designs presented in this paper. The nuclear ships were all designed with reactor cores capable of powering the ships for their full service life without refueling, reducing life cycle costs.

Professor Femenia identifies the benefit of developing a series of discreet sized primary nuclear plants that could be shared between multiple ships. The AP study used this approach adopting current Navy

Figure 1A: Plant Sizing for Small Surface Combatant

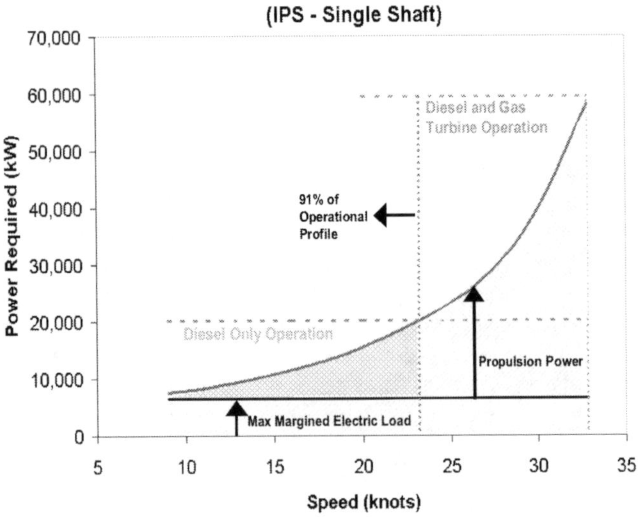

Figure 2A: Normalized Small Combatant Fuel Consumption

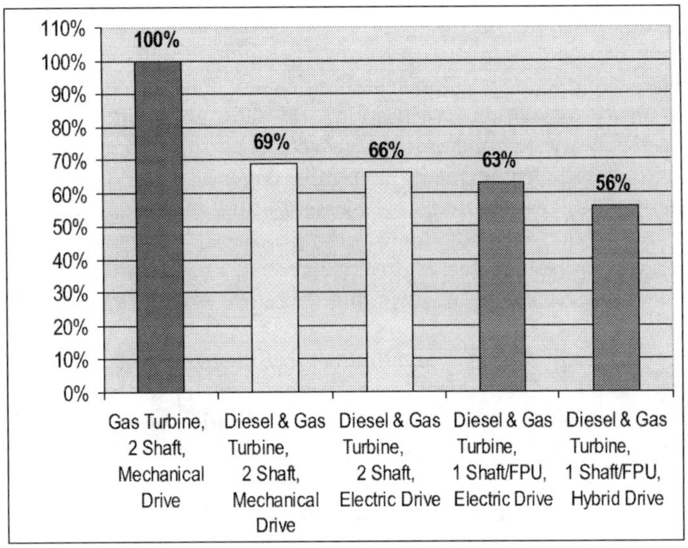

Figure 3A: Compartment Deactivation Approach Sample

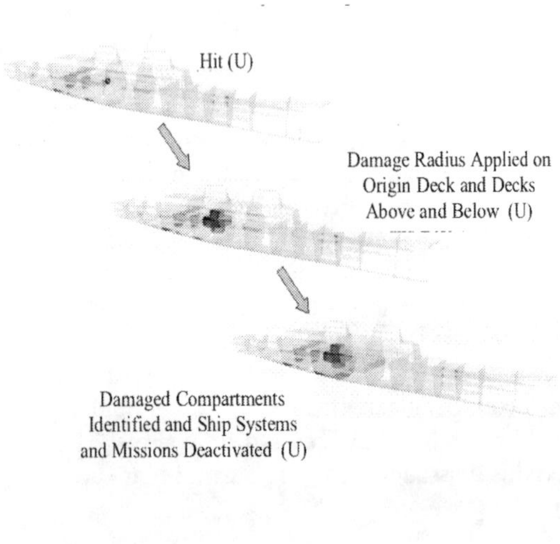

primary nuclear plants in each ship type. The nuclear reactor and primary propulsion plant components evaluated in this study are common with the CVN 78 Ford Class aircraft carrier for the MSC and amphibious warfare ship. The reactor and primary plant of the Seawolf Class submarine were integrated into the small surface combatant. The fossil fuel propulsion plants were designed to be efficient in the most probable speed range (5-20 knots) and electric load power levels. High efficiency is achieved by using a small "cruise" plant that is efficient at the lower speeds/power levels that constitute most Navy ship operations. A second high-power density "boost" plant is used to satisfy peak power demand. Figure 1A illustrates how this methodology was applied to a combined diesel and gas turbine powered small combatant with electric drive. Below about 23 knots this ship runs only on its efficient 20 MW diesel generator plant. When high speed is required, the lightweight gas turbine boost generators make the additional 40 MW required for its 30+ knot maximum speed.

It was found that the use of diesel engines instead of gas turbines as the prime movers in the "cruise" plant resulted in a 30% reduction in annual fuel consumption for both the small and medium surface combatants.

Mr. Bebar requested additional discussion of the benefits and limitations associated with a single shaft configuration incorporating forward propulsion units (FPU), specifically how they compare to a conventional twin shaft configuration. Several variants of the Small Combatant were designed with an FPU in lieu of a second shaft line as the redundant propulsion source. This forward propulsion unit is sized to provide a mission speed of 15 knots, which encompasses more than 85% of the small combatant's projected operational profile. The elimination of the second shaft line allows the single shaft line to run through the centerline skeg, significantly reducing appendage drag and improving hydrodynamic efficiency.

The single shaft with FPU configuration and diesel cruise engines resulted in a 44% reduction in annual fuel use over a traditional gas turbine powered twin shaft small combatant. Figure 2A shows the efficiency gains that can be realized in a small combatant. The single shaft/FPU ship design exhibited slightly lower acquisition cost and the lowest life cycle cost of the fossil fueled Small Combatants. The single shaft/FPU configuration was used for the Small nuclear combatants (medium and amphibious nuclear combatants have a twin shaft configuration). The efficient single shaft configuration allowed that small nuclear combatant to achieve a 30-knot sustained speed and a 25-year core life with a small nuclear power plant.

The longitudinal separation of the propulsion equipment (propeller aft and FPU forward) reduced vulnerability in the mobility mission area when compared to the twin shaft ship (all propulsors aft). This was determined using a model that combines a probabilistic hit distribution with a compartment deactivation methodology. Figures 3A illustrates this approach.

Mr. Kiss commented on the use of service life allowance, stating that it may lead to increased cost of naval ships. The authors concur that there may be some additional cost up front. That said, service life allowances ensure that ships can be modernized at their midlife point to prevent obsolescence and early retirement. This also ties into Mr. Karafiath's concern that the nuclear ships, although they have life of ship cores, may be obsolete before all of the nuclear fuel is consumed. Future Navy ships, especially those that have expected service lives longer than 20 years, should employ sufficient service life allowances, modular mission systems, and open architecture to allow the ships to be reconfigured to maintain relevancy in a changing world.

Mr. Bebar raises the issue of balancing fuel efficiency with signatures performance. A number of NATO member and partner nations have successfully balanced radiated electromagnetic and acoustic performance with fuel efficiency in their coastal defense littoral warfare patrol craft, corvettes, frigates and destroyers. The US Navy employs a blend of diesel and gas turbine prime movers on FFG-7 and LCS class surface combatants. The CODOG and CODAG architectures can isolate machinery noise and heat sources providing a balance between operational availability, mobility and survivability capability and cost.

Dr. Amy addresses the modeling limitation encountered by the AP study team in capturing all elements of total life cycle cost. Nonrecurring engineering (NRE) research and development and design costs were omitted from the cost estimate, as Dr. Amy points out. NRE design cost models are insufficiently sensitive to capture differences in system alternatives within common weight. The study team attempted to mitigate differences between NRE research and development costs of propulsion system alternatives by including only mature technology components; either demonstrated at full scale or fielded systems. Design complexity associated with total ship computing, integrated electric propulsion and nuclear power that are expected to create differences in design cost are acknowledged as a future challenge for warship NRE cost estimation.

Mr. Bennett identifies warfare mission system energy system loads as design opportunities rather than fixed outcomes effecting ship energy management system architecting and ship synthesis. The authors agree that the design of experiments approach can be used to determine the influence of warfare mission subsystem design and operational employment on energy management system acquisition cost and ship life cycle cost.

Further, to answer Mr. Bennett, a low operating profile (zero major combat operations plus normal training and presence type operations) was developed; however, it was not presented in the final analysis, as history does not bear it out. From the end of World War II through present day, the US Navy has been engaged in major combat operations or equivalent levels of military deployment within any particular seven-year period modeled in the study.

Mr Bennett further comments on cost estimation uncertainty, suggesting that engineering methods used to model performance result in higher confidence predictions. The authors acknowledge the impact of cost uncertainty on the inferiority/superiority of the technologies modeled in the study. Equipment and architectures that have not been integrated into a ship system will be burdened with high cost margins, reflective of the low level of confidence associated with technology maturation and undefined complexity of design integration. The cost uncertainty alone will make new technologies uncompetitive when compared to existing systems of lower capability but where high confidence in cost estimation is available from return data. The point is well taken and the authors concur that additional sources of uncertainty should be captured in the systems engineering processes used to holistically evaluate cost versus operational sufficiency. The methodology used in the AP study exposes requirements sensitivity as well as the impact of technical modeling uncertainty. Risk-based cost estimation processes being adopted by the NAVSEA cost engineering group capture individual sources of cost risk and their impact on overall system cost. The risk-based methodology is being used to develop investment strategies that balance capability improvements with R&D investment, thus getting to the core of Mr. Bennett's issue.

Mr. Garzke provides important historical prospective in the areas of cost, operability, strategic versus tactical decisions, and lessons learned.

Dr. Genalis discusses the inherent challenges of the need to characterize the entire fleet to be able to reach conclusions on specific costs associated with a member of that fleet. The authors concur and are participating in the Naval Sea Systems Command initiative to develop assessment tools that evaluate the impact of individual ship class capabilities on fleet groups to the determine optimum cost versus operational sufficiency.

A paper accepted for publication in the Journal of Ships and Offshore Structures (Koenig, Czapiewski, and Hootman, 2007) describes an in-house, interdisciplinary study of alternative fleet architectures with a goal of addressing the high cost and extended duration of naval vessel design and construction. This paper describes the approach used and variables studied in the naval architectural and force architecture components of the study, i.e., the fleet synthesis and analysis part. The methods and findings of other recent studies in this subject were reviewed, and an approach was developed, which includes capabilities requirements, ship design and cost, plus quantitative tracking of the long-term evolution of the fleet as ships are introduced and old ones are retired.

With acquisition cost reduction in mind, a systematic, quantitative approach to synthesizing an alternative naval fleet was developed that takes account of the main drivers: naval architecture, shipboard warfare systems, ship service life, new construction program planning, and naval operations in peacetime and wartime. Two key features of this study were: (1) explicit checking of ship designs and fleet architectures for peacetime presence and warfighting sufficiency; and (2) tracking of the retirement of older ships and their replacement with future concepts. This highlighted the lag between initial ship design, and realization of a fleet in which that design is not only present, but is the dominant component in its type. These two features are absent from many other fleet architecture studies in the literature, and failure to consider them can encourage an inadequate appreciation of the magnitude of effort required to affect large changes in naval force architectures, given a large existing navy.

The alternate propulsion study built off of this effort, going the next step to marry the force structure and ship design with cost and performance trade offs. Doing so has allowed the study team to evaluate warfighting performance characteristics as well as determine potential energy demand signals of individual ships in a total force construct.

Mr. Harrington provides the prospective of the design agent in identifying complexities identified during later stages of design that are often overlooked in sensitivity studies performed at the concept design level. Ship synthesis modeling performed using ASSET provides a balanced design that relies on user interaction to verify that system, subsystem and component relationships are valid. Sensitivities to vulnerability and integrated power and propulsion system architecture implementations were performed external to ASSET after a balanced design had been achieved; a converged synthesis loop had been achieved. The latest version of ASSET has migrated to an object oriented design environment permitting increased resolution models to be integrated into the synthesis process. Studies being conducted by NAVSEA are automating vulnerability model variable interfaces with IPS design code providing a holistic assessment of Power Quality of Service within the synthesis loop. Additional Next Generation IPS program planned efforts being executed by the Electric Ship Office of PEO Ships will integrate deactivation diagram logic into compartment arrangements and distributive system service. As recommended by Mr. Harrington, an independent design assessment is required to validate automated design decisions made by coded optimization processes.

A number of discussers raised questions concerning the potential applicability of a net present value type analysis to this study. Net present value (NPV) is a commonly used method to measure alternate investment strategies, by discounting the effects of inflation and the potential of alternative investments, such as Treasury bills and U.S. bonds. OMB Circular A-94 provides guidance for these calculations. In turn, OSD provides specific interpretation for defense weapons systems. Shipbuilding has been granted relief from OSD/OMB inflation guidance owing to its unique inflationary pressures. Hence, a single NPV analysis can become complicated. The alternate propulsion study uses a shipbuilding-unique interpretation of OMB Circular A-94.

OMB Circular A-94 Definition

NPV analyses use a process that discounts future cash flows in order to transform costs occurring in different time periods to a common dollar-year basis for comparison. OMB Circular A-94 "Guidelines and Discount Rates for Benefit-Cost Analysis of Federal Programs" defines Net Present Value (NPV) as "the discounted monetized value of expected net benefits (i.e., benefits minus costs)." The NPV is computed by assigning monetary values to benefits and costs, discounting future benefits and costs using an appropriate discount rate and subtracting the sum total of discounted costs from the sum total of discounted benefits. Discount rates are based on the interest rates for Treasury notes and bonds (government's cost of borrowing).

DoD Guidance for Net Present Value

DoD Instruction 7041.3 "Economic Analysis for Decisionmaking" details the policy for conducting costeffectiveness economic analysis used to evaluate decisions concerning acquisition investment alternatives. The results of the alternate propulsion study were not intended to be used as the basis for determining final investment decisions between alternative ship platforms, but instead to present a series of breakeven analyses comparing the recurring life cycle cost of nuclear versus fossil fuel powered ships over a continuum of dollars per barrel of crude oil. More detailed review would be required to support specific NPV analysis, hence conclusions that could be drawn from a partial NPV analysis may be misleading without these assessments.

Alternate Propulsion Study

By example, a full NPV analysis requires a complete life cycle cost analysis of competing alternatives. The alternate propulsion study trades off comparable recurring life cycle costs of specific nuclear and fossil fuel powered follow ship concepts in constant dollars. The scope of the study did not assess procurement quantities, development costs of nonpropulsion technologies, potential beneficial impacts to the nuclear industrial base, nor the cost and time required to reestablish a nuclear surface ship construction capability. In addition, costs were not normalized to reflect the quantified benefits for the nuclear variant of increased operational sustainability when on station and improved timeliness in surge operations; this could potentially lead to fewer ships needed within a ship class to support fleet requirements. The future availability of crude oil and potential impacts on fleet operations were also not assessed. The results of the study are meant to indicate conditions where nuclear propulsion could be considered a viable alternative for future analysis, rather than determine specific recommendations concerning future investment decisions.

Net Present Value at 0%: Unique Shipbuilding Inflation

It has been recognized that annual OMB inflation guidance markedly underestimates shipbuilding inflation. In 2004, ASN(RDA) Mr. John Young directed that shipbuilding programs budget for realistic inflation projections rather than OMB guidance. Congress has subsequently appropriated ships to these inflation projections, and the Congressional Budget Office uses this index for their projections of ship costs. Per OMB Circular A-94, real discount rates (assessing the time value of an investment) can be approximated by subtracting expected or actual inflation from a nominal interest rate. OMB Circular A-94 Appendix C, revised January 2007, states a nominal discount rate of approximately 5% for periods of 3-30 years and a real rate of 3% for internal government investments (5% less the 2% annual inflation guidance). Actual shipbuilding inflation is 4.5%. When adjusting the nominal interest rate of 5% by subtracting for actual shipbuilding inflation, this effectively equates to a real discount rate of zero percent. Thus, the alternate propulsion study has reflected this zero real discount rate.

NPV Results at 3% Discount Rate

While widely accepted shipbuilding indices support a lower real discount rate than standard policy, NPV analysis using a real discount rate of 3% can be used for evaluating DoD investment decisions. DoDI 7043.1 directs the use of discount rates contained in Appendix C of OMB Circular A-94 for determining NPV of DoD investment alternatives. The AP Study estimated costs in Constant Year 2007 dollars, thereby removing inflationary bias over the project life. Analyses that involve constant dollar flows are discounted using real discount rates. From OMB Circular A-94 Appendix C revised January 2007, the current real discount rate for projects with durations greater than 20 years (as those in the AP Study) is 3%.

Medium surface combatants were identified in the AP Study as the most compelling case for more detailed study. A life cycle cost premium of 0-10% was identified for nuclear propulsion follow ships compared to fossil fuel propulsion. Using a 3% discount rate for NPV, this premium is 4-13%, well within the estimating error for concept designs.

NPV Results at 7% Discount Rate

OMB Circular A-94 specifies the use of a real discount rate of 7% to determine NPV for public investments that benefit society as a whole, such as Federal highway projects. Per DoDI 7043.1, "Societal costs and benefits outside the Federal Government are usually not included in a DoD analyses." Therefore, the standard 7% discount rate specified in OMB Circular A-94 for public investments providing benefits to the general public would not be appropriate in DoD analyses such as the AP Study.

Mr. Karafiath also asks about any other promising alternate power and propulsion technologies; we would refer him to a complementary paper to this one presented at ASNE Day 2007 (Webster, et al. 2007).

Mr. Keane identifies the value of learning ship design through practice in multi-disciplinary teams with experienced designers, referring to key foundation al capabilities for designing cost effective warships. Lamb et al. (2007) describes the education of naval ship designers in the U.S. Other efforts underway by NAVSEA include the University of Michigan Summer Naval Surface Ship Design Program (SNSSDP) and MIT Professional Summer Program. The SNSSDP and Professional Summer Program effectively meets defense and commercial market human capital needs for personnel development and career skills refreshment. The course curricula are aimed at providing the warship design work force with the capabilities needed to design cost effective and innovative naval ships.

Building further off of Keane's discussion, a particular system architecture consists of a selection of the types of system elements, their characteristics, and their arrangement (International Council on Systems Engineering 2004). Systems architecting is the synthesis of selected elements within the bounds of design objectives. In the case of alternate propulsion systems, energy storage elements and subsystems, energy conversion systems, energy distribution systems, and mobility systems were synthesized within power quality, vulnerability, system efficiency, and cost objectives. Systems architecting tools used in the AP studies relied upon deterministic performance objectives in each operational objective area. The integrated power and propulsion system architecting tools balanced warfare mission systems and propulsion system energy loads using US Navy deterministic power quality and power quality of service criteria. Alternative energy management system architectures perturbed segregation, redundancy and power quality within a common set of energy management system requirements.

A design of experiments (DOE) methodology was used in alternate propulsion studies to quantify cost, performance, and operational sufficiency differences between peer alternate propulsion system architectures and components integrated into small and medium surface combatants and amphibious warfare ships. A multivariate approach perturbed vulnerability and mobility capabilities holding warfare mission systems constant across a ship-type baseline. Dominant solutions suggested by the DOE response variables were synthesized using the same propulsion and power system architecting and ship synthesis tools used to develop DOE data set elements to validate the cost and operational sufficiency response.

A number of discussers raised questions concerning build strategy. This has been topic has been a topic of much interest, and was discussed at length in March 1, 2007, by the House Armed Services Subcommittee on Seapower and Expeditionary Forces Hearing on Integrated Nuclear Power Systems. An excerpt from the transcript of that hearing is included as follows:

Rep. Taylor:

Keeping in mind that we have five surfacecombatant shipyards, or capable shipyards at this moment. To what extent could the other three participate in bidding on these? To what extent could a very large portion of the hull be built and then towed to either one of the two nuclear yards for the nuclear components? How expensive is it for a yard that is not at the moment nuclear-certified to become nuclear-certified? How big a hurdle is that?

ADM Donald:

I will go ahead and address the piece of that about what it takes to certify and to reestablish or establish another nuclear-capability yard, and then I will ask that Secretary Etter and probably Admiral Sullivan discuss the second piece of it. Just the basics of what it takes to have a nuclear-certified yard, to build one from scratch, or even if one existed once upon a time as it did at Pasacagoula, and we shut it down, first and foremost you have to have the facilities to do that. What that includes, and I have just some notes here, but such things as you have to have the docks and the dry-docks and the pier capability to support nuclear ships, whatever that would entail. You would have to have lifting and handling equipment, cranes, that type of thing; construction facilities to build the special nuclear components, and to store those components and protect them in the way that would be required.

The construction facilities would be necessary for handling fuel and doing the fueling operations that would be necessary on the ship -- those types of things. And then the second piece is, and probably the harder piece other than just kind of the brick-and- mortar type, is building the structures, the organizations in place to do that work, for instance, nuclear testing, specialized nuclear engineering, nuclear production work. If you look, for instance, at Northrop Grumman Newport News, right now, just to give you a perspective of the people you are talking about in those departments, it is on the order of 769 people in nuclear engineering; 308 people in the major lines of control department; 225 in nuclear quality assurance; and then almost 2,500 people who do nuclear production work. So all of those would have to be, you would have to find that workforce, certify and qualify them, to be able to do that.

And then finally you have to train them, obviously, and then qualify them to do the work. So my view of this is we have some additional capacity at both Electric Boat and at Northrop Grumman Newport News. My primary concern is if we are serious about building another nuclear-powered warship, a new class of warship, cost is obviously going to be some degree of concern, and certainly this additional costs, which would be -- and I don't have a number to give you right now, but I think you can see it would be substantial to do it even if you could. It probably doesn't help our case to move down the path toward building another nuclear-powered case, when we have the capability existing already in those existing yards. And then I would turn it over to...Admiral Sullivan....

VADM Sullivan:

Sir, we are building warships in modular sections now. So if we were going to, could you assemble this, could you build modules of this ship in different yards and put it together in a nuclear-certified yard, the answer is yes, definitely, and we do that today with the Virginia Class. As you know, we are barging modules of submarine up and down the coast.

What I would want is, and sort of following along

with what Admiral Donald said, you would want the delivering yard to be the yard where the reactor plant was built, tooled, and tested, because they have the expertise to run through all of that nuclear work and test and certify the ship and take it out on sea trials. But the modules of the non-reactor plant, which is the rest of the ship, could be built theoretically at other yards and barged or transported in other fashion to the delivering shipyard. If I had to do it ideally, that is where I would probably start talking to my industry partners, because although we have six shipyards, it is really two corporations, and those two corporations each own what is now a surface combatant shipyard and they each own a nuclear-capable shipyard. I would say if we were going to go do this, we would sit down with them and say, you know, from a corporation standpoint, what would be the best work flow? What would be the best place to construct modules? And how would you do the final assembly and testing of a nuclear-powered warship?

Mr. Kiss requests additional definition of the cost comparison between AC and DC medium voltage systems when installed in the medium surface combatant. This particular comparison was not part of the study, however AC and DC low voltage systems were compared for the medium surface combatant.

The notional DC ZEDS zonal architecture developed for this study is shown for illustrative purposes in Figure 4A and illustrates the level of fidelity reflected in the low voltage distribution system architecting.

The DC low voltage system had a weight and cost premium for as yet unquantified benefits in Quality of Service (QOS) and vulnerability. Also, the reductions in weight and cost in the mission system equipment possible due to the enhanced QOS and vulnerability performance of the power system were not accounted for. Future efforts will provide quantification of performance improvements and attempt to capture load equipment impacts.

Mr. Kiss requested a clarification of Charts 23, 24, and 25. Unfortunately the original text was incorrect as he noted. The text published in conjunction with this closure has been corrected.

Figure 4A: DC ZEDS Centralized Voltage Conversion

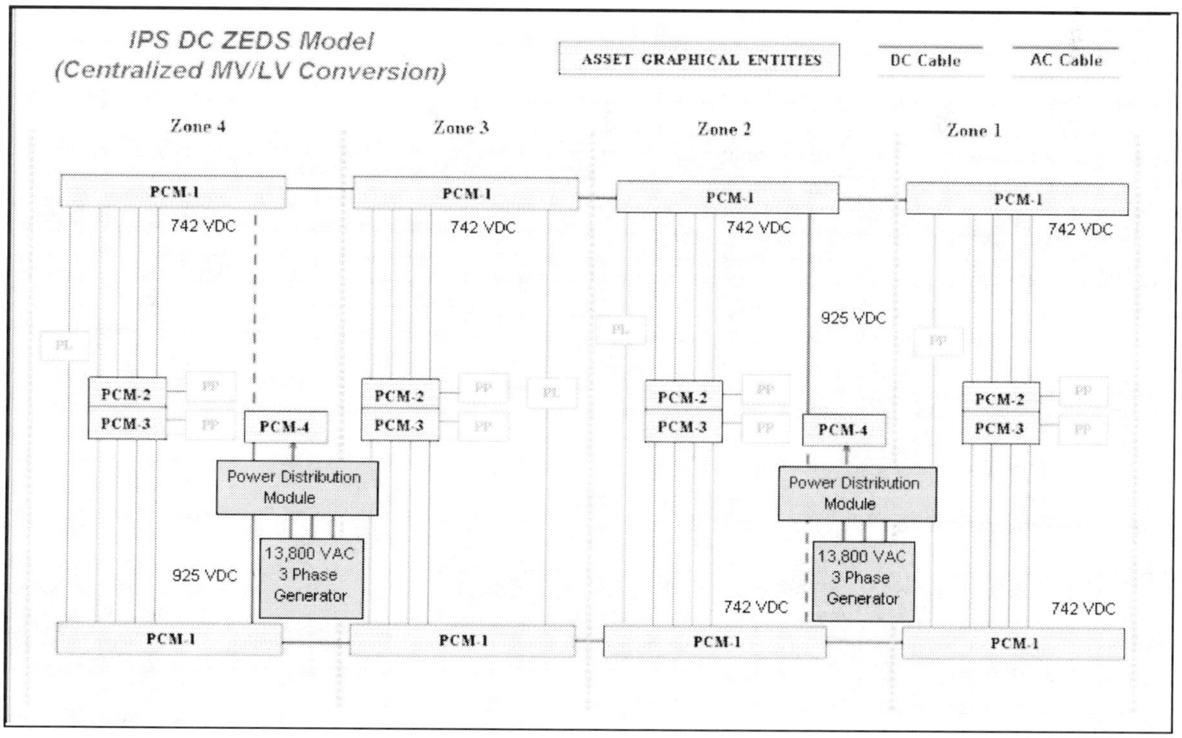

Dr. Palmer suggests the addition of a speed versus efficiency graph to assist in propulsion system architecting. The authors concur that this is a good approach that when integrated into the AP study methodology will provide an evaluation of the time weighted efficiency versus mission system load.

We thank Captain Tibbitts for catching an oversight in our definition of life-cycle cost, which was the exclusion of acquisition costs in the text only. Acquisition costs included actual cost return data or vendor quotes for power and propulsion system material, when possible, due to the maturity of power and propulsion system equipment. The LCC estimates incorporated acquisition cost, inactivation, defueling (nuclear variants only), disposal, burdened fuel costs, manpower costs, and maintenance.

The Report to Congress presents a breakeven analysis using a fifth ship life cycle cost perspective. As such, the results are meant to merely indicate conditions where nuclear propulsion should be considered in future analyses. These future analyses will trade off nonrecurring costs with overall force structure requirements, such as quantity, to determine specific breakeven points. Nonrecurring costs were not specified as they are dependent on capability growth (both military and industrial) which is outside the scope of this study. Further, a breakeven analysis amortizes non-recurring costs over a given quantity over an acquisition timeline. Because the scope of the study did not assess quantity, the development cost of non-propulsion technologies, the beneficial impacts to the nuclear industrial base, nor the costs and time required to re-establish a nuclear surface ship construction capability, the results presented herein are meant to merely indicate conditions where nuclear propulsion can compete in future analyses. More detailed review would be required for specific tradeoffs.

Further to Captain (ret) Vergara comment on alternative liquid hydrocarbon fuel sources, a product generated by Fischer Tropsch processes has been successfully tested by the Air Force in gas turbine prime movers and, when blended with fossil derived fuel sources by the Army, on diesel prime movers. Bio fuels have been evaluated by the Navy and determined to present challenges in the marine environment, such as resistance to mixing with seawater. Koenig, Czapiewski, and Hootman (2007) evaluates of a broad range of energy sources and energy conversion systems, their technology readiness level and applicability to current and future US Navy ships. Breakeven cost studies did not evaluate the source of liquid hydrocarbon.

The introduction of synthetic and other fossil derived fuels is reflected in driving total fuel supply and will be market driven.

In closing, the authors would like to thank the many discussers for their insightful and valuable comments, as well as the Society for the opportunity to discuss these timely topics.

Reference

Koenig, Czapiewski, and Hootman 2007. "Synthesis and Analysis of Future Naval Fleets," Journal of Ships and Offshore Structures, accepted for publication, Fall.

Webster, Fireman, Allen, Mackenna, and Hootman 2007. "US Navy Studies on Alternative Fuel Sources and Power and Propulsion Methods for Surface Combatants and Amphibious Warfare Ships," ASNE Day.

Lamb, Doerry, Webster, and Hootman 2007. "The Education of Naval Ship Designers in the U.S.A.," 2nd World Scientific and Engineering Academy and Society's International Symposium on Advances in Naval Science, Research and Education, Venice, Italy, November 21-23.

2004. International Council on Systems Engineering Systems Engineering Handbook, INCOSE-TP-2003-016-02, Version 2a, June 1.

Tradition and Innovation: The Design of the Ferry *Island Home*

Douglas M. Wolff, (M)

Island Home *is a new ferry operating between Woods Hole, Massachusetts and Martha's Vineyard Island and represents a leap in capabilities in The Steamship Authority's (SSA) fleet. This new double-ended ferry is significantly larger than other vessels in the SSA fleet, and features Americans with Disabilities Act compliance, lift decks, bow and stern thrusters, weathertight doors at both ends of the freight deck and a slide evacuation system. The design process started with a seakeeping analysis for the numerical modeling of realistic wave shapes and periods, which were recreated in a model basin. Tests were performed to predict ship motions and to confirm powering predictions. Numerical flow analysis was performed to identify hull form improvements. Studies assisted with decisions regarding styling, propulsion system, and the number of pilothouses. This paper follows the design process, construction period and sea trials.*

INTRODUCTION

Martha's Vineyard Island, lying 4.5 miles south of Cape Cod, is a study in contrasts. The winter population of 12,200 swells to 72,600 in the summer, when the rich and famous return to their multi-million dollar cottages and day-trippers fill the streets and stores. Islanders place high value on their privacy but depend on summer tourist dollars to carry them through the lean months of winter.

Proud of their centuries-old traditions, island residents are also quick to take advantage of new technologies. They are fiercely independent in spirit yet dependent upon the ferries for delivery of food, mail, building supplies and even medical care.

Year-round ferry service to Martha's Vineyard is provided by The Woods Hole, Martha's Vineyard and Nantucket Steamship Authority (SSA), an organization chartered, but not supported financially, by the Commonwealth of Massachusetts. SSA operates a fleet of vessels on the following routes:

- Woods Hole to Vineyard Haven, Martha's Vineyard
- Woods Hole to Oak Bluffs, Martha's Vineyard
- Hyannis to Nantucket

Particulars of the vehicle/passenger ferries operated by SSA are provided in Table 1. Note that the *Islander* was retired in March, 2007 when the *Island Home* entered service on the Woods Hole to Vineyard Haven run.

NOMENCLATURE

ABS	American Bureau of Shipping
ADA	American with Disabilities Act
Alpha	Coefficient dependent on fetch and wind speed
AMR	Auxiliary machinery room
CFD	Computational fluid dynamics
CFR	Code of Federal Regulations
CL	Centerline
CORD	Cape Organization for the Rights of the Disabled
CR	Residuary resistance coefficient
EBDG	Elliott Bay Design Group
EOS	Engineer operating station
FEA	Finite element analysis
G	Gravitational constant
Gamma	Wave shape parameter
GM	Vertical distance from the vessel center of gravity to the metacenter
HP	Horsepower
Hs	Significant wave height
HVAC	Heating, ventilation and air conditioning
IMO	International Maritime Organization
JONSWAP	A specific sea spectrum typically used for near shore applications
LULA	Limited Use Limited Availability
LT	Long ton, equal to 2240 pounds
MARSEC	Maritime Security
MCR	Mean continuous rating
NCE	Noise Control Engineering
NOAA	National Oceanographic and Atmospheric Administration
OCMI	Office in Charge of Marine Inspection
P_D	Developed power
PE	Effective power
PH	Pilothouse
RMS	Standard deviation (root mean square)
RPM	Revolutions per minute
SSA	The Woods Hole, Martha's Vineyard and Nantucket Steamship Authority
T_P	Wave period
USCG	United States Coast Guard
VTHM	VT Halter Marine

Table 1: Existing Vehicle/Passenger Vessels in the SSA Fleet

Vessel	Year	LOA	Beam	Vehicles /Pax	Speed, kts	HP
Islander	1950	201'	58'-1"	48/788	11.5	1800
Governor	1954	242'	64'-1"	65/241	12	1200
Nantucket	1974	230'	60'-0"	52/789	14	3000
Eagle	1987	233'	61'-6"	55/800	14	3000
Martha's Vineyard	1993	230'	60'-0"	55/1387	14	3000
Gay Head	1981	235'	40'-0"	/138	13	3050
Katama	1982	235'	40'-0"	/142	13.5	3050
Sankaty	1981	197'	40'-0"	/292	12.5	2520

While not sister vessels, the *Martha's Vineyard* and *Islander* (Figure 1) have been paired in recent years to provide regular service over the 6.2 mile Woods Hole / Vineyard Haven route and the 7.5 mile Woods Hole / Oak Bluffs route. With daytime departures every 75 minutes, these two workhorses make about 9700 trips annually, totaling 63,000 miles. Their regularity and dependability have exemplified the SSA motto "Lifeline to the Islands."

Fig. 1: *Islander*

Islander was built in 1950 and has served SSA and its patrons dependably for these past 57 years. *Islander* is a double-ended vessel with a capacity of 788 passengers and 48 vehicles. Hinged doors at each end protect against boarding seas and freezing spray. Unfortunately, obsolete equipment and degraded materials have driven up operation and maintenance costs so that replacement became an obvious requirement.

SSA began the replacement process in 1997, working with a consulting naval architect to produce a concept design for a new single-ended vessel based on the designs of the *Nantucket* and *Martha's Vineyard.* This replacement design featured a capacity of 1400 passengers and 55 vehicles. However, the political dynamics within SSA changed, leading to the shelving of this first design when the island residents demanded a double-ended design.

As time passed and the *Islander* became increasingly expensive to operate, the urgency to restart the design and acquisition process became acute. Accordingly, SSA issued a Request for Proposal (RFP) to several naval architecture consulting firms that had been researched and judged capable of designing the replacement vessel. Each firm was invited to ride the SSA ferries so as to acquire a basic understanding of SSA's mission, as well as its opportunities and challenges, in providing its services.

Competing firms each submitted a proposal in response to the RFP. After verifying references, SSA conducted interviews and based their selection on a number of factors, including:

- Experience in ferry design
- Technical approach outlined in the proposal
- Demonstrated understanding of SSA requirements
- Firm size and depth of expertise
- Firm location
- Experience of proposed design team members.

Elliott Bay Design Group (EBDG) was pleased to be selected from the group of well-qualified naval architectural firms. As the design team formed and the lines of communication between SSA in Woods Hole and EBDG in Seattle were established, project managers on both ends reviewed, discussed and refined the fundamental design criteria for the new vessel. The established criteria were:

- Capacity for 1200 passengers with 600 inside seats
- Capacity for 70 vehicles
- Double-ended
- Designed in general compliance with American Bureau of Shipping rules
- Compliant with ADA guidelines for passenger vessels, insofar as possible
- 16 knots top speed and 14 knots cruising speed
- Hull shape to fit existing terminals
- Exceptional maneuverability
- Excellent security for the protection of passengers and crew
- 10'-6" loaded draft
- Excellent seakeeping characteristics in the steep choppy seas of Vineyard Sound

Because the existing terminals have single lane ramps, relatively long loading and unloading times are an accepted fact of operation. SSA did specify, however, that the freight deck arrangements needed to be as open as possible to accommodate tractor trailer units and to simplify car loading and parking.

One unusual specification was that the new vessel carry no more than 10,000 gallons of petroleum products so as to avoid USCG regulations pertaining to fuel transfer and loading. Because SSA vessels are dockside every night, refueling from trucks is simple to schedule and poses no operational difficulties.

Americans with Disabilities (ADA) accessibility played an important role during the development of the arrangements and was driven not only by the obvious need to provide reasonable accommodation, but also by an agreement between SSA and the Cape Organization for the Rights of the Disabled (CORD) (Reference 1). In this agreement, SSA committed to design and operate any new vessel in conformity with ADA standards, insofar as possible. SSA was fully aware that CORD would hold them to the agreement, and would be inspecting the finished vessel carefully to verify compliance.

The *Islander*'s 11-knot speed results in a one-way trip time of 75 minutes (45 minutes in transit and 30 minutes for loading and unloading), resulting in departures at seemingly random times throughout the day. A cruising speed of 14 knots, if matched by the other vessel on the Vineyard Haven run, would permit hourly sailings, increasing the number of trips daily while making departure times easy to remember.

Although exceptional maneuverability and excellent seakeeping characteristics were requirements for the new vessel, no objective goals were provided. Instead, SSA elected to rely upon the experience of the design team, augmented by computational fluid dynamics calculations and model tank testing, both of which will be addressed later.

Recent heavy investments in terminal pilings, dolphins and fenders, coupled with the fact that other SSA vessels are not scheduled for replacement for some years, mandated that the new vessel fit into the existing slips with single lane boarding ramps. SSA arranged for an accurate survey of each slip so that the shape of the vessel ends could be precisely specified.

Description of Terminal Harbors

Vineyard Haven Harbor (Figure 2) is well protected by a breakwater and offers no challenges from current or prevailing wind. The channel has at least 14 ft water depth; however, it is very congested in the summer, with expensive yachts crowding the channel leading to the terminal. Wake wash is a concern, and the desire to eliminate the need to turn around in the harbor led to the demand from the island community that the vessel be double-ended. In addition to eliminating the risk involved with the turning operation, several minutes of voyage time would be eliminated.

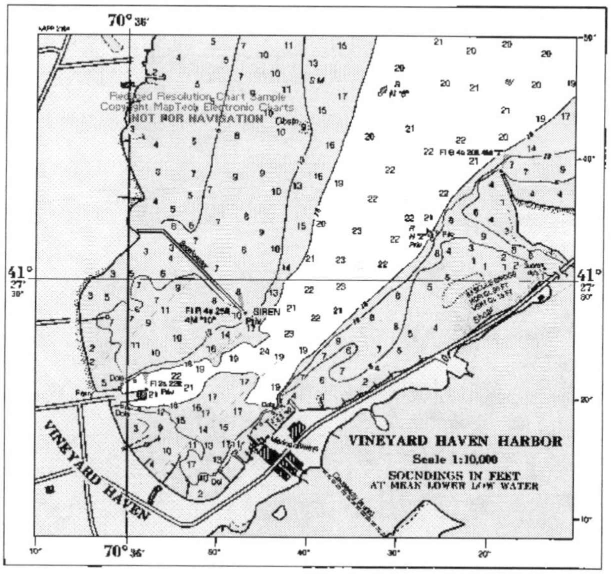

Fig. 2: Vineyard Haven Harbor

The Oak Bluffs terminal (Figure 3), located on an exposed shoreline, is subject to strong currents and cross winds. In the worst conditions, ferries are diverted to the Vineyard Haven terminal. At normal low tides, however, the maximum safe draft at Oak Bluffs is 10'-6"; hence the design draft requirement.

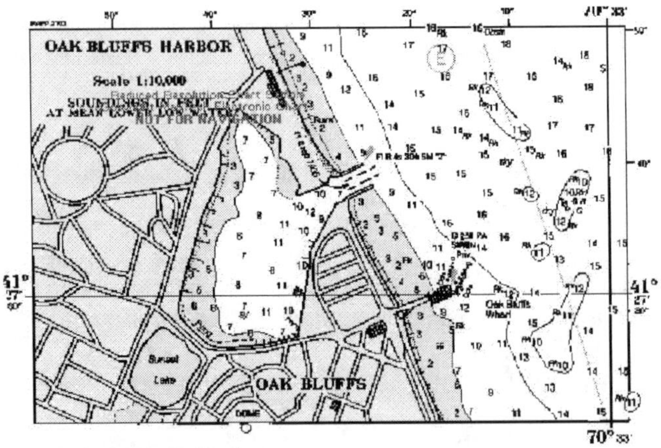

Fig. 3: Oak Bluffs Harbor

The terminal at Woods Hole (Figure 4) is commonly subject to gale force winds in the winter and to tidal cross currents approaching 4 knots. Both wind and current directions are typically perpendicular to the line of approach during docking, leading to what has been described by SSA insiders as a "controlled crash" method of docking the *Islander*. That this is not exaggeration is verified by the regular need for repairs to the fenders, dolphins and to the dock itself. Drift ice from Buzzards Bay presents yet another source of concern during winter months.

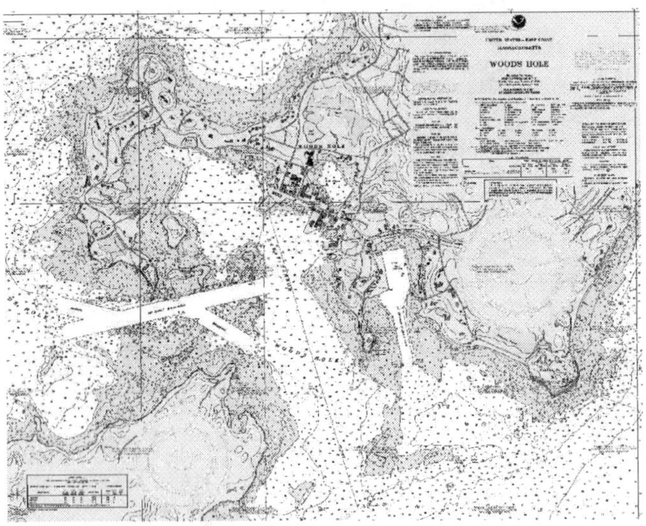
Fig. 4: Woods Hole Harbor

Vineyard Sound is shallow, with an average mean low water depth of about 25 feet. As mentioned above, the Sound is also subject to strong tidal currents and to strong winds. This combination of factors frequently leads to steep, choppy wave conditions which can cause severe discomfort among the riders. As double-ended vessels are not generally noted for their seakeeping performance, extra effort would be required to marry the propulsion and seakeeping requirements. This effort is described later in this paper.

Styling Study

Generations of island residents have grown up with the *Islander* as one of the primary means of transportation to and from the mainland. As the ferry has aged, its quirks and flaws have become in the public mind lovable idiosyncrasies, and SSA staff heard from many quarters that the new ferry should be "just like the *Islander*." Desired or not, a defacto gold standard had been established by the ridership, and SSA was determined to achieve that standard while still acquiring a vessel that would comply with current regulations and serve well for the next 40 years or more.

Understanding that both the operators and the ridership had a keen interest in the new vessel, SSA decided to include both groups in the design process by soliciting comments during the concept design phase. In addition to postings on the SSA web site, three public meetings were held on Martha's Vineyard at one-month intervals. At each meeting, the EBDG project manager presented a slide show highlighting design features expected to be of general interest, and then fielded questions from the very interested attendees. SSA personnel observed at each meeting and noted trends in the comments. As appropriate, design modifications were incorporated to reflect public concerns and desires.

To accommodate the larger traffic volumes, the new ferry would necessarily have to be larger than the *Islander*. The requirement for the new vessel to be double ended was included to eliminate the additional time and perceived risk incurred while turning the vessel around in the congested Vineyard Haven Harbor, but forced an entirely different look which, it was feared, could displease island residents.

It was therefore decided to produce a styling study and to solicit input regarding the appearance. Four outboard profiles were produced (Figure 5):

- Modern styling with a pilothouse at each end
- Modern styling with a single pilothouse amidships
- Traditional styling with a pilothouse at each end
- Traditional styling with a single pilothouse amidships

It is apparent from the four profiles that SSA was grappling not only with the need to appeal to those who were keenly interested in appearance, but also with the issues of pilothouse visibility and security. A single pilothouse amidships would offer the advantages of improved operational control (no need for the Master to walk from one end of the vessel to the other when changing directions in an emergency) and security (Master is not exposed during the transition). A pilothouse at each end would offer the advantages of traditional appearance and improved visibility forward during the notorious Vineyard Sound foggy conditions, as well as when approaching the slip.

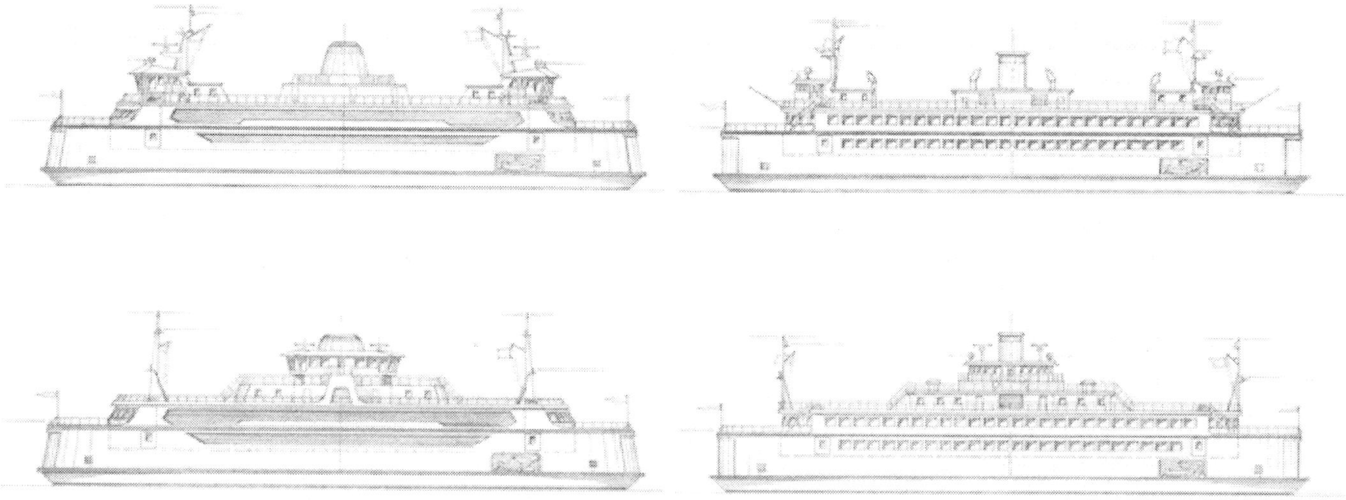

Figure 5: Styling Study Outboard Profiles

A simple comparison of the visibility "shadow" is shown in Figure 6. While both the single- and two-pilothouse versions have adequately small shadows, the two-pilothouse version has a much shorter distance to the water and so will offer better visibility in fog conditions.

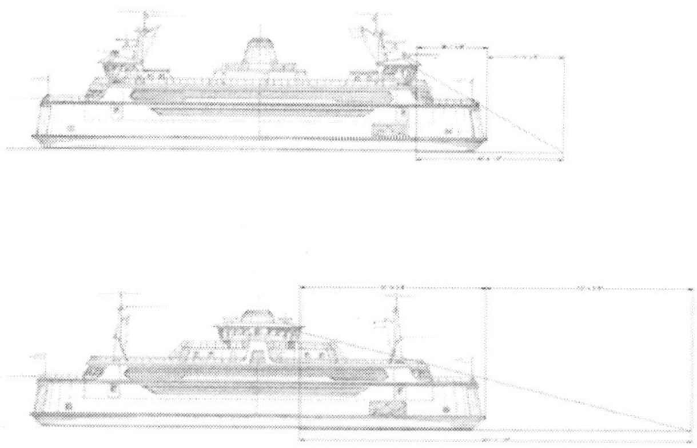

Fig. 6: Visibility Sketches

Table 2 was produced to compare visibility from the pilothouse of the existing ferry *Nantucket* with the single- and two-pilothouse options. The *Nantucket* was selected for comparison as the vessel operators generally agreed that its visibility was adequate in both dark and foggy conditions. The results of Table 2 strengthened the argument that an amidships pilothouse offered acceptable visibility.

Table 2: Visibility Comparison

	PH aft of stem	Shadow distance to target on water	PH to target on water
Nantucket	81'	149'	230'
Fore & aft PH's	39'	42'	81'
Amidships PH	101'	113'	214'

A new issue, pilothouse security, also needed to be considered. At all Maritime Security (MARSEC) levels, it is necessary to control access to the bridge. This requirement is easily met through the use of pass-key or cipher-lock doors. At MARSEC levels 2 and 3, the bridge team must be able to move freely and securely from one pilothouse to another (in the two-pilothouse version) without direct passenger interaction.

If the two-pilothouse version were selected, there are two obvious means of compliance with the security requirement:

1. Deny passenger access to the weather deck between the pilothouses.
2. Provide an enclosed crew passageway between the pilothouses on or above the weather deck.

The first of these options was deemed to be the better of two bad choices, as the enclosed passageway would divide the 03 Deck in half longitudinally, restricting access to the port side stair towers. However, it was decided to avoid both problems by selecting the single pilothouse, which also offered the potential of reducing the quantity of electronic devices through the use of repeaters.

Public forums were held to gauge public reaction, and the selected profile – modern styling with an amidships pilothouse, appeared to satisfy all concerned and so was selected. As the design process developed, however, it became evident that advocates of the two-pilothouse version, including some of the masters and pilots, were merely biding their time and marshalling resources.

Propulsion Study

While the public viewed and discussed the outboard profiles, SSA and EBDG personnel were evaluating the propulsion options. Key factors included:

- Double-ended vessel design
- Fleet commonality of main engines
- Draft limitation
- Maneuverability and control requirements
- 16-knot speed requirement

The following propulsion system options were evaluated and scored:

1. Two (one at each end) main engines with fixed-pitch propeller, high-lift rudder and two directional control pump-jet thrusters (one at each end)
2. Two (one at each end) main engines and two azimuthing thrusters (one at each end)
3. Four (two at each end) main engines and two azimuthing thrusters (one at each end)
4. Three DC generators serving two (one at each end) fixed-pitch propellers, two high-lift rudders and two directional control pump-jet thrusters
5. Three DC generators serving two (one at each end) azimuthing thrusters
6. Three DC generators serving four (two at each end) azimuthing thrusters

A scoring matrix was created and included the following factors:

Maneuverability
- o At speed
- o While docking
- o Seakeeping
- o Crash stop performance

Availability
- o Reliability
- o Serviceability

Cost
- o Capital cost
- o Operating cost
- o Maintenance cost

Construction
- o Arrangement impacts
- o Weight
- o Passenger/crew comfort

Support
- o Installation history
- o Regulatory compliance
- o Fleet similarity
- o Training
- o Helmsman stress

Environmental emissions

SSA and EBDG assigned individual weighting factors and scored the six propulsion system options independently with remarkably similar results. The first, second and third choices, identical for both scorers, were Options 4, 1 and 5, respectively.

Accordingly, Option 4 was selected and the design proceeded for a diesel-electric propulsion system with high-lift rudders and DC motors driving fixed-pitch propellers and pump-jet thrusters. Significant progress in the propulsion system design was necessary before a detailed weight analysis could be performed. The results of this analysis showed a large weight increase as compared to the baseline design, Option 1. Application of the Option 4 weight into the vessel weight and stability studies revealed major problems in maintaining the maximum draft requirement of 10'-6". This issue was significant enough to force two changes in the vessel design:

- Propulsion system Option 1 was selected in lieu of Option 4 to reduce weight.
- The vessel hull was lengthened five feet to increase buoyancy.

Concept Arrangements

As these initial studies were completed and decisions made, the concept design began to take shape. It had become evident that weight would be a problem that could not be resolved by increasing the hull size, as the beam was restricted by the terminal shape and length was restricted by master and pilot concerns regarding maneuverability, and by public concerns regarding visual impact in Vineyard Haven Harbor. As discussed above, the weight issue eliminated from consideration the diesel-electric propulsion system. It also led to the decision to fabricate the pilothouse, deckhouse and upper superstructure from aluminum.

An enclosed freight deck is a standard feature in the SSA fleet, providing protection from the freezing spray common in the winter months. Doors are operated hydraulically; a variety of geometric configurations have been installed on the various vessels, with varying degrees of success. The most common problem with the currently installed doors is cracking. This issue raised concerns for the new design as well, since the decision had been made to fabricate the freight deck doors of aluminum in order to minimize weight. Looking to reduce operations-induced stresses in the doors, it was decided to develop a new design which would roll rather than lift out of the way, and the concept design incorporated this feature.

Vehicle capacity is an obvious topic of concern as the fundamental mission of the vessel is to transport vehicles over water. Less intuitive, however, is the issue of maneuvering and parking vehicles, both automobiles and tractor-trailer rigs, on the freight deck (Figure 22). One or more longitudinal "island" structures on the freight deck serve to provide passenger access to the upper decks, and to support the structure of the decks above. Location of the island or islands has a major impact not only on how many vehicles can fit on the freight deck, but also how long those vehicles will take to load and unload. As expected, narrower lanes and increased number of turns both increase the time it will take for the driver to negotiate the vehicle to the designated parking location. This is especially true for tractor-trailer rigs, which have limited ability to maneuver in tight quarters. It should be noted here that the SSA loading ramps have only a single lane, making it difficult to maneuver large rigs too far from the vessel centerline.

Ferries in the SSA fleet have a variety of island configurations including a single centerline island as well as port and starboard islands. For the new ferry, SSA required that the ship centerline be clear of obstructions so as to facilitate rapid loading and unloading of tractor-trailers. Passenger loading mezzanines at the 01 Deck port and starboard precluded the option of locating stair towers and elevators adjacent to the side shell, so the only logical choice for the island location was at the inboard side of a mezzanine, as two islands, one at each mezzanine, would occupy too much valuable traffic lane space on the freight deck. A single island maximizes lane width, facilitating loading and unloading, as well as providing space for a wheelchair-access lane running the full length of the freight deck.

One drawback to a single off-centerline island is the long span for the transverse racking structure (Figure 7). Lengthened spans translate to deeper members which in turn dictate the height of the 02 Deck, which must allow for clearance for 13'-6" maximum legal height tractor-trailers. The height of the 02 Deck was also increased when it was decided to provide additional parking for automobiles on port and starboard lift decks. These decks, to be fabricated of aluminum to minimize weight, stow in the freight deck overhead when not needed or when a full load of tractor-trailers is loaded.

All of these factors played key roles in determining the layout of the freight deck and all the decks above. We were fortunate that none of these decisions had to be revisited as a result of the ongoing public presentations and meetings.

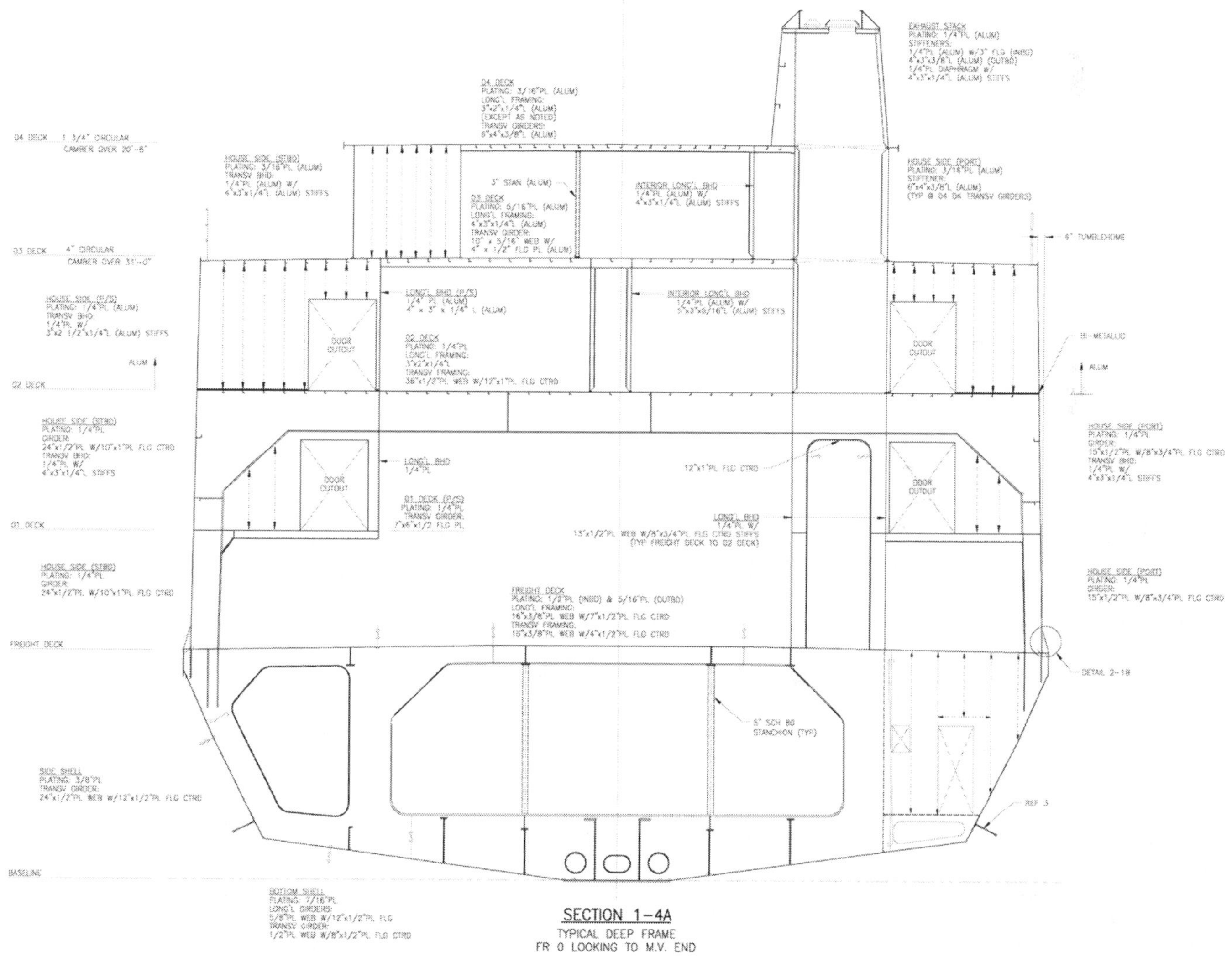

Fig. 7: Structural Midship Section

Structural Arrangements

In keeping with typical practice for double ended ferries, it was decided that the hull would be transversely framed. A frame spacing of 28 inches was selected, with half frames installed at the freight deck to minimize deck plating deformation resulting from vehicle traffic.

Also typical for double-ended ferries, Frame 0 was located at amidships, with frame numbers increasing towards each end of the vessel. To differentiate vessel ends, one end was designated the Woods Hole end and frame numbers were given the suffix "W" (1W, 2W, 3W, etc.) The opposite end was designated the Martha's Vineyard end and frame numbers received the suffix "M."

Transverse racking structure consisting of deep frames and bulkheads was installed at intervals of five frames (11.67 feet).

To reduce weight, longitudinal framing was specified at the 02 Deck and above. In addition, structure above the 02 Deck was fabricated from aluminum. Bimetallic welding strips were used to connect the aluminum structure to the steel 02 Deck.

Truck Maneuvering Study

Because the loading ramps are only a single lane wide, maneuvering long trucks into and out of parking lanes is problematic. To ensure that the proposed truck-parking lane arrangement would work, tests were performed with trucks on a full size mockup. A large vacant lot on the Seattle waterfront was located, and chalk, cones and lumber were used to mark the lanes and interferences. Loading and unloading ramps were also bordered with lumber.

Fig. 8: Truck Maneuverability Study

Three tractor-trailer rigs, each 66 ft-5 inches long, were rented and the professional drivers demonstrated no difficulty in maneuvering to any of the three center lanes, and thence to the unloading ramp (Figure 8). The successful conclusion of this test provided great confidence that no significant parking problems could be expected when the ferry enters service.

Concept Hull Form

The two highest priorities in creating the hull form (Figure 9) were that the deck plan precisely match the existing slipways, and that the full load draft not exceed 10'-10". The bearding lines at the vessel ends are uncommonly long and flat to accommodate the rudders and thrusters, but the sections transition rapidly to a V-shape for improved seakeeping. A hard chine form was selected for ease of construction and reduced cost._

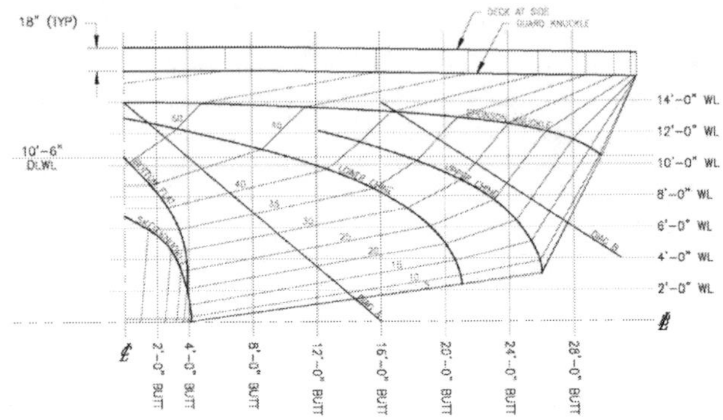

Length overall	254'-8"
Length, waterline	238'-8"
Beam, molded	64'-0"
Depth at amidships, molded	17'-6"
Design draft	10'-6"
Block coefficient	0.415
Midship coefficient	0.791
Prismatic coefficient	0.526
Deadrise angle	7.5°
Molded displacement	1928 LT

Fig. 9: Body Plan and Particulars

Computational Fluid Dynamics Study

Force Technology of Brøndby, Denmark, was directed to perform computational fluid dynamics (CFD) calculations at even keel with a design draft of 10'-6" and speed of 16 knots ($Fn(L_{pp}) = 0.321$).

Potential flow CFD calculations were performed using SHIPFLOW 2.4 (XPAN). Non-linear free surface boundary conditions and free dynamic sinkage/trim were specified; appendages were not included in the calculations. Results were evaluated by means of:

- Axial velocity distribution
- Tufts (streamlines)
- Wave pattern and profile

The wave system and profile were found to be consistent and typical for the vessel lines and Froude number. The wave system exhibited a pronounced bow wave, a wave trough at about 40% L_{pp} aft of the forward perpendicular, and a stern wave system with a peak at 110% L_{pp}. The calculated wave profile was found to be similar or better than that for in-house double-ended ferry CFD data.

It was recommended that the bottom flat forward of the skeg be eliminated to reduce the probability high wave impact loads in a seaway. This recommendation was not incorporated due to the need for a flat bottom for the omnidirectional thruster discharge ring.

It was also recommended that the upper and lower chines be better aligned with the calculated flow lines to reduce drag due to cross-flow over the knuckles; see Figure 10 which shows the calculated flow lines in plan view – note the flow discontinuity at the leading edge of the skeg about one third of the way from the left side of the diagram. Alignment of chines with the calculated flow lines was accomplished by raising the chines towards the ends of the vessel, creating more of a V-section. This change partially satisfied the previous recommendation regarding wave impact loads as well.

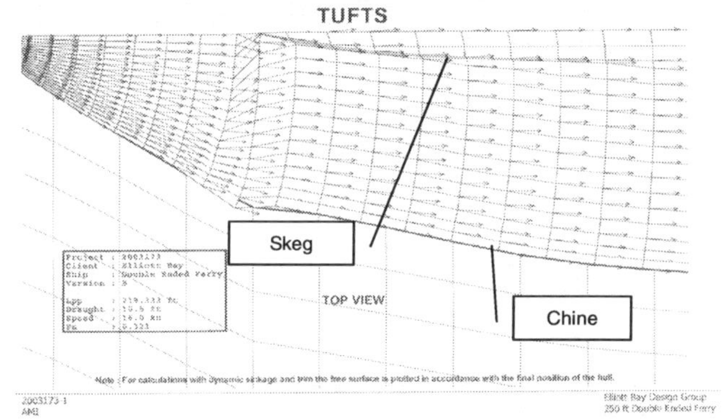

Fig. 10: Calculated Flow Lines

Some difficulty in course-keeping was predicted after consideration of the hull form and ratios, particularly the L/B ratio. This issue was later examined by self-propelled zig-zag tests and the concerns proved to be unfounded.

The calculated dynamic sinkage values at 16 knots were 3.1 inches and 12.7 inches at the forward and aft perpendiculars, respectively.

Model Test Protocol

Chine modifications recommended by the CFD study were incorporated, and Force Technology proceeded to create a self-propelled model at a scale of 1:14.69; this 17.34-foot model is shown in Figure 11. The model included the propellers, rudders and thrusters at each end. Test propellers were 5.7 inches in diameter, fixed-pitch, and four-bladed with the following full scale characteristics:

Diameter	6.99 feet
Pitch/diameter ratio	1.00
Blade area ratio	0.7

Draft, roll period and stability tests were conducted on the model to confirm that hydrostatic and stability properties of the model matched the predicted values. Roll and pitch mass moment distributions were corrected to provide the calculated natural periods, and added mass and damping were accounted for.

Fig. 11: Test Tank Model

Still water tests, all at a draft of 10'-6", were specified as follows:

- Propeller wake survey – bow and stern propellers operating with the vessel proceeding at 16 knots.
- Stream paint tests – determine the flow in way of the bilge keels and appendages with the vessel operating at 16 knots.
- Calm water resistance with appendages – measure resistance in two knot increments from 8 to 18 knots.
- Self propulsion test – determine optimum power loading distribution between bow and stern propellers at 16 knots with five different power ratios.
- Self propulsion test – measure power requirements (rpm, torque, power and thrust) in two knot increments from 8 to 18 knots.

Optimization of Power Distribution

Because the propellers are fixed pitch, the forward propeller operates in the reverse direction and the forward propeller blades meet the oncoming flow with their trailing edges. This is obviously far from ideal, and leads to a low efficiency for the forward propeller. In an effort to determine the optimum power ratio between forward and aft propellers, self-propelled tests were performed at 16 knots with varying power ratios. Results are provided in Table 3.

Table 3: Power Distribution Measurements

RPM_{fwd}/RPM_{aft}	PD_{fwd} (HP)	PD_{aft} (HP)	PD_{fwd}/PD_{aft}	PD total (HP)
.80	695	3160	.22	3855
.85	939	3014	.31	3953
.90	1176	2802	.42	3978
.95	1476	2643	.56	4119
1.00	1743	2423	.72	4166

Not surprisingly, the lowest total power requirement occurs when the aft propeller is most heavily loaded. The total power required to propel the vessel at 16 knots with a 0.80 RPM ratio still exceeds the available power for the aft propeller, however, so a 42% forward/aft power ratio was selected for the subsequent self-propelled testing.

Speed & Powering Prediction

Resistance and self-propulsion tests were performed at the design draft with the forward propeller running at 90% of the aft propeller speed, absorbing approximately 40% of the power of the aft propeller. Table 4 shows the propeller speed and shaft power predictions for the design speed of 16.10 knots. Table 4 and Figure 12 are reproduced from Reference 2.

Table 4: Speed and Power Prediction Summary

Propeller	Forward	Aft
Shaft Power (HP)	1196	2880
Propeller RPM	308	340

The five-bladed propellers were designed and manufactured by Rolls Royce Naval Marine, optimized for 16 knots. Propellers are fixed pitch, 90 inches diameter with 93.5 inches (nominal) pitch, 12 degrees of skew and a developed area ratio of 0.80. Slight modifications were made to the standard pitch, skew, camber and chord distributions to improve performance in the backing direction.

3-D Wake Measurements

A five-hole pitot tube was used to perform 3-D wake measurements at the design draft and vessel speed of 16 knots. Measurements were taken at the plane of the aft propeller; the aft rudder was removed for this test to provide good access to all of the propeller disk points. It was decided not to measure the forward propeller wake as the forward propeller operates in the shadow of the forward rudder and because the forward propeller is operated in reverse, an off-design condition.

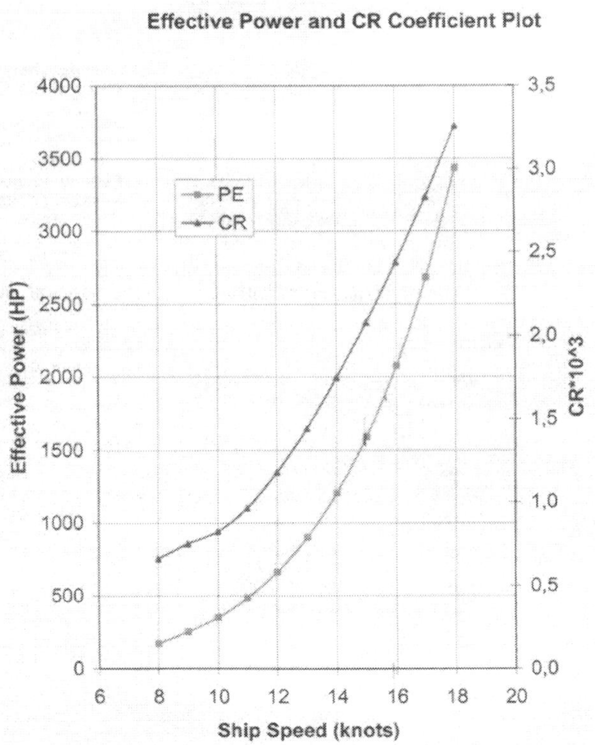

Fig. 12: Speed and Horsepower Curve

Objectives of the wake measurement test were to obtain information for adequate propeller design and to measure cavitation excitation performance. Test results determined that the axial wake is distributed evenly in both the radial and circumferential directions. The axial velocity component varies between 80% and 100% of the undisturbed flow velocity, with a mean value over the entire disk of 91%. Calculated inflow angles showed very low fluctuations, suggesting an evenly distributed circumferential axial wake, with a low risk of blade cavitation.

Transverse velocity flow components (Figure 13) demonstrated a typical buttock flow, with a very weak vortex at the 180 degree position, probably generated by the skeg. It was accordingly recommended to reduce the skeg width if possible and to round the lower edges of the skeg. This second recommendation was incorporated into the design, but it was not practical to reduce the skeg width for structural reasons.

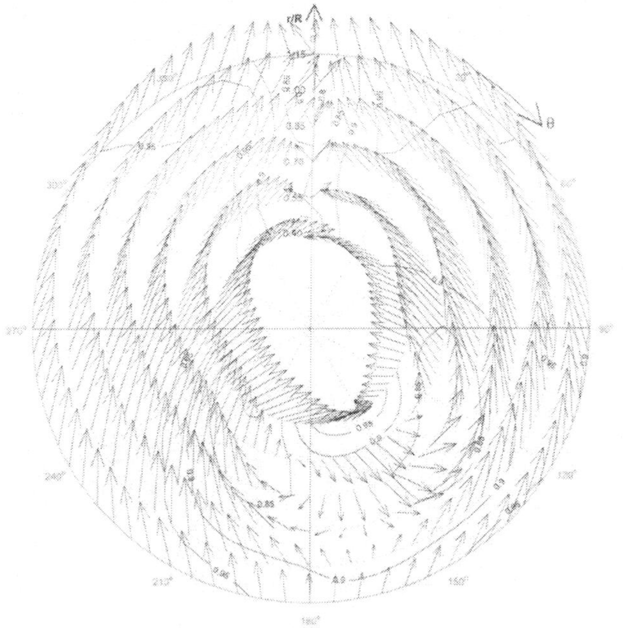

Fig. 13: Propeller Wake Vectors of Transverse Velocities

Streamline Test

Streamline tests were performed with the vessel self-propelled (both propellers operating) at 16 knots and a draft of 10.50 feet. Resulting streamlines, shown in Figure 14, reveal typical buttock-type flow with no areas of pronounced flow separation. Streamlines aligned well with the chines in the amidships areas, but did cross the chines at a small angle toward the ends of the model. It is interesting to note that this result was predicted by the CFD modeling described above. Possible flow separation at the lower skeg edges was also observed and it was accordingly recommended to round those edges if possible. As discussed above, this change was accomplished.

Fig. 14: Streamline Test

Seakeeping Study – Wave Design

Data from a_NOAA weather buoy at the entrance to Buzzards Bay (the closest buoy to the actual operating area) were analyzed and sea spectra wave parameters, provided in Table 5, were determined for the seakeeping tests.

Table 5: Sea Spectra Wave Parameters

	Sea State 4	Sea State 5
	JONSWAP	JONSWAP
Hs, feet	6.33	8.71
Tp, seconds	8	8.5
Gamma	3.2	3.2
Alpha	0.01	0.01

Seakeeping Tests

Roll and pitch decay tests were performed at the design displacement and with a mass distribution that mimicked the estimated vessel radius of gyration and metacentric height. Because of the high natural restoring forces for roll and pitch motions, a weight of about 180 pounds (a staff member) was applied and released to initiate motions. Natural periods were then calculated to be:

$T_{Roll} = 5.8$ seconds
$T_{Pitch} = 5.37$ seconds

Attempts to perform a heave decay test were unsuccessful due to the huge heave restoring force.

Seakeeping tests were performed with the vessel at a draft of 10.50 feet for the two sea states specified in Table 5 and at three headings (beam, following and head seas.) Tests in beam seas were conducted at zero speed. Tests in head seas and following seas were conducted at three speeds corresponding approximately to 100% MCR, 75% MCR and 50% MCR.

Vertical accelerations were measured by accelerometers at the locations listed in Table 6.

Table 6: Accelerometer Locations

Location	X, ft	Y, ft	Z, ft
Pilot house	20	0	60
Passenger deck	72	26	38
Fwd bicycle area	95	21	19
Aft bicycle area	-95	21	19

X, Y and Z coordinates are referenced to amidships, centerline and baseline, respectively.

Table 7: Summary of Motions (RMS) in Head and Beam Seas

Speed	Wave height (Hs)	Roll	Pitch	Heave	Vert Accel. Pilot house	Vert Accel Pass space	Vert Accel Fwd Bike	Vert Accel Aft Bike
Knots	Feet	Deg	Deg	Feet	G	G	G	G
11.14	6.33		1.16	0.36	0.06	0.09	0.11	0.08
12.60	6.33		1.15	0.35	0.06	0.10	0.12	0.09
14.21	6.33		1.03	0.33	0.06	0.09	0.12	0.08
0	6.33	3.17		0.47	0.03	0.06	0.05	0.06
10.13	8.71		1.63	0.52	0.07	0.11	0.14	0.10
11.64	8.71		1.60	0.52	0.08	0.12	0.15	0.11
12.97	8.71		1.55	0.52	0.08	0.12	0.15	0.11
0	8.71	3.70		0.61	0.04	0.07	0.07	0.07

The measured roll angles and vertical accelerations, summarized in Table 7, are higher than would likely be comfortable for passengers on voyages longer than two hours. However, the master does have the option of slowing down, and the length of exposure to rough seas on the intended route is well under one hour, so these values are considered to be acceptable.

Speed Loss Calculation

Based upon the maximum predicted speed in calm water of 16.1 knots, and assuming constant power output, speed loss was calculated as the difference between the speed/power curves in the corresponding sea state and the still water reference; see Table 8. Speed / power curves were calculated from the measured propeller torque and revolutions. Predictions included wind resistance for wind speeds of 31 ft/sec and 37 ft/sec for the two sea states, respectively.

Table 8: Speed Loss (knots)

Sea State	Head sea Max speed	Head sea Speed loss	Foll. sea Max speed	Foll. sea Speed loss
Hs = 6.33'	14.1	2.0	15.7	0.4
Hs = 8.71'	12.3	3.8	15.1	1.0

Course Stability

Self-powered maneuvering tests (zig-zag maneuvers) were conducted in still water with the vessel operating at the design draft with a GM of 26.05 ft. and 16 knots initial speed. Three series of zig-zags were performed, at rudder yaw angles of 4 -4 degrees, 6 – 6 degrees and 10 – 10 degrees (Reference 3).

One key measurement of course stability is the "overshoot angle," measured in degrees and defined as the heading deviation from the moment the rudder is reversed to the moment that rate of change of heading is zero. For ships longer than 100 meters, IMO Maneuvering Resolution A.751(18) specifies that for the 10-10 degree zig-zag criterion, the vessel maintains good course stability when the first and second overshoot angles are less than 10 and 25 degrees, respectively. The procedure for the 10-10 degree zig-zag maneuver is to set the aft rudder 10 degrees to port (for example) until the vessel achieves a heading of 10 degrees to port, at which time the rudder is reversed to 10 degrees starboard until the vessel course is 10 degrees starboard. The forward rudder was fixed on centerline for all maneuvering tests.

Table 9: Overshoot Angles

Rudder Angle	First Overshoot	Second Overshoot
IMO 10/10 Zig-zag	10° max	25° max
4°	4.0°	3.6°
6°	5.8°	7.0°
10°	7.5°	12.0°

Zig-zag test results are provided in Table 9 and Figure 15. Both the first and second overshoot angles were found to be well within the IMO criterion; it was therefore concluded that course stability would not be a problem.

It should be noted that low speed maneuvering was not tested as the combination of a thruster and a rudder at each end of the hull was expected to provide exceptional control.

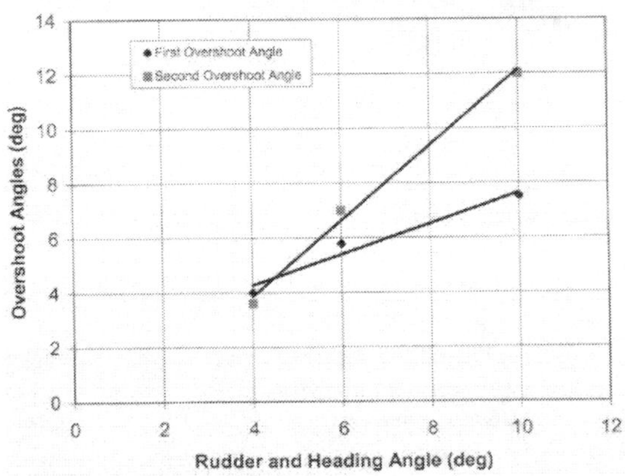

Fig. 15: Plot of Zig-zag Overshoot Angles

Pilothouse Issue Revisited

As mentioned earlier, the decision to design the vessel with a single pilothouse amidships was well-considered but not unanimous. Over a period of several months, smoldering disagreement evolved into active dissent and letters expressing concern about visibility in foggy conditions began to appear in the local newspapers. SSA personnel took notice and held public forums to determine the strength of the dissent.

After much discussion and careful consideration of the consequences of changing the design, it was finally decided to redesign the vessel with two pilothouses. Unfortunately, the structural design had been completed some time earlier and much additional design work was required to reconcile the new arrangements and structure. The final cost of this change, including engineering and additional materials, labor and equipment, was estimated at about one million dollars.

Disabilities Accommodations

As stated earlier, it was the intention of SSA to comply insofar as practical with the standards of the Americans with Disabilities Act (ADA) (Reference 4).

Many of these standards are relatively simple to achieve, but all require the designer to maintain focus on the standards throughout the design process so as not to compromise the disabilities accommodations. Typical of the standards for accommodations are:

- Clear height through doorways of 80 inches
- Clear passage widths of 32 inches
- Signage in Braille
- Automated door openers
- Provision for wheelchair seating throughout the passenger areas
- Flashing light alarm signals

Another accommodation for disabilities required more careful planning to include in the design. The most obvious need was for an elevator providing access to upper decks for the mobility-impaired. The Steamship Authority's experience indicated that when a single elevator is installed on a vessel, passengers and food service contractors frequently compete for its use. Early in the design process, therefore, SSA made the decision to provide two elevators with one being for the exclusive use of passengers.

Elevator size was constrained by the available length and width of the freight deck island (Figure 22), which in turn was constrained by the need to maximize traffic lane width on a deck with limited beam and length. Careful design of the island made it possible to fit the uptakes, three stair towers, utility trunks and two elevators, but there was inadequate deck space for fully ADA-compliant elevators. ADA standards anticipated this situation by offering a reduced size elevator, termed Limited Use Limited Availability (LULA). LULA elevators are required to have clear inside dimensions of 42 inches width × 54 inches depth with a door clear opening width of 32_inches. Both elevators comply with LULA standards.

An elevator would do little to provide access for the mobility-impaired unless convenient access is to the elevator is provided. This was accomplished by providing a pedestrian path for the full length of the freight deck. The pedestrian path is 36_inch wide, adequate for wheelchair passage, and painted high-visibility yellow to assist drivers in keeping clear.

Evacuation Arrangements

Given the USCG requirement to evacuate the full passenger load of 1200 people within thirty minutes, efficiency in every aspect of the evacuation process was essential. Early in the design process, SSA decided that passengers would evacuate from the 01 Deck rather than from the freight deck, reasoning that the freight deck was much more likely to be involved in a fire than the 01 Deck passenger spaces.

Evacuation from the 01 Deck, located 15 ft-6 inches feet above the full load waterline, would require either a chute or a slide from the 01 Deck to the escape platform. SSA carefully considered these two options and selected a slide arrangement, believing that it would be easier to coax passengers, including the elderly and small children, onto a slide.

At the lower end of each slide, evacuees land on the escape platform, a floating raft which serves as a queue while the evacuees are directed by a crew member into the adjoining life rafts. Two evacuation systems are provided, one located at the Woods Hole end of the vessel, port side, and the other at the Martha's Vineyard end, starboard side. Each slide, platform and raft assemblage is intended to serve 600 passengers.

Before the *Island Home* entered service, an evacuation drill was performed under the watchful eye of the USCG in January, 2007, at Fairhaven, MA. Frigid conditions (16°F with 20 knot winds) necessitated the support of two small craft to push the sea ice away from the evacuation system deployment area. The sample size of two crew members and ten hardy volunteers was able to evacuate in 20 seconds. Extrapolating the evacuation time to 600 passengers, and allowing for the 80 second slide and raft deployment time, it was determined that a full load of 1200 passengers could evacuate in 22 minutes.

Rescue Boats

The two rescue boats are SOLAS-approved, each 4.3 meters long with a 6-person capacity and fitted with a 30 HP outboard motor. A SOLAS-approved davit was furnished for launching and retrieving each boat.

As with the evacuation systems, SSA wanted the rescue boats to be located in the most efficient locations. Past experience with launching rescue boats in high wind conditions convinced SSA that the boats should be located as close to the waterline as practical. Therefore, space on the freight deck was allocated for the rescue boats, which are located at the Woods Hole end, starboard side and Martha's Vineyard end, port side, as shown in Figure 16.

Numerous drills conducted during crew training confirmed the wisdom of this decision, as quick launches and rapid recoveries were performed in a variety of weather conditions.

Lift Decks

The original design concepts did not include lift decks, but after inspecting successful installations of lift decks in other ferry operations, SSA elected to include them in the new design. The rationale for the lift decks was to increase automobile capacity without reducing truck capacity. Increased automobile capacity would reduce waiting time for drivers during the busy summer months while simultaneously increasing revenue.

To avoid reducing truck capacity, the lift decks needed to stow in the overhead of the freight deck, maintaining 14.5 feet clear underneath for the tallest road-legal trucks in Massachusetts. A variety of mechanisms to accomplish this were evaluated and some consideration was given to providing the lift deck design as part of the contract package. In the end, it was decided to leave responsibility for the design with the shipyard so as to provide the best opportunity for integration of the lift decks with the vessel structure and systems.

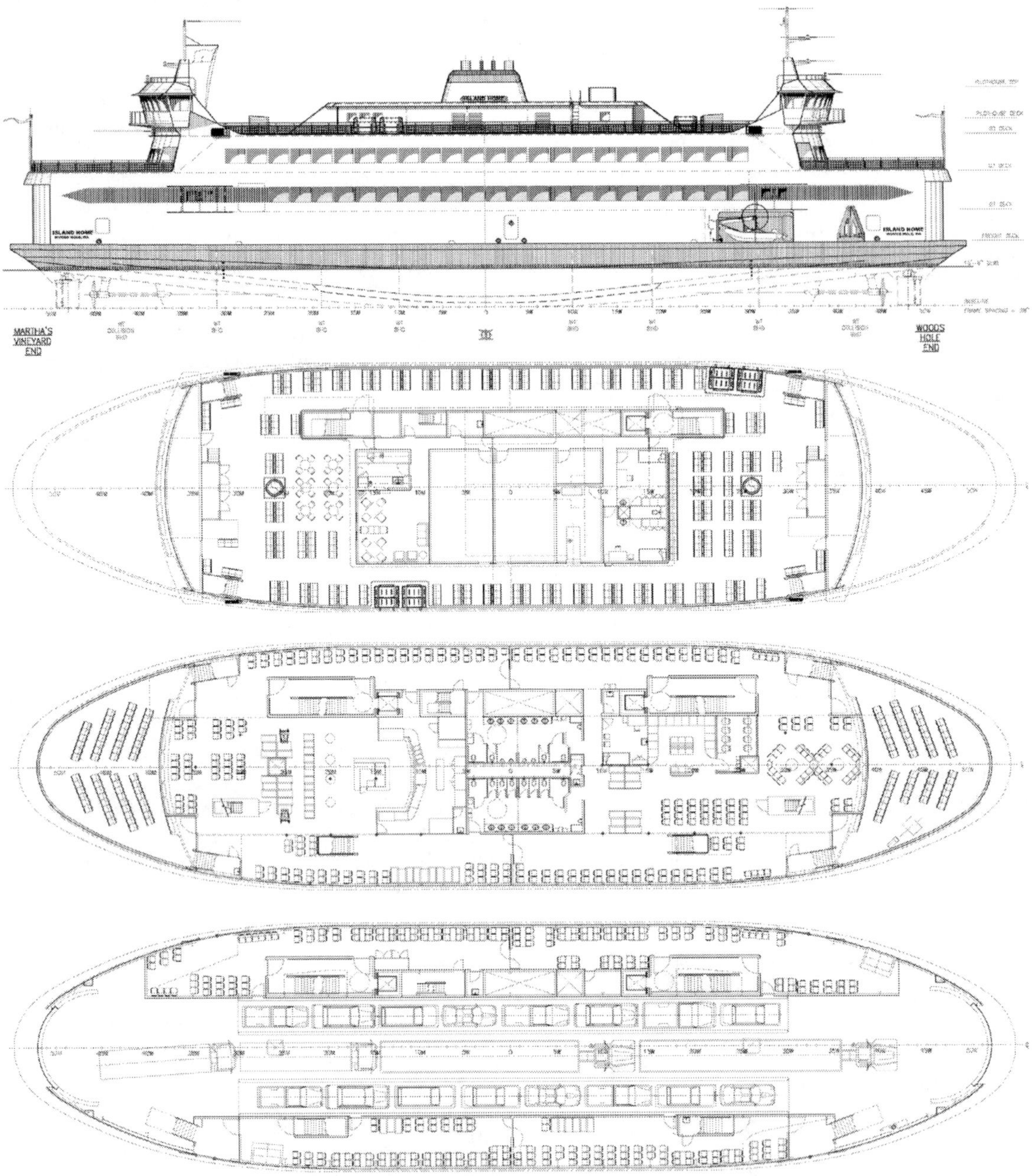

***Island Home* Particulars**	
Length overall	254'-8"
Length, waterline	238'-8"
Beam, molded	64'-0"
Depth at amidships, molded	17'-6"
Design draft	10'-10"
Vehicle capacity	76
Passenger capacity	1200
Fuel oil capacity	8700 gallons
Fresh water capacity	1000 gallons

Figure 16: *Island Home* Profile and Arrangements

Huber, Inc. of New Orleans was awarded the design of the lift decks and produced the successful results seen in Figure 17. Each of the two lift decks is 140 feet long and has a capacity of eight typical automobiles. In addition to the complex geometry involved in the scissors-lift for the center sections and the hydraulic cylinders to operate the ramps, it was necessary to develop practical details for folding handrails, safety interlocks for the access doors at the 01 Deck, and sprinkler piping and heads on the underside of the ramps.

Fig. 17: Lift Deck

Security Features

Security concerns typical of present day operations were addressed in the design phase of this project. In addition to the omnipresent security cameras and cipher locks on doors accessing secure spaces, the design includes a crew-only stair tower providing secure access from the engine room to the crew spaces on the 03 Deck.

The original design had an amidships pilothouse; this stair tower originally terminated inside the pilothouse. However, the change to the installation of two pilothouses eliminated the possibility of accessing pilothouses without passing through passenger spaces. The final design configuration requires that passengers be prohibited from the 03 Deck when heightened security measures are in force, so that the master and pilot can transit between pilothouses without passenger contact.

Machinery Arrangements

The engine room arrangement is shown in Figure 18. An Engineer Operating Station (EOS) is located at the engine room port side, with direct access to the crew stair tower. The EOS provides good visibility of the engine room and also has several monitors to permit the engineers to view activity in the steering gear rooms, reduction gear rooms, and auxiliary machinery room. The main switchboard and motor control center are also located within the EOS.

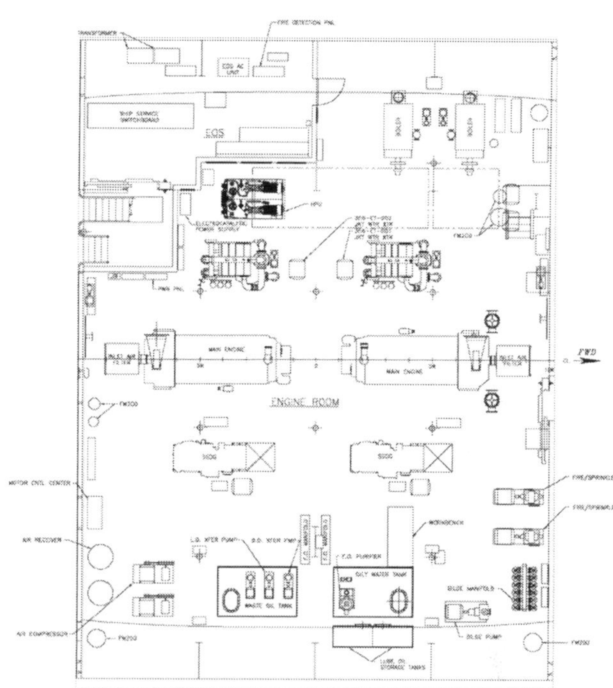

Fig. 18: Engine Room Arrangement

Perhaps the most obvious feature of the engine room is the location of the two main engines. Also obvious is the lack of reduction gears, which are installed in the reduction gear rooms close to the ends of the vessel. The reason for the separation of engines and reduction gears is that the maximum compartment length, as determined by the floodable length regulations, does not allow all four pieces of machinery to fit in the engine room. A benefit of separation of engines and reduction gears is the ability to use smaller diameter, high speed shafting to transmit power from the engines to the reduction gears, saving weight and cost.

The Auxiliary Machinery Room (AMR) is adjacent to the engine room with access through an automatic watertight door. Primary machinery in the AMR includes the sewage treatment system and refrigeration compressors. The AMR arrangement can be seen in Figure 19.

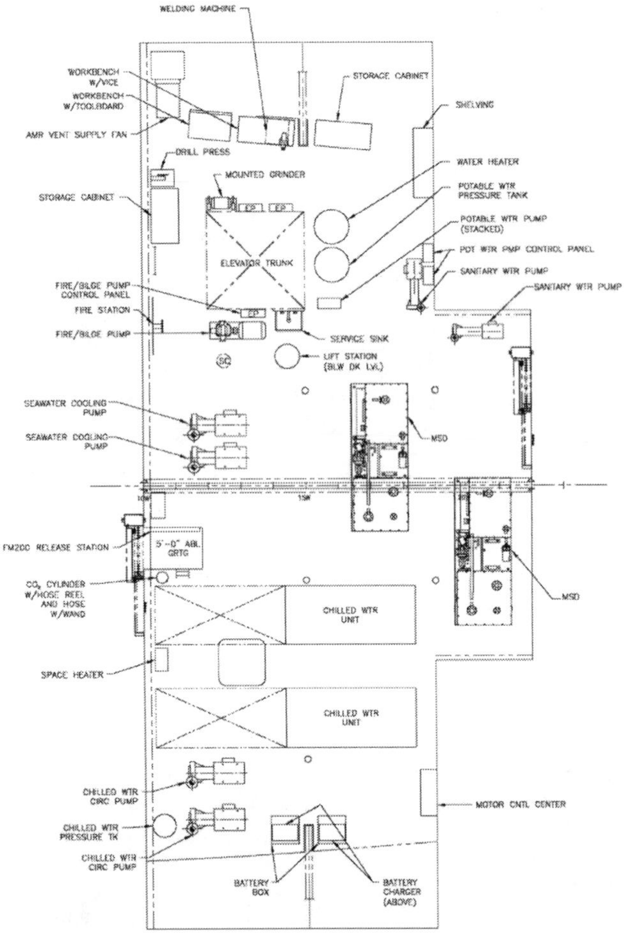

Fig. 19: Auxiliary Machinery Room Arrangement

Rudders

One measure of success for the design of the *Island Home* would be its ability to accomplish controlled dockings under the adverse conditions so common at the Woods Hole terminal. High efficiency rudders were therefore very appealing, and Becker articulating rudders were selected for this application. So as to keep steering torque requirements (and therefore steering gear capacity) at reasonable values, it was decided that the forward rudder would not be operable at vessel speeds exceeding 5 knots.

The rudder installation is shown in Figure 20; note the ice knife located at the forward end of the rudder, designed to protect the rudder from ice flows.

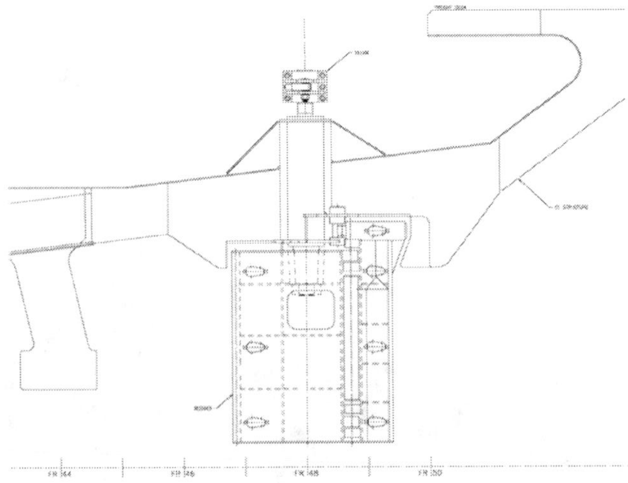

Fig. 20: Rudder Arrangement

Thrusters

SSA has had successful experience with the use of Eliot Whitegill (subsequently named Tees) jet thrusters, and elected to supplement the articulating rudders with a 400 HP electric motor-driven Tees thruster at each end of the vessel. This decision, which was made early in the design process, had a significant impact on the hull shape and location of the rudders.

It was necessary to locate the pump inlet as far below the design waterline as possible so as to minimize ingestion of ice or other floating debris. To maximize the effectiveness of the water jet, it was necessary to locate the pump discharge as close to the end of the vessel as possible while trying to minimize systemic losses resulting from friction and turbulence between the inlet and the pump. These two requirements can normally be reconciled easily in a bow thruster installation on a conventional hull, but seem at odds when trying to locate them around the propeller and rudder. The end result can be seen in Figure 21.

Shipyard Selection

The size and complexity of the vessel limited the number of qualified and interested shipyards, as did the requirement for construction bonding. VT Halter Marine (VTHM) of Pascagoula, MS was ultimately awarded the construction contract and immediately proceeded with the detail design engineering.

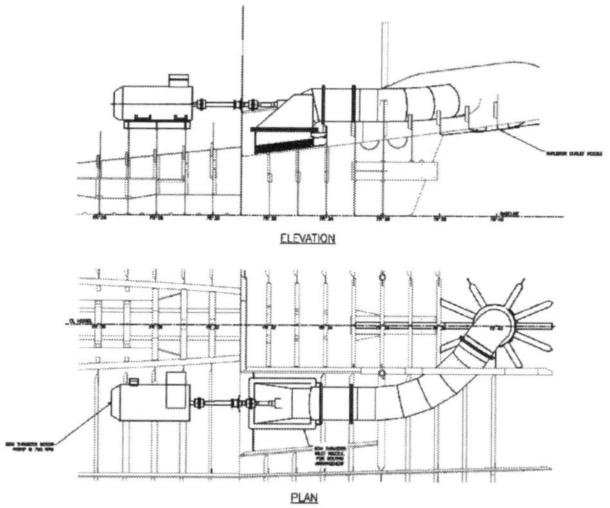

Fig. 21: Thruster Installation Arrangement

Vibration Predictions

VTHM was required to produce a vibration analysis to predict natural frequencies, global mode shapes and forced vibration response in the hull structure. Once the structural design was largely completed, Noise Control Engineering performed this analysis using a marine-based FEA code to model the hull and superstructure, and a general purpose FEA code to conduct the forced response analysis. Outfitting weights and liquids in tanks were included in the model, but the weights of passengers and vehicles were not. Forcing frequencies, excerpted from Reference 5, are provided in Table 10.

Table 10: Forcing Frequencies for Engines, Shafts and Propellers, Hz

Vessel speed (knots)	8	10	12	14	15	16
Engine speed (RPM)	400	500	600	700	800	900
Shaft speed (RPM)	131.8	164.7	197.6	230.6	263.5	296.4
Engine rotation frequency (Hz)	6.7	8.3	10.0	11.7	13.3	15.0
Shaft rate frequency (Hz)	2.2	2.7	3.3	3.8	4.4	4.9
Blade passing frequency (Hz)	11.0	13.7	16.5	19.2	22.0	24.7

Reference 4 predicted "resonant coincidences between the propeller shaft rotation rate primary and second harmonics and natural frequencies at 3.9 and 7.8 Hertz. The first resonant condition is within 10% of the first vertical mode and the second coincidence is within 10% of the fourth vertical mode/third torsional mode." It was recommended that these frequency coincidences be avoided by "ensuring proper balance, shaft concentricity and shaft coupling alignment." These precautions were taken to heart, and the line shaft material was changed from steel to carbon fiber to further alter the natural frequency of the shafting systems.

Serious excesses to the American Bureau of Shipping "COMF" Class vibration criteria (Reference 6) were also predicted, caused by vibration from the propulsion diesel engines and the shafting between the gearboxes and propellers. The precautions adopted as described above should also address the shafting vibration issue, but the diesel engine vibration issue would require additional measures.

The two practical options for reducing main engine-induced vibration were:

- Increasing structural mass to alter the natural frequency
- Resiliently mounting the main engines

Draft limitations precluded serious consideration of the first option, and the builder strongly advocated the use of resilient mounts. Accordingly, SSA chose to proceed with the installation of resilient mounts under the main propulsion engines.

Design Changes

Every significant engineering design could have been improved with the advantage of 20/20 hindsight, and this effort was not an exception. Despite extensive experience in the design of passenger/vehicle ferries, the designers did not always interpret the regulations correctly, did not anticipate some new USCG interpretations, and did not avoid all of the opportunities for what can only be called mistakes. Resolving unexpected challenges can be true test of a team's cohesiveness, and it is a testament to the team of SSA, EBDG and VTHM that technical challenges were overcome professionally, without rancor and without schedule delay.

The first technical surprise involved stability, downflooding, and the joiner doors on the outboard (port) side of the island. Several of these doors provide access to companions leading below the freight deck and so are potential downflooding points. Calculations performed during the contract design indicated that the vessel met the applicable stability criteria without the need for weathertight doors at these locations. It is an unfortunate quirk of the stability software used for the calculations that the software defaults to starboard heel when calculating righting energy. As the doors are located on the port side, it is necessary to either mirror the downflood points to the starboard side, or to direct the software to heel both port and starboard. This need was overlooked; fortunately VTHM discovered the error before the doors were installed, and arranged to install weathertight rather than joiner doors.

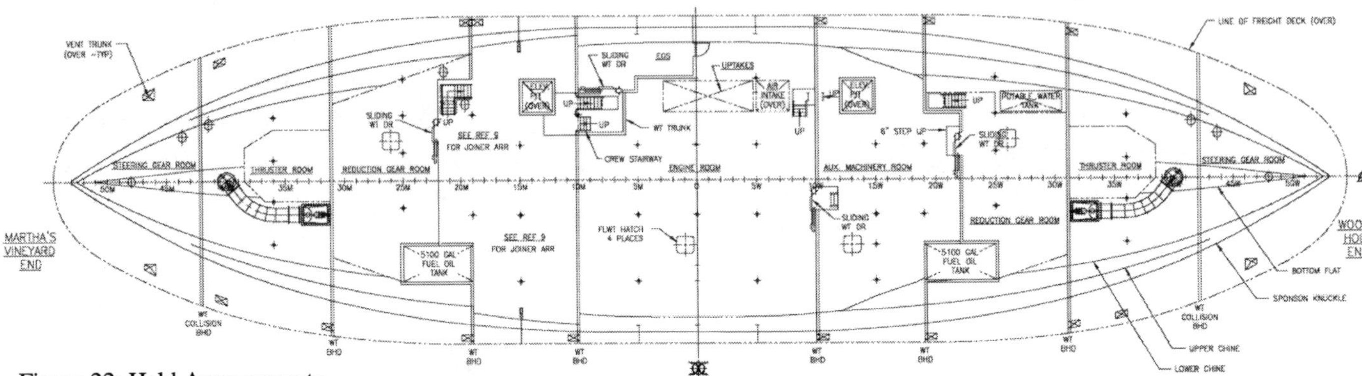

Figure 22: Hold Arrangements

The second technical surprise also involved stability but resulted from a carefully considered decision rather than an oversight. 46 CFR 171.072 establishes permeabilities for different types of spaces, specifically 0.85 for machinery spaces and 0.95 for "other" spaces. During the contract design phase, the reduction gear rooms were considered to be a machinery spaces with a defined permeability of 0.85, a decision based on numerous USCG approvals of nearly identical arrangements. Unfortunately for this project, the precedent of former interpretations was deemed no longer appropriate by the USCG, which applied a permeability of 0.95. Equally unfortunate, the greater permeability more accurately reflected reality, eliminating the opportunity for using a calculated value.

The solution to this issue was expensive and difficult; it was necessary to relocate the inboard portions of the already-fabricated transverse bulkheads at frames 19W and 19M, creating stepped bulkheads (Figure 22) to reduce the volume of the reduction gear rooms. A positive outcome of this development was the opportunity to expand the crew quarters and auxiliary machinery room.

Two regulatory issues related to the aluminum structure caught the designers by surprise. The first was the need for steel stairs within the A-60-insulated stair towers, and the other was the need to insulate both sides of aluminum structure in order to achieve A-60-level protection for stair tower bulkheads.

Each project has several features that would be changed on the next hull built to the same design, and the *Island Home* is again no exception to this truism. Given the opportunity, we would locate the fresh water tank on the starboard side of the Martha's Vineyard end rather than the port side of the Woods Hole end so as to counterbalance the port side island structure. An even more minor improvement would be to delete the tonnage opening in the crew lounge on the 03 Deck – this opening proved to be unnecessary insurance.

The last significant technical issue to be raised occurred just days before the *Island Home* departed on the delivery voyage. 46 CFR 112.05(d) requires "Each compartment containing this equipment [the emergency power source] must be readily accessible from the open deck ..." This requirement was met in the concept design, but concerns over vessel security led to a decision to move the access to a passageway in the secure crew accommodations. The regulations were not revisited during the contract design, and USCG approved the arrangements without comment, leaving this violation to be discovered by OCMI. The solution was to add a watertight hatch for access from the exterior, a solution which meets the letter of the law while still discouraging passengers from unwanted exploration.

One final OCMI requirement delayed the delivery voyage departure by a day when it became necessary to replace the plastic tubes serving the soft-drink dispenser with stainless steel tubes.

Hurricane Katrina

Hurricane Katrina arrived right in the middle of the construction program. Several of the amidships hull modules had been erected on the building ways, but the end modules and upper levels were at ground level still undergoing fitting and welding.

At the peak of the storm surge about ten feet of sea water flooded the shipyard. One large vessel moored and anchored at the yard broke free and floated inland, requiring a half mile channel to be excavated to extract the vessel from the marshes.

The *Island Home* was much more fortunate – the erected modules suffered very little damage and the units on the ground were undamaged, although a variety of snakes and other reptiles inhabited the nooks and crannies when the yard workers were able to return and begin the cleaning up operations.

The shipyard suffered extensive damage to its buildings and shops. All ground level offices and facilities were completely destroyed, including machine shops, prefab shops and the Owner's trailer office. The warehouse was flooded, destroying the ship's generator sets and other equipment. Fortunately, the main engines had been blocked high and appeared to be undamaged, but were returned to the supplier for inspection as a precaution and to validate the warranty.

VT Halter was able to reopen in remarkably fast time, but restoration of full capabilities was long delayed by the inability to obtain equipment, fuel and supplies as every remaining viable business competed for scarce resources. Probably the hardest resource to get back on line was labor, as many of the production workers lost their homes and had to either relocate out of the area or spend months rebuilding.

The net result for the *Island Home* project was a delay of about nine months, a remarkably short period when the devastation to the shipyard and surrounding towns and homes is considered.

Sea Trials Results

The sea trials agenda was based on Reference 7. Sea trials were conducted in the Gulf of Mexico south of Pascagoula, MS on January 12, 2007. Test conditions were as follows:

Water depth:	45 feet
Wave size	4 – 6 feet
Wave direction	NW
Wind velocity	8 – 12 knots
Wind direction	SE
Draft forward	10'-0"
Draft aft	9'-10"

Speed trials were conducted at a shaft speed of 300 RPM, resulting in an average speed of 16.25 knots. The highest recorded speed for a 10 minute run was 17.0 knots.

Whole Body Vibration Survey

A whole body vibration survey was conducted to confirm compliance with the contract requirement that the vessel comply with the American Bureau of Shipping "COMF" level criteria, Reference 6. The "COMF" requirement is an acceleration level of not more than 0.315 meters/second^2 through the frequency range of 0.5 to 80 Hertz.

Testing was performed by Noise Control Engineering, Ltd. (NCE). Test instruments were configured to measure one-third octave band vibration from the 1 Hertz to the 80 Hertz one-third octave bands with an rms averaging period of 60 seconds. Note that this measurement range does not match the "COMF" range due to limitations of the test instruments and because vibrations in the range under 2 Hertz tend to be controlled by sea state rather than equipment vibration.

Table 11. Whole Body Measurement Results

Deck	Space	Frame	Side	Vibration m/sec2
Freight	Vehicle	30M	CL	0.227
Freight	Vehicle	0	CL	0.169
Freight	Vehicle	30W	CL	0.160
01	Passenger	15M	Port	0.335
01	Passenger	15M	Stbd	0.220
01	Passenger	15W	Port	0.292
01	Passenger	15W	Stbd	0.247
02	Dining	30M	CL	0.358
02	Passenger	30W	CL	0.289
03	Passenger	30M	CL	0.403
03	Passenger	30W	CL	0.310
04	Pilothouse	36M	CL	0.338
04	Pilothouse	36W	CL	0.443

Vibration measurements were taken at the locations indicated in Table 11 above using a triaxial accelerometer affixed to the freight deck with magnetic mounts and to locations above the freight deck with bees wax. Vibration readings were in decibels relative to micro-g (dB//μg) and later converted to linear accelerations in meters/second^2.

Both main engines were operated at full power during the whole body vibration survey. It can be seen in Table 11 that several locations exhibited vibrations in excess of the 0.315m/sec^2 limit. It is believed that these excessive values are due in large part to the fact that both propellers were operating at full power, with the forward propeller cavitating heavily. Because the forward propeller will not normally be operated above 75% power, it is believed that vibrations will be reduced to an acceptable level during normal operations.

Engine Vibration and Displacement Survey

The purpose of the engine vibration and displacement survey was to confirm that the main engine vibration isolators were properly selected and installed so that the engines, EMD model 12-710G7B, exhibited reasonable displacements throughout the full power range. The maximum allowable static displacement of the engines, limited by the flexible coupling on the turbo-charger intake, is 22mm (0.867 inches).

Vibration measurements were performed by NCE using a 16-channel data collection system. Data was collected in the 10 Hertz to the 10,000 Hertz one-third octave bands using tri-axial accelerometers with rms averaging for 30 second periods. Readings were taken in decibels relative to micro-g (dB//μg). Accelerometers were located at the back engine foot (above the vibration isolator), the front engine foot (above the vibration isolator), above and at the side of the air intake. Table 12 is a sample of the results obtained from the measurements.

Table 12. Static Transverse Displacement

Engine Speed (rpm)	Ship Speed (knots)	Minimum Displacement (inches)	Maximum Displacement (inches)
350	3.8	-0.005	+0.007
400	5.3	0.000	+0.008
450	6.5	0.000	+0.010
500	6.4	-0.001	+0.020
550	7.5	-0.001	+0.028
600	8.0	-0.010	+0.030
650	8.4	-0.002	+0.030
700	10.2	+0.020	+0.040
750	12.2	+0.020	+0.050
800	12.6	+0.030	+0.060
850	14.0	+0.040	+0.070
900	14.2	+0.050	+0.070

A temporal variation in the data of 0.02 inches with a 5 second period was observed and was ascribed to vessel motions. Trials were conducted in high Sea State 1 conditions and the vessel was experiencing "moderate" rolling conditions.

No resonant conditions were observed. The greatest recorded transverse-direction motion was less than 0.01 inches and well below the allowable 0.867 inches as determined by the turbo-charger flexible coupling. Measurements taken above and below the isolation mounts exhibited vibration attenuation in excess of 20 decibels.

Airborne Noise Survey

The initial contract specification established desired and maximum allowable noise levels during normal conditions with the main engines operating at full power and the HVAC system operating. During contract negotiations, SSA agreed to change the maximum values to "goals" rather than firm requirements. All noise levels were specified and measured in a-weighted decibels (dB(A)).

Sound pressure level measurements were performed using a hand-held sound level meter with a ½ inch microphone shielded with an acoustically transparent windscreen. Integration response was set to "slow" (one second rise-time) and RMS averaging was performed for at least ten seconds.

Excluding the freight deck, one measurement was taken per compartment. Measurements were taken at typically occupied locations but normally at least three feet from large flat surfaces. Desired values, "goal" values and measured values, all in dB(A), are included in Table 13.

Table 13. Airborne Noise Measurements

Space	Desired Value	Goal Value	Measured Value
02 Dk Pass Area	60	65	62
02 Dk Outside			77
01 Dk Pass Area	60	65	68
Dining Area	65	70	63
Pilothouses	55	65	68
03 Dk Crew Spaces	65	70	58
Hold Crew Spaces	65	70	69
EOS	70	75	
Engine Room	100	108	108
Aux Machy Room	85	95	96
Red Gear Rooms	85	95	94
Strg Gear Rooms	85	95	91
Freight Deck	75	75	84

Excessive noise values were in most cases ascribed to cavitation noise from the forward propeller; actual operating noise levels will normally be lower as the forward propeller will not be operated at full power.

Unfortunately and surprisingly, no measurements were recorded in the Engineers Operating Station.

Illumination Survey

The VTHM survey procedure specified the grid system data point method for the illumination survey. Measurements were taken at a distance approximately thirty inches above the deck and after fluorescent fixtures had been lighted for at least thirty minutes. In high traffic areas such as passageways and stairways, measurements were taken at the deck level, at both the top and bottom of stairs. Compartments were divided into 3 foot square grids from the center out and illumination levels were measured at each grid intersection. Readings were averaged and compared with the standard established by Reference 8.

Typical results are included in Table 14. Note that the dining area measured 8% below the required value; SSA elected to accept this minor shortcoming.

Table 14. Illumination Levels

Space	Required Footcandles	Measured Footcandles
Strg Gear Room	30	53.7
Thruster Room	30	39.7
Red Gear Room	30	34.7
Aux Machy Rm	30	53.1
Engine Room	30	34.8
EOS	30	30.5
Crew Stateroom	10	14.9
Crew Passage	8	41.2
Freight Deck	15	19.3
Elevator	15	29.8
01 Dk Pass Area	15	22.8
Restrooms	20	20.2
Ship's Office	20	21.9
02 Dk Pass Area	15	16.6
Dining Area	20	18.3
Food Prep Area	20	24.8
Crew Pantry	15	29.6
HVAC Room	15	33.4
Emerg. Gen Room	30	60.7

Delivery Trip Results

The delivery trip from Pascagoula to Woods Hole was largely uneventful. Most of the trip was accomplished with a stern propeller RPM of 750 and a bow propeller RPM of 550, resulting in a fuel consumption of about 110 gallons per hour. Although waves up to 15 feet and wind gusts to 50 knots were encountered, there were no injuries and no damage to the vessel.

Commissioning and First Months of Service

The *Island Home* was dedicated in a ceremony at Vineyard Haven on Martha's Vineyard, Massachusetts on March 3, 2007, and entered full service on March 5. With far fewer than expected start-up issues to be resolved, passenger reaction has been almost unanimously positive. Although the initial impressions seemed to focus on the substantial difference in size between the *Island Home* and the *Islander*, the tremendous improvement in interior accommodations has won over virtually all of the skeptics and traditionalists, and SSA can look forward to the start of a new tradition.

ACKNOWLEDGEMENTS

It is safe to say that Capt. Ed Jackson, recently retired from SSA, deserves most of the credit for this highly successful project. His patience, diplomacy and untiring efforts were instrumental in negotiating compromises and maintaining production under uncommonly adverse conditions.

REFERENCES

1. Cape Organization for the Rights of the Disabled, 1019 Iyannough Road #4, Hyannis, MA 02601.
2. Force Technology, "Still Water Performance Tests", DMI 2003173, Rev. 2.0, 2003.
3. Force Technology, "Seakeeping and Zig-Zag Tests", DMI 2003173, Rev. 2.0, 2004.
4. United States Access Board, "Draft Accessibility Guidelines for Passenger Vessels", 2006.
5. Noise Control Engineering, Report 2005-002, "Martha's Vineyard Ferry FEA Results", 2005.
6. American Bureau of Shipping "Guide for Passenger Comfort on Ships," 2001.
7. SNAME, Technical and Research Bulletin No. 3-47 "Code for Sea Trials", 1989.
8. Journal of the Illuminating Engineering Society, "Recommended Practice for Marine Lighting" RP-12-97.

Discussion

Bruce O. Whittemore, Member

First, I want to congratulate Doug Wolff for his excellent presentation of this subject. He leads us very clearly through the design stages and restrictions encountered to complete this project. I particularly appreciate the frank discussion of problems that raised their ugly heads and how they were resolved, apparently to everyone's satisfaction.

I have four comments /questions about the design of this vessel:

1. The prismatic coefficient selected for this project was Cp = 0.526. Why was such a low prismatic coefficient selected? The design speed of 16 knots with a waterline length of 238.7 feet is a speed-length ratio of 1.036 (V/(LWL)1/2 English units). All the studies I am familiar with indicate that the least wave making resistance occurs at a prismatic coefficient of about 0.58 to 0.60 when the speed-length ratio is about 1.05.

This was also proved during model testing a series of proposed hull forms for the new Super Pacific class tuna seiners built by Marco Seattle-owned Campbell Industries of San Diego. Three models were built with very similar hull shapes including very deep "U" shaped bow sections, but with the main difference among each being the prismatic coefficients, namely about 0.57, 0.59, and 0.64. A fourth model was also built to suit rougher northern waters with more "V" sections forward and a prismatic coefficient of about 0.60. All the models had the same waterline length and displacement. The model that had the least wave making resistance at the design speed-length ratio of 1.05 had a prismatic coefficient of 0.59 followed by the model at Cp =0.60 with the "V" sections forward. The model with the Cp = 0.57 had a small increase in resistance over the 0.59 and 0.60 models and the model with Cp = 0.64 had the most resistance of the group. However, as the speed was increased significantly above the design speed-length ratio of 1.05, the lowest wave making resistance occured at higher prismatic coefficient values such as 0.64.

The big differences between the subject ferry and the tuna seiners are that the displacement and displacement-length ratio of the ferry are about one half of that of the tuna seiner and the beam-draft ratio of the ferry is about 2.5 times that of the tuna seiner.

The above comments indicate that the subject ferry could have a prismatic coefficient of about 0.58 for lower resistance. With a similar midship section at the same 1928 ton displacement, the loaded draft would be reduced. Or, the displacement could have been increased by 190 tons at the same draft. This would allow the entire superstructure to be built of steel, making it easier to meet structural fire protection requirements.

2. I agree that simple chine structure is an economical method for building such a vessel. However, the body plan for this vessel shows a single chine at the midship part of the vessel, which changes to a double chine toward the ends of the vessel. My recommendation for such a vessel is to use a double chine form amidship that continues to the vessel ends. The reduced angle at the chines would reduce the hull resistance closer to that of a round bilge vessel. The chine plate, which may be about 3 feet wide, makes it easier to place the bilge keel in the flow stream lines for lower resistance.

3. As stated in the paper, doors were installed at each end of the vessel to prevent sea spray and ice spray from damaging the cars and trucks on the car deck in heavy weather. The photo of the predecessor ferry *Islander* had similar doors. What means have been employed to prevent an *Estonia*-type accident from occurring due to free surface flooding of the car deck after damage to the bow doors in heavy weather? The ferry *Estonia* had a double layer bow door system, but even that did not protect her from bow door damage in heavy seas, causing the flooding of the car deck and the sudden capsize with very high loss of life. Some possibilities that come to mind include: bow door open sensors and/or car deck sumps with level alarms so the captain is warned and can reverse the bow end, double locking devices on the bow doors, very large hinged freeing ports but with heaters to prevent freezing closed. There must be other solutions.

4. I am unfamiliar with the Tees jet thruster. Most units of this type are notorious for their low efficiency at near 0 vessel velocity as compared with traditional transverse bow thrusters. However, the traditional transverse bow thruster rapidly loses side thrust with increased vessel speed. What were the results of any dock side thruster tests of this ferry? And what evaluation was made with the vessel under way, such as up to 5 knots?

Again, congratulations Doug, for your excellent paper.

I. Hilary Rolih, Member, **Allen Chin**, Member, and **Chi-Cheng Yang**, Member

1. Wake wash is a concern with the design. Does the Steamship Authority (SSA) have any wake wash criteria to define the acceptable level (wave height and location)? CFD calculates wave pattern and profile,

which indicate that waves trough at 40% of LBP aft of fore peak and peak at 110% of LBP. Could you show the wave profile and wave height? How far from the centerline of the ship is this calculated wave profile located?

2. The bearding line at the vessel ends are uncommonly long and flat to accommodate the rudder and thruster. Naval architects understand the weakness of this hull form, which will cause a bigger entrance angle, significant bow wave. and less directional stability. In the paper, the author tried to keep the V-section bow or cut-wave bow. At Section 50, it seems close to V-shaped, but at Section 45 it is closer to U-shaped. Could you advise what the entrance angle is at design water line of 10′-6″? What is the impact on wave resistance and directional stability?

3. It is noted that the hull form has lower chine under water instead of a round bilge. Is this selection purely based on construction cost and was the vessel's performance considered? Based on Sharp's experience, the round bilge will have 2-3% lower resistance and less induced wave. Please comment on this.

The upper chine varies above the waterline from 0 feet to about 3 feet. In the model test, does the wave impact on the bottom plate of sponson?Please comment.

4. CFR §72.05-10 indicates that the hull superstructure and deck house shall be divided by suitable structural steel or other equivalent bulkhead material into main vertical zones. This main vertical zone shall not, in general, exceed 131 feet on any one deck. The length of 01 and 02 decks exceed 131 feet. Are there A-60 doors at Fr. 0? If not, how did you overcome this issue? Could you share your experience on resolving stair tower issues?

5. Is this vessel considered operating in open waters or partially protected waters?

6. Regarding optimization of power distribution, Sharp did similar tests for optimization of power distribution for the new *Molinari* Class Staten Island ferries. The lowest total power is at 8.3% bow and 91.7% stern power distribution. At this power distribution, the bow propeller produces a negative thrust with a positive power supply. To avoid this negative thrust, the power distribution of 15% bow and 85% stern was selected, which is different from *Islander Home*'s selection of 29.5% bow and 70.5% stern. The *Islander Home*'s selection is limited by the power available at the stern.

7. It is noted that the *Islander Home* uses the rudder combined with 400 HP electric motor-driven Tee thruster to improve the maneuverability. What is the rudder area, percentage of lateral area under water, maximum rudder turning angle and deflection rate?

8. The vibration level at 02 deck and above appears to be excessive. At what frequency does excess vibration occur? Does the design locate the vibration source? In contract design, were any criteria or guidance utilized to design the local structural members, global structural members, and the main engine foundations to avoid the excess vibration? VTHM's 3-D Finite Element Study predicted some vibration problems, and the shipyard resolved them. We reviewed the vibration measurement results in Table 11 and noted that all excess vibration occurs at the 02 level and above, which is aluminum structures. The Young's modulus (E) for aluminum is 10,000,000 PSI, which is one-third of 30,000,000 PSI for steel. The aluminum structures have less structural stiffness than steel structures. These soft aluminum structures will induce relatively large deflection and vibration. What kind of aluminum was used in the subject vessel? Do you think this contributed to the excess vibration above 02 deck?

9. The forward propeller when fully powered is cavitating heavily. It is known that with the propeller in pull mode, it will be very difficult to avoid cavitation at full power. In your propulsion test, what is the maximum delivery power for the forward propeller in pull mode. Also, at what percentage of full power will the cavitation disappear?

10. Did you measure the vibration level and noise level at the steering gear flat which is directly above the propeller? What is the propeller tip clearance for this ferry?

11. We note that in the propulsion study, options 4, 5 and 6, include three DC generators. We assume that one unit is a standby. With only two DC generators, if one of the two is out of service, the ferry could still operate at a cruising speed of 14 knots (see Fig. 12). We question, if only two 12-710G7B DC generators had been selected for these options, if the 10′-6″ draft could have been retained after the elimination of all propulsion shafts and, if used, reduction gears. In this case, an installation with electric power available for a series of services could complement the supply from the ship's service generators.

12. During the MOLINARI design, we had problems with USCG interpretation of fireload requirements, when the USCG waived the requirements for stairtowers. We noticed there was no mention of fireload and/or vertical fire zone bulkheads. We were wondering what the author's experience with the USCG was with these fire prevention requirements?

13. It is interesting to note the author's comments on his experience with maneuverability using both ends of the propellers. During trials, we found that we could move sideways at about 2-3 knots. Did you try to move sideways during the trials?

14. The author indicates the ferry is, "in general compliance with ABS rules." Is the vessel ABS certificated or is it not?

15. Any design could encounter numerous obstacles and problems. In the final analysis, the vessel has been constructed and is running successfully. The author should be congratulated for this achievement and valuable information in the paper.

Author's Closure

Many thanks to Messers Whittemore, Rolih, Chin and Yang for the effort to review this paper and offer your keen insights. I have attempted below to provide replies in the same order as the questions are listed above.

Bruce O. Whittemore

1. It would be very interesting to investigate the affect on resistance of an increased prismatic coefficient on a double-ended hull form, and the benefits Mr. Whittemore suggests would certainly be welcomed. No such investigation was performed in this instance; however, careful attention was paid to the design of the ends of the vessel. It was particularly challenging to fit in the omnidirectional thruster discharges as close as possible to the ends of the hull while keeping the thruster intakes as low as possible so as to avoid ingestion of ice. This unusual requirement, combined with the need to design the ends to minimize slamming in the short, steep waves of Vineyard Sound, drove the design team to the shape of the ends and resultant prismatic coefficient.

2. In addition to simplifying hull construction, it was expected that the single chine hull form would provide some measure of additional roll reduction. The design team and the owner were willing to accept the consequential increased resistance, although the computational fluid dynamics studies showed minimal affect of cross-chine flow, leading us to believe that the effect on resistance was not significant. That being said, the author now believes that a shipyard constructing a ship of this size and complexity should have the ability to fabricate a round bilge hull form with minimal cost differential and would design a molded hull form if he were starting the design all over.

3. The "enclosed" freight deck is not a watertight structure and freeing ports are provided for the length of the vessel. In addition, the vessel is operating on Vineyard Sound, on a far more protected route than the Baltic Sea. The Steamship Authority does not operate its fleet of vessels during extreme weather conditions.

4. The owner specified thrusters based on previous satisfactory experience with this type of thruster on single-ended vessels. To the author's knowledge, the thrusters were only tested to ensure that they operated correctly; systematic testing was not performed. Vessel operators have by now undoubtedly each developed a personal preference for using the thruster, the bow rudder, or a combination of both to assist with maneuvering into the slip under adverse conditions.

Hilary Rolih, Allen Chin, and Chi-Cheng Yang

1. Wake wash was indeed a concern, but the issue has been and continues to be addressed by the simple expedient of operating at slow speed through the congested Vineyard Haven harbor. This is prudent not only from the wake wash standpoint, but also because of the need to avoid the many small craft that operate in and adjacent to the channel. As a result, wave profiles at speed were not a particular concern and analysis of this data was never performed.

2. The entrance half-angle at the 10'-6" design waterline is about 18°. The impact of the end shape on wave resistance cannot be quantified as no other shapes were investigated; however, the test basin report indicates that resistance and wave making for this hull form were well within the performance envelope of other double-ended designs. As discussed in the paper, our concerns regarding directional stability proved unfounded as the model performed very well in the zig-zag tests. Vessel operations have confirmed that the Island Home has excellent maneuverability as well as excellent directional stability. Perhaps the bow rudder, locked in the fore-and-aft position, enhances directional stability by acting as a skeg.

3. As mentioned in the response to an earlier comment, the single chine was selected largely to reduce construction cost, with the intended additional benefit of roll reduction. The author has since concluded that the minimal additional cost would be a good tradeoff for the resistance reduction. Model tests predicted some minor impacting of waves on the sponson bottoms — this prediction was certainly expected and sponson deadrise was maximized to reduce slamming forces. In practice, the Island Home has not yet experienced significant slamming, although the vessel is just entering its first winter service period at the time of this writing.

4. A main vertical zone, complete with A-60 doors, is indeed fitted at Frame 0. The stair towers were designed with capacity to comply with the requirements of CFR 72.05-10.

5. This vessel operates on a route considered by USCG to be partially protected waters.

6. The author concurs with the discussers' conclusions.

7. Each rudder has a lateral area of about 42 square feet, combining to provide about 4.2% of the underwater lateral area of the hull. Rudder stops are set

at 45 degrees; the flaps operate at twice the angle of the rudder stock.

8. Excessive vibration is caused by propeller cavitation when the leading propeller is operated at high power levels. This situation was performed during trials but is not expected under normal operating conditions. The author concurs that the use of aluminum is likely to exacerbate vibration issues. This tendency was overcome for practical power levels by changing the natural frequency of the shafting system and by installing vibration isolation for the main engines. The use of different aluminum alloys was not investigated as the specified alloys, 5086-H116 and 6061-T6, are in common use by commercial shipyards.

9. The maximum delivery power for the forward propeller in the pull mode during the propulsion tests was 1150 HP, which was 29% of the total driving horsepower at 16 knots. No measurements were taken to determine the point at which cavitation disappears.

10. No measurements of vibration and noise were made in the steering gear flat. Propeller tip clearance is 15 inches, or 16.7% of the propeller diameter.

11. The use of three generators for the diesel electric configuration was an owner's requirement. It is unfortunate that the schedule and budget for the design effort did not permit investigation of options such as the commenter has proposed. It is very possible that the elimination of the third generator, reduction gears and the redundant boiler, among other weight control measures, would have permitted the use of diesel electric propulsion while maintaining the maximum permissible draft.

12. This vessel design required no waivers from USCG and so did not have the same issues to resolve as the *Molinari* design apparently did. That being said, all materials used for insulation, sheathing, linings and furniture were non-combustible.

13. The author is unaware of any tests on this vessel to measure sideways speed.

14. The Island Home is not certificated by the American Bureau of Shipping.

Accident Precursors and Safety Nets: Initial Results from the Leading Indicators of Safety Project

Martha R. Grabowski (V), **Premnath Ayyalasomayajula** (V), **Haiyuan Wang** (V), **Jason R. Merrick** (V), **Denise McCafferty** (V), **Martin L. Meador** (V), **Craig Kinney** (V)

ABSTRACT

Leading indicators, one type of accident precursor, are conditions, events or measures that precede an undesirable event, and that have some value in predicting the arrival of the event, whether it is an accident, incident, near miss, or undesirable safety state. This paper describes initial results from the ABS Leading Indicators of Safety project, which was initiated in response to the need for empirically validated guidance on safety management principles, indicators and measures. We first discuss the concept of leading indicators and safety factors, followed by a description of the data analysis undertaken with our industry partner, a global leader in energy transportation. The results of the analysis are then presented, followed by a discussion of the usefulness of the leading indicators. Implications of the analysis for future work conclude the report.

KEY WORDS: Safety; leading indicators; marine transportation; tanker operations; predictive indicators; lagging indicators; accident precursors; safety management; safety culture.

INTRODUCTION

Safety performance has traditionally been measured by 'after the loss' type of measurements such as accident and injury rates, incidents and dollar costs. However, there is a growing consensus among safety professionals and researchers that these "lagging" indicators, which means that an accident must occur or a person must get injured before a measure can be made, may not provide the necessary insights for avoiding future accidents. A low reported accident rate, even over a period of years, is no guarantee that risks are being effectively controlled, nor will it ensure the absence of injuries or accidents in the future (Lindsay, 1992). Moreover, in many safety-critical settings, the likelihood of catastrophic events or accidents is low; thus, the absence of unlikely events is not, of itself, an indicator of good management (Van Steen, 1996). Recognizing signals before an accident occurs offers the potential for improving safety, and many organizations have sought to develop programs to identify and benefit from alerts, signals and other types of prior indicators.

Leading indicators, one type of accident precursor, are conditions, events or measures that precede an undesirable event, and that have some value in predicting the arrival of the event, whether it is an accident, incident, near miss, or undesirable safety state. Leading indicators are associated with proactive activities that identify hazards and assess, eliminate, minimize and control risk. Lagging indicators, in contrast, are measures of a system that are taken after events and assess outcomes and occurrences. Examples of leading indicator measurement programs include near hit reporting in anesthesia management (Pate-Cornell, 2003), accident precursor assessment programs in nuclear safety (Sewell, Khatib-Rahbar & Erikson, 1999; Sattison, 2003), and hazard identification and analyses for offshore oil and gas in the United Kingdom (Step Change in Safety, 2004). Examples of lagging indicator measurements include recordable injury frequencies, lost time frequencies, lost time severity, vehicle accident frequencies, workers' compensation losses, property damage costs, and numbers and frequency of accident investigations (Gross, Ayres, Wreathall, Merritt & Moloi, 2001).

This paper reports on the preliminary results of the American Bureau of Shipping (ABS) Leading Indicators of Safety project, undertaken for ABS by Rensselaer Polytechnic Institute, with assistance from Virginia Commonwealth University. This project, which began in 2003, focuses on identifying, analyzing and evaluating a set of leading safety indicators in marine transportation. Initially, two industry partners—one a domestic U.S. tanker organization, and one an international tanker organization--joined the project, and data were collected from those organizations in 2006. This report describes the results of the data analysis conducted with the domestic U.S. tanker organization. A third industry partner, a domestic U.S. container and government shipping organization, joined the project in 2006; data collection with the third industry partner is underway in 2007. Additional research partners in 2007 and 2008 include the international parent companies for the domestic U.S. tanker and container operators, as well as an international cruise ship line.

The ABS Leading Indicators of Safety project was initiated in response to the need for empirically validated guidance on safety management principles, indicators and measures. In this paper, we first discuss the concept of leading indicators and safety factors, followed by a description of the data analysis undertaken. The results of the analysis are then presented, followed by a discussion of the usefulness of the leading indicators. Implications of the analysis for future work conclude the report.

LEADING INDICATORS

Leading indicators have been studied in many types of systems, with widely varying results. Many economic systems, including the U.S. economy, use composite indexes and economic series with leading, coincident, and lagging indicators of economic performance (Conference Board, 2004). In economic systems, leading indicators are those indicators that tend to shift direction in advance of a business cycle. Coincident economic indicators, such as employment and production, are broad series that measure aggregate economic activity, and thus define the business cycle. Lagging indicators tend to change direction after the coincident series. In economic systems, lagging indicators are used to confirm turning points and to warn of structural imbalances in the economy.

Over the past thirty years, the medical community has developed increasingly sophisticated leading indicators of health in the United States. Initially, these efforts focused on identifying predictors of individual mortality; recently, the focus has shifted to include broader leading indicators of the nation's health (Chrvala & Bulger, 1999). The nuclear power industry has also evaluated the predictive validity of leading indicators of individual and group safety and performance in nuclear power plants (Gross, et al., 2001; Ayres & Gross, 2002; Lehtinen & Wahlstrom, 2002).

Some industries, such as aviation, have a relatively long history of seeking to identify leading indicators; others organizations, such as blood banks and hospitals, are relative newcomers to the field. Nevertheless, each field uses similar information-gathering processes and weighs common design choices (Tamuz, 2003). Some of these industries discovered accident precursors based on their common experiences, such as having to draw on small samples of accidents (March, Sproull & Tamuz, 1991), while other industries developed signal detection programs as a result of learning by imitation (Levitt & March, 1988), such as medicine's Patient Safety Reporting System, which drew on aviation's experience with its Aviation Safety Reporting System (Tamuz, 2003). Recent work in leading indicators has focused on the use of leading indicators in resilience engineering as a way to assist organizations in recognizing and adapting to change and surprise (Woods & Hollnagel, 2006).

Safety Factors and Leading Indicators

Research to identify leading indicators of safety has often begun with a search for safety factors, elements or conditions that can be linked to high levels of organizational safety performance. Beginning with Shafai-Sahrai's 1971 dissertation, which examined 11 matched pairs of low- and high-injury rate companies in order to identify common factors in low injury rate companies, researchers have sought to identify factors in organizations that are associated with high levels of safety performance (Cohen, 1977; Cohen & Cleveland, 1983; Chew, 1988; Shannon, Walters, Lewchuk, Richardson, Moran, Haines & Verma, 1996; Shannon, Mayr & Haines, 1997; Mearns, Flin, Gordon, & Fleming, 2001; Mearns, Whitaker & Flin, 2001; 2003; DeJoy, Schaffer, Wilson, Vandenberg & Butts, 2004).

Review of the safety factors identified in several major studies undertaken over the past 30 years shows that there is general agreement about the factors that influence organizational safety (Dufort & Infante-Rivard, 1998; Zimolong & Elke, 2006):

- *Genuine and consistent management commitment to safety*, including prioritization of safety over production; maintaining a high profile for safety in meetings; personal attendance of managers in safety meetings and in walk-abouts; face-to-face meetings with employees that feature safety as a topic; and jobs descriptions that include safety contracts.
- *Communication about safety issues*, including pervasive channels of formal and informal communication and regular communications between management, supervisors and the workforce, and
- *Involvement of employees, including empowerment,* delegation of responsibility for safety, and encouraging commitment to the organization (Zimolong & Elke, 2006, pp. 7-8).

One of the difficulties with previous work in leading indicators and safety factors, however, has been the lack of predictive validity in the safety factors and the leading indicators. Thus, one of the goals of the ABS Leading Indicators of Safety project is to develop a set of predictive, empirically-validated leading indicators of safety in marine transportation, and to benchmark the leading indicators over time. In the next section, we describe the steps taken to develop the leading indicators with the first industry partner.

LEADING INDICATORS OF SAFETY IN A DOMESTIC U.S. ENERGY TRANSPORTATION ORGANIZATION

The first industry partner was the domestic U.S. affiliate of a large international energy transportation company. A total of 77 shipboard and 22 shoreside individuals participated in the research project. Shipboard, the participants included all employees working on board the vessels, officers--captains, chief engineers, mates, and engineers—as well as crewmembers from all departments aboard the vessel. Shoreside, the participants included employees holding managerial positions as well as administrative and individual workers. The average age of the participants was 45.5 years, with a mean of 20 years' experience in the maritime industry, and a mean of 16 years of employment with the industry partner.

As set forth by the ABS Guidance Note (2006), six steps were followed to identify and evaluate leading indicators of safety when working with the first industry partner:

- Identify the safety decisions in the organization,
- Identify the fundamental objectives for achieving safety in the organization,
- Identify safety factors to achieve the fundamental objectives,
- Measure the organization's safety performance,
- Measure the safety factors, and
- Determine the candidate leading indicators.

Step 1: Identify Safety Decisions in the Organization

The first step in the Leading Indicators project was to identify the safety decisions in the industry partner organization. Central to the organization's safety culture was a commitment to minimizing accidents. Thus,

Organization's Strategic Objective = Minimize Accidents.

The members of the organization, including senior management; vessel managers; safety, health and environmental management; and vetting managers then worked to identify the decisions that were made about procedures and operations in the company that could either avoid accidents altogether or ensure that the correct actions were taken when exposure occurred. Once the key safety decision in the organization—Minimize Accidents—was identified, the fundamental objectives of safety for the organization were identified.

Step 2: Identify Fundamental Objectives for Achieving Safety

In order to determine the fundamental objectives, interview sessions were arranged with experts with different decision contexts within the industry partner organization. Vessel crew management; safety, health and environmental personnel; senior management; and vetting personnel were identified as critical decision makers whose fundamental objectives were critical to the development of leading indicators.

The interviews began by defining the decisions to be analyzed and the alternatives that might be chosen in each decision. From these discussions, the fundamental safety objectives for the industry partner were identified, as shown in Figure 1. These fundamental safety objectives were:

- Improving the Organizational Safety Culture,
- Improving the Shipboard Safety Culture, and
- Improving Each Individual's Safety Attitude.

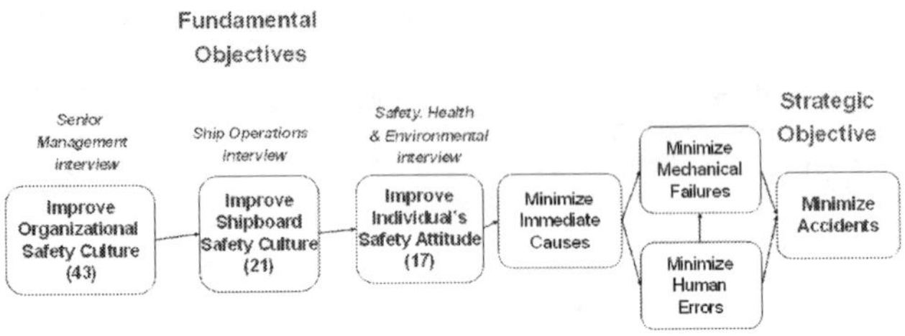

Figure 1. Fundamental Objectives for Improving Safety

Step 3: Identify Safety Factors to Achieve the Fundamental Objectives

The next phase of the project was to identify safety factors for each fundamental objective using the same group of experts. Four groups of 3-4 people comprised the expert groups. Participants represented the shipboard operations group; the safety, health and environmental group; the vetting group; and the senior management team. A total of thirteen personnel participated in four group exercises, which took place over a period of two days in the organization's offices.

Figure 2 summarizes the safety factors identified in the sessions as subsets of the fundamental objectives of Figure 1. Review of previous safety factor studies showed the items to be common with safety factors identified in other studies (Shafai-Sahrai, 1971; Mearns, et al., 2001; 2003; Zimolong & Elke, 2006). Once the safety factors were identified, metrics to measure organizational safety and for the safety factors were determined.

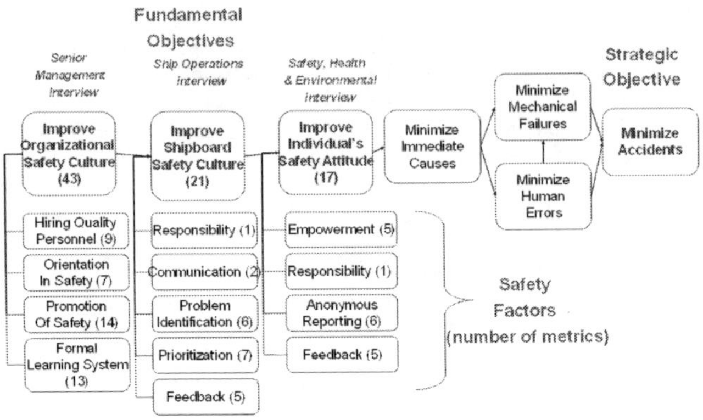

Figure 2. Fundamental Objectives, Safety Factors

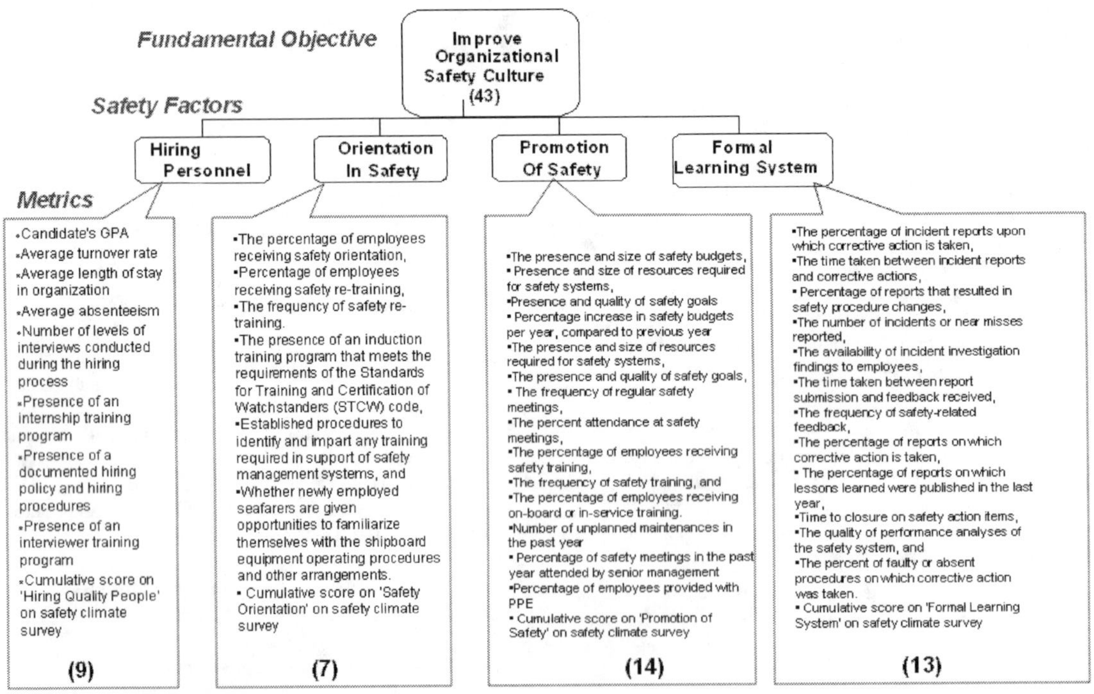

Figure 3. Leading Indicator Metrics for Organizational Safety

Step 4: Measure the Organization's Safety Performance

Official safety statistics and survey data from each of the participating vessels were utilized to develop the organization's safety performance data. The safety performance metrics were traditional lagging indicator data—number of accidents per year, number of near losses per year, number of long term injuries requiring 3 or more days absence from work (LTI>=3) per year, number of port state deficiencies per year, and number of conditions of class per year. However, since data for each of the safety performance measures suffered from a small sample size, only near losses could be used as a safety performance variable to support statistically significant analyses.

Surveys were also administered between January and March 2006 to gather safety climate data from individuals and vessels. This data was utilized to evaluate the significance of the safety factors in the industry partner organization.

Table 1. Summary of Correlation Analysis of Shipboard Safety Factors (* = small sample size, n = 7 vessels)

Vessel Safety Factor	Spearman Correlation with Normalized Near Losses	p-value
Prioritization of Safety	-0.74 (n=7) *	0.07
Problem Identification	-0.85 (n=7) *	0.02
Vessel Feedback	-0.75 (n=7) *	0.07

Table 2. Summary of Correlation Analysis of Individual Safety Factors

Individual Safety Factor	Correlation with Perceived Safety	p-value
Individual Empowerment	0.52	<0.0001
Anonymous Reporting	0.59	<0.0001
Individual Feedback	0.67	<0.0001
Individual Responsibility	0.60	<0.0001

Table 3. Candidate Leading Indicators and Metrics

Leading Indicator Categories	Standard Metrics for Assessing the Leading Indicators
Vessel	*Normalized near losses = dependent variable*
Prioritization of Safety	1. Employees' perceptions of whether officers on their vessel would stop work due to safety concerns even if it meant losing money.
	2: Employees' perceptions of whether a consistent message that that work pressures must not compromise safety is communicated by the vessel management to the work force.
Problem Identification	3: Employees' perceptions of whether all jobs have safety procedures, instructions and rules.
	4: Employees' perceptions of the presence of rules on the vessel that make it easy for employees to identify procedures that are not safe.
Vessel Feedback	5: Employee satisfaction with regard to follow-up and measures taken after accidents, injuries and near losses have taken place on the vessel.
Individual	*Perceived safety = dependent variable*
Empowerment	**6: Employees' perceptions of their influence on vessel health and safety performance.**
	7: Employees' perceptions of their control over job safety outcomes.
	8: Employees' satisfaction with their involvement in vessel safety.
	9: Employees' perceptions of their involvement when health and safety procedures are developed and reviewed.
	10: Employees' perceptions of their influence on safety decisions being made by superiors.
Anonymous Reporting	**11: Employees' perceptions of the effectiveness of anonymous reporting systems in the organization.**
	12: Employees' perceptions of the presence of a 'no-blame' culture in the organization.
	13: Employees' perceptions of the ability of the organization to correct mistakes without punishment and to treat them as a learning opportunity.
	14: Employees' perceptions of their willingness to report a co-worker's failure.
Individual Feedback	**15: Employee satisfaction with the feedback given on accidents, incidents, near losses and injuries that occur aboard the vessel.**
	5: Employee satisfaction with regard to follow-up and measures taken after accidents, near losses and injuries have taken place.
Individual Responsibility	**16: Employees' perceptions of safety as a top priority when performing duties.**
	17: Employees' perceptions of their willingness to point out safety regulations being broken.
	18: Employees' perceptions of their involvement in informing management of important safety issues.
	19: Employees' perceptions of their ability to hold themselves accountable for reporting safety violations—actual or potential—on the vessel.

Step 5: Measure the Safety Factors

Metrics for the safety factors were then identified. Preference was given to metrics that were available, such as ones for which operating companies or government entities were already collecting data so as to facilitate future data gathering and leading indicator analysis. Figure 2 shows the number of metrics identified for each of the safety factors in parentheses. Figure 3 gives an example of the metrics identified for the organizational safety factors.

Figure 2 shows that for some safety factors, constructed or proxy indicators were required, such as for Improving Vessel Responsibility, Improving Vessel Communication, and Improving Individual Responsibility, as few metrics were identified for those safety factors. In addition, Figure 2 shows that more than twice as many organizational metrics were identified, compared to the vessel and individual metrics. This suggests that data collection and measurement systems in the industry partner organization focused on broad organizational measurements, rather than on individual metrics. This distribution of metrics may change in the future, as metrics are developed for some of the safety factors without metrics, and as measurement and control systems in the organization change over time.

Step 6: Determine the Candidate Leading Indicators

Significant safety factors were identified by identifying statistically significant correlations between the safety factor metrics and the organization's safety performance data. Tables 1 and 2 present the correlations of the safety factors with the safety performance data (normalized near losses for the vessel safety factors; perceived safety for the individual safety factors). Near loss data was normalized by the number of vessels in the fleet to facilitate comparison with other companies. Caution is advised with the Table 1 vessel safety factor results, as the sample size is small (only 7 vessels). As a result, p-values within 0.10 are reported as significant in Table 1.

Once the significant safety factors were identified, the metrics corresponding to the safety factors were examined for correlation with the organizational safety data. The significant safety factors and metrics were then identified as candidate leading indicators, as shown in Table 3.

Discussion

Seven categories of leading indicators were identified as significantly correlated with safety performance in the first industry partner organization. At the vessel level, Prioritization of Safety, Problem Identification and Vessel Feedback were significantly correlated with normalized near losses; at the individual level, Empowerment, Anonymous Reporting, Individual Feedback and Individual Responsibility were significantly correlated with perceived safety. Organizational level safety factors were not evaluated in this study, as there was only one research partner at the time. The number of leading indicators roughly corresponds to the number of leading indicators identified in other accident precursor and leading indicator studies (Chrvala & Bulger, 1999; Sattison, 2003; Step Change in Safety, 2004; Zimolong & Elke, 2006).

The distribution of leading vs. lagging indicator results is also consistent with earlier work. In general, leading and lagging indicators can be expected to differ by granularity and focus, as seen in Figure 4 (van den Bergh, 2003). Leading indicators are primarily focused at the individual and departmental levels. In contrast, lagging indicators are broader in scope and generally focus on organizational measures. The results of this study are consistent with van der Bergh, 2003, as leading indicators were found at the individual and department (vessel) level. No organizational leading indicators were able to be evaluated because there was only one industry partner participating at the time of this research.

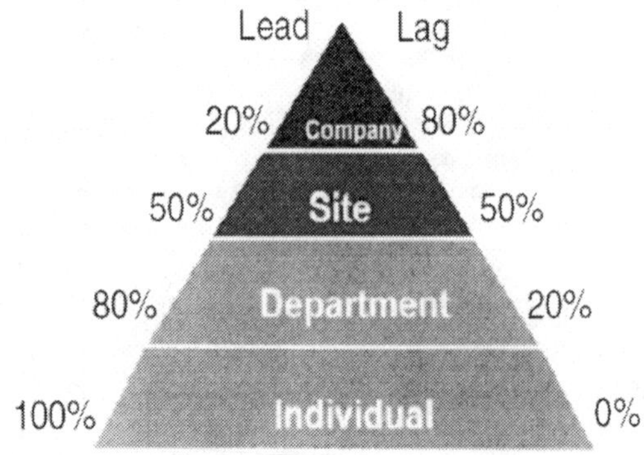

Figure 4. Leading and Lagging Indicators (van der Bergh, 2003)

Over the past twenty years, the leading energy transportation firms have developed a number of analytical tools to identify risk in operations, operators, and in vessels. Among these tools are the Oil Companies International Marine Forum's (OCIMF's) Tanker Management and Self Assessment program (TMSA) (OCIMF, 2004), which provides a tool to assess safety-management systems against key performance indicator categories, as shown in Table 4 (OCIMF, 2004), and OCIMF's Ship Inspection Report (SIRE) program (OCIMF, 2007), which provides a data-based tanker risk assessment tool that assesses performance in the areas identified in Table 4 (OCIMF, 2007).

Table 4 shows that although there is some overlap between the TMSA, SIRE and leading indicator categories, there are also some differences. For instance, the TMSA and SIRE program elements are primarily focused at the organizational level, assessing broad categories of organizational performance, while the leading indicators in Table 4 are focused at the individual and department (vessel) levels of the organization. Inspection of the three research instruments also reveals a qualitative difference between the information gathered with the TMSA and SIRES programs and in the leading indicator program: the TMSA and SIRES program elements determine whether the different systems identified in Table 4 are in place or not. In contrast, the leading indicator data verifies the effectiveness of the systems or their components. Thus, use of leading indicator data in concert with the TMSA and SIRE programs can provide insight as to the effectiveness of existing systems, as well as insight and warnings of impending events at the individual and department (vessel) level.

There are other differences in the leading indicators identified in this research. As can be seen from Table 3, the leading indicators are primarily subjective measures, in contrast to more traditional quantitative metrics of safety performance. Thus, the use of leading indicators, in conjunction with traditional safety and performance measuring systems, can provide a balance between quantitative, organizational, lagging metrics of safety performance and qualitative individual and vessel leading indicators. The subjective leading indicator data might serve as an effective complement to existing data collection and management systems in place in industry. However, new data gathering, analysis and monitoring processes may be required with a leading indicator program, since much safety data analysis is currently focused on quantitative, organizational lagging metrics, rather than qualitative, individual and departmental leading indicators.

Table 4 Leading Indicator Comparisons with TMSA and SIRE Program Elements

Organizational Level	Leading Indicator Categories (Significant Safety Factors)	OCIMF Tanker Management Safety Assessment (TMSA) Program Elements	OCIMF Ship Inspection Report Program (SIRE) Program Elements
Organizational Level		Management Leadership, and Accountability	
		Recruitment and Management of Shore-based Personnel	
		Recruitment and Management of Ships' Personnel	Crew Management
		Reliability and Maintenance Standards	
		Management of Change	
		Environmental Management	
		Emergency Preparedness and Contingency Planning	
		Measurement, Analysis and Improvement	
			Certification and Documentation
		Navigational Safety	Navigation
		Safety Management	Safety Management
			Pollution Prevention
			Structural Condition
		Cargo Ballasting and Mooring Operations	Cargo and Ballast Systems
			Mooring
			Engine and Steering Compartments
			General Appearance and Condition
Vessel Level	Prioritization of Safety		
	Problem Identification		
	Vessel Feedback		Communication
Individual Level	Empowerment		
	Anonymous Reporting	Incident Investigation and Analysis	
	Individual Feedback		
	Individual Responsibility		

Standardized metrics for measuring the leading indicators are provided in Table 3. Many of these metrics were derived from validated survey instruments developed by the aviation, railway, and marine transportation communities along with the aerospace and nuclear communities over the past 40 years. The root of these surveys is the Flight Management Attitudes Questionnaire (FMAQ), an aviation safety culture and safety climate survey (Sexton, Helmreich, Neilands, Bowen, Vella, Boyden, Roberts & Thomas, 2006; Helmreich, Merritt, Sherman, Gregorich & Weiner, 1993). The Table 3 metrics were also derived from validated safety climate surveys from the U.K. oil and gas industry, including the Health and Safety Climate Tool (HSCST), the Offshore Safety Climate Questionnaire (OSQ), the Computerized Safety Climate Questionnaire (CSCQ), the Loughborough Safety Climate Assessment Toolkit (LSCAT), the Quest Safety Climate Questionnaire (QSCQ), the Ship Management Attitudes Questionnaire (SMAQ), and the Safety Attitudes Questionnaire (Health & Safety Executive, 1997a; b; 1999; Mearns, Flin, Gordon, and Fleming, 1998; Flin, R., Mearns, O'Connor, and Bryden, 2000; Mearns, Whitaker, and Flin, 2003; Andersen, Garay, & Itoh, 1999; Sexton, Helmreich et al. (2006), Sexton & Thomas, 2003). The Table 3 metrics can be helpful in standardizing administration and data collection of qualitative leading indicator survey data across the marine transportation system.

The seven categories of leading indicators identified in Table 4 are preliminary in nature, and are in the process of review by the industry partner organizations. These results require assessment over time in order to establish their validity, particularly through a continuous improvement process that identifies and monitors leading indicators, to determine their predictive validity.

USE OF LEADING INDICATORS

The purpose of leading indicators is to identify gaps in the current environment and to guide actions for improving future performance. Thus, one of the important characteristics of leading indicators is that they should facilitate continuous improvement, and be seen as a part of a process for improving future performance (Chrvala & Bulger, 1999; Step Change in Safety, 2004).

Ship operators, owners and charterers can utilize leading indicators as part of their on-going screening, review, monitoring and control processes. Leading indicators can serve as a critical component of effective planning, monitoring and control systems, and of a continuous learning and improvement processes, within and between shipping organizations, owners and charterers. Following Table 4 and Figure 5, an extension of the Oil Companies International Marine Forum Vessel Inspection Questionnaire for bulk, oil, chemical and gas carriers (OCIMF, 2000) might be envisioned to incorporate leading indicator metrics, as might extensions to the Tanker Management Safety Assessment (TMSA) system.

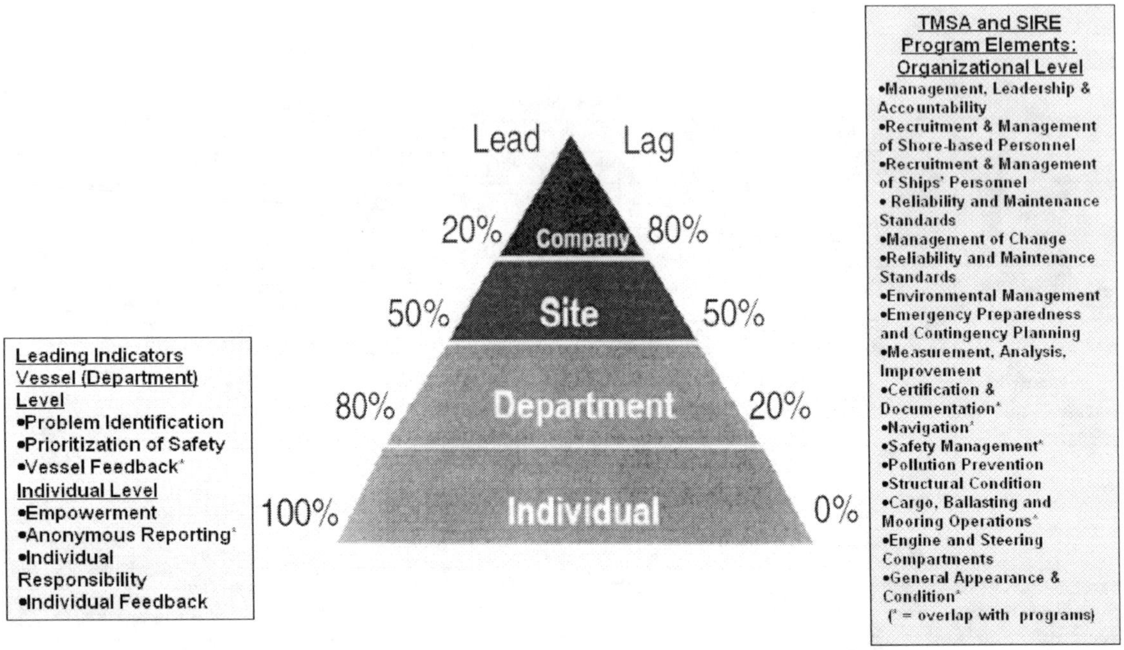

Figure 5. TMSA, SIRE and Leading Indicators (following van der Bergh, 2003)

Regulatory organizations might utilize leading indicators as part of their oversight, compliance, and review responsibilities. Understanding which facets of an organization, culture, operation or vessel to monitor, observe or highlight as models for others in industry is an important benefit of an established and empirically validated leading indicator program. More importantly, understanding leading indicators of safety within different subcultures in an organization might significantly enhance regulatory oversight, understanding and compliance activities. Classification societies might also use leading indicator results as benchmarking metrics within and between class societies, as well as to provide services to clients. Thus, a common leading indicator framework used by industry, regulators, classification societies, and others, could provide the foundation for a strengthened safety culture and precursor measurement/management program.

Despite the promising nature of these findings, however, there are several cautions that should be heeded with these initial results. First, as described, the results reported in this paper are preliminary, the results of a cross-sectional study, and represent one data point in one organization, in what must be an on-going series of organizational measurements. No longitudinal data has been collected at this point from any of the research partners, although discussions about such data collection have begun. Analysis of safety culture and objective data from the industry partner over time can be helpful in understanding how safety culture perceptions fluctuate over time as well as how relationships between safety factors and safety performance metrics change over time. It will be important to establish the validity of any leading indicators over sufficiently long periods of time in order to determine their validity. In both industry and regulatory use, strong longitudinal correlations between the indicators and safety performance must be established in order to prove the indicators' reliability.

Integration of diverse data sets from different operators will also be critical to developing a generalized leading indicator model that reaches across the marine transportation system. Longitudinal results that reflect many organizations' performance and perceptions over time will be important in establishing effective and predictive leading indicators--for industry, for regulators, for different sectors of the marine transportation system, and for the marine transportation system itself.

There were a number of research design factors that limited the strength of these results. First, the results in this study relied on correlations, which is not causation, and only reveals bivariate associations between variables. Higher order models examining the causal relationships between safety factors and safety performance metrics (i.e., structural equation modeling (SEM)) will be considered in future research. In addition, the current study was limited by small sample sizes. In this research, data was collected in only one organization, with a small fleet (n = 7 vessels), which meant that the significance of any organizational level safety factors could not be evaluated, and the vessel safety factors were hampered by small sample sizes. Corroboration of the trends noted in the vessel results with larger and more diverse samples is an important next step in this research.

Finally, the data required to establish a leading indicators of safety program is a mix of existing quantitative safety performance data (already collected by industry, government, non-profit and environmental groups) as well as new qualitative safety perception data (not collected on a regular basis by any of the groups just noted). In order to establish an effective leading indicators program, new data collection and analysis methods, processes and requirements need to be developed, tested, promulgated and revised over time. The size and scope of this effort should not be underestimated. Industry might voluntarily agree to comply with new suggestions for data collection, reporting, and maintenance, and regulators might develop reporting, analysis and monitoring mechanisms for such precursor data, but the costs and infrastructure required to establish such a program industry-wide would be significant and warrant careful investigation (Chrvala & Bulger, 1999; Phimister, Bier & Kunreuther, 2003).

It will be important to analyze the effectiveness of leading indicators over time as safety efforts change, as goals are achieved, and as new areas of improvement are identified. Thus, in addition to initial validation of the leading indicators, a periodic analysis of leading indicators should be undertaken in order to reveal which indicators are not effective and need replacement. Analysis of the effectiveness of leading indicators can also be used to communicate the achievement of safety goals and to maintain momentum on safety efforts.

SUMMARY AND NEXT STEPS

Using the methodology described in this paper, seven categories of leading indicators for one domestic U.S. energy transportation company were identified. The methodology was described, safety performance and safety climate data were collected, and an analysis of the data was undertaken. Metrics operationalizing the leading indicator categories were presented, and the leading indicator results were compared to safety management concepts at use within the industry.

Data analysis from the first two industry partners in the ABS Leading Indicators of Safety project was completed in 2007. Data collection with the third industry partner, a domestic U.S. container and government shipping organization, began in April 2007; analysis of this data will also be completed in 2007. Additional research partners in 2007 and 2008 include the international parent companies for the domestic U.S. tanker and container operators. In addition, a second round of data collection with the original industry partners is anticipated in 2007 so as to provide early trend analyses. Following analysis of this data, a generalized model of leading indicators of safety in marine transportation will be proposed and empirically evaluated.

REFERENCES

Obituaries

James L. Albright

Affiliate

James L. Albright, of Flower Mound, Texas, passed away on April 7, 2006, in Plano, Texas. He was born in Columbus, Ohio, on June 25, 1938 and received a bachelor of science degree in civil engineering from Ohio State University in 1962.

He entered the employ of the Caterpillar Tractor Company in mid-1962 as a trainee in the Marketing Division. He remained with the Caterpillar organization for 38 years, advancing regularly throughout his career. In the late 1990s he was regional manager at Plano, Texas, for Caterpillar, Inc. He retired in September 2000.

Mr. Albright joined the Society as an Affiliate in 1976.

Merle W. Allen

Member

Captain Merle W. Allen, of Toms River, New Jersey, passed away on April 16, 2007. He was born at Fairport, New York, on March 11, 1915, and was a graduate of the New York State Maritime College in 1938. He was one of the youngest men command a merchant ship during World War II, having been promoted to captain in 1943.

After a seagoing career during the period from 1938 to 1947, he became a marine surveyor, and in 1953 formed Marine Surveys Co., Inc., based in Staten Island, New York. Later he formed Marine Container Equipment Certification Co. He retired in 1999.

Captain Allen joined the Society as a Member in 1970.

Theodore C. Ballew

Member

Theodore C. Ballew, of Federal Way, Washington, passed away on April 16, 2007. He was born at Port Orchard, Washington, on December 5, 1925, and lived in the Puget Sound region for his entire life. He attended Wabash College, the University of Washington and Olympic College, and was a torpedoman in the Navy during World War II.

His career took him to many of the companies in marine design and construction in the Seattle-Tacoma area, including Todd Shipyards Corp., Puget Sound Bridge and Dock (later Lockheed Shipbuilding), Tacoma Boatbuilding Co., Morris Guralnick Associates and others.

Later, he formed Ballew Marine Designers, Inc, and was active in the design and construction of small craft. At one time, he bought a fishing boat and gill-netted for salmon in Alaska — with his wife and three cats.

Ted Ballew joined the Society as a Member in 1969.

Ralph R. Brown

Member

Ralph R. Brown, of Dunedin, Florida, passed away on May 15, 2007. He was born at Milwaukee, Wisconsin, on November 2, 1927, and attended Wagner College in New York, and The College of William and Mary at Norfolk, Virginia.

His early career found him a seagoing engineer, first with the U. S. Army Transportation Corps and later with the Military Sea Transportation Service. In 1959, he joined the American Bureau of Shipping as a surveyor and was stationed at many of the Bureau's offices, including Yokohama, Kure and Cadiz. His later Bureau career found him as principal surveyor at Beaumont and Hoboken. He retired in the early 1990s.

Mr. Browm joined the Society as a Member in 1962.

Joseph D. Bryan

Life Member

Joseph D. Bryan, of Houston, Texas, passed away on November 13, 2005. He was born at Montgomery, Alabama, on October 23, 1930. He attended the University of Alabama and other schools and was employed during the late 1950s as a marine surveyor. He later worked for a number of companies mainly in the offshore petroleum industry.

Mr. Bryan joined the Society as a Member in 1962 and became a Life Member in 1982.

Roger A. Burger

Member

Roger A. Burger, of Katy, Texas, passed away on June 9, 2007. He was born on October 19, 1956. He was a graduate of Mississippi State University in 1979 with a bachelor's degree in marine engineering technology.

At that time he commenced his career as a naval architect at McDermott Incorporated, at New Orleans, specializing in projects related to the growing offshore industry. In 1983, he transferred to Hudson Engineering Corp., a McDermott subsidiary at Houston.

Following a short period as a consultant, he joined Ultramarine, Inc. in 1991, where he remained until his passing.

Mr. Burger joined the Society as a Student Member in 1977 and became a Member in 1996.

William C. Carvill

Life Member

William C. Carvill, of Old Saybrook, Connecticut, passed away on April 28, 2004. He was born on January 5, 1918. His career began in 1938 at the piping design section at Gibbs & Cox in New York, followed by seven years from 1941with George B. Drake. After a period as an independent marine surveyor, he joined Sparkman & Stevens as assistant head of the machinery design department.

From 1952 onward, he was with National Bulk Carriers, Inc. and Hess, Inc., and later was with C. P. Steuber and Company, United Tanker Corp. and Pittston Marine Corp. His later activities were mainly as a consultant under the name Maradvice.

Mr. Carvill joined the Society as a Life Member in 1974.

Barton B. Cook, Jr.

Fellow

Barton B. Cook, Jr., of Sun City Center, Florida, passed away on May 7, 2007. He was born at Brooklyn, New York, on August 29, 1920, and received his maritime education at the University of Michigan, which awarded him a bachelor of science degree in naval architecture and marine engineering in 1943.

He joined Gibbs & Cox, Inc., at New York following graduation and remained with that firm until 1955. He was the head of the Machinery Scientific Section at Gibbs & Cox.

A change of employer came in 1955, when he joined the De Laval Steam Turbine Company (later De Laval Turbine, Inc.) as a marine engineer. In January 1961, he was promoted to manager of the Marine Department. He became vice president, marketing, of De Laval Turbine, Inc. in 1968 and in 1971 assumed corporate management of the company's broad interest in the energy field as vice president and assistant group manager, energy products. The year 1979 brought his promotion to executive vice president of Transamerica De Laval, Inc. Retirement came in 1988.

Mr. Cook was the author of "Automation Applied to Steam Turbine Driven Marine Propulsion Plants," presented before the Great Lakes-Great Rivers Section in 1963. He was also the author, with W. I. H. Budd, of a 1966 Spring Meeting paper entitled "Where Are Marine Steam Power Plants Headed?"

His contributions to the operation of SNAME were impressive. Among committee and panel memberships on this long list may be found the following: several Ships' Machinery Committee panels and chairman of Panel M-20 (1964-1967); the Banquet Committee (1966-1971); the T&R Steering Committee (1968-1973); and the Scholarships Committee (1979-1986). In addition, he was chairman of the Philadelphia Section (1971-1972) and was a vice president of SNAME (1974-1975).

Mr. Cook joined the Society as a Junior Member in 1947. He upgraded to Associate member in 1949 and to Member in 1961. He was elected a Fellow in 1978 and received his Fifty-Year Membership Certificate in 1998.

Wilson Lyall Creighton

Member

Wilson Lyall Creighton, of Vancouver, British Columbia, passed away on October 15, 2006. He was born at Ottawa, Canada, on August 3, 1926, attended Carleton University and was a graduate of the Indiana Institute of Technology in 1951, where he earned a bachelor of science degree in mechanical engineering. Having been born at Ottawa, he liked to refer to himself as a "capitalist."

Upon graduation, he entered the employ of Brown Boveri Canada Ltd., where, during a career of about 28 years, he rose to the position of president. In 1980 he became president of Atomic Energy of Canada International and from 1982 was president of the Vancouver consulting firm Dynamic Engineering Inc.

Mr. Creighton joined the Society as a Member in 1986.

Cecil D. Davies

Life Member

Captain Cecil D. Davies, of Charlottesville, Virginia, passed away on February 2, 2005. He was born at Reading, Pennsylvania, on August 27, 1918. He was a graduate of the Pennsylvania Nautical School (aboard the ex-USS Annapolis) in 1938.

His seagoing career with American Export Lines, from 1939 to 1946, resulted in wide experience in ship repair under wartime conditions. His experience with three Liberty ships stands out: a deep tank explosion on John L. Motley, a collision involving Charles Carroll and the mining and beaching of Horace Binney. The difficulty of making repairs to these vessels is difficult to imagine at this time.

He came ashore in 1946 and held a number of positions until 1953, when he became manager of marine personnel for Calmar Steamship Corporation. In 1962 he joined Continental Oil Company as assistant manager, marine operations and was also vice president of two Conoco subsidiary ship-operating companies. He was later associated with Keystone Shipping Company and retired in the early 1990s.

Captain Davies joined the Society as an Associate Member in 1954 and transferred to the grade of Member in 1966. He became a Life member in 1984.

Horacio Garcia

Associate Member

Horacio Garcia, of Buenos Aires, Argentina, passed away on June 13, 2006. He was born on May 27, 1953, and received a degree in naval architecture and marine engineering from the University of Buenos Aires in 1977.

He was active in the Argentinian maritime industry throughout his career, at one time with AFNE-SA and most recently as chairman of Hermes C81 S.A.

Sr. Garcia joined the Society as a Student Member in 1977 and transferred during the following year to Associate Member.

Robert P. Giblon

Honorary Member, Fellow

Robert P. Giblon, of Shrewsbury, New Jersey, passed away on January 31, 2007. He was born at Maplewood, New Jersey, on November 4, 1914, and attended Stevens Institute of Technology, where he earned a bachelor of science degree in mechanical engineering in 1936. This was followed by employment for four years as a machinery draftsman and later as an engineer in the Engineering Scientific Department of Federal Shipbuilding and Dry Dock Company.

In 1938, while employed at the Federal yard, he returned to evening classes at Stevens Institute, and in 1943 received a master of science degree in mechanical engineering.

In December 1940, he was employed by George G. Sharp, Inc. in the Engineering Design and Scientific Department. His expertise in the design of propulsion plant design led to an ever upward trajectory within the Sharp organization. He became Sharp's chief marine engineer in 1951, vice president in 1963 and president in 1969. He retired in 1984. His lifetime accomplishments covered a wide range of marine engineering, including steam, gas turbine, diesel and nuclear power plant design, along with important aspects of design such as plant efficiency and economics. His expertise led to the design and supervision of propulsion plants of more than 130 ships.

This expertise was brought to the Society's attention in many papers, three of which were especially notable. In 1953, he presented, with William Elmer, Jr. and John F. Nace, "Machinery Design of the Barrett Class MSTS Transports" at the New York Metropolitan Section. This paper won The President's Award (later the Vice Admiral E. L. Cochrane Award). The 1961 Spring Meeting produced "The Effect of Steam Conditions and Cycle Arrangement on Marine Power Plant Performance as Determined by The Electronic Computer," co-authored with C. W. Stott, and at the 1966 Spring Meeting, he and I. H. Rolih presented "Effect of Different Types of Propulsion Power Plants on the Ship Power Requirement for Dry Cargo Ships."

His other SNAME activities included chairmanship of the New York Metropolitan Section in 1955-1956, membership on the T&R Steering Committee and the Awards Committee. He was chairman of the Ships' Machinery Committee and of Panel M-15 (Heat balances) and was a member of Panel M-21 (Automation of Ship Propulsion Plants).

Mr. Giblon was elected an Honorary Member for Life in 1983. In 1994 he was awarded the David W. Taylor Medal, in recognition of his many innovative ideas, hard work, dedication and determination. As President Bill Kime said in the medal's presentation, "Bob, you've done it all."

Bob Giblon joined the Society as an Associate Member in 1943, upgrading to Member in 1947. He became a Life Member in 1981 and received his Fifty-Year Membership Certificate in 1993. He was elected a Fellow in 1994.

John T. Gilbride

Life Fellow

John T. Gilbride, of Stamford, Connecticut, the retired chairman and chief executive officer of Todd Shipyards Corporation, passed away on March 17, 2007. His father was president of Robins Dry Dock and Repair Company, later acquired by Todd Shipyards Corp. He was born at Brooklyn, New York, on May 29, 1916, and his shipyard career began when he became a pipefitter's helper at the Robins yard in Brooklyn around 1930 — at the age of fourteen. Summers while in college were spent in various departments of the Robins yard, which was renamed Todd Erie Basin Dry Docks, Inc. in 1942.

He was a graduate of the University of Pennsylvania with a bachelor of arts degree in 1938 and later attended Pratt Institute. In 1939, he was employed as a draftsman in the Robins yard and two years later he was a ship superintendent. Throughout the years 1942-1945, he was superintendent in charge of Navy work, and in the latter year he became the yard's production manager. A transfer to the Los Angeles yard as assistant general manager came in 1946 and in the same year he was promoted to general manager.

Mr. Gilbride's upward voyage continued in 1958, when he was named president of Todd Shipyards Corporation, and again in 1975, when he became the corporation's chairman and chief executive officer. He retired from that position in 1986.

His activities and responsibilities at SNAME were no less impressive. He was a Member of Council Representing Members and Affiliates (1961-1962); Vice President (1962-1964); member of the Committee on Finance and Audit (1966-1973); member of the Executive Committee (1969-1971); member of the Awards Committee (1972-1973); and chairman of the T&R Finance and Administration Committee (1974-1983). In addition, he was an Honorary Vice President for Life from 1968 onward.

He was awarded the prestigious Vice Admiral "Jerry" Land Medal in 1966 for outstanding accomplishments in the marine field. Throughout his career he was a staunch and forceful supporter of the American merchant fleet, and his efforts were recognized in the 1966 Land Medal presentation as follows:

Today shipbuilding in the United States and, indeed, ship operations under the American flag face a crisis. Although this fact is widely recognized, improvement and ultimate correction seem a long way off. With enthusiasm and pride in the industry, which are truly hallmark characteristics of Jack Gilbride, he has been supporting, and continues to press for, new ways to invigorate the total marine industry. That the fortunes of the industry have further deteriorated during the intervening forty-one years is a clarion call for remedial action of a kind that would please Mr. Gilbride.

Jack Gilbride joined the Society as a Junior Member in 1940. He upgraded to Associate in 1944 and to Member in 1946. He was elected a Fellow in 1974 and became a Life Member in 1982. He was presented with his Fifty-Year Membership Certificate in 1990.

Jon R. Gribskov

Member

Jon R. Gribskov, of Washington, DC, passed away on November 21, 1966. He was born at Eugene, Oregon, on September 25, 1950.

Among the positions he filled during his career was that of project manager for Intergraph Corporation in the late 1990s.

He served on the Journal of Ship Production Committee in 1999 and 2000.

Mr. Gribskov joined the Society as a Member in 1993.

Norman A. Hamlin

Life Member

Professor Norman A. Hamlin, of Brunswick, Maine, passed away on October 29, 2007. He was born in New York City on November 23, 1922. He was a graduate of Webb Institute of Naval Architecture in 1944 with a bachelor of science degree in naval architecture and marine engineering. A period as an ensign in the Naval Reserve followed, with assignments at Puget Sound Naval Shipyard and David Taylor Model Basin. He then received a master of science degree in naval architecture and marine engineering at Massachusetts Institute of Technology, awarded in 1951.

He then was employed by Bethlehem Steel Corporation at the Central Technical Division in Quincy, Massachusetts. In 1964, he joined the faculty at Webb Institute of Naval Architecture as professor of naval architecture, a position he held until he retired in 1987.

Professor Hamlin was lead author of "Liquid Sloshing in Slack Ship Tanks — Theory, Observations and Experiments," at the 1986 Annual Meeting. His other papers included "The In-Servicxe Roughness Allowance: Effects of Drydocking, Recoating and the Passage of Time" at the 1980 Spring Meeting; "Evaluating the Seakeeping Performance of Destroyer-Type Ships in the North Atlantic" presented before the New England Section in 1968;

and "Assessment of Seakeepability" in Marine Technology in October 1966. He was also the author of "Chapter One — Ship Geometry" in the 1988 edition of Principles of Naval Architecture.

He was a longtime member of the Scholarships Committee (1970-1987) and was a member of other groups, including the Flow Studies Panel (1957-1960), the Hydrodynamics Committee (1958-1961 and 1965-1967) and the Resistance and Propulsion Panel (1960-1972 or later).

He was active in the New England Section around 1960, serving on its Papers and Executive Committees, and was the section's vice chairman and chairman.

Professor Hamlin joined the Society as a Junior Member in 1945, transferred to Associate Member in 1951 and to Member in 1959. He bcame a Life Member in 1989.

Norm Hamlin was the naval architect's naval architect — one hundred percent of the time. During a relaxing cruise aboard a well-known vessel, he was observed crouching motionlessly at the corner of the ship's swimming pool, a position he maintained for many minutes. When questioned about this later, he stated that he was observing the pattern of water flow at the pool's corner as the ship rolled and pitched ever so slightly. Could it have been a nostalgic reaffirmation of his keen interest in ship motion and liquid sloshing?

Bent K. Jakobsen

Member

Bent K. Jakobsen, of Silver Spring, Maryland, passed away on September 26, 2007. He was born in Denmark on April 5, 1940, and received a master of science degree in naval architecture and mechanical engineering from the Technical University of Copenhagen in 1965.

From 1966 through 1983, he was employed at the Danish Ship Research Laboratory in research and engineering. He then came to the United States and was originally with Tracor Hydronautics, Inc. Later he joined Advanced Marine Enterprises, which later came part of Computer Sciences Corporation as CSC Advanced Marine Center. Mr. Jakobsen was with CSC-AME when he died.

He served as a member of Panel H-10 (Ship Controllability) of the Hydrodynamics Committee from 1995.

Mr. Jakobsen joined the Society as a Member in 1985.

William S. Lagen

Associate Member

William S. Lagen, of Bellevue, Washington, passed away on April 15, 2007. He was born on July 29, 1923, and received a bachelor of science degree in engineering from the U. S. Naval Academy in 1946.

He was the owner of the Meydenbauer Bay Marina, on Lake Washington at Bellevue.

Mr. Lagen joined the Society as an Associate Member in 2002.

Duane H. Laible

Life Fellow

Duane H. Laible, of Hansville, Washington, passed away on September 3, 2007. He was born at Beulah, North Dakota, on June 5, 1940. He was a graduate of Webb Institute of Naval Architecture in 1962, earning a bachelor of science degree in naval architecture and marine engineering. In addition, he completed the Management Program at the University Of Washington Graduate School of Business Administration in 1985.

Upon graduation from Webb, he was employed by the Marine Systems Division of The Boeing Company, but in 1964 joined Marine Construction and Design Company, of Seattle, as a naval architect, and three years later was employed in the same capacity by another Seattle firm, W. C. Nickum & Sons.

In 1971, he joined L. R. Glosten & Associates, Inc., of Seattle, as vice president, advanced to president in 1981, and became the firm's chairman in 2000. He held this position at the time of his passing.

While at Glosten, he was active in the design of many ship types, including oceanographic research vessels, tanker-escort tugs, tractor tugs with cycloidal propulsion, and ferry vessels for Washington and Alaska. He served on the Academy of Sciences Spill Response Committee, the United States Arctic Research Commission, and on the board of the Ballard Maritime Academy, a program of the Seattle Public Schools to attract students to the marine industry and prepare them for further maritime education.

Mr. Laible was a quiet, highly motivated individual who enjoyed his profession, the Pacific Northwest and the need to operate ships in a responsible manner.

Duane Laible joined the Society as an Associate member in 1963, transferred to the grade of Member in 1977, became a Life Member in 1990 and was elected a Fellow in 2003.

C. William Lapworth, Jr.

Member

C. William Lapworth, Jr., of Laguna Beach, California, passed away on April 3, 2006. He was born at Detroit, Michigan, on December 12, 1919. He was a graduate of the University of Michigan in 1941, earning a bachelor of science degree in naval architecture and marine engineering.

Upon graduation, he entered the employ of the American Shipbuilding Company at its Cleveland yard. In 1944, he entered the U. S. Navy as an ensign and was assigned initially to Hingham, Massachusetts, and later to San Diego, California.

From 1946 to 1951, he was engaged in a private practice specializing in small yachts and commercial craft, but returned to the Navy as a docking officer at the Long Beach Naval Shipyard in 1951 for a two-year assignment.

He returned to his small craft design business in 1953. His Cal 40 class sloop was one of the first production fiberglass yachts constructed.

Mr. Lapworth joined the Society as a Junior Member in 1943. He was upgraded to Associate Member in 1949 and to Member in 1956.

Edward M. MacCutcheon

Life Fellow

Edward M. MacCutcheon, of Cornwell, Pennsylvania, passed away on May 17, 2007. He was born at Bridgeport, Connecticut, on November 12, 1915, and was a graduate of Webb Institute of Naval Architecture in 1937, at which time he earned a bachelor of science degree in naval architecture and marine engineering. He later was awarded a master of science degree in engineering administration by George Washington University.

His long career in the maritime field began with a short period at the New York Shipbuilding Corporation, where he was a welding test engineer and a naval architect. In 1938, he moved to the Bureau of Marine Inspection and Navigation of the Department of Commerce, later absorbed by the United States Coast Guard. It was during this period that he was a member of a group that investigated the structural failures that plagued the Liberty ship and the T2 tanker in the early 1940s, events that threatened our ability to supply wartime material to the Allies.

From the close of the war until 1962, he held several positions with the U. S. Navy, initially at David Taylor Model Basin, followed by the Bureau of Ships, the Office of Naval Research and as technical director of the Naval Civil Engineering Laboratory at Port Hueneme, California. In 1962 he became Chief of the Office of Research and Development of the Maritime Administration, where, inter alia, he was involved with the operation of NS Savannah and the hydrofoil Dennison. Later he was assistant director of the Coast and Geodetic Survey in charge of the Office of Systems Development. He retired in 1972.

He was the author of two papers at Society annual meetings. "Rivet Slip, Stress Distribution and the Deflection of Ships' Hulls" was presented in 1949, and "Maritime Administration Research and Development" followed in 1963. In addition, he was a co-author of two papers presented before the Chesapeake Section in 1946, "Stress Relief of Welded Hulls" and "Structural Alterations on Liberty Ships." The latter papers reflected upon the experience with hull structure failures during World War II which led to the formation and early work of the Ship Structure Committee.

His assignments with Society Committees and Panels were manifold. His activities included membership on the Education Committee (1987-1994, the T &R Steering Committee (1962-2003), the Ship Repair and Conversion Panel (1970-1995) and the Ocean Engineering Advisory Group (1966-1971). He was a member of council (1964-1966 and 1988-1990) and was chairman of the Chesapeake Section (1956-1957).

Ed MacCutcheon joined the Society as a Junior Member in 1938. He became an Associate Member in 1945, a Member in 1947, a Life Member in 1981 and was elected a Fellow in 1984. He was awarded a Distinguished Service Award in 1988 and also received his Fifty-Year Golden Membership Certificate in 1988.

"Mac" was the consummate naval architect, but he had a humorous side. At the age of 65, he stated in a letter to the Society, "I am preparing a report on geriatrics and naval architecture, but it is not yet complete." That did not occur until some 27 years later.

James W. Mavor

Life Associate Member

James W. Mavor, Jr., of Woods Hole, Massachusetts, passed away on August 29, 2006. He was born at Schenectady, New York, on January 18, 1923, and received a bachelor of science degree in naval architecture and marine engineering at Massachusetts Institute of Technology in 1944.

He was then commissioned as an ensign in the Naval Reserve and was assigned as a docking and repair officer in the Pacific, followed by two years at D. W. Taylor Model Basin as a naval architect and one year at the Portsmouth Naval Shipyard. He received a master of science degree from M.I.T. in 1950 and was an assistant professor of marine engineering at the U. S. Naval Academy from 1950 to 1953. Other professorial assignments through 1959 followed.

He became associated with Woods Hole Oceanographic Institution in 1957 and became a full-time employee in 1961. As a research specialist, he was one of the pioneers in underwater exploration and contributed to the development of the submarine Alvin, one of the earliest successful vehicles for that highly specialized branch of naval architecture. One notable task assigned to Alvin that is little remembered was the recovery of the nuclear bomb lost in an aircraft accident off the coast of Spain.

He was one of the co-authors of a memorable 1966 SNAME Annual Meeting paper entitled "Alvin, 6000-Ft Submergence Research Vehicle."

Mr. Mavor joined the Society as a Junior Member in 1950 and later advanced to Associate Member. He became a Life Associate Member in 1989.

Donald N. McCallum

Associate Member

Donald N. McCallum, of Silver Spring, Maryland, passed away on June 2, 2007. He was born in Scotland on October 29, 1937, and was a graduate of the Royal College of Science and Technology, in Glasgow, Scotland, in 1959, at which time he received a bachelor of science degree in naval architecture.

Following his arrival in North America in 1962, he was a naval architect with Davie Shipbuilding, Philip F. Spaulding & Associates, and Todd Shipyards Corp., and in 1971 he joined Naval Sea Systems Command, where stack design became one of his several fields of expertise.

Mr. McCallum was co-author, with Gary J. Baham, of "Stack Design Technology for Naval and Merchant Ships," presented at the 1977 Annual Meeting, and one of the authors of a 1982 Annual Meeting paper, "Causes and Corrections for Propeller-Excited Airborne Noise on a Naval Auxiliary Oiler."

Mr. McCallum joined the Society as an Associate Member in 1972.

William L. McCarthy

Member

William L. McCarthy, of Essex, Connecticut, passed away on January 6, 2007. He was born at Cohasset, Massachusetts, on May 8, 1947, and was a graduate of the U. S. Merchant Marine Academy in 1969 with a bachelor of science degree in marine engineering.

After a short period at sea as a third and second assistant engineer, he joined the National Maritime Research Center in 1971 as an Information Specialist. From 1975 to 1977, he was a self-employed consultant in marine engineering (while receiving a master of science degree in Ocean Engineering from Long Island University) and was employed by Ocean Environmental Systems, Ltd.

In 1977 he joined Seaworthy Engine Systems (later Seaworthy Systems, Inc.), where he held the positions of manager, power plant performance analysis, and, from 1980, vice president. At the time of his death, he was executive vice president of engineering.

He was a member of several SNAME panels relating to his fields of expertise, including Panel M-15 (Power Plant Economics), Panel O-21 (Ship Operating Efficiency), and Panel M-26 (Boiler Operating Efficiency).

Mr. McCarthy joined the Society as an Associate Member in 1972 and became a Member in 1978.

William B. Mitchell

Life Associate Member

Captain William B. Mitchell, USNR (Ret), of Cottage Grove, Oregon, passed away prior to June 2007. He was born at Detroit, Michigan, on September 12, 1913, and was a graduate of the University of Michigan in 1941, earning a bachelor of science degree in naval architecture and marine engineering.

Shortly after receiving his degree, he entered the U. S. Navy and was attached to many naval facilities, including the Mare Island Naval Shipyard and the Supervisor of Shipbuilding at San Pedro. Later he was Head of the Marine Engineering Branch of the Naval Ship Engineering Center at Port Hueneme, California. In addition, he was a naval architect with Joshua Henry Corporation and an engineer at the Pacific Missile Range, Point Mugu, California. He retired to Oregon in the mid 1980s.

Captain Mitchell joined the Society as an Associate Member in 1941 and became a Life Associate Member in 1981. He received his Golden Award Fifty-Year Membership certificate in 1991.

Edward K. Moll

Fifty-Year Member

Edward K. Moll, of Bath, Maine, died on September 11, 2007. He was born at Toledo, Ohio, on January 30, 1933, and received a bachelor of arts degree in applied science from Harvard University — unusual for a SNAME member — in 1955, and then attended Massachusetts Institute of Technology, which granted him a degree in naval architecture and marine engineering the following year.

He began his career with the Maritime Administration, and from 1963 was employed by Bath Iron Works, initially as chief hull designer and later as director of production design. He retired in 1997.

He served as a member of Panel O-34 (Computers) from 1969 to 1973.

Mr. Moll joined the Society as an Associate Member in 1957 and transferred to the grade of Member in 1969. He earned his Fifty-Year Membership Certificate in January 2007.

Waldemar S. Nelson

Member

Waldemar S. Nelson, of New Orleans, Louisiana, passed away on November 15, 2005. He was born on July 8, 1916, and was a graduate of Tulane University in mechanical engineering and electrical engineering in 1936. Civil engineering was added to his wide expertise, and he was the holder of professional engineering licenses in forty-four states.

After a short period as a field engineer, he opened his practice as a consulting engineer in 1939 and was a partner in Bedell & Nelson Engineers after World War II. In 1956, the firm of Waldemar S. Nelson & Company was founded.

Mr. Nelson's expertise in maritime-related projects was primarily in the offshore petroleum industry, where the company developed technology in tension-leg platforms and many other aspects of offshore activities. Another notable project was the installation of 230kv and 115kv submarine transmission lines across the Mississippi River to replace the previous overhead lines.

Mr. Nelson, who was known as the "engineer's engineer" to his colleagues, joined the Society as a Member in 1996.

William B. Prechtl

Life Member

William B. Prechtl, of Bridge City, Texas, passed away on an unknown date prior to November 2007. He was born at Elmira, New York, on October 13, 1923. He was a consulting engineer specializing in electrical systems.

Among his many associations over the years were with Metropolitan Shipping Ltd., of Athens, Greece, and Applied Systems Technology, of Springfield, Virginia.

Mr. Prechtl joined the Society as a Member in 1976.

Kris Ramakrishna

Member

Kris Ramakrishna, of Virginia Beach, Virginia, passed away on July 3, 2006. He was born at Madras, India, on October 6, 1924, and received a first class certificate as an engineer while aboard the merchant marine training ship Dufferin.

He was a seagoing engineer with British India Steam Navigation Co., Ltd., from 1946 to 1951, at which time he became an engineer and ship surveyor for the Indian Ministry of Transport. In 1956 he joined Hooghly Docking & Engineering Co., of Calcutta and remained there in various managerial positions until he emigrated to the United States in 1969.

He was then a port engineer for Allied Towing Corp. and later an estimator with Moon Engineering Co. He retired in the mid-1990s.

Mr. Ramakrishna joined the Society as a Member in 1969.

Virgil W. Rinehart

Member

Captain Virgil W. Rinehart, USCG (Ret), of Williamsburg, Virginia, passed away on September 23, 2006. He was born on August 1, 1925 at Chillicothe, Ohio. He attended the Coast Guard Academy, graduating in the class of 1948. In 1954, he completed a course in naval engineering at Massachusetts Institute of Technology, and he later studied a number of technical subjects at the University of Washington.

His varied Coast Guard career, which came to an end with his retirement in October 1971, included tours at sea as engineering officer of USCGC Eastwind and executive officer of USCGC Winona. Landside assignments included four years as an assistant professor at the Coast Guard Academy; as chief of the Naval Engineering Branch of the 13th Coast Guard District; and as design chief for replacement ice breakers. A short period with NOAA followed his retirement.

In 1973, he joined the Maritime Administration as a Research and Development Program Manager. Among his later assignments were as director of Advanced Ship Operations and director of Maritime Technology. He retired from Marad around 1995.

During his Society membership, he was a member of the Journal of Ship Production Committee and of Panels MS-5 and MS-9 of the Marine Systems Committee

Few Society members will know that he was a dedicated choral singer and a member of the Choral Arts Society of Washington.

Virge Rinehart joined the Society as an Associate Member in 1954 and became a Member in 1968.

Thomas A. Santobianco

Member

Thomas A. Santobianco, of Ocean Springs, Mississippi, passed away on July 27, 2007. He was born on January 24, 1930, and earned a bachelor of science degree in mechanical engineering at Penn Military College in 1962.

After a number of years involved with hydraulic and mechanical systems relating to non-marine applications, he joined Ingalls Shipbuilding as Director of Material Operations and Control in 1972, remaining in that position until he retired in the mid-1990s.

He was a member of the 1989 Spring Meeting/ STAR Symposium Steering Committee and of the Exhibitions Committee in 1993-1994.

Mr. Santobianco joined the Society as a Member in 1980.

Fred George Speidel

Life Member

Fred George Speidel, of Sarasota, Florida, passed away peacefully on February 7, 2007. He was born on March 2, 1918 at Brooklyn, New York, and began his career in 1940 with Franque A. Dickins as a chargeman in the design of shipboard power systems. He studied at the Polytechnic Institute of Brooklyn.

After serving in the United States Navy during World War II, he found employment as an electrical designer at Federal Shipbuilding & Dry Dock Co., followed by short periods back at the Dickins firm and with George G. Sharp, Inc.

In 1949 he joined Smith-Meeker Engineering Co., and in 1961 moved to Modern Metal Manufacturing Co., in each case maintaining his interest in the design of marine electrical systems. Later, he was self-employed as a consultant, and upon retirement in 1981 moved to Florida.

Mr. Speidel joined the Society as a Member in 1964. He became a Life Member in 1990.

Edward V. Stewart

Life Member

Edward V. Stewart, of Redlands, California, passed away on April 20, 2007. He was born at Jamesville, Virginia, on May 7, 1917, and served as a deck officer aboard tankers of Atlantic Refining Company from 1941 to 1946. He then attended the University of Michigan and graduated in 1952 with a bachelor of science degree in naval architecture and marine engineering.

After graduation, he was employed by Newport News Shipbuilding for about six months, after which he joined Atlantic Refining Company as a technical assistant, beginning a longtime association with that company and its successors. By 1968, he had advanced to the position of manager of construction and repair. In 1969 the company became Atlantic Richfield Company, and Mr. Stewart retained his title until 1975, when he was promoted to manager, operations. A year later he was manager, technical services, for Arco Transportation Company, and from 1980 until his retirement several years later he was manager of engineering for Arco Marine, Inc.

Mr. Stewart was a member of the Ships' Machinery Committee from 1974 to 1976, and a member and later chairman of Panel M-17 (Disposal of Shipboard Wastes) during the same period. He was chairman of the Los Angeles Metropolitan Section in 1980.

Mr. Stewart joined the Society as a Student Member in 1951, transferred to Associate Member in 1952, became a Life Associate member in 1958, and a Life Member in 1980. He received his Fifty-Year Membership Certificate in 2001.

Richard H. Suehrstedt

Life Member

Richard H. Suehrstedt, of Berea, Ohio, passed away on January 15, 2007. He was born on December 15, 1928. He received his bachelor of science degree in naval architecture and marine engineering in 1951.

He was successively chief naval architect, vice president and president of Marine Consultants and Designers, of Cleveland, Ohio, and contributed significantly over the years to the design of vessels for operation on the Great Lakes.

He served as chairman of the Great Lakes-Great Rivers Section in 1971-1972 and was the author of an Eastern Canadian Section paper in 1973 entitled "Recent and Future Design Trends for Great Lakes Bulk Carriers."

Mr. Suehrstedt joined the Society as a Student Member in 1950 and upgraded successively to Associate Member and Member. He became a Life Member in 1981 and received his Fifty-Year Membership Certificate in 2000.

Rashid J. Tamboowalla

Associate Member

Rashid J. Tamboowalla, of Mumbai, India, passed away on April 27, 2006. He was born on July 5, 1969, and was a senior technical manager for Silver Ship Management, of Mumbai.

He was a first class engineer (motor), as recognized by the Indian Ministry of Transport.

Mr. Tamboowalla became an Associate Member of the Society in 1997.

Robert L. Tapscott

Life Member

Robert L. Tapscott, of Westfield, New Jersey, passed away on October 23, 2006. He was born in Greenfield, Indiana, on March 20, 1916, was a graduate of the U. S. Naval Academy in 1937, and received a bachelor of science degree in naval architecture and marine engineering in 1939.

After a short period as a technical assistant in the Hull Design Department at Newport News Shipbuilding & Dry Dock Co., he joined George G. Sharp, Inc, in November 1939 as a naval architect in the Basic Design Department. After frequent promotions, he was appointed head of the Hull Design Department in September 1948. One of his interesting assignments during World War II was the development of a program to move 18 Great Lakes cargo vessels to salt water via the Chicago Drainage Canal and the Illinois and Mississippi Rivers. He was to remain with George G. Sharp, Inc. — in latter years as Vice President — until he retired in March 1980.

He was a member of the Ship Technical Operations Committee from 1970 to 1981 and a member of the committee's Panel O-31 (Cargo Handling) from 1970 to 1978. In addition, he was a member of the Marine Technology Committee from 1969 to 1970 and a co-author of the chapter "General Arrangement and Interior Design" of the 1955 edition of Design and Construction of Steel Merchant Ships and was the sole author the chapter "General Arrangement" of the 1969 edition of Ship Design and Construction. He was a member of the Ship Design and Construction Control Committee from 1975 to 1980.

Mr. Tapscott joined the Society as a Junior Member in 1940. He became an Associate Member in 1944, a Member in 1952, and a Life Member in 1969. He received a Golden Award Fifty-Year Certificate in 1990.

William C. Thompson

Member

William C. Thompson, of Freeport, New York, passed away on July 28, 2005. He was born on January 19, 1919, and received a bachelor of mechanical engineering degree from Pratt Institute in 1943. In 1947 there followed a master of mechanical engineering (marine option) degree from New York University.

Upon graduation from Pratt Institute, he joined Columbian Bronze Corporation and devoted his entire career to propeller design and manufacture. He became chief engineer of Columbian, and latterly he held the same position with Ferguson Propeller, Inc. He retired in the late 1990s.

Mr. Thompson joined the Society as a Member in 1971.

Larry P. Toups

Member

Larry P. Toups, of Houma, Louisiana, passed away in April of 2006. He was born on July 24, 1950, and was educated at Tulane University, where he received a bachelor of science degree in mechanical engineering in May 1972.

After a two-year period with C. F. Bean Corp., a Gulf Coast-based dredging contractor, he joined J. Ray McDermott, Inc., where he remained until 1991. With McDermott, he was responsible for maintaining the company's extensive fleet of offshore vessels. There followed short periods of employment with other companies, the last of which was at North American Shipbuilding, in Larose, Louisiana, which he joined in 1997.

In 1999, he joined Bollinger Shipyards, Inc., at Lockport, Louisiana, as a mechanical engineer. He remained at Bollinger until his passing.

A professional engineer licensed in his native state, he was well versed in preventive and predictive maintenance of machinery — an important aspect of the maritime and offshore worlds.

Mr. Toups became a Member of the Society in 1999.

In 1999, he joined Bollinger Shipyards, Inc., at Lockport, Louisiana, as a mechanical engineer. He remained at Bollinger until his passing.

A professional engineer licensed in his native state, he was well versed in preventive and predictive maintenance of machinery — an important aspect of the maritime and offshore worlds.

Mr. Toups became a Member of the Society in 1999.

Charles W. Wilson

Honorary Member, Life Fellow

Charles W. Wilson, of Pompton Plains, New Jersey, passed away on August 24, 2007. He was born at New York City on August 4, 1921. After serving in the Army as a Captain in the Corps of Engineers from 1942 to 1946, he attended New York University evening courses and earned both a bachelor of science and a master of science degree in mechanical engineering.

In 1946, he began a long association with The Babcock & Wilcox Company Boiler Division. By 1968, he was promoted to manager of Marine Sales for the company in its Industrial and Marine Division. Early in 1982, after taking early retirement from B&W, he joined Seaworthy Systems, Inc., as vice president and manager of its New York office. He remained with Seaworthy until 1985. After his retirement, he spent a number of years assisting the SNAME staff part-time as a technical advisor.

His contributions to the Society as a volunteer were many. He was a member of the Dinner Dance Committee (1961-1967); Membership Committee (1964-1968); Awards Committee (1981-1984); Applications Committee (1982-1990); Ships' Machinery Committee (1986-1988); Panel M-24 Marine Fuel Systems and Panel M-26 Boiler Operation (1973-1975). He was chairman of the New York Metropolitan Section (1972-1973) and also served in other offices and activities of the section.

He was a Member of Council and a Vice President representing Members and Affiliates (1987-1989).

Charlie Wilson joined the Society as an Associate Member in 1949, and advanced to Member in 1955. He became a Life Member in 1989, was presented with a Distinguished Service Award in 1993, and was elected a Fellow in 1998 and an Honorary Member in 1999.

Sigmund V. Wroblewski

Member

Sigmund V. Wroblewski, Captain, USNR, Retired, of Greenland, New Hampshire, passed away on January 8, 2007. He was born at New York City on January 18, 1919, and received a bachelor of science degree in naval architecture and marine engineering from the University of Michigan in January 1942. He was commissioned an ensign in the Naval Reserve, with his initial assignment being at the Portsmouth Naval Shipyard. In 1947, he became a design officer with the Supervisor of Shipbuilding at Groton, Connecticut.

Later Navy assignments included sea time as an engineering officer aboard the cruiser Des Moines, as design superintendent at Boston Naval Shipyard, planning officer at the First Naval District, and many other assignments before his retirement.

Captain Wroblewski joined the Society as a Member in 1957.

Memorial Resolution

ROBERT P. GIBLON

THE SOCIETY OF NAVAL ARCHITECTS AND MARINE ENGINEERS
unanimously adopted, at a meeting of its Executive Committee, the following memorial resolutions:

WHEREAS, The Society of Naval Architects and Marine Engineers and the maritime industry have suffered a great loss in the death of ROBERT P. GIBLON, on January 31, 2007, a greatly respected, highly honored and deeply beloved Honorary Member and Fellow of the Society, it desires to record these memorial resolutions.

Mr. Giblon first joined the Society as an Associate Member in 1943, upgrading to Member in 1947. He became a Life Member in 1981, and was elected an Honorary Member for Life in 1983. Mr. Giblon received his Fifty-Year Membership Certificate in 1993 and was elected a Fellow in 1994. His other SNAME activities included chairmanship of the New York Metropolitan Section in 1955-1956, membership on the T&R Steering Committee and the Awards Committee. He was Chairman of the Ships' Machinery Committee and of Panel M-15 (Heat Balances), and was a member of Panel M-21 (Automation of Ship Propulsion Plants).

Mr. Giblon was awarded the David W. Taylor Medal in 1994, in recognition of his many innovative ideas, hard work, dedication and determination. As President Bill Kime said in the medal's presentation, "Bob, you've done it all."

Mr. Giblon was born at Maplewood, New Jersey, on November 4, 1914, and attended Stevens Institute of Technology, where he earned a Bachelor of Science degree in Mechanical Engineering in 1936. This was followed by employment for four years as a machinery draftsman, and later as an engineer in the Engineering Scientific Department of Federal Shipbuilding and Dry Dock Company. In 1938, while employed at the Federal yard, he returned to evening classes at Stevens Institute, and in 1943 received a Master of Science degree in Mechanical Engineering. In December 1940, he was employed by George G. Sharp, Inc. in the Engineering Design and Scientific Department. His expertise in the design of propulsion plant design led to an ever upward trajectory within the Sharp organization. He became Sharp's Chief Marine Engineer in 1951, Vice President in 1963 and President in 1969. He retired in 1984. His lifetime accomplishments covered a wide range of marine engineering, including steam, gas turbine, diesel and nuclear power plant design, along with important aspects of design such as plant efficiency and economics. His expertise led to the design and supervision of propulsion plants of more than 130 ships.

Mr. Giblon's expertise was brought to the Society's attention in many papers, three of which were especially notable. In 1953, he presented, with William Elmer, Jr. and John F. Nace, "Machinery Design of the Barrett Class MSTS Transports" at the New York Metropolitan Section. This paper won The President's Award (later the Vice Admiral E. L. Cochrane Award). The 1961 Spring Meeting produced "The Effect of Steam Conditions and Cycle Arrangement on Marine Power Plant Performance as Determined by the Electronic Computer," co-authored with C. W. Stott, and at the 1966 Spring Meeting, he and I. H. Rolih presented "Effect of Different Types of Propulsion Power Plants on the Ship Power Requirement for Dry Cargo Ships."

In tribute to his long, distinguished, and invaluable service to the Society, the maritime industry and his fellow man, hereby be it

RESOLVED, That The Society of Naval Architects and Marine Engineers, and the maritime industry, have lost, in the death of ROBERT P. GIBLON, a beloved member of the Society, a distinguished, capable and fine gentleman and a true friend.

Further, be it

RESOLVED, That these resolutions be recorded in the TRANSACTIONS of the Society and that a copy be sent to his family as a token of the Society's deep regard for ROBERT P. GIBLON.

JOHN T. GILBRIDE

THE SOCIETY OF NAVAL ARCHITECTS AND MARINE ENGINEERS
unanimously adopted, at a meeting of its Executive Committee, the following Memorial Resolutions:

WHEREAS, The Society of Naval Architects and Marine Engineers and the maritime industry have suffered a great loss in the death of JOHN T. GILBRIDE on March 17, 2007, a greatly respected, highly honored and deeply beloved Life Member and Fellow of the Society, it desires to record these memorial resolutions.

Mr. Gilbride first joined the Society as a Junior Member in 1940. He upgraded to Associate in 1944 and to Member in 1946. He was elected a Fellow in 1974 and became a Life Member in 1982. He was presented with his Fifty-Year Membership Certificate in 1990. He was a Member of Council Representing Members and Affiliates (1961-1962); Vice President (1962-1964); Member of the Committee on Finance and Audit (1966-1973); Member of the Executive Committee (1969-1971); Member of the Awards Committee (1972-1973); and Chairman of the T&R Finance and Administration Committee (1974-1983). In addition, he was an Honorary Vice President for Life from 1968 onward.

Mr. Gilbride was awarded the prestigious Vice Admiral "Jerry" Land Medal in 1966 for outstanding accomplishments in the marine field. Throughout his career he was a staunch and forceful supporter of the American merchant fleet, and his efforts were recognized in the 1966 Land Medal presentation as follows:

Today shipbuilding in the United States and, indeed, ship operations under the American flag face a crisis. Although this fact is widely recognized, improvement and ultimate correction seem a long way off. With enthusiasm and pride in the industry, which are truly hallmark characteristics of Jack Gilbride, he has been supporting, and continues to press for, new ways to invigorate the total marine industry.

That the fortunes of the industry have further deteriorated during the intervening forty-one years is a clarion call for remedial action of a kind that would please Mr. Gilbride.

Born in Brooklyn, New York, on May 29, 1916, Mr. Gilbride graduated from the University of Pennsylvania with a Bachelor of Arts degree in 1938, and later attended Pratt Institute. His father was president of Robins Dry Dock and Repair Company, later acquired by Todd Shipyards Corp. Mr. Gilbride's shipyard career began when he became a pipefitter's helper at the Robins yard in Brooklyn around 1930 --aged fourteen. Summers while in college were spent in various departments of the Robins yard, which was renamed Todd Erie Basin Dry Docks, Inc. in 1942. In 1939, he was employed as a draftsman in the Robins yard and two years later he was a ship superintendent. Throughout the years 1942-1945, he was superintendent in charge of Navy work, and in the latter year he became the yard's production manager. A transfer to the Los Angeles yard as Assistant General Manager came in 1946 and in the same year he was promoted to General Manager. Mr. Gilbride's upward voyage continued in 1958, when he was named President of Todd Shipyards Corporation, and again in 1975, when he became the corporation's Chairman and Chief Executive Officer. He retired from that position in 1986.

In tribute to his long, distinguished, and invaluable service to the Society, the maritime industry and his fellow man, hereby be it

RESOLVED, That The Society of Naval Architects and Marine Engineers, and the maritime industry, have lost, in the death o f JOHN T. GILBRIDE, a beloved member of the Society, a distinguished, capable and fine gentleman and a true friend.

Further, be it

RESOLVED, That these resolutions be recorded in the TRANSACTIONS of the Society and that a copy be sent to his family as a token of the Society's deep regard for JOHN T. GILBRIDE.

EDWARD M. MacCUTCHEON

THE SOCIETY OF NAVAL ARCHITECTS AND MARINE ENGINEERS
unanimously adopted, at a meeting of its Executive Committee, the following memorial resolutions:

WHEREAS, The Society of Naval Architects and Marine Engineers and the maritime industry have suffered a great loss in the death of EDWARD M. MacCUTCHEON on May 17, 2007, a greatly respected, highly honored and deeply beloved Life Member and Fellow of the Society, it desires to record these memorial resolutions.

Mr. MacCutcheon first joined the Society as a Junior Member in 1938. He became an Associate Member in 1945, a Member in 1947, a Life Member in 1981 and was elected a Fellow in 1984. He was awarded a Distinguished Service Award in 1988, and also received his Fifty-Year Golden Membership Certificate in 1988. Mr. MacCutcheon's assignments with Society Committees and Panels were manifold. His activities included membership on the Education Committee (1987-1994, the T &R Steering Committee (1962-2003), the Ship Repair and Conversion Panel (1970-1995) and the Ocean Engineering Advisory Group (1966-1971). He was a Member of Council (1964-1966 and 1988-1990) and was Chairman of the Chesapeake Section (1956-1957).

Mr. MacCutcheon was the author of two papers at Society annual meetings. "Rivet Slip, Stress Distribution and the Deflection of Ships' Hulls" was presented in 1949, and "Maritime Administration Research and Development" followed in 1963. In addition, he was a co-author of two papers presented before the Chesapeake Section in 1946 -"Stress Relief of Welded Hulls" and "Structural Alterations on Liberty Ships." The latter papers reflected upon the experience with hull structure failures during World War II which led to the formation and early work of the Ship Structure Committee.

Mr. MacCutcheon was born at Bridgeport, Connecticut, on November 12, 1915, and was a graduate of Webb Institute of Naval Architecture in 1937, at which time he earned a Bachelor of Science degree in Naval Architecture and Marine Engineering. He later was awarded a Master of Science degree in Engineering Administration by George Washington University. Mr. MacCutcheon's long career in the maritime field began with a short period at the New York Shipbuilding Corporation, where he was a welding test engineer and a naval architect. In 1938, he moved to the Bureau of Marine Inspection and Navigation of the Department of Commerce, later absorbed by the United States Coast Guard. It was during this period that he was a member of a group that investigated the structural failures that plagued the Liberty ship and the T2 tanker in the early 1940s, events that threatened our ability to supply wartime material to the Allies.

From the close of the war until 1962, he held several positions with the U. S. Navy, initially at David Taylor Model Basin, followed by the Bureau of Ships, the Office of Naval Research and as Technical Director of the Naval Civil Engineering Laboratory at Port Hueneme, California. In 1962 he became Chief of the Office of Research and Development of the Maritime Administration, where, inter alia, he was involved with the operation of NS Savannah and the hydrofoil Dennison. Later he was Assistant Director of the Coast and Geodetic Survey in charge of the Office of Systems Development. He retired in 1972.

In tribute to his long, distinguished, and invaluable service to the Society, the maritime industry and his fellow man, hereby be it

RESOLVED, That The Society of Naval Architects and Marine Engineers, and the maritime industry, have lost, in the death of EDWARD MacCUTCHEON, a beloved member of the Society, a distinguished, capable and fine gentleman and a true friend.

Further, be it

RESOLVED, That these resolutions be recorded in the TRANSACTIONS of the Society and that a copy be sent to his family as a token of the Society's deep regard for EDWARD M. MacCUTCHEON.